Science of Gel

Yoshihito Osada
Editor

Science of Gel

Springer

Editor
Yoshihito Osada
RIKEN
Wako, Saitama, Japan

ISBN 978-981-95-0300-1 ISBN 978-981-95-0301-8 (eBook)
https://doi.org/10.1007/978-981-95-0301-8

Translation from the Japanese language edition: "Geru no Kagaku" by Yoshihito Osada,

This Springer imprint is published by the registered company Springer Nature Singapore Pte Ltd.
The registered company address is: 152 Beach Road, #21-01/04 Gateway East, Singapore 189721, Singapore

Preface to Japanese Edition

Gel science, while a significant area of research within polymer science, is a multidisciplinary field that requires a wide range of academic knowledge. To deeply and comprehensively understand it, knowledge is needed not only in the fields of polymer chemistry and physics, but also in areas such as reaction kinetics, solution theory, thermodynamics, and other branches of physical chemistry. Additionally, mathematical sciences, including statistical mathematics and probability theory, mechanics and materials science, including rheology, and even biology, medicine, and pharmacology are essential. In today's highly specialized academic system, achieving a comprehensive understanding of gels as a substance is no easy task.

A gel is a two-component system composed of an insoluble polymer network and solvent, and it is a unique substance with an inhomogeneous three-dimensional network structure. Its behavior is not clearly defined—whether it behaves like a liquid, a solid, or some intermediate state. Even though the term "gel" was proposed by the Scottish chemist Thomas Graham in 1861, the reason it remained largely neglected and did not become a research focus for so long seems to lie in these ambiguities.

In the present day, where there is a demand to integrate various established academic frameworks into a new multidisciplinary field, research on materials like gels, which are multiphase complex systems, may be one of the most suitable research topics. The description of the process of polymer network formation through repeated branching reactions shares common conceptual frameworks with nuclear reactions, molecular aggregation, infectious diseases, cell division, and genetic evolution. Additionally, the fact that polymer networks exhibit discrete characteristics and phase transition phenomena despite having multi-element and inhomogeneous structures can be interpreted as phenomena that correspond to natural or societal group systems. Furthermore, the assertion that gels possess dynamic motion and local diversity while creating ordered or hierarchical structures provides a powerful approach to understanding the dynamism and emergence in biological tissues. These are all contemporary challenges that need to be addressed.

Given this academic background, challenges, and potential, there has been no book that comprehensively and integratively explains the nature of gels—covering their formation process, structure, thermodynamics, rheology, and materials science. This gap in literature became the motivation for this book's publication. When I shared this situation with Mr. Kenji Gomi of Kodansha Scientific Co., he immediately understood and readily agreed to publish the book. Moreover, he generously allowed for an unusually long preparation period of four to five years. This occurred during the 22nd Polymer Network Group Meeting and the 10th Gel Symposium held at the University of Tokyo in 2014.

However, the editing process proved more challenging than anticipated. The issue was how to accurately describe the development of theories and understanding of this difficult material, considering the historical progression of gel science up to the present. This task was vital because the book's mission is to clarify unresolved issues in gel science and to contribute to the future development of the field. It was a challenging endeavor. After discussions among the authors, the following structure was decided upon:

Chapter 1: "What is a Polymer Gel?"(Yoshihito Osada)
Definition of gels, their history, and classification in gel science.
Chapter 2: "Basic Properties of Gels" (Yoshihito Osada)
An overview of the general properties and characteristics of gels.
Chapter 3: "Formation of Gels" (Yoshihito Osada and Karel Dušek)
Classification of gels and general methods for network formation, as well as the theory of inhomogeneity and branching formation.
Chapter 4: "Thermodynamics of Gels" (Karel Dušek)
The foundational swelling theory and thermodynamics, followed by the development of improved models. Theory of volume phase transitions.
Chapter 5: "Kinetics of Swelling and Shrinkage" (Mitsuhiro Shibayama)
The swelling kinetics based on solvent diffusion and network expansion. A discussion on various gel forms.
Chapter 6: "Mechanical Properties and Rheology of Gels" (Kenji Urayama)
Deformation and mechanical responses of gels, with theoretical explanations based on experimental results on sol-gel transitions.
Chapter 7: "Structural Analysis of Gels"(Mitsuhiro Shibayama)
Theoretical and experimental methods for structural analysis, primarily using light and neutron scattering.
Chapter 8: "Gels in Biology" (Yoshihito Osada)
A hierarchical discussion of biological tissues from the perspective of gels.
Chapter 9: "Biomimetic Functions and Applications of Gels" (Yoshihito Osada)
Results and trends in the application of synthetic gels that aim to mimic biological functions.

Chapters 1–3 serve as introductory sections, providing a bridge to Chapters 4 and beyond, without using too many mathematical formulas. In Sect. 3.4, the theoretical models of branching formation are described. Chapter 4 covers how thermodynamic approaches to gels have developed and constructed theories, starting from the Flory-Huggins solution theory. It is recommended to review the basics of polymer solution theory before reading this chapter. This chapter may differ from usual textbooks in that it does not omit a large number of references, as understanding the development process and the evolution of ideas in gel theory is important for future theoretical construction. Similarly, in Chaps. 5 and 6, physical descriptions and mathematical expressions are not avoided, as these chapters are also intended to explain important concepts in a way that is comprehensive. These chapters do not necessarily need to be read in sequence; one could choose to read Chaps. 5 and 6 before Chaps. 3 and 4.

Gels are a major constituent of biological organisms, and their structure, properties, and functions share numerous commonalities with biological tissues. Understanding biological tissues from the perspective of gels is one of the crucial challenges for future gel science and will have significant contributions to fields such as medicine and pharmacology. For this reason, Chaps. 8 and 9 were included.

I would like to briefly introduce Dr. Karel Dušek, who wrote Sects. 3.2, 3.4 and Chap. 4. Dr. Karel Dušek has long been involved in the theoretical and experimental research on the formation of branched and cross-linked polymers. In 1967, he predicted for the first time in the world the volume phase transition of gels, which led to empirical verification by Dr. Toyoichi Tanaka and others about ten years later, marking the beginning of the flourishing of modern gel science. His detailed and precise descriptions on reaction kinetics and thermodynamics are essential in this field. The number of references he cites is extensive and includes a vast array of materials. While including all these references was somewhat challenging, it would be both his tribute to the work of previous researchers and a historical record reflecting the dynamic progress of the field. I, as an editor, decided, with the understanding of Kodansha Scientific, to maintain this structure. I hope readers will appreciate this intention.

In the modern era of information overload, the role and mission of books must change. While one approach is to provide a comprehensive, easy-to-understand overview, this book takes a different approach. As a result, it may not cover every topic in an easily accessible manner, but I believe it successfully incorporates the most essential and core aspects of gel science.

Finally, I would like to express my deepest gratitude to Mr. Kenji Gomi of Kodansha Scientific Co. for making the publication of this book a reality. He not only provided more than five years for research and writing but also personally took on the laborious task of translating Dr. Karel Dušek's manuscript to Japanese. This work, both lengthy and highly advanced, could not have been published without his dedicated efforts, which were backed by his expertise in polymer science obtained during his doctoral studies. On behalf of all the authors, I would like to express my heartfelt thanks to Mr. Gomi.

Wako, Saitama, Japan — Yoshihito Osada
February 2020

Preface to English Edition

It has been nearly five years since the Japanese edition of *The Science of Gel* was published in 2020. From the initial planning stages, the intention was to publish this work in English to widely convey the importance of gel science to the world. However, due to changing circumstances, the decision was made to first publish the Japanese edition and then follow it with the English edition. As a result, while the English edition was initially planned to be published shortly after the Japanese one, the unexpected situation of the COVID-19 pandemic has led to the delay.

As I revisited each chapter in preparation for the English edition, I reflected on the Japanese edition's preface, which states: "Gels, with their polymer networks having multi-faceted and heterogeneous structures, exhibit discrete characteristics and phase transition phenomena that can be seen as analogous to phenomena in natural or social collective systems. Moreover, the proposition that gels, while having dynamic mobility and local diversity, form ordered or hierarchical structures which provide a powerful approach for understanding the dynamism and emergence of biological tissues. These are all pressing, contemporary challenges that require resolutions." This insight into the importance of gel science remains as valid as ever, and I have come to feel that the significance of publishing this book has only grown stronger over time.

Therefore, I did not request the authors to update their chapters with more recent findings when preparing the English edition. This is because the preface of the Japanese edition emphasizes that the mission of this book is "to accurately describe the historical development of gel science, from which we can highlight unsolved problems and contribute to the advancement of future gel science." As this book is intended to provide an overview of gel science over a long historical span, it is meant to have lasting value. This might naturally be the outcome of such a perspective.

What is most regrettable is that, during this period, one of the great pioneers of gel science and a co-author of this book, Prof. Karel Dušek, passed away. I deeply regret that we were unable to present the book to him while he was alive. During the writing of this book, I had many discussions with him, and his passion for contributing to this work was beyond words. It was truly inspiring to witness his unwavering commitment to this project, and I will always remember this as a profound example of what it means to be a true scientist. Though this book has, unintentionally, become a posthumous work, I am grateful that we can share a part of his legacy and passion for gel science through it.

In the publication of this book, I would like to express my heartfelt gratitude to Dr. Shinichi Koizumi (Senior Editor, Springer Nature Japan), who has been of immense help. When the plan for an English edition seemed uncertain, a mutual acquaintance introduced me to Mr. Koizumi. Upon hearing the circumstances, he immediately understood the significance of the book and graciously agreed to support the publication. Throughout the process, his quick and careful guidance made this publication possible. Without his deep understanding and collaboration, this book may not have come to fruition. On behalf of all the authors, I would like to extend our deepest thanks to him.

Wako, Japan
April 2025

Yoshihito Osada
Editor-in-Chief

Contents

Editor and Contributors

About the Editor

Yoshihito Osada graduated from Waseda University in 1966 and completed his doctoral course (Macromolecular Chemistry) at Moscow University in 1971. After serving as a professor at Ibaraki University, he became a professor at Hokkaido University in 1992. He has served as Dean, School of Science, and Vice President of Hokkaido University. He was invited as Deputy Director of Advanced Science Institute of RIKEN (The Institute of Physics and Chemistry), Japan. Now, he is a senior visiting scientist at Riken and a professor emeritus of Hokkaido University.

Contributors

Karel Dušek Institute of Macromolecular Chemistry, Academy of Sciences of the Czech Republic, Prague 6, Czech Republic

Yoshihito Osada RIKEN, Wako, Saitama, Japan

Mitsuhiro Shibayama Neutron Science and Technology Center, Comprehensive Research Organization for Science and Society (CROSS), Ibaraki, Japan

Kenji Urayama Department of Material Chemistry, Kyoto University, Kyoto, Japan

Chapter 1
Gel is a Comprehensive State of Matters

Yoshihito Osada

Abstract Materials consisting of polymer network and fluid is called polymer gel(in case of water: hydrogel). They are swollen like jelly or agar and soft and wet. Living organisms are largely consisting of hydrogels. They show a variety of unique properties and behaviors which cannot be observed by solid and liquid. In this chapter, unique properties of the gel are introduced and their basic mechanism is explained. Brief history of gel science and the classification of gels are also described.

Keywords Polymer network · Hydrogel · Swelling · Deborah number · Superabsorbent · Chemical gel · Physical gel · Biological gel · Xerogel · Gelatine · Agar · Colloid

1.1 Introduction

1.2 What is Polymer Gel?

Materials consisting of cross-linked polymer network with fluid filling the interstitial spaces of the network is called polymer gel [1]. The polymer network sustains the fluid (liquid) and prevents it from flowing out through the strong interaction with the fluid. Thus, the polymer network plays the role of a container that holds a large amount of fluid. Once the polymer chains are cross-linked, either chemically or physically, they form three-dimensional (3D) network with huge molecular size and lose their macroscopic (translational) mobility. The network behaves as a solid matter despite they are totally surrounded with a mass of fluid. Although the translational motion of polymer chains are restricted, the network itself can spread into the fluid trapping the fluid and swell to form highly swollen 3D conformation (Fig. 1.1). Instead of losing the translational motion as whole, the swollen network obtains vigorous local mobility of vibration to give the viscoelastic property. Thus,

Y. Osada (✉)
RIKEN, Wako, Saitama, Japan
e-mail: osadayoshi@riken.jp

Y. Osada (ed.), *Science of Gel*, https://doi.org/10.1007/978-981-95-0301-8_1

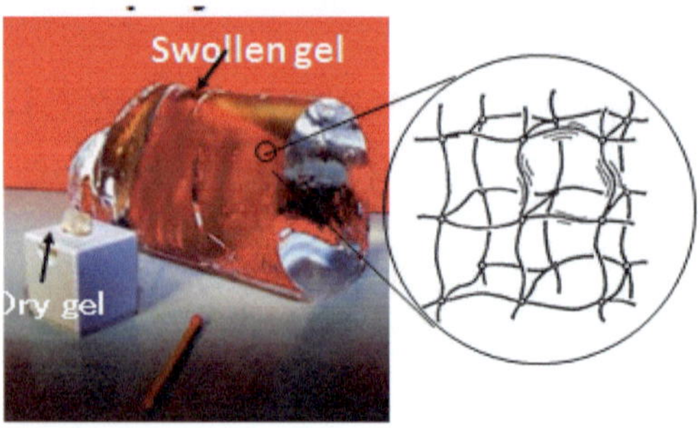

Fig. 1.1 Dry polymer network can absorb water more than thousand times in its weight and forms highly swollen hydrogel. Insert indicates the schematic illustration of polymer network

the viscoelastic property is one of basic properties of swollen "polymer gels" originated from dynamic nature of elastic cross-linked network and the fluid filling the interstitial spaces. Therefore, the viscoelastic property is strongly influenced by the amount of the fluid incorporated, which is closely related to the crosslinking density of the polymer network. The lower the crosslinking density, the higher the amount of the fluid incorporated and the swelling degree increases.

The swelling degree depends not only on the network density, but on the thermodynamic properties of the penetrating fluid. When the crosslinking density is same, the higher the affinity of fluid to the network, the higher the swelling degree of the gel. If the fluid in the network is water which is a highly polarized containing ions and the network is hydrophilic, water would be strongly attracted to the network by the electrostatic interaction to give water-swollen gel called hydrogel. As a result, water molecules in the gel, especially those closely located to the network are strongly bound through electrostatic interactions to it and change the intrinsic local properties as viscosity, density, partial volume, ionization degree, and melting temperature. Water in this state is called bound water or non-freezable water (details refer to 2.7). Thus, the swelling is determined by the competitive balance between the fluid affinity which tend to expand the volume and the rubber elasticity that keeps the network in a state of moderate expansion. Flory explained the water-retention capability of crosslinked polyelectrolytes by employing the ionic network theory [2]. Details will be described in Chaps. 4 and 5. Thus, gel swells and shrinks incorporating and releasing the fluid.

From a macroscopic viewpoint, the swelling of gel seems to be different from the cases of the paper and cotton when are wet. Paper and cloth can absorb water, but the amount of water absorption is limited since they absorb water mainly by the capillary effect which is weak. There, water can easily be excluded by press or squeeze. In contrast, polymer gel absorbs the fluid substantially when immersed in the fluid and swells until no more water penetrates to the network and attains equilibrium state. The gel holds the water strongly and does not exclude it even under the pressure.

If the amount of fluid in the gel is very low, the gel would behave as solid-like but if it is very high, it would show liquid-like properties. Here, the term of solid-like or liquid-like behavior of the matters should carefully be understood because it is rather difficult to define easily since it is the matter of time and space. In order to define the fluidity or solidity of the matter, Markus Reiner proposed Deborah number [3] in 1964 as the ratio of the observation time and relaxation time: De $= t_c/t_p$,

where t_c refers to the relaxation time that the matter required to recover the original state after deformed by the stresses. t_p refers to the characteristic time of the observation of the process. Since *Deborah number* is defined as their ratio, the material behaves as Newtonian fluid when De is low. When De is high, it behaves as non-Newtonian reaches or solid which are mainly dominated by viscoelastic or elastic properties. Since polymer gel can cover a wide range of *Deborah number* depending on the amount of fluid incorporated, one can say that polymer gel is a comprehensive state of matters characterizing both liquid-like and solid-like properties.

As understandable from the definition of De, it is different from traditional understanding of solid and liquid. Important feature is that it is the matter of the scale of time and space [4]. Gel behaves as solids on the macroscopic level: they have definite shapes even under the load. At the same time, gel behaves as liquid on the microscopic level: a drop of colored liquid soaks through the fluid in the gel. This is the crucial reason why the body of living organism consists of hydrogel. It can transport the physiologically necessary compounds for the organism behaving as a liquid, while keeps the three-dimensional shape behaving as a solid, thus sustain the vital activity. The importance of gel to keep the vital activity will be described in detail in Chap. 8.

The history of gel as a state of matter goes back for many centuries down to even prehistorical era since our daily life is fully surrounded with gels. Animal meat, thick white egg, sweet jelly-cakes, and many other foods are gels of biological origins. At the same time, nowadays a tremendous amounts of gels are mass-produced in industry and play important roles in our daily life. Water absorbing polymer hydrogels are widely used as diaper, sanitary goods, and soil treatment materials. Ion-exchange resins and reverse osmosis membranes are very important to supply sweet water. Gels are widely used for biomedical applications such as soft contact lenses, artificial vitreous, artificial kidney, and electrophoresis scaffold. Their functions and applications will be described in Chap. 9.

1.3 Living Organism Consists of Hydrogel

Human body consists of hard tissue (bone, tooth, horn) and soft tissue. Apart from the question whether hard tissue is the gel state or not [5], soft tissue which is definitely gel which further can be divided into two categories—connective tissue and organ. There, the various kinds of proteins and polysaccharides entangled, penetrated, physically aggregated through molecular interactions and chemically cross-linked each other to build up intricate network structure. Particularly, connective tissue such as ligaments, tendons, cartilage and others build up topological networks with highly hierarchical structure and due to this topological nature, it can exert excellent mechanical toughness to protect the body from mechanical stress outside. At the same time, it has highly viscoelastic nature and sustains the body robust behaving as "shock absorber" under load. Organs such as stomach, gut, vein, nerve, muscle, cornea, are all in gel state filled with large amount of water.

Living cells which consist organs are also typical gels showing viscoelastic property. Although the cell cytosol contains majority of cell fluid, it can be treated as constituted of two phases: a viscous fluid and an elastic solid. It composed of liquid phase containing proteins, DNA, RNA in which suspended cytoskeletal filaments as well as organelles to give viscous cytosol. The liquid component itself has a high compression modulus, meaning it is very fluid (less viscous) while the solid phase has a lower compression modulus, exhibiting more elastic properties. Thus, cells show viscoelastic characteristics: viscosity, elasticity, and creep compliance. The viscosity of the cytoplasm dominates the transport and movement of subcellular macromolecules such as DNA and proteins, while elasticity controls the response of the cytoplasm to mechanical stresses at shorter timescales (seconds), organelles and cytoskeleton elements over long timescales (minutes), and the contractile mechanical regime governs responses at even longer time scales (minutes to hours).

Cytoskeletal filaments which mainly respond to mechanical functions of the cells are composed of three types of self-assembled protein filaments–actin, tubulin and intermediate. They are the mechanical system responsible to determine the shape of the cell's surface, whole-cell locomotion, mechanical strength and the positions of membrane-enclosed organelles and intracellular transport. Viscoelasticity of whole cells is the result of combined contribution of these cellular components of biological polymers network and macromolecules. Recently, various methods have been developed to measure the associated parameters which may lead to understand the relationship between mechanical properties and function of whole cell and deformations of cytoskeletal network. Nevertheless, study of the mechanical behavior of cells to associate with the component biopolymers remains challenging. This would certainly be the future task of gel scientist.

It should also be mentioned that sub-organs which constitute cells such as organelle, endoplasmic reticulum, mitochondrion, Golgi body are all in the gel state. Cells and these sub-organs are surrounded with membranes with liquid crystalline-like lipid layers to separate from outside. There, the membranes simultaneously allow to selectively penetrate water molecules, ions and molecules through channels with various diffusion constants. Peptidoglycan associated lipoprotein is one of the most abundant outer membrane proteins in *E. coli* cells and makes a major contribution to viscoelasticity.

1.4 Gels in Our Life

Polymer gels have numerous applications and widely used in our common life. Figure 1.2 shows some examples.

The most successful and popular example of gels for biomedical application would be the soft contact lenses developed by Czech scientists Wichterle and Lím. They together. Succeeded in preparing a cross-linked and transparent hydrogel using

Fig. 1.2 hydrogels in our life **a**, Jelly-cake, **b**, Water absorbing gel for diaper, **c**, Ion exchange resin, **d**, Artificial kidney for dialysis

poly(2-hydroxyethyl methacrylate) (pHEMA). This gel can absorb water up to 40% which have made more comfortable contact with eyes with suitable mechanical properties. Since that time polymer hydrogels for soft lenses have been manufactured over a half century improving primarily oxygen permeability. In 1998, the first silicone hydrogel contact lenses were released which has extremely high oxygen permeability. Later, various hydrogels have been proposed for soft contact lenses. Attempts to replace vitreous materials are also under progress.

Artificial kidney is another important contribution in the medical science. Artificial kidney can be produced from various synthetic and natural polymers, but the important characteristics is to design the network structure capable of permeating small molecules selectively such as urine, other exhausts, and water to outside and taking necessary minerals from outside while keeping larger molecules such as proteins and cells circulating within the system.

Ion-exchange resins have long history and they are massively produced in industries. Sulfonated polystyrene gel behaves as cation exchanging resin and quaternary amino polystyrene gel behaves as anion exchanging resin. They have long been employed to produce sweet water from sea water. In addition, they are used as supporting matrix for immobilization of enzyme and bio-active substances. Nowadays, reverse osmosis (RO) membranes and fibers are widely used for desalination of sea water. RO membranes are relatively highly crosslinked polymer gels made from aromatic polyamide and other crosslinked polymers. They should be slightly swollen in sea water to selectively permeate water molecules under high pressure.

Typical superabsorbent materials are made of sodium salt of polyacrylic acid gel. The high rate of swelling and high swelling ability to absorb hundred times of water in their weight has made them useful as water-retaining media in diapers and water supplier for soil. They are also used as soil conditioning media (water-retaining and soil-aggregating agents) through their ion-exchange abilities as well as various nutrients releasing ability to plants.

Various polymer gels are important in chemical and biological researches and widely used as separation techniques. The most widely used physical method in the biological science would be gel electrophoresis using polyacrylamide gels. The basic idea of electrophoresis is to separate molecules based on their intrinsic electrical charge and molecular size through polymer network. Here, the water-swollen polymer network plays as a sieve filtering selectively molecules with different molecular size through interstitial spaces of the swollen network. DNA is often separated into different sizes using agarose gel which is the polysaccharide extracted from seaweed. (refer to Chap. 3). The pore size of the agarose gel is suitable for separating DNA consisting of several hundred nucleotides (around 200–50 k base pair). DNA of more smaller size fragment (5–1000 base pair) and proteins are separated by synthetic polymer gel of polyacrylamide since the pore size is smaller.

Various kinds of chromatographic techniques are used to selectively separate organic and inorganic solutes by using amphiphilic polymer gels where hydrophobic and hydrophilic proportions are carefully selected. Novel uses for crosslinked gels can be found in regenerative medicine and pharmaceutical science. Various type of synthetic and biological gels have been developed as scaffolds to cultivate and grow cells eventually to make 3 D organs. Development of drug delivery system (DDS) which can release drug and other bio-active compounds at required time to required place is very important subject of gel science. Various novel hydrogels for DDS have been synthesized by controlling and designing nano-meter ordered texture of the network. In this way, hydrogel is becoming more and more important for our daily life. Details will be described in Chap. 9.

1.5 History and Definition of Gel

Although the history of gel goes back to even prehistorical era and it was mainly familiar sources of food, scientific approach of gel is more recent dating in the nineteen century from the work of Thomas Graham (1805–1869) (Fig. 1.3) and it has unbelievably difficult history. Thomas Graham was a Scottish chemist and known as the founder of colloid chemistry [6]. He was the person created the word "*gel*" in 1861 making the word gelatin shorten [7]. He said *"gel" is the solid, jelly-like material that can have properties ranging from soft and weak to hard and tough."* Gelatin was really a good example of the gel which was extracted from collagen. Other examples in those days are agar derived from seaweed and thick white egg. It was easy to imagine from their characters that they include the non-flowing soft materials of biological origin, but the scientific definition of gel was very difficult due to its unique and obscure behaviors.

The first appearance of the term "gel" in the literature was in 1894 in the journal *Z. Anorg. Chem.* as "Der Hydrogel und das Kristallinische Hydrat des Kupferoxydes," [8]. The terms "gel" and "sol" were used there to distinguish between two types of colloids. It is written that *"a sol has the properties of a liquid; it pours, whereas a gel has some of the properties of a solid; it does not pour readily."* Note that a gel has "some of the properties of a solid." This indicates how difficult to scientifically define the term "gel" in those days. Indeed, Lloyd pointed out correctly in 1926 that the gel state is easier to recognize than to define [9].

When you refer to encyclopedia *BRITANNICA* 14th edition (1929 through 1940), you can find an independent column of *"Gel,"* but it describes only two lines without

Fig. 1.3 Scottish chemist, Thomas Graham (1805–1869)

definition as *"the name given by Graham to the coagulated precipitate from a sol or colloidal solution (see Colloid)."* When you open the column of *"Colloid"* (which occupies one and half pages!) of *BRITANNICA* 14th edition, you can find the term "gel" as sub-column describing as *"Reference has already been made to the products of-reversible or irreversible-transformation of sols, to which Graham gave the name of gels."* But it surprisingly continues full description with almost correct today's understanding as follows;

> The gels of the lyophilic sols, to which the term is now more or less confined, are of particular interest and importance; one of the typical forms is the gel of gelatin, familiar to everyone in the shape of table jellies. These contain 5–6% of the dry substance and thus exhibit at once the most striking characteristic of gels: that they retain their shape and behave up to a point like solids, although consisting very largely of liquid.

Further it describes the melting behavior of gelatin and introduces the presence of swelling pressure of gelatin together with its method of measurement. It continues that "*This swelling property is sensitive to the electrolyte content similar to those of tissues of peas and living organism.*" "*Gel combines in a curious way the properties of the solid and liquid state. The volume remains almost exactly constant during deformation; if the deformation is maintained beyond a short time, the gel yields and the force required to maintain a given deformation becomes less and less with time.*"

The encyclopedia even describes that *"when the stress is applied for a considerable time, and especially if the gel is dries while stressed, the double refraction remains permanent; it is a curious and significant fact that many structural elements of organisms, which consist of gels or mixtures of such, always exhibit double refraction. Gelatinizing properties of agar is more marked than those of gelatin and because of that the agar has been much used for adulterating jams and jellies."*

BRITANNICA introduces even lyophilic gels describing that *"cellulose acetate and benzyl alcohol behave exactly like gelatin and water: forming a sol at 40 °C which on cooling sets to a jelly resembling gelatin gel in appearance and elastic properties. Both natural and vulcanized rubber swell in hydrocarbons and the volumes of liquid taken up and the pressure generated are much greater than with gelatin."*

Description on silica gel was also made as *"formed from silicic acid was called rigid gel in contradiction to the type represented by the gelatin and rubber, which are known as elastic gels"*.

It should be noted that in this 14th edition (1929) BRITANNICA has a sub-column of "Diffusion in Gels" in the column of *"Colloid"* describing that *"Graham, who carried out the first quantitative investigations on the rate of diffusion of number of substances, found that it was not much lower in dilute gels than in the pure solvent, but the rate of diffusion is markedly reduced in the more concentrated gels. Many measurements have been carried out in such, as the difficulties arising from vibration and convection are eliminated."*

The BRITANNICA of 1929 edition has a sub-column of *"The Sol–Gel Transformation"* in the column of *"Colloid"* and describes as *"The formation of a gel, with some of the properties of a solid, from a sol which may contain a little as 0.5%of dry matter (agar) is an extremely striking change, which has very naturally been the subject of much speculation. Most gels are not differentiated in the ultra-microscope, so that theory must rely on inferences from their known properties. The most generally received opinion is that the particles somehow join up to form chains and a network, or "ramifying aggregates," which some investigators (McBain) assume to be, to some extent, present already in the sol and account for its viscosity of the particles. It is also probable that such linking-up is accompanied by some change in the hydration. In any case, there must almost certainly be continuous liquid paths through a gel, to account for the case with which diffusion proceeds through it."*

In this way, although the unique behaviors of the gel were well understood in those days, the term *"gel"* apparently seemed to be one of characteristic behaviors of *"colloid"* and no independent column was used till the BRITANNICA 1961 edition. The historically available characterization referred to a gel would be the description by Alexander in 1949 [10]. Alexander and Johnson exampled in the quoted reference some cellulose derivatives which transfer from sol state to gel state on warming and described as "a two components system of a semi-solid nature, rich in liquid". There they pointed that the gels in these cases are formed from solutions or suspensions by forming three-dimensional structure forming cross-linkages between the solute molecules.

At the same time, Hermans assigned the term *gel* to the systems characterizing the following aspects in 1949 [11]:

(a) They are coherent colloidal disperse systems of at least two components;
(b) They exhibit mechanical properties characteristic of the solid state;
(c) Both the dispersed component and the dispersed medium extent themselves continuously throughout the whole system.

The first independent column of *"gel"* appeared only in 1961 in BRITANNICA 15th edition. However, used somehow only 1/4 page which is much less than the sub-column in the *"colloid"* in 1929 edition. It describes *"A gel is an elastic coherent mass consisting of a liquid in which ultramicroscopic particles are either dispersed or arranged in a fine network extending throughout the mass. The particles may be, for example, large molecules, such as proteins; or small crystals, such as bentonite;*

or polymer particles, such as polystyrene. Extensive swelling (that is, the increase in gel volume) results in a gradual transformation of the gel into a colloidal solution. The rheological properties of a gel vary between those of viscus or elastic liquid and those of solid."

After this description, explanation of the terms of melt, xerogel, aerogel, rubber, lyo-gel, and many examples of synthetic and natural polymer gels are followed.

Nowadays, one can find the term *"gel"* in the column of BRITANNICA (The New Encyclopedia (15th, Edition, 1990, Volume 5, p165), as *"coherent mass consisting of a liquid in which particles too small to be seen in an ordinary optical microscope are either dispersed or arranged in a fine network throughout the mass. A gel may be notably elastic and jellylike (as gelatin or fruit jelly), or quite solid and rigid (as silica gel). Gels are colloids (aggregates of fine particles, dispersed in a continuous medium) in which the liquid medium has become viscous enough to behave more or less as a solid."*

The term "gel" can also be found in the column of *"gelatin"* in the same encyclopedia (The New Encyclopedia (15th, Edition, 1990, Volume 5, p165) *explaining as "gelatin solution solidifies to form a gel as it cools. The gel state is reversible to a sol state at higher temperatures and the sol can be changed back to a gel by cooling."*

Here, it should be noted that the term *gel* created by Graham and later discussed for many time are concerned with the gels consisting of both macromolecules and inorganic compounds. Scientists tried to understand the gels from their unique behaviors and not from their origins of the constituents. Because of this history there appeared various and independent definitions about gels and each of them has different definition meaning different state of materials which leads to confusion up to now. For example, the term "gel" are widely used for low molecular and inorganic compounds. When a concentrated sodium silicate solution is neutralized and kept for a while, a coherent transparent homogeneous mass called gel is formed. Upon drying it, a hard transparent glassy-like solid is obtained which is still called gel (aerogel).

Sol–gel reaction is typical and famous method to form inorganic three-dimensional gel through chemical reactions. As early as 1820, formation of silica-alumina gel was reported as a result of the reaction of hydrofluorosilicic acid and aluminum hydroxide [12].

Inorganic gels were also prepared by mixing two concentrated solutions of barium sulfate and manganese sulfate. Similar inorganic gels are also formed from aqueous suspensions of aluminum hydroxide, ferric hydroxide, and bentonite by adding various metallic salts. Nowadays, a various kind of silicone, titanium, and other inorganic metal containing gels are prepared by the corresponding organometallic compounds through hydrolysis. Sol–gel transition can also be observed by cooling organic solutions of organo-metal compounds such as aluminum stearate or oleate.

The term *gel formation* or *gelation* is traditionally used in polymer science when the solute simply comes out from the solution as a precipitate and the process is called as *gelation* and the precipitate is called *gel*. Therefore, an addition of incompatible solvents and precipitating agent is the simple way to get the gel. For example, when poor solvent such as ethyl acetate is added to the polystyrene solutions of toluene, polystyrene gel is produced as a precipitate. When chain polymerization such as radical polymerization proceeds in the concentrated viscous solution, the elementally processes of the polymerization are substantially influenced and the overall polymerization scheme is changed from that of the homogeneous reaction, for example acceleration of polymerization and the effect is called "gel-effect."

Different definition meaning different state of materials seems to have happened because the term "gel" is used just for convenience and easiness by different communities to describe different matters and states and systems. There seems to be no clear-cut bordering lines between them. Thus, in addition to difficulty of the definition, the situation became more complex as referred to Sect. 1.5 (classification). On these confusing background, IUPAC definition of gel has been introduced and proposed as IUPAC recommendations in 2007 [13]. This was made by IUPAC Inorganic chemistry Division and Polymer Division, Commission on Macromolecular Nomenclature. For instance, gel is defined as non-fluid colloidal network or polymer network that is expanded throughout its whole volume by a fluid. Refer to the whole description of IUPAC definition of gel in the appendix.

1.6 Classification of Polymer Gel

As early as 1950s P.J. Flory proposed the following classification of polymer gels in his article [14].

I *Well-ordered lamellar structures, including gel mesophases.*
II *Covalent polymeric networks; completely disordered.*
III *Polymer networks formed through physical aggregation, predominantly disordered, but with regions of order.*
IV *Particulate disordered structures.*

The synthetic polymer gels are mostly categorized by II classification. These gels consist of chemically cross-linked network with infinite molecular weight and consequently they swell when immersed in a good solvent. In this book definitions II, III are mainly covered.

Nowadays, there are a variety of approaches to classify the polymer gels and some of them are more popular and easier to understand. Followings are typical and widely accepted classification of polymer gels Table 1.1.

1.6.1 By the Fluid Filling the Space

One simple way is to classify by the fluids that fill network polymers. In principle, any fluid can fill the interstitial spaces of the network to form the gel if it has a good affinity with the network. When it is filled with water, it is called hydrogel and when it is filled with organic fluid, it is called organic gel (or organogel). We can find many examples of hydrogels in foods like jelly, gelatin and various water absorbing

Table 1.1 Classification of polymer gels

By fluid		
	Water organic fluid air	Hydrogel organogel, organic gel xerogel、aerogel
By crosslinking	Covalent bonding	Chemical gel (irreversible gel)
	Interaction	Physical gel (reversible gel)
By network	Polyelectrolyte	Polyelectrolyte gel, ionic gel
	Biopolymer	Biogel, biological gel, natural gel

materials. The organic liquid such as mineral oil, silicon oil, or vegetable oil can be component of organic gel and they are found as cosmetics and drugs.

Xerogel or aerogel is the special and traditional name in which no solvent is present in the pores. Silica gel is a typical example of xerogel and is usually formed by drying the silicate containing solution under keeping the size constant. Therefore, silica gel is highly porous (15–50%) with small pore size (1–10 nm) and numerous surface area (150–900 m^2/g). Although they have the term "gel" we do not treat in this book unless otherwise noted.

1.6.2 By the Crosslinking Characteristics

Another popular way is to classify the gel by the characteristics of crosslinking. In this category polymer gels can roughly be classified into two classes; chemically cross-linked gels and physically cross-linked gels.

"Chemical gel" is made of chemical crosslinking of covalent bonding. The typical network of synthetic polymer is obtained by the radical polymerization of vinyl monomers in the presence of di-vinyl monomers which act as a crosslinking agent. Crosslinking is also available by radiation of polymers already prepared. Other polymerization method is poly-condensation of multifunctional monomers or poly-addition reaction with precursors. Details will be described in Chap. 3.

"Physical gel" is formed by cooperative molecular interactions between polymer chains. Physical gel is formed by hydrogen bonding, electrostatic ionic bonding, hydrophobic interaction, van der Waals and dipole–dipole interactions and others. Physical gel is also formed by the extensive entanglement of polymer chains or by inter-penetrating polymer chains. In this case increases of chain length and the concentration enhance the gel formation.

Generally, chemical gels are not solubilized due to crosslinked by covalent bonding and called irreversible gel. On the other hand, physical gels are solubilized by dissociation by changes in concentration, temperature, composition of solvent, pH, and others and they are called as "reversible gel."

1.6.3 By the Network Structure

When polymer network consists of polyelectrolyte network (cross-linked polymers containing with ions) and low molecular counter ions in the fluid, it is called polyelectrolyte gel or ionic gel. Polyelectrolyte gel can adsorb a significant amount of water within its network, they are produced in industry as diaper and other water absorbing materials. Polymer network without ions on the network is sometimes called neutral gel.

Gels prepared from (biology-originated) natural polymers are called Biological Gel or Natural Gel and gels prepared from polysaccharides, proteins and DNA are included into this category. Gelatin is protein gel prepared from collagen. Agar and carrageenan are made from seaweed polysaccharides. Sephadex is the gel used for chromatography and produced in industry from cellulose. Biological gel can be prepared both by physical interaction and chemical crosslinking. Some of biological gels have ions in the network and play important biological functions in the organism.

Biological gels often form non-covalent crosslinking area called "junction zones." They are consisting of bundles of macromolecules to give ordered structure such as helical structure. Gelatin gel is the typical example with this structure. Some polysaccharides form ion mediated junction zone called *"egg box."* These biological gels undergo thermo-or chemical-reversible sol–gel transition. Detailed mechanism and structure is described in Chap. 3.

Some globular proteins such as serum albumin and β-lactoglobulin form insoluble and irreversible gels by heating, changes of pH and the presence of ions or chemical substances. The gel formation is due to the disordering or unfolding of natural helical structure and this process is called denaturation categorized in Flory's definition IV.

Chemical crosslinking of bio-and natural polymers are often performed to obtain stable biological gels for specific biomedical applications. A variety of crosslinking agent has recently been developed to obtain chemical bio-gels without losing the original ordered structure and some examples are introduced in Chap. 3.

1.7 Why Gel Science is Important?

Science of gel has long been suffered from two serious difficulties for investigation. One is insoluble property of the gel and the other is its inhomogeneity, both are inherent properties of the gel.

When scientists investigate the structures and properties of the matters, they usually make a solution and try to isolate every molecule independently to avoid any interactions between them. However, the gel swells but is not dissolved due to the crosslinked network structure. This brings about difficulties extremely to investigate the matters on molecular level.

The other and even more serious subject is the fact that the gel is essentially an inhomogeneous system consisting of two-phase; network and fluid or ordered region and disordered region. As will be described in Chap. 7, the distance between crosslinking junctions of the network is widely distributed (crosslink density distribution) and the structure is quite heterogeneous. This random-distributed heterogeneity of the network size gives inhomogeneous interstitial spaces and gives rise to highly fluctuated dynamic motion of the network. Due to this, the amount of the surrounding fluid near the network is widely distributed to give very complicated behaviors and properties bringing about extremely broad and multiple-overlapped spectra of characteristics of the fluid. Furthermore, many biological gels consist of two different regions; crystalline area and disordered amorphous area. This inhomogeneity of the structure results in entirely different dynamics of the network and the fluid and substantially changes their characteristics and behaviors. This heterogeneity of the network also gives extremely wide and complicated dynamic spectra to give non-linear viscoelastic, mechanical, optical, electrical responses as whole. These behaviors cannot be understood by the classical solid and fluid theories and left for further studies of gel scientists. Gel science is important not for modern materials science, but for future life science to drive the comprehensive understanding of mechanism of living organism by bridging them.

Appendix

Pure Appl. Chem., Vol. 79, No. 10, pp. 1801–1829, 2007.

doi: 10. 1351/pac200779101801

INTERNATIONAL UNION FOR PURE AND APPLIED CHEMISTRY

INORGANIC CHEMISTRY DIVISION

and

POLYMER DIVISION

COMMISSION ON MACROMOLECULAR NOMENCLATURE*

SUBCOMMITTEE ON MACROMOLECULAR TERMINOLOGY and

SUBCOMMITTEE ON POLYMER TERMINOLOGY

(IUPAC Recommendations 2007)

GELS

Gel

Non-fluid *colloidal network* or *polymer network* that is expanded throughout its whole volume by a fluid.

Note 1: A gel has a finite, usually rather small, yield stress.

Note 2: A gel can contain

a. a covalent *polymer network*, e.g., a network formed by crosslinking polymer chains or by non-linear polymerization;
b. a *polymer network* formed through the physical *aggregation* of polymer chains, caused by hydrogen bonds, crystallization, helix formation, complexation, etc., that results in regions of local order acting as the network *junction points*. The resulting swollen network may be termed a *thermos-reversible gel* if the regions of local order are thermally reversible;
c. a polymer network formed through glassy *junction points*, e.g., one based on block copolymers. If the *junction points* are thermally reversible glassy domains, the resulting swollen network may also be termed a *thermos-reversible gel*;
d. lamellar structures including mesophases, e.g., soap gels, phospholipids, and clays;
e. particulate disordered structures, e.g., a flocculent precipitate usually consisting of particles with large geometrical anisotropy, such as in V2O5 gels and globular or fibrillar protein gels. Corrected from [2], where the definition is via the property identified in Note 1 (above) rather than of the structural characteristics that describe a gel.

Aerogel

Gel comprised of a microporous solid in which the dispersed phase is a gas.

Note: Microporous silica, microporous glass, and zeolites are common examples of aerogels.

Corrected from [2], where the definition is a repetition of the incorrect definition of a gel followed by an inexplicit reference to the porosity of the structure.

R. G. JONES

c 2007 IUPAC, *Pure and Applied Chemistry* 79, 1801–1829

1806

Alcogel

Gel in which the swelling agent consists predominantly of an alcohol or a mixture of alcohols.

Aquagel

Hydrogel in which the *network* component is a *colloidal network.*

Colloidal Gel

Gel in which the *network* component comprises particles of colloidal dimensions.

See also *colloidal network.*

Gel Microparticle

See *microgel.*

Gel Nanoparticle

See *nanogel.*

Humming Gel

See *ringing gel.*

Hydrogel

Gel in which the swelling agent is water.

Note 1: The *network* component of a hydrogel is usually a *polymer network.*

Note 2: A hydrogel in which the network component is a *colloidal network* may be referred to as an *aquagel.*

Microgel

Gel microparticle

Particle of *gel* of any shape with an equivalent diameter of approximately 0.1 to 100 m.

Modified from [2]. The definition proposed here is recommended for its precision and because it distinguishes between a microgel and a *nanogel*.

Nanogel

Gel nanoparticle

Particle of *gel* of any shape with an equivalent diameter of approximately 1 to 100°*nm*.

Neutralized Gel

Gel containing acidic or basic groups that have been neutralized.

Particulate Gel

Gel in which the *network* component comprises solid particles.

c 2007 IUPAC, *Pure and Applied Chemistry* 79, 1801–1829

Inorganic and polymeric gels, networks, and materials 1807

Polyelectrolyte Gel

Polymer gel in which the *polymer network* contains ionic or ionizable groups in a significant fraction of its constitutional units.

Polymer Gel

Gel in which the *network* component is a polymer.

Responsive Gel

Gel that responds to external electrical, mechanical, thermal, light-induced, or chemical stimulation.

Note: The use of the term *intelligent gel* is discouraged.

Rheopexic Gel

Gel for which the time of solidification after discontinuation of a relatively high shear rate, is reduced by applying a small shear rate.

Rheotropic Gel

See *rheopexic gel.*

Ringing Gel

Humming gel

Gel with energy dissipation in the acoustic frequency range.

Note: A ringing gel is often a *hydrogel* with a surfactant as a third component and has a composition within an isotropic, one phase region of its ternary phase diagram.

Sonogel

Colloidal gel produced by the action of ultrasonically induced cavitation.

Thermoreversible Gel

Swollen *network* in which the *junction points* are thermally reversible.

See also *gel.*

Thixotropic Gel

Gel which has a reduced viscosity on the application of a finite shear but which recovers its original viscosity when the shear is discontinued.

Xerogel

Open *network* formed by the removal of all *swelling agents* from a *gel.*

Note: Examples of xerogels include silica gel and dried out, compact macromolecular structures such as gelatin or rubber.

Modified from [2]. The definition proposed here is recommended as being more explicit.

See also *swelling*.

R. G. JONES

c 2007 IUPAC, *Pure and Applied Chemistry* 79, 1801–1829

1808

Drying Control Chemical Additive (DCCA)

Cosolvent included to facilitate the rapid drying of *gels* without cracking.

Gel Point

Gelation point

Point of incipient *network* formation in a process forming a chemical or physical *polymer network*.

Note 1: In both network-forming polymerization and the *crosslinking* of polymer chains, the gel point is expressed as an extent of chemical reaction (cf. gel time).

Note 2: At the gel point, a solid (network) material spanning the entire system is formed. See also *gel fraction.*

Note 3: The gel point is often detected using rheological methods. Different methods can give different gel points because viscosity is tending to infinity at the gel point and a unique value cannot be measured directly.

Gel Temperature

See *gelation temperature*.

Gel Time

Gelation time

Time interval from the start of a *network*-forming process to the *gel point*.

Gelation Point

See *gel point*.

Gelation Temperature

Gel temperature

Temperature threshold for the formation of a *thermos-reversible gel*.

Note 1: A thermos-reversible gel is usually formed by cooling a polymer solution. In these cases, the gel temperature is a maximum temperature at which the presence of *network* is observed.

Note 2: Since gel temperatures depend on the method of determination, this should always be indicated.

Gelation Time

See *gel time*.

Swelling Agent

Fluid used to swell a *gel*, *network*, or solid.

See also *swelling*.

c 2007 IUPAC, *Pure and Applied Chemistry* 79, 1801–1829

References

1. Followings are suitable textbooks and reviews. P.J. Flory, *Principles of Polymer Chemistry, Chapter 9* (Cornell University Press, Ithaca, New York, 1953). O. Wichterle, Encyclopedia of polymer science and technology. Interscience **15**, 273 (1977). *Gels Handbook*, vol. 1–4 (Academic Press, 2000). D. de Rossi, D. Kajiwara, K. Osada, A. Yamauchi, *Polymer Gels-Fundamentals and Biomedical Applications* (Plenum Press, New York, 1991)
2. P.J. Flory, *Principles of Polymer Chemistry, Chapter 13* (Cornell University Press, Ithaca, New York, 1953)
3. M. Reiner, The Deborah number. Phys. Today **17**(1), 62 (1964)
4. P.G. De Gennes, *Scaling Concepts in Polymer Physics* (Cornell University Press, New York, 1979)
5. Y.O. Author, Considers the hard tissue is also categorized in the gel state. Bone consists of mainly hydroxyapatite and collagens and the total water content of the bone is around 11% and strengthen the mechanical properties
6. Graham firstly made the description on *colloid chemistry* in the journal *"Philosophical Transactions"* (1861) entitled "Liquid Diffusion applied to Analysis". Therefore, the term *"COLLOID"* cannot yet be found in the encyclopedia **BRITANNICA 8thedition** (1854) and **9thedition** (1877)
7. Gelatin (from Latin gelatius meaning "stiff", "frozen") is the partially hydrolyzed product of collagen extracted from animal skin, bone, and connective tissues. Gelatin is soluble in hot water, but forms the gel on cooling when the concentration is high enough. Gelatin is widely used as a glue and various gelling agents in food, photography, pharmaceutical and cosmetic industries. Refer to Chapter 2 in detail. The encyclopedia **BRITANNICA** 8th edition (1853) spells "GELATINE" and describes through half page. 9th edition (1875) somehow spells "GELATIN" using more than three pages. 14th edition (1929) used one and half page and 15th edition (1961) used only a quarter page,
8. J.V. Van Bemmelen, Z. Anorg. Chem. **5**, S.466 (1894)
9. D.J. Lloyd, The problem of gel structure, in *Colloid Chemistry*, vol.1, ed. by J. Alexander (Chemical Catalogue Co., New York 1926), pp. 766–782
10. A.E. Alexander, P. Johnson, *Colloid Science*, vol. 2.1 (Clarendon Press, Oxford, 1949)
11. P.H. Hermans, *Colloid Science*, vol. 2, ed by H.R. Kruyt (Amsterdam, Elsevier 1949)
12. refer to E.A. Hauser, *Silica Science*, (Van Nostrand, Princeton, New Jersey, 2.1, 2.2 1952)
13. Pure Appl. Chem. **79**(10), 1801–1829 (2007)
14. P.J. Flory, Introductory lecture. Disc. Faraday Soc. **57**, 7 (1974)

Chapter 2
Basic Properties of Gel

Yoshihito Osada

Abstract Gel shows a variety of unique properties and behaviors which are not observed both in solid and liquid. Rubber elasticity, viscoelastic response, abrupt collapse of the volume, presence of non-freezing water are the examples. These properties and behaviors are arisen by the strong interactions between polymer network and the fluid. The process and the mechanism of interaction in the gel are introduced.

Keywords Gel point · Viscoelastic property · Storage modulus · Loss modulus · Volume phase transition · Non-freezing water · Bound water · Diffusion constant

2.1 Introduction

2.2 Gel Point

The time or temperature at which the sol loses liquidity and the gel is formed is called **gel point (GP)** [1]. In the contrary, the time or temperature at which the gel starts flowing is called **melting point (MP)** [2]. Some methods have been developed to experimentally determine the GP and MP such as tube "tilting" method (TTM), inverted test tube method(ITT) [3], and ball drop method (BDM) [4]. Although they are qualitative and phenomenological natures, they have widely been used to determine the gel point.

TTM would be the simplest method to determine GP. In this method, the tube containing the sample solution is cooled slowly and the temperature at which the sample loses the fluidity is the gelation temperature. If the sample is quenched in advance at low temperature and then gradually warms it up, the temperature at which the sample gel starts to flow is T_m [5].

In case of BDM, a steel ball is placed on the top of the gel and the gel is slowly heated. The temperature at which the gel melts and the ball descends is T_m.

Y. Osada (✉)
RIKEN, Wako, Saitama, Japan
e-mail: osadayoshi@riken.jp

Y. Osada (ed.), *Science of Gel*, https://doi.org/10.1007/978-981-95-0301-8_2

One can easily notice that GP determined by these methods is qualitative. GP is a transition state between sol and gel, and the boundary is characterized by the molecular motions which should be correlated with space and time of the system. In fact, MP should not necessarily the same as GP and hysteresis is often observed. These parameters strongly depend on the reaction conditions of gel formation such as concentration, temperature, and reaction procedures of the sample. Therefore, discussions to determine where the exact borderline between sols and gels lies have been repeated. With respect to the concentration, for example, the higher the concentration, the easier the gelation and gave higher GP than that of lower concentration. Therefore, one of the methods to determine GP more accurately is to prepare the gel under minimized concentration of solute molecules necessary to form gel. In fact, Ferry [3] found linear logarithmic relation between c and $1/T_m$ as log c ~ $1/T_m$ where c is the gelatin concentration and T_m is melting temperature. In the similar manner, a linear relation was obtained between log M and $1/T_m$, where M is the molecular weight of gel forming compound. Here, the extrapolated values of c and M are characterized as the minimum concentration and minimum molecular weight, respectively, that sustains the infinite network. In the case of chemical gels which usually obtained by the polymerization of multifunctional monomers or the crosslinking reaction between linear polymers, GP is determined by the time when the insoluble gel starts forming after the reaction started. Again, the gel point is not independent but mutually related in terms of solution concentration, reaction conditions, and temperature. Although these methods lack generality as described, they are often used to qualitatively know the condition of gelation.

The **viscoelastic measurement** is now standard method to scientifically determine the GP and MP. When a small deformation oscillatory shear strain of frequency ω is applied, the resulting stress wave of an elastic solid material is in phase with the strain wave (i.e., phase angle, $\delta = 0°$) and the stress is proportional to strain, while in the case of a Newtonian liquid the stress and strain waves are exactly 90°out of phase (refer to Chap. 6 in detail). Since gel has both solid-like and liquid-like characters with long relaxation time, δ lies between 0°and 90°indicating a viscoelastic material [6]. When a sinusoidal strain is applied, modulus which is defined as the ratio of stress to strain is characterized by two factors: the stress in phase called storage modulus (*G'*) and the stress 90°out of phase called loss modulus (*G"*). *G'* and *G"* correspond to the real and imaginary parts of the complex share modulus. *G'* is a measure of the energy stored in the material and recovered from it indicating the elastic character of the material, and *G"* represents a measure of the energy dissipated indicating liquid character. Therefore, gel is considered as the **soft and wet matter** exhibiting the viscoelasticity that consist of two components: liquid and solid.

As described before, gelation is a kinetic process and therefore, *G'* and *G"* are functions of time, i.e., the rheological character must be made by changing the frequencies. The typical ω-dependence of *G'* and *G"* of physical gel has the following characteristics: In the range of low frequencies, *G"* is larger than *G'* indicating the material is fluid (Fig. 2.1). Both of G' and C" increase with increase of frequency and becomes viscous fluid. However, since *G'* increases as a power of two ($\alpha\omega^2$, slope = 2 in Fig. 2.1a) and *G"* as a power of one ($\alpha\omega^1$ slope = 1) with increase in frequency, there appears a cross-over point at which *G'* becomes larger than *G"* and this is the GP. With increasing frequencies, elastic response starts to dominate and the material behaves like more solid-like. At higher ω, *G'* and *G"* are parallel keeping *G'* > *G"* and frequency-insensitive, i.e., both *G'* and *G"* are $\alpha\omega^0$, giving an equilibrium modulus (slope ~0) referred to as the plateau region (Fig. 2.1b). When the frequency decreased, *G"* becomes larger than *G'* at low frequencies through "cross-over" point and the material is no more solid-like and flow again as viscous liquid.

Thus, one of the characteristic features of gel from their mechanical viewpoint is its modulus and gels can be defined as "gels are viscoelastic solids that possess the plateau of the real *G'* (larger modulus) of complex modulus G* in a wide frequency range" [7].

According to this definition, it is not necessary for gel state that the relaxation time is to be infinite if *G'* is flat on the order of seconds and much larger than *G"*. In this case, the time scale of observation becomes important from the dynamic point of view. In the case of physical gel in which the network is formed by the entanglement of penetrating polymer chains, the system appears as gel when the frequency applied for measurement exceeds the lifetime of the entanglement. That is why the physically

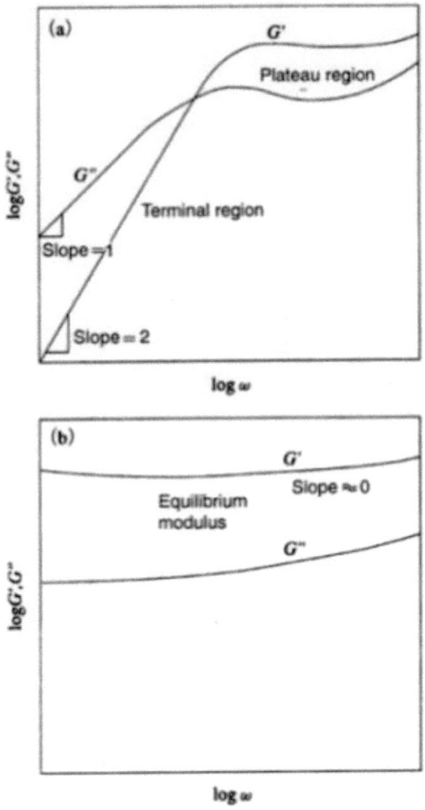

Fig. 2.1 Dynamic mechanical spectrum of (**a**) entanglement networks (physical gel) and (**b**) covalently crosslinked networks (chemical gel) (from ref. [6])

Fig. 2.2 Effect of molecular weight of linear polymer on *G'* as a function of frequency, ω. (from ref. [6])

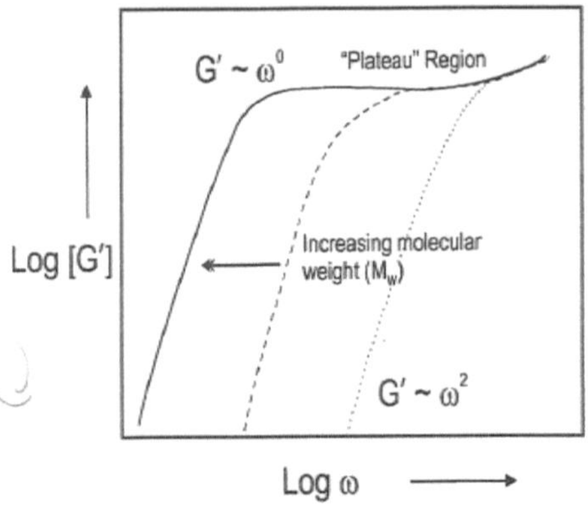

entangled polymer network is also a gel state (refer to Chap. 3). Entangled networks show the relationships as $G' \propto \omega^2$; and $G'' \propto \omega^1$ which indicates that, in the low-frequency region, it behaves as high-viscosity liquid. Since the entanglement occurs when the product of molecular weight and the concentration exceeds a certain value, the plateau region of the physical gel will move to lower frequencies as the molecular weight increases and the gelation time becomes less as the concentration of polymer is increased (Fig. 2.2).

According to the reputation theory [8], the longest relaxation time of entangled polymers τ_1 is proportional to the third power of the molecular weight M

$$\tau_1 \propto M^3 \tag{2.1}$$

Therefore, having the longest relaxation time is equivalent to the molecular weight being infinite. Therefore, it does not contradict the Flory–Stockmayer definition (refer to the reference [4] in Chap. 3).

The entangled polymer gels dissolve by dilution, whereas chemical gels do not dissolve but only swell. In the case of chemical gels, the gel point corresponds to the point where the longest relaxation time becomes infinite and they do not show such effect.

As shown in Fig. 2.2, *G'* and *G"* of chemical gels show little frequency dependence for a small dynamic shear deformation; $G\prime \propto \omega^0$; $G'' \propto \omega^0$.

Therefore, chemical gels and physical gels can easily be distinguished by measuring their mechanical properties. When the storage modulus *G'* and loss modulus *G"* are measured as a function of dynamic shear deformation of the frequency ω, the spectra of the chemical and physical gels show similar mechanical behavior in the high-frequency region while they should show significant difference at low ω as a result of the difference of the crosslinking nature.

2.3 Rubber Elasticity

As described earlier, rubber elasticity is one of the most important properties characterizing polymer gels. Rubber elasticity has attracted strong attention since the early era of research of polymer science and has been developed through statistical mechanics. The basis of rubber elasticity is the micro-Brownian motion of the polymer chains originated from entropy and it is essentially different from that of the rigid sold which is based, in principle, on energetic elasticity. The *ideal* relationship between stress and strain of rubber elasticity is given by

$$\sigma = E\left(\lambda - \frac{1}{\lambda^2}\right) \tag{2.2}$$

where λ is the elongation ratio along the stretching direction, and E is Young's modulus. Here, the word *ideal* means that the chains obey Gaussian statistics where there is no chain-to-chain interaction (they pass through as though they are ghosts), the network have a complete homogeneous structure, there should not be any crosslink density distribution and finally, the network growth is under the assumption of the dendritic structure ([9], refer to 3.1).

When a single polymer chain with molecular weight M and density ρ is stretched, the following relationship is obtained by calculating the entropy change of deformation:

$$E = \frac{3\rho \mathrm{RT}}{M} \tag{2.3}$$

In this calculation, both ends of the polymer chain are fixed and there is no micro-Brownian motion of the molecule as whole. The physical cause of fixation of both ends is the crosslinking. When there are free spaces to allow micro-Brownian motion between the crosslinks, the rubber elasticity based on the entropy appears.

Supposing networks with regular structure deform, the front factor of Eq. (2.3) differs by two orders of magnitude depending following two models. Affine model assumes that the crosslink point moves proportionately to the deformation of the whole body [10], while they are spatially fixed, or Ghost model assumes that the networks are fixed to the frame while the internal crosslink points can move freely [11] (details refer to Chap. 6).

In any case, if Eq. (2.3) can be rewritten as

$$E = \frac{3g\mathrm{RTNe}}{V\mathrm{mol}} \tag{2.4}$$

and Eq. (2.2) can be used. Here, g is the so-called rubber front factor, V_{mol} is the molar volume of the chain (it is approximately l/ρ in bulk), and Ne denotes the chains in a unit volume that contribute effectively to elasticity and can be calculated from the degree of crosslinking where those chains in the networks that do not contribute to elasticity should be excluded.

In the Affine model, the crosslink points are fixed in the space and they do not contribute to the entropy of the system. Therefore, the chains that constitute the networks move independently of each other. In other words, the chemical structure of the networks, how many chains come out from the crosslink point has no relation to the elasticity, but the numbers of free chains that can move independently in the network contribute to the elasticity. In this case $g = 1$. In the Ghost model, the crosslink points can move along with the deformation and this part of free motion of the crosslink points can contribute to the entropy of the system. However, the motion of the chains constituting the networks is connected by the crosslink points and it is not independent of each other. Therefore, in this model, the elasticity will substantially depend on the network structure. In this case, $g = 1–2/f$, where f is number of chains that come out from the crosslink point [12].

Equations (2.3) and (2.4) indicate that the modulus E is directly proportional to the absolute temperature. This is because the modulus of a gel is assumed to depend only on the entropy rather than enthalpy. However, the modulus often reduces as the temperature increases particularly, in the case of physical gels since they are reversible gel and change the gel structure due to the deformation arose by the enthalpic contribution. In practice, the contribution of enthalpy rarely is taken into account, and the modulus of a gel is usually discussed considering only the topology of the networks assuming that the gels are completely homogeneous as derived from Eqs. (2.3) and (2.4). Viscoelastic property and rubber elasticity of gels are described in detail in Chap. 6.

2.4 Solvent Absorption and Retention

One of the specific features of gel is that it can substantially absorb the fluid and sustain it through the affinity of polymer network with the adsorbed fluid. The affinity is realized by various interactions such as electrostatic force, hydrogen bonding, hydrophobic interaction, or van der Waals forces. The retention capability of gels depends on crosslink density as described earlier, but it also depends on the affinity between polymer network and the fluid. Among them, polymer gel which has charges on the crosslinked chain and the counter ion in the fluid is called as **polyelectrolyte gel** and it is the special class that can absorb large amount of water and used as water absorbent materials.

As a result of swelling a gel produces substantial space and the elastic force, which in turn, produces the driving force to transfer the fluid and the molecules in it. Depending upon the interactions between the network and the molecules in the fluid, certain molecules would be adsorbed strongly and other would pass within the gel. This feature allows to use the gel as material capable of transporting, exchanging, separating, releasing, and up-taking solutes in the fluid. Examples for these properties include ion exchange, electrophoresis, chromatographic, drug delivery, and

Table 2.1 Functions and applications of gels

Function	Application
Water absorption, water retention	Superabsorbent material, diapers, feminine products, oil–water separation materials
Moisture absorption	Drying agents, anti-frost agents
Sustained release	Drug substrates, horticultural water holding agents, air fresheners
Absorption and permeation	Impurity removal material, selective separation membranes, chromatographic column materials, ion exchange resins
Increase of viscosity	Foods, cosmetics, culture substrates
Transparency, light transmission	Contact lens, optical fiber, and device
Electric properties	Sensors, soft electrodes
Vibration damping, sound absorption	Sound barrier walls, insulators, impact-damping material artificial joints, artificial tissues
Flexibility, plastic deformation, elasticity	Shape-deformable materials, slurring agents for soil, packing materials
Solidification of water	Digestive aids, fertilizer additives
Depression of freezing point	Coolants, cool pillows
Swelling	Toys, packing materials, ceiling materials

dialysis substances. These separating functions of gels are commercially produced as selective separation membranes, dialysis membranes, ion exchange resins, chromatography column packing materials, and electrophoretic gels. Selective interaction based on compatibility or solubility with a penetrant is not limited in the fluid. An example of gels in which gas permeability is important is soft contact lenses with biocompatibility. The oxygen permeability of contact lenses is known to increase as the water content of the gel increases. Functions and applications of the gel will be reviewed in detail in Chap. 9. Here, the typical characteristics of gels and applications that employ those properties are summarizes in Table 2.1.

2.5 Swelling and the Phase Transition

When the dry polymer network is placed in the fluid with good affinity of the network, it absorbs the fluid and gradually swells in proportion to the amount of the solvent incorporated and continues swelling until the equilibrium between elasticity and amount of the absorbed fluid is attained.

In general, the process of swelling of a dried polymer network can roughly be classified into three consecutive procedures: the diffusion of fluid ("solvent" is commonly used) molecules into polymer networks (step 1); relaxation of polymer chains from a glassy state to a rubbery state by the interaction with the solvent (step 2); and diffusion of polymer networks into the solvent (step 3). Depending on which process is the rate-limiting step, the swelling behavior of gels can be characterized. When step 1 is the limiting factor, the amount of solvent to penetrate to the gel is limited and relaxation of the polymer chain hardly occurs. However, if the polymer chain relaxes quickly, the swelling of the gel is controlled by the diffusion of solvent molecules into the networks (Fickian or case-l diffusion). When the relaxation of polymer chains dominates, that is, when step 2 is the limiting factor, the swelling behavior deviates from the Fickian diffusion (non-Fickian or anomalous diffusion). When step 3 is the limiting process, the swelling of the gel is controlled by the diffusion of polymer chains in the solvent and the swelling behavior of the gel is explained by cooperative diffusion.

However, certain gels show abrupt change of their size when the property of the solvent is slightly changed. For example, when ethanol is added slowly to water in which a partially ionized polyacrylamide gel is swollen, the gel starts to gradually shrink in water–ethanol mixture since the ethanol is a poor solvent for the gel. However, when ethanol concentration reaches to a certain extent, the volume of the gel suddenly and quickly collapses to several percent of the initial volume. This phenomenon is called **volume phase transition** and it was found later that it occurs not only by change in chemical composition of the solvent but also by a variety of external environmental changes such as temperature, pH, electric fields, and light. Thus, gel behaves as stimuli-responsible materials responding to the external environment.

Phase transition in polymer gels was theoretically predicted by Dusek, the author of this book, and Patterson in 1968 on the base of the coil-globule transition phenomena [13]. Refer to Chap. 4 in which theoretical deriving process is described in detail. This prediction was later experimentally confirmed by Tanaka using partially ionized gel [14–16]. The discrete volume phase transition is understood as a manifestation of the competition among three osmotic pressures acting on the polymer network: osmotic pressure due to the affinity among polymers, the rubber elasticity that keeps the network in a state of moderate expansion, and osmotic pressure of counter ions. The combined effect of these three pressures determines the equilibrium volume. Temperature, pH, and ions affect both positive and negative pressures, whereas solvent composition influences only the negative pressure. Changing one of these factors creates an excess nonzero pressure, which results in a new equilibrium volume. On the basis of Flory–Huggins theory [12], Eq. 2.7 was theoretically derived by Tanaka et al. [14].

Since overall free energy of polymer gel (ΔG) is the sum of three free energies; network-solvent mixture, ΔG_{mix} which is derived from Flory–Huggins equation, 1st term in Eq. (2.5), network elasticity, $\Delta G_{el,}$ 2nd term in Eq. (2.5), and counter ions ($\Delta G_{ion,}$ 3rd term in Eq.).

$$\begin{aligned}\Delta G &= \Delta G_{mix} + \Delta G_{el} + \Delta G_{ion} \\ &= -\kappa T[(1-\Phi)\ln(1-\Phi) + \chi\Phi(1-\Phi)]\end{aligned}$$

$$+ 3\nu_e\kappa T/2(\lambda^2 - 1 - \ln\lambda) - \nu_e f\kappa T\ln(V_0\lambda^3/nv_i) \tag{2.5}$$

where κ is the Boltzmann constant, T is the absolute temperature, Φ is the volume fraction of the polymer network, ν_e is effective number of polymer network, χ is an interaction parameter, λ is elongation ratio after swelling ($= d/d_0$), f is number of ions between junctions, V_0 is initial volume of gel before swelling, n is number of solvent, and v_i is the partial molar volume of solvent.

Osmotic pressure **Π** is expressed as follows with ΔG,

$$\Pi = -N/v_i(\Delta G/dn_1)_{t,p} = -N\kappa T/v_i[\ln(1-\Phi) + \Phi^2]$$
$$+ \nu_e\kappa T/2(\Phi/2\Phi_0 - (\Phi/\Phi_0)^{1/3})1 + \nu_e f\kappa T(\Phi/\Phi_0) \tag{2.6}$$

where N is Avogadro number, Φ_0 is volume fraction of polymer before swelling, and swelling ratio $\boldsymbol{\alpha}$ is expressed as

$$\alpha = \lambda^3 = {}^{V}/_{V_0} = {}^{\Phi}/_{\Phi_0}$$

Since $\Pi = 0$ at equilibrium swelling, Eq. 2.6 becomes as follows

$$\tau = 1 - 2\chi = -\nu_e v_i/N\Phi 2[(2f+1)((\Phi/\Phi_0) - 2(\Phi/\Phi_0)^{1/3})$$
$$+ 1 + 2/\Phi + 2\ln(1-\Phi)/\Phi^2 \tag{2.7}$$

Thus, the equilibrium network volume fraction is determined as a function of the reduced temperature, i.e., one can get theoretical swelling curve when $\boldsymbol{\tau}$ is plotted as a function of the swelling ratio Φ/Φ_0.

For certain values of reduced temperature, the above equation is satisfied by three values of polymer network volume fraction Φ corresponding to two minima and one maximum of the free energy. The value of Φ corresponding to the lower minimum represents the equilibrium value. A discrete volume transition occurs when the two free-energy minima have the same value. Figure 2.3 shows the theoretical phase diagram of gels. $\boldsymbol{\tau}$ is third ordered functional curve which increases continuously if $f < 0.659$.

2.6 Polyelectrolyte Gel

A polyelectrolyte gel is a polymer network having charges on the network chain and swollen state in the fluid. Since polyelectrolyte gels are capable to absorb a significant amount of water (even up to~3000 times the dry polymer weight) within its network structure, they are widely used for water-absorbing materials as diaper and sanitary products. The localized high density of macro-ions on the network brings about high electrical potential which enables to exchange their micro-counter ions in the fluid

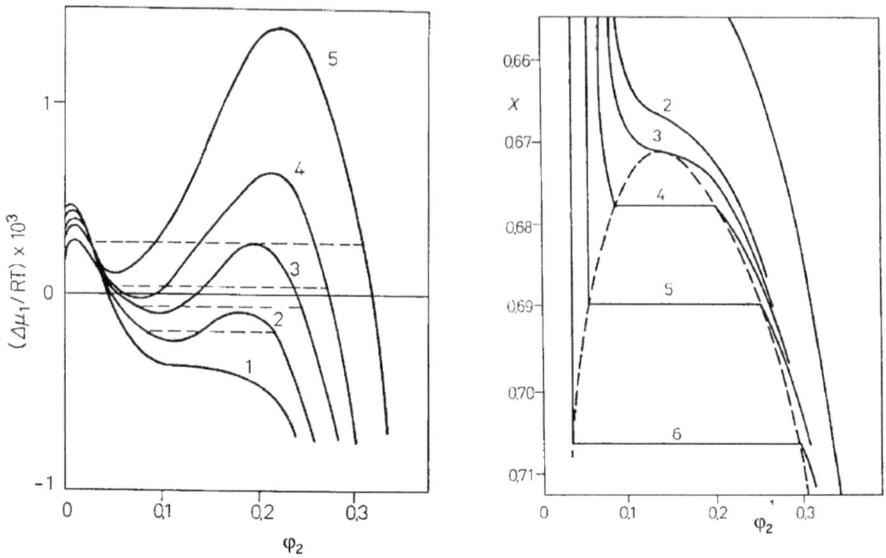

Fig. 2.3 Theoretical phase diagram of gels. **a**, Dependence of the chemical potential of solvent on polymer concentration for gel exhibiting phase transition; **b**, phase equilibria as a function of the interaction parameter (e.g., temperature) for different concentration of elastically active network chains [13]

and they are widely used as ion exchange resins to produce sweet water from sea water.

Biological gels made from proteins and polysaccharides and contain ionic groups are also categorized as polyelectrolyte gel. Soft tissues of animals consist of well-aligned assemblies of highly charged polysaccharide and filamentous proteins and their important properties to keep their vital activities are largely originated from the basic properties as polyelectrolytes. Articular cartilage, for example, exhibits excellent elasticity, low surface friction, and physical repulsive forces to withstand stress. The unique features are closely related to the electrostatic forces and large water content, both of which is induced by the negative charges on the gel matrix of polysaccharide (aggrecan–hyaluronic acid) which is connected with proteins. This will be described in detail in Chap. 8.

The swelling ability—the ability to absorb water is determined by the balance of three parameters: affinity of polyelectrolyte with water, i.e., electrostatic interaction, the osmotic pressure generated by a high ionic concentration within the gel, and rubber elasticity based on the network dynamics which closely related to the crosslinking density. Flory explained the water-retention capability of crosslinked polyelectrolytes by employing the ionic network theory [17].

$$\begin{aligned}\text{Swelling ability} &= \text{osmotic pressure of ions}\\ &\quad + \text{affinity of polyelectrolyte with water /crosslink density}\end{aligned}$$

Among these parameters, the strong electrostatic interaction of charged polymer network with water and the counter ions dominates the fundamental properties of polyelectrolyte gel. To quantitatively elucidate the mechanism of water adsorbing and ion exchanging abilities of polyelectrolyte gel, numerical approach of electrostatic potential distribution of the gel has been made on the base of the Poisson-Boltzmann equation [18]. It was calculated that there exist deep electrostatic potential valleys along the polymer chains and deeper potential wells exist at every crosslink point (Fig. 2.4). Figures 2.2 and 2.4 show a spatial profile of electrostatic potential energy in the unit of eV on the planes of mesh-like networks. It is seen that the potential energy wells are formed at every crosslinking point where the valleys along the polymer chains cross each other. Since water is a polarized molecule with large dipole moment and therefore, they are preferentially attracted to the deep electrostatic potential valleys and wells through electrostatic forces and give the gel high water adsorption ability.

Counter-ion distribution in the gel was also calculated using the Boltzmann distribution and it was found that counter ions are mostly localized in the deep potential wells and valleys around polymer network. The deep potential wells and high counter ion densities consequently bring about an instability to result in counter ion condensation in a similar as predicted by Oosawa [13] and Manning [14] for the linear polyelectrolyte solution. Counter ions located in the potential valley ($\gg \kappa_B T$) decreases its density very sharply with an increase in the distance from the polymer chain.

Thus, the more counter ion is condensed, the depth of the valley and the well become less and the gel swells less. This is why the degree of swelling decreases by the addition of salt. Similarly, increase of crosslinking density increases the number of bound counter ions and decreases the swelling degree. The high ion exchange ability of counter ions of the ion exchange resins can be explained by the same reasoning.

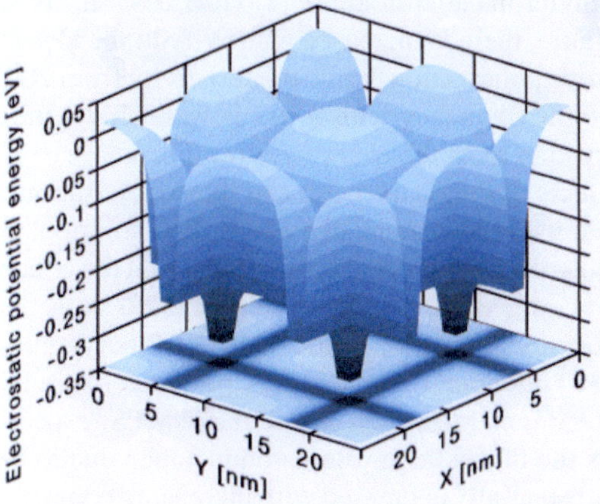

Fig. 2.4 Spatial profile of electrostatic potential energy for the plane within the mesh-like network. *X* and *Y* axes are in unit of r_i (= 0.6 nm) [18]

As described before, the polyelectrolyte gels have high charge density and due to this, ions close to the network are less mobile and condensed near the network. According to Oosawa-Manning condensation theory [19, 20] the portion of apparent free counter ions f is expressed as $f = 0.866\zeta^{-1}$ where $\zeta = \lambda/\mathrm{b}$ (b: distance between charges and 0.25 nm for polyelectrolytes prepared from vinyl monomers, λ : distance where electrostatic repulsion energy and thermal energy becomes equivalent called Bjerrum length and expressed as $\lambda = \mathrm{e}^2/4\pi\varepsilon\kappa\mathrm{T}$). In an aqueous solution, f is calculated as roughly 0.3, meaning 1/3 of total counter ions are free and contribute to the electric conductivity, while 2/3 of them are condensed near the network with apparent decreased potential due to the neutralization with charges on the network. The amount of condensed counter ions increases with increase of ionic strength.

2.6.1 Electric Conductivity

Associating with the electric responses in the living organism such as muscle contraction and nerve transformation where specific ion transfer processes are performed, basic study of the electrical properties of gels is critically important. However, it is poorly studied comparing with that of solutions due to various methodological problems. Only recently it has become possible to accurately measure the electrical conductivity and the reproducible experimental data have been obtained. Electrical conductivity of gels was measured by using alternating current dipolar method (the variable frequency method) to avoid electrolysis and the polarization effect of solvent molecules and LCZ meter with black-covered platinum was used as electrodes. The conductivity was measured by applying alternative voltage of 1 V through 1 cm^3 cubic gel sandwiched between a pair of electrode plates of 1 cm^2 and changing the frequency from 50 to 100 Hz under constant temperature. The result of the measurement for fully ionized polyelectrolyte gel–poly (2-acrylamide 2-methylpropane sulfonic acid) (PAMPS) gel is shown in Fig. 2.5. As shown in this figure, there is increased resistivity in the low-frequency region < 200 Hz due to electrode polarization, while the resistivity from 200 Hz to 1 kHz increases in proportion to the frequency with − 1/2 power. This is the common frequency dependence observed in most the electrolyte solutions.

Effects of the counter-ion and the condensation of ions on the conductivity were studied using low-molecular electrolyte solution, polyelectrolyte solution [19, 20], and polyelectrolyte gel [18] and found that the concentration dependence of between them is quite different. Figures 2.2 and 2.6 shows the concentration dependences of the equivalent (molar) conductance of 2-acrylamido-2-methylpropanesulfonic acid (AMPS) monomer (a), PAMPS polymer (b), and PAMPS gel (c) with different counterv ions [20]. Low-molecular AMPS, especially the proton as a counter ion shows the highest equivalent conductance due to high mobility, but the conductivities substantially decreased with increasing concentration due to decrease of mobility by the strong interactions of the ionized ions. Conductivity of electrolytes with other counter ion was much lower. In contrast, the conductance of polymer (PAMPS)

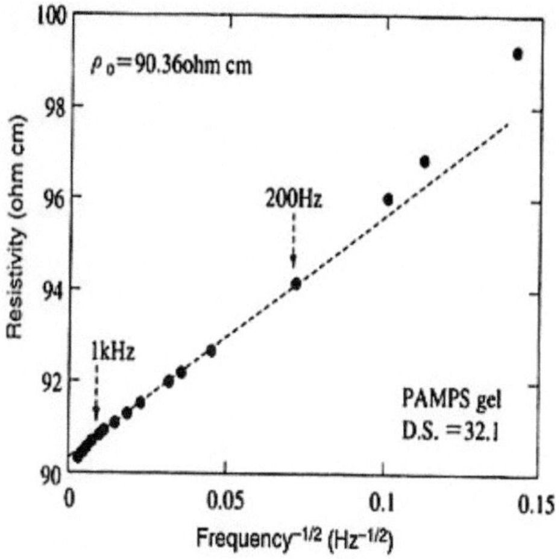

Fig. 2.5 Frequency characteristics of electrical resistance of PAMPS gel (swelling degree: 32) showing − 1/2 dependence on the frequency

solution with proton as a counter ion increased with increase of concentration. This phenomenon should contribute to increase of free (more mobile) counter ions liberated from the condensed state by the conformational change of the polyelectrolyte chains from extended to coiled compaction. Important fact is that such compaction of does not occur for the polyelectrolyte gels and the gel shows almost equal equivalent conductance with almost no concentration dependence as shown in Fig. 2.6c.

2.6.2 *Low-Frequency Dielectric Relaxation*

The dielectric relaxation spectrum gives the information about ions that move under alternative electric field. The high-frequency relaxation in MHz frequency region reflects the fluctuation of free counter ions which exist far from the polyelectrolyte. The low-frequency relaxation arises from the fluctuation of bound counter ions near the polyelectrolyte. Therefore, the measurement of the conductivity does not sufficiently contribute to know the behaviors of bound counter ions since highly condensed counter ions do not effectively contribute to the electric conduction, but they can migrate along the network. Thus, the experiment of the low-frequency relaxation is important to know the behaviors of bound counter ions. However, due to the difficulty of the experiment that originates from their high conductivity and large electrode polarization effect, there were no report of the experiment. Recently, the low-frequency relaxation experiment was made in the range of 50 ~ 100 kHz using full-ionized anionic poly (sodium 2-acrylamido-2-methylpropanesulfonate) (PNaAMPS) gels of various cross-lining density and compared with those of the linear polymer solutions [22]. By measuring the complex dielectric constant, dielectric relaxation time of the ions in a lower frequency region was calculated. It was found that mean relaxation time of the linear polymer solution which shows constant relaxation time

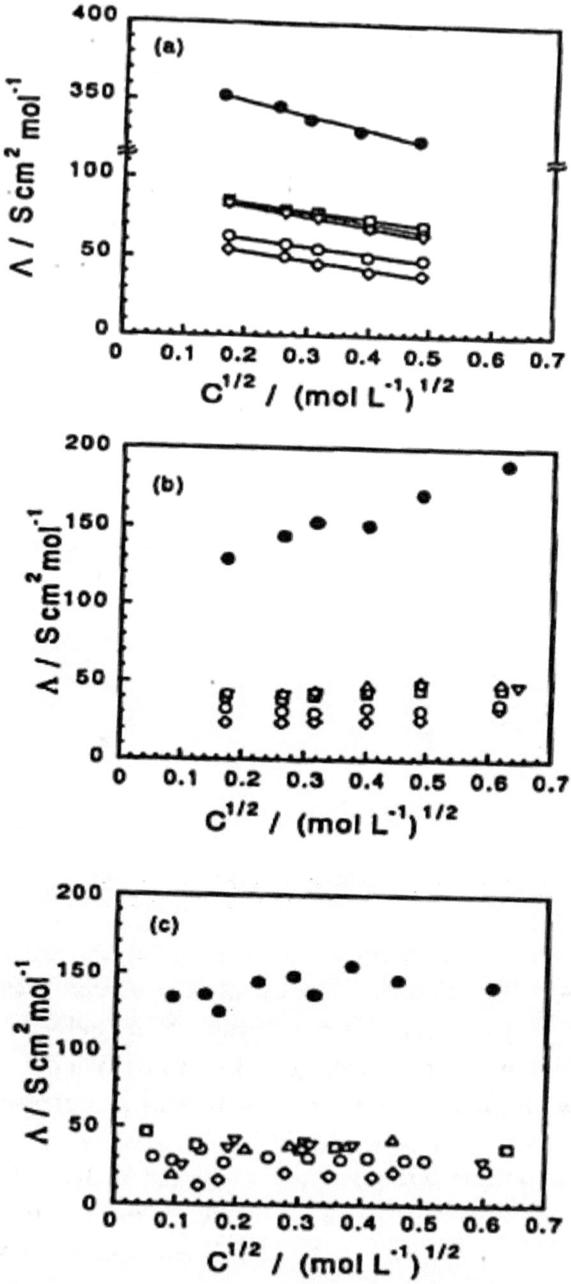

Fig. 2.6 Equivalent conductance of **a**, AMPS monomer solution, **b**, PAMPS polymer solution, and **c**, PAMPS gel. Counter ions: (●): H; (◇) :: Li; (0): Na; (□): K; (△): Rb; (▽) : Cs, measured at 25 °C [21]

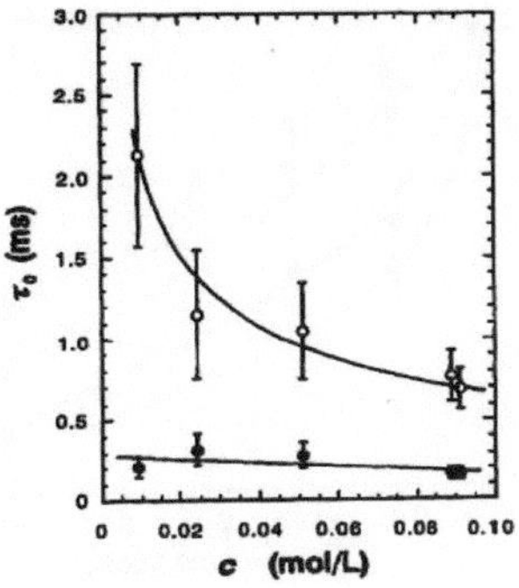

Fig. 2.7 Concentration dependence of the mean relaxation time τ_0 of PNaAMPS gel and polymer solutions: (○): polyelectrolyte gels; (●): linear polymer solutions [22]

over the wide polymer concentration (Fig. 2.7). Observed almost no concentration dependence of low-frequency relaxation of the polymer solution suggests the tightly bound counter ions fluctuation along the polymer chains. In contrast, low-frequency relaxation of the gel is quite different. It shows much longer relaxation time which substantially decreased with increase of the crosslinking density (or decrease in the distance of the crosslinked junction). This result could be explained in terms of the crossing of the bound counter ions through the crosslinking points [23]. As shown in Fig. 2.4, the polyelectrolyte gel has deep electro-potential valleys at the crosslinking points and the bound counter ions which fluctuate and migrate along polymer chain of the network have to cross the deep potential valley where results in a low-relaxation time. According to this assumption the diffusion constant at the crosslinking points was calculated as three orders of magnitude lower than that along the network chain.

2.7 Water in the Gel

Water in gels is known to exist in several different forms. The presence of deep potential wells at the crosslinking joints Fig. 2.4 allows the gel not only to adsorb a large amount of water, but also to induce the polarization of water molecules in the network cavities and confine their motion. Thus, independent motion of water molecules is substantially restricted comparing with that of common water molecules (often called bulk water) and give rise certain preferred orientation or more organized structure of water molecules at the water/network interface. Water existing close to the network is called **bound water**. There, strong interaction induces entropy and enthalpy changes of water molecules to give unfavorable arrangement to crystalize and decrease in the freezing temperature of water occurs. Water in the cavities of gel does not freeze below the freezing point of the bulk water. Among bound water that does not freeze below 0 °C is called non-freezable or un-freezable water. The presence of bound water is observed in the protein and polysaccharide gels. Decrease in freezing temperature is also observed for organic fluid in organic gels.

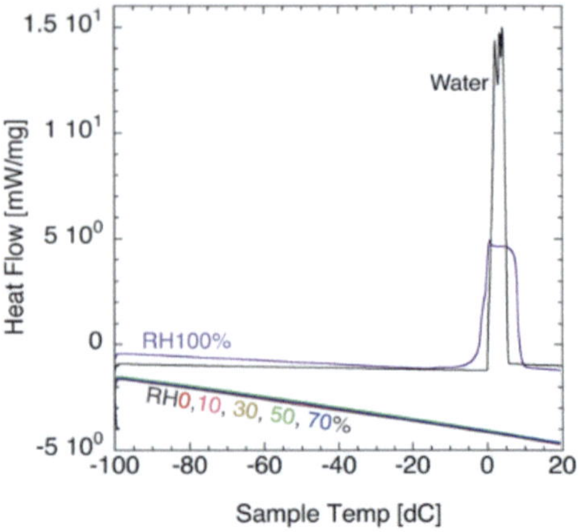

Fig. 2.8 DSC curves of DN gels equilibrated under the chamber of various RH indicate relative humidity of the chamber where the gels were equilibrated [24]

A differential scanning calorimeter (DSC) is usually used to experimentally prove the presence of "non-freezable water". DSC is the equipment to measure the amount of heat flowing in and out of the sample changing temperature continuously. Figure 2.8 shows DSC curves of the double network (DN) gel (refer to Chap. 9) equilibrated at various relative humidity (RH) [24]. The DN gel containing 16% water (equilibrated in the chamber at 70% RH) or less shows no peak corresponding to freezing or melting point, while DN gel containing 90% shows a strong exothermal peak at around 0 °C. This result indicates that water in the DN gel containing water less than 16% consists of all non-freezing water (un-freezable water), but contains both non-freezing water and normal water if the gel contains 90% water.

As described, the molecular motion of non-freezable water should be different from the common bulk water. In general, the relaxation time of a water molecule adsorbed near the ionized polymer is must be longer than that of bulk water since the ionized polymer in the gel could affect to the inertia of water molecules within the hydrated layers, as a result, should give a wide peak. An attempt to determine the relaxation time of non-freezing water in DN gels was made using quadrupole solid-echo NMR method. Figure 2.9 shows wide-line deuterium NMR spectra of DN gels after immersing in D_2O gas for 5 h. As shown in Fig. 2.9, water in the DN gel shows very wide peak and the profile is also different both from crystalline D_2O (ice) [22]. This figure indicates that non-freezing water (D_2O) in the DN gel has very slow relaxation time due to thermal fluctuation.

Described results obtained from DSC and NMR demonstrate that the interaction between polymer network and the water molecules is strong enough to restrict the mobility of the water and form non-freezing water. Since the mobility of non-freezing water in the DN gel is restricted, the mechanical toughness (tensile stress) of DN containing water less than 16% is increased [24].

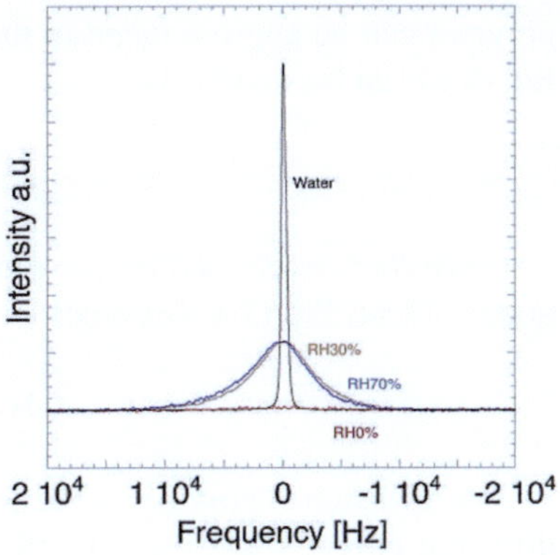

Fig. 2.9 Heavy water solid-echo NMR spectrum. RH indicates relative humidity of the chamber where the gels were equilibrated [24]

2.8 Diffusion and Transport in the Gel

Diffusion of fluid and solvated molecules in the polymer gels is one of the basic functions characterizing structure and dynamics of polymer network and experimental and theoretical investigations have been continuously made. As already mentioned, extra cellular matrix (ECM) in the living organism controls the transportation of water, ions, hormones, and nutrition for cellular viability. Therefore, study of diffusion of a solvent, molecules, and macromolecules in the swollen network gives not only important information for these transport phenomena but also gives important knowledge on the interaction dynamics with the network chains.

The classical and easy method to measure the diffusion of solute in a gel is to measure the change of amount of the solute permeated through the gel with time. Permeation of the solute by assaying time course of the concentration change through a gel is a typical example. There, one can obtain the intrinsic diffusion coefficient by extrapolating the time profile to zero concentration. Dynamic light scattering technique which allows to determine the translational diffusion coefficient is also common method. The self-diffusion coefficient can be obtained by the pulsed magnetic gradient spin echo (PGSE)-NMR technique. The principle of this technique is based on measurement of the time dependence of average distance of the nuclear magnetic probe.

The diffusion of macroscopic rigid sphere in a fluid follows Stokes—Einstein Equation and the diffusion coefficient of the rigid sphere D_0 is expressed as follows

$$D_0 = kT/\zeta = kT/6\pi R\eta \tag{2.8}$$

where k is the Boltzmann constant, T is the absolute temperature, ζ is the friction coefficient of the sphere during transport, R is the radius of the rigid sphere, and η is the viscosity of the fluid.

In the case of diffusion of polymer in a random coil state, the solvent within the coil is assumed to travel together with the coil. In that case, the diffusion of the polymer can be approximated to the behavior of a rigid sphere with radius R_h, and Eq. (2.8) can be written as

$$D_0 = kT/6\pi R_h \eta \tag{2.9}$$

Since R_h is expressed by the molecular weight M *as* $R_h = M^{\nu}$ and ν is 0.6 in a good solvent, Eq. (2.8) becomes

$$D_0 = kT/6\pi M^{0.6}\eta \tag{2.10}$$

The diffusion of polymer in the gel is as follows. If any specific interaction between diffusing polymer and the network is not exists, the diffusion is largely dominated by the effect of exclusion volume. When the concentration (density) of polymer network is low, i.e., the gel is highly swollen, the diffusion of permeating polymer would mainly be dominated by the interaction with the solvent in the gel, similarly to the case of diffusion in solution. In contrast, when the concentration of polymer network is high (the gel is moderately swollen), the diffusion is substantially be restricted by the network itself. In general, diffusion coefficient of molecules in the gel D is expressed as a function of volume fraction of the gel, more exactly volume fraction of network fibers φ as follows

$$D/D_0 \sim \Phi^{-\alpha} \tag{2.11}$$

where D_0 is the diffusion coefficient in the solvent and α is determined by the ratio of the size of diffusing molecule and the network size. When the size of diffusing molecule R is much smaller than the network size, $\alpha = 0.5$ and the diffusion coefficient D of the particle in the gel is expressed as

$$D = D_0 \exp(-\kappa R) \tag{2.12}$$

where κ^{-1} is the fluid dynamic network size and is called as dynamic screening distance and this equation indicates that the diffusion is controlled by the ratio of the particle and network size. When the diffusing polymer is in a random coil conformation and R_h is smaller than κ^{-1} the equation $R_h = M^{\nu}$ could be held. However, when R_h is large enough, the transport in the transverse direction of the diffusing polymer might be highly restricted by the network chains and the movement along the chain direction would become feasible. This is so-called reputation region and $\alpha = 1.75$. The diffusion in this case is expressed as Eq. (2.13) that is similar case to the diffusion in a molten polymer.

$$D = M_e M^{-2} \tag{2.13}$$

Here, M_e. is the molecular weight between the entanglement points, i.e., between crosslinking junctions [25]. Thus, the diffusion of polymers in the highly dense

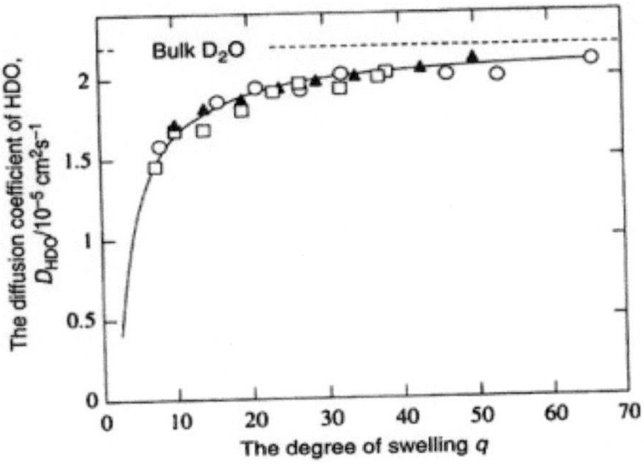

Fig. 2.10 Diffusion coefficient of HDO (D_{HDO}) as a function of swelling degree q of poly (N,N-dimethylacrylamide) gel. Molecular weight of polyethylene glycol included in the gel as probe molecule: (□) 4250 (▲) 10,890 (•) 20,000, measured at 30 °C [28]

network gel is reciprocally proportional to the second power of the molecular weight of the diffusing polymer [26].

2.8.1 *Diffusion of Solvent*

The diffusion of HDO (included in a minute amount in heavy water) in a poly (N,N-dimethylacrylamide) gel was investigated using (PGSE) NMR technique [27]. The gel was swelled to equilibrium in heavy water and polyethylene glycol was included as the probe molecule. As shown in Fig. 2.10, the solvent (HDO) diffusion in the gel substantially decreased with decrease in the swelling degree regardless of the molecular weight of the probe molecule. The relationship between the degree of swelling q and the diffusion coefficient D of HDO was obtained [28]. The diffusion coefficient D_{H2O} of water in the poly (methacrylic acid) gel was determined by (PGSE) 1 H NMR techniques [29]. $D_{H2O}$was proportional to the 3rd power of the degree of swelling which indicates that the mobility of water increases in one dimensional with the network size.

2.8.2 *Diffusion of Macromolecules*

If the size of diffusing polymer is smaller than that of the network junction, the diffusion of the probe molecule can be expressed by Eq. (2.5) where the fluid dynamic radius R_h is used instead of the size of the particle *R*. Thus, diffusion coefficient of poly(ethylene glycol) (PEG) in highly swollen poly(N,N-dimethyl-acrylamide) gel equilibrated in heavy water was determined by using the (PGSE)-1 H NMR technique [28]. When the diffusing polymer is larger than the network size, the

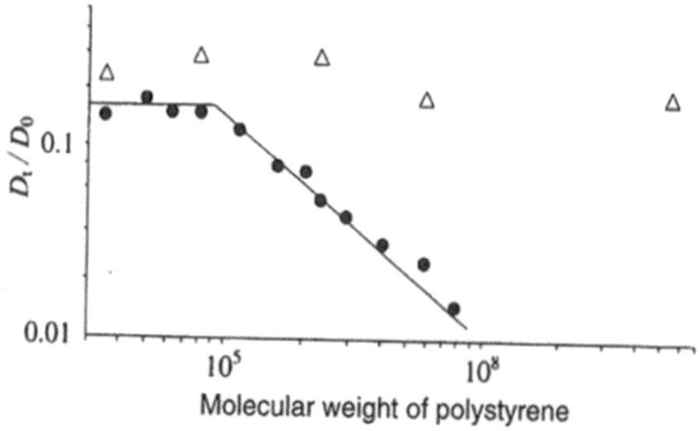

Fig. 2.11 Dependence of the molecular weight of polystyrene on the diffusion coefficient (D_t) in the poly (methyl methacrylate) gel (•) D_0 is the diffusion coefficient of polystyrene in a toluene solution. (△) is the diffusion coefficient of polystyrene in the poly (methyl methacrylate) solution. Poly (methyl methacrylate) gel was swelled in toluene. Degree of crosslinking 12.5%, measured at 4.6 °C [26]

Stokes Einstein (S-E)-type diffusion in Eq. (2.9) cannot be used and the polymer diffuses by reputation and the diffusion coefficient is expressed by Eq. (2.13).

Diffusion coefficient of polystyrene (PS) *Dt* in the poly (methyl methacrylate) (PMMA) gel swelled in toluene was measured by using dynamic light scattering [26]. As shown in Fig. 2.11, Dt/D_0 (D_0 is diffusion coefficient of PS in toluene solution) in the range of molecular weight (M) below 80,000 shows no molecular weight dependency and *Dt* followed $Dt \sim \mathrm{M}^{-0.6}$ in toluene. In this region PS is considered to follow the Stokes—Einstein type diffusion with radius R_h as in toluene solution. However, when molecular weight of PS becomes higher than 100,000, the diffusion is substantially restricted with increase of molecular weight and showed the slope of $Dt/D_0 \sim M^{-1.2}$ and the molecular weight dependence of *Dt was* $Dt \sim \mathrm{M}^{-1.8}$. This is the strong contrast with the diffusion of PS in the PMMA solution where molecular weight dependence was not seen as shown in Fig. 2.9. The slope of the molecular weight dependence (− 1.8) is close to the parameter—2 in Eq. (2.9) and suggests the diffusion takes place by the reputation mechanism, i.e., the polymer diffuses by the molecular motion along the main chain direction. These results also suggest that the disentanglement of PMMA chains can locally be occurred in the solution, while it does not happen in chemical crosslinked PMMA chain of the gel.

2.8.3 *Diffusion of Proteins*

The study of diffusion of proteins through gels is especially important to understand the processes in living organism a well as separation processes by electrophoretic and chromatographic methods. The diffusion of proteins should be depended not only the size, but also the charges and the shape. Diffusion of proteins in the gels has been

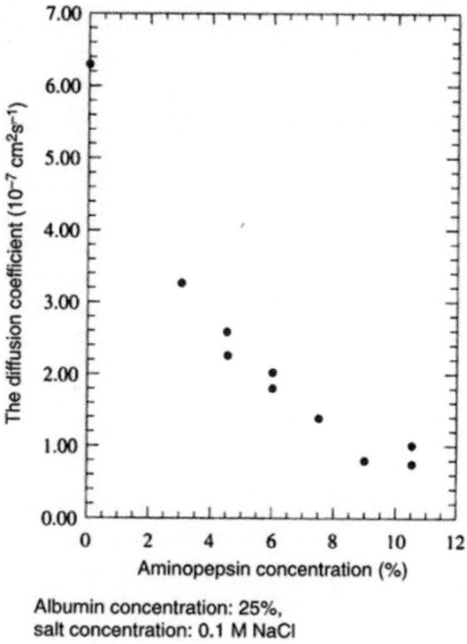

Fig. 2.12 Dependence of the diffusion coefficient of albumin on the concentration (density) of amylopectin gel albumin concentration: 25%, NaCl concentration: 0.1 mol [31]

studied by various methods including dynamic light scattering, pulsed field-gradient nuclear magnetic resonance, and holographic interferometry. Diffusion of globular protein in agarose gel was investigated and the experimental data of globular protein agreed with the combined model proposed by Clague and Philips [30].

The effect of amylopectin concentration (density of amylopectin gel) on the diffusion coefficient of albumin was studied by FTIR measurement. As shown in Fig. 2.12, the diffusion coefficient of albumin substantially decreases with increase of network concentration of amylopectin [31].

In general, dynamic screening distance κ^{-1} of the network in Eq. (2.12) is expressed with the network concentration *c as*:

$$\kappa^{-1} = C^U \tag{2.14}$$

u is the constant and takes a value of 0 5–1.0 depending on the property of the network polymer. From Eqs. (2.8) and (2.10), we obtain

$$D = D_0 \exp\left(-R_C^U\right) \tag{2.15}$$

From the slope of logarithmic plot of Fig. 2.12, one can obtain the dependence of the albumin diffusion on the amylopectin concentration as $u = -0.93$.

It is important to know how strongly the charges and shapes of proteins affect to the diffusion behavior in the charged gels. To clear this problem, two kinds of polysaccharide gels, neutral agarose gel and negatively charged carrageenan gel, were used and diffusion of myoglobin which is a globular protein was investigated. The advantage to use polysaccharide gels is that they do not swell or shrink over a wide range of pH and ionic strength and can keep the volume fraction of network constant. A newly developed electronic speckle pattern interferometry (ESPI) method [32] was used in this study.

It was found that diffusion coefficient of myoglobin in the neutral agarose gel is not affected either by changes in pH and in ionic strength, while that of myoglobin substantially decreased in the anionic λ-carrageenan gel in the pH range higher than the isoelectric point (pI). This indicates that the diffusion of myoglobin is extensively hindered in the charged gel by the electrostatic repulsion. This diffusion of protein in λ-carrageenan gel well agrees with the Tsai and Strieder model and an apparent radius of the myoglobin at various pH were estimated [33].

In addition of charges, the shape of the diffusing macromolecules, in particular, the aspect ratio of the molecules would be another important factor affecting the diffusion process. The effect of the aspect ratio on the diffusion coefficients was studied using the charged rod-like protein: tropomyosin (aspect ratio $R = 26$) in agarose gel and negatively charged λ-carrageenan gel and compared with those of the globular protein: myoglobin ($R = 1.6$). As shown in Fig. 2.13, the diffusion coefficients of the tropomyosin and myoglobin showed a similar behavior in the neutral agarose gel, i.e., both decreased with increase of the volume fraction of the gel (network volume fraction Φ) with a scaling exponent $\alpha = -0.3$ to -0.4 in Eq. 2.11. This value is close to the Rouse model of $\alpha = -0.5$.

In contrast, the diffusion of tropomyosin in the λ-carrageenan gel was quite different from that of myoglobin. It consisted of two regions: When Φ is low, D was proportional to $\Phi^{-0.3}$, which is close to the Rouse model, while when Φ is high, D was proportional to $\Phi^{-1.8}$, which is close to the reputation model of $\alpha = -1.75$ [23, 34]. Thus, the diffusion of the tropomyosin in λ-carrageenan gel decreases significantly with increase in Φ in a power law and transfers from Rouse-like to the reputation-like diffusions. From these results a pore size of the gels was also estimated. Thus, the diffusion of rod-like protein with large aspect ratio was confirmed to follow reputation-like behavior in the condensed hydrogel.

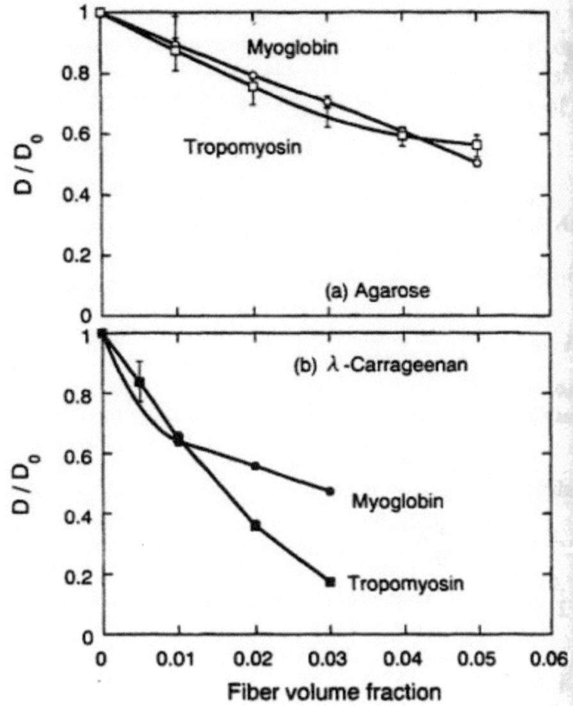

Fig. 2.13 Effect of fiber volume fraction of the polysaccharide gels on the diffusion coefficient of proteins. (**a**) In agarose gel, (**b**) in λ-carrageenan gel; (○, □, ·) myoglobin; (□, □, ■) tropomyosin, solution: pH 6.8, 0.5 M KCl, measured at 20 °C [33]

References

1. C.O. Edwards, L. Mandelkern, Crystallization–gelation process of homopolymers and copolymers from solution. J. Polymer Sci. Polym. Lett. **22**, 366 (1982)
2. C.M. Tseng et al., Uniform polymer particles by dispersion polymerization in alcohol. J. Polym. Sci. Polym. Chem. **24**, 2995 (1986)
3. J.E. Eldridge, J.D. Ferry, Studies of the cross-linking process in gelatin gels. III. Dependence of melting point on concentration and molecular weight. J. Phys. Chem. **58**, 992 (1954)
4. M.A. Harrison, P.H. Morgan, G.S. Park, A simple method for determining sol—gel transition temperatures in weak polymer jellies. Br. Polymer J. **3**, 154 (1971)
5. R.C. Domszy, R. Alamo, C.O. Edwards, L. Mandelkern, Thermoreversible gelation and crystallization of homopolymers and copolymers. Macromolecules **19**, 310.1–3 (1986)
6. A.H. Clark, S.B. Ross-Murphy, Adv. Polymer Sci. **83**, 60 (1987)
7. W. Burchard, S.B. Ross-Murphy (eds.), *Physical Networks: Polymers and Gels* (Elsevier Applied Science, London, 1990)
8. M. Doi, S.F. Edwards, *The Theory of Polymer Dynamics* (Oxford University Press, Oxford, 1986)
9. G.R. Dobson, M. Gordon, J. Chem. Phys. **43**, 705 (1965)
10. W. Kuhn, F. Grun, Relationships between elastic constants and stretching double refraction of highly elastic substances. Kolloid-Z. **101**, 248 (1942)
11. H.M. James, E. Guth. Theory of the elastic properties of rubber. J Chem. Phys. **11**, 455 (1943)
12. P.J. Flory, *Principles of Polymer Chemistry Chapter 11* (Cornell University Press, Ithaca, New York, 1953)

13. K. Dusek, D. Patterson, Transition in Swollen Polymer Networks Induced by Intramolecular Condensation. J. Polym. Sci. **Pt A2 6**, 1209 (1968)
14. T. Tanaka, Collapse of gels and the critical endpoint. Phys. Rev. Lett. **40**, 820 (1978)
15. T. Tanaka, Gels. Sci. Am. **244**, I 10 (1981)
16. T. Tanaka, D.J. Fillmore, S.T. Sun, L. Nishio, G. Swislov, S. Shah, Phase transitions in ionic gels. Phys. Rev. Lett. **45**(20), 1636 (1980)
17. P.J. Flory, *Principles of Polymer Chemistry Chapter 13* (Cornell University Press, Ithaca, New York, 1953)
18. J.P. Gong, Osada Y., Chem. Lett. **449** (1995)
19. F. Oosawa, A simple theory of thermodynamic properties of polyelectrolyte solutions. J. Polym. Sci. **23**, 43 (1957)
20. G.S. Manning, Limiting laws and counterion condensation in polyelectrolyte solutions. III. An analysis based on the Mayer ionic solution theory. J. Chem. Phys. **51**, 924 (1969)
21. J.P. Gong, N. Komatsu, T. Nitta, Y. Osada, Electrical conductance of polyelectrolyte gels. J. Phys. Chem. B **101**, 740 (1997)
22. T. Mitsumata, J.P. Gong, K. Ikeda, Y. Osada, Low-frequency dielectric relaxation of polyelectrolyte gels. J. Phys. Chem. B **102**, 5246 (1998)
23. P.G. de Gennes, *Scaling Concepts in Polymers Physics* (Cornell Univers1ty Press, Ithaca, New York, 1979)
24. T. Tominaga, K. Sano, K. Ijiro, Y. Osada et al., Hydrophilic double-network polymers that sustain high mechanical modulus under 80% humidity. ACS Macro Lett. **1**, 432 (2012)
25. P.G. de Gennes, Conjectures on the transport of a melt through a gel. Macromolecules **19**, 1245 (1986)
26. S. Pajevic, R. Bansil, C. Konak, Diffusion of linear polymer chains in methyl methacrylate gels. Macromolecules **26**, 305 (1993)
27. E.D. von Meerwall, E.J. Ames, J.D. Ferry, Macromolecules **18**, 260 (1985)
28. H. Yasunaga, M. Kobayashi, S. Matsukawa, H. Kurosu, I. Ando, *Annual Reports on NMR Spectroscopy,* vol. 34, p. 39, ed by G.A. Webb, I. Ando (Academic Press, London, 1997)
29. H. Yasunaga, I. Ando, Dynamic behavior of water in hydro-swollen crosslinked polymer gel as studied by PGSE 1H NMR and pulse 1H NMR. Polym. Gels and Networks **1**, 83 (1993)
30. D.S. Clague, R.J. Phillips, Hindered diffusion of spherical macromolecules through dilute fibrous media. Phys. Fluids **8**(1720), 7 (1996)
31. R.E. Cameron, M.A. Jalil, A.M. Donald, Diffusion of bovine serum albumin in amylopectin gels measured using Fourier transform infrared microspectroscopy. Macromolecules **2**, 2708 (1994)
32. X. Zhang, N. Hirota, T. Narita, J.P. Gong, Y. Osada, Investigation of molecular diffusion in hydrogel by electronic speckle pattern interferometry. J. Phys. Chem. B **103**, 6069 (1999)
33. J.P. Gong, N. Hirota, A. Kakugo, T. Narita, Y. Osada, Effect of aspect ratio on protein diffusion in hydrogels. J. Phys. Chem. B **104**, 9904 (2000)
34. M. Doi, S.F. Edwards, *The Theory of Polymer Dynamics*·(Oxford University Press, Oxford, 1986)

Chapter 3
Network Formation of Gels

Yoshihito Osada and Karel Dušek

Abstract Polymer gel is classified by two crosslinking methods: chemical gel and physical gel. Radical crosslinking copolymerization gives covalently crosslinked, irreversible gel, and the formation process is described. Interactions between polymer chains give reversible physical gel which undergoes sol–gel transition. Formation mechanism and the structure of physical gel are explained. Theoretical repeated branching models and simulation methods of network structure buildup are explained in detail.

Keywords Chemical gel · Physical gel · Crosslinking · Network · Percolation · Diffusion control · Trommdorff effect · Cyclization · Egg box junction · Gellan gum · Sol–gel transition

3.1 What Is Crosslinking?

According to IUPAC definition [1], a crosslink is defined as a bond that links one polymer chain to another either by chemical bonding or physical bonding to give polymer network. Crosslinking is used both for synthetic polymers and for the natural (biological) polymers. There, a crosslink is characterized as: a small region in a macromolecule from which at least four chains emanate and formed by reactions involving sites or groups on existing macromolecules or by interactions between existing macromolecules.

Notes (1) The small region may be an atom, a group of atoms, or number of branch points connected by bonds, groups of atoms, or oligomeric chains.

Y. Osada (✉)
RIKEN, Wako, Saitama, Japan
e-mail: osadayoshi@riken.jp

K. Dušek
Institute of Macromolecular Chemistry, Academy of Sciences of the Czech Republic, Prague 6, Czech Republic

Y. Osada (ed.), *Science of Gel*, https://doi.org/10.1007/978-981-95-0301-8_3

(2) In the majority of cases, a crosslink is a covalent structure, but the term is also used to describe sites of weaker chemical interactions, portions of crystallites, and even physical interactions and entanglements.

As explained in the previous chapters and defined by IUPAC, the gel is classified by two crosslinking methods: chemically crosslinked gel (**chemical gel**) and physically crosslinked gel(**physical gel**). Chemical gel consists of polymer chains with both ends connected by covalent bonding and therefore usually leads to irreversible gel formation. The chemical gel is further classified by the crosslinking procedure of chemical reaction at the chain end; by chain reactions of unsaturated monomers or by stepwise branching reactions of multifunctional monomers or precursors.

On the other hand, physical gel is formed by the molecular interactions between polymer chains and therefore undergoes a reversible sol–gel transition by changes in temperature, pH, and ionic concentration and sometimes by the presence of metallic ions. Therefore, some physical gels can be solubilized reversibly beyond the gel point.

The physical gels can also be formed by polymer entanglement either in the solution or melt even without specific interactions. In this case, gel formation is characterized by product of concentration and the molecular weight exceeding a certain critical molecular weight [2, 3]. In this case the longest relaxation time of entangle polymers depends on M^3. In other words, the principal difference between chemical gel and physical gel can be understood in terms of their life time of the network junctions. Crosslinking junctions of chemical gel are permanent while junction zones of physical gels have infinite lifetime which depends on the species of interactions.

According to the Flory–Stockmayer gelation theory [4], the gel point is defined as the point at which the apparent molecular weight becomes infinite by connecting structural units. In this model, multifunctional structural units (monomers) react sequentially through covalent bonding in case of chemical gel and grow like tree branches. When the branches fill the entire space, the gelation is completed.

Using the cascade theory Gordon [5], proposed a process of f-functional monomers connected randomly by condensation polymerization. As shown in Fig. 3.1, the tree that expresses gelation and the equations that are 1:1are given in a dummy variable θ. The branching point of the tree (●) in Fig. 3.1 corresponds to the dummy variable θ and the probability of the tree growing, and α is the probability of further branching meaning.

This formulation indicates that f-functional random condensation polymerization system forms homogeneous gels without forming intramolecular cyclization and side reaction. These are definitions of gelation since it is not necessary to be two components. More details will be described in Sect. 3.4.

From structural viewpoint of polymer gels, various attempts have been made to characterize the specificity of gel by branching fraction [6] or correlation length by scaling laws [7]. The scaling concept described the structure of random fractal systems in terms of scaling laws trying to illustrate general properties.

In order to obtain the covalently connected together with continuous network, the network should develop from one edge of the gel to the other by going through

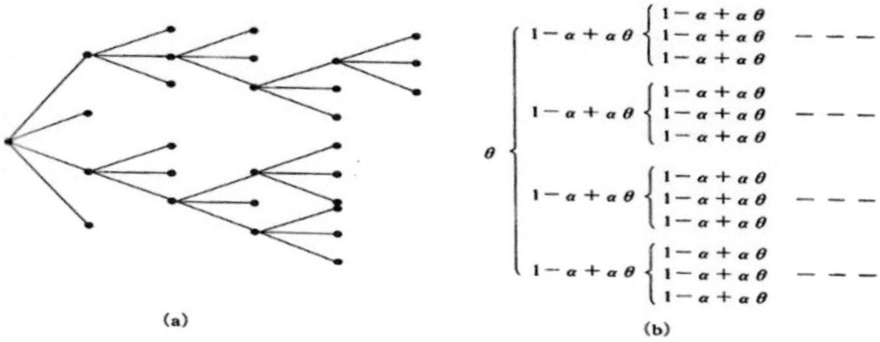

Fig. 3.1 Tree model (f −4) of functional condensation polymerization and its corresponding probability processes

a crosslinking junction, following another polymer chain going through a different crosslinking joint [7, 8]. This is called **percolation**. The probability Ω of finding a crosslinkpoint of the polymer chain to which another polymer chain is connected through crosslinking is given as

$$\Omega = \varepsilon(\chi - 1) = \varepsilon\chi \tag{3.1}$$

where χ is degree of polymerization, and ε is the concentration of the crosslinked chemical repeat unit. Generally, the critical branching index αC is given by $\alpha C = 1/(F-1)$, where F is the number of polymer chains that pass through a crosslink point or region. In this situation, problems of polymer concentration and loop formation have not been considered. When these issues are considered, the theory is known as percolation theory, and will be discussed more in detail in Sect. 3.4.

3.2 Chemical Gel

A chemically crosslinked polymer network is composed of two parts: crosslinking junction and chain backbone. Crosslinking junction has four-armed branches covalently connecting with other junction point through chain backbone. Chain backbone is linear portion of polymer chain between two junction points. Thus, crosslinked polymer network is considered as whole macroscopic matter consisting of one huge "giga" molecule. Polymer network can be formed by any chemical reaction to form covalent bonding and is largely classified by two procedures.

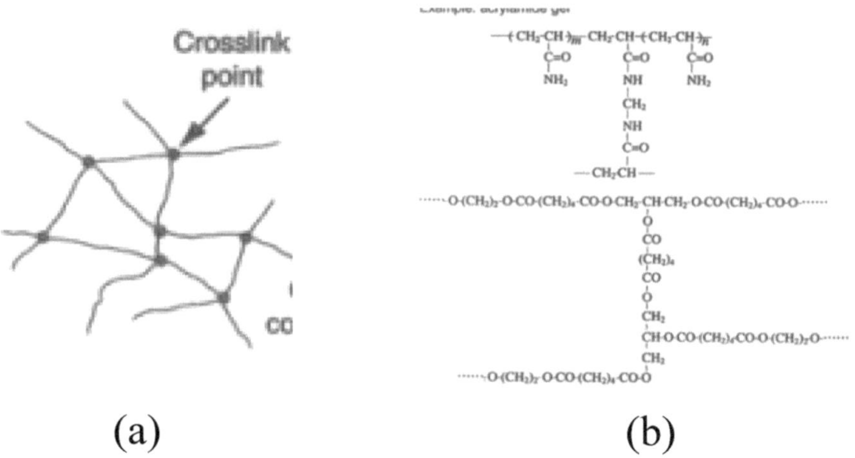

Fig. 3.2 **a** Typical network structure of chemical gel showing crosslink point and chain backbone, **b** chemical structure of polyacrylamide gel crosslinked by N,N-methylenbis acrylamide

3.2.1 Crosslinking During Polymerization Reaction

(1) **Crosslinking during polymerization reactions** (simultaneous polymerization and crosslinking):
(2) **Crosslinking of preformed polymer chains** (crosslinking of preformed linear polymer chains or biopolymers by chemical reaction)
(3) Can further be divided into two major groups by the polymerization process.

   1. Chain polymerization (radical polymerization, ionic polymerization, etc.) in the presence of divinyl or other crosslinking compounds
   2. Step growth reactions (polycondensation, polyaddition, etc., in the presence of multifunctional compounds)

3.2.1.1 Chain Polymerization

Radical Crosslinking Copolymerization (RCC)

Among chain polymerization, free-radical crosslinking copolymerization (RCC) of unsaturated monomers using divinyl group as a crosslinking agent is the most frequently used to prepare polymer networks (Fig. 3.2). Ion exchange resins (both anionic and cationic) are typically produced in industry by RCC of styrene derivatives and divinyl benzene.

In the case of RCC, chemical initiators are usually used, but UV light, gamma ray, and electron beam radiation can also be used. The most typical polymerization procedure is as follows: RCC to give chemically crosslinked polymers gel is performed by polymerizing an unsaturated vinyl monomer in the presence of appropriate amount (usually a few %) of divinyl compounds. If RCC performed in a good solvent, the gel is obtained in swollen state [6, 9]. On the other hand, if the polymerization is performed in a poor solvent, the gel is obtained as a particle or as a precipitate. Peroxides or azobis-based compounds are often used as a radical initiator. Typical crosslinking divinyl compounds for organic media are ethylene glycol dimethacrylate, divinyl benzene, and different bisacrylamide derivatives (N,N-hexamethylene dimethacrylamide, N,N-decamethylene dimethacrylamide), and dimethacrylic esters of 1, 12-(*P*-oxyphenyl) dodecane are used. Typical crosslinking agents in an aqueous medium is methylene bisacrylamide. These crosslinking agents are used at elevated temperature (50 °C and higher) to decompose the initiator to produce initiating radicals.

A low temperature (room temperature or below) RCC can be carried out by irradiating UV light, γ-ray, or electron beam. In case of UV irradiation there are two methods. One is to irradiate UV light that matches the vinyl monomer's absorption wavelength. The other is to shine UV (or even visible light) in the presence of photosensitizer that produces free radicals by absorbing the light. Hydrogels from water-soluble monomers such as acrylic acid, vinylpyrrolidone, and N-isopropylacrylamide are often prepared by γ-ray irradiation since water used as a solvent produces hydroxyl and other radicals which can additionally initiate the polymerization. Low-temperature RCC can also be carried out by using redox initiators, such as hydrogen peroxide-Fe^{2+} and persulfate-sulfite. Low-temperature RCC has advantages in creating gels containing bioactive substances to immobilize them. However, some monomers do not polymerize or crosslinked by γ-ray irradiation.

It should be noted that polymer network prepared by RCC has strong irregular structures, i.e., the distribution of crosslinking junction is quite random giving different network size (size of segment between two crosslinking points). This comes from (1) the complicated statistical nature of the elementary processes of radical polymerization, (2) concentration change of reacting component accompanied with the reaction progress, and (3) surface effect due to insolubility of formed network. Detailed process and mechanism is described in [9].

RCC contains various types of side reactions in its elementary processes and is far from simple propagating and crosslinking reactions. Chain transfer reactions occur to initiator, polymers backbone, solvent, and others to give chain termination. These side reactions bring about numerous structural defaults such as loops, dangling chains, large fluctuation of chain length between branching points, and heterogeneous composition from chain-to-chain. For instance, when the monomer concentration is high, it gives highly condensed polymers and enhances the intermolecular crosslinking reactions by attacking the chain backbone (chain transfer to polymer chain) and leads to form the irregularly distributed network. Therefore, RCC in bulk (without solvent) or under high monomer concentration is generally not recommended. Oppositely, when monomer concentration is very low, the intramolecular crosslinking reaction predominates at the early stage of polymerization and produces highly heterogeneous structure where loops and branching points are preferentially distributed. If these side reactions take place, highly branched microgels

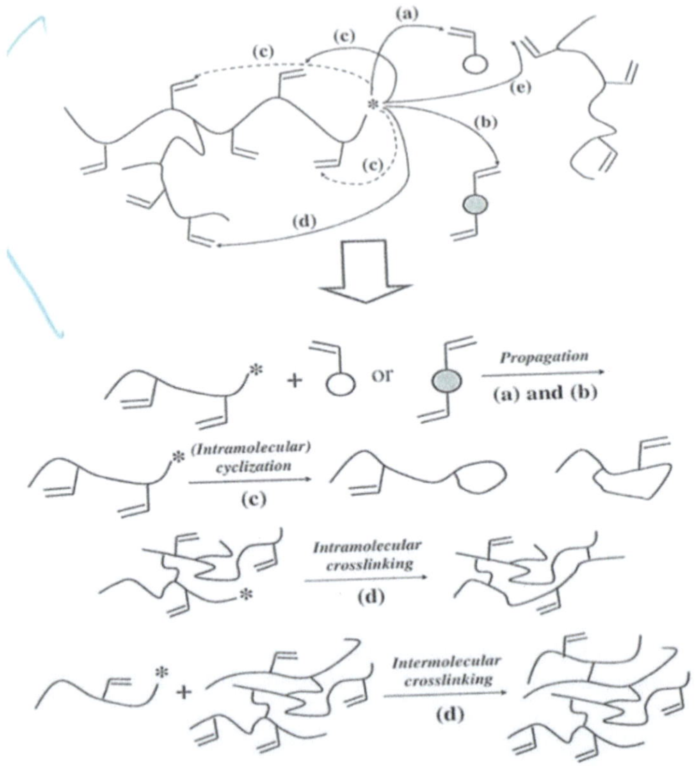

Fig. 3.3 Schematic representation of elementary processes of radical crosslinking copolymerization. These side reactions lead to the irregular network structure

are formed in an early stage that subsequently bind together and eventually form the macroscopic gelation with highly irregular network distribution.

Figure 3.3 shows possible elementary processes during RCC. The active center at the growing chain end has various possibilities to react with: (a) monomer, (b) divinyl monomer (crosslinker), (c) pendant double bonds of crosslinker on the same chain (lead to cyclization), (d) pendant double bonds of the same macromolecule but located on other chains (intramolecular crosslinking), (e) double bond of another macromolecule (inter-macromolecular crosslinking).

Another feature that makes the crosslinking process more complex is that one of two double bonds of crosslinking agent has different reactivity after the first double bond has been reacted. When the first double bond is reacted, the crosslinking agent is incorporated in a part of growing polymer chain as a pendant vinyl group. Therefore, the apparent reactivity of the second double bond becomes lower due to steric hindrance as well as the change in its electronic state of double bond. In order to minimize the inhomogeneity of the network structure, it is important to carefully select a combination where reactivity ratio of the monomer (r_1) and crosslinking divinyl monomer (r_2) are similar. If the copolymer reactivity ratio is equal, i.e., $r_1 = r_2 = 1$, this combination of RCC gives, in principle, network with regular-distributed homogeneous structure. However, such a combination of reactivity ratio is practically difficult to find when the copolymer reactivity ratio, r_1 and r_2, is different values.

One should note that different values of r_1 and r_2 bring about a large difference of the consumption rates of monomer and crosslinking agent and result in a substantial heterogeneity of the network which becomes prominent with the progress of the polymerization. If $r_1 < r_2$, the major crosslinking reaction will be completed in the early stage of polymerization, and if $r_1 > r_2$, the crosslinking may preferentially occur at the later polymerization stage.

Difference of the apparent reactivity to give heterogeneous network structure occurs not only by the change of chemical reactivity, but by the spatial kinetic inhomogeneity of reactions due to the different molecular mobility of reacting species. The pendant double bond is usually occluded as a part of growing chain in the internally crosslinked macromolecules, monomer molecules become less accessible to react (excluded volume effect), and the apparent reactivity significantly decreases. This has been verified by the presence of residual double bonds in the network when obtained at high conversion degree but the exact reactivity of the second double bond is not satisfactorily evaluated. This process is called "gel effect" and described in detail at Sect. 3.2.2.

Microgel consisting of limited number of networks is characterized with decreased heterogeneity of the network structure and as will be described in Chap. 9 and has a variety of special applications in industry. In that case RCC in dilute solution or precipitated condition is employed since it favors intramolecular crosslinking reactions.

Emulsion polymerization is another popular method to form microgel. There, growing chains within micelles are isolated from those of other micelles and prevent to develop macro-gelation. It yields smaller particles of narrower size distribution due to enhanced internal crosslinking.

To control the network structure or minimize the heterogeneity of the network, **controlled radical polymerization (CRP)** [10] has been developed, and the irregular network structure has significantly improved. They include atom transfer radical polymerization (ATRP), reversible addition fragmentation chain transfer polymerization (RAFT), stable free-radical-mediated polymerization (SFRP), and nitroxide-mediated polymerization (NMP). They are often called living radical polymerization. When CRP is performed in the presence of a divinyl compound, a crosslinking reaction of divinyl monomer occurs much later than RCC. In case of CRP, short chains are formed at the early stages, and numbers of pendant double bonds in each chain remain unreacted since the reactivity of these pendant double bonds is much lower than conventional radical polymerization [11]. These unreacted crosslinkable pendant moieties make possible to crosslink polymer chains subsequently to give regular network structure.

Among these controlled radical polymerization methods, ATRP has most actively studied and significantly improved network with microgel cores was prepared via formation of star polymers [12]. For example, ATRP polymerization of methyl

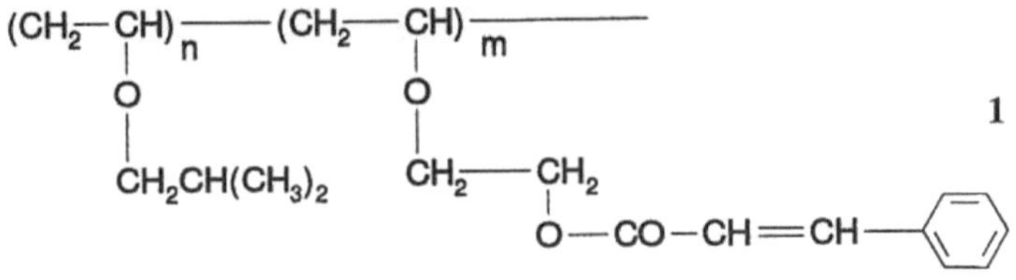

1

Fig. 3.4 Chemical structure of poly(iso-butyl vinyl ether)-block-poly[2-(vinyloxy)ethyl cinnamate] (PIBVE-b-PVEC)

methacrylate was carried out in the presence of ethylene glycol dimethacrylate used as a crosslinker. There, the intra-chain crosslinking reaction to form cyclic structure was minimized, and the gel point of the obtained gel well fitted with Flory–Stockmayer theory suggesting more homogeneous network structure. A hipper-branched polymer which subsequently forms nano-ordered gel was also obtained by ATRP.

Although many polymer gels with controlled network structure have been synthesized by CRP [12], the strictly controlled network size has not been made at present, and important problem has left for further studies of gel science.

Ionic Polymerization

Anionic and cationic polymerizations are characterized by the methods to prepare polymers with stereo-regular structure and mono-or narrow distribution of molecular size. They are widely used to prepare block and graft copolymers. In principle, polymer gel could also be obtained by the ionic copolymerization if monomers are polymerizable by ionic polymerization. However, the reactivity of ionic polymerization is sensitive to chemical structure of monomers, particularly to the electronic density, and therefore, hydrophilic monomers to give hydrogel have not directly been polymerized unless special after reaction is followed.

The networks were first prepared block copolymer as a chain backbone by ionic polymerization and crosslinked afterward by radical polymerization. A block copolymer poly(iso-butyl vinyl ether)-block-poly[2-(vinyloxy) ethyl cinnamate] (PIBVE-b-PVEC) (Fig. 3.4) was first prepared by sequential anionic and cationic polymerizations, and the obtained di-block copolymers (i.e., polymer brushes with crosslinked anchoring layers) were made radical crosslinking through vinyl groups by UV irradiation [13].

In the similar way, the polystyrene-block-poly(2-hydroxyethyl methacrylate) (PS-b- PHEMA) was synthesized by anionic polymerization as a precursor. In the presence of this precursor, radical polymerization of cinnamoyethyl methacrylate was made to give polystyrene-b-poly (cinnamoyethyl methacrylate) (PS-b-PCEMA) network. Poly(ε-caprolactone) containing OH groups at both chain ends was synthesized by living cationic ring-opening polymerization of ε-caprolactone. Here, diol-containing initiator was used. Acrylate monomer was attached to both chain ends to give a macromolecular crosslinker. This macromolecular divinyl compound was polymerized by living polymerization to obtain a polymer gel with regular crosslinking network.

Examples described here may not exactly be included in the category of gel with controlled network since the crosslinking is made by radical polymerization onto the precursors prepared by ionic polymerization, but described here since whole process of synthesis was performed in the continuous processes and suitable to synthesize multifunctional gels with telechelic chemical structure.

3.2.1.2 Step Growth Reaction

Step growth polymerization refers to a polymerization in which two kinds of bifunctional monomers capable of reacting each other bind together and form polymers in stepwise. In case of condensation polymerization to form polyesters or polyamide, low molecular product as H_2O is released (Fig. 3.5), while addition polymerization does not produce any low-molecular side product. In order obtain high-molecular-weight polymers, a high extent of reaction of strictly equimolar mixture of bifunctional monomers is required. When monomer with three functionalities or more is mixed with bifunctional monomers, branching reaction and subsequently a crosslinking reaction take place in the course polymerization and ultimately form network.

The point at which this three-dimensional structure is formed is known as the **gel point** and can easily be observed by an abrupt formation of insoluble gel (refer to Chap. 2). The crosslinking density is proportional to the amount of multi(three or more) functional monomer added, and the higher the concentration of multifunctional monomer, the less it swells. Step growth reactions include esterification, transesterification, reactions of phenols with formaldehyde, epoxide groups, isocyanate groups to form polyurethanes, hydrolysis and condensation of alkoxysilanes, and others. Typical methods of network formation are shown in Table 3.3.

There exist several side reactions in the step growth polymerization and introduce irregular network structure with large polydispersity. They are:

1. Cyclization reactions.
2. Branch formation due to topological limit of conversion of functional groups.
3. Change in apparent chemical reactivity due to diffusion.

Fig. 3.5 Example of polycondensation to give polyester bifunctional p-terephthalic acid and ethylene glycol react to give poly(ethylene-p-terephthalate)

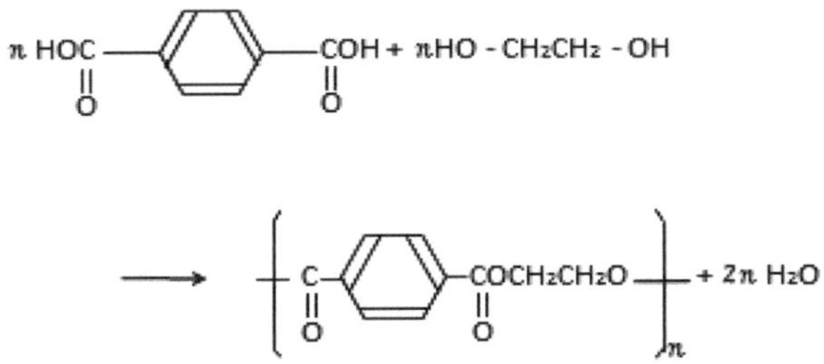

Condensation Polymerization

The condensation polymerization of bifunctional monomers, for example, ethylene glycol and p-terephthalic acid, will form linear polyesters, but in the presence of an appropriate amount of trifunctional monomer such as glycerin, it gives 3D network structure.

In case of condensation polymerization, chain length grows stepwise with the progress of reaction, and the gel is formed when the molecular weight becomes infinite as a result of network formation. Flory [6], Stockmayer [14], and Case [15] have studied in detail on this type of gelation. According to Flory [6], the critical condition to obtain 3D network structure is suggested to be

$$\alpha_c = 1/(f - 1) \tag{3.2}$$

where α_c is the critical branching index for gelation, and f is the number of the monomer's functional group. The relationship between α, reaction rate, and reactant ratio has been reported and shown as

$$\alpha = \rho / \left\{ R - P^2 (1 - \rho) \right\} \tag{3.3}$$

where P is the reaction rate of reacting group, R is the reactant ratio, and ρ is the ratio of monomer in branch units against the total number of monomers. These two equations can predict the extent of reaction at the gel point in the condensation polymerization of bifunctional monomers with multifunctional monomers once the feed ratio of monomers is given.

There is a class of networks formed by condensation polymerization which possess relatively high crosslinking density. Some types of reverse osmosis (RO) membranes are made by reacting aromatic diamine and aromatic dicarboxylic acid anhydride to form polyamide where an appropriate amount of tricarboxylic acid hydride is mixed as a crosslinking agent. The RO membrane and fibers do not show substantial swelling and swell only enough to permeate water. Another example of highly network formation by condensation reactions to give polyester in industry is known as alkyd resin which is used mainly as a coating material. It is stable but does not swell.

Silicon-based microgels is also an example prepared by condensation polymerization. They are synthesized by so-called sol–gel reaction, i.e., direct emulsion polycondensation of trimethoxymethylsiloxane to give polydimethylsiloxane network where trimethoxy group extends the chains and branches to give network.

Addition Polymerization

A polymer containing isocyanate (R–N=C=O) undergoes polyaddition reaction with diols or diamines resulting in urethane and urea bonds, respectively. When a simultaneous crosslinking is made using three functional compounds, a crosslinked gel is formed. This crosslinking method can produce gels with

Table 3.1 Typical crosslinking agents and the partner polymers with functional group

Crosslinking agent	Type of polymers
Dialdehyde compounds	Polymers with amino or hydroxyl group
Amines	Polymers with halogen, carboxylic ester, isocyanate, epoxy, metyrol group
Aziridine	Polymers with carboxyl group
Di or polymetyrol phenolic resin	Polymers with nitril, mercapto, carboxyl group
Halogen compound	Polymers with amine, diene group
Di or polyisocyanate compound	Polymers with active hydrogen as hydroxyl, thiol, amino. carboxyl group
Diol, polyol, bisphenol	Polymers with chlorosulfonate, isocyanate group
Diepoxy compound	Polymers with carboxyl, hydroxide, mercapto, chlorosulfonate group

relatively regular network structure compared to that of simultaneous chain crosslinking copolymerization. (see Table 3.1). There is a few examples of microgels obtained by polyaddition reaction. Polyurethane gels were synthesized from α,ω-dihydroxytelechelic polyethyleneoxide (PEO) and bis-(cyclohexylmethane-4,4-diisocyanate) using 1,2,6-hexanetriol as crosslinker [8]. Microgels based on polyurethanes were obtained where PEO plays the role of a steric stabilizer to prevent macrogel formation.

3.2.2 Diffusion Control of Network Formation (Dusek)

Covalent bond formation is controlled either by reactivity of functional groups or their diffusivity which depends on the mobility of segments they are bonded to. The reactivity also includes steric and microenvironment effects. In low-molecular-weight chemistry, the control by diffusivity is relatively rare and becomes effective only for very fast reactions. In macromolecular systems, the microenvironment effect becomes more important due to much stronger thermodynamic exclusion and selective interactions. Also, the diffusion control is much more frequent and important, and for a number of macromolecular systems, the diffusivity of large molecules or segments is much lower, and the diffusion control determines the application potential of such systems. Especially in crosslinked systems, the mesh structure of the network hinders movement of molecule, and the hindrance becomes very size-specific—small molecules hardly feel it but large molecules are effectively hindered from moving. At higher crosslink densities, it may be difficult to distinguish between hindrance in moving and full obstruction of the reaction. In the next few paragraphs, we will consider the diffusion-controlled kinetics where the diffusion process can be relatively well defined. We will distinguish between *overall* and *specific* diffusion control. The overall diffusion control affects mobility of all reactive groups;

it is determined by increasing viscosity and is typical for transition of molten or rubbery state into the glassy state. The specific control is concerned with mobility of specific chains or structures and reactive groups attached to it, distinguished, for instance, by their degree of polymerization. The Trommdorff or gel effect is associated with reduction of termination rate in free-radical polymerization and results in acceleration of polymerization.

3.2.2.1 Chemical Versus Diffusion Control

The reactivity and diffusion control kinetics are described by the Eyring transition state theory [16] formulated for bimolecular reaction as

$$-\mathrm{A}+\mathrm{B}-\underset{k_{\mathrm{AB},-\mathrm{D}}}{\overset{k_{\mathrm{AB,D}}}{\longleftrightarrow}}\langle-\mathrm{AB}-\rangle^{*}\overset{k_{\mathrm{AB,C}}}{\longrightarrow}-\mathrm{AB}- \qquad (3.4)$$

The group A attached to a chemical structure reacts with group B, and a transition (or activated) complex $\langle-\mathrm{AB}-\rangle^{*}$ is formed. The rate of complex formation is determined by rate constant $k_{\mathrm{AB,D}}$ (D means diffusion control). The formed activated complex is unstable and dissociates again back into the initial reactants with a rate controlled by the rate constant $k_{\mathrm{AB,-D}}$. The other choice for the complex is the transformation into a stable product, the bond –AB–. This happens with a rate controlled by the rate constant $k_{\mathrm{AB,C}}$.(C means **c**hemical control). When $k_{\mathrm{AB,D}}$ and $k_{\mathrm{AB,-D}}$. >> $k_{\mathrm{AB,C}}$, i.e., the reaction partners form the complex many times and dissociate again before they form a stable product, and the reaction outcome is controlled by $k_{\mathrm{AB,C}}$., i.e., by chemical reactivity. However, when the diffusion is slow and $k_{\mathrm{AB,D}}$., $k_{\mathrm{AB,-D}}$. < $k_{\mathrm{AB,C}}$., the stable product is formed almost each time whenever the activated complex is formed by a rate controlled by $k_{\mathrm{AB,D}}$ (diffusion or encounter control). Equation 3.4 can be solved, and using some simplifying assumptions, one arrives at Rabinowitch equation in the form

$$\frac{\mathrm{d}[-\mathrm{AB}-]}{\mathrm{d}t}=\frac{k_{\mathrm{AB,C}}k_{\mathrm{AB,D}}}{k_{\mathrm{AB,C}}+k_{\mathrm{AB,-D}}}[\mathrm{A}][\mathrm{B}]=k_{\mathrm{app}}[\mathrm{A}][\mathrm{B}] \qquad (3.5)$$

from which is equivalent to the usual form of Rabinowitch's equation [17]

$$\frac{1}{(k_{\mathrm{AB}})_{\mathrm{app}}}=\frac{1}{k'_{\mathrm{AB,C}}}+\frac{1}{k_{\mathrm{AB,D}}},\ \text{where}\ \frac{1}{k'_{\mathrm{AB,C}}}=\frac{1}{k_{\mathrm{AB,C}}}\frac{k_{\mathrm{AB,D}}}{k_{\mathrm{AB,D}}} \qquad (3.6)$$

The constant $k'_{\mathrm{AB,C}}$ rate constant is controlled by Arrhenius dependence on temperature (different from $k_{\mathrm{AB,D}}$ dependence on temperature)

$$k'_{\mathrm{AB,C}}(T)=A\exp(-\Delta F_{\mathrm{act}}/RT);\quad k_{\mathrm{AB,D}}(T)=\phi_{\mathrm{mob}}(T) \qquad (3.7)$$

The constant $k'_{AB,C}$ is independent of conversion but $k_{AB,D}$ does depend on conversion because the segmental mobility depends on the structure; for instance, the glass transition temperature changes (increases) as the structure develops, or chain displacement time increases when it is longer.

Thus, for a bimolecular reaction of groups bonded to polymer chains one can write

$$\frac{d[-AB-]}{dt} = k_{app}[-A][B-] \tag{3.8}$$

Whether k_{app} is controlled by reactivity and is independent of conversion or whether it depends on conversion and is controlled by groups diffusivities depend on relative rates of chemical transformation and diffusion. For many macromolecular systems in which bonds are formed by step reactions ("slow" kinetics), the groups reenter the activated state many times before a stable bond is formed. The kinetic laws valid for low-molecular-weight systems are obeyed, although the numerical values of the rate constants need not be the same. No change in functional group kinetics is observed in the vicinity of the gel point. Only at high conversions of functional groups, the difference between low-molecular-mass and high-molecular mass kinetics gets noticeable, and a topological limit of reaction extent exists [18].

The mobility of molecules decreases with increasing size and increasing viscosity of the medium. Trommdorff effect is a demonstration of diffusion control in linear or crosslinking chain polymerization. It is manifested by acceleration of polymerization due to the onset of diffusion controlled termination by free-radical recombination.

According to this reasoning, the gelation phenomenon should always be connected with diffusion control because it involves the reactions between very large molecules embedded in medium of large viscosity. Although it is not necessary that the huge molecular clusters are displaced as a whole, the study of critical exponents near the gel point shows that they are different [19] than the mean-field model on which the mass action law is based. However, we do not see any break on the conversion-time dependence near the gel point. This is due to that that an overwhelming majority of groups is bonded to "small" molecules which move fast enough [20].

3.2.2.2 Overall Diffusion Control (Glass Transition Control)

Many crosslinking systems forming eventually glassy materials pass during their formation the stage of transformation of melt or rubber states to the glass state. As the viscosity increases and the glass transition region is approached, the reaction rate becomes diffusion controlled. In terms of Eq. 3.7, the kinetics change from initial $k_{AB,D} >> k'_{AB,C}$ to $k_{AB,D} \approx k'_{AB,C}$ and eventually to $k_{AB,D} << k'_{AB,C}$", the transition being governed by the Rabinowitch's equation (Eq. 3.7).

While the temperature dependence of the "chemical" rate constant is governed by the Arrhenius equation (Eq. 3.7), the dependence of k_D is different. The constant k_D is related to the segmental diffusion coefficient

$$k_D = k_{D0} D_s \tag{3.9}$$

which is a function of temperature, conversion, and volume fraction of solvent, if present. The diffusion coefficient D_s or k_D is related to T_g through the change of the free volume (volume fraction v_f) or through the configurational entropy. Most treatments of diffusion-controlled kinetics are based on free volume change as described by the WLF (Williams, Landell, Ferry) [21] or by Doolittle viscosity [22], (see also [23]) theories. They have been applied to a number of thermosetting systems controlled by step or chain reactions. To use the theory of Adam and Gibbs [24] based on configurational entropy is an alternative approach. More details can be found in Refs. [25–37].

According to the free-volume theory

$$D_s = D_{s0} \exp\left(\frac{b_D}{v_f}\right) \tag{3.10}$$

where the fraction of free volume as a function of temperature is expressed as

$$v_f = v_{fg} + \alpha_f (T_c - T_g) \tag{3.11}$$

where v_{fg} ($\approx$0.025) is the fraction of free volume at T_g. This relation gives the dependence of k_D on temperature

$$k_D = k_{D0} \exp\left(-\frac{b}{v_{fg} + \alpha_f (T_c - T_g)}\right) \tag{3.12}$$

Here, T_c is the reaction (cure) temperature, and α_f is the expansion coefficient of free volume for the reacting system. If the monomer greatly differs in size from polymer, their expansion coefficients differ as well, and like in free-radical polymerization, a mixing rule is applied [36].

$$v_f = [0.025 + \alpha_P (T - T_{gP}]\alpha + [0.025 + \alpha_M (T - T_{gM}](1 - \alpha) \tag{3.13}$$

where α_P and α_M are expansion coefficient of the polymer and monomer, respectively, and α is reaction conversion. For non-isothermal cure, Simon and Gillham [34] took into account of the Arrhenius temperature dependence of the constant k_{D0}

$$\ln k_D = \ln A_D - \frac{E_D}{RT} + \frac{C_1 (T_c - T_g)}{C_2 + (T_c - T_g)} \tag{3.14}$$

where C_1 and C_2 are the WLF constants, and A_D and E_D the pre-exponential factor and activation energy of k_{D0}, respectively.

The Adam–Gibbs theory [24] relates the diffusion coefficient to the configurational entropy S_c

$$D_s = D_{s0} \exp\left(-\frac{K}{TS_c}\right) \quad (3.15)$$

which eventually gives [35]

$$D_s = D_{s0} \exp\left(-\frac{m}{T_c \ln\left(T_c/\left(T_g - T'\right)\right)}\right) \quad (3.16)$$

where m and T' are constants; $T' \approx 50$ K

Equations 3.9–3.12 or 3.14 need the dependence of T_g on conversion, α. Several semi-empirical equations are available such as

$$\frac{1}{T_g} = \frac{1-\alpha}{T_{g0}} + \frac{\alpha}{T_{g\infty}} + \frac{(1-\alpha)\alpha}{C} \quad (3.17)$$

where T_{g0} and $T_{g\infty}$ are T_g values at conversion $\alpha = 0$ and $\alpha = 1$, respectively. Alternatively, Di Benedetto equation reads

$$\frac{T_g - T_{g0}}{T_{g\infty} - T_{g0}} = \frac{\lambda\alpha}{1-(1-\lambda)\alpha} \quad (3.18)$$

where λ is an adjustable parameter

Fraga et al. [28] analyzed various forms of equations relating k_D to $T - T_g$ difference with respect to the value of the diffusion controlled constant at T_g if expressed through the kernel of the Smoluchowski coagulation equation

$$k_{Dg} = 4\pi N_A r D_g \quad (3.19)$$

For plausible values of the parameter, they came to a surprising conclusion that the values of the constant are very different, varying in the range of almost six decades. Nevertheless, about 5 degrees below T_g, k_D values and their decrease were almost the same. It means that the change of the denominator is the controlling factor, only the temperatures at which the diffusion control becomes noticeable, namely the $T_c - T_g$ differences, are different.

Below, an example is shown of overall diffusion control of the curing reaction of a diamine with a di-epoxide. Slowing down or full arrest of the reaction is due to increase of viscosity of the reaction medium and transition of the reaction system into the glassy state. Figure 3.6 shows the effect of the diffusion control on the reaction kinetics for reaction taking place at different temperatures. It is evident that at lower temperatures, the reaction does not go to completion.

Because the glass transition temperature, T_g, increases with increasing conversion (Fig. 3.7), at a certain conversion T_g can become larger than the curing temperature T_c (Fig. 3.7).

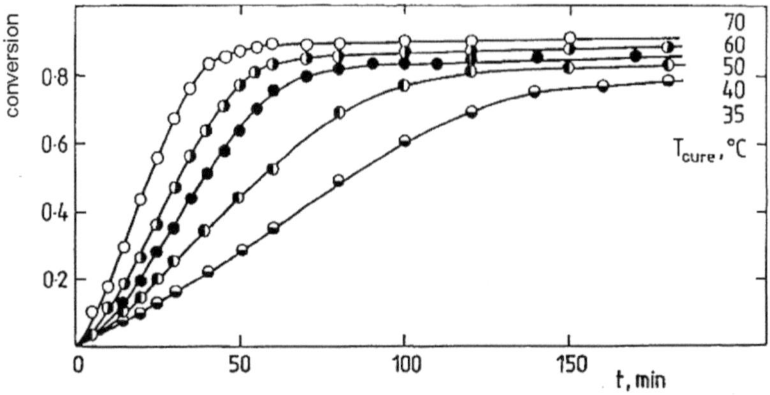

Fig. 3.6 Curing of diglycidyl ether of bisphenol A (DGEBA, 2,2-bis(4-glycidyloxyphenyl) propane) with 1,3-diminopropane in stoichiometric ratio at different temperature. Reprinted with permission from Prog. Org. Coatings 1993, 22,145–159. ©2000 Elsevier Science Ltd. [35]

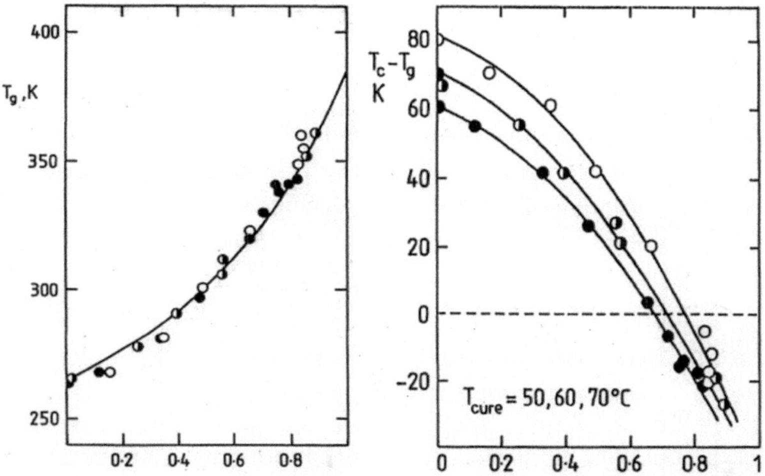

Fig. 3.7 Dependences of T_g on conversion of epoxide groups (left figure) and difference between the temperature of cure T_c and T_g (right figure) on conversion for the reaction of epoxy groups of diglycidyl ether of bisphenol A with 1,3-diaminopropane. Reprinted with permission from Prog Org Coatings 1993, 22,145–159. ©2000 Elsevier Science Ltd. [35]

In Fig. 3.8 the apparent rate constants are compared with values calculated ones using the configurational entropy model (Eqs. 3.7, 3.10, 3.15, 3.16) and the dependence of T_g on conversion of Eq. 3.17.

It is seen that the model describes well the sharp fall of the apparent rate constant starting in the vicinity of T_g. Hypothetically, the reaction should stop when the denominator of Eqs. 3.12 or 3.16 become zero (which is about 50 K below T_g), but

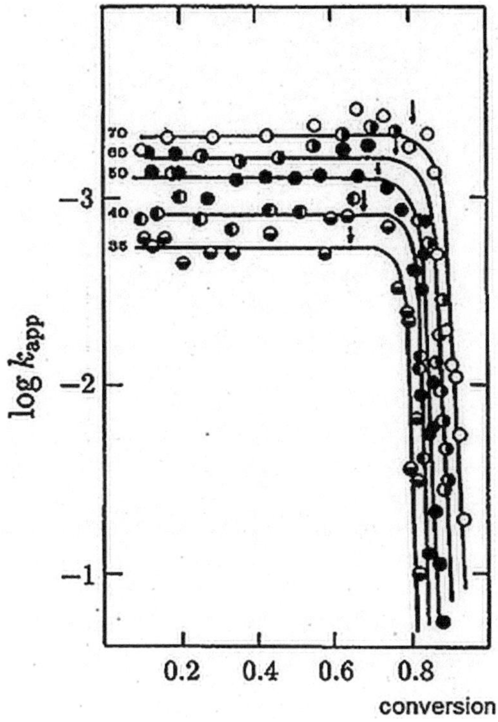

Fig. 3.8 Experimental and calculated (—) dependences of the apparent catalytic rate constant for the reaction of epoxy groups of diglycidyl ether of bisphenol A with 1,3-diaminopropane on conversion of epoxide groups at temperatures indicated; T_g is marked by an arrow. Reprinted with permission from Prog Org Coatings 1993; 22:145–59. © 2000 Elsevier Science Ltd. [35]

the free-volume theory based on WLF concept does not hold any longer and the volume relaxation of glass (physical aging of glass) rather follows another dependence on temperature. The relaxation process is described by stretched exponential (Kohlrausch–Williams–Watts equation)

$$\phi(t) = \exp\left(-\frac{t}{\tau}\right)^{\beta} \tag{3.20}$$

rather than simple exponential

$$\phi(t) = \exp\left(-\frac{t}{\tau}\right) \tag{3.21}$$

Lee and McKenna [38] investigated the applicability of a time-aging time–temperature-conversion superposition principle to describe the influence of crosslink density on the physical aging behavior and found that as the aging time increased with increasing crosslink density. The aging time effects on glass relaxation were recently reviewed [39], and some relations for glass aging depending on glass formation history have been proposed. In the case of a system reacting in the glassy state, bond formation is connected with increase of free volume due to reaction contraction, so that the reaction rate is expected to be higher than would correspond to relaxation of a non-reacting glass. Definitely, the volume relaxation due to aging should be considered for long-lasting application in all cases where large free volume is generated by early glass transition of not fully reacted systems and/or containing volatile solvents.

Similarly, as in stepwise reacting thermosetting resins, the overall diffusion control effect is observed in chain free-radical polymerizations in the later stages of reaction if T_g of the final polymer exceeds the reaction temperature. This is often the case of ambient temperature crosslinking, but also elevated temperature crosslinking. For instance, conventional monomer like poly(methyl methacrylate) or polystyrene has T_g around 100 °C, while thermal polymerizations are carried out at about 60–80 °C; the same problem is with hydrophilic monomers line poly(2-hydroxyethyl methacrylate) or polyacrylamide. Then, in the final stages, all reaction steps are hindered by transition of the reacting system into the glassy state. Figure 3.4 shows an example of such case for dimethacrylate dental resin [40] (Fig. 3.9).

Obtained by simulations [36], the net deceleration effect of overall diffusion control on initiation, propagation, deactivation, and termination rate constants is shown in Fig. 3.10.

The decrease of rate constants is the net effect; otherwise, the system exhibits the acceleration by termination controlled by specific diffusion effect which will be discussed in the next paragraph.

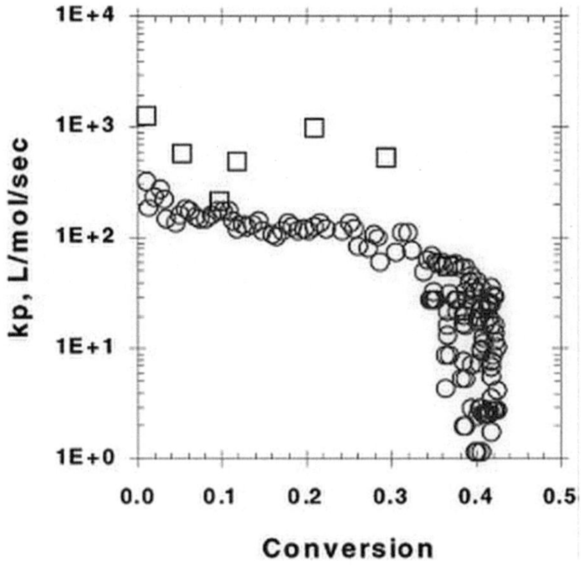

Fig. 3.9 Propagation rate constant k_p as a function of conversion for photopolymerization 75/25 bisGMA/TEGMA (bisGMA = 2,2bis[p-(2-hydroxy-3-methacryloyloxyprop-1-oxy)-phenyl]propane, TEGMA = tetraethyleglycol methacrylate); results by NIR (□) and EPR (o). Reprinted from ref. [40] (Fig. 3.3) with permission

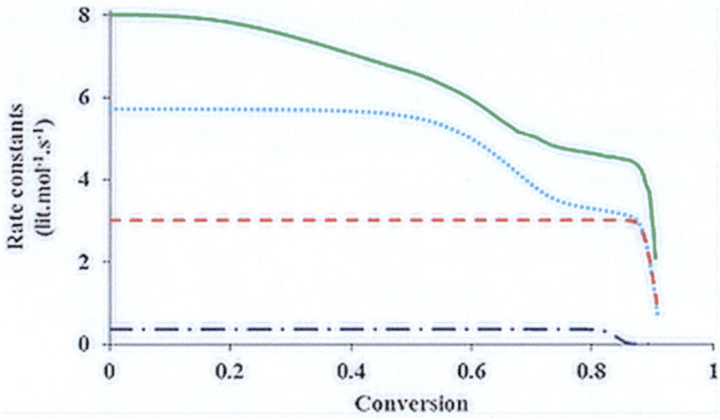

Fig. 3.10 Rate coefficients in logarithmic scale of termination (——), deactivation (····) and propagation (– – –), as well as activation (–· –), as a function of monomer conversion. Reprinted from ref. [36] with permission

3.2.2.3 Specific Diffusion Control Trommdorff or Gel Effect

In contrast to the overall diffusion control where all reactions are slowed down by increasing viscosity, the specific diffusion control is concerned with specific molecules or substructures; the movement of these structures is retarded differently. There are many such instances of diffusion-controlled associations in colloid and macromolecular science, and they are traditionally described by Smoluchowski coagulation equations. In its simplest form, the rate by which particles i and j merge is given by

$$\frac{\mathrm{d}c_{i+j}}{\mathrm{d}t} = k_{\mathrm{t}}^{ij} c_i c_j$$
$$k_{\mathrm{t}}^{ij} = 4\pi b\big(D_i + D_j\big)N_{\mathrm{A}} \tag{3.22}$$

where D_i and D_j are diffusion constants, and b is capture radius. The coefficient k_{ij} depends on temperature, molecular weight, architecture of the polymer, and solvent concentration.

Here, we will remind only one kind of specific diffusion control typical for linear and crosslinking (co)polymerization—the Trommdorff or gel effect. In the course of free-radical polymerization, acceleration of polymerization is observed which is caused by an increase of the free-radical concentration due to retardation of termination by macroradicals coupling. The retardation is caused by decreased mobility of macroradicals when their length is big enough, and the viscosity of the medium they move through is high enough. The typical conversion versus time dependence for polymerization of methyl methacrylate is displayed in Fig. 3.11.

Crosslinking contributes to viscosity, beyond the gel point the free radical attached to the gel becomes less mobile, and the gel effect is operative earlier than in linear

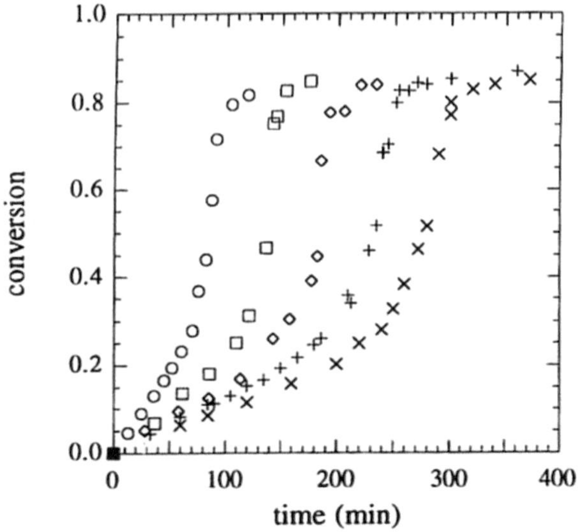

Fig. 3.11 Conversion as a function of time for MMA polymerizations at 70 °C and five different initiator (AIBN) concentrations: (o) 6 wt%; (□) 2 wt%; (◇) 1 wt%; (+) 0.5 wt%; (x) 0.3 wt%. Reprinted from [41] (Fig. 3.6) with permission copyright American Chemical Society

polymerization (Fig. 3.11). In the figure, also the overall diffusion control by entering the glass transition region is well seen. Because of the dependence of *D* on molecular weight, the dispersity in degree of polymerization is important [42] (Fig. 3.12).

The Trommdorff (also called Trommdorff–Norrish) effect was reviewed by Barner–Kowollik and Russel [43]. Much more light was shed on the mechanism of diffusion-controlled macroradicals recombination based on analysis of the results of controlled radical polymerization such as reversible addition fragmentation chain transfer (RAFT) and application of sophisticated spectroscopic approaches,

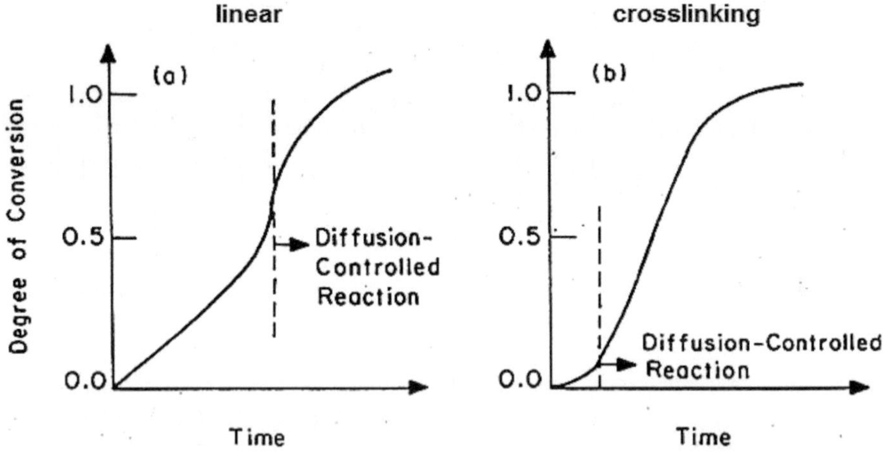

Fig. 3.12 Comparison of the gel effect (acceleration of polymerization) in linear and crosslinking chain polymerization

including time-resolved electron paramagnetic resonance spectroscopy, to monitor the concentration of radicals after a laser pulse. Studies of the rate of free-radical recombination in non-polymerizing system have shown that recombination rate follows exactly the Smoluchowski equation.

Smith et al. [44] proposed a composite model to express the dependence of the termination constant in the region of diffusion control on the degrees of polymerization of reacting macroradicals as

$$k_t^{ij} = \frac{1}{2} k_t^{11} \left(i^{-\alpha} + j^{-\alpha} \right) \tag{3.23}$$

Johnstone–Hall and Monteiro [45] found for linear polymerization that the matrix architecture had little effect on the Trommdorff acceleration and concluded that it was the chain overlap that was the main cause of the observed acceleration. Entanglements are not believed to have any specific effect on diffusion control and reptation can be operative only at very high conversions. In most cases, Rouse friction model serves well. The dependence of the termination constant for reaction between two macroradicals of polymerization degree i varies with i as shown in Fig. 3.13.

Below α_S the chains are short and move freely, and above α_S the chain end radicals start to be buried inside the coils and have to reorient to be able to react. When the polymerization degree i reaches the value i_{gel}, the overlap limit is reached.

Trommdorff effect is operative also in preparation of hydrogels but when diluted with a good solvent (water), its effect is reduced, and the typical acceleration is only

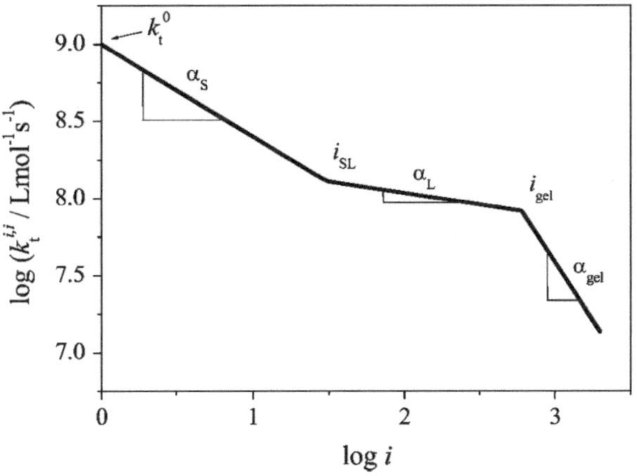

Fig. 3.13 Generalized composite model description for termination is illustrated in a log–log plot of the chain length-dependent termination rate coefficient, $k_t^{i_1}$, versus chain length, i. Also showed are the chain length-dependent exponents (Eq. 3.21) for the termination of short chains in dilute solution, α_S, and long chains in dilute solution, α_L, along with the crossover chain length between short- and long-chain behavior, i_{SL}, the chain length at onset of the gel effect, i_{gel}, and the chain length-dependent exponent for termination in the gel regime. Reprinted from ref. [45] Fig. 3.12 with permission

weak or at high dilution it is not observed at all. Other problems may be encountered in highly does not gel.

Diluted system: (a) the reactive group concentration is very low, and reaction rate is very slow, so that full conversions are not reached, (b) cyclization overrides intermolecular crosslinking and the system.

3.2.3 *Crosslinking of Preformed Polymer Chains*

3.2.3.1 Synthetic Polymers

Any synthetic and natural polymers can be crosslinked if they have chemically reactive groups such as hydroxyl, carboxyl, and amine on the pendant side chain or the main chain ends (Table 3.1). They can react with partner polymers containing the aldehyde, hydroxyl, carboxylic acid, and epoxy groups. These reactions trigger the crosslinking of linear polymers at the inter or intramolecular level. The starting material can have a wide selection from various kinds of already prepared synthetic macromolecules as precursors. One example is vinyl polymers containing glycidyl ether moiety in part, and it can crosslink by reacting with partner polymer-containing amines.

Reaction between macromolecules containing aldehyde such as poly (aldehyde gluronate) and adipic dihydrazide gives biodegradable and injectable crosslinked polymer gel by Sciff base reaction.

One of the advantages of the crosslinking of preformed polymers is that the distribution of crosslinking joint has essentially a statistical nature since no preference domains exist. This is different from the case of RCC where distribution of the crosslinking joint is determined by the chemical reactivity ratio of monomer and crosslinking agent and physical conditions which often lead to the heterogeneous crosslinking (refer to this chapter). Another advantage of this method is one can use many practically available materials under "monomer-free" condition. The third additionally important perspective is one can get materials with block-order networks which give own special properties.

The crosslinking reaction on macromolecular level has a long history. Bevington, as early as 1948, has reported that a crosslinked polymers such as polyvinyl chloride and polyvinyl acetate can be obtained by using Friedel–Crafts catalysts [46]. Crosslinking of polystyrene was also recognized in1949 by the reaction between the halogen containing solvent (dichloroethane) and styrene units [47]. Later, this crosslinking reaction was called as the chloromethylation and became a common route to obtain anion-exchange resins of polystyrene in industry. Chloromethylation of polystyrene is usually carried out using chloromethyl methyl ether, or formaldehyde-hydrochloric acid mixtures, in the presence of Friedel–Crafts catalysts. A crosslinking reaction occurs simultaneously following to the primary chloromethylation in early stage. Figure 3.14a shows the crosslinking structures derived from side reactions of chloromethylation of polystyrene. Cation-exchange

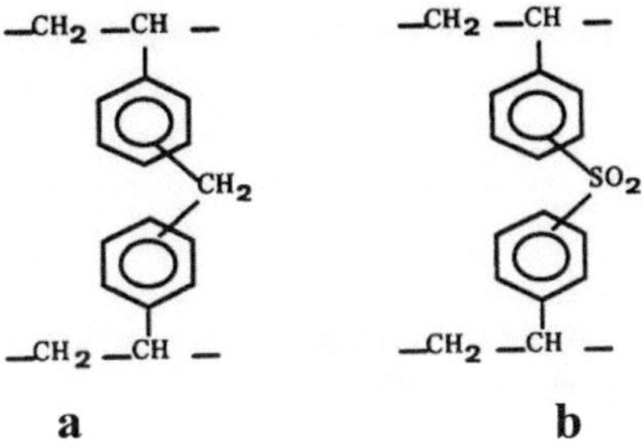

Fig. 3.14 Crosslinking reaction derived from parallel reactions during chloromethylation (**a**) and sulfonation (**b**)

resin is usually prepared by using sulfonating agents such as concentrated sulfuric acid, chlorosulfuric (chlorosulfonic) acid, and sulfur trioxide solutions. Among them chlorosulfuric acid is the most popular to introduce sulfonic group. Figure 3.14b shows the crosslinking structures derived from side reactions of sulfonation of polystyrene.

Crosslinking between polymers are often made, and the vulcanization of sulfur-containing compounds is the famous example widely used in the rubber industry.

Polymers containing unsaturated group can make crosslinking with peroxide units.

Another way to obtain crosslinked network from preformed polymers is to use chemical precursors. A precursor is a chemical compound in which functional groups capable of branching reactions are attached. These precursors with different molecular architecture are introduced as a component of the polymers to form the gel through chemical reaction or physical irradiation. Here, hyperbranched dendrimer and highly branched functional polymers are also used to attach the desired function to the gel. One advantage to use precursor with functional reactive groups is its easiness to synthesize both macrogel and microgel.

When the microgel is required to obtain, the reaction under dilute conditions, i.e., at a concentration generally much lower than the critical overlap concentration (C*) of the polymer is generally used to enhance the intramolecular chemical crosslinking.

However, when the concentration is increased, the pendant vinyl-groups present in the microgel may react with propagating radical species of a neighboring particle and may lead to inter-particular crosslinking to give microgel with increased size and eventually a macrogel. In contrast, when the formation of a macroscopic gel is required, the reaction could be made under increased concentration to develop the crosslinking continuously. Here, one should note that the substantial increase of the viscosity of the reacting medium takes place with progress of the polymerization, and the reaction under highly viscous medium is dominated by the diffusion process of reacting species to give heterogeneous network structure as described in detail in Sect. 3.2.2.

Interesting and new method to form polymer gels from preformed polymer is to use so-called ***Crick reaction*** which was developed by Professor Sharpless—the 2001 Novel Prize winner. Typical Crick reaction is the addition between azide and alkyne to give triazol in the presence of Cu(1) as a catalyst. Using Crick reaction homogeneously crosslinked polyethylene glycol (PEG) gel was obtained by reacting PEG-attached acetylene and PEG-attached tetra azide at their chain ends at room temperature. In the similar manner, poly (t-butyl acrylate) which contains azide groups at the chain ends was polymerized by ATRP method (refer to Sect. 3.2) and subsequently crosslinked by reacting with tetra-acetylene derivatives to give homogeneous crosslinked network.

Another type of Crick reaction is between thiol and unsaturated bond. Block copolymer of polylactic acid—PEG containing two acrylate groups at the chain ends was reacted with tetrathiol to give biodegradable gel with PLA-PEG-PLA block by illuminating visible light (365 nm) for ten minutes. Similarly, branched PEG containing double bond at each chain end formed homogeneously crosslinked PEG gel by reacting branched PEG containing thiol groups attached at the chain ends.

Crosslinking of the preformed polymers and precursors by exposing radiations such as UV light, gamma ray, and electron beam is widely used method. One advantage of these methods is capable of obtaining gels containing no monomer or catalysis (radical initiator) residues which are required for especial uses in medicine.

Photo-crosslinking is a method especially suitable to prepare large amounts of crosslinking and low swelling polymer gels. This technique is divided into photo-addition and photo-dimerization reactions. Examples of photo-addition reactions include tetraazonium salts and bisazide compounds. Water-soluble polymers, such as poly(vinyl alcohol) and poly(N-vinyl pyrrolidine), are often photo-crosslinked via diazo resins, bisazides, and chromic acid. When light is shone on a multifunctional aromatic diazonium salt (such as tetraazonium salt), free radicals are formed which subsequently crosslink the polymers. Bisazide compounds are decomposed by light to produce nitrane which is a reaction intermediate to crosslink polymers by addition, insertion, or hydrogen extraction reactions [47].

The crosslinking of polymers by photo-dimerization is achieved by introducing groups capable of photo-dimerization such as styrilbazorium and styrilbenzorium salt (Fig. 3.15). Water-soluble polymers having photosensitive group such as styrilbazorium salt are photo-dimerized to make crosslinking. Similarly, anthracene moiety and thymine derivatives were crosslinked with the partner polymer chains by photo-dimerization (Fig. 3.16).

The characteristics of photo-crosslinking are summarized as: (1) crosslinking is relatively homogeneous; (2) crosslinking takes place at low temperature and therefore suitable for enzymes and cells immobilization; and (3) patterning of gels similar to

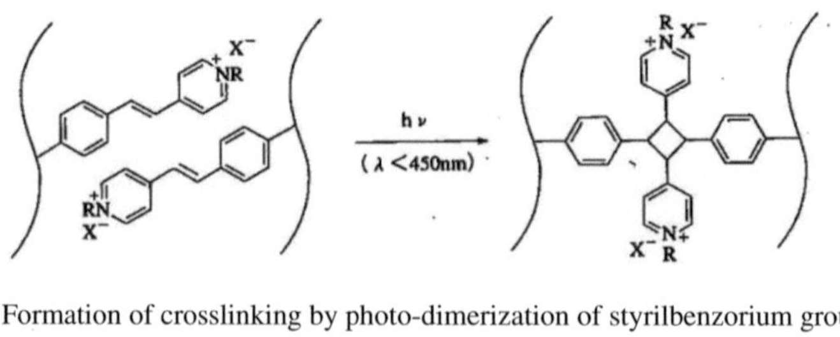

Fig. 3.15 Formation of crosslinking by photo-dimerization of styrilbenzorium group

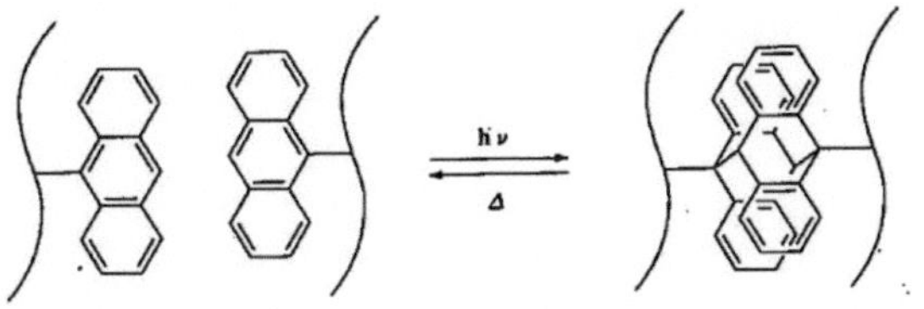

Fig. 3.16 Formation of crosslinking by photo-dimerization of anthranyl group

photoresist reaction is possible. The degree of crosslinking is controlled by the density of functional groups, the polymer concentration, and the wavelength, intensity, and duration of the light. The radiation crosslinking of performed polymers in the solid state is widely used in industry to produce electric wire, cable, and tire, polymer films, and others. If these crosslinked polymers are swollen in an appropriate solvent, a variety of crosslinked gels would be obtained.

Radiation crosslinking can also be carried out in solution. When water is used as a solvent, hydrogen abstraction by splitting water molecules and extraction of hydrogen from the main polymer chain occurs to form free radicals. The effective crosslinking is followed by the coupling reactions of the radicals. The rate of crosslinking in solution is much higher than in solids due to a high mobility of the polymer free radicals formed.

Crosslinking by irradiating gamma rays is performed in water using poly(methyl vinyl ether), poly(vinylpyrrolidone), polyethylene glycol, polyethylene, polystyrene, polyacrylate, and natural polymers. Figure 3.17 is the schematic process of crosslinking when a poly (vinyl alcohol) is irradiated by gamma rays in an aqueous solution.

Exposing electron beam is also widely used in industry to crosslink polyolefins such as polyethylene and polypropylene mainly to improve mechanical and thermal properties and not to get the gel at present. Radiation crosslinking is affected by the total dose, rate of irradiation, properties of the polymer (presence of multiple bonding, aromatic ring, etc.), and others. One should note in these radiation reactions that the chain scission of polymers take place competitively with crosslinking. The number of polymers that can be used for crosslinking is limited. In general, the reactivity against radiation is closely related to the chemical structure and is believed that the polymers with acrylic group [$-CH_2-CHR-$]$_n$ are crosslinked, while polymers with methacrylic group [$-CH_2-C(CH_3)R-$]$_n$, are decomposed.

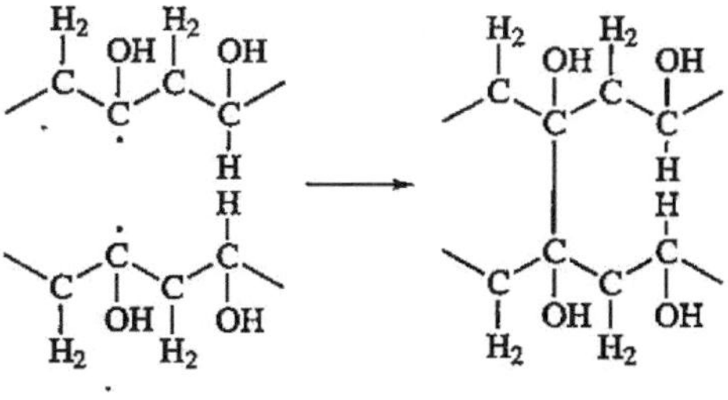

Fig. 3.17 Schematic process of crosslinking of poly(vinyl alcohol) by gamma rays in an aqueous solution

3.2.3.2 Biopolymers

Protein

Chemically crosslinked protein networks widely exist in our body, where the crosslinking is performed by enzyme-catalyzed reactions. Examples can be found as hair, skin, and cartilage where crosslinking by disulfide bonding or isopeptide bonding is most common. Transglutaminase (TGase), tyrosinase, and peroxidase are the typical enzyme-catalyzed crosslinker. Transglutaminase is the enzyme which forms isopeptide bonding between glutamine and lysin. Tyrosinase first oxidizes phenol group of tyrosine to form quinone and then binds amino residue through Michael addition reaction. Peroxidase binds two phenol moieties of tyrosine in the presence of hydroxy peroxide.

Many protein gels have been prepared by enzyme-catalyzed crosslinking. Gelatin and several oligopeptide gels were prepared by crosslinking with TGase, and rheological and mechanical properties were characterized to examine the possibility for tissue engineering and surgical adhesive applications. A gelatin-catechin conjugate was obtained by the lactase catalyzed oxidation of catechin in the presence of gelatin. F-Actin fibers can chemically be crosslinked by using transglutaminase.

Since many biological polymers possess chemically functional group as hydroxyl, carboxyl, and amino groups on the side chain or the chain ends, they can be crosslinked by using functional groups such as aldehyde, hydroxyl, carboxylic acid, and epoxy groups of partner chains. For example, poly(γ-benzyl-L-glutamate (PBLG)) gel with a cholesteric liquid crystalline structure was synthesized by crosslinking with ethylene diamine or triethylenetetramine. This type of gel showed nematic liquid crystalline structure with swelling anisotropy under a magnetic field [48]

For a long time, hydroxy peroxide was the common crosslinker used to obtain chemically crosslinked gelatin gel. Now the popular crosslinkers are the imidoester, dimethyl suberimidate, N-hydroxysuccinimidyl ester (NHS ester), and glutaraldehyde. Each of these crosslinkers undergoes nucleophilic attack the amino group of lysine and subsequent covalent bonding via the crosslinker. The carbodiimide crosslinker1-ethyl-3-(3-dimethylaminopropyl)carbodiimide hydrochloride(EDC) also binds primary amines of the protein. Isocyanate, succinimide anhydride, maleimide attack thiol group of cysteine, diazo-compounds, and transglutaminase bind carboxylate functions by converting carboxyls to amine-reactive iso-urea intermediates that bind to lysine residues or other primary amines. Succinimidyl-4-(*N*-maleimidomethyl)cyclohexane-1-carboxylate (SMCC) or its water-soluble analog, sulfosuccinimidyl-4-(*N*-maleimidomethyl) cyclohexane-1-carboxylate (sulfo-SMCC), is the common crosslinker to prepare antibody-hapten conjugates for antibody candidates.

There are multifunctional polymeric crosslinkers in which crosslinking groups are attached at the chain ends. Cytoskeletal proteins such as actin, myosin, and microtubule were crosslinked by using bis-maleimide PEG through cysteine residues. Protein is also crosslinked by bis-NHS-polyethylene glycol(bis-NHS-PEG) through binding lysine, respectively. The storage modulus G' of such chemically crosslinked actin gels was higher than 10^3 Pa which substantially exceeded those of cytoplasmic actin gels formed by actin-binding proteins [49]. A particular interest is that the crosslinked F-actin gel crosslinked by bis-NHS-PEG undergoes reversible sol–gel transition in vitro by changing the ionic concentration or by applying oscillatory share strain. In the similar manner, a mixture of tubulin dimer and tetra-NHS-PEG-tubulin makes conjugate spontaneous polymerization to form chemically crosslinked microtubule (MT) hydrogel when temperature is raised higher than 25 °C [49]. This process is reversible and the de-polymerization of crosslinked MT hydrogel occurs to give tubulin dimer by lowering temperature.

Peptide ligation is a chemical method to form amide(peptide) bond by reacting thiol group of cysteine with thioester of another peptide. This method was applied to form peptide-containing hydrogel by reacting peptide-dendrimer containing PEG to which aldehyde is attached at the chain end.

It should be noted that the "chemical crosslinking" is sometimes used to study the protein–protein interactions or to confirm the geometrical proximity of two proteins. When two proteins are assumed to be close to each other, a chemical crosslinker is used to prove the adjacent protein–protein interaction by making chemical bonding them together. Since the interactions of these proteins are usually weak or transient to be detected, the chemical crosslinking reinforces and stabilizes to make treating easier.

Polysaccharide

Polysaccharides can be crosslinked by reacting hydroxyl, carboxyl, and amine group in the chains with aldehyde, hydroxyl, carboxylic acid, and epoxy groups which are chemically introduced into the partner chains. Aldehyde group in the alginic acid reacts with gelatin to give biocompatible polysaccharide-protein hybrid gel in the presence of tetraboric acid.

Polysaccharide gels of hyarulonic acid or alginic acid can be prepared by using horse-radish peroxidase as a crosslinking catalyst. Chitin is the natural polysaccharide which composes the exoskeleton or outer body covering crab, shrimp, and insects. Although chitin is practically insoluble in water, its partial de-acetylated product, chitosan is water soluble. Since chitosan is biodegradable and biocompatible, a various types of hydrogels have been obtained by using glutaraldehyde or isocyanates through reactions with amino-residue. Many attempts to apply biodegradable and biocompatible chitosan hydrogels for medical and analytical applications are made.

Epichlorohydrin and ethylene glycol diglycidyl ether are often used to get polysaccharide gels containing hydroxyl groups. Poly(γ-methyl glycomannan and pullulan gels have been obtained by using these compounds. Cellulose and chitin could also be crosslinked by 1,2,3,4 butantetracarboxylic acid dianhydride.

3.3 Physical Gel

Physical gels can be formed by interaction forces such as hydrogen bonding, Coulombic force (electrostatic ionic force), van der Waals, hydrophobic, dipole–dipole, and others of intra- and inter-macromolecules. Gelatin gel from which the term "gel" was originated would be the most popular example of physical gel, and gelatin gel is formed by decreasing the temperature of the aqueous solution lower than, for example 25 °C and transfer to sol (sol is a colloidal dispersion in solvents having fluidity)when the temperature is increased up to 40 °C. The temperature of transition is sensitively changed by molecular weight, specie of gelatin, and reaction conditions such as concentration and rate of temperature change. The sol–gel transition refers to the status of the sol and gel being reversibly transferred to each other. The sol–gel transition occurs not only by change in temperature, but by ionic strength, pH, and others.

The physical gels may be classified by their interaction schemas, and typical examples are shown in Fig. 3.18. Mechanism of the formation of various physical gel and the structure will be followed in the next sections.

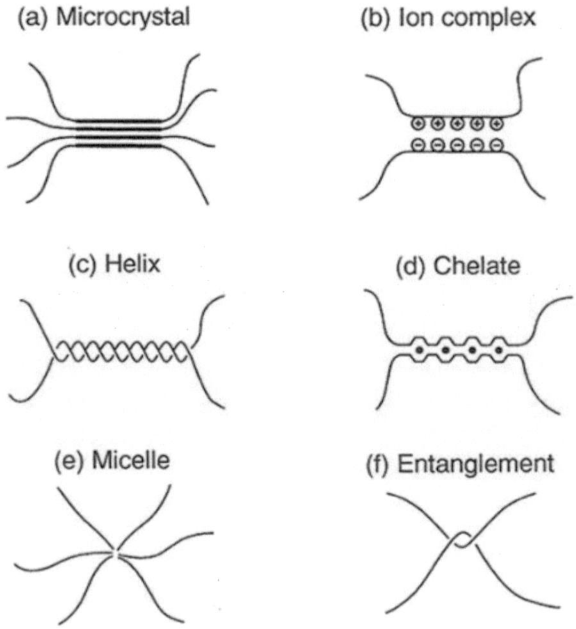

Fig. 3.18 Formation of crosslinking by intermolecular linkages

3.3.1 Synthetic Polymers

Polyelectrolytes can easily be crosslinked to give gels by mixing oppositely charged di-, tri-, and multi-valent counter ions (salt-bridged polyelectrolyte gels). For example, highly charged polyelectrolytes such as sodium salt of poly(styrene sulfonic acid) form insoluble gel by simply mixing with alkaline earth metal such as calcium, magnesium, and aluminum ions. Similarly, poly (vinyl trimethylbenzyl ammonium chloride) forms gels by mixing with disulfate or phosphate anions. Weakly charged polyelectrolytes such as poly(acrylic acid)(PAA) can form crosslinked gel by mixing multi-valent counter ions under increased pH condition where PAA is ionized (Fig. 3.19a).

This type of crosslinking occurs by inter-macromolecular electrostatic bridging of multi-ions with ionized polymers. There is a relation between the radius of the hydrating of multi-valent ions and bridging strength: the smaller the radius of hydration of multi-valent ions, the easier the gel is formed due to increased static interaction. Since it is an inter-macromolecular reaction, the gelation also depends on the molecular weight and concentrations of polyelectrolytes. When the polymer concentration or the molecular weight of polymer is not high enough, the microgel is dominantly formed due to intramolecular interaction, and sometimes they dissolved by additional dilution. The structure of obtaining gel is related to the charge density, and the higher the charge density, the lower the swelling degree since the crosslinking density is increased. These salt-bridged polyelectrolyte gels are dissolved by addition of neutral salt since it weakens the salt bridging interaction.

Polyelectrolytes gels can form ***polyelectrolyte complex or inter-polyelectrolyte complex*** or more widely ***polymer–polymer complex*** by mixing oppositely charged polyelectrolytes [50] (Fig. 3.18b). Polyelectrolyte complex is a special class polymer

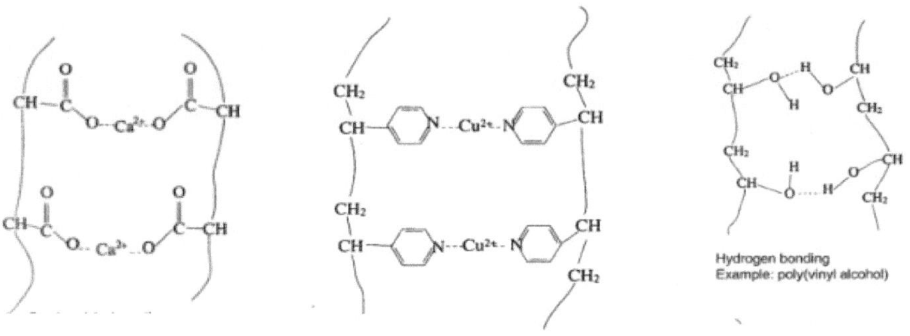

a b c

Fig. 3.19 Examples of physical gel of synthetic polymers, **a** salt-bridged polyelectrolyte gels by Coulombic force, PAA, and Ca^{2+}, **b** polyelectrolyte gels by coordination bonding, poly4-vinyl pyridine, and Cu^{2+}, **c** polymer–polymer complex by cooperative hydrogen bonding, unionized PAA compounds which is formed by the interaction between oppositely charged polyelectrolytes or by so-called matrix polymerization where ionic monomer is polymerized on the opposite-charged polymer matrix. For example, mixing of sodium salt of poly(styrenesulfonic acid) and poly(vinyl trimethylbenzyl ammonium chloride) forms insoluble gel, and the structure and the property are changed by their mixing ratio. The structure and the condition of gel formation depend on pH (in case of weak polyelectrolytes), ionic strength, and polymer concentration. It is possible to control the degree of crosslinking, i.e., the degree of swelling and modulus of the obtaining gels by adjusting these variables [51]. If full-ionized poly-cation and poly-anion are mixed with the ratio of 1:1, fully neutralized gel is, in principle, formed, and it is a highly crosslinked ladder-like bridging structure and less swelling. If charges of one of polyelectrolytes are excess, the complex gel still has ionic character and more swelling character. These polyelectrolyte complexes become soluble in an appropriate ratio of the ternary mixture of water, organic solvent, and salt which allow to give polymer films by coating. Similar complexes are formed not only by the combination of synthetic polymers, but by a variety of combination of proteins, polysaccharides, and DNA [52].

These polymer–polymer complexes can also be prepared via cooperative hydrogen bonding between two different polymers such as poly(methacrylic acid)–poly(ethylene glycol) or poly(acrylic acid)–poly-(vinyl alcohol) (Fig. 3.19c) [53]. The repeated freezing and thawing process of poly (vinyl alcohol) forms gels with excellent mechanical and water uptake properties due to their enhanced hydrogen bonding between polymer chains. Mixing syndiotactic and isotactic poly(methyl methacrylate) solutions forms stereo-complex gel which is considered to form crystalline structure through van der Waals interaction (Fig. 3.18a).

Polymer–polymer complexes are obtained by the self-assembly process through hydrophobic interaction. For example, amphiphilic block copolymers constituting hydrophobic and hydrophilic blocks form micro and macrogels depending on the concentration when temperature is lowered. Poly(vinylether) (PVE) is usually synthesized by ring-opening cationic polymerization, and it shows lower critical solution temperature (LCST) at which PVE dissolves or upper critical solution temperature (UCST) at which PMV precipitates. Therefore, it forms gels by changing temperature.

Ordered block copolymers consisting of hydrophilic and hydrophobic segments, for example, a poly(styrene-b-ethyleneglycol-b-styrene) triblock copolymer, form three-dimensional network to give physical gels in selective solvents. The gel is formed by the hydrophobic interaction between styrene segments in water (Fig. 3.18e). Polyethylene oxide (PEO) and PVE copolymers or polymer complexes containing PEO and PVM as a constituent polymer form physical gels when the temperature of solution is raised and dissolved when temperature is decrease. This phenomenon is occurred by the increased hydrophobic interaction between component polymer units under elevated temperatures.

Since RAFT polymerization (refer to Sect. 3.2.1.1) gives various ABA-type block copolymers with regulated molecular size, many physical gels are prepared from polymers showing lower critical solution temperature (LCST). An example is the physical gel prepared from N-isopropyl-acrylamide and N,N-dimethyl acrylamide which undergo sol–gel transition at around 35 °C.

3.3.2 *Biopolymers*

3.3.2.1 Protein

It is widely known that proteins such as egg albumen and serum albumin form the gel when heated or changed pH. This irreversible type of gelation occurred due to denaturation, i.e., the globular structure is destroyed. In this case, hydrophobic parts of the protein chain are exposed on the water surface as a result of denaturation and coagulated through the hydrophobic interaction in water. Another well-known example of irreversible gel formation of protein is Tofu—one of the typical Japanese foods known as bean curd. Tofu is made by from soy milk coagulating by adding two valent magnesium or calcium ion.

In this way, the gel formation of protein is historically familiar in our common life, but the exact understanding of the mechanism of the gel formation and the structure is still under the way.

Collagen is the protein with high oriented triple-stranded helix structure which constitutes soft tissue as ligament, joint, and others. Gelatin is the product of thermal or denaturation of collagen and is soluble to be used as glue, cosmetics, and others.

Crosslinking of gelatin gel is formed by helix formation via cooperative hydrogen bonding between the NH group of the proline and its neighboring CO group of hydroxyproline (Fig. 3.18c). A part of the chain of this helix is connected each other to form network. Thus, the crosslink point to form the network is no longer a “point”, but the helix domain which is considered as the phase-separated ordered aggregated phase bridging over the definite areas or zone with an ordered structure. Fig. 3.20 shows the schematic illustration of thermal denaturation of collagen and reversible sol–gel transition of gelatin.

It should be noted that immobilized enzyme gels are produced via crosslinking of the physical enzyme aggregates and used as in biochemical industry.

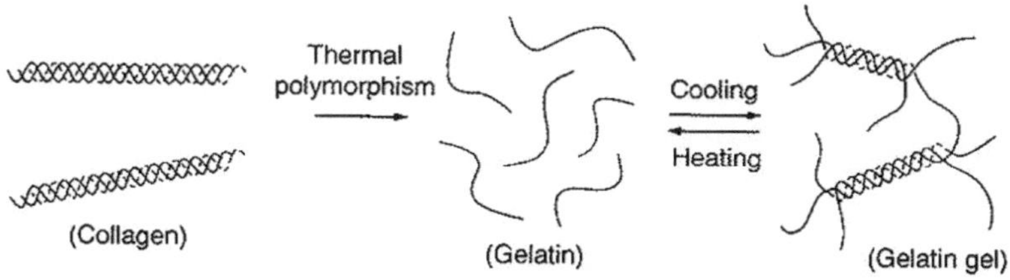

Fig. 3.20 Thermal denaturation of collagen to gelatin and sol–gel transition of gelatin

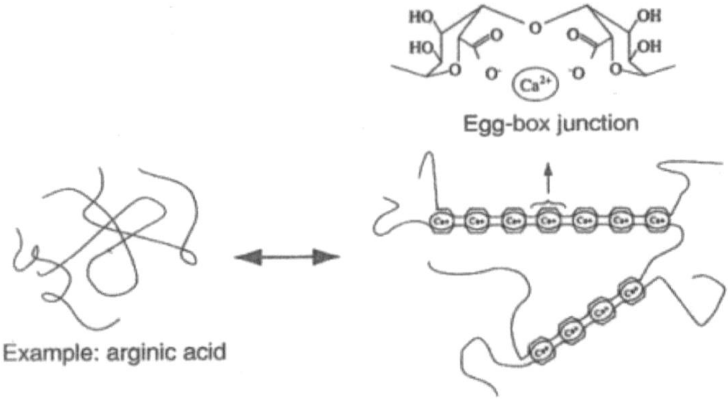

Fig. 3.21 Schematic model of gel formation of alginic acid in the presence of calcium ion

3.3.2.2 Polysaccharide

Polysaccharides such as agar–agar, salt of alginic acid, pectin, and konjak (konjac, konnyaku) mannan form gels when divalent metallic ions are added. For example, a solution of sodium alginic acid is added dropwise into a calcium salt solution, and a spherical gel is readily formed. Many other polysaccharides such as agar, agarose, and carrageenan form reversible gels through usual formation processes. All these gels are formed through cooperative helix formation of polysaccharide chains although the structure and the mechanism are different each other (Fig. 3.18c).

Many years discussion has been made on the mechanism of gel formation, particularly on the formation of "junction zones" of polysaccharide on a molecular level. Now it is considered that the junction zone is formed through relatively strong and cooperative interactions to give helical structure which is additionally stabilized by the metal-ion chelation-dependent crystallites (alginates) or by the hydrophobic interactions (carboxymethyl celluloses).

Alginic Acid

Alginic acid is a negative charged linear polysaccharide consisting of a block copolymer, (1–4)-linked β-D-mannuronate (M), and its C-5 epimer, α-L-guluronate (G) residues and obtained as a sodium salt of alginate. They are covalently linked together in different sequences or blocks with three structural units. The gel of alginate is obtained by binding calcium ions to form bridges between two polysaccharide chains (Fig. 3.18d) The bridge or junction formed by calcium ions is often called as "egg box junction" that is schematically shown in Fig. 3.21.

The egg box-type crosslinked junction zone is formed by incorporating calcium ions by the carboxylic groups of alginic acid. If the calcium ion is removed by using a chelating agent such as ethylene diamine tetraacetate, the gel transfers to the sol state, and the process is reversibly.

Agar (Agar–Agar)

Agar (agar–agar) [54] is extracted from sea weeds, and the main component is a polysaccharide, agarose. It consists of D-galactose and 3,6-anhydro-L-galactose. Agar–agar gel is formed when the agarose solution is cooled below 32–40 °C and melts when heated higher than 85 °C [55]. The GP and MP exhibits notable hysteresis and varies depending on the types of agalose. For example, agarose derived from *Gelidium* forms a gel between 34 and 38 °C and melts at the temperature of 90–95 °C, while other agarose derived from *Gracilaria* forms the gel in the range of 40–52 °C and melts at 85–90 °C. As described in Chap. 2, the GP and MP also depend on the concentration, molecular weight, and the condition of the gel preparation.

The process of formation of agarose gel is considered as follows:

(1) Agarose, for example, consisting of ~800 molecules of galactose chain transfers from random coil to single or double helices via cooperative hydrogen bonding when temperature is lowered.
(2) These helices aggregate each other to form an ordered parallel arrangement or supercoiled structure with a radius of 20–30 nm. This is the crystalline junction zone to form 3D network connecting flexible chains of agarose.

The aggregates are quasi-rigid and have a wide range of size depending on the agarose concentration. In the gel state, the fibers form a three-dimensional network of a diameter ranging from 50 to >200 nm depending on the concentration of agarose used. The higher concentration of agarose yields lower average pore diameters. However, detailed analysis on the structure of crosslinked domains has not been done. The melting and gelling temperature of agarose can be modified by chemical modifications, most commonly by hydroxyethylation because it reduces the number of intra-strand hydrogen bonds, resulting in lower melting. Since the 3D network structure is formed by the hydrogen bonds, it can be disrupted by heating back to a liquid sol state.

Carrageenan

κ-Carrageenan is a polysaccharide derived from red seaweed consisting of D-galactose 4-sulfate and 3,6-anhydo-D- galactose. It is divided into three groups according to its content, ι-,κ-or λ-carrageenan, and the properties also differ. Generally, κ-carrageenan is easiest to form the gel.

The gelation process is considered as follows (Fig. 3.21).

(1) By lowering temperature, double helices are formed between carrageenan chains from random coils.

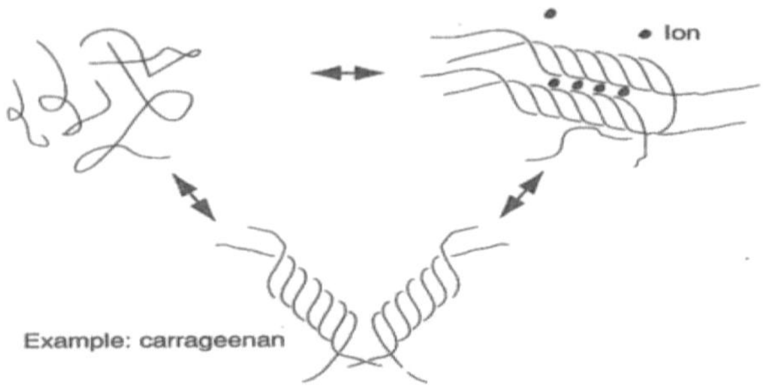

Fig. 3.22 Schematic process of gel formation of carrageenan in the presence of metallic ions

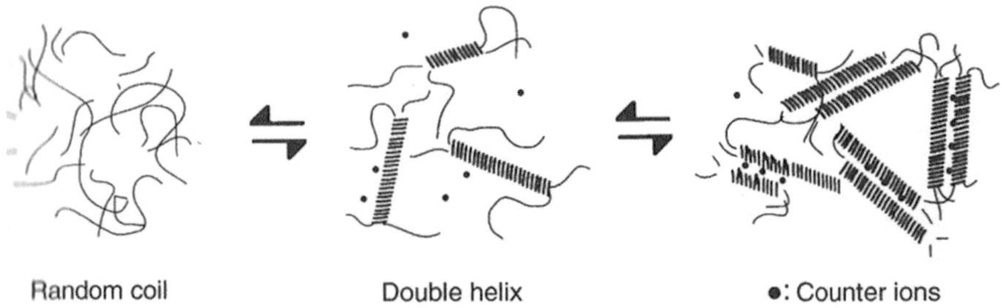

Fig. 3.23 Schematic process of gel formation of gellan gum by lowering temperature

(2) Further cooling promotes the aggregation of these helices to form crosslinked network through counter ions such as the K^+ ion with junction zones (double-helix association).

Gellan Gum

Gellan is a water-soluble anionic polysaccharide produced from the bacterium *Sphingomonas elodea*. The repeating unit of the gellan is a saccharide units consisting of two residues of D-glucose and two residues of L-rhamnose and D-glucuronic acid (Fig. 3.22).

The gelation of gellan gum initiates by formation of double helices in the aqueous solution when temperature is lowered, followed by double helices association between them in a way to form in parallel packing to form a crystalline junction (Fig. 3.23).

Cellulose Derivatives

Aqueous solutions of methyl cellulose and hydroxypropyl cellulose aggregate when heated and form gels. Gelation accompanies with micro and macrophase separation which occurs continuously. The gel exhibits no specific order structure at the early stage, but gradual ordering takes place with time to give crystalline gel.

When hydrophobic alkyl side chains with 6, 12, and 16 carbons are chemically introduced to hydroxyethyl cellulose, the viscosity of the solution increases due to hydrophobic interactions between side chains and give gels. The concentration of the alkyl side chain to give aggregate decreases with increase in the chain length since the hydrophobic interaction increases. As well established, hydrophobic interaction takes place due to increase in entropy by the liberation of water molecules which surrounded the hydrophobic alkyl side chains, and the mobility was restricted (low entropy). Hydrogel from carboxymethyl cellulose (CMC) can also be obtained by adding divalent acid like citric acid and oxalic acid. Native chitin is insoluble in water, but it dissolves in N,N-dimethyl acetamide containing 5% LiCl and shows sol–gel transition when heated up to 110 °C.

3.4 Theoretical Models of Network Formation

This chapter is focused on modeling of structure development of branched and crosslinked polymers by formation of covalent bonds. The results can also be applied to physical networks in the case when the physical bonds are strong enough to be considered stable within the experimental observation time window.

The network buildup process can be characterized by several phases:

- evolution of the molecular weight distribution before the gel point, characteristic by its broadening and divergence of the second and higher moments of the degree of polymerization (molecular weight) distributions diverging at the gel point
- critical time and conversion of functional groups at the gel point when the infinite structure is formed for the first time; the gel fraction is zero at the gel point
- transformation of finite (soluble) part of the system (sol) into the "infinite" gel is characterized by the decrease of mass fraction of sol and increase of the mass fraction of the gel and by the differences in composition of sol and gel. Also by changes of the molecular weight distribution of the sol, decrease of its molecular weight averages
- increase of connectivity of the gel, resulting in changes of amounts or molecular weights of various substructures (dangling chains, elastically active network chains, etc.)

The network formation process can proceed by various mechanisms such as step growth or chain growth, from small molecules or larger building blocks by bond formation of different types, or by crosslinking (vulcanization) of polymer chains of different lengths.

These changes depend on the structure of the building units (monomer units or multifunctional network precursor units), such as their mass, numbers, types, and reactivities of functional groups, and on the rules of forming bonds between reactive groups (reaction mechanism) and types of specific bonds populations as they develop in time. Chemistry information is vital for understanding, modeling, and simulation of network formation ("chemistry is about making and breaking bonds"), particularly information about reaction mechanism and kinetics of all participating groups as developing in time is necessary. According to the original Flory's assumption, the groups' reactivities were considered independent of the size of molecules they are bonded to. This assumption has been found valid experimentally for some systems but it is not of general validity. On contrary, there is enough evidence of steric hindrance and thermodynamic exclusion which affect the apparent reactivity of the functional groups. Among these factors, the specific and overall diffusion control discussed in Sect. 2.5 is of special importance.

The ring-closing reactions (cyclization, intramolecular reaction) which affect the buildup process are another important issue. Before the point, a ring-closing reaction does not affect the mass of the cluster but contributes to the conversion degree of functional groups. Beyond the gel point, any reaction between functional groups bonded to the gel (gel–gel reaction) is a ring-closing reaction. However, with respect to the elastic response of chains between multifunctional junctions, one should distinguish between (small) loops and elastically active network chains (EANCs). The former are believed not contributing to the elastic response, while the other do contribute. However, this division is somewhat nebulous because the contribution to the number of EANCs by a loop forming chain can range from zero to one depending on the number of units between ends of ring-closing chain and its properties and affinity or non-affinity of their displacement relative to external macroscopic strain.

Some of the characteristic changes during crosslinking are illustrated by Fig. 3.1 for the case of crosslinking of a functional copolymer (groups A) of number-average functionality $f_n = 7.3$, with a trifunctional crosslinker (B).

For various systems differing in functionality and bond formation mechanism, the critical point (critical conversion) critical fraction of crosslinked units) extends from 10^{-3} (vulcanization of long chains, crosslinking polymerization) to 1 (100%) (low-functionality step polyaddition, alternating off-stoichiometric systems). This categorization is only relative, since the critical point position depends very much on dilution of the system during network formation: increasing dilution shifts the gel point to higher conversion, and beyond critical dilution, the system does not gel even at 100% conversion of reactive groups, and the majority of reactive groups being transformed into ring-forming bonds.Likewise, for $A + B \rightarrow AB$ type reactions, a similar shift is caused under non-stoichiometric condition of the reagents even at high or 100 % conversion. there remains unreacted reagent of the majority component; there remain unreacted many groups of the majority component; the is very loose many of its chains are terminated by unreacted groups or gelation does not occur at all. These two cases of subcritical final structure are shown in Fig. 3.2.

Along with structural changes characterizing network formation discussed above, changes of several other physical properties are associated with network formation such as changes of viscosity and radiation scattering. Combination with branching theory makes the structural interpretation possible.

3.4.1 Models of Network Buildup

The models describing the buildup of branched and crosslinked structures can be grouped in several categories.

- Statistical methods: Generation of branched and crosslinked structures from distribution of *building units in different reaction states*, i.e., such that contain variable numbers of unreacted functional groups and reacted groups engaged in bonds (possibly, of several types). The distribution of units in different reaction states is determined by equilibrium constant or by chemical kinetics. In these models, the distributions of molecules composed of these units are generated by statistical (stochastic) methods. These methods are recursive, which means that a population of units in a molecule (molecular cluster) is calculated systematically in a finite number of steps; the type and number of units in the n-th generation are derived from numbers and types of units in the previous generation. The theoretical assemblage into larger clusters is controlled by transition probabilities of Markov statistics. The ability of two unreacted functional groups of different units to form a bond can be made dependent on the state of other functional groups in the molecule under consideration (first-order Markov statistics) or on the states of topologically more distant units (higher-order Markov statistics). As building units, monomer units or preprepared larger precursor units are used [56, 57]. The buildup process is controlled by transition probabilities which are obtained from concentrations of dyads, triad, etc. Their concentrations are determined from the corresponding sets of differential equations of chemical kinetics in which the rate constants are obtained either experimentally or can be calculated by quantum chemistry.
- Kinetics methods: Evolution of distributions of the whole molecules is described by solving systems of kinetic differential equations as a function time. Molecules are formed by bond formation between any pair of groups on already existing molecules and allowed by reaction mechanism. The intensity of the reaction is a function of their concentration and reaction ability. The reaction ability of a molecular cluster depends on the number and reactivity of functional groups they bear and, possibly, on their physical properties like group diffusivity or steric hindrance. The differential equations are of the type of the Smoluchowski coagulation equations describing aggregation (including branching); the aggregation process is embedded in space; for application in polymer branching, see, e.g., [58–60]).
- Simulation in 3D space. This category includes modeling and simulation of the structure evolution process in 3D space. Various methods have been used ranging from percolation on rigid lattice to molecular dynamic simulation of space-filling segments. These ways of simulation are not intended to be discussed in this section, and only a brief account of this category is given. Stauffer et al. [61] introduced the percolation techniques on static rigid lattices [62]. Bonds were introduced between neighboring lattice sites occupied by monomer units. In off-lattice simulations, the distributed precursor chains were randomly distributed in 3D space,

and a capture radius was ascribed to the reactive groups; a bond was formed when the capture spheres came into contact and overlapped. With increasing conversion, the capture radii increased [63–65]. The artificial static picture of bond formation was relaxed by introduction of the bond fluctuation model: after formation of a bond, the structure was allowed to relax by a multiple change of segment positions keeping the connectivity intact [66–68]. This method is being currently used in network formation simulations [69, 70]. The methods listed above offer topological information on the crosslinking system but do not address the physical (thermomechanical) properties of the systems. Information on these properties and the topology can be obtained by combining the topology evolution with molecular dynamics simulation. These methods have been applied so far to thermoset systems—cured epoxy or cyanate resins to predict the glass transition temperature, thermal expansion coefficients, and elastic properties of each of the crosslinked systems. The simulation is multiscale and consists in the following steps: (a) mapping of the polymerizing monomers onto a coarse-grained model, (b) crosslinking the monomers at mesoscale by applying dissipative dynamics and a probabilistic weighting, (c) reverse mapping of the coarse-grained model to a fully atomistic representation, and (d) simulation of the atomistic model through standard molecular dynamics technique. Examples can be found in refs. [71–80]. There are many other instances where 3D Monte Carlo simulations of structure growth are employed, partly on small systems serving as input for the perturbation methods of graph-like theories. Since MC simulations are time-consuming, the simulations method driven by adaptive sampling of networks called Markov state models (MSMs) can yield tremendous time and resource savings [81].

In this section, only the models of the first category (stochastic buildup from units) will be discussed, and briefly the second category (kinetics generation) will be mentioned. The multiscale modeling involving the molecular dynamics is a promising step forward but it is computationally more demanding and needs wider validation by experiment.

3.4.2 Stochastic Generation of Branched and Crosslinked Structures

The generation of branched structures is viewed as branching or multiplicative process characteristic for population growth and in biology for cell cloning, transfer of genetic information, spreading of diseases and epidemic, and cascades of atom transformations produced by cosmic rays or fission of nuclei by neutrons. Branching processes are typical for structure growth in macromolecular science either by covalent bond formation of or by physical association. In mathematics, the term "branching process" was coined by Kolmogorov and Dmitriev [82]. Several monographs have been published on modeling of various branching processes (see, e.g., refs. [83–85]). Stochastic methods are very important in biology to treat multiplicative (branching) processes such as cell cycle kinetics and cell division, drug resistance and chemotherapy, evolution theory, gene amplification, molecules aggregation, section, mutations, or polymerase chain reaction. Several book series have been published on this topic for decades; two Springer series *Lecture Notes in Biomathematics* and *Branching Processes in Biology* serve as examples. In two of them, articles are collected that are related to branching in macromolecules [86, 87].

3.4.2.1 Branching: Historical Connotations

The statistical methods used in contemporary network buildup theories have interesting roots in history: they belong to the category of so-called *Galton–Watson processes*. Francis Galton, a great Victorian scientist interested in inheritance problems [87] and collaborator of Darwin, called biometrician by his peers, [88, 89] was also interested in the probability of extinction of family names of important English families due to decreasing fertility. In 1873, he published in the journal Educational Times a problem called Problem 4001 [90].

Problem 4001: A large nation, of whom we will only concern ourselves with adult males, N in number, and who each bear separate surnames, colonize a district. Their law of population is such that, in each generation, a_0 percent of the adult males have no male children who reach adult time; a_1 have one such child, a_2 have two, and so on up to a_5, who have five.

Find (1) what proportion of surnames will have become extinct after r generations; and (2) how many instances there will be of the same surname being held by m persons (source [84]).

Galton received only one answer which was incorrect and asked his friend, clergyman and mathematician Reverend Henry William Watson to take up the matter. Watson attacked the problem, recognized the importance of functional iteration, and used the formalism of *probability generating functions* (pgf) for solution. He recognized the recursive character of the process solved by substitution of pgf variable by another pgf. However, in analyzing the *extinction probability*, he made a mistake by considering also the trivial root equal to one. The answer was published in the same year [91], and the problem of extinction of family names was treated in a joint publication [92]. In the decades that followed, correct solution was found and extended to multitype processes. The formalism of the Galton–Watson process provides insight into one of the fundamental problems of actual populations, the extinction problem, and its complement, the question of size stabilization: If a freely reproducing population does not die out, can it stabilize or must it grow beyond bounds? The answer was that there existed no freely reproducing populations with stable number of members and population size stability must result of forces other than individual reproduction. More on historical development of the Galton–Watson can be found in monographs [83, 84].

The Galton–Watson formalism of branching or multiplicative processes was further developed and introduced into the macromolecular science by Irving John

Good and broadly developed by Manfred Gordon and his collaborators. Irving John Good (1916–2009) was a British mathematician who during WWII worked as a cryptologist at Bletchley Park with Alan Turing. He was born to a Polish Jewish family Gudak after they emigrated to Great Britain and changed his name to Good. He was working on breaking the Enigma code used by the German Kriegsmarine which determined the victory of Allied fleet against Kriegsmarine.

Alan Turing had caught Good sleeping on the floor while on duty during his first night shift (and after that several times). At first, Turing thought Good was ill, but he was cross when Good explained that he was just taking a short nap because he was tired. Good was well known for his work on Bayesian statistics. He published a number of books on probability theory. Good got interested in the family name transfer work and later—talking to colleagues—he recognized that, what he called cascade process, is also relevant to certain chemical processes which in certain aspects are simpler than birth and death processes in living nature. This shift of interest occurred while working for Admiralty Research Laboratory in Teddington [93–97]. It is well possible that his interest in branching of polymers had arisen from studies of high energy radiation effect on polymers (crosslinking and degradation in parallel to birth and death)—in U. K. at that time intensively studied by Brigadier-General Professor Arthur Charlesby (1915–1996) in Atomic Energy Research Establishment (AERE) at Harwell and later at the Royal Military College of Science at Shrivenham.

Manfred Gordon was born in Leipzig, and he and his family fled to UK via Switzerland when living in Hitler's Germany became too dangerous for Jews. After finishing university studies, he started working with Dunlop Company on rubber vulcanization and continued his research and teaching at Royal Technical College in Glasgow; passing through several places in UK including Imperial College London, he returned to Scotland to become Professor of Physical Chemistry at the University of Strathclyde at Glasgow. After three years, he mowed to, then new, University of Essex in Colchester to head the Institute of Polymer Science. Almost all his students at Strathclyde moved with him. After his retirement, Manfred moved to his beloved Statistical Laboratory of Mathematical Department of the University of Cambridge for which he served before for many years in the rank of Adjunct Professor and which offered him an office [98]. He was a friend of the renowned Professor Peter Whittle, for some time Director of Statistical Laboratory; he also got involved into Good–Gordon problems of branching [99]. One of his later books [100] may offer inspiration for future development of polymer gels and networks. While at Imperial College, Manfred discovered Good's paper on cascade theory and immediately recognized tremendous potential of the method for description of branching and crosslinking in chemistry and biosystems and of statistical properties of such systems. With his many students from various countries and collaborators visiting him for shorter time, he was able to replace treatments based on ad hoc probabilistic arguments which soon became cumbersome with exact defined expression enabling treatment of systems of high complexity, as contemporary systems of practical importance are. It is not known to me whether Gordon and Good knew they shared the same habit—stretching on floor aside for a while when getting tired. Manfred did not mention Good when he looked for a proper corner of lecture room during one of the Prague Macro Meetings when the others left for lunch. Manfred was a very knowledgeable person not only in science but also in philosophy a medieval history—he visited Vatican Library to study in original what Dante Alighieri wrote about gels and their functions. Mentioning of name of his inspiratory in the title of Manfred's first paper on cascade theory "Good's Theory of Cascade Processes Applied to the Statistics of Polymer Distributions" [101] reflects his modesty.

Let me add at this place a few personal recollections: I met first Manfred at the UIPAC Macro 1965 in Prague where he presented results of his "cascade" theory. At that time, I understood practically nothing of the theory (generating functions, cascade substitution, and so forth) but I immediately realized the great potential of the method making possible to treat many real systems of practical importance. Two years later, when Willem Prins of TH Delft invited me to come and write a joint review on polymer networks [102], I decided that Manfred's "cascade" theory must be explained there. After Manfred's short visit to Delft and my short visit to Colchester in early 1968, I got sufficient theoretical background. Our contacts continued through my one-month visiting professorship at Colchester in 1975 and then through several decades through our meetings at conferences and Polymer Networks Group meetings, as well as several private visits of Manfred and his wife Blanche in Czechoslovakia.

The Galton–Watson process is the oldest, simplest, and best-known branching process. Any particle existing in the process, except for the ancestor of the process, can be assigned to a subprocess traceable to a particular first-generation offspring of the ancestor. In other words, the process can be represented as a union of the subprocesses initiated by the first-generation offspring of the ancestor particle. The formalism of the Galton–Watson process provides insight into one of the fundamental problems of actual populations, the extinction problem, and its complement, and the question of size stabilization or size limitation is a result of forces other than individual reproduction, of the interplay between populations and their environment. Such interactions are important also for branching of polymers whenever they are strong enough.

The generality of the *cascade* method and detailed information on topology of branched and crosslinked structure on one side, and the use of, for non-mathematicians strangely sounding *generating functions* as transform method, are horrifying many synthetic and physical chemists. On the other side, this notion of this feeling inspired polymer scientist Chris Macosko and mathematician Doug Miller to offer a simpler approach. Also, Good–Gordon's first papers written in a somewhat mathematical style being decorated by a few lemmas and corollaries had a detracting effect on pedestrian readers. The Miller–Macosko recursive theory is simple in that for some quantities the mean values are formulated as expectancies known better (by pedestrians) from the probability theory. In the *cascade* method, the *distributions* (units, molecules, and gel substructures) are formulated first *implicitly,* and the means are obtained by differentiation. The models—basically ring-free and graph-like—are the same as well as the final results. The advantage of the *cascade* method can be found in that distributions available implicitly can be made explicit by Lagrange expansion and can be weighted by specific factors which make it possible to calculate structure factors in scattering functions; also, transforms between various moments of distributions can be made routinely and easily. The required knowledge of mathematics is really minimum—just differentiation and rarely integration—on the level taught in basic course of math at any technically oriented university. The main advantage of the use of probability generating functions (pgf) is keeping track of composition of any structures generated by the branching process in terms of numbers of building units of various types and bonds connecting them. The meaning of the coefficients in front of the pgf variable is then fully clear. Even Miller and Macosko resorted to the use of pgfs when the complexity of the system increased [103]. At present, description of generating function has penetrated also through other theoretical approaches which will be mentioned below.

In a broader view on development of branching theory, we have to consider the Galton–Watson–Good–Gordon contribution as tributary to already stable stream initiated by Flory and Stockmayer, and based on experimental evidence of gelation and reversible elasticity stemming from technologically important discoveries of rubber vulcanization by Charles Goodyear in 1844 and phenol–formaldehyde thermosets formation by Leo Baekeland in 1907 as well as systematic work of the father of macromolecular chemistry Hermann Staudinger (1881–1965) [104] was awarded in 1953 the Nobel Prize in Chemistry. The main theoretical stream based on probabilities of states and their combinations was enriched by a kinetic look where the development of whole molecules (not units in different reaction states) was considered. The theoretical foundation of such processes was layed by Marian von Smoluchowski working mainly in Polish Krakow. The Smoluchowski coagulation equation is a population balance equation describing the time evolution of the number density of particles as they coagulate. This kinetic view was later applied to polymer network formation because it is able to reflect the efficiency of various physical and chemical interactions through modification of the equation kernel. The Smoluchowski coagulation equation can be formulated in discrete or continuous forms (integro-differential equations). A great impetus to development of theoretical description of crosslinking of polymers (combined with chain scission) gave the studies of ionizing radiation effect on polymers. Pioneering work was done by the group of Brigadier-General A. Charlesby in U.K. already mentioned above (his work is summarized in the monograph [105, 106]) and a group of Japanese scientists (Saito, Amemyia, and Inokuti) who worked independently, but later some of them published their results jointly. Osamu Saito worked at Department of Physics, Faculty of Science and Engineering, Chuo University, Tokyo, Korakuen campus, Ayao Amemyia was a member of Department of Applied Physics, Faculty of Engineering, University of Tokyo, and Mitio Inokuti (1933–2009), an eminent atomic physicist and boyhood friend of the emperor of Japan, was first affiliated with the Department of Applied Physics, Faculty of Engineering, University of Tokyo. In 1963, he moved to Argonne National Laboratory, Argonne, Illinois, USA. They always considered simultaneously occurring crosslinking and degradation. The changes in the degree of polymerization, gelation sol $\rightarrow$ gel and gel $\rightarrow$ sol transitions were described by continuous integro-differential equations describing variations of molecular weight distributions, gel and sol fractions caused by irradiation. The main papers related to polymers can be found under the following references [107–121]. Their work inspired the treatment of crosslinking and degradation by the cascade method which made possible to consider various types of degree of polymerization distributions and gave access to postgel properties [122, 123].

The development of the branching theories is visualized in Fig. 3.1—Flory and Stockmayer can be considered as fathers of this branch of polymer science by developing the statistical approach to distributions and gelation. Their motivation came from experimental observations made by chemists and technologists.

This approach was given a great impetus by introduction elements of mathematical methodology dealing with multiplicative processes in human society. Along with this statistical approach, the branching theory was being enriched by the kinetics approach stemming from Smoluchowski aggregation/coagulation kinetics treatment. This approach plays now an irreplaceable role in braching theory applied to polymers.

3.4.2.2 Application of Statistical Methods to Branching and Network Buildup

Statistical methods for polymer chains buildup were first used by Flory and Stockmayer for the case when all functional groups reacted independently of each other (Bernoulli statistics) [124]. The degrees of polymerization/molecular weights averages were obtained by summations of discrete distributions. The theory was then extended beyond the gel point to describe the sol–gel transition and evolution of connectivity of the gel component [124]. However, certain simplifications had to be made especially those concerned with the network defects which in some cases represent too severe approximations. When the approach was applied to systems with unequal reactivities of functional groups, heavy and difficult tractable expressions were obtained [125]. Inspired by the work of Good [93–97], Gordon et al. started describing branching and crosslinking as a multivariate Galton–Watson process to deal with various phenomena that characterize formation of branched and crosslinked structures [125–150]. In this approach, the distributions (bond types, reaction states of units, degrees of polymerization, molecular weights, etc.) are expressed implicitly in the form of probability generating functions (pgf), as already was done by Watson [91], by a simple transform which allows us to obtain easily various averages. Moreover, the pgf formalism allows us to label various numerical coefficients which define the physical (probabilistic) meaning of the particular coefficient even after complicated transformations. This approach—expression of distributions by probability generating functions and transform of distribution by manipulating with pgfs—was called *cascade theory*, although later Gordon preferred to call this approach *theory of branching processes*, to avoid possible identification of the crosslinking process with cascades of particles produced by cosmic rays. Since the latter designation is too broad, we will continue using *cascade theory* as an acronym for the branching theory that uses the pgf formalism.

The cascade theory has been developed in the past decades by several groups. Perhaps the most important work was done by groups of Burchard [138], Kajiwara and Gordon [139], Burchard [142], Kajiwara and Burchard [143], Burchard et al. [144] in German, Trappe et al. [148], Kajiwara [149, 150], Kajiwara and Ribeiro [151], Tanaka [152–154], Tanaka and Stockmayer [155], Tanaka and Ishida [156], Nakao et al. [157] in Japan, Nakao et al. [158], Irzhak [159], Kuchanov and Povolotskaya [160], Irzhak and Tai [161], Kuchanov and Zharnikov [162], Kuchanov et al. [163, 164], Deberdeev et al. [165], Irzhak and Irzhak [166] and Irzhak [167], Irzhak and Tai [161], Kuchanov and Zharnikov [162], Kuchanov et al. [163] in Russia, Irzhak et al. [168], Galina and Szustalewicz [169], Galina and Lechowicz [170, 171] in Poland, and the Prague group (K. Dušek) in Czechia [172–201]. The references shown here are examples characterizing the work of the groups, and there have been other papers published in which the cascade theory was used.

The rigor and wide applicability of the cascade theory on one side but too "mathematical" language of the early papers as well as the use of, for many chemists unknown and deterring, probability generating functions inspired Šomvársky et al. [202], Macosko and Miller [203] to develop a simpler approach, sometimes called *recursive theory* or *In–Out recursive modeling* when only averages of the distribution were required. Both theories are based on the same reactivity models and probability arguments, both are recursive in nature, and, for the same assumptions, the results are identical. The formal differences are in that in the cascade theory the distributions (for instance, of molecular weight) are formulated implicitly in the form of pgf and then transformed by simple differentiation into averages of whatever kind one wishes. Also, one can get information about composition of generations at a certain topological distance from the selected unit which is useful for evaluation of certain physical properties. The recursive theory calculates the weight averages directly from average probability (expectation) of structure continuation (similar to Flory [124] branching coefficient) bypassing the formulation of the (implicit) distribution function and its transformation to average. There are small formal differences in that the extinction probability used in cascade theory is conditional given a bond exist, whereas in the finite "In–Out" method the unreacted group is included into consideration. Indeed, if one is satisfied with averages and the starting components as well as the crosslinking mechanism are simple, the M&M recursive theory is less demanding. However, for more complicated systems, the authors also used the formalism of probability generating functions, e.g. [103, 204]. The basic development of the M&M recursive theory can be found in refs. [202–211]. Further development of the *in–out recursive modeling* shifted to Argentina (Bahia Blanca) and has been headed by Claudia Sarmoria who collaborated initially with deceased Douglas Miller (cf. refs. [212–220]). The Argentine group is now an ardent propagator of the use of probability generating functions (cf., additionally, e.g., [221–225]). Another Argentine group headed by Fortunatti et a. [226], Aranguren et al. [227], Vásquez et al. [228], Riccardi and Williams [229], Williams et al. [230], Williams [231], Williams and Vallo [232] adapted the M&M recursive theory for application to various system of practical importance. One should not forget important contribution to the statistical theory of network formation by Japanese scientists Fukui and Yamabe [233].

In the part of this Chapter that follows, we will be using the formalism of probability generating functions (pgf) and the cascade method; first, we briefly explain what a pgf is and the basic operations with it to transform the distribution and to get averages.

3.4.2.3 Distributions and Their Manipulations—Averages

A characteristic property of the world of macromolecules is their dispersity (often used prefix *poly*dispersity is redundant according to IUPAC) described by distribution functions. Each polymer chain can adopt various shapes in space (conformations) described by conformation distribution. Chemically, a polymer is composed of many molecules differing in degree of polymerization; in a copolymer, each mer of a given degree of polymerization distribution has a distribution in composition. If one or more monomer units carry one or more functional groups capable of forming bonds, there is on top of the former distributions a functionality distribution. Each of the functional groups can occur in unreacted state or reacted states, and each reacted group participates in a bond; such distributions are called distributions of reacted states. The distributions are described by number, weight (mass), or other fractions of units in the specified state. In the branching theory, these fractions are related to probabilities of finding such state in the respective probability space. As we have seen above, already Watson [91] in the nineteenth century described distributions in the problem of transfer of family names by a probability generating function.

Probability generating function (pgf) function $A(Z)$

$$A(Z) = \sum_i a_i Z^i \tag{3.24}$$

is defined by a power series of Z in which the coefficients a_i of a sequence $\{a_i\}$ have a meaning of the probability of an event, $P(u_i)$, to which a random variable i is assigned. Since the sequence $\{a_i\}$ fills in the probability space

$$A(Z = 1) = \sum_i a_i = 1 \tag{3.25}$$

Z is a dummy (auxiliary) variable which serves only as operator for the distribution $\{a_i\}$; later, it disappears by putting it equal to unity or zero ($Z = 1,\ Z = 0$).

The average value of i, $\bar{i}$, is obtained by differentiation of $A(Z)$ and putting $Z = 1$

$$\frac{\partial A(Z)}{\partial Z} \equiv A'(Z) = \sum_i i a_i Z^{i-1}, \quad A'(Z = 1) \equiv A'(1) = \sum_i i a_i = \bar{i} \tag{3.26}$$

Similarly, by differentiating $A(Z) k$ times and putting $Z = 0$, the a_k term is obtained: for the k-th differentiation of $A(Z) = \sum_{i=0}^{m} a_i Z^i$

$$\frac{\partial A^k(Z)}{\partial Z^k} \equiv A^{(k)}(Z) = \sum_{i=k}^{m} m \cdot (m-1) \cdot ...(k-i)a_i Z^{m-k}$$
$$== k!a_k + \frac{k!}{2!}a_{k+1}Z + ... + \frac{m!}{(k-1)!}a_m Z^{m-k} \quad (3.27)$$

so that

$$\left[\frac{\partial A^k(Z)}{\partial Z^k}\right]_{Z=0} \equiv A^{(k)}(0) = k!a_k, \quad a_k = A^{(k)}(0)/k! \quad (3.28)$$

More concretely, for a polymer composed of monomer units (mers) of degree of polymerization i, the i-mer chain (exponent of Z^i, i is the value of the number of monomer units) can be found among other chains with probability n_i (n_i is number fraction of the i-mer), the number fraction generating function $N(Z)$ reads

$$N(Z) = \sum_i n_i Z^i, \quad N(Z=1) \equiv N(1) = \sum_i n_i = 1 \quad (3.29)$$

and number-average degree of polymerization P_n is obtained by differentiation

$$\frac{\partial N(Z)}{\partial Z} = \sum_{i=1}^{\infty} in_i Z^{i-1}, \quad \left[\frac{\partial N(Z)}{\partial Z}\right]_{Z=1} = N'(Z=1) \equiv N'(1) = \sum_{i=1}^{\infty} in_i = P_{\mathrm{n}} \quad (3.30)$$

Equation 3.7 represents the first moment of the distribution; a simple manipulation with the pgf leads to the distribution for the second moment and to what is called weight-average degree of polymerization.[1]

$$Z\frac{\partial N(Z)}{\partial Z} = \sum_i \mathrm{in}_i Z^i = P_n \sum_i \frac{\mathrm{in}_i}{\sum \mathrm{in}_i} Z^i = P_n \sum_i w_i Z^i \quad (3.31)$$

and

$$\left[\frac{\partial\left(Z\frac{\partial N(Z)}{\partial Z}\right)}{\partial Z}\right]_{Z=1} = \sum_i i^2 n_i = P_w P_n \quad (3.32)$$

In this way, we can go to higher averages. The opposite way, i.e., e.g., from weight to number distribution, one can proceed by integration

[1] This distribution is identical with weight (mass) distribution only if the weight of all monomer units are the same; otherwise, the different mass of units must be respected as shown below.

$$P_n \int_0^Z \frac{1}{Z} W(Z) = \int_0^Z \frac{1}{Z} \sum_i \mathrm{in}_i Z^i \mathrm{d}Z = \int_0^Z \sum_i \mathrm{in}_i Z^{i-1} \mathrm{d}Z = \sum_i \frac{\mathrm{in}_i}{i} Z^i = N(Z) \tag{3.33}$$

In homopolymers, the contributions of numbers of all mers to the molecular weight are described by Eq. 3.10. To get the molecular weight averages, we put the mass of the monomer unit into the exponent

$$N_{\mathrm{M}}(Z) = \sum_i n_i Z^{iM_0}, \quad \left[\frac{\partial N_{\mathrm{M}}(Z)}{\partial Z}\right]_{Z=1} = \sum_i \mathrm{in}_i M_0 = M_{\mathrm{n}} \tag{3.34}$$

The distribution of copolymers composed of units of different mass can be described by multivariate pgf of more auxiliary variables. Thus, for binary copolymers composed of units A and B, as an example,

$$\begin{aligned} N_{\mathrm{M}}(Z_{\mathrm{A}}, Z_{\mathrm{B}}) &= \sum_{i,j} n_{ij} Z_{\mathrm{A}}^{iM_{\mathrm{A}}} Z_{\mathrm{B}}^{jM_{\mathrm{B}}}, \\ M_{\mathrm{n}} &= \left[\frac{\partial N_{\mathrm{M}}(Z)}{\partial Z_{\mathrm{A}}} + \frac{\partial N_{\mathrm{M}}(Z)}{Z_{\mathrm{B}}}\right]_{Z_{\mathrm{A}}=Z_{\mathrm{B}}=1} \\ &= \sum_{i,j} n_{ij}(iM_{\mathrm{A}} + jM_{\mathrm{B}}) \end{aligned} \tag{3.35}$$

Similarly, one can proceed with multicomponent system and to higher averages where mixed second moments exists—for the example here related to i^2, j^2, and ij. In applications, the degree of polymerization distribution and compositional distribution have specific closed forms (cf., e.g., [132, 138, 199]). For instance, for the Flory's most-probable distribution

$$N_{\mathrm{M}}(Z) = \frac{(1-q)Z}{1-qZ} = \sum_{i=1}^{\infty} (1-q) q^{i-1} Z^i, \quad q = 1 - \frac{1}{P_{\mathrm{n}}} \tag{3.36}$$

The formalism of the pgfs makes it possible to identify the exact meaning of the probability coefficient appearing in front of the variables.

Units and bonds: While we were so far concerned with counting units, for the branched and crosslinked systems, the bonds connecting units are important. It is the bond through which the multiplicative process occurs—the structure propagation takes place. To distinguish between information about units property and bonds connecting units, we have proposed some time ago to use upper case (capital) letter (Z) for the unit property as auxiliary variable with subscript denoting type of unit X (Z_{X}) and lower case letter (z) for identification of bonds extending from unit X to unit Y (z_{XY}) or, when necessary, from unit X via group a to group b of unit Y (z_{XabY}). The bonds in the branching theory are always oriented: for a bond looking out of the unit X, X-a → b-Y; the same bond, looking out of the Y unit, it denoted as Y-b → a-X.

3.4.2.4 Basic Information on the Branching System

Basic information on the composition of the systems in terms of distribution of reaction states including their participation in bonds is described by the pgf $F_0(\mathbf{Z}, \mathbf{z})$. This function is also called a pgf for the units in the root of a tree [126, 127]; all reaction states of all units of the system occur in (serve as) the root; therefore, the function $F_0(\mathbf{Z}, \mathbf{z})$ describes the composition of the whole system. Its correct formulation is of utmost importance and requires knowledge of the reaction mechanism and kinetics (or equilibrium) to obtain the molar fractions of reaction states at time t or conversion α. The conversions can be calculated by analytical or numerical solution of sets of equations of chemical kinetics or equilibrium constants.

The basic information about the units and reaction states is stored in the pgfs F_0 which are used in several forms such as

$$
\begin{aligned}
F_{0\mathrm{n}}(\mathbf{Z}, \mathbf{z}) &= \sum_R n_R F_{0\mathrm{R}}(Z_R, \mathbf{z}) \\
F_{0\mathrm{m}}(\mathbf{Z}, \mathbf{z}) &= \sum_R m_R F_{0\mathrm{R}}(Z_R, \mathbf{z}) \\
F_{0\mathrm{R}}(Z_R, \mathbf{z}) &= Z_R^P \sum_{\mathrm{p,q,S,k}} a_{\mathrm{RpqS},k} z_{\mathrm{RpqS}}^k
\end{aligned}
\tag{3.37}
$$

If the crosslinking system is composed of several components, the pgf of the system is represented by sums weighted by the number n_R or mass m_R fractions of pgfs of components (Eq. 3.15); the pgf of component R displays the value of a property P (e.g., contribution to the weight characterized by the molecular weight of the component M). The pgf displays further the contribution of the unit R to bonds extending from R via its group to group q attached to unit S; $a_{\mathrm{RpqS},k}$ is the probability of finding k of such bonds normalized to all groups of the unit of type R.

Alternatively, the link probability generating function (lpgf) used in Gordon's et al. and other lists only outgoing bonds; it is obtained by putting $\mathbf{Z} = \mathbf{1}$ in Eq. 3.14.

$$
\begin{aligned}
F_{l,0\mathrm{n}}(\mathbf{z}) &= \sum_R n_R F_{0\mathrm{nR}}(\mathbf{z}) \\
F_{l,0\mathrm{m}}(\mathbf{z}) &= \sum_R m_R F_{0\mathrm{nR}}(\mathbf{z}) \\
F_{1,0\mathrm{R}}(\mathbf{z}) &= \sum_{\mathrm{p,q,S,k}} a_{\mathrm{RpqS},k} z_{\mathrm{RpqS}}^k
\end{aligned}
\tag{3.38}
$$

In the literature, the type of bonds RpqS is sometimes denoted differently, as RpSq, especially when p and q are numerals (e.g., A1B2), or by upper case letters (capitals), if each component contains a single type group (AabB = AB)

$$\begin{aligned} \mathrm{XstY} &\equiv \mathrm{XsYt} \\ \mathrm{AstA} &\equiv \mathrm{st} \\ \mathrm{AabB} &\equiv \mathrm{AB} \end{aligned}$$

Correlation depth: Bond formation and existence may depend not only on the types of reacting functional groups and their reference reactivities but also on the chemical enviroment of the unit the group is a part of. Such correlation is usually called *first-shell substitution effect* (*fsse*). Although not precisely defined, *fsse* takes into account the state of the unit defined by the numbers and types of issuing bonds irrespective of the reaction states of nearest beighbor units. If the fond formation and population balance depend on reaction states of nearest neighbors, such correlation is called *second shell substitution effect* (*ssse*). The difference between fsse and ssse is illustrated in Fig. 3.2.

Generation of larger structures occurs by transition probabilities of Markov process of first and second order, respectively.

Examples—Single Type of Units

Next, some examples will be shown:

(a) single unit and single type groups of the same and independent reactivity (each group reacts independently of the other—convolution of probabilities of states for each group)

$$\begin{aligned} F_0(Z_\mathrm{A}, z_\mathrm{AA}) &= Z_\mathrm{A}^{M_\mathrm{A}}(1-\alpha_\mathrm{A}+\alpha_\mathrm{A} z_\mathrm{AA})^{f_\mathrm{A}} \\ &\equiv Z_\mathrm{A}^{M_\mathrm{A}} \sum_{i=0}^{f_\mathrm{A}} \frac{f_\mathrm{A}!}{i!(f_\mathrm{A}-i)!}(1-\alpha_\mathrm{A})^{(f_\mathrm{A}-i)} \alpha_\mathrm{A}^i z_\mathrm{AA}^i \\ &\equiv Z_\mathrm{A}^{M_\mathrm{A}} \sum_{i=0}^{f_\mathrm{A}} a_i z_\mathrm{AA}^i \end{aligned} \tag{3.39}$$

Figure 3.3 shows the distribution of units of a tetrafunctional (star-shaped) monomer in different reaction states; probabilities of number fractions, a_i, are assigned.

The units are represented in 2D for convenience; in fact, the structures are just graphs without any correlation with dimensionality of the space. As shown in the bottom part of Fig. 3.4, in this representation only the number of reacted and unreacted group matters, their geometrical arrangements are irrelevant. It should be remarked that in the literature the units are represented in different geometrical forms, dependent on whether their represents a root or part of a branch of a tree in which the bonds are oriented (Fig. 3.4)

(b) groups of the same type but of different reactivity A1 and A2

Reacting groups	Bond	Rate constant
A1 + A1	A1A1	k_{11}
A1 + A2	A1A2 = A2A1	k_{12}
A2 + A2	A2A2	k_{22}

By solving the respective kinetic differential equations, the concentrations of bonds [A1A1], [A1A2] = [A2A1], and [A2A2] are obtained. Then, the pgf F_0 reads

$$\begin{aligned} F_0(Z_A, \mathbf{z}) &= Z_A^{M_A}(1 - \alpha_{A1} + \alpha_{A1}(p_{11}z_{A1A1} + p_{12}z_{A1A2}))^{f_{A1}} \\ &\quad (1 - \alpha_{A2} + \alpha_{A2}(p_{21}z_{A2A1} + p_{22}z_{A2A2}))^{f_{A2}} \\ \mathbf{z} &= (z_{A1A1}, z_{A1A2}, z_{A2A1}, z_{A2A2}) \end{aligned} \tag{3.40}$$

In Eq. 3.40, the coefficients p are Markovian transition probabilities determined by the concentration of bonds, either calculated from the respective reaction scheme using rate constants, or determined experimentally as a function of conversions α_{A1}, or α_{A2}, or total conversion $\alpha_{A1} + \alpha_{A2}$

$$\begin{aligned} p_{11} &= \frac{[A1A1]}{[A1A1] + [A1A2]}, \quad p_{12} = \frac{[A1A2]}{[A1A1] + [A1A2]}, \quad p_{11} + p_{12} = 1 \\ p_{22} &= \frac{[A2A2]}{[A2A1] + [A2A2]}, \quad p_{21} = \frac{[A2A1]}{[A2A1] + [A2A2]}, \quad p_{21} + p_{22} = 1 \end{aligned} \tag{3.41}$$

where [XY} are concentration of dyads XY

The formation of bonds of several types for the $f_{A1} = 2$ and $f_{A2} = 2; f_A = 4$ case is shown in Fig. 3.5

(c) substitution effect—the reactivity of a groups in a building unit depends the state (unreacted versus reacted) of other units. Such an effect is expressed by different rate constants for the groups. For instance, for an A_4 unit, the transformation scheme in which the number of reacted groups is indicated by subscript x and that of unreacted groups by subscript y, A_{xy} can be described as shown in Fig. 3.3

One can see in Fig. 3.6 that various reaction paths progressing faster and some slower. Here, the rate constants $k_{ij} = k_{ji}$.

Schemes for the bond formation are shown in Fig. 3.6; the concentrations of dyads are obtained by numerical solution of kinetics equation for bimolecular (or other order) reactions based on mass action law. Their solution determines the transition probabilities.

$$4[\mathrm{A04}] + 4[\mathrm{A04}] \xrightarrow{k_{08}} [\mathrm{A13}-\mathrm{A13}]\ p11 = \frac{[\mathrm{A13-A13}]}{[\mathrm{A13-A13}]+[\mathrm{A13-A22}]+[\mathrm{A13-A13}]+[\mathrm{A13-A40}]}$$
$$4[\mathrm{A04}] + 3[\mathrm{A13}] \xrightarrow{k_{01}} [\mathrm{A13}-\mathrm{A22}] \qquad .$$
$$4[\mathrm{A04}] + 2[\mathrm{A22}] \xrightarrow{k_{02}} [\mathrm{A13}-\mathrm{A31}] \qquad .$$
$$4[\mathrm{A04}] + [\mathrm{A31}] \xrightarrow{k_{03}} [\mathrm{A13}-\mathrm{A40}]\ \ p14 = \frac{[\mathrm{A13-A40}]}{[\mathrm{A13-A13}]+[\mathrm{A13-A22}]+[\mathrm{A13-A13}]+[\mathrm{A13-A40}]}$$

$$\vdots$$

$$[\mathrm{A31}] + 4[\mathrm{A04}] \xrightarrow{k_{30}} [\mathrm{A40}-\mathrm{A13}]\ p41 = \frac{[\mathrm{A40-A13}]}{[\mathrm{A40-A13}]+[\mathrm{A40-A22}]+[\mathrm{A40-A13}]+[\mathrm{A40-A40}]}$$
$$[\mathrm{A31}] + 3[\mathrm{A13}] \xrightarrow{k_{31}} [\mathrm{A40}-\mathrm{A22}] \qquad .$$
$$[\mathrm{A31}] + 2[\mathrm{A22}] \xrightarrow{k_{32}} [\mathrm{A40}-\mathrm{A13}] \qquad .$$
$$[\mathrm{A31}] + [\mathrm{A31}] \xrightarrow{k_{33}} [\mathrm{A40}-\mathrm{A40}]\ \ p44 = \frac{[\mathrm{A40-A40}]}{[\mathrm{A40-A13}]+[\mathrm{A40-A22}]+[\mathrm{A40-A13}]+[\mathrm{A40-A40}]}$$

The pgf F0 which balances units with respect to the number and type of issuing bonds reads

$$\begin{aligned}
F_{0n}(\mathbf{z}) &= a_0 + a_1\zeta_1 + a_2\zeta_2 + a_3\zeta_3 + a_4\zeta_4 \\
\zeta_1 &= p_{11}z_{11} + p_{12}z_{12} + p_{13}z_{13} + p_{14}z_{14} \\
\zeta_2 &= p_{21}z_{21}^2 + p_{22}z_{22}^2 + p_{23}z_{23}^2 + p_{24}z_{24}^2 \\
\zeta_3 &= p_{31}z_{31}^3 + p_{32}z_{32}^3 + p_{33}z_{33}^3 + p_{34}z_{34}^3 \\
\zeta_4 &= p_{41}z_{41}^4 + p_{42}z_{42}^4 + p_{43}z_{33}^4 + p_{44}z_{34}^4
\end{aligned} \tag{3.42}$$

$$F_{0\mathrm{n}}(Z_{\mathrm{A}}, \mathbf{z}) = Z_{\mathrm{A}}^{M_{\mathrm{A}}} F_{0\mathrm{n}}(\mathbf{z})$$

where

$$a_i = \frac{\sum_j [\mathrm{A_{ij}}]}{\sum_i \sum_j [\mathrm{A_{ij}}]} \tag{3.43}$$

are molar fractions of units with i reacted functional groups.

Schematically, the network buildup in the case of homopolyaddition of monomer of functionality equal to 4, respecting differences in reaction paths, is shown in Fig. 3.8.

Experimental determination of all rate constants would be quite difficult, but for predictive pursposes, the rate constants can be calculated with the aid of quantum chemistry.

The lowest level of correlation in system with substitution effect corresponding to Bernoulli statistics is disregarding differences in bond sequences between groups of different reactivity: comparing to the random case, it is only the distribution of the number of reacted groups A per monomer unit which is not random and is not equal to coefficients of binomial expansion. If the reaction time correlations are neglected and reacted groups are connected to form bonds AA randomly, the corresponding pgf reads

$$F_0(Z_A, z_{AA}) = Z_A^{M_A}[a_0 + a_1 z_{AA} + a_2 z_{AA}^2 + a_3 z_{AA}^3 + a_4 z_{AA}^4] \tag{3.44}$$

For irreversible reactions or non-equilibrium states of reversible reactions, Eq. 3.21 can be a very poor approximation; in the case of milder differences in reactivity, approximation by Eq. 3.21 can be quite good.

(d) In contrast to the case (b), groups of the same type are placed on different monomers of different functionality, reactivity, and mass $A1_{fA1} + A2_{fA2}$ but their reactivity is independent of the reaction conversion. Then, the pgf F_0 reads

$$\begin{aligned} F_0(Z_A, \mathbf{z}) &= n_{A1} Z_{A1}^{M_{A1}}(1 - \alpha_{A1} + \alpha_{A1}(p_{11} z_{A1A1} + p_{12} z_{A1A2}))^{f_{A1}} \\ &\quad + n_{A2} Z_{A2}^{M_{A2}}(1 - \alpha_{A2} + \alpha_{A2}(p_{21} z_{A2A1} + p_{22} z_{A2A2}))^{f_{A2}}, \\ \mathbf{z} &= (z_{A1A1}, z_{A1A2}, z_{A2A1}, z_{A2A2}) \end{aligned} \tag{3.45}$$

where n_{A1} and n_{A2} are molar fractions of monomers $A1_{fA1}$ and $A2_{fA2}$, respectively.

Examples—Various Types of Groups and Units—Selective Reactions

Systems with selectively reacting groups are in practice much more important than those, where all groups react with each other (Sect. 3.4.2.1). Classical is the A_{fA} + B_{fB} system where only the A + B → AB reaction occurs. If the reactivities of A groups and B groups in $A_{fA} + B_{fB}$, respectively, are equal and independent, the pgf F_0 version according to Eq. 3.14 has a simple form

$$\begin{aligned} F_0(Z_A, \mathbf{z}) &= n_A Z_A^{M_A}(1 - \alpha_A + \alpha_A z_{AB})^{f_A} \\ &\quad + n_B Z_B^{M_B}(1 - \alpha_B + \alpha_B z_{BA})^{f_B}, \\ \mathbf{z} &= (z_{AB}, z_{BA}) \end{aligned} \tag{3.46}$$

The bond-only version of Eq. 3.15 is given by lpgf

$$\begin{aligned} F_{1,0}(Z_A, \mathbf{z}) &= n_A(1 - \alpha_A + \alpha_A z_{AB})^{f_A} \\ &\quad + n_B(1 - \alpha_B + \alpha_B z_{BA})^{f_B}, \\ \mathbf{z} &= (z_{AB}, z_{BA}) \end{aligned} \tag{3.46a}$$

where α_A and α_B are conversions of functional groups A and B, respectively, n_A and n_B, molar fractions, respectively, of monomers A_{fA} and B_{fB}, and M_A and M_B molecular weights of monomers A_{fA} and B_{fB}; z_{AB} and z_{BA} denominate bonds looking out of the unit A_{fA} and B_{fB}, respectively.

Again, one or both components can have groups of different reactivity, or there can exist a substitution effect in one or both components. If both components exhibit substitution effect, the distribution of reaction states in terms of bond types is quite extensive, and it is determined by quite a number of rate (or equilibrium) constants, like in the case c).

Considerable simplification is achieved if one of the components has groups of independent reactivity for which the probability to find the group, for instance, in the unreacted state is $1 - \alpha_B$ and the reaction state α_B. The distinguishable oriented bonds are:

$$\text{A1-B, A2-B, A3-B, A4-B, B-A1, B-A2, B-A3, B-A4}$$

and the respective pgf F_0 reads

$$\begin{aligned} F_0(Z_A, Z_B, \mathbf{z}) = {} & n_A Z_A^{M_A}(a_{A0} + a_{A1} z_{A1B} + a_{A2} z_{A2B}^2 + a_{A3} z_{A3B}^3 + a_{A4} z_{A4B}^4) \\ & + n_B Z_B^{M_B}\left[1 - \alpha_B + \alpha_B(p_{BA1} z_{BA1} + p_{BA2} z_{BA2} \right. \\ & \left. + p_{BA3} z_{BA3} + p_{BA4} z_{BA4})\right]^{f_B} \end{aligned} \quad (3.47)$$

where

$$p_{BAj} = \frac{j a_j}{\sum_j j a_j}, \text{ i.e.,} p_{BA1} = \frac{a_1}{a_1 + 2a_2 + 3a_3 + 4a_4}, \text{ ....etc.} \quad (3.48)$$

The latter case (Eqs. 3.49 and 3.50) as well as the random case (Eq. 3.47) are relatively frequent.

The distribution of reaction states of units and bonds going out of the units is shown in Fig. 3.32.

Examples—Crosslinking of Primary Chains—Vulcanization Reactions

First, first crosslinking of uniform collection of primary chains will be considered. In the basic model, each monomer unit is considered to be crosslinkable by the same probability, independent of the state of neighbor unit. There is a single type of bonds connecting the polymer segments (monomer units); each crosslinked unit can participate in a bond. Since the bond is a single type, one pgf variable z is sufficient for designation. The reaction states of building units (primary chains) are distinguished by the number of crosslinked segments i out of the total number of segments x:

$$\begin{aligned} F_0(Z, \mathbf{z}) &= \left[Z^{M_0}(1 - \alpha + \alpha Z^{M_v/2} z)\right]^x \\ &\equiv Z^{xM_0} \sum_{i=0}^{x} \frac{x!}{i!(x-i)!}(1-\alpha)^{x-i}\alpha^i z^i Z^{i(M_v/2)} \end{aligned} \quad (3.49)$$

In this equation, M_0 is the molecular weight of the monomer unit, and α is the fraction of crosslinked monomer units. The inclusion of $Z^{M_v/2}$ takes care of a (small) increase of molecular weight by the crosslinker of molecular weight M_ν (half of each is assigned to primary chain). This contribution to the weight is usually omitted, but it should be taken into account when the crosslinking (vulcanization) agent is large.

For certain purposes (radii of gyration, structure factors) [139], or crosslinking and random degradation [122], it is preferable to root a primary chain (or each primary chain of a distribution) by its monomer unit.

The two branches issuing from a selected unit (called, for instance, the right, p. and left q branch) have only a finite continuation to add to the total number of mers, x. For random degradation of the primary chain, $\beta = 1 - \gamma$, the weight fraction generating function

$$\begin{aligned}
W_{xi}(Z) &= Z(1-\gamma+\gamma u_{\mathrm{q},1})(1-\gamma+\gamma u_{\mathrm{p},1}) \\
u_{\mathrm{q},1} &= Z(1-\gamma+\gamma u_{\mathrm{q},2}) \\
u_{\mathrm{q},2} &= Z(1-\gamma+\gamma u_{\mathrm{q},3}) \\
&\bullet \\
&\bullet \\
u_{\mathrm{q,i-2}} &= Z(1-\gamma+\gamma u_{\mathrm{q,i-1}}) \\
u_{\mathrm{q,i-1}} &= Z \\
\\
u_{\mathrm{p},1} &= Z(1-\gamma+\gamma u_{\mathrm{p},2}) \\
u_{\mathrm{p},2} &= Z(1-\gamma+\gamma u_{\mathrm{p},3}) \\
&\bullet \\
&\bullet \\
u_{\mathrm{p,x-i-1}} &= Z(1-\gamma+\gamma u_{\mathrm{p,x-i}}) \\
u_{\mathrm{p,ix-i}} &= Z
\end{aligned} \tag{3.50}$$

If degradation proceeds simultaneously with degradation, each monomer unit can become crosslinked with probability α, and for an x-mer it holds

$$
\begin{aligned}
W_{xi}(Z) &= Z(1-\gamma+\gamma u_{\mathrm{q},1})(1-\gamma+\gamma u_{\mathrm{p},1})(1-\alpha+\alpha u)\\
u_{\mathrm{q},1} &= Z(1-\gamma+\gamma u_{\mathrm{q},2})(1-\alpha+\alpha u)\\
u_{\mathrm{q},2} &= Z(1-\gamma+\gamma u_{\mathrm{q},3})(1-\alpha+\alpha u)\\
&\bullet\\
&\bullet\\
u_{\mathrm{q,i-2}} &= Z(1-\gamma+\gamma u_{\mathrm{q,i-1}})(1-\alpha+\alpha u)\\
u_{\mathrm{q,i-1}} &= Z(1-\alpha+\alpha u)\\
\\
u_{\mathrm{p},1} &= Z(1-\gamma+\gamma u_{\mathrm{p},2})(1-\alpha+\alpha u)\\
u_{\mathrm{p},2} &= Z(1-\gamma+\gamma u_{\mathrm{p},3})(1-\alpha+\alpha u)\\
&\bullet\\
&\bullet\\
u_{\mathrm{p,x-i-1}} &= Z(1-\gamma+\gamma u_{\mathrm{p,x-i}})(1-\alpha+\alpha u)\\
u_{\mathrm{p,ix-i}} &= Z(1-\alpha+\alpha u)\\
u &= x^{-1}\sum_{i=1}^{x} Z(1-\gamma+\gamma u_{\mathrm{p},1})(1-\gamma+\gamma u_{\mathrm{q},1})
\end{aligned}
\tag{3.51}
$$

In the absence of degradation, $\gamma = 1$.

Preprepared Precursors with Distributions of Molecular Weights, Number of Groups, and Bond Types

In an overwhelming majority of cases of practical interest, the network precursors are mixtures of compounds characterized by distributions of molecular weights and number of functional groups. Often, several types of functional groups are involved [57, 200]. These distributions are incorporated into $F_0(\mathbf{Z}, \mathbf{z})$ in forms of corres-dponding probability generating functions. The resulting $F_0(\mathbf{Z}, \mathbf{z})$ can be further routinely transformed to get the gel point and characteristic properties of sol and gel in the same way as simple systems. Even more straightforward is the use of precursors preprepared by polymerization process, possibly including branching. When such branched precursor is used, the preexisting branch points become gradually activated which means that they are transferred from the elastically inactive to the state of elastically active junctions. This transfer can be well described theoretically [193]. The functional groups that have not reacted in this first stage and are designed for reaction in the second (crosslinking) stage are marked. In the second stage, the bonds of the same type formed in the first and second stages must be distinguished by assigning to them different pgf variables z. In the next paragraph, we will show some closed forms pgfs for degree of polymerization/molecular weight distributions, compositional distributions, and a typical example of precursor as network formation intermediate.

Most typical *degrees of polymerization/molecular weight distributions* of linear polymers in the form of pgfs are displayed in Table 3.2 (cf. [57]).

Table 3.2 Number fraction probability generating functions $N(Z) = \sum_{i=1}^{\infty} n_i Z^i$ for the most typical degrees of polymerization distributions of linear homopolymers

Distribution	n_i	Parameter	$N(Z)$	P_n	P_w
Most probable	$(1-q)q^{i-1}$	$q = 1 - \frac{1}{P_n}$	$\frac{(1-q)Z}{1-qZ}$	$\frac{1}{1-q}$	$\frac{1+q}{1-q}$
SCMP*	$(1-q)^2 q^{i-1}$	$q = 1 - \frac{2}{P_n}$	$\frac{(1-q)^2 Z^2}{(1-qZ)^2}$	$\frac{2}{1-q}$	$\frac{2+q}{1-q}$
Schulz-Zimm	$(P_w - P_n)^{\frac{P_n}{P_w-P_n}} \cdot i^{\frac{2P_n-P_w}{P_w-P_n}} \exp\left(\frac{-\frac{i}{P_w-P_n}}{\Gamma\left(\frac{P_n}{P_w-P_n}\right)}\right)$	P_n, P_w	$\left(\frac{1}{1-(P_w-P_n)\ln Z}\right)^{\frac{P_n}{P_w-P_n}}$	P_n	P_w
Poisson	$\frac{\lambda^{i-1}}{(i-1)!}\exp(-\lambda)$	$\lambda + 1 = P_n$	$Z\exp[\lambda(Z-1)]$	$\lambda + 1$	$\frac{\lambda^2+3\lambda+1}{\lambda+1}$

*Self-convoluted most-probable distribution

Table 3.3 Types of substructures in the network classified according to the number of bonds with infinite continuation

Type of substructure	Number of bonds having infinite continuation $(1-v)$ type	Number of bonds having finite continuation (v type) and unreacted groups
Sol	0	All (f)
Dangling chain	1	$f-1$
Elastically active network chain	2	$f-2$
Elasticallyactive crosslink (junction)	3 and more	$f-3, f-4, \ldots$

Thus, for random crosslinking of primary chains disperse in degree of polymerization, Eq. 3.26 is transformed into

$$F_0(Z,z) = N(\zeta); \quad \zeta = Z^{M_0}(1-\alpha+\alpha Z^{M_v/2}z) \tag{3.52}$$

For the most-probable distribution of primary chains, as an example, Eq. 3.27 gives

$$F_0(Z,z) = \frac{(1-q)\zeta}{1-q\zeta}; \quad \zeta = Z^{M_0}(1-\alpha+\alpha Z^{M_v/2}z) \tag{3.53}$$

for Poisson distribution, it reads

$$F_0(Z,z) = \zeta \exp[\lambda(\zeta-1)]; \quad \zeta = Z^{M_0}(1-\alpha+\alpha Z^{M_v/2}z) \tag{3.54}$$

where M_0 is the molecular weight of the monomer unit, and M_v is the molecular weight of the crosslinker.

The *compositional distribution* of a copolymer described by compositional probability generation function for a binary copolymer x-mer

$$N_{c,x}(Z_A, Z_B) = \sum_{i=0}^{x} c_{i,x-i} Z_A^i Z_B^{x-i} \tag{3.55}$$

depends on the way copolymers (copolyadducts) are prepared. For free-radical copolymers prepared batchwise, the conversion drift of copolymer composition is characteristic; the distribution $c_{i,x-i}$ is complicated and depends on time (conversion). On the other hand, the composition of copolymers can be kept constant, if copolymerization is carried out under starving conditions—the mixture of monomers is fed continuously and is transformed into the copolymer of the same composition. For such copolymers, the distribution can be considered as random, i.e., it is equal to coefficients of binomial expansion as shown in Eq. 3.31

$$N_{c,x}(Z_A, Z_B) = (n_A Z_A + n_B Z_B)^x = \sum_{i=0}^{x} \frac{x!}{i!(x-i)!} n_A^i (1-n_A)^{x-i} Z_A^i Z_B^{x-i} \quad (3.56)$$

The coefficient in front of $Z_A^i Z_B^{x-i}$ is the probability of finding an x-mer chain composed i units A and $x - i$ units B irrespective of their sequential arrangements. If one needs to consider molecular weights of the units, Eq. 3.31 is modified

$$N_{cM}(Z_A, Z_B) = (n_A Z_A^{M_A} + n_B Z_B^{M_B})^x = \sum_{i=0}^{x} \frac{x!}{i!(x-i)!} n_A^i (1-n_A)^{x-i} Z_A^{iM_A} Z_B^{(x-i)M_B} \quad (3.57)$$

For a copolymer of A and B monomers of which monomer A carries a functional group A which subsequently reacts with group C of a f_C-functional crosslinker bearing C groups of independent reactivity (AC bonds are formed)

$$F_0(\mathbf{Z}, \mathbf{z}) = n_{AB} N_P(N_{cM}(Z_A \xi_{AC}(z_{AC}), Z_B)) + n_C Z_C^{M_C}(1 - \alpha_C + \alpha_C z_{CA})^{f_C} \quad (3.58)$$

where N_P and N_{cM} describe degree of polymerization and compositional distributions; for the most-probable DP distribution and random compositional distributions

$$N_P(N_{cM}(Z_A \xi_{AC}(z_{AC}), Z_B)) = \frac{(1-q)(n_A Z_A^{M_A} \xi_{AC} + n_B Z_B^{M_B})}{1 - q(n_A Z_A^{M_A} \xi_{AC} + n_B Z_B^{M_B})},$$
$$\xi_{AC} = (1 - \alpha_A + \alpha_A z_{AC}) \quad (3.59)$$

where n_A and n_B are molar fractions of units A and B, respectively, in the copolymer AB, M_A and M_B are the respective molecular weights of monomer units, and α_A is the conversion of A groups in the reaction with C groups. Further examples related to experimentally studied systems including those related to various applications especially in coatings, and thermosets areas can be found in Refs. [57, 199, 200].

Basic Information—Conclusions

The few examples shown here illustrate the transformation of all necessary input information (given by composition of the crosslinking system and mechanism and kinetics of bond forming reactions) about the building units and their ability to form bonds with other units into the form of a basic probability generation function, denoted as $F_0(\mathbf{Z}, \mathbf{z})$, where $\mathbf{Z} = (Z_A, Z_B, ...)$ and $\mathbf{z}$ are vectors of units properties and bonds connecting units via reacted groups, respectively. The function $F_0(\mathbf{Z}, \mathbf{z})$ in the Galton–Watson-type branching theory based on rooted trees (Good–Gordon and others) is called the probability generating function *for the root*, because every unit of the forest is chosen as a root while not all units are able to occur in the branches (cf. next section). Correct formulation of $F_0(\mathbf{Z}, \mathbf{z})$ is of utmost importance because it not only balances all building units, but also bonds according to their types and numbers. This set of information is needed to describe the structure growth to form larger molecules or transformation of sol into "infinite" molecule—the gel. No additional input information is necessary.

3.4.2.5 Pregel State: Buildup of Larger Structures (Molecular Trees) and Their Average Mass

So far, we were concerned with structure of building units and their branching properties; now, we will find how to join them in order to generate larger branched molecules. At this point, we will consider tree-like structures generated by the Galton–Watson process. This is how common statistical theories (including Flory–Stockmayer, cascade, Miller–Macrosko recursive) are working. Let us repeat that the initial information we have available is the distribution of building units in different reaction states. The particular reaction state is characterized by types and numbers of bonds extending from the building unit to another building unit of a specified type. In the cascade version, that information is stored in $F_0(\mathbf{Z}, \mathbf{z})$ or $F_{l,0}(\mathbf{z})$ for so-called roots of probability trees. This function characterizes the numbers and types of bonded neighbors. In contrast to root unit, the attached neighbor unit must have at least one reacted functional group through which it is bonded. While $F_0(\mathbf{Z}, \mathbf{z})$ characterizes the *roots* of the trees or *generation 0* and includes all existing units and their reaction states, the distribution of reaction states in *generation* 1, characterized by pgf $F_1(\mathbf{Z}, \mathbf{z})$, excludes reaction states which cannot exist in generations $i \geq 1$ (i.e., those with all groups in unreacted states) and makes the probability of their attachment to preceding generation proportional to the number of bonds of the specified type. One can say that the unit in generation $i \geq 1$ is rooted on generation $i - 1$ by one of its bonds. Let it be, for instance, the bond BA extending from unit B (in $g = i$) to unit A (in $g = i - 1$); from the point of view of unit A in generation$i - 1$ it is the bond AB. Thus, there are so many functions $F_1(\mathbf{Z}, \mathbf{z})$ as many are components of the vector $\mathbf{z}$. It is the basic property of the Galton–Watson statistics that $F_i(\mathbf{Z}, \mathbf{z})$ is independent of i for $i \geq 1$ therefore, one can write $F_1(\mathbf{Z}, \mathbf{z}) = F_2(\mathbf{Z}, \mathbf{z}) = F_3(\mathbf{Z}, \mathbf{z}) = \cdots \equiv F(\mathbf{Z}, \mathbf{z})$. The function F is marked by the type of bond by which it is rooted, and the bond is considered as *incoming*.

The functions $F(\mathbf{Z}, \mathbf{z})$ are obtained from the basic function $F_0(\mathbf{Z}, \mathbf{z})$ by differentiation with respect to the bond labeling variables and subsequent normalization. Similarly to Eqs. 3.14 and 3.15, it can be formulated in two forms: as bearing information on molecular weight of the component or as link probability generating function (subscript l) (Eqs. 3.35 and 3.36, respectively)

$$F_0(\mathbf{Z}, \mathbf{z}) = \sum_X n_X F_{0X}(Z_X, \mathbf{z}),$$

$$F_{0X}(Z_X, \mathbf{z}) = Z_X^{M_X} \sum_{\mathrm{s,t,Y}} \sum_{i=0}^{f_X} a_{\mathrm{XstY},i} z_{\mathrm{XstY}}^i$$

$$F_{\text{YtsX}}(Z_{\text{X}}, \mathbf{z}) = \frac{\partial F_{0\text{X}}(Z_{\text{X}}, \mathbf{z})}{\partial z_{\text{XstY}}} \Big/ \left[\frac{\partial F_{0\text{X}}(Z_{\text{X}}, \mathbf{z})}{\partial z_{\text{XstY}}} \right]_{Z_{\text{X}}=1, z=1} \tag{3.60}$$

$$F_{1,0}(\mathbf{z}) = \sum_{\text{X}} n_{\text{X}} F_{1,0\text{X}}(\mathbf{z}),$$

$$F_{1,0\text{X}}(\mathbf{z}) = \sum_{\text{s,t,Y}} \sum_{i=0}^{f_{\text{X}}} a_{\text{XstY},i} z_{\text{XstY}}^{i}$$

$$F_{\text{YtsX}}(\mathbf{z}) = \frac{\partial F_{1,0\text{X}}(\mathbf{z})}{\partial z_{\text{XstY}}} \Big/ \left[\frac{\partial F_{1,0\text{X}}(\mathbf{z})}{\partial z_{\text{XstY}}} \right]_{z=1} \tag{3.61}$$

where $a_{\text{XstY},i}$ is the probability that a X-unit participates in i XstY bonds in direction looking out of a X unit.

Examples—Single Type of Bonds

For single type of units and groups A and bonds AA and equal and independent reactivity of groups A (the right-hand-side expressions refer to the random case)

$$F_0(\mathbf{Z}, \mathbf{z}) = Z_{\text{A}}^{M_{\text{A}}} \sum_{i=0}^{f_{\text{A}}} a_i z_{\text{AA}}^{i} = Z_{\text{A}}^{M_{\text{A}}} (1 - \alpha_{\text{A}} + \alpha_{\text{A}} z_{\text{AA}})^{f_{\text{A}}}$$

$$F(\mathbf{Z}, \mathbf{z}) = Z_{\text{A}}^{M_{\text{A}}} \sum_{i=1}^{f_{\text{A}}} i a_i z_{\text{AA}}^{i-1} \Big/ \sum_{i}^{f_{\text{A}}} i a_i = Z_{\text{A}}^{M_{\text{A}}} (1 - \alpha_{\text{A}} + \alpha_{\text{A}} z_{\text{AA}})^{f_{\text{A}}-1}$$

$$F_{\ell}(\mathbf{Z}, \mathbf{z}) = \sum_{i=1}^{f_{\text{A}}} i a_i z_{\text{AA}}^{i-1} \Big/ \sum_{i}^{f_{\text{A}}} i a_i = (1 - \alpha_{\text{A}} + \alpha_{\text{A}} z_{\text{AA}})^{f_{\text{A}}-1} \tag{3.62}$$

In the next paragraph, we will show how to formulate pgf for the number of units in the generation i, $G_i(Z)$ and for the number of units in generation i, or its weight contribution to the weight of the whole tree, $W_i(Z)$ starting from generation 0 up to generation i. Such distributions are obtained by substitution of the pgf variable by another pgf; this substitution is been done repeatedly. It is called *cascade substitution.* To make it clearer, let us consider the simplest case of homopolyaddition of A groups forming single bonds AA and formulate pgfs $G_i(Z)$ and $W_i(Z)$

$$\begin{aligned}
G_0(Z_{\text{A}}) &= Z_{\text{A}} & W_0(Z_{\text{A}}) &= Z_{\text{A}}^{M_{\text{A}}} \\
G_1(Z_{\text{A}}) &= (1 - \alpha_{\text{A}} + \alpha_{\text{A}} Z_{\text{A}})^{f_{\text{A}}} & W_1(Z_{\text{A}}) &= Z_{\text{A}}^{M_{\text{A}}} (1 - \alpha_{\text{A}} + \alpha_{\text{A}} Z_{\text{A}}^{M_{\text{A}}})^{f_{\text{A}}} \\
G_2(Z_{\text{A}}) &= (1 - \alpha_{\text{A}} + \alpha_{\text{A}} (1 - \alpha_{\text{A}} + \alpha_{\text{A}} Z_{\text{A}})^{f_{\text{A}}-1})^{f_{\text{A}}} & W_2(Z_{\text{A}}) &= Z_{\text{A}}^{M_{\text{A}}} (1 - \alpha_{\text{A}} + \alpha_{\text{A}} Z_{\text{A}}^{M_{\text{A}}} \\
& & &\quad (1 - \alpha_{\text{A}} + \alpha_{\text{A}} Z_{\text{A}}^{M_{\text{A}}})^{f_{\text{A}}-1})^{f_{\text{A}}} \\
&\vdots & &\vdots
\end{aligned} \tag{3.63}$$

The last substitution in the W-series is just equal to $Z_{\mathrm{A}}^{M_{\mathrm{A}}}$ because all units exceeding this generation are cut off (not counted).

Let us further define functions up to generation k, $G_k(Z)$ and $W_k(Z)$ (dropping the unit and bond identification subscripts A, AA) which are a function of the recursive function $U_k(Z)$ describing the structure continuation for $k > 0$.

$$\begin{aligned}
U_{\ell,k}(Z) &= F_\ell(U_{\ell,k-1}(Z)) = F_\ell(F_\ell...(F_\ell(Z)...)) \\
&= (1-\alpha_{\mathrm{A}} + \alpha_{\mathrm{A}}(1-\alpha_{\mathrm{A}}+\alpha_{\mathrm{A}}Z)^{f-1}...)^{f-1} \\
G_k(Z) &= F_{1,0}(U_{1,k-1}(Z)) = F_{\ell,0}(F_\ell...(F_\ell(Z)...)) \\
&= (1-\alpha_{\mathrm{A}} + \alpha_{\mathrm{A}}(... + \alpha_{\mathrm{A}}(1-\alpha_{\mathrm{A}}+\alpha_{\mathrm{A}}Z)^{f-1}...)^{f-1}) \\
W_k(Z) &= Z^M F_{\ell,0}(Z^M U_{\ell,k-1}(Z)) = Z^M F_{\ell,0}(Z^M F_\ell...(Z^M F_\ell(Z))...) \\
&= Z^M(1-\alpha_{\mathrm{A}} + Z^M\alpha_{\mathrm{A}}(1-\alpha_{\mathrm{A}}+\alpha_{\mathrm{A}}Z^M)^{f-1}...)^{f-1})^f \\
&= F_0(F...(F(Z)...)) = Z^M(1-\alpha_{\mathrm{A}} + Z^M\alpha_{\mathrm{A}}(1-\alpha_{\mathrm{A}}+\alpha_{\mathrm{A}}Z^M)^{f-1}...)^{f-1})^f
\end{aligned} \tag{3.64}$$

The set of Eq. 3.40 shows the correlation between quantities with subscript l (=link) and quantities including M—the weight of the contributing unit

In the limit of $k \to \infty$, i.e., including trees of all possible sizes, the recurrent Eq. 3.41 reads

$$\begin{aligned}
u(Z) &= Z^M F_\ell(u(Z)) = F(u(Z)) = Z^M(1-\alpha_{\mathrm{A}}+\alpha_{\mathrm{A}}u(Z))^{f-1} \\
\frac{\partial u(Z)}{\partial Z} &= MZ^{M-1}F_\ell(u(Z)) + Z^M\frac{\partial F_\ell(u(Z))}{\partial u(Z)}\frac{\partial u(Z)}{\partial Z}
\end{aligned} \tag{3.65}$$

For $Z = 1$, $u(Z) = 1$, $\left[\partial F_\ell(u(Z))/\partial u(Z)\right]_{Z=1} \equiv F_\ell'(1)$, $\left[\partial u(Z)/\partial Z\right]_{Z=1} \equiv u'(1)$

$$u'(1) = \frac{M}{1-F_\ell'(1)} = \frac{M}{1-(f-1)\alpha_{\mathrm{A}}} \tag{3.66}$$

Equation 3.42 shows the contribution to the weight by units of the tree except of the root and its nearest neighbors. The total weight is obtained from $W(u(Z))$.

As was stated in the introduction to Sect. 3.2, differentiation of a pgf with respect to its variable and putting the variable(s) equal to 1 gives the average value.

$$\begin{aligned}
\left[\frac{\partial G_k(Z)}{\partial Z}\right]_{Z=1} &= F_0'(1)[F'(1)]^{k-1} \\
\left[\frac{\partial W_k(Z)}{\partial Z}\right]_{Z=1} &= M\left(1+F_0'(1)\left(1+F'(1)\left(1+F'(1)\left(...\left(1+F'(1)\right)...\right)\right)\right)\right) \\
&= MF_0'(1)\sum_{j=1}^{k-1}F'(1)^{j-1}
\end{aligned} \tag{3.67}$$

When k grows above all bounds, $[\partial G_k(Z)/\partial Z]_{Z=1}$ is finite only if $F'(1)$ does not exceed unity. The pgf $W_k(Z)$ transforms to weight fraction generation function $W(Z)$ and the power series of Eq. 3.39 converging to zero to that of Eq. 3.44

$$\lim_{k\to\infty} W_k(Z) = W(k) \equiv \sum_{i=1}^{\infty} w_i Z^i,$$
$$\left[\frac{\partial W(Z)}{\partial Z}\right]_{Z=1} \equiv M_w = M + \frac{MF_0'(1)}{1-F'(1)}$$
$$= M_A + \frac{M_A f_A \alpha_A}{1-(f_A-1)\alpha_A} \tag{3.68}$$

$W(Z)$ can also be formulated with the aid of $u(Z)$ (Eq. 3.41)

$$W(Z) = F_0(u(Z)), \quad u(Z) = F(u(Z)) = Z^M(1-\alpha_A+\alpha_A u(Z))^{f-1} \tag{3.69}$$

The righ-hand-side expression is valid for equal and independent reactivity case of polyaddition of f_A monomer. For a similar system with *first-shell substitution effect* (fsse, the functional groups reactivities in the building unit are initially equal but when one or more of them react, the reactivity of remaining groups changes) a more general form (Eq. 3.70, 3.71) was sometimes used.

$$F_0(Z_A, z_{AA}) = Z_A^{M_A}\sum_{i=0}^{f_A} a_i z_{AA}^i, \quad F(z_{AA}) = Z_A^{M_A}\sum_{i=0}^{f_A} i a_i z_{AA}^{i-1} \Big/ \sum_{i=0}^{f_A} i a_i,$$
$$\left[\frac{\partial F(z_{AA})}{\partial z_{AA}}\right]_{Z_A=z_A=1} \equiv F'(1) = \sum_{i=0}^{f_A} i(i-1)a_i \Big/ \sum_{i=0}^{f_A} i a_i \tag{3.70}$$

The weight-average molecular weight is given by the relation

$$M_w = M_A\left(1 + \frac{F_0'(1)}{1-F'(1)}\right)$$
$$= M_A\left(1 + \frac{\sum_{i=1}^{f_A} i a_i}{1 - \sum_{i=1}^{f_A} i(i-1)a_i \Big/ \sum_{i=1}^{f_A} i a_i}\right)$$
$$= M_A\left(1 + \frac{(f_A\alpha_A)^2}{2f_A\alpha_A - \sum_{i=1}^{f_A} i^2 a_i}\right)$$

$$\sum_{i=`1}^{f_A} ia_i = f_A\alpha_A \tag{3.71}$$

Such formulation is based on the implicit assumption that kinetically generated paths of bonds at a certain conversion are rearranged randomly. Compared with the random case, the system differs only in the populations of units in different reaction states, a_i. The populations of states is obtained by solving the system of kinetic differential equations corresponding reaction scheme between the respective A groups in a way as for instance in ref. [178] when additivity of activation free energies is assumed.

$$\begin{aligned}
-\frac{da_0}{dt} &= 3k_0a_0S \\
-\frac{da_1}{dt} &= 2k_1a_1S - 3k_0a_0S \\
-\frac{da_2}{dt} &= k_2a_2S - 2k_1a_1S \\
-\frac{da_3}{dt} &= -k_2a_2S \\
S &= k^{-1}c_0(3k_0a_0 + 2k_1a_1 + k_2a_2)
\end{aligned} \tag{3.72}$$

where k is an arbitrary reference, rate constant a, and c_0 the reference group concentration at which the rate constants k were determined. The Eq. (3.48) can be transformed into $-\,da_0/dt = K_{00}a_0^2 + K_{01}a_0a_1 + K_{02}a_0a_2$, etc., where $K_{ij} = (3-i)(3-j)k_ik_jc_0/k$. An analytical solution for a_i as a function of a_0 and in terms reactivity ratios is shown in Eq. 3.49

$$\begin{aligned}
&2k_1/3k_0 = \kappa_1,\ k_2/3k_0 = \kappa_2 \\
&a_1 = \frac{a_0 - a_0^{\kappa_2}}{1-\kappa_1} \\
&a_2 = \frac{\kappa_1}{1-\kappa_1}\left[\frac{\kappa_2 a_0 - a_0^{\kappa_2}}{1-\kappa_2} - \frac{a_0^{\kappa_1} - a_0^{\kappa_2}}{\kappa_1 - \kappa_2}\right] \\
&a_3 = 1 - \frac{\kappa_1}{1-\kappa_1}\left[\frac{\kappa_2 a_0 - a_0^{\kappa_2}}{1-\kappa_2} - \frac{\kappa_2 a_0^{\kappa_1} - \kappa_1 a_0^{\kappa_2}}{\kappa_2^2 - \kappa_1\kappa_2}\right] \\
&a_1 + 2a_2 + 3a_3 = 3\alpha_A
\end{aligned} \tag{3.73}$$

Graphically, the assembly process is visualized in Fig. 3.13.

The method of treating substitution effect discussed above (random rearrangement of bonds among non-random populations of reaction states) has been commonly used, cf., e.g., refs. [123, 127, 131, 168, 172, 173, 178, 193, 205, 207, 233]. The same approach has been used for treating competitions between groups of different reactivity described by irreversible chemical kinetics. In contrast, if kinetically generated paths between units with the substitution effect should remain preserved, distinguishing of AA bonds according to the reaction states of units the bond is connecting (Sect. "Carrageenan") should be preserved. Then, the results are much closer to exact kinetics solution.

Examples—Multiple Types of Units and Bonds, Substitution Effect

The bonds A–B formed by reaction of groups A and B are considered either as bonds A → B (AB) or bond B → A (BA) and are labeled by variables z_{AB} and z_{BA}, respectively (Eq. 3.23). Passage to higher generations occurs by units A or B which are rooted on groups B and A, respectively. Using the recipe (3.34) one gets

$$F_{BA}(Z_A, z_{AB}) = Z_A^{M_A} \sum_{i=1}^{f_A} ia_i z_{AB}^{i-1} \Big/ \sum_{i=1}^{f_A} ia_i = Z_A^{M_A}(1 - \alpha_A + \alpha_A z_{AB})^{f_A - 1}$$

$$F_{AB}(Z_B, z_{BA}) = Z_B^{M_B} \sum_{i=1}^{f_B} ib_i z_{BA}^{i-1} \Big/ \sum_{i=1}^{f_A} ib_i = Z_B^{M_B}(1 - \alpha_B + \alpha_B z_{BA})^{f_B - 1} \quad (3.74)$$

where the right-hand-side corresponds with the random case without substation effect.

By cascade substitution, the functions u_{BA} and u_{AB} are constructed, resulting in the weight fraction gf, $W(\mathbf{Z})$

$$u_{BA}(Z_A, u_{AB}) = Z_A^{M_A} \sum_{i=1}^{f_A} ia_i u_{AB}^{i-1} \Big/ \sum_{i=1}^{f_A} ia_i = Z_A^{M_A}(1 - \alpha_A + \alpha_A u_{AB})^{f_A - 1}$$

$$u_{AB}(Z_B, u_{BA}) = Z_B^{M_B} \sum_{i=1}^{f_B} ib_i u_{AB}^{i-1} \Big/ \sum_{i=1}^{f_A} ib_i = Z_B^{M_B}(1 - \alpha_B + \alpha_B u_{BA})^{f_B - 1}$$

$$\begin{aligned} W(\mathbf{Z}) &= m_A W_{0A}(Z_A, u_{AB}) + m_B W_{0A}(Z_A, u_{AB}) \\ &= m_A F_{0A}(Z_A, u_{AB}) + m_B F_{0b}(Z_B, u_{BA}) \end{aligned} \quad (3.75)$$

After substitutions, the weight fraction generating function $W(Z)$ depends only on the variables Z_A and Z_B; after differentiation with respect to Z_A and Z_B, one obtains the weight-average molecular weight. Differentiation with respect to these two variables offers the desired M_w. This is illustrated by the random case expressed by the right-hand side of Eq. 3.52. In Eq. 3.53 and below, a short-hand notation for values of derivatives is used in the form

$$y_{XY}^{J} \equiv \left[\frac{\partial y_{XY}(\mathbf{Z}, \mathbf{z})}{\partial Z_J} \right]_{Z=z=1}$$

Thus, M_w is given by Eq. 3.51

$$M_w = W^{Z_A,Z_B} = m_A[M_A + f_A\alpha_A(u^A_{AB} + u^B_{AB})] + m_B[M_B + f_B\alpha_B(u^A_{BA} + u^B_{BA})] \tag{3.76}$$

For the random case, the values of derivatives of components of the vector u are listed below

$$\begin{array}{lll} u^A_{BA} = M_A + (f_A - 1)\alpha_A u^A_{AB} & u^A_{BA} = M_A/D & D = 1 - (f_A - 1)\alpha_A(f_B - 1)\alpha_B \\ u^B_{BA} = (f_A - 1)\alpha_A u^B_{AB} & u^B_{BA} = (f_A - 1)\alpha_A M_B/D & \\ u^A_{AB} = (f_B - 1)\alpha_B u^A_{BA} & u^A_{AB} = (f_B - 1)\alpha_B M_A/D & \\ u^B_{AB} = M_B + (f_B - 1)\alpha_B u^B_{BA} & u^B_{AB} = M_B/D & \end{array} \tag{3.77}$$

so that the explicit form for M_w reads

$$\begin{aligned} M_w = m_A\left[M_A + \frac{1}{D}f_A\alpha_A[(f_B - 1)\alpha_B M_A + M_B]\right] \\ + m_B\left[M_B + \frac{1}{D}f_B\alpha_B[M_A + (f_A - 1)\alpha_A M_B]\right] \end{aligned} \tag{3.78}$$

where m_A and m_B are weight fractions of monomer units.

A few examples in Figs. 3.14 and 3.15 illustrate the dependence of the number- and weight-average molecular weights as a function of the functionality of monomer units and initial molar ratio of A to B groups, $r_A = [A0]/[B0]$. The figures display also the values of number-average molecular weight M_n. The dependence of M_n on conversion is obtained from a simple reasoning valid universally (also for the substitution effect) that the number-average degree of polymerization P_n is equal to the number of monomer units per molecule, and likewise M_n is equal to the weight of monomer units per molecule. The number of molecules is equal to the number of monomer units minus number of bonds between them. The number of bonds (in fact, half-bonds) per unit is given by the value of derivative of the function F_0, or $F_0'(1)$ which is equal to $f\alpha$ for a single-component system, or to $\sum_i n_i f_i \alpha_i$ for a multicomponent system. Thus

$$M_n = \frac{\sum_i n_i M_{0i}}{1 - \sum_i F_0^{i\prime}(1)/2} = \frac{\sum_i n_i M_{0i}}{1 - \sum_i n_i f_i \alpha_i/2} \tag{3.79}$$

For the random alternating system $A_{f_A} + B_{f_B}$

$$M_n = \frac{n_A M_{A0} + n_B M_{B0}}{1 - (n_A f_A \alpha_A + n_B f_B \alpha_B)/2} \tag{3.80}$$

for this alternating system $n_A f_A \alpha_A = n_B f_B \alpha_B$ (number of bonds A–B is equal to the number of bonds B–A).

The weight-averages, M_w, diverge at the gel point while M_n reaches a finite value; M_n would diverge at higher conversion when the denominator of Eqs. 3.54 or 3.55 approaches zero, but the values beyond the gel point have no physical meaning since the units are non-randomly distributed between sol and gel components. Also, from Figs. 3.14 and 3.15, it can be seen that off-stoichiometry has a large effect on molecular weights.

In Sect. 3.4.1 (c), the homopolyaddition'of a tetrafunctional monomer with substitution effect was discussed where the rate of formation of a new AA bond was considered dependent on how many groups of the partner molecules have already reacted. This case is described by

$$\begin{aligned} F_0(Z_A, \mathbf{z}) = Z_A^{M_A}[&a_0 + a_1(p_{11}z_{11} + p_{12}z_{12} + p_{13}z_{13} + p_{14}z_{14}) \\ &+ a_2(p_{21}z_{21} + p_{22}z_{22} + p_{23}z_{23} + p_{24}z_{24})^2 \\ &+ a_3(p_{31}z_{31} + p_{32}z_{32} + p_{33}z_{33} + p_{34}z_{34})^3 \\ &+ a_4(p_{41}z_{41} + p_{42}z_{32} + p_{43}z_{33} + p_{44}z_{34})^4] \end{aligned} \tag{3.81}$$

According to Eq. 3.34,

$$\begin{aligned} F_{11}(Z_A, \mathbf{z}) &= Z_A^{M_A}, \quad F_{21}(Z_A, \mathbf{z}) = Z_A^{M_A}, \\ F_{31}(Z_A, \mathbf{z}) &= Z_A^{M_A}, \quad F_{41}(Z_A, \mathbf{z}) = Z_A^{M_A} \\ F_{12}(Z_A, \mathbf{z}) &= F_{22}(Z_A, \mathbf{z}) = F_{32}(Z_A, \mathbf{z}) \\ &= F_{42}(Z_A, \mathbf{z}) = Z_A^{M_A}(p_{21}z_{21} + p_{22}z_{22} + p_{23}z_{23} + p_{24}z_{24}) \\ F_{13}(Z_A, \mathbf{z}) &= F_{23}(Z_A, \mathbf{z}) = F_{33}(Z_A, \mathbf{z}) \\ &= F_{43}(Z_A, \mathbf{z}) = Z_A^{M_A}(p_{31}z_{31} + p_{32}z_{32} + p_{33}z_{33} + p_{34}z_{34})^2 \\ F_{14}(Z_A, \mathbf{z}) &= F_{24}(Z_A, \mathbf{z}) = F_{34}(Z_A, \mathbf{z}) \\ &= F_{44}(Z_A, \mathbf{z}) = Z_A^{M_A}(p_{41}z_{41} + p_{42}z_{42} + p_{43}z_{43} + p_{44}z_{44})^3 \end{aligned} \tag{3.82}$$

The recurrent equations for u_{ij} and values of their derivatives read

$$\begin{aligned} u_{11} &= u_{21} = u_{31} = u_{41} = Z_A^{M_A} \\ u_{12} &= u_{22} = u_{32} = u_{42} = Z_A^{M_A}(p_{21}u_{21} + p_{22}u_{22} + p_{23}u_{23} + p_{24}u_{24}) \\ u_{13} &= u_{23} = u_{33} = u_{43} = Z_A^{M_A}(p_{31}z_{31} + p_{32}z_{32} + p_{33}z_{33} + p_{34}z_{34})^2 \\ u_{14} &= u_{24} = u_{34} = u_{44} = Z_A^{M_A}(p_{41}u_{41} + p_{42}u_{42} + p_{43}u_{43} + p_{44}u_{44})^3 \end{aligned} \tag{3.83}$$

Altogether, there are 16 values of differentiation of variables **u** determined by 16 equations. The values of transition probabilities p_{ij} are determined by (numerical) solution of kinetic equations as shown in Sect. 3.4.1 (c).

Passing to multicomponent systems composed of units with different types of groups, like that of alternating addition of A_4–B_3 (Fig. 3.7), one can visualize graphically the branching process as shown in Fig. 3.13. The structure growth is usually visualized in the vertical direction as trees use to grow—from the roots to the branches (Fig. 3.13a). The continuation of structure is marked by tripledot.

Difference in Reactivities. When monomer units contain group of different reactivity or substitution effect exists, formation of specific paths of bonds occurs with different probability like in the case of homopolyaddition with substitution effect, and the transport of this information over generations is possible. When treating systems composed of monomers bearing groups of different reactivity, we have two options differing in the level of reaction correlations. Either the selection for bond formation involving two specific groups (chracterized by transition probability) is assumed proportional to their extents of reactions (conversion model), or the transition probability is obtained from irreversible kinetics involving all participating groups (kinetic model). The transition probabilities are calculated from populations of the respective dyads. If the differences in reactivities are small to moderate, the molecular weight averages and the gel points are almost the same, but if the reactivity differences grow larger, the differences in the gel point conversion become inportant.

One of such systems was reported by Huybrechts and Dušek [234], Bruchmann et al. [235], Bruchmann and Schrepp [236]. The system is composed of diisopropanolamine and diisocyanate bearing different kinds of isocyanate groups (Fig. 3.17). The ratios of reaction constants for urea and urethane formation were about 150 in the case of isophorone diisocyanate and about 15,000 in the case of 2,4-toluylene diisocyanate. Such systems in fact simulate an in situ formed first monomer of hyperbranched polymer—reaction of A1 with B1, and then, depending on stoichiometry, a hyperbranched polymer or a crosslinked product is formed.

Such systems were treated theoretically [193], and great difference in molecular weight development and gelation was shown to be caused by differences in the correlation depth. In the case of the "conversion model", the transition probabilities p were calculated as shown below

$$p_{A1\to B1} = p_{A1B1} = \frac{n_{B1}\alpha_{B1}}{n_{B1}\alpha_{B1} + n_{B2}\alpha_{B2}}$$

$$p_{A1\to B2} = p_{A1B2} = \frac{n_{B2}\alpha_{B2}}{n_{B1}\alpha_{B1} + n_{B2}\alpha_{B2}}$$

$$p_{A2\to B1} = p_{A2B1} = \frac{n_{B1}\alpha_{B1}}{n_{B1}\alpha_{B1} + n_{B2}\alpha_{B2}}$$

$$p_{A2\to B2} = p_{A1B2} = \frac{n_{B2}\alpha_{B2}}{n_{B1}\alpha_{B1} + n_{B2}\alpha_{B2}}$$

$$p_{B1\to A1} = p_{B1A1} = \frac{n_{A1}\alpha_{A1}}{n_{A1}\alpha_{A1} + n_{A2}\alpha_{A2}}$$

$$p_{B1\to A2} = p_{B1A2} = \frac{n_{A2}\alpha_{A2}}{n_{A1}\alpha_{A1} + n_{A2}\alpha_{A2}}$$

$$p_{B2\to A1} = p_{B2A1} = \frac{n_{A1}\alpha_{A1}}{n_{A1}\alpha_{A1} + n_{A2}\alpha_{A2}}$$

$$p_{\mathrm{B2}\to\mathrm{A2}} = p_{\mathrm{B2A2}} = \frac{n_{\mathrm{A2}}\alpha_{\mathrm{A2}}}{n_{\mathrm{A1}}\alpha_{\mathrm{A1}} + n_{\mathrm{A2}}\alpha_{\mathrm{A2}}}$$
$$\text{calculated from conversions } \alpha_{\mathrm{A1}}, \alpha_{\mathrm{A2}}, \alpha_{\mathrm{B1}}, \alpha_{\mathrm{B2}} \quad (3.84)$$

In the case of the "kinetic" model, the transition probabilities are generated by irreversible kinetics for formation of dyads

$$\begin{aligned}
p_{\mathrm{A1}\to\mathrm{B1}} &= p_{\mathrm{A1B1}} = \frac{[\mathrm{A1B1}]}{[\mathrm{A1B1}]+[\mathrm{A1B2}]} & p_{\mathrm{A1}\to\mathrm{B2}} &= p_{\mathrm{A1B2}} = \frac{[\mathrm{A1B2}]}{[\mathrm{A1B1}]+[\mathrm{A1B2}]}\\
p_{\mathrm{A2}\to\mathrm{B1}} &= p_{\mathrm{A2B1}} = \frac{[\mathrm{A2B1}]}{[\mathrm{A2B1}]+[\mathrm{A2B2}]} & p_{\mathrm{A2}\to\mathrm{B2}} &= p_{\mathrm{A2B2}} = \frac{[\mathrm{A2B2}]}{[\mathrm{A2B1}]+[\mathrm{A2B2}]}\\
p_{\mathrm{B1}\to\mathrm{A1}} &= p_{\mathrm{B1A1}} = \frac{[\mathrm{B1A1}]}{[\mathrm{B1A1}]+[\mathrm{B1A2}]} & p_{\mathrm{B1}\to\mathrm{A2}} &= p_{\mathrm{B1A2}} = \frac{[\mathrm{B1A2}]}{[\mathrm{B1A1}]+[\mathrm{B1A2}]}\\
p_{\mathrm{B2}\to\mathrm{A1}} &= p_{\mathrm{B2A1}} = \frac{[\mathrm{B2A1}]}{[\mathrm{B2A1}]+[\mathrm{B2A2}]} & p_{\mathrm{B2}\to\mathrm{A2}} &= p_{\mathrm{B2A2}} = \frac{[\mathrm{B2A2}]}{[\mathrm{B2A1}]+[\mathrm{B2A2}]}
\end{aligned} \quad (3.85)$$
$$\text{calculated from dyads } [\mathrm{A2B1}] = [\mathrm{B1A2}], \text{ etc.}$$

$$\begin{aligned}
&\frac{\mathrm{d}[\mathrm{A1B1}]}{\mathrm{d}t} = k_{\mathrm{A1B1}}[\mathrm{A1}][\mathrm{B1}]; \quad \frac{\mathrm{d}[\mathrm{A1B2}]}{\mathrm{d}t} = \cdots \text{ etc..}\\
&\frac{\mathrm{d}[\mathrm{A1}]}{\mathrm{d}t} = -[\mathrm{A1}](k_{\mathrm{A1B1}}[\mathrm{B1}] + k_{\mathrm{A1B2}}[\mathrm{B1}]; \quad \frac{\mathrm{d}[\mathrm{A2}]}{\mathrm{d}t} = \cdots \text{ etc.}\\
&t = 0 \quad [\mathrm{A1}] = [\mathrm{A1}]_0,\ [\mathrm{A2}] = [\mathrm{A2}]_0,\ [\mathrm{B1}] = [\mathrm{B1}]_0,\ [\mathrm{B1}] = [\mathrm{B1}]_0,
\end{aligned} \quad (3.86)$$

The basic pgf F_0 containing all information necessary for pregel and postgel description of network buildup is formulated as

$$\begin{aligned}
F_{0\mathrm{n}}(\mathbf{z}) &= n_{\mathrm{A}}(1 - \alpha_{\mathrm{A2}} + \alpha_{\mathrm{A2}}\zeta_{\mathrm{A2}})^2(1 - \alpha_{\mathrm{A1}} + \alpha_{\mathrm{A1}}\zeta_{\mathrm{A1}})\\
&\quad + n_{\mathrm{B}}(1 - \alpha_{\mathrm{B1}} + \alpha_{\mathrm{B1}}\zeta_{\mathrm{B1}})(1 - \alpha_{\mathrm{B1}} + \alpha_{\mathrm{B1}}\zeta_{\mathrm{B2}})\\
\zeta_{\mathrm{A1}} &= p_{\mathrm{A1B1}}z_{\mathrm{A1B1}} + p_{\mathrm{A1B2}}z_{\mathrm{A1B2}};\\
\zeta_{\mathrm{A2}} &= p_{\mathrm{A2B1}}z_{\mathrm{A2B1}} + p_{\mathrm{A2B2}}z_{\mathrm{A2B2}};\\
\zeta_{\mathrm{B1}} &= p_{\mathrm{B1A1}}z_{\mathrm{B1A1}} + p_{\mathrm{B1A2}}z_{\mathrm{B1A2}};\\
\zeta_{\mathrm{B2}} &= p_{\mathrm{B2A1}}z_{\mathrm{B2A1}} + p_{\mathrm{B2A2}}z_{\mathrm{B2A2}};
\end{aligned} \quad (3.87)$$

This figure clearly demonstrates that the "conversion" and "kinetics" formulations of transition probabilities give different results but also that in a certain range ($k = 1 - 10$) the difference is well below experimental error.

Other cases shown in examples of Sects. 3.4.1 and 3.4.2. can be treated similarly.

Examples—Crosslinking of Primary Chains

This is the case when reactive groups exist on preexisting chains (see Sect. 3.4.3). Vulcanization of rubbers is a typical example of introducing bonds between like reactive moities. The primary chains considered here contain only one type of reactive site which is a part of each monomer unit. However, the dispersity in the degrees of polymerization (molecular weights) is a typical property of such systems. Various polydienes and their vulcanization can serve as examples.

The building units are represented by primary chains having various numbers of units in crosslinked state. The basic information on such collection of uniform (monodisperse) primary chains is given by Eq. 3.64. Here, for simplicity reasons, the crosslinks are represented by immaterial connections between monomer units; α is the fraction of crosslinked monomer units. M_0 is the molecular weight of monomer unit

$$F_0(Z,\mathbf{z}) = \left[Z^{M_0}(1-\alpha+\alpha z)\right]^x \equiv Z^{xM_0}\sum_{i=0}^{x}\frac{x!}{i!(x-i)!}(1-\alpha)^{x-i}\alpha^i z^i \tag{3.88}$$

Passing to collection of primary chains of different degrees of polymerization, we get

$$\begin{aligned} F_{0\mathrm{n}}(Z,\mathbf{z}) &= \sum n_x\left[Z^{M_0}(1-\alpha+\alpha z)\right]^x; \\ F_{0\mathrm{w}}(Z,\mathbf{z}) &= \sum w_x\left[Z^{M_0}(1-\alpha+\alpha z)\right]^x \end{aligned} \tag{3.89}$$

for the number and weight distributions of primary chains.

The function $F(\mathbf{Z},\mathbf{z})$ is obtained by differentiation from $F_{0\mathrm{n}}(\mathbf{Z},\mathbf{z})$

$$F(Z,\mathbf{z}) = \sum_{x=1}^{\infty} xn_x Z^{xM_0}(1-\alpha+\alpha z)^{x-1}\Big/\sum_{x=1}^{\infty} xn_x = \sum_{x=1‘}^{\infty} w_x\left[Z^{xM_0}(1-\alpha+\alpha z)\right]^{x-1} \tag{3.90}$$

Equations (3.65) and (3.66) give direct access to number- and weight-average degrees of polymerization and molecular weights. For uniform collection of primary chains

$$M_\mathrm{n} = \frac{M_0 x}{1-x\alpha/2} \tag{3.91}$$

and for a distribution of primary chains

$$M_\mathrm{n} = \frac{M_0 P_{0\mathrm{n}}}{1-P_{0\mathrm{n}}\alpha/2} \tag{3.92}$$

The weight and higher averages of molecular weights are calculated from the weight fraction gf which is obtained by cascade substitution and recurrent equation for the variable u. Thus, pro uniform (monodisperse) polymer

$$\begin{aligned} W(Z,u) &= Z^{xM_0}(1-\alpha+\alpha u)^x, \quad u(Z) = Z^{xM_0}(1-\alpha+\alpha u)^{x-1} \\ M_\mathrm{w} &= xM_0\left(1+\frac{x\alpha}{1-(x-1)\alpha}\right) \end{aligned} \tag{3.93}$$

For random crosslinking of a polymer disperse in the degree of polymerization

$$W(Z,u) = \sum_{x=1}^{\infty} w_x Z^{xM_0}(1-\alpha+\alpha u)^x, \quad u(Z) = \sum_{x=1}^{\infty} w_x Z^{xM_0}(1-\alpha+\alpha u)^{x-1}$$

$$M_w = P_w^0 M_0 \left(1 + \frac{P_w^0 \alpha}{1-(P_w^0-1)\alpha}\right) \tag{3.94}$$

The denominator $1-(P_{0w}-1)\alpha$ turns to zero at the gel point when P_w and M_w diverge.

In practical applications, the summation over the degree of polymerization distribution is replaced by an analytical DP distribution in the form of generating function; the most frequently used distribution functions and the respective pgfs are listed in Table 3.1. For instance, for the most-probable distribution, the weight fraction distribution function is given by

$$W(Z,u) = \frac{(1-q)^2 Z}{(1-qZ)^2} \tag{3.95}$$

For crosslinking of the distribution of chains, Z is replaced by

$$Z \to Z^{M_0}(1-\alpha+\alpha u)$$

and the recurrent variable u reads

$$u(Z) = \frac{(1-q)^2 Z^{M_0}}{(1-qZ^{M_0}(1-\alpha+\alpha u(Z))^2} \tag{3.96}$$

and

$$M_w = M_0 \frac{1+q}{1-q}\left(1+\frac{(1+q)\alpha}{1-q-2q\alpha}\right) = P_{0w} M_0 \left(1+\frac{P_{0w}\alpha}{1-(P_{0w}-1)\alpha}\right)$$

$$M_n = \frac{M_0}{1-q-\alpha/2} = \frac{M_0 P_{0n}}{1-P_{0n}\alpha/2} \tag{3.97}$$

This route gives the same expression as Eq. 3.70 when $P_w^0 = (1+q)/(1-q)$, $P_n^0 = 1/(1-q)$. Other distributions of primary chains, like Poisson or Schulz-Zimm, are handled in the same way.

Figure 3.19 shows the dependences of M_n and M_w for a polymer composed of monomer units of molecular weight $M_0 = 100$ Dalton for various values of q.

While the weight average and averages related to higher moments of distribution diverge at the gel point, the number average remains finite until $\alpha = 2/P_{0n}$ but the values beyond the gel point have no physical meaning. Therefore, the Mn dependences in Fig. 3.19 are cut off at the gel point. For the other distributions (Table 3.1), a similar shape is observed, but the dependences differ in their steepness.

Distributions: It is the advantage of the cascade approach that the generating functions contain implicitly information about the degrees of polymerization or molecular weights distributions. Formally, one can get the fractions of mers by succesive differentiations of $W(Z)$ or $N(Z)$ and putting the variable $Z = 0$, but it is cumbersome and impractical. Spouge [237, 238] suggested to use Lagrange expansion; more efficient is the numerical inversion of the pgfs using Stehfest or Gavor algorithms as performed by Dušek [177] and Sarmoria's group [216, 223, 239]. A basic analysis of the inversion problem was offered by Sarmoria et al. [240].

Examples—Compositional/Molecular Weight Distributions

Often, the molecular weight distribution is combined with the compositional distribution. In solving many practical applications, functional copolymers bearing functional groups are needed and applied. The units bearing functional groups participate in crosslinking and gelation and network formation depend on the distribution of the functional groups in types and their numbers. This distribution is determined by three kinds of distributions: degree of polymerization distribution, compositional distribution, and distribution of the number of reacted functional groups. All these distributions are described by probabilitity generating functions.

The degree of polymerization distribution $N(Z)$ was already discussed above

$$N_{\mathrm{P}}(Z) = \sum_{x=1}^{\infty} n_x Z^x. \tag{3.98}$$

The composition of an x-mer copolymer of units A, B, C, ... is defined by a pgf $C_x(\mathbf{Z})$

$$C_x(\mathbf{Z}) = \sum_i c^{(x)}_{i,j,k,\ldots} Z_{\mathrm{A}}^{iM_{\mathrm{A}}} Z_{\mathrm{B}}^{jM_{\mathrm{A}}} Z_{\mathrm{C}}^{kM_{\mathrm{A}}} \ldots \tag{3.99}$$

where $c^{(x)}_{i,j,k,\ldots}$ is the molar concentration of an x-mer composed of $i, j, k (i + j + k + \ldots = x)$ units A, B, C, ..., respectively. The compound distribution (number fraction generating function) $N_{P,C}(\mathbf{Z})$ distinguishing the type of units is obtained by substituting

$$N_{\mathrm{P,C}}(\mathbf{Z}) = N(C_x(\mathbf{Z})) = \sum_{x=1}^{\infty} n_x \sum_{\substack{i,j,k \\ i+j+k=x}} c_{(x),i,j,k\ldots} Z_{\mathrm{A}}^i Z_{\mathrm{B}}^j Z_{\mathrm{C}}^k \ldots \tag{3.100}$$

For random compositional distribution

$$c_{(x),i,j,k..} Z_{\mathrm{A}}^i Z_{\mathrm{B}}^j Z_{\mathrm{C}}^k = (x_{\mathrm{A}} Z_{\mathrm{A}} + x_{\mathrm{B}} Z_{\mathrm{B}} + x_{\mathrm{C}} Z_{\mathrm{C}})^x \tag{3.101}$$

If one or more groups (e.g., A and B) are engaged in crosslinking with one or more components of the system (e.g., R or S), the forming bonds are marked by variables z_{AR} and z_{BS}, and the corresponding distribution is amended by distribution of numbers of bonds of AR and BS type

$$N_{P,C,\alpha}(\mathbf{Z}) = N(\zeta_{Cx}(F_{0,1}(Z,\mathbf{z}))) = \sum_{x=1}^{\infty} n_x (x_A Z_A F_{0A}(z_{AR}) + x_B Z_B F_{0B}(z_{BS}) + x_C Z_C)^x. \tag{3.102}$$

Let us take, as an example, a binary copolymer of functional monomer A with a non-functional monomer B. The degree of polymerization is of the Flory's most-probable type, and the compositional distribution is random, given by molar fractions (in practice, prepared under starving conditions, with a continuous monomer feed). The functional groups A then react with a crosslinker X, and the group reactivity of the same type is equal and independent. Then, the pgf F_0 reads

$$\begin{aligned} F_0(\mathbf{Z},\mathbf{z}) &= n_{AB}\frac{(1-q)C}{1-qC} + n_X Z_X^{M_X}(1-\alpha_X+\alpha_X z_{XA})^{f_C} \\ C &= x_A Z_A^{M_A}(1-\alpha_A+\alpha_A z_{AX}) + x_B Z_B^{M_B} \end{aligned} \tag{3.103}$$

The functions F for first and higher generations read

$$\begin{aligned} F_{XA}(\mathbf{Z},\mathbf{z}) &= \frac{\partial F_0}{\partial z_{AX}} N = \frac{(1-q)^2}{(1-qC)^2} Z_A^{M_A} \\ C &= x_A Z_A^{M_A}(1-\alpha_A+\alpha_A z_{AX}) + x_B Z_B^{M_B} \\ F_{AX}(\mathbf{Z},\mathbf{z}) &= Z_X^{M_X}(1-\alpha_X+\alpha_X z_{XA})^{f_X-1} \end{aligned} \tag{3.104}$$

The recurrent equations of the function $u(\mathbf{Z})$

$$\begin{aligned} u_{CA} &= \frac{(1-q)^2}{(1-q\left[x_A Z_A^{M_A}(1-\alpha_A+\alpha_A u_{AC}) + x_B Z_B^{M_B}\right])^2} \\ u_{AC} &= Z_C^{M_C}(1-\alpha_C+\alpha_C u_{CA})^{f_C-1} \end{aligned} \tag{3.105}$$

The values of the derivatives of Eq. 3.81 with respect to Z_A, Z_B, and Z_C are listed below as Eq. 3.82

$$\begin{aligned} u_{XA}^{A} &= x_A M_A + \frac{2q x_A M_A}{(1-q) - 2q x_A \alpha_A \alpha_X (f_X-1)} \\ u_{XA}^{B} &= x_B M_B + \frac{2q x_B M_B}{(1-q) - 2q x_A \alpha_A \alpha_X (f_X-1)} \\ u_{XA}^{X} &= \frac{2q x_A \alpha_A M_X}{(1-q) - 2q x_A \alpha_A \alpha_X (f_X-1)} \end{aligned}$$

$$u_{\mathrm{AX}}^{\mathrm{A}} = \frac{2qx_{\mathrm{A}}M_{\mathrm{A}}(f_{\mathrm{X}}-1)\alpha_{\mathrm{X}}}{(1-q)-2qx_{\mathrm{A}}\alpha_{\mathrm{A}}\alpha_{\mathrm{X}}(f_{\mathrm{X}}-1)}$$
$$u_{\mathrm{AX}}^{\mathrm{B}} = \frac{2qx_{\mathrm{B}}M_{\mathrm{B}}(f_{\mathrm{X}}-1)\alpha_{\mathrm{X}}}{(1-q)-2qx_{\mathrm{A}}\alpha_{\mathrm{A}}\alpha_{\mathrm{X}}(f_{\mathrm{X}}-1)}$$
$$u_{\mathrm{AX}}^{\mathrm{X}} = M_{\mathrm{X}} + \frac{2qx_{\mathrm{A}}\alpha_{\mathrm{A}}M_{\mathrm{X}}(f_{\mathrm{X}}-1)\alpha_{\mathrm{X}}}{(1-q)-2qx_{\mathrm{A}}\alpha_{\mathrm{A}}\alpha_{\mathrm{X}}(f_{\mathrm{X}}-1)} \tag{3.106}$$

The weight fraction generating function $W(\mathbf{Z})$ is a function of Z_{A}, Z_{B}, and Z_{X} and can be obtained as

$$W(\mathbf{Z}) = m_{\mathrm{A}}W_{\mathrm{A}}(\mathbf{u}(\mathbf{Z})) + m_{\mathrm{X}}W_{\mathrm{X}}(\mathbf{u}(\mathbf{Z}))$$
$$W_{\mathrm{A}}(\mathbf{u}(\mathbf{Z})) = Z_{\mathrm{A}}^{M_{\mathrm{A}}}F_{0\mathrm{A}}(\mathbf{u}_{\mathrm{A}}(\mathbf{Z})), \quad W_{\mathrm{X}}(\mathbf{u}(\mathbf{Z})) = Z_{\mathrm{X}}^{M_{\mathrm{X}}}F_{0\mathrm{X}}(\mathbf{u}_{\mathrm{X}}(\mathbf{Z})) \tag{3.107}$$

and the average M_{w} as the value of a sum of its derivatives with respect Z_{A}, Z_{B}, and Z_{X}

Path—Weighting Generating Functions—Structure Factors and Scattering Functions

The probability generating functions F store information on the ability of units to contribute to structure propagation. They represent a simple genetic code for structure growth through formation of branched paths of bonds. The process is recursive, and the way of its expression is the cascade substitution, which means substituting a pgf variable by another pgf. We have already seen that we can describe not only the growth of sequences of bonds but also the distributions of numbers and weights of units bonded together in the sequences. Also, changes in other properties can be described even if their contribution is not directly proportional to the number of units in the trees and subtrees. Such forms of pgfs were called by Kajiwara, Gordon et al. *path-weighting generating functions.*

Unlike the usual cascade substitution, counted in numbers of units

$$u(Z) = Zu(Z) \quad u(Z) = ZF_{\ell}(Z) \equiv F(Z) \tag{3.108}$$

with a constant exponent at Z equal to 1, or to the unit's weight M_0

$$u(Z) = Z^{M_0}u(Z) \quad u(Z) = Z^{M_0}F_{\ell}(Z)$$

valid for all generations, in the case of path-weighting gf this exponent may be generation number dependent [144, 241]

$$
\begin{aligned}
u_0(Z) &= Z^{\Phi_0} F_0(u_1) \\
u_1(Z) &= Z^{\Phi_1} F_2(u_2) \\
u_2(Z) &= Z^{\Phi_2} F_3(u_3) \\
&\vdots \\
u_k(Z) &= Z^{\Phi_k} F_k(u_{k+1}) \\
&\vdots
\end{aligned}
\tag{3.109}
$$

By differentiation and successive substitution for u_k to express u_{k+1}, one obtains [for the values of derivatives ($Z = 1$)]

$$
u_0'(1) = \Phi_0 + F_{\ell 0}'(1) \sum_{k=1}^{\infty} [F_{\ell 0}'(1)]^{k-1} \Phi_k \tag{3.110}
$$

This expression differs from Eqs. 3.37–3.39 (cf., also Sect. 3.5.7) by the presence of Φ_k: $F_{\ell 0}'(1)[F_{1\ell}'(1)]^{k-1}$ is the average number of units in the k-th generation. This average is just weighted by Φ_k, and $u_0'(1)$ is equal to the z-average of the property $\langle \Phi \rangle_{\mathrm{z}}$

$$
u_0'(1) = P_{\mathrm{w}} \langle \Phi \rangle_{\mathrm{z}} \tag{3.111}
$$

For instance, the average distance between chain ends of N-th power is given by

$$
\left\langle r_n^N \right\rangle = A_N n^{N/2} \tag{3.112}
$$

where n is the number of Kuhn's segments; for linear chains, the root-mean-square end-to-end distance $\left(\left\langle r_n^2 \right\rangle\right)^{1/2}$ is equal to

$$
\left(\left\langle r_n^2 \right\rangle\right)^{1/2} = \mathrm{l} n^{1/2} \tag{3.113}
$$

Equations 3.86 and 3.87 give access to average distances between end groups of branched chains in a collection of randomly branched chains which is related to their average radius of gyration. For random f-functional polyadducts, Kajiwara [149] found for the weight-average radius of gyration $\left\langle S^2 \right\rangle_{\mathrm{w}}$

$$
\left\langle S^2 \right\rangle_{\mathrm{w}} = \frac{\sigma^2}{2(f-2)} \left[\frac{f(f-1)(1-\alpha)^2}{(f-2)\alpha} \left(\ln \frac{1-\alpha}{1-(f-1)\alpha} \right) - \frac{f(f-1)(1-\alpha)}{f-2} - \frac{f\alpha}{2} \right] \tag{3.114}
$$

where σ^2 is the mean-square effective bond length, and α is conversion of functional groups. Also, the translational diffusion coefficient of the collection of randomly branched chains can be calculated [149].

The path-weighting gf was used for evaluation of scattering experiments applied to solutions of branched polymers. In the case of static scattering, the scattering intensity $R(q)$ as a function of the scattering vector q is given by

$$R(q) = KcM_{\mathrm{w}}P_{\mathrm{z}}(q) \tag{3.115}$$

where $P_{\mathrm{z}}(q)$ is the z-average of the particle scattering factor, K is the contrast factor, and c polymer concentration. For randomly branched systems, the particle scattering factor can be calculated using the cascade branching theory. The result is illustrated by the dependence of $P_{\mathrm{z}}(q)$ on conversion for a random f-functional polyadduct expressed, for instance, through its average radius of gyration $\langle S^2 \rangle_{\mathrm{z}}$

$$\begin{aligned} P_{\mathrm{z}}(q) &= \left(1 + q^2\langle S^2\rangle_{\mathrm{z}}/3\right) \\ \langle S^2\rangle_{\mathrm{z}} &= \frac{f-1}{2f}P_{\mathrm{w}} = \frac{(f-1)(1+\alpha)}{2f[1-(f-1)\alpha]} \end{aligned} \tag{3.116}$$

Relations for $P_{\mathrm{z}}(q)$ and $\langle S^2 \rangle_{\mathrm{z}}$ for branched polymers of various architecture and collection of branched molecules formed by various branching reactions are tabulated in Burchard's review [144].

Similarly, the path-weighting gf formalism can be used for calculation of the diffusion coefficient effective in dynamic light scattering D_{z} or the hydrodynamic factor R_{h}. The advantage of the use of the path-weighting gf combined with the cascade substitution can be found in that conformational properties are calculated in one step, instead, as done before, in two steps—calculating first the number of units in the given generation, weighting the result by the factor characterizing the property at that generation, and performing the summation of the results. The topics related to radiation scattering covering several kinds of branching systems are most comprehensively treated in the review by Burchard [144] which also summarizes earlier work on this subject (cf., e.g.,. [126–128, 138, 144–150, 241].

Gel Points

The gel point or the gel time are milestones in the process of nertwork formation. At this point, one of the paths of bonds in the largest molecule becomes infinite. In Sect. 3.5.2, we saw that at the gel point the weight-average molecular weight became infinite; more specifically, the denominator of M_{w} became equal to zero

$$1 - F'(1) = 0 \quad \text{or} \quad 1 - \mathbf{F}'(1) = 0 \tag{3.117}$$

where $F'(1)$ is the value of derivative of the function F for $Z = 1$ in the case of single type of bonds, or a matrix of derivatives in the case of multiple types of bonds. Alternatively, one can ask what the probability is that one finds a unit in generation g when $g \to \infty$. The distribution for the number of units in generation k is described by the pgf $G_k(Z)$, Eq. 3.64,

$$\begin{aligned} G_k(Z) &= F_{\ell,0}(u_{\ell,k-1}(Z)) = F_{\ell,0}(F_\ell...(F_\ell(Z)...)) \\ &= (1-\alpha_A+\alpha_A(...+\alpha_A(1-\alpha_A+\alpha_A Z)^{f-1}...)^{f-1}) \end{aligned} \tag{3.118}$$

and the value of the derivative for $Z = 1$

$$\begin{aligned} G_{gel} &= \lim_{k\to\infty} G_k'(1) = F_{\ell,0}'(1)(F_\ell'(1))^{k-1} = f\alpha[(f-1)\alpha]^{k-1} \\ &\text{for } F_{\ell,0}'(1) < 1,\ G_{gel} = 0;\quad \text{for } F_{\ell,0}'(1) > 1,\ G_{gel} = \infty \end{aligned} \tag{3.119}$$

It follows that infinite number of units is reached after the value of $F_\ell'(1)$ surpasses 1.

For several types of bonds, described by a vector of $\mathbf{F}(\mathbf{z}) = \left[F_{ij}(z_{jk})\right]$, the gel point is determined by the determinant D of the matrix of derivatives being equal to zero

$$D = \begin{vmatrix} 1-F_{11}^{11} & F_{12}^{21} & \cdots & -F_{1,k-1}^{k-1,1} & -F_{1,k}^{k,1} \\ -F_{21}^{12} & 1-F_{22}^{22} & \cdots & -F_{2,k-1}^{k-1,2} & -F_{2,k}^{k,1} \\ \vdots & \vdots & \vdots & \vdots & \vdots \\ -F_{k-1,1}^{1,k-1} & -F_{k-1,2}^{2,k-1} & \cdots & 1-F_{k-1,k-1}^{k-1,k-1} & -F_{k-1,k}^{k,k-1} \\ -F_{k,k}^{k,1} & -F_{2,k-1}^{k-1,2} & \cdots & -F_{k,k}^{k,1} & 1-F_{k,k}^{k,k} \end{vmatrix} = 0 \tag{3.120}$$

For some of the examples given in Sects. 3.4.1.4 and 3.5.1.4, the gel point conversions of selected examples for ring-free systems are summarized below. Next to the more general form, the explicit form for random systems is displayed.

Single type of groups of equal reactivity

$$1 - F_A^A = 1 - (f-1)\alpha_A = 0,\quad \alpha_A = \frac{1}{f-1} \tag{3.121}$$

Groups of unequal reactivity

$$1 - F_A^A = 1 - \frac{\sum_{i=1}^{f} i(i-1)a_i}{\sum_{i=1}^{f} ia_i} = 0,\quad 1 - \frac{\sum_{i=1}^{f} i(i-1)a_i}{\sum_{i=1}^{f} ia_i} = 2 - \frac{\sum_{i=1}^{f} i^2 a_i}{f\alpha_A} = 0 \tag{3.122}$$

which, in terms of ratios of rate constants, can be transformed into a dependence on conversion through Eq. 3.49.

For mixtures of monomers bearing different number of functional groups of equal reactivity

$$1-(f_{\mathrm{w}}-1)\alpha_{\mathrm{A}}=0,\quad \alpha_{\mathrm{A}}=\frac{1}{f_{\mathrm{w}}-1},\quad f_{\mathrm{w}}=\frac{\sum_{1}^{f}n_if_i^2}{\sum_{i}^{f}n_if_i} \tag{3.123}$$

Simple alternating system A + B = AB: for groups of equal reactivity:

$$\begin{aligned}1-F_{\mathrm{AB}}^{\mathrm{BA}}F_{\mathrm{AB}}^{\mathrm{BA}}&=1-(f_{\mathrm{A}}-1)(f_{\mathrm{B}}-1)\alpha_{\mathrm{A}}\alpha_{\mathrm{B}}=0,\\ \alpha_{\mathrm{A}}&=\frac{n_{\mathrm{B}}\alpha_{\mathrm{B}}f_{\mathrm{B}}}{n_{\mathrm{A}}f_{\mathrm{A}}}=r_{\mathrm{B}}\alpha_{\mathrm{B}},\quad \alpha_{\mathrm{B}}=r_{\mathrm{A}}\alpha_{\mathrm{A}}\\ \alpha_{\mathrm{A}}^2&=\frac{r_{\mathrm{B}}}{(f_{\mathrm{A}}-1)(f_{\mathrm{B}}-1)},\quad \alpha_{\mathrm{B}}^2=\frac{1}{(f_{\mathrm{A}}-1)(f_{\mathrm{B}}-1)r_{\mathrm{B}}}\end{aligned} \tag{3.124}$$

for groups of unequal reactivities

$$1-F_{\mathrm{AB}}^{\mathrm{BA}}F_{\mathrm{AB}}^{\mathrm{BA}}=1-\frac{\sum_{i=1}^{f_{\mathrm{A}}}i(i-1)a_i\sum_{j=1}^{f_{\mathrm{B}}}j(j-1)b_i}{f_{\mathrm{A}}\alpha_{\mathrm{A}}f_{\mathrm{B}}\alpha_{\mathrm{B}}}=0 \tag{3.125}$$

Gelation resulting from crosslinking of primary chains (Sect. 3.5.3) is equivalent to crosslinking of a mixture monomers of different functionality and equal reactivity

$$1-(P_{\mathrm{w}0}-1)\alpha_{\mathrm{A}}=0,\quad \alpha_{\mathrm{A}}=\frac{1}{P_{\mathrm{w}0}-1} \tag{3.126}$$

In the case of degree of polymerization and composition distributions, the calculation of the gel point condition is simpler than calculation of molecular weights. It is only the path of bonds and not its mass contributions that matter. To show an example of the more complicated cases, let us consider crosslinking of a functional copolymer AB composed of units A bearing each reactive groups A and units B inactive in crosslinking. Groups A can react with groups C of a f_{C}-functional crosslinker C. Such case is described by the pgf

$$\begin{aligned}F_0(\mathbf{z})&=n_{\mathrm{AB}}\frac{(1-q)\zeta}{1-q\zeta}+n_{\mathrm{C}}(1-\alpha_{\mathrm{C}}+\alpha_{\mathrm{C}}z_{\mathrm{CA}})^{f_{\mathrm{C}}};\\ \zeta&=x_{\mathrm{A}}(1-\alpha_{\mathrm{A}}+\alpha_{\mathrm{A}}z_{\mathrm{AC}})+x_{\mathrm{B}}\end{aligned} \tag{3.127}$$

The pgfs F_{CA} and F_{AC} are obtained by differentiation, and the gel point is given by the gel point condition $1-F_{\mathrm{CA}}^{\mathrm{AC}}F_{\mathrm{AC}}^{\mathrm{CA}}=0$. For the case of Eq. (3.103)

$$1-\frac{2q}{1-q}x_{\mathrm{A}}\alpha_{\mathrm{A}}(f_{\mathrm{C}}-1)\alpha_{\mathrm{C}}=0 \tag{3.128}$$

For stoichiometric ratio of groups A and C, $\alpha_A = \alpha_C = \alpha$, and the gel point condition reads

$$\alpha = \sqrt{\frac{1-q}{2qx_A\alpha_A(f_C-1)}} = \sqrt{\frac{1}{(P_{0w}-1)x_A\alpha_A(f_C-1)}} \tag{3.129}$$

Further examples can be found in Refs. [57, 200]. The relations given here are so-called ring-free, i.e., the models are based on a fully tree-like structure of the molecules. In reality, a part of bonds close cycles, and thus, the gel point is shifted to higher conversions than would correspond to the ring-free case. This means that the real, experimental critical conversion is somewhat higher than the ring-free value discussed here. What the magnitude of this shift is and how it depends on the precursor structure and the crosslinking process will be discussed in Sects. 3.8, 4.1, and 4.5.

3.4.2.6 Postgel State

The postgel region is delimited by the gel point on one side and reaction completion (full conversion of reactive groups, or minority reactive groups) on the other side. After surpassing the gel point, bonds continue to be formed

- between groups on finite molecules (sol–sol reaction) by which the molecular weight of the sol increases; if cycle formation occurs before the gel point, it can continue within the sol molecules,
- between groups of finite molecules and the gel (sol–gel reaction) by which the molecular weight of the sol decreases and the gel mass increases,
- between groups residing on the gel units (gel–gel reactions). Any new bond formed between groups of the gel closes a cycle. The cycles are of different sizes and may contain branch points issuing different number of bonds of infinite and finite continuation.

Whereas sol–sol reactions can be distinguished as intermolecular and intramolecular irrespective of the ring size, the sol–gel reactions are always intermolecular, and gel–gel reactions are always intramolecular. Yet, there is a difference in the gel–gel reactions with respect to their size and their effect on equilibrium elasticity.

The current graph theories based on the recursive multiplicative process ignore the space-filling problem. A pure mathematical graph has no metrical information, which means that any information on the embedding of the graph in three-dimensional space will be impossible to infer from the graph alone. At conversions above the gel point, more than one bond extends from units of generation g to units of generation $g+1$, i.e., the value of the derivative $F' > 1((f-1)a > 1$ for the random case). The number of units to be found in generation g is equal to $N_g = F_0'(F')^{g-1}$ whereas the space available to accommodate them increases as $\Delta v \propto g^3 - (g-1)^3$; it is clear that the ratio $N_g / \Delta v$ diverges for $g \to \infty$ for any $F' > 1$. This space overcrowding effect generated by the recursive model in contrast to the uniformly space-filling model (excess bonds close cycles) is shown in Fig. 3.20

One should realize that any tree-like structure (a) of any finite size must also contain unreacted groups: the minimum number of bonds to connect N units into a tree-like graph without cycles (spanning tree) is $N - 1$; for maintaining tree-like structure, the limit $\alpha = 2/f$ should not be surpassed (Stockmayer case). However, the equal reactivity assumption (independent of conversion), found valid for many experimental systems, does not make difference between intermolecular bond formation or uncorrelated ring forming between groups of the gel. Then, α is allowed to go to completion, and $N(f\alpha/2-1)$ is the cycle rank.

The overcrowding problem gave rise to some doubts about applicability of the tree-like models to real systems. Indeed, generation of paths of bonds in the gel by the standard recursive method would be impossible. Path weighting should be applied securing that $F'_{g=1} = (f-1)\alpha$ and $F'_{g\to\infty} = 1$. Yet, the comparison of the tree-like model with experiment over decades was in general successful and has been used to explain successfully effects of small changes in reaction mechanism, a kinetics of bond formation or scission. How is it possible?

There exist arguments that the overcrowding effect becomes operative only at some distance (in number of generations) from the unit under consideration and does not (or does only marginally) influence the quantities characterizing the network structure (fraction or molecular weight of dangling chains, elastically active network chains, and elastically active crosslinks).

(a) Using the recursive method, the characteristics are derived from the distribution of states of building units distinguishing not only types of bonds but also distinguishing whether these bonds have finite or infinite continuation and paths through the gel are not generated; the issuing bonds are distinguished by whether they have or do not have continuation to infinity. The distinguishing quantity is the *extinction probability* (conditional probability that provided a bond exist it has finite continuation; designated as scalar v but it is usually a vector $\mathbf{v}$). Its complement—probability that the continuation is infinite—is obtained by weighting the probability of bond existence by this complemen $(1 - v)$; its application in long-range cascade substitution would lead to overcrowding. Contrary to this, it is argued that the extinction probability is derived from the condition of finite continuation and is not affected at all by possible overcrowding in the gel. Also, initial parts of the long sequences are not affected by overcrowding. Circuit closing in the gel adds additional paths to infinity in excess to that existing in its spanning and does not change the determination of the path to reach "infinity". What remains to be proved is whether uncorrelated combination of bonds with finite and infinite continuations issuing from a building unit, in the form of binomial expansion, is a reasonable Ansatz.

(b) For the random case, there is one-to-one correspondence between the cascade generation and the kinetics method (discussed in Sect. 3.4) described by Smoluchowski coagulation equations [176]. According to mass action law, the rate constants are disrectly proportional to the concentrations of molecules and number of unreacted groups they bear. No recurrent cascade substitution generation is applied. The solution for the distribution of the sol phase offers identical equations in the form of a pgf as those obtained by cascade substitution. The equivalence of statistical (cascade) and kinetics solution was further analyzed by Kajiwara et al. [242].

Several directions can be investigated to solve the problem of overcrowding when cascade substitution is applied to the gel component. Ziff et al. [243] suggested that the principle of recursive methods that there is just one incoming bond for a unit on generation $g > 0$ should be relaxed, and more incoming bonds per unit should be allowed (up to their number equal to f). These incoming bonds in addition to the necessary one are engaged in the uncorrelated ring closing in the gel. It is an interesting idea, but it should be found how to control the distribution of incoming and outgoing bonds per monomer unit in order to prevent overcrowding. There is a vast literature available on percolation models [244] which could offer inspiration for formulation of a path-weighting generating function capable to handle the overcrowding problem when cascade substitution is applied.

Next section will deal with the key quantity of the postgel state—extinction probability.

Extinction Probabilities

The way of dealing with bonds in the postgel state is to distinguish between those that, looking out of the unit, are part of finite sequences (finite subtree) or those that connect to gel structure (which is considered infinite). The probability that a bond has only finite continuation is called *extinction probability*. It is a conditional probability that, given the bond exists, it has only finite continuation. It is equivalent to the probability of extinction of family name as formulated by Watson and Galton (cf. Sect. 3.1). The probability that a bond does have continuation to infinity (to the gel) is complementary to extinction probability.[2]

The set of extinction probabilities is equal in number to the set of probability generating functions **F** for the number of bonds issuing from a unit on generation $g > 0$.

The calculation of the extinction probability (symbol v) is based on the equalities of probabilities for incoming and outgoing bonds: the incoming bond exists with probability 1 but the probability that it is a bond with finite continuation is v; in that case, all outgoing bonds (issuing from reacted functional groups) must also have finite continuation. Thus, for single type of bonds and equal and independent reactivity

$$v = F(v); \quad v = (1 - \alpha + \alpha v)^{f-1} \tag{3.130}$$

[2] In Miller–Macosko's version of the statistical theory, the probability that the outgoing bond has finite continuation, $W(\text{out})$ includes also the state, where all $f-1$ functional groups are in unreacted state, whereas the estinction probability v applies only to existing bonds.

Similarly, for unequal or dependent reactivity of single type bonds (AA)

$$v = \sum_{i=0}^{f_A} i a_i v_{AA}^{i-1} \bigg/ \sum_{i=0}^{f_A} i a_i, \tag{3.131}$$

For several types of bonds, more extinction probabilities exist, and their number is equal to the number of auxiliary variables z_{ij} and generating functions F_{ij} (technically, in $F_{ij}(z_{jk})$ z_{ij} is replaced by v_{ij} and the equation or set of equations are solved) (for the sets of z and F, consult Sect. 3.4.1)

$$v_{ij} = F_{ij}(v_{jk}) \tag{3.132}$$

or if one needs more detailed classification of bonds

$$v_{JyxI} = F_{JyxI}(\mathbf{v}) \tag{3.133}$$

Because F are probability generation functions, the equations are always satisfied for $v_{ij} = 1$, which is called trivial root. Before the gel point, it is the only root in the interval [1, 0]. Beyond the gel point there is another root in that range $0 \leq v_{ij} < 1$ which is the relevant one. For simple, low-functionality system, the trivial root can be eliminated analytically; otherwise, a numerical solution is used and in the interval excluding $v = 1$.

For alternating A + B → AB reaction, the extinction probabilities v_{BA} and v_{AB} can be calculated from the following equations, in which the right-hand side shows the result for the random case

$$\begin{aligned} v_{BA} &= \sum_{i=1}^{f_A} i a_i v_{AB}^{i-1} \bigg/ \sum_{i=1}^{f_A} i a_i = (1 - \alpha_A + \alpha_A v_{AB})^{f_A - 1} \\ v_{AB} &= \sum_{i=1}^{f_B} i b_i v_{AB}^{i-1} \bigg/ \sum_{i=1}^{f_A} i b_i = (1 - \alpha_B + \alpha_B v_{BA})^{f_B - 1} \end{aligned} \tag{3.134}$$

For crosslinking of primary chains

$$v = \sum_{x=1}^{\infty} m_{1x}(1 - \alpha + \alpha v)^{x-1} \tag{3.135}$$

for the case of precursors having molecular weight and functionality distribution

$$\begin{aligned} v_{XA} &= \frac{(1-q)^2}{(1-qC)^2}, \quad C = x_A(1 - \alpha_A + \alpha_A v_{AX}) + x_B \\ v_{AX} &= (1 - \alpha_X + \alpha_X v_{XA})^{f_X - 1} \end{aligned} \tag{3.136}$$

Classification of Units in the Postgel Stage

It is generally assumed that the probability for a bond to have an infinite continuation is complementary to v, i.e., it is defined as $1 - v$. Thus, for single type of bonds and equal and independent group reactivities the pgfs describing the whole system, $F_0(\mathbf{Z}, \mathbf{z})$, the sol, $F_{0,\text{sol}}(\mathbf{Z}, \mathbf{z})$, and the gel, $F_{0,\text{gel}}(\mathbf{Z}, \mathbf{z})$ read

$$\begin{aligned}
F_0(\mathbf{Z}, \mathbf{z}) &= Z_{\text{A}}^{M_{\text{A}}}(1 - \alpha_{\text{A}} + \alpha_{\text{A}}(v z_{\text{AA},v} + (1 - v) z_{\text{AA},1-v})^{f_{\text{A}}} \\
&\equiv Z_{\text{A}}^{M_{\text{A}}} \sum h_{ij} z^i_{\text{AA},v} z^j_{\text{AA},1-v} \\
F_{0,\text{sol}}(\mathbf{Z}, \mathbf{z}) &= Z_{\text{A}}^{M_{\text{A}}} \frac{(1 - \alpha_{\text{A}} + \alpha_{\text{A}} v z_{\text{AA},v})^{f_{\text{A}}-1}}{(1 - \alpha_{\text{A}} + \alpha_{\text{A}} v)^{f_{\text{A}}-1}} \\
F_{0,\text{gel}}(\mathbf{Z}, \mathbf{z}) &= Z_{\text{A}}^{M_{\text{A}}} \frac{(1 - \alpha_{\text{A}} + \alpha_{\text{A}}(v z_{\text{AA},v} + (1 - v) z_{\text{AA},1-v})^{f_{\text{A}}-1} - (1 - \alpha_{\text{A}} + \alpha_{\text{A}} v z_{\text{AA},v})^{f_{\text{A}}-1}}{1 - (1 - \alpha_{\text{A}} + \alpha_{\text{A}} v)^{f_{\text{A}}-1}}
\end{aligned} \tag{3.137}$$

The coefficient $h_{ij} z^i_{\text{AA},v} z^j_{\text{AA},1-v}$ at $F_0(\mathbf{Z}, \mathbf{z})$ is the probability of finding a unit engaged in i bonds with finite and j bonds with infinite continuation per unit in the whole system.

For the random A + B alternating case, the corresponding equations read

$$\begin{aligned}
F_0(\mathbf{Z}, \mathbf{z}) = {} & n_{\text{A}} Z_{\text{A}}^{M_{\text{A}}}(1 - \alpha_{\text{A}} + \alpha_{\text{A}}(v_{\text{AB}} z_{\text{AB},v} + (1 - v_{\text{AB}}) z_{\text{AB},1-v})^{f_{\text{A}}} \\
& + n_{\text{B}} Z_{\text{B}}^{M_{\text{B}}}(1 - \alpha_{\text{B}} + \alpha_{\text{B}}(v_{\text{BA}} z_{\text{BA},v} + (1 - v_{\text{BA}}) z_{\text{BA},1-v})^{f_{\text{B}}}
\end{aligned} \tag{3.138}$$

For the sol and gel, the corresponding single component—single bond Eq. 3.113 is reformulated accordingly. For outgoing bonds from units on generations $g > 0$ branch has finite or infinite continuation,

$$\begin{aligned}
F_{\text{BA, sol}}(\mathbf{Z}, \mathbf{z}) &= Z_{\text{A}}^{M_{\text{A}}} \frac{(1 - \alpha_{\text{A}} + \alpha_{\text{A}} v_{\text{AB}} z_{\text{AB},v})^{f_{\text{A}}-1}}{(1 - \alpha_{\text{A}} + \alpha_{\text{A}} v_{\text{AB}})^{f_{\text{A}}-1}}, \\
F_{\text{AB, sol}}(\mathbf{Z}, \mathbf{z}) &= Z_{\text{B}}^{M_{\text{B}}} \frac{(1 - \alpha_{\text{B}} + \alpha_V v_{\text{BA}} z_{\text{BA},v})^{f_B-1}}{(1 - \alpha_{\text{B}} + \alpha_V v_{\text{BA}})^{f_{\text{B}}-1}} \\
F_{\text{BA, gel}}(\mathbf{Z}, \mathbf{z}) &= Z_{\text{A}}^{M_{\text{A}}} \frac{(1 - \alpha_{\text{A}} + \alpha_{\text{A}}(v_{\text{AB}} z_{\text{AB},v} + (1 - v_{\text{BA}}) z_{\text{BA},1-v})^{f_{\text{A}}-1} - (1 - \alpha_{\text{A}} + \alpha_{\text{A}} v_{\text{AB}} z_{\text{AB},v})^{f_{\text{A}}-1}}{1 - (1 - \alpha_{\text{A}} + \alpha_{\text{A}} v_{\text{AB}})^{f_{\text{A}}-1}} \\
F_{\text{AB, gel}}(\mathbf{Z}, \mathbf{z}) &= Z_{\text{B}}^{M_{\text{B}}} \frac{(1 - \alpha_{\text{B}} + \alpha_{\text{B}}(v_{\text{BA}} z_{\text{BA},v} + (1 - v_{\text{BA}}) z_{\text{BA},1-v})^{f_{\text{B}}-1} - (1 - \alpha_{\text{B}} + \alpha_{\text{B}} v_{\text{BA}} z_{\text{BA},v})^{f_{\text{B}}-1}}{1 - (1 - \alpha_{\text{B}} + \alpha_{\text{B}} v_{\text{BA}})^{f_{\text{B}}-1}}
\end{aligned} \tag{3.139}$$

In all these cases, the variable z_{XY} in $F_0(\mathbf{Z}, \mathbf{z})$ is expanded into $v_{\text{XY}} z_{\text{XY},v} + (1 - v_{\text{XY}}) z_{\text{XY},1-v}$ which implies that no correlations between numbers of v-type of $1 - v$ type bonds are assumed. In Eq. 3.115, the generation functions are normalized by division by the number fractions of units in sol and gel, respectively. The numbers of $1 - v$ type bonds determine the type of substructures in the network (Table 3.3).

The substructures are graphically illustrated in Fig. 3.22.

Sol and Gel Fractions

In the sol, all units issue bonds with finite continuation. Therefore, the probability that a bond exists (α) is multiplied by the extinction probability v. The sol fraction is usually considered as weight fraction of the whole system. Thus, for multicomponent systems, the weight fractions of components are considered. For the random alternating $A_{fA} + B_{fB}$ system, the sol w_s and gel w_g fractions are given by

$$\begin{aligned} w_s &= m_A(1 - \alpha_A + \alpha_A v_{AB})^{f_A} + m_B(1 - \alpha_B + \alpha_B v_{BA})^{f_B} \\ w_g &= 1 - w_s \end{aligned} \tag{3.140}$$

The sol and gel fraction are important quantities chracterizing the crosslinking reaction beyond the gel point particularly its degree of completion. A few examples of the dependence of w_g on conversion are shown in Fig. 3.23.

Figure 3.23 demostrates the strong suppression of network buildup by off-stoichiometry; even at 100% conversion, the gel fraction does not exceed about 30%. On the other hand, a "low" sol fraction, e.g., 5%, does not guarantee high conversion; in the case of system 1, the conversion would not exceed 60–70%.

Substructures in gel: One can single out a specific substructure and formulate the respective pgf for it. For the sol fraction, one puts $z_{XY,1-v} = 0$ in Eq. 3.104, and the expression is renormalized. For instance, for the alternating random case (subscripts for sol)

$$F_{0,s}(\mathbf{Z}, \mathbf{z}) = \frac{n_A Z_A^{M_A}(1 - \alpha_A + \alpha_A v_{AB} z_{AB,v})^{f_A} + n_B Z_B^{M_B}(1 - \alpha_B + \alpha_B v_{BA} z_{BA,v})^{f_B}}{n_A(1 - \alpha_A + \alpha_A v_{AB})^{f_A} + n_B(1 - \alpha_B + \alpha_B v_{BA})^{f_B}} \tag{3.141}$$

Similarly, to get the pgf for the gel, $F_{0,g}(\mathbf{Z}, \mathbf{z})$ the part related to sol (numerator of Eq. 3.117) is subtracted, followed by normalization.

Alternatively, Eq. 3.113 for $F_{0,s}$ and $F_{0,g}$ can be reformulated by using conversions of groups on units sol and gel, and also the molar fractions of units A and B in sol and gel, which are not the same but different

$$\begin{aligned} F_{0,s}(\mathbf{Z}, \mathbf{z}) &= n_{A,s} Z_A^{M_A}(1 - \alpha_{A,s} + \alpha_{A,s} z_{AB,v})^{f_A} + n_{B,s} Z_B^{M_B}(1 - \alpha_{B,s} + \alpha_{B,s} z_{BA,v})^{f_B} \\ F_{0,g}(\mathbf{Z}, \mathbf{z}) &= n_{Ag} Z_A^{M_A}(1 - \alpha_{A,g} + \alpha_{Ag} \zeta_{AB})^{f_A} + n_{B,g} Z_B^{M_B}(1 - \alpha_{B,g} + \alpha_{B,g} \zeta_{BA})^{f_B} \\ \zeta_{AB} &= (v_{AB} z_{AB,v} + (1 - v_{AB}) z_{BA,1-v}) \\ \zeta_{BA} &= (v_{BA} z_{BA,v} + (1 - v_{BA}) z_{BA,1-v}) \end{aligned} \tag{3.142}$$

From simple balances of numbers of units and reacted functional groups in sol and gel, one gets expressions for compositions and conversions in sol and gel

$$n_{A,s} = \frac{n_A(1 - \alpha_A + \alpha_A v_{AB})^{f_A}}{n_A(1 - \alpha_A + \alpha_A v_{AB})^{f_A} + n_B(1 - \alpha_B + \alpha_B v_{BA})^{f_B}}$$

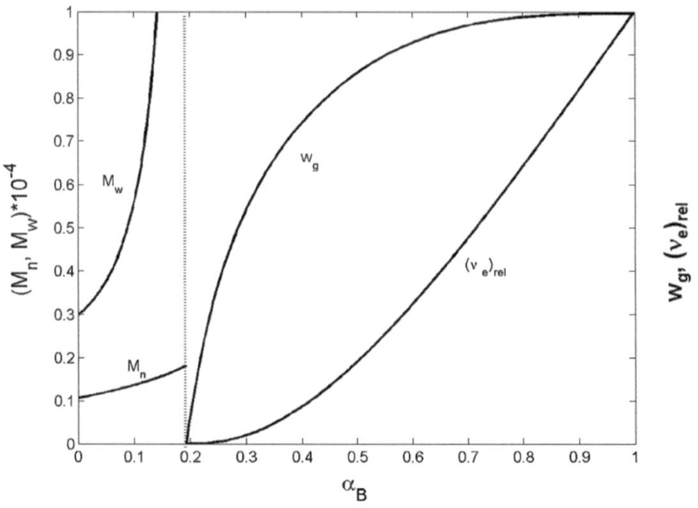

Fig. 3.24 Changes of molecular weight averages, M_n and M_w, gel fraction, w_g, and relative concentration of EANC's for a statistical copolymer bearing groups A, $(f_A)_n = 7.3$, crosslinked with a trifunctional monomer bearing groups B; reduced concentration of elastically active network chains (EANC), $(\nu_e)_{\mathrm{rel}} = \nu_e/(\nu_e)_{\max}$

$$n_{\mathrm{B,s}} = \frac{n_{\mathrm{B}}(1-\alpha_{\mathrm{B}}+\alpha_{\mathrm{B}}\nu_{\mathrm{BA}})^{f_{\mathrm{B}}}}{n_{\mathrm{A}}(1-\alpha_{\mathrm{A}}+\alpha_{\mathrm{A}}\nu_{\mathrm{AB}})^{f_{\mathrm{A}}}+n_{\mathrm{B}}(1-\alpha_{\mathrm{B}}+\alpha_{\mathrm{B}}\nu_{\mathrm{BA}})^{f_{\mathrm{B}}}}$$
$$n_{\mathrm{A,g}} = \frac{n_{\mathrm{A}}(1-(1-\alpha_{\mathrm{A}}+\alpha_{\mathrm{A}}\nu_{\mathrm{AB}})^{f_{\mathrm{A}}})}{n_{\mathrm{A}}(1-(1-\alpha_{\mathrm{A}}+\alpha_{\mathrm{A}}\nu_{\mathrm{AB}})^{f_{\mathrm{A}}})+n_{\mathrm{B}}(1-(1-\alpha_{\mathrm{B}}+\alpha_{\mathrm{B}}\nu_{\mathrm{BA}})^{f_{\mathrm{B}}})}$$
$$n_{\mathrm{B,g}} = \frac{n_{\mathrm{B}}(1-(1-\alpha_{\mathrm{B}}+\alpha_{\mathrm{B}}\nu_{\mathrm{BA}})^{f_{\mathrm{B}}}}{n_{\mathrm{A}}(1-(1-\alpha_{\mathrm{A}}+\alpha_{\mathrm{A}}\nu_{\mathrm{AB}})^{f_{\mathrm{A}}})+n_{\mathrm{B}}(1-(1-\alpha_{\mathrm{B}}+\alpha_{\mathrm{B}}\nu_{\mathrm{BA}})^{f_{\mathrm{B}}})} \quad (3.143)$$

and

$$\alpha_{\mathrm{A,s}} = \frac{\alpha_{\mathrm{A}}\nu_{\mathrm{AB}}}{1-\alpha_{\mathrm{A}}+\alpha_{\mathrm{A}}\nu_{\mathrm{AB}}}; \quad \alpha_{\mathrm{B,s}} = \frac{\alpha_{\mathrm{B}}\nu_{\mathrm{BA}}}{1-\alpha_{\mathrm{B}}+\alpha_{\mathrm{B}}\nu_{\mathrm{BA}}}$$
$$\alpha_{\mathrm{A,g}} = \frac{\alpha_{\mathrm{A}}(1-\nu_{\mathrm{AB}}\nu_{\mathrm{BA}})}{1-(1-\alpha_{\mathrm{A}}+\alpha_{\mathrm{A}}\nu_{\mathrm{AB}})^{f_{\mathrm{A}}}}; \quad \alpha_{\mathrm{B,g}} = \frac{\alpha_{\mathrm{B}}(1-\nu_{\mathrm{BA}}\nu_{\mathrm{AB}})}{1-(1-\alpha_{\mathrm{B}}+\alpha_{\mathrm{B}}\nu_{\mathrm{BA}})^{f_{\mathrm{B}}}} \quad (3.144)$$

Figures 3.24, 3.25, 3.26, 3.27, 3.28, 3.29, 3.30, and 3.31 offer examples of such dependences.

These examples show that the compositions of sol and gel and conversions of functional groups can be quite different. It is generally accepted that the molecular weights of the sol decrease with increasing conversion, but not so much attention is paid to the fact that the reacting system splits at the gel point into two chemically different parts. From the examples analyzed here, one can make the following conclusions:

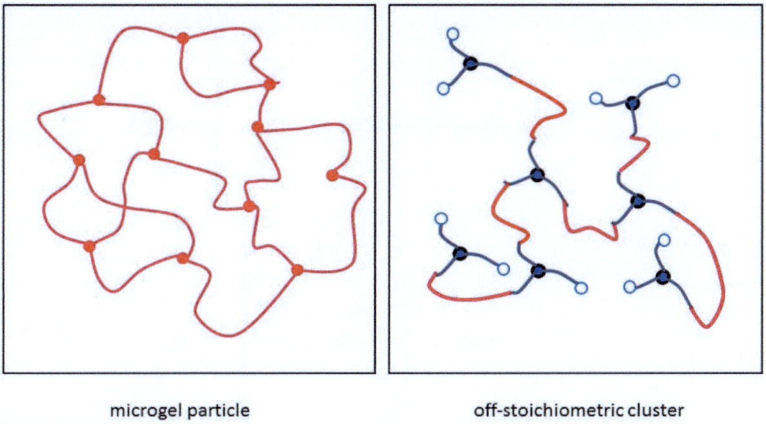

Fig. 3.25 Microgel particle obtained by self-condensation of *A*3 monomer and an off-stoichiometric particle rom A2 + B3 monomers; excess of B groups. In both cases, conversion of A groups has reached 100%

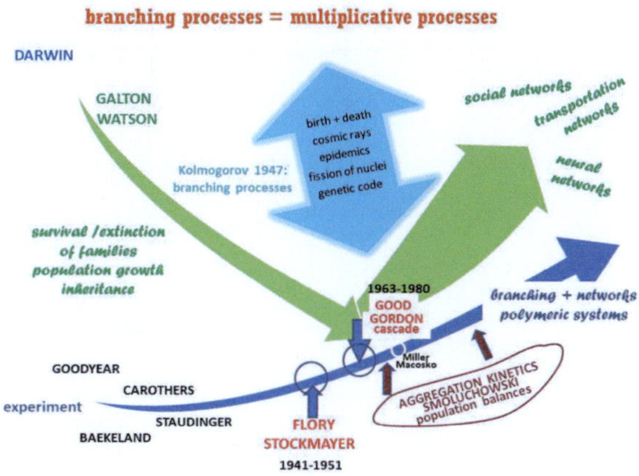

Fig. 3.26 Various branching processes

(a) The difference in *composition* of the stoichiometric systems is small at the gel point but widens with increasing conversion, and the sol becomes richer in the lower-functional monomer (here B) in composition approaching pure monomer (but its amount is very small) when the reaction approaches its completion. For off-stoichiometric systems, the difference at the gel point is larger but the change with conversion is milder.

(b) The split of *conversions* of functional groups A and B at the gel point is larger, and, for stoichiometric systems, conversions in the sol for both components converge to zero when full conversion is approached. Thus, the last traces of sol are composed exclusively of lower-functional monomer (B); i.e., the conversion

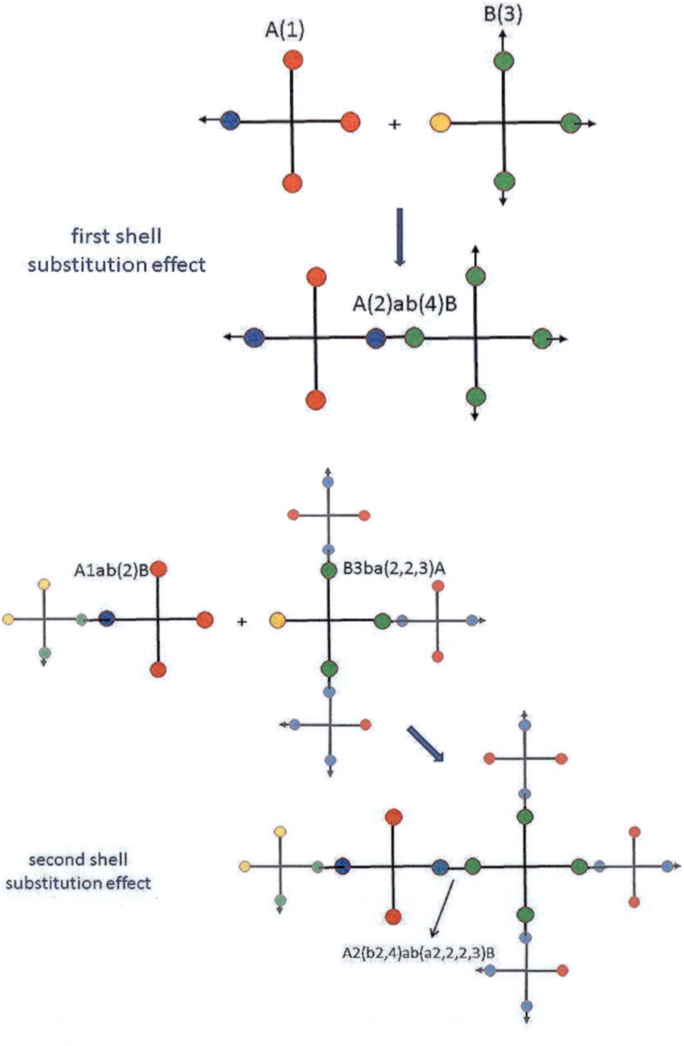

Fig. 3.27 Formation of bond between A groups and B groups of A4 and B4: upper part—under the reactivity control by *fsse*; lower part—under the control by *ssse*

is zero. Again, for off-stoichiometric systems, the differences are larger at the gel point but the changes with conversion are weaker.

As a result, the large differences between compositions and group conversions in sol and gel can lead to a change in types and numbers of interacting sites and possibly to thermodynamic intability and phase separation. The critical regions around the gel point are very susceptible to development of instabilities.

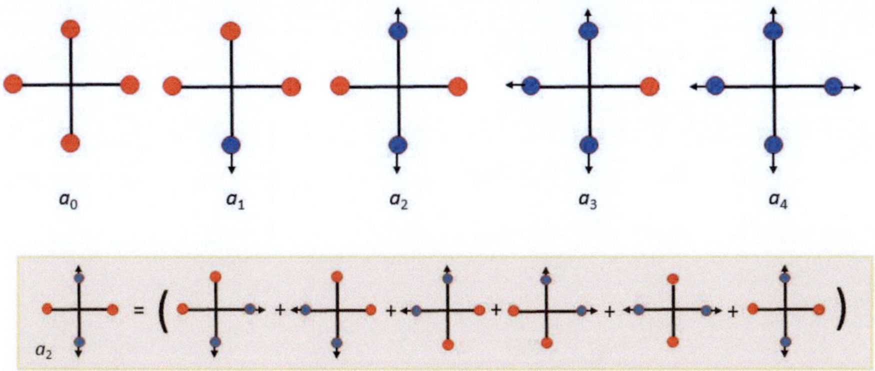

Fig. 3.28 Reaction states of a tetrafunctional monomer unit. Red discs represent unreacted functional groups, blue discs with an arrow represent reacted functional groups participating in a bond

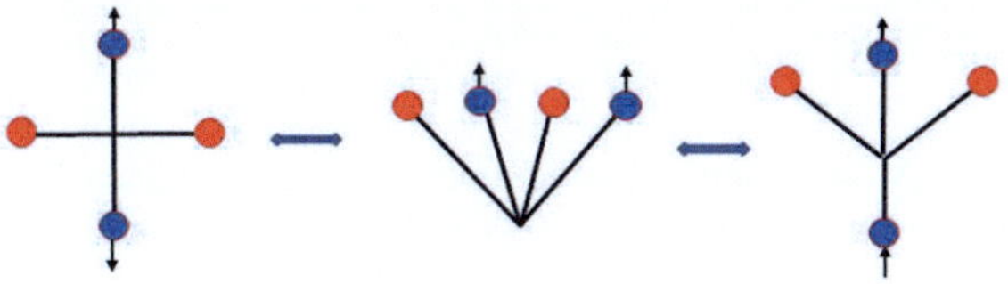

Fig. 3.29 Various representations of a tetrafunctional unit with two reacted functional groups

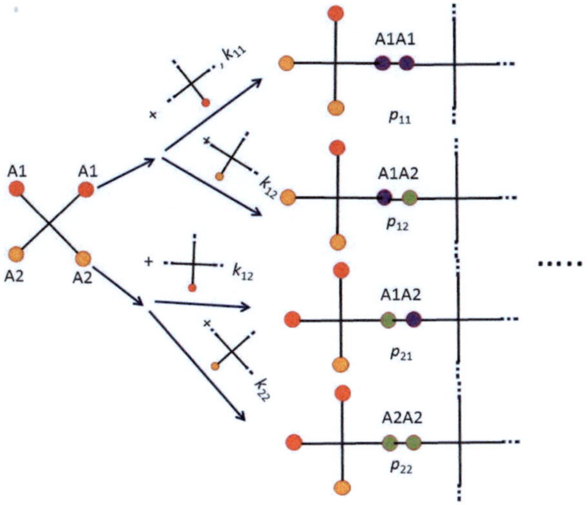

Fig. 3.30 Formation of branched or crosslinked products by formation of bonds of various types: A_{fA1fA2} monomer with $f_{A1} = 2$ and $f_{A2} = 2$

Molecular Weights of Sol

$$A_{04} + \begin{cases} A_{04} \xrightarrow{k_{00}} A_{13} + A_{13} \\ A_{13} \xrightarrow{k_{01}} A_{13} + A_{22} \\ A_{22} \xrightarrow{k_{02}} A_{13} + A_{31} \\ A_{31} \xrightarrow{k_{03}} A_{13} + A_{40} \end{cases} \qquad A_{13} + \begin{cases} A_{04} \xrightarrow{k_{10}} A_{22} + A_{13} \\ A_{13} \xrightarrow{k_{11}} A_{22} + A_{22} \\ A_{22} \xrightarrow{k_{12}} A_{22} + A_{31} \\ A_{31} \xrightarrow{k_{13}} A_{22} + A_{40} \end{cases}$$

$$A_{22} + \begin{cases} A_{04} \xrightarrow{k_{20}} A_{31} + A_{13} \\ A_{13} \xrightarrow{k_{21}} A_{31} + A_{22} \\ A_{22} \xrightarrow{k_{22}} A_{31} + A_{31} \\ A_{31} \xrightarrow{k_{23}} A_{31} + A_{40} \end{cases} \qquad A_{31} + \begin{cases} A_{04} \xrightarrow{k_{30}} A_{40} + A_{13} \\ A_{13} \xrightarrow{k_{31}} A_{40} + A_{22} \\ A_{22} \xrightarrow{k_{32}} A_{40} + A_{31} \\ A_{31} \xrightarrow{k_{33}} A_{40} + A_{40} \end{cases}$$

Fig. 3.31 Reaction scheme for time/conversion evolution of units in different reaction states for an A4 monomers with substitution effect; all reaction paths taken into account

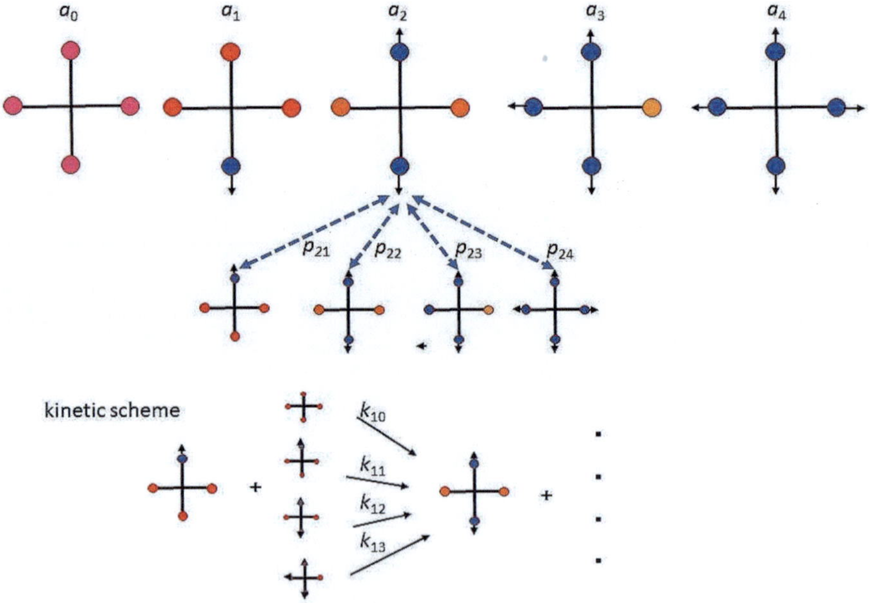

Fig. 3.32 Scheme illustrating network formation by homopolyaddition of a tetrafunctional monomer with substitution effect, respecting differences of bonds connecting units in different reaction states

The calculation of average molecular weights of the sol is analogous to the calculation of M_n in the pregel state

$$M_{n,s} = \frac{n_{A,s}M_A + n_{B,s}M_B}{1-(n_{A,s}f_A\alpha_{A,s} + n_{B,s}f_B\alpha_{B,s})/2}, \quad n_{A,s}f_A\alpha_{A,s} = n_{B,s}f_B\alpha_{B,s} \tag{3.145}$$

Figure 3.32 shows the change of number-average molecular weight of the sol, $M_{n,s}$, for two stoichiometric systems.

Whereas the weight average diverges at the gel point, $M_{n,s}$ remains finite.

The pgf for the sol units in generations $g > 0$ is given by Eq. 3.115. In the same way as in the pregel state, one get from Eq. 3.105 the expression for the weight-average molecular weight

$$\begin{aligned} M_w = {} & \frac{n_{A,s}M_A}{n_{A,s}M_A + n_{B,s}M_B}\left[M_A + \frac{1}{D_s}f_A\alpha_{A,s}[(f_B-1)\alpha_{B,s}M_A + M_B]\right] + \\ & \frac{n_{B,s}M_B}{n_{A,s}M_A + n_{B,s}M_B}\left[M_B + \frac{1}{D_s}f_B\alpha_{B,s}[M_A + (f_A-1)\alpha_{A,s}M_B]\right] \\ D_s = {} & (f_A-1)(f_B-1)\alpha_{A,s}\alpha_{B,s} \end{aligned} \tag{3.146}$$

Substructures in Gel

All units forming gel must issue at least one bond with infinite continuation, and the other $f - 1$ bonds are either in the unreacted state or have reacted and issue bonds with finite or infinite continuation. The commonly discussed substructures of the gel include *dangling chains* (composed of units issuing 1 and only 1 with infinite continuation), *elastically active network chains* (composed of units issuing 2 and only 2 with infinite continuation), and *elastically active crosslinks (junctions)* (composed of units issuing 3 or more bonds with infinite continuation). For a simple $A + A \rightarrow AA$ system, the balance equation reads

$$F_0(z_v, z_{1-v}, z_\omega) = \sum_{\substack{i=0 \\ i+j+k=f}}^{f}\sum_{j=0}^{f}\sum_{k=0}^{f} a_{ijk} z_v^i z_{1-v}^j z_\omega^k \rightarrow ((1-\alpha)z_\omega + \alpha(v z_v + (1-v)z_{1-v})^f \tag{3.147}$$

where a_{ijk} is the molar fraction of units having k groups in the unreacted state (variable z_ω), i groups are engaged in v-type bonds (z_v), and j groups in $1 - v$ type bonds (z_{1-v}). The r.h.s expression corresponds to the random case. Since we are at present not interested in distribution of the number of unreacted groups, we can put $z_\omega = 1$ and transform a_{ijk} into $a_{ij} = \sum_k a_{ijk}$. The respective pgf then reads

$$F_0(z_v, z_{1-v}) = \sum_{i=0}^{f}\sum_{j=0}^{f} a_{ijk} z_v^i z_{1-v}^j \rightarrow (1-\alpha+\alpha(vz_v+(1-v)z_{1-v})^f \quad (3.148)$$

and can be transformed into a sum of distributions for the numbers of bonds with finite continuation at a fixed number of bonds with infinite continuation

$$\begin{aligned} F_0(z_v, z_{1-v}) &= \sum_{i=0}^{f}\sum_{j=0}^{f} a_{ij} z_v^i z_{1-v}^j = \sum_{i=0}^{f} a_{i0} z_v^i + \sum_{i=0}^{f} a_{i1} z_v^i z_{1-v} \\ &+ \sum_{i=0}^{f} a_{i2} z_v^i z_{1-v}^2 + \sum_{i=0}^{f} a_{i3} z_v^i z_{1-v}^3 + \ldots \end{aligned} \quad (3.149)$$

The sums on the r.h.s. express the fractions of units in the sol, dangling chains, elastically active networtk chains, and elastically active crosslinks, respectively. The fractions of units of the respective category and the average numbers of issuing (outgoing) bonds with finite continuation are obtained by differentiations. Thus

$$\begin{aligned} \sum_{i=0}^{f} a_{i0} &= [F_0(z_v, z_{1-v})]_{z_v=1, z_{1-v}=0} \quad \mathrm{j}=0 \\ \sum_{i=0}^{f} a_{i\mathrm{j}} &= \left[\frac{1}{\mathrm{j}!}\frac{\partial F_0^{\mathrm{j}}(z_v, z_{1-v})}{\partial z_{1-v}^{\mathrm{j}}}\right]_{z_v=1, z_{1-v}=0} \quad \mathrm{j} \geq 1 \\ \sum_{i=0}^{f} i a_{i\mathrm{j}} &= \left[\partial \frac{1}{\mathrm{j}!}\frac{\partial F_0^{\mathrm{j}}(z_v, z_{1-v})}{\partial z_{1-v}^{\mathrm{j}}} \Big/ \partial z_v\right]_{z_v=1, z_{1-v}=0} \quad \mathrm{j} \geq 1 \end{aligned} \quad (3.150)$$

For the example of single-component system, the pgf $F_0(z_v, z_{1-v}) = (1-\alpha+\alpha(vz_v+(1-v)z_{1-v})^f$ can be used. The molar fraction of units in dangling chains (per unit of the whole system) reads

$$\sum_{i=0}^{f} a_{i1} = f\alpha(1-v)(1-\alpha+\alpha v)^{f-1} \quad (3.151)$$

the molar fraction of units in EANCs

$$\sum_{i=0}^{f} a_{i2} = \frac{1}{2} f(f-1)\alpha^2(1-v)^2(1-\alpha+\alpha v)^{f-2} \quad (3.152)$$

The number (molar) fraction of all EACs, N_{eac}, is given by

$$N_{\text{eac}} = \sum_{i=0}^{f} a_{i3} + \sum_{i=0}^{f} a_{i4} + \sum_{i=0}^{f} a_{i5} + \cdots = 1 - \sum_{i=0}^{f} a_{i0} - \sum_{i=0}^{f} a_{i1} - \sum_{i=0}^{f} a_{i2};$$

$$N_{\text{eac}} = 1 - (1-\alpha+\alpha v)^f - f\alpha(1-v)(1-\alpha+\alpha v)^{f-1} - \frac{1}{2}f(f-1)\alpha^2(1-v)^2(1-\alpha+\alpha v)^{f-2} \quad (3.153)$$

and the number of EANCs, N_e, can be expressed as

$$N_{\text{eanc}} = \frac{1}{2}\left(\sum_{i=0}^{f} 3a_{i3} + \sum_{i=0}^{f} 4a_{i4} + \sum_{i=0}^{f} 5a_{i5} + \cdots\right) = \frac{1}{2}\left(\sum_{j=1}^{f}\sum_{i=0}^{f} ja_{ij} - \sum_{i=0}^{f} a_{i1} - \sum_{i=0}^{f} 2a_{i2}\right); \quad (3.154)$$

which gives

$$\begin{aligned} N_{\text{eanc}} &= \frac{1}{2}(f\alpha(1-v) - f\alpha(1-v)(1-\alpha+\alpha v)^{f-1} \\ &\quad - f(f-1)\alpha^2(1-v)^2(1-\alpha+\alpha v)^{f-2}) \\ &= \frac{1}{2}(f\alpha(1-v)^2[1-(f-1)\alpha(1-\alpha+\alpha v)^{f-2}]) \\ &= \frac{1}{2}\left(f\alpha(1-v)^2\left[1-\frac{(f-1)\alpha v}{1-\alpha+\alpha v}\right]\right) \end{aligned} \quad (3.155)$$

Another quantity useful for characterization of the gel is the number of bonds with *finite* continuation issuing from units of EANCs and, possibly from EACs, or EACs + EANCs. For EANCs, the number of issuing v-type bonds is equal to the number of dangling chains

$$N_{\text{DC,eanc}} = \sum_{i=0}^{f} ia_{i2} = \frac{1}{2}\left[\partial\frac{\partial F_0^2(z_v, z_{1-v})}{\partial z_{1-v}^2} \Big/ \partial z_v\right]_{z_v=1, z_{1-v}=0}$$

$$N_{\text{DC,eanc}} = \sum_{i=0}^{f} ia_{i2} = \frac{1}{2}f(f-1)(f-2)\alpha^3(1-v)^2 v(1-\alpha+\alpha v)^{f-3} \quad (3.156)$$

whereas the number of dangling chains (= number of v-type bonds) issuing from an EAC, $N_{\text{DC,eac}}$, is defined as

$$N_{\text{DC,eac}} = \sum_{j=3}^{f}\sum_{i=1}^{f} ia_{ij} \quad (3.157)$$

This number is calculated after subtracting from the pgf (3.115) terms pertaining to units issuing 0, 1, and 2 bonds with infinite continuation

$$\begin{aligned}N_{\mathrm{DC,eac}} =& f\alpha v\big(1-(1-\alpha+\alpha vz)^{f-1}-(f-1)\alpha(1-v)(1-\alpha+\alpha vz)^{f-2}\\&-(f-1)(f-2)\alpha^2(1-v)^2(1-\alpha+\alpha vz)^{f-3}\big)\end{aligned} \tag{3.158}$$

The values of moments of the pgf F_0 (Eq. 3.115) can be used for calculation of various averages of gel substructures. In the following paragraphs, number-average values for random alternating step reaction of f_A- with f_B-functional monomers are shown, without a detailed derivation of the corresponding expression. The basic pgf will be employed.

Dangling Chains in Gel

Dangling chains in gel are composed of units issuing one bond with infinite continuation, whereas the remaining bonds in DC have only finite continuations.

The total number of units contained in dangling chains (per unit of the system), N_{DC}, is equal to

$$\begin{aligned}N_{\mathrm{DCU}} =& n_A\sum_{i=1}^{f_A} a_{i1} + n_B\sum_{i=1}^{f_B} b_{i1} = n_A f_A \alpha_A(1-v_{AB})(1-\alpha_A+\alpha_A v_{AB})^{f_A-1}\\&+n_B f_B\alpha_B(1-v_{BA})(1-\alpha_B+\alpha_B v_{BA})^{f_B-1}\end{aligned} \tag{3.159}$$

These units are attached to the elastically active network chains through bonds with finite continuation. Finite subtrees (dangling chains) issue also from elastically active crosslinks. They are of the same length, but their number is different than the number of those issuing from units of EANCs.

The number of sites on which dangling chain units are rooted is obtained by counting the number of bonds with *finite* continuation issuing from units already participating in two or more bonds with infinite continuation. This number, $N_{\mathrm{DC,eanc}}$, is given by

$$\begin{aligned}N_{\mathrm{DC,eanc}} =& n_A\sum_{i=1}^{f_A} i a_{i2} + n_B\sum_{i=1}^{f_B} i b_{i2}\\ =& \frac{1}{2} n_A f_A(f_A-1)(f_A-2)\alpha_A^3(1-v_{AB})^2 v_{BA}(1-\alpha+\alpha v_{AB})^{f_A-3}\\&+\frac{1}{2} n_B f_B(f_B-1)(f_B-2)\alpha_B^3(1-v_{BA})^2 v_{AB}(1-\alpha+\alpha v_{BA})^{f_B-3}\end{aligned} \tag{3.160}$$

The number of dangling chains issuing from EACs

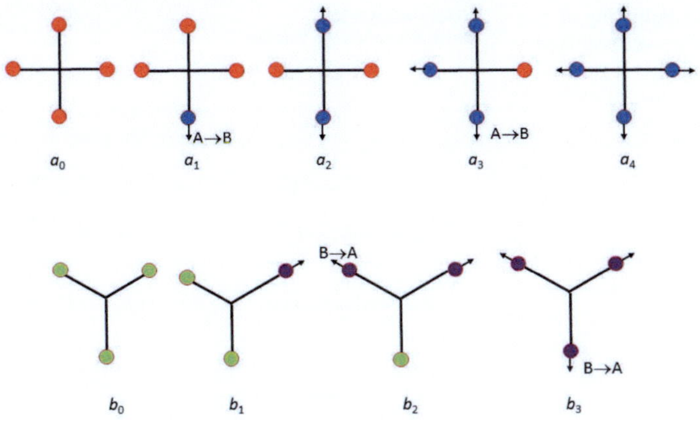

Fig. 3.33 Reaction states of units of A4 and B3 monomers bonded by alternating A + B reaction

$$\begin{aligned} N_{\mathrm{DC,eacA}} =& n_{\mathrm{A}}\alpha_{\mathrm{A}} f_{\mathrm{A}} \nu_{\mathrm{AB}} \Big(1 - (1-\alpha_{\mathrm{A}} + \alpha_{\mathrm{A}}\nu_{\mathrm{AB}})^{f_{\mathrm{A}}-1} \\ & - (f_{\mathrm{A}}-1)\alpha_{\mathrm{A}}(1-\nu_{\mathrm{AB}})(1-\alpha_{\mathrm{A}}+\alpha_{\mathrm{A}}\nu_{\mathrm{AB}})^{f_{\mathrm{A}}-2} \\ & - \frac{1}{2}(f_{\mathrm{A}}-1)(f_{\mathrm{A}}-2)\alpha_{\mathrm{A}}^2(1-\nu_{\mathrm{AB}})^2(1-\alpha_{\mathrm{A}}+\alpha_{\mathrm{A}}\nu_{\mathrm{AB}})^{f_{\mathrm{A}}-3}\Big) \end{aligned} \quad (3.161)$$

and an analogous expression describes the contribution by the B component.

The balances of Eqs. 3.125–3.127 store all information for characterization of dangling chains—their total number $N_{\mathrm{DC}} = N_{\mathrm{DC,eanc}} + N_{\mathrm{DC,eac}}$ and weight, $w_{\mathrm{g,DC}}$, which gives access to their weight fraction w_{DC} and number-average molecular weight of dangling chains, $M_{\mathrm{DC,n}}$

$$w_{\mathrm{g,DC}} = \frac{m_{\mathrm{DCU}}}{n_{\mathrm{A}}M_{\mathrm{A}} + n_{\mathrm{B}}M_{\mathrm{B}}}, \quad M_{\mathrm{DC,n}} = \frac{m_{\mathrm{DCU}}}{N_{\mathrm{DC,eanc}} + N_{\mathrm{DC,eac}}} \quad (3.162)$$

Dependences of $w_{\mathrm{g,DC}}$ together with weight fraction of gel, w_{g}, are displayed in Figs. 3.33, 3.34, 3.35, and 3.36.

In stoichiometric systems, the weight fraction of dangling chains passes through a maximum (Figs. 3.33 and 3.35), and for off-stoichiometric systems, the maximum does not appear. For strong off-stoichiometry, the mass of dangling chain is almost equal to the mass of the gel (Figs. 3.34 and 3.36).

Elastically Active Chains

Elastically active network chains (EANC) are strands of units extending between elastically active crosslinks (junctions) (EAC). As defined in Sect. 3.6.3, an EAC is represented by a unit issuing three or more bonds with infinite continuation. Two such issuing bonds (in fact half-bonds) from EACs contribute by one to the number

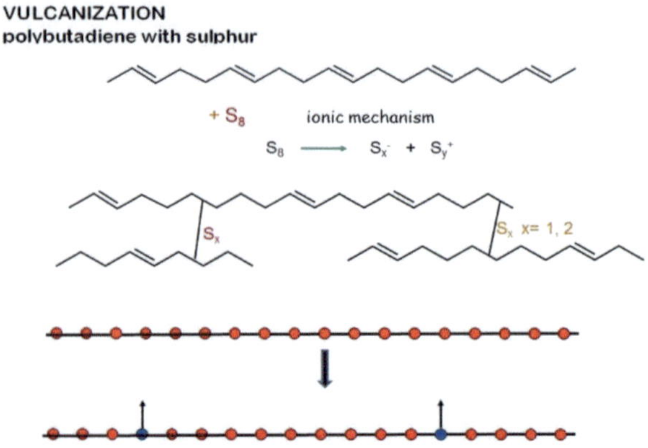

Fig. 3.34 Crosslinking of primary chains—crosslinked and uncrosslinked segments. The whole chain as a root

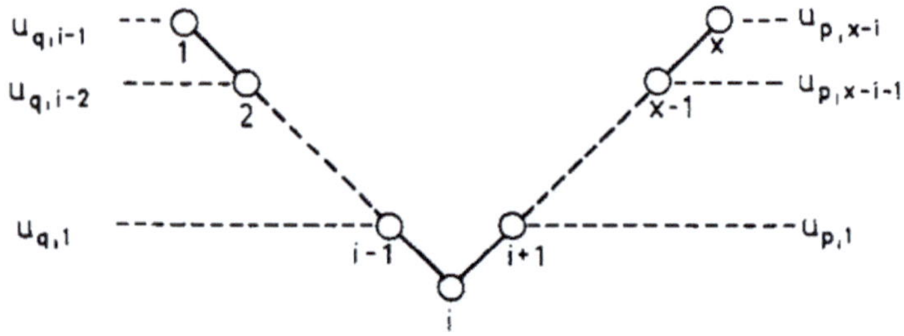

Fig. 3.35 Rooting of a primary chain by one of its segments

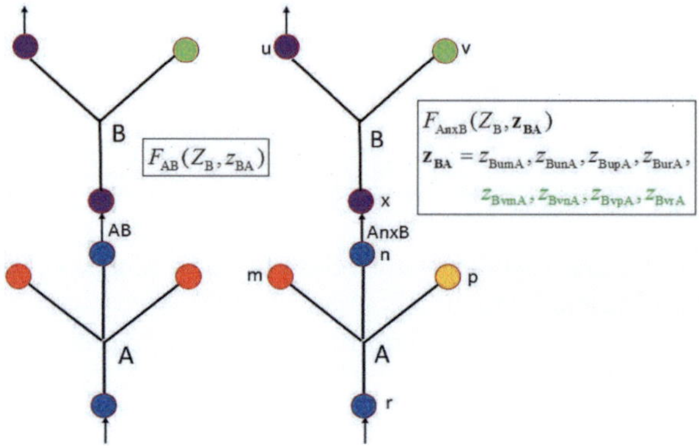

Fig. 3.36 Formulation of the pgf for the number of bonds issuing from a unit B in generation $i \geq 1$ to an A unit in generation $i + 1$ and rooted by a bond B–A, looking out of the B unit, which is an incoming bond A–B looking in the B unit. Left scheme, all B groups and all A groups, respectively, are of the same reactivity; right schcme, A groups are distinguishable (m, n, p, r) and B groups as well (u, v, x)

of elastically active network chains, N_e. Relating N_e to the whole system (sol + gel), its value can be calculated from $F_0(\mathbf{Z}, \mathbf{z}) \to F_0(\mathbf{Z}, \mathbf{v}\mathbf{z}_v + (1-\mathbf{v})\mathbf{z}_{1-v})$ expressed by a power series

$$F_0(\mathbf{Z} = \mathbf{1}, \mathbf{v}\mathbf{z}_\mathbf{v} + (\mathbf{1} - \mathbf{v})\mathbf{z}_{\mathbf{1}-\mathbf{v}}) \equiv \sum_{i,j=0}^{f_A} t_{ij} z_v^i z_{1-v}^j \tag{3.163}$$

and in terms of the coefficients t_{ij}

$$N_e = \left(\frac{1}{2}\right) \sum_{i=0,j=3}^{i+j=f_A} j t_{ij}$$

This sum can be expressed by values of the pgf F_0 and its derivatives

$$\begin{aligned} N_e &= \left(\frac{1}{2}\right) \sum_{i=0,j=3}^{i+j=f_A} j t_{ij} = \left(\frac{1}{2}\right)\left(\sum_i t_{i1} + 2t_{i2} + 3t_{i3} + 4t_{i4} \ldots - t_{i1} - 2t_{i2}\right) \\ &= \left(\frac{1}{2}\right)\left(\left[\frac{\partial F_0}{\partial z_{1-v}}\right]_{z=1} - \left[\frac{\partial F_0}{\partial z_{1-v}}\right]_{z_v=1, z_{1-v}=0} - \left[\frac{\partial^2 F_0}{\partial z_{1-v}^2}\right]_{z_v=1, z_{1-v}=0}\right) \end{aligned} \tag{3.164}$$

When applied to the random alternating $A_{fA} + B_{fB}$ case the contribution by the A component N_{eA} and B component by its sum N_e read (after transformation)

$$\begin{aligned} N_e &= (n_A N_{eA} + n_B N_{eB}) \\ N_{eA} &= \frac{1}{2} f_A \alpha_A (1 - v_{AB})\left(1 - v_{BA} - \frac{(f_A - 1)\alpha_A (1 - v_{AB}) v_{BA}}{1 - \alpha_A + \alpha_A v_{AB}}\right) \\ N_{eB} &= \frac{1}{2} f_B \alpha_B (1 - v_{BA})\left(1 - v_{AB} - \frac{(f_B - 1)\alpha_B (1 - v_{BA}) v_{AB}}{1 - \alpha_B + \alpha_B v_{BA}}\right) \end{aligned} \tag{3.165}$$

To calculate the molecular weight of EANCs including backbone and dangling chains, one needs the mass of EANC units, m_e. The number of units in EANC backbone N_{eU} according to Eqs. (3.116) and (3.119) reads

$$\begin{aligned} N_{eU} &= n_A \sum_{i=0}^{f} a_{i2} + n_B \sum_{i=0}^{f} b_{i2} = n_A N_{eUA} + n_B N_{eUB} \\ N_{eUA} &= \frac{1}{2} f_A (f_A - 1) \alpha_A^2 (1 - v_{AB})^2 (1 - \alpha_A + \alpha_A v_{AB})^{f_A - 2} \\ N_{eUB} &= \frac{1}{2} f_B (f_B - 1) \alpha_B^2 (1 - v_{BA})^2 (1 - \alpha_B + \alpha_B v_{BA})^{f_B - 2} \end{aligned} \tag{3.166}$$

and the weight of these units is

$$m_{\mathrm{eU}} = n_{\mathrm{A}} M_{\mathrm{A}} N_{\mathrm{eUA}} + n_{\mathrm{B}} M_{\mathrm{A}} N_{\mathrm{eUB}} \tag{3.167}$$

The number of dangling chains rooted on elastically active network chains, $N_{\mathrm{DC,e}}$, is somewhat different from that rooted on elastically active crosslinks can be obtained from the term $\sum_{i=0}^{f} a_{i2} z_v^i$ by differentiation

$$\begin{aligned}
N_{\mathrm{DC,e}} &= n_{\mathrm{A}} N_{\mathrm{DC,eA}} + n_{\mathrm{B}} N_{\mathrm{DC,eB}} \\
N_{\mathrm{DC,eA}} &= \frac{1}{2} f_{\mathrm{A}} (f_{\mathrm{A}} - 1)(f_{\mathrm{A}} - 2)\alpha_{\mathrm{A}}^3 (1 - v_{\mathrm{AB}})^2 v_{\mathrm{AB}} (1 - \alpha_{\mathrm{A}} + \alpha_{\mathrm{A}} v_{\mathrm{AB}})^{f_{\mathrm{A}}-3} \\
N_{\mathrm{DC,eB}} &= \frac{1}{2} f_{\mathrm{B}} (f_{\mathrm{B}} - 1)(f_{\mathrm{B}} - 2)\alpha_{\mathrm{B}}^3 (1 - v_{\mathrm{BA}})^2 v_{\mathrm{BA}} (1 - \alpha_{\mathrm{B}} + \alpha_{\mathrm{B}} v_{\mathrm{BA}})^{f_{\mathrm{B}}-3}
\end{aligned} \tag{3.168}$$

The concentration of EANCs, ν_{e}, a quantity obtained, e.g., from equilibrium stress–strain measurements, is related to N_{e} as

$$\nu_{\mathrm{e}} = \frac{N_{\mathrm{e}} \rho_2}{M_{0,\mathrm{n}}} \tag{3.169}$$

where ρ_2 is the specific gravity of the dry polymer, and $M_{0,\mathrm{n}}$ is the number-average molecular weight of the building (monomer) units $(n_{\mathrm{A}} M_{\mathrm{A}} + n_{\mathrm{B}} M_{\mathrm{B}})$. The quantity ν_{e} depends on functionality of the components and increases with increasing conversion. For the alternating A + B system, it strongly depends on the initial ratio of functional groups, expressed by r_{A} or r_{B} because the gel contains unreacted groups of the monomer in excess and also by dangling chains. A few examples of the change of ν_{e} with conversion are displayed in Fig. 3.37.

The concentration of EANCs is very low when the gel point is surpassed and initially increases steeply but closer to full conversion state the increase almost directly proportional to conversion (depending on monomer functionality). For off-stoichiometric systems, one can prepare chemically stable systems (at full conversion of minority groups) of very low crosslink density which is challenging for a number of applications. The functional groups of component in excess (here A) can be deactivated by reaction with monofunctional monomer if necessary, so that the product is fully stable.

Often, in the literature and conference discussions, the crosslink density of polymer networks is characterized by the quantity M_{c}—molecular weight of chain connecting two adjacent (elastically active) crosslinks. It is the usual habit to calculate M_{c} from experimental ν_{e} using the relation

$$M_{\mathrm{c}} = M_{\mathrm{c,tot}} = \frac{\rho_2}{\nu_{\mathrm{e}}} \tag{3.170}$$

Except of the perfect networks, this quantity has little physical meaning because it includes sol. Sol can be removed from consideration just by division of ν_{e} by w_{g}

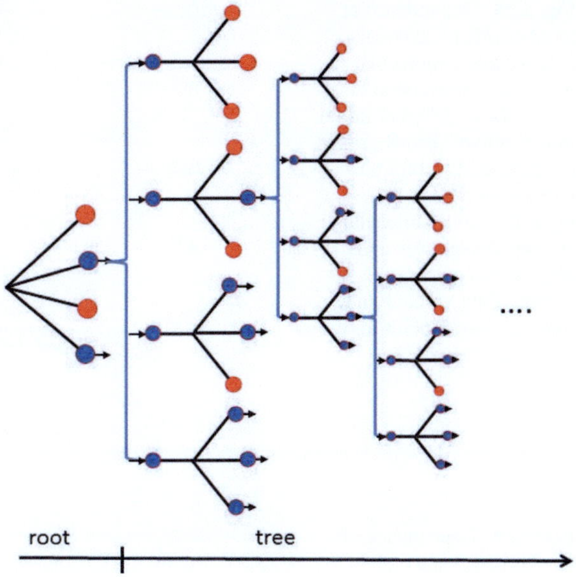

Fig. 3.37 Branching tree of units of A4 monomer listed in Fig. 3.3; each reacted group (blue circle) extends a bond to another reacted groups of a unit with at least one reacted group. An arrow issuing from a reacted group can generate tree structures like those shown in this picture

$$M_{\mathrm{c,gel}} = \frac{w_{\mathrm{g}}\rho_2}{\nu_{\mathrm{e}}} \tag{3.171}$$

Yet, there is a certain ambiguity in the imagination of an EANC. Perhaps, closer to user's imagination is consideration of an EANC as a sequence of units connecting two neighboring EACs and possibly adding to it the corresponding two arms of EAC (if it is star-shaped—Fig. 3.38) or another arbitrary part of EAC. In such perception, let such sequence call *backbone EANC,* and let it characterize by its molecular weight $M_{\mathrm{c,b}}$. Another variant is to count the backbone and dangling chains issuing from it. The corresponding molecular weight is designated by $M_{\mathrm{c,BBDC}}$. The compositions of various versions of EANCs are illustrated by Fig. 3.38.

The molecular weight of the backbone EANC, $M_{\mathrm{c,BB}}$, is calculated from the weight of EANCs units m_{eU} divided by its number, N_{e}, and contribution by two EANC arms, M_{arm} is added (Figs. 3.39, 3.40, 3.41, and 3.42)

$$M_{\mathrm{c,BB}} = \frac{m_{\mathrm{eU}}}{N_{\mathrm{e}}} + 2M_{\mathrm{arm}} \tag{3.172}$$

When also contribution to dangling chains issuing from EANC units is considered, the formula for $M_{\mathrm{c,bdc}}$ reads

$$M_{\mathrm{c,BBDC}} = \frac{m_{\mathrm{eU}} + N_{\mathrm{DC,e}}M_{\mathrm{DC}}}{N_{\mathrm{e}}} + 2M_{\mathrm{arm}} \tag{3.173}$$

Comparison of the variants of M_{c} is shown in Figs for A3 + B2 and A4 + B3 systems Fig.3.46 and Fig.3.47, respectively.

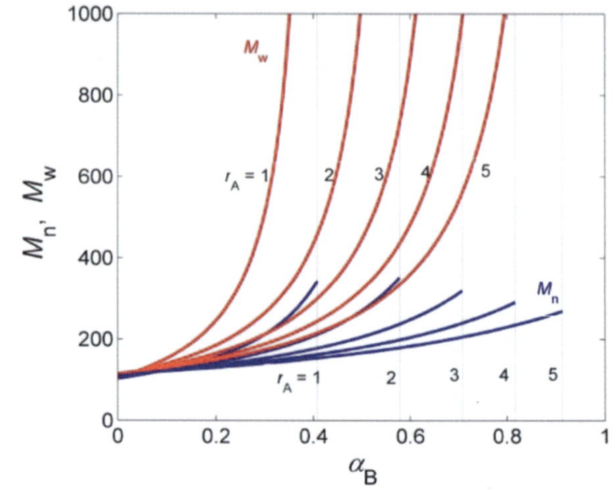

Fig. 3.38 Dependence of number-(M_n) and weight (M_w)-average molecular weights on conversion of groups B for different initial molar ratio of A to B groups, r_A, indicated and for an alternating random $f_A = 4$, $f_B = 3$ system. Molecular weight of monomer units $M_{A0} = 120$ and $M_{B0} = 90$ Dalton. Thin vertical lines indicate the gel point for the respective value of r_A

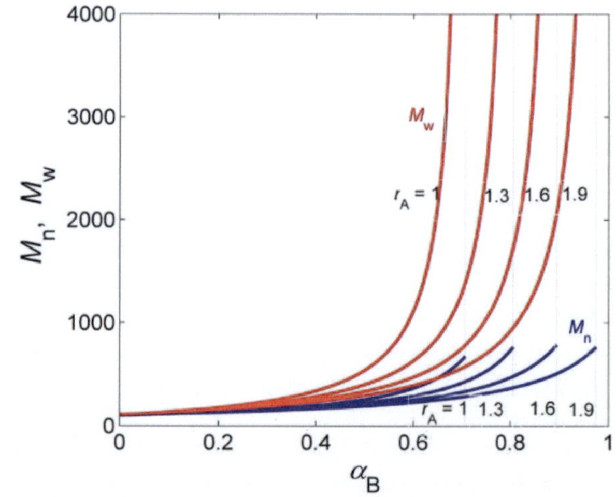

Fig. 3.39 Dependence of number-(M_n) and weight (M_w)-average molecular weights on conversion of groups B for different initial molar ratio of A to B groups, r_A, indicated, for an alternating random $f_A = 3$, $f_B = 2$ system. Molecular weight of monomer units $M_{A0} = 120$ and $M_{B0} = 90$. Thin vertical lines indicate the gel point for the respective value of r_A

$M_{c,tot}$ is unphysical and should not be used at all because this value contains also sol molecules if sol is present. The values of $M_{c,gel}$ and $M_{c,BBDC}$ are almost the same in all cases. $M_{c,BBDC}$ does not include DC issuing from the elastic active junctions. In the case of stoichiometric A4 + B3, a certain difference can be noticed, but it does not exceed 60 Dalton or g/mol. In the case of the A3 + B2 system, no dangling chain can issue from elastically active crosslinks because $f_A = 3$. There is a relatively large difference between $M_{c,BB}$, and $M_{c,BBDC}$ especially closer to the gel point and for off-stoichiometric systems. For stoichiometric systems, all values converge to one point because there are no dangling chains, and all functional groups have reacted and formed bonds.

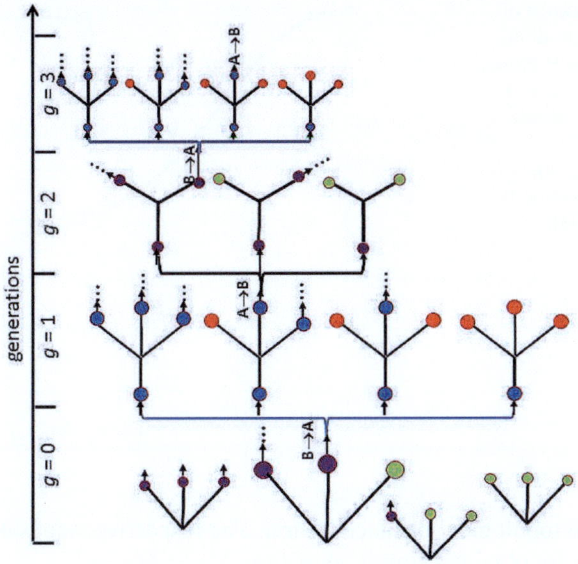

Fig. 3.40 Branching tree structure generated by alternating reaction A + B → AB of A_4 with B_3; the distribution of units of A_4 is added to each reacted B group and of units of B_3 to each reacted A group of A_4, respectively[3]

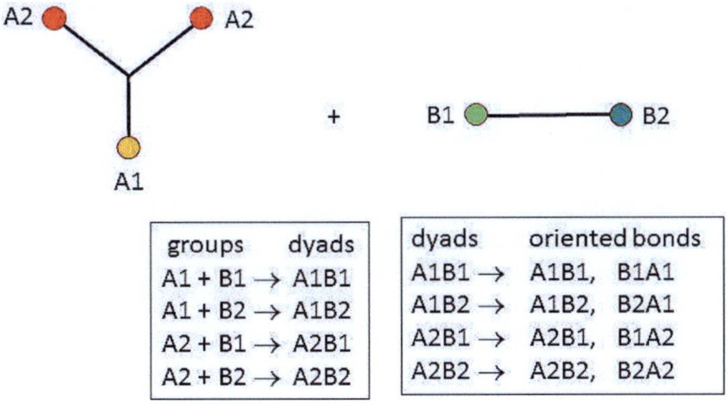

Fig. 3.41 Scheme showing a system with large difference in reactivities of functional groups A1 and A2, and B1 and B2. Alternating reaction A + B → AB

Should M_c be used to characterize the network buildup instead of ν_e, it should be the molecular weight between crosslink $M_{c,BBDC}$ or $M_{c,BB}$. In the latter case, the use of the value of the degree of polymerization of the backbone would perhaps

[3] The probability tree extends verically from the root (generation 0) to branches situated in generations higher than 0, as trees use to grow.

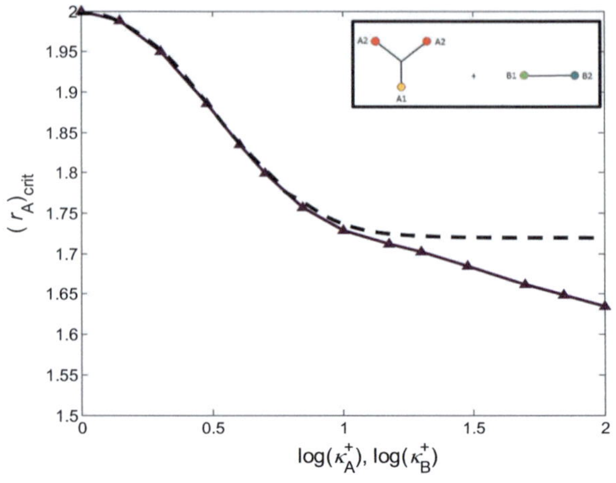

Fig. 3.42 Dependence of the critical value of initial molar ratio of A to B groups, for the system $A1(A2)_2$ + B1B2 on the ratio of rate constants $k_A = k_{A1}/k_{A2}$, $k_B = k_{B1}/k_{B2}$ (here, $\kappa_A = \kappa_B$, assumed). Data from ref. [193]

characterize the topological distance clearer. The respective values of the degree of polymerization are obtained simply by putting $M_A = M_B = 1$.

In all cases, adding M_{arm} for the contribution by two arms is optional and can be modified; 2 M_{arm} is the shortest possible connection between two junctions. It can make some difference for short EANCs but becomes negligible when they are long.

Elastically active network chains in networks formed from more complicated precursors as primary chains, functional copolymers, or precursors containing branch points in its architecture are defined in the same way as before, i.e., contributed by junctions issuing 3 or more paths of bonds with infinite continuation. Yet, there may be certain specific differences in calculating the concentration of EANCs. For vulcanization reactions, the crosslink is usually short (e.g., a–S–S– bond), and a tetrafunctional crosslink is formed from two crosslinked units. Crosslinked *end* and crosslinked *inner* units issuing bonds with infinite continuation contribute differently to the number of EANCs (cf., Fig. 3.43).

An inner unit always merges into an elastically active junction and contributes to the number of EANC by 2*(1/2) = 1. An end unit contributes to the number of EANC only if its partner in the crosslink is an inner unit. Then, it contributes by ½ each. If two end units are connected, the crosslink issues only two paths with infinite continuation and does not contribute to the number of EANCs at all because such crosslink does not qualify for an elastically active junction—it becomes only a part of an EANC and possibly issues two dangling chains. In each chain composed of 2 and more units active crosslinked units, there are $j - 2$ inner units and 2 end units. Quantification is made by the use of Eqs. 3.159 or 3.162, putting $z_v = 1$ and $t_{ij} = t_j$, $F_0(z_{1-v}) \equiv \sum_{j=0}^{\infty} t_j z_{1-v}^j$. The number of EANCs per primary chain is then equal to $N_e = \sum_{j=2}^{\infty} \left((j-2)t_j + t_j p_{inn}\right)$, where $p_{inn,j}$ is the probability that an active end junction meats an inner junction of the partner molecule $p_{inn} = \frac{\sum (j-2)t_j}{\sum j t_j}$. N_e is expressed in

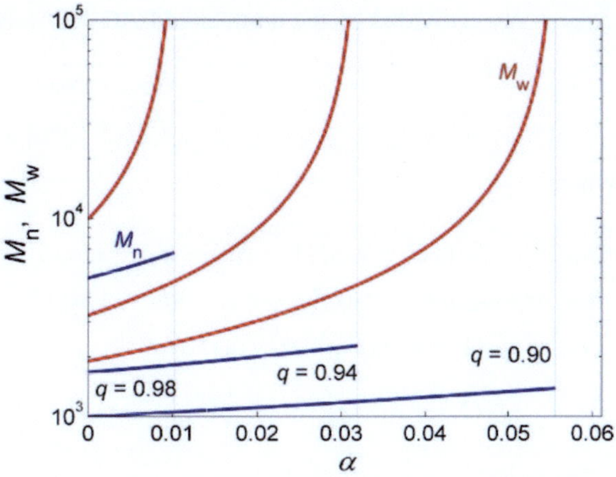

Fig. 3.43 Dependence of number (M_n) and weight (M_w) average molecular weights on fraction of crosslinked units, α, of randomly crosslinked primary chains of most-probable degree-of-polymerization distribution. The values of the parameter q indicated; $P_{0n} = 1/(1-q)$; $P_{0n} = 10, 16.7, 50$, $M_0 = 100$ Dalton M_n dependences are cut off at the gel point value of α

number of EANCs per building unit, which is in this case equal to number-average degree of polymerization of primary chain. More detailed treatment is offered in ref. [245] and Appendix to [246]. The concentration of EANC contributed by preexisting branch point in the precursors occurs by activation process described and quantified in [57, 192] and demonstrated for crosslinking of dendrimers and hyperbranched polymers as examples. This approach can be further elaborated to cover other substructures of the gel.

Effective Functionality of Elastically Active Crosslink f_e

In the mixing part of the swelling equation discussed in Chap. 4.5

$$\frac{\Delta\mu_{1,\text{mix,n}}}{RT} = \left(\frac{1}{m_1}\ln\phi_1 + \frac{\phi_2}{m_1} + \frac{f_e-2}{f_e}\nu_e V_{\text{mol},1}\phi_2 + \left(g(\phi_2) - \frac{\partial g(\phi_2)}{\partial\phi_2}\phi_1\right)\phi_2^2\right)$$

and equations that follow, there appears the quantity f_e called *effective functionality*. The quantity has been defined as number-average number of bonds with infinite continuation issuing from an elastically active crosslink (EAC). In terms of the coefficients t_{ij}

$$f_e = \frac{\sum\limits_{i=0,j=3}^{i+j=f_A} j t_{ij}}{\sum\limits_{i=0,j=3}^{i+j=f_A} t_{ij}} = \frac{\sum\limits_i (t_{i1} + 2t_{i2} + 3t_{i3} + 4t_{i4} + \ldots - t_{i1} - 2t_{i2})}{\sum\limits_i (t_{i0} + t_{i1} + t_{i2} + t_{i3} + \ldots - t_{i0} - t_{i1} - t_{i2})} \tag{3.173a}$$

When applied to the random alternating $A_{fA} + B_{fB}$ case, f_e is equal to number of paths with infinite continuation issuing from elastically active units (EAC) A and B,

$N_{\mathrm{fAe}} + N_{\mathrm{fBe}}$, divided by the number of EAC $N_{\mathrm{Ae}} + N_{\mathrm{Be}}$

$$f_{\mathrm{e}} = \frac{N_{\mathrm{fAe}} + N_{\mathrm{fBe}}}{N_{\mathrm{Ae}} + N_{\mathrm{Be}}} \tag{3.173b}$$

where

$$\begin{aligned}
N_{\mathrm{fAe}} &= n_{\mathrm{A}} f_{\mathrm{A}} \alpha_{\mathrm{A}} (1 - \nu_{\mathrm{AB}}) \big(1 - \nu_{\mathrm{BA}} + (f_{\mathrm{A}} - 1)\alpha_{\mathrm{A}}(1 - \nu_{\mathrm{AB}})(1 - \alpha_{\mathrm{A}} + \alpha_{\mathrm{A}}\nu_{\mathrm{AB}})^{f_{\mathrm{A}}-2}\big) \\
N_{\mathrm{fBe}} &= n_{\mathrm{B}} f_{\mathrm{B}} \alpha_{\mathrm{B}} (1 - \nu_{\mathrm{BA}}) \big(1 - \nu_{\mathrm{AB}} - (f_{\mathrm{B}} - 1)\alpha_{\mathrm{B}}(1 - \nu_{\mathrm{BA}})(1 - \alpha_{\mathrm{B}} + \alpha_{\mathrm{B}}\nu_{\mathrm{BA}})^{f_{\mathrm{B}}-2}\big) \\
N_{\mathrm{Ae}} &= n_{\mathrm{A}} \big(1 - (1 - \alpha_{\mathrm{A}} + \alpha_{\mathrm{A}}\nu_{\mathrm{AB}})^{f_{\mathrm{A}}} \\
&\quad - f_{\mathrm{A}}(1 - \nu_{\mathrm{AB}})\alpha_{\mathrm{A}} \left(\nu_{\mathrm{BA}} + \frac{1}{2}(f_{\mathrm{A}} - 1)\alpha_{\mathrm{A}}(1 - \nu_{\mathrm{AB}})(1 - \alpha_{\mathrm{A}} + \alpha_{\mathrm{A}}\nu_{\mathrm{AB}})^{f_{\mathrm{A}}-2} \right)\bigg) \\
N_{\mathrm{Be}} &= n_{\mathrm{B}} \big(1 - (1 - \alpha_{\mathrm{B}} + \alpha_{\mathrm{B}}\nu_{\mathrm{BA}})^{f_{\mathrm{B}}} \\
&\quad - f_{\mathrm{B}}(1 - \nu_{\mathrm{BA}})\alpha_{\mathrm{B}} \left(\nu_{\mathrm{AB}} + \frac{1}{2}(f_{\mathrm{B}} - 1)\alpha_{\mathrm{B}}(1 - \nu_{\mathrm{BA}})(1 - \alpha_{\mathrm{B}} + \alpha_{\mathrm{B}}\nu_{\mathrm{BA}})^{f_{\mathrm{B}}-2} \right)\bigg)
\end{aligned} \tag{3.173c}$$

Figure 3.42a and b demonstrates the changes of f_{e} with conversion for various combination of monomer functionalities and stoichiometric as well as off-stoichiometry cases.

By definition of an EAC, the minimum value of f_{e} is 3. This is the value of f_{e} irrespective of the chemical functionality of monomers X provided $f_{\mathrm{X}} \geq 3$. Figure 3.42a and b shows that it is so. With increasing conversion, the effective functionality increases and reaches the maximum value at full conversion (value 1 for stoichiometric systems had the maximum value for the majority groups for off-stoichiometric systems). It may be interesting to note that (a) for $f_{\mathrm{A}} = 3, f_{\mathrm{B}} < 3, f_{\mathrm{e}} = 3$ for all conversions beyond the gel point, (b) f_{e} for $f_{\mathrm{A}} = 4, f_{\mathrm{B}} = 3$ is lower than for the system $f_{\mathrm{A}} = 4, f_{\mathrm{B}} = 2$; this is due to the fact that $f_{\mathrm{B}} = 3$ contribute to the average but $f_{\mathrm{B}} = 2$ does not; (c) for highly off-stoichiometric networks f_{e} remains close to $f_{\mathrm{e}} = 3$ irrespective of chemical functionality of the components. Off-stoichiometric networks are important in applications, for instance, as various (very) soft hydrogels or pressure sensitive adhesives. It is quite usual practice that the difference between f_{e} and f_{X} is not being taken into account in evaluation of swelling experiments which may lead to more or less false conclusions.

3.4.2.7 Possible Use of the Statistical Theory for Chain Crosslinking Copolymerization

Networks are often prepared by chain polymerization of multivinyl monomers or by their copolymerization with monovinyl monomer. By its nature—buildup of a network from building (monomer) units in different reaction states—the statistical theory is not suitable for modeling systems determined long memory such as free-radical polymerization or copolymerization where the degree of polymerization distribution depends on several irreversible steps (e.g., initiation, propagation, chain transfer, and termination). For two-stage processes, the problem can be overcome by assigning different pgf variables to bond formation in each of the stages, and, for irreversible reactions, the probabilities for different reaction paths determined by Markov transition probabilities calculated from populations of dyads. To explain the situation more explicitly: statistical generation of linear chains gives always the most-probable distribution for their DP, or the so-called pseudo-most-probable distribution. The "pseudo most probable" means that, in the distribution, the first n terms are different from those of the most-probable distribution but the tail from $(n + 1)$th one to infinity satisfies the most-probable distribution.

Yet, the application of the cascade theory in some special cases can be successful—for instance, if the polymerization is fully controlled by chain transfer by disproportionation when the chain growth is terminated randomly followed by growth of a new chain. Such a case was considered by Gordon et al. in their early paper [128]. To build up a network from primary chains, distinguishing reacted and unreacted vinyls of the divinyl monomer molecule is another option. The weak point is the necessity to know the corresponding distribution of the primary chains and to guarantee that this distribution does not change. This is generally not the case. Only a minor improvement was brought by the *fragment method* by Williams [230, 231], who distinguished between chain propagating units and various end units formed by initiation or several kinds of chain transfer but assumed a random combination of them. More substantial progress was made by application of the *combined method* (combination of kinetic and statistical approach) see, e.g., Refs. [181, 183, 190]. The combined method consists in kinetics generation of subcritical fragments from small building units where the crosslinking units (e.g., a bisunsaturated monomer unit) are cut into two halves and the points of cut labeled. Then, the kinetics method is applied to build up the distribution of still finite fragments (e.g., primary chains) bearing the labels. In the second step, the labels are recombined into crosslinks, and crosslinked gel is generated. If the double bonds in the crosslinker are of the same reactivity, the labels are recombined randomly; if not, more types of labels are used and recombined as follows from the kinetics. The advantage of the combined method consists in the easy passage over the gel point and in the possibility to characterize substructure of the gel as was done in the case of statistical treatment of the step growth systems. Disadvantage and simplification of the combined method consists in "random mixing" of primary chains formed at different reaction times which differ in the degree of polymerization and number of labels. Splitting the total reaction time into several stages could be way for approximate solution of the problem.

Let us note at this place that so far we have considered only intermolecular reactions and no ring forming. We will see in the next section that this assumption does not hold for majority of free-radical polymerizations.

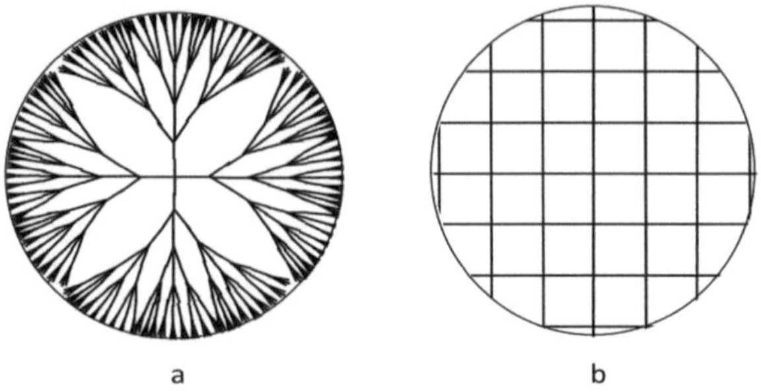

Fig. 3.44 Comparison in 2D of crosslinked structures at high conversion of functional groups; **a** tree-like model generated by recursive model, **b** model allowing uncorrelated ring closing and allowing uniform space filling

3.4.2.8 Treatment of Cyclization Within the Statistical Method

Pregel State and Gelation

Ring closing, or cyclization, or intramolecular reaction means formation of a bond connecting two building units already connected by a path of bonds. Intramolecular reaction always accompanies intermolecular reaction. In linear polymerizations, cyclization is relatively weak [247]. In branching systems, intra- and inter-reactions can be distinguished well below the gel point, because by an intermolecular reactions by which pairs of molecules are connected the molecular weight increases. By an intramolecular reaction, a bond is formed between groups of the same molecule, and its molecular weight does not change. Intermolecular reaction is usually bimolecular and the cyclization reaction monomolecular. Essentially, in both cases the two reacting groups must meet within the reaction volume, but there is a difference in the effect of concentration on inter and intramolecular reaction. Whereas the probability of entering of reactive group residing on another molecule into the reaction volume around a randomly selected group decreases with increasing dilution, the probability of entering of group of the same molecule is essentially independent of dilution. Therefore, the ratio of intra to intermolecular reactions increases with decreasing concentration. These considerations are valid up to the gel point. Change of the number-average molecular weight and shift of critical conversion at the gel point are conveniently used for characterization of cyclization extent (Fig. 3.44). The dependence of critical conversion, α_{crit}, on 1/c (c is concentration) extrapolated to fictive $1/c = 0$ gives the ring-free value. This value should be equal to that obtained by a ring-free statistical theory (e.g., cascade, Miller–Macosko, etc.). Extrapolation spans the physically unrealistic region of densities exceeding the density of bulk, undiluted system.

The difference between real (measured) value of α_g and the extrapolated value is ascribed to the fraction of bonds consumed by ring closing. It is often expressed as a fraction s of bonds consumed in ring closing

$$s = \frac{\alpha_g(\text{exptl}) - \alpha_g(\text{inter})}{\alpha_g(\text{exptl})} \tag{3.174}$$

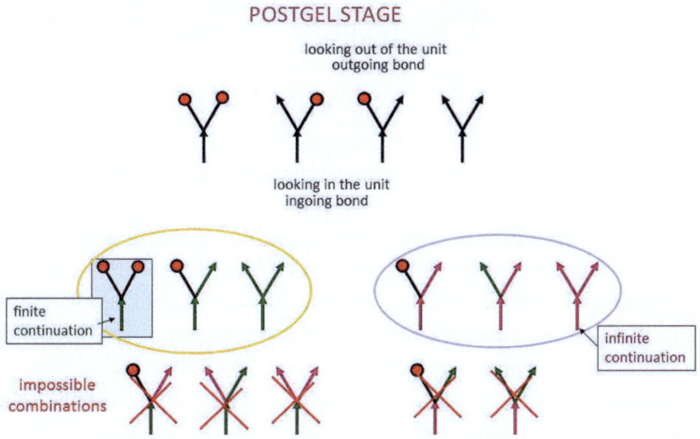

Fig. 3.45 Bonds with finite and infinite continuation. The upper row shows distribution of units according to the number of outgoing bonds; the lower part shows that beyond the gel point bonds are considered to have finite or infinite continuations. Combinations within the ovals are possible, those crossed are logically impossible. For definition of extinction probability (i.e., a conditional probability that the paths of bonds becomes extinct), only states in the left oval are considered.

In the literature, various data can be found on cyclization determined by this method—fewer data is offered based on the shift of number-average molecular weight. Data of Semlyen [248], Stepto [249], Stepto et al. [250], Cail and Stepto [251], Dutton et al. [252], Rolfes and Stepto [253] are probably the most extensive. Also, cyclization involving crosslinking of more complicated structures was characterized through the shift of the gel point [176, 179, 197, 198, 254–256]. In bulk aromatic epoxy-amine systems, cyclization was found to be negligible [174], and in undiluted polyurethane networks, it amounted to a few percent as follows from the works of Stepto, Ilavský, and others. As the functionality of the monomer increases, also the degree of cyclization increases. The degree of cyclization increases with dilution, and eventually, at limiting dilution, the system does gel at all [255], and only a distribution of branched and internally crosslinked molecules is formed possibly assisted by phase separation (microgels and microgel-like systems). To make application of many crosslinking systems possible, the systems have to contain solvent in the application stages—in order to keep the system in soft (rubbery) state—gels, or to adjust the fluidity of the system before gelation—protective organic coatings. In these cases, the shift of the gel point to higher conversion is noticeable. In Fig. 3.45a and b, the degrees of cyclization found for some systems of practical interest are displayed.

Statistical models of cyclization in the pregel state are based on the difference in probability that the reacting group chooses as reaction partner a group already connected by a sequence of bonds or a group of another molecules.

This particular situation can be kinetically described by applying mass action law

$$v_{\text{inter}} = \frac{\mathrm{d}[\mathrm{A}]}{\mathrm{d}t} = k[\mathrm{A}][\mathrm{B}]$$
$$v_{\text{intra}} = \frac{\mathrm{d}[\mathrm{A}]}{\mathrm{d}t} = k_{\text{intra}}[\mathrm{A}][\mathrm{B}]_{\text{intra}} = [\mathrm{A}]k\Lambda_{\mathrm{B}} \tag{3.175}$$

where v_{inter} corresponds to a bimolecular reaction, and [A] and [B] are average concentrations over the whole volume. This is not true for v_{intra}, where $[\mathrm{B}]_{\text{intra}}$ is dependent on the probability that the corresponding group occurs in the reaction volume; thus, it depends on the conformational properties of the paths of bonds connecting groups under consideration. Often, this sequence is considered to behave as a freely jointed chain characterized by Gaussian function for the distribution of end-to-end distances. Jacobson and Stockmayer [259] showed that the probability of forming a ring of size n Kuhn's segments is proportional to $n^{-3/2}$. Thus, Λ in Eq. 3.140 for formation of ring of size n

$$\Lambda_n = \left(\frac{3}{2\pi n l^2}\right)^{3/2} \tag{3.176}$$

Considering all possible ring sizes and arrangements, the cyclization parameter Λ in Eq. 3.140 is a sum of products of the number of unreacted groups on units in generation g,$N_{\omega g}$, and the Jacobson–Stockmayer weighting factor Λ_{n} for the conformation distance of the unit on generation g.

$$\Lambda = \sum_{g=g_{\min}}^{\infty} N_{\omega g}\left(\frac{3}{2\pi n_0 g l^2}\right)^{3/2} \Big/ \mathrm{N_A} \quad N_{\omega g} = [(f-1)\alpha]^g n_{\omega 1} \tag{3.177}$$

Each generation contributes by n_0 Kuhn's segments to the ring size, and in each generation, there is $N_{\omega g}$ unreacted groups; $\mathrm{N_A}$ is Avogadro number, and $n_{\omega 1}$ is the number of unreacted groups per unit on generation $g > 0$. The sum converges well before and at the gel point $(f-1)\alpha \leq 1$.

Such reasoning has been used in the development of *spanning-tree approximation* by Gordon et al. [129], Gordon and Scantlebury [130], and Jacobson and Stockmayer [259]. In contrast to the ring-free system considered so far, only a part of bonds formed were considered intermolecular (α), another part was engaged in closing cycles (σ), and the unreacted groups were marked as ω. Therefore, the basic pgf F_0 was transformed as

$$F_0(z_\alpha, z_\sigma, z_\omega) = \sum_{i=0}^{f}\sum_{j=0}^{f}\sum_{k=0}^{f} a_{ijk} z_\alpha^i z_\sigma^j z_\sigma^k \tag{3.178}$$

In building-up larger structures, F for higher generations is obtained by differentiation with respect to z_α. Consideration of random distribution of intermolecular and ring-closing bonds is the simplest but crudest approximation

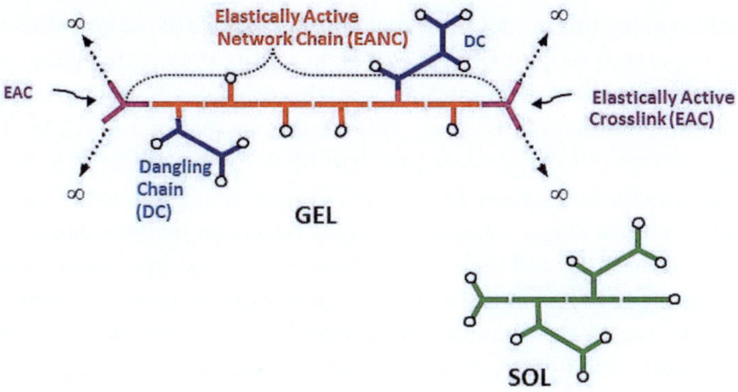

Fig. 3.46 Network units classified according to the number of bonds with infinite continuation: sol (green)—all bonds have only finite continuation; dangling chains (blue)—one bond has infnite continuation, all other bonds have finite continuation; elastically active network chain (red)—two bonds have infinite continuation, all other bonds have finite continuation; elastically active crosslinks (EAC, violet)—each EAC unit issues at least three bonds with infinite continuation, and the other bonds can have finite or infinite continuation

Fig. 3.47 Dependence of gel fraction on conversion of *B* groups for alternating copolymerization of A4 + B3 and A3 + B2 monomers; (1) $f_A = 4, f_B = 3, r_A = [A]_0/[B]_0 = 1$; (2) $f_A = 3, f_B = 2, r_A = 1$; (3) $f_A = 4, f_B = 3, r_A = 5$; (4) $f_A = 3, f_B = 2, r_A = 1.9$

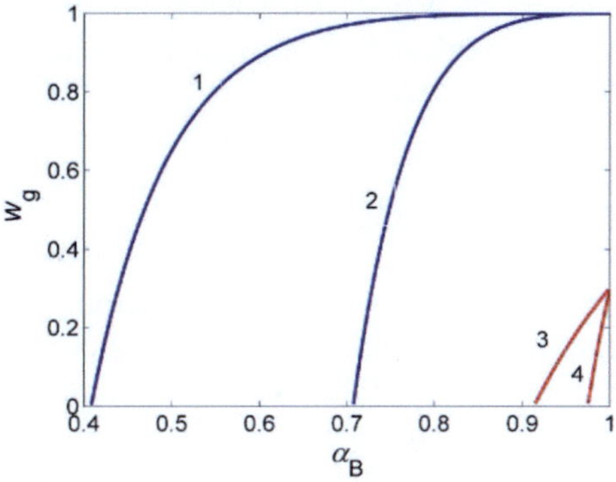

$$F_0(z_\alpha, z_\sigma, z_\omega) = (1 - (\alpha + \sigma)z_\omega + \alpha z_\alpha + \sigma z_\sigma)^f \quad (3.179)$$

Its effect is just equal to that of lowering the conversion in a ring-free case. However, this lowering is conversion-dependent.

In the more general case, the coefficients a_{ijk} of Eq. 3.144 representing the distribution of building units and the fraction σ are calculated by solving the system of the respective kinetic differential equations. If the components are long-chain telechelics (DP of arms larger than about 10–20), control by cyclization probability Λ of Gaussian form (Eq. 3.143) can be a good approximation. Otherwise, the probability of forming the smallest cycles will be different and is to be formulated.

Figure 3.46 shows a reaction scheme for time evolution of distribution of building units (Fig. 3.47).

Calculations using this model show that the dependence of the gel point conversion on dilution expressed as plot of $\alpha + \sigma$ versus $1/c$ is practically linear which can justify the analysis of experimental data in this way. Such generation avoids the artifical random combination of α, σ, and ω states as that of Eq. 3.144. In order to take into account specific ring-closing probabilities for the smallest rings, the size of the building unit can be increased to include fragments composed of two (or more) units and the smallest rings. Closing of larger rings can be considered to occur by the Gaussian probability, and their size distribution can be generated recursively.

Sarmoria et al. [214] analyzed the effect of approximations inherent to rate and the original version of the spanning-tree models. They came to a conclusion that these approximate models may give reasonable results for concentrated systems but fail at high dilutions.

Application of the Gaussian form of the parameter Λ of Eqs. 3.142 and 3.143 as well as of the function $N_{\omega g}$ equal to the number of unreacted groups in generation g is another problem. For highly branched system (hyperbranched polymers), the decrease of $N_{\omega g}(3/2\pi n_0 g l^2)^{3/2}$ was found too mild to satisfy experimental data [260]. In general, it is difficult to calculate Λ and extent of cyclization for different precursor architectures, for instance, as those shown in Fig. 3.45. Therefore, we suggest the way of determining the fraction s (or σ) experimentally and using this value for calculation of molecular weights backwards—moving from the gel point to zero conversion.

Postgel Cyclization

The postgel state differs from the pregel state in that intramolecular bond formation between pairs of functional groups residing at the gel chains is intramolecular in nature by which cyclic structures—circuits—are formed. This circuit closing accompanies the intermolecular reaction between functional groups on sol molecules, cyclization reaction in sol molecules, and intermolecular reaction between sol and gel by which the gel fraction increases. The circuit closing is considered uncorrelated with the conformational properties of the connecting path. Flory described the rate this circuit closing by the mass action law using average concentrations, and the same approach to circuit closing has been applied in other theories. Therefore, this circuit closing is of second order despite of existence of bond paths connecting the reacting groups. Comparison with experiments over decades revealed that this is a relatively good approximation for concentrated systems. However, the equilibrium elastic modulus of the networks decreases much more than is predicted by rubber elasticity theories. This is another general observation. There is no reason to assume that the conformationally determined ring closing of first order stops when the gel point is surpassed. Elastically active network chains (EANC) span two different elastically active crosslinks (EAC). On contrary, the sequence of units satisfying the condition that it has two bonds with infinite continuation returns to the same EAC where it started cannot be considered as elastically active; it is sometimes called *loop*. Loops can be formed by reaction of groups residing both on dangling chains, or two

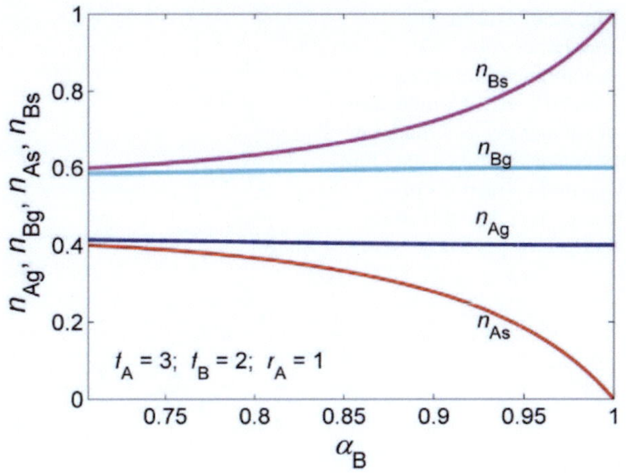

Fig. 3.48 Alternating copolymer A3 + B2; stoichiometric case, $r_A = [A]_0/[B]_0 = 1$. Composition of sol and gel as a function of conversion: n_{As}, n_{Ag}, n_{Bs}, n_{Bg} molar fractions of component A and B in sol and gel, respectively

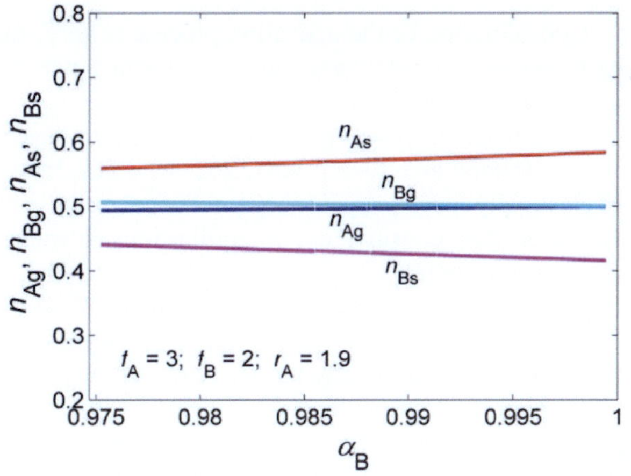

Fig. 3.49 Alternating copolymer A3 + B2; off-stoichiometric case, $r_A = [A]_0/[B]_0 = 1.9$. Composition of sol and gel as a function of conversion: n_{As}, n_{Ag}, n_{Bs}, n_{Bg} molar fractions of component A and B in sol and gel, respectively

groups both on the same backbone EANC. Figure 3.48 shows example of formation of such rings.

Also, there exist structures in the gel where a pair of EAC is connected by more than one sequence of units as in Fig. 3.48, backbone-dangling. The elastic response of such two-chain connection is different and depends on their relative length. If it is very asymmetric, the longer chain can be considered as loop.

There is another important process changing the status of chains in an elastically inactive cycle. It has been called *activation* of cycles (loops). When units in an elastically inactive cycle bear unreacted groups and one of them happens to react intermolecularly with a group residing on another EANC or its dangling chain, elastically active strands are formed as shown in Fig. 3.49 and the former cyclic structure looses the status of loop.

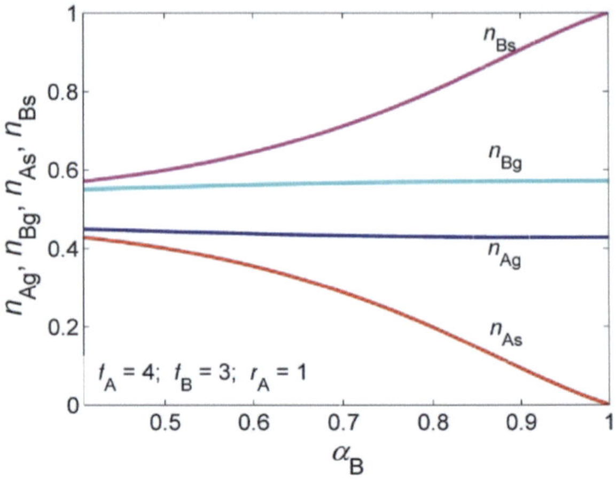

Fig. 3.50 Alternating copolymer A4 + B3; stoichiometric case, $r_A = [A]_0/[B]_0 = 1$. Composition of sol and gel as a function of conversion: n_{As}, n_{Ag}, n_{Bs}, n_{Bg} molar fractions of component A and B in sol and gel, respectively

Consideration of the activation process is very important because in pregel and early postgel regions large rings are formed which "disappear" later due to the activation process.

This mechanism was given some thought in development of theoretical models. It was Helfand and Tonelli [261, 262] who developed the first model for formation of elastically inactive cycles beyond the gel point in crosslinking of primary chains. However, they considered it as equilibrium between conformationally determined ring-forming bonds and intermolecular bonds. Dušek, Gordon, and Ross-Murphy improved this model by formulating it kinetically (by solving systems of differential equations) with inclusion of the rate of activation. The change of the fractions of bonds wasted in elastically inactive cycles [263] is shown in Fig. 3.50.

Formation of elastically inactive loops at low crosslink densities occurs beyond the gel point almost with the same intensity as before the gel point and there is not much difference between sol and gel. Contribution to cyclization by the sol diminishes as its fraction falls beyond the gel point. However, the activation process going on only in the gel changes the situation: The number of elastically inactive cycles goes down to a limit determined by cycles that cannot be activated because they do not contain any crosslinkable unit (at the given value of ν).

An analogous procedure was developed for alternating Ax + By systems [175] and later applied to polyurethane networks [176, 179]. These stepwise systems of low functionality are quite different with respect to network formation compared to crosslinking of primary chains: gelation starts at high conversions, and there is a narrow range of conversions available for network build. A narrow space also exists for development of inactive cycles and their activation. Two Figs. 3.51 and 3.52 show how the fraction s is expected to change with conversion and what the effect of inactive ring formation can be appeared.

The fraction of bonds lost in inactive cycles is relatively low (a few percent) compared to crosslinking of primary chains. This is due to the high molecular weight of the smallest possible cycle. Such low values were also found experimentally.

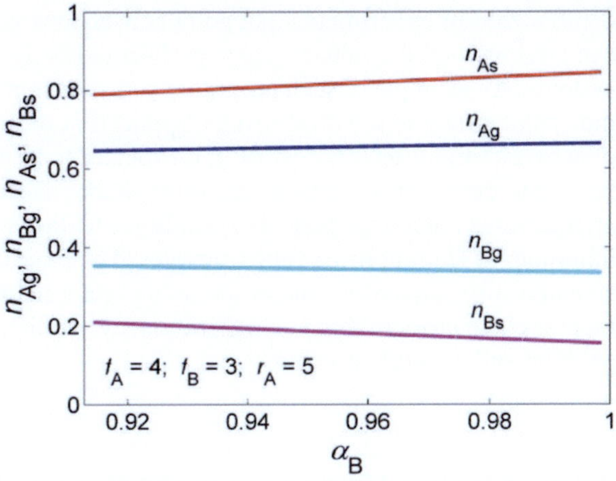

Fig. 3.51 Alternating copolymer A4 + B3; off-stoichiometric case, $r_A = [A]_0/[B]_0 = 5$. Composition of sol and gel as a function of conversion: n_{As}, n_{Ag}, n_{Bs}, n_{Bg} molar fractions of component A and B in sol and gel, respectively

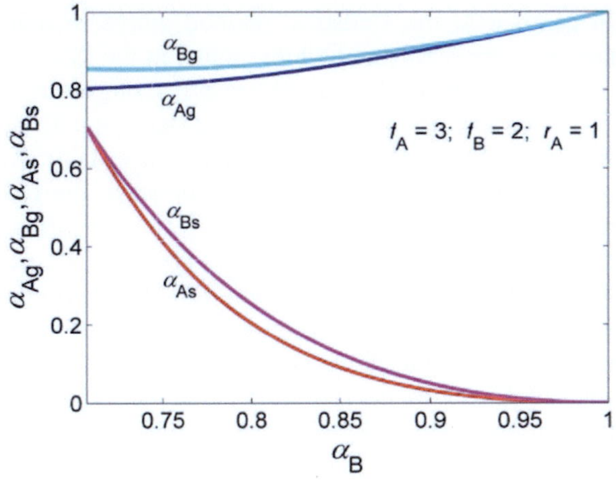

Fig. 3.52 Alternating copolymer A3 + B2; stoichiometric case, $r_A = [A]_0/[B]_0 = 1$. Conversions of groups A and B in sol and gel as a function of conversion: α_{As}, α_{Ag}, α_{Bs}, α_{Bg} conversions of groups A and B in sol and gel, respectively

Figure 3.52 shows that the effect of wastage of this small fraction of bonds causing relatively larger drop in ν_{eg}.

In conclusion: The definition of cycles in the structures is clear, and the models are relatively simple and can be further developed. Classification of cyclic structure in gels with respect to its role in equilibrium elasticity is difficult. Categorization of chains, formally satisfying the condition for a unit to be part of an EANC (two bonds with infinite continuation), into only two states—active and inactive—is possible, but apparently insufficient. In loop units, the two bond continuations to infinity are not independent because they merge at one point. The two-chain EANCs have variable elastic response depending on the disparity in chain lengths.

Rescaling of the ring-free dependences against the experimental gel point helps to approach a single master curve, but superposition of theoretical dependences is satisfactory only in the range of low crosslink densities [175]. Nevertheless, proper combination of experimental gel point conversions with ring-free recursive models can offer predictions valuable for experimentalists developing particular crosslinked systems. An example of such procedure was shown in ref. [197]. For obtaining more molecular structure sensitive results, application of molecular dynamics simulation of network formation is necessary. It is expected to be more predictive and also would be able to deal with systems where short chains, steric obstructions, or complicated architectures come into play. The equilibrium elastic response should be related to minimum of Helmholtz or Gibbs energies. The issue of cyclization is important and should not be ignored as one of the unimportant network flaws. We will see below that for chain crosslinking polymerization, cyclization may have a decisive effect on gelation and network structure.

3.4.3 Kinetics Description of the Branching Process

In contrast to the statistical method where branched and crosslinked structures are reconstructed from building units, in the kinetics (kinetic) approach the whole molecules (species) are considered as reacting entities. The transformations—formation and transformation of molecules—are considered as a function of time; the time variable can be converted into conversion variable. As in chemical kinetics, the rate of transformation is directly proportional to the product of concentrations of the reacting species, but the rate constant can have various forms and possibly be dependent on conversion. In the chemist's view, the reactivity of a species is proportional to the number and reaction ability of unreacted functional groups of each species. Each species is characterized by the number of unreacted functional groups by the number of building units (x) and number of unreacted functional groups (l). For single type groups and reactions taking place only intermolecularly, the reaction can be described by a set of differential equations

$$\frac{\mathrm{d}c(x)}{\mathrm{d}t} = \frac{1}{2}\sum_{i=1}^{x-1} k_{\mathrm{inter}}(i, x-i)[c(x-i)c(i)] - c(x)\sum_{i=1}^{\infty} k_{\mathrm{inter}}(x)c(i) \tag{3.180}$$

by which the product composed of x-units is formed by formation of a bond between i and x − i mer; two functional groups are consumed in this reaction. In an acyclic molecule formed from f-functional monomer, the number of unreacted functional groups l_x is related to x as

$$l_x = x(f-2) + 2$$

The rate constant is a function of the product of the numbers of unreacted functional groups

$$k_{\mathrm{inter}}(i, x-i) = kl_i l_{x-i} \tag{3.181}$$

Equation 3.146 with rate constant proportional to numbers of unreacted functional groups (as in Eq. 3.148) is one of the forms of Smoluchowski coagulation equations proposed by the Polish scientist in the time of WW1 [264, 265]. The Smoluchowski equation describes aggregation processes occurring by various mechanisms controlled by chemical reactions (reaction limed aggregation (RLA) or controlled by species diffusivities (diffusion limited aggregation, DLA). The proportionality of k_{inter} to l_x and to x is important. The x-proportionality instead of l_x-proportionality suppresses the reaction ability of small molecules in comparison with the larger ones [266].

The rate constant of Eq. 3.146 can have various forms of dependence on x, also different from that of Eq. 3.147, is called *Kernel* $K(x)$ of the coagulation equation. There is a vast literature on Smoluchowski coagulation equations—perhaps the most close and relevant to the covalent crosslinking topic is the review by Galina and Lechowicz [58]. In many real cases of crosslinking, factors other than chemical reactivity are operative in bond formation. Relevant information and inspirations for advanced modeling and for description of the gelling systems can be found, for instance, in Refs. [162, 171, 267–276]. Two types of Kernels are distinguished. Those proportional to the product of i and$x - i$ and to the sum of functions of i and$x - i$

$$K_{\text{prod}} \propto f(i(x-i)); \quad K_{\text{sum}} \propto F(f(i) + f(x-i)) \tag{3.182}$$

For A + B type reactions

$$K_{\text{prod}} \propto f(x_A x_B); \quad K_{\text{sum}} \propto F(f(x_A) + f(x_B)) \tag{3.183}$$

Equation 3.146 represents an infinite set of differential equations for evolution of distribution of species according to the number of units x at time t up to the gel point. Smoluchowski-type equations work with average concentrations of species and belong to the category of mean-field theories. However, the critical exponents are equal to those obtained by statistical theories (and called mean-field) only for the random step growth polyaddition; in general, they differ from those (cf., Refs. [277, 278]). There is abundant literature on solution of the infinite sets of Smoluchowski-type differential equations. For some types of Kernels, analytical solution can be found as a function of time or function of conversion. For some other systems, the solutions are numerical, or the distribution is obtained by Monte Carlo simulation of the process.

Method of moments is a usual way of characterization of the distributions. The distribution is characterized by a few moments of the distribution which determine its shape and the gel point. The moment methods are explained in sufficient detail in the review by Galina and Lechowicz [58]; for application of the moment method, the tutorial article by Mastan and Zhu [279] can be helpful. Briefly, the polymer chain formation is described by several sets of kinetic differential equations characterizing evolution of various species in time. By manipulations with sets of time-dependent sums of concentrations of species, for instance, transient and "dead" molecules can be obtained. From these sums, moments of the distribution of dead distribution can be obtained. The moments, $M_0, M_1, M_2 \ldots M_j$, for degrees of polymerization distribution are defined in terms of concentrations of x-mers as

$$M_j = \sum_{x=0}^{\infty} x^j c_x \tag{3.184}$$

The moment $M_0 = \sum c_x$ is equal to the molar concentration of all molecules and gives the number-average degree of polymerization P_n. M_1 gives the concentration of monomer units; the averages of degrees of polymerization are related to the moments M_j as

$$P_n = M_1/M_0; \quad P_w = M_2/M_1; \quad P_z = M_3/M_2, \text{ etc.} \tag{3.185}$$

Mastan et al. [280] demonstrated the application of the moments method on an example of simple free-radical polymerization. The reaction steps considered are summarized in Table 3.4.

The solution in terms of moments, M_j, of dead molecules and moments of the radical species, $Y_j = \sum x^j c_x^*$. By summations of kinetic equations for evolution of distribution of species, a system of kinetic differential equations for the moments is obtained

$$\begin{aligned}
\frac{dY_0}{dt} &= 2fk_d c_I - k_t Y_0^2 \\
\frac{dY_1}{dt} &= k_p c_m Y_0 - k_t Y_1 Y_0 + k_p^* Y_0 M_2 \\
\frac{dY_2}{dt} &= k_p c_m (Y_0 + 2Y_1) - k_t Y_0 Y_2 + k_p^* (2Y_1 M_2 + Y_0 M_3) \\
\frac{dM_0}{dt} &= \left(k_{td} + \frac{k_{tc}}{2}\right) Y_0^2 - k_p^* Y_0 M_1 \\
\frac{dM_1}{dt} &= k_t Y_0 Y_1 - k_p^* Y_0 M_2 \\
\frac{dM_2}{dt} &= k_t Y_0 Y_2 + k_{tc} Y_1^2 - k_p^* Y_0 M_3 \\
k_t &= k_{td} + k_{tc}
\end{aligned} \tag{3.186}$$

Table 3.4 Reaction steps in free-radical polymerization of a vinyl monomer considered [279]

Initiation	$I \xrightarrow{k_d} 2P_0^*$
Propagation	$P_x^* + m \xrightarrow{k_p} P_{x+1}^*$
Termination by disproportionation by combination	$P_x^* + P_s^* \xrightarrow{k_{td}} P_x + P_s$ $P_x^* + P_s^* \xrightarrow{k_{tc}} P_{x+s}$

The system of Eq. 3.152 is solved numerically, and average degrees of polymerization are calculated. The summations of terms of the differential equations for distributions to get the time change of the moments are sometimes tedious especially if the polymerization mechanism is more complicated.

Several other groups have used the formalism of generating functions by which obtaining of the equations for the moments becomes simpler because summations are replaced by differentiations. This direction was pursued by Galina and Lechowicz [58] in their review. In some cases, it is not necessary to use the moment methods, and the distribution can be expressed through a generating function explicitly. This is illustrated by a simple example of living polymerization [183].

Classical living polymerization can be described by two different steps—initiation and propagation—yielding distribution of chains with living end

$$\begin{aligned} & \mathrm{m} + \mathrm{I}^{\otimes} \xrightarrow{k_{\mathrm{i}}} \mathrm{P}_1^{\otimes} \\ & \mathrm{P}_1^{\otimes} + \mathrm{m} \xrightarrow{k_{\mathrm{tp}}} \mathrm{P}_2^{\otimes} \\ & \ldots \\ & \mathrm{P}_x^{\otimes} + \mathrm{m} \xrightarrow{k_{\mathrm{tc}}} \mathrm{P}_{x+1}^{\otimes} \\ & \ldots \end{aligned} \qquad (3.187)$$

Each chain is terminated by a living end which is then deactivated by a blocking agent b to get a distribution of inactive polymer species

$$\mathrm{P}_x^{\otimes} + \mathrm{b} = \mathrm{P}_x$$

The corresponding kinetic equations read

$$\begin{aligned} & \mathrm{d}c_{\mathrm{I}}/\mathrm{d}t = -k_{\mathrm{I}} c_{\mathrm{I}} c_{\mathrm{m}} \\ & \mathrm{d}c_1/\mathrm{d}t = k_{\mathrm{I}} c_{\mathrm{I}} c_{\mathrm{m}} - k_{\mathrm{p}} c_{\mathrm{m}} c_1 \\ & \ldots \\ & \mathrm{d}c_x/\mathrm{d}t = k_{\mathrm{p}} c_{\mathrm{m}} c_{x-1} - k_{\mathrm{p}} c_{\mathrm{m}} c_x \\ & \ldots \end{aligned} \qquad (3.188)$$

where $c_{\mathrm{I}},\ c_{\mathrm{m}},\ c_x$ are concentrations of initiator monomer, and x-mer, respectively.

Defining a generating function $g(Z)$ as

$$g(Z) \equiv \sum_{x=1}^{\infty} c_x Z^x \qquad (3.189)$$

and multiplying each equation of the set 153 by the corresponding power of the variable Z, and summing left and right hand sides, one obtains

$$g(Z) \equiv \sum_{x=1}^{\infty} c_x Z^x$$

$$\frac{\partial g(Z)}{\partial t} = c_m g(Z)(Z-1) + \kappa c_m c_I Z \tag{3.190}$$

where $\kappa = k_I/k_p$, $\tau = k_p t$ and $\partial c_I/\partial\tau = -\kappa c_m c_I$, $\int c_m d\tau = -\ln(x)/\kappa$. This gives $g(Z)$ in the following form which no longer explicitly dependent on time but only through the time-dependent concentration of initiator, c_I

$$g(Z) = \frac{\kappa c_{I0} Z}{Z - 1 + \kappa}\left[\left(\frac{c_I}{c_{I0}}\right)^{-(Z-1)/\kappa} - \frac{c_I}{c_{I0}}\right] \tag{3.191}$$

where c_{I0} is the initial concentration of the initiator. This dependence can be normalized by the factor $c_{I0} - c_I$ and transformed into the dependence on the concentration of the monomer, c_m, using the balance

$$c_{m0} - c_m = ((\kappa - 1)/\kappa)(c_{I0} - c_I) + (c_{I0}/\kappa)\ln(c_{I0}/c_I) \tag{3.192}$$

The normalized function 156 offers a direct access to averages of degree of polymerization or to the moments of distribution if needed. For systems where transformation into an equation for the generating is not possible, numerical transform methods exist which will be briefly discussed in the section that follows.

Let us now return to comparison of the results obtained by kinetics and statistical methods. Flory [124] was the first who formulated kinetic solution of distribution evolution and showed that the solution was identical with that of Flory derived statistically. The kinetic approach to the step growth reactions involving groups of unequal and/or dependent (substitution effect) reactivities was pursued by several groups (see, e.g., [161, 168, 168, 171, 178, 183, 281]). It was found that in the case of substitution effect the distributions and the gel point conversions were not the same but for usual systems the differences were small and varied within experimental errors. When the substitution effect is strong and nonlinear, the differences can be substantial [171, 178]. Zhou and Szabo [281] found very good agreement in the kinetics of polymerization described by the Emoluchoweski equation and simulatrion in 3D space. Thus, it might seem that the statistical approach is satisfactory for modeling of a number of conventional systems and could assist well in optimizing real systems. However, it is not so in the case of chain growth and other history-dependent systems. Such dependence is well documented by the living polymerization example discussed in the preceding paragraph. When the initiation is fast with respect to propagation, all chains start growing at the same time, and the fluctuations in chain length are minimal, whereas when the initiation rate decreases, the distribution gets broader. For the case of much faster initiation than propagation ($k_I >> k_p$), the distribution converges to the Poisson distribution described by the probability generating function

$$g(Z) = Z\exp((Z-1)(P_n - 1)) \tag{3.193}$$

Such distribution can never be simulated by applying the statistical method. Such statistically generated distribution of degrees of polymerization would be of the most-probable or pseudo-most-probable type. The prefix pseudo means that a few first terms of distribution are calculated exactly and deviate from those of most-probable distribution, but the tails of the distribution are of the most-probable type [183].

3.4.3.1 Cyclization

Formally, there should no problem to include a cyclization term into the Smoluchowski-type equations by considering formation and consumption of molecules with bonds closing cyclic paths. The problem is how the Kernel for cyclization should depend on the degree of polymerization and the number of formed cycles. Usually, Smoluchowski equation counts molecules built-up from units (expressed as $c(x)$), but does not distinguish between isomers. If cyclization is allowed, each isomer will be characterized by a distribution of cycles of various sizes because conformationally determined cyclization depends on topological distances of groups. It seems manageable to retain the form of Eq. 3.146 and modify k_{inter} to be proportional to the number of unreacted functional groups (which is lower than in the case of tree-like isomers)

$$l_x = [x(f-2)+2](1-s_x) \tag{3.194}$$

where s_x is a fraction of groups consumed to form bonds closing cycles; s_x could be obtained from probability for a reacted group to close a cycle (taken as average over x-mer isomers with all possible ring sizes), $p_{s,x}$, relative to the probability that the group reacts intermolecularly, $p_{\text{int,x}}$, $s_x. = p_{s,x/}\left(p_{s,x}+p_{\text{int,x}}\right)$. For calculating $p_{s,x}$, the conformational probability of encounter of two unreacted groups would be considered. Because of decreasing probability of ring closure between more distant functional groups, s_x converges rapidly to a constant value with increasing x. This suggested approximation may remind the spanning-tree approximation explained here for statistical treatment of step growth reactions. The inherent smearing of effects of all isomeric shapes and ring sizes seems to be the simplest way to simulate the effect of the pregel cyclization on x-distribution. So far, the possibility of intramolecular reaction was considered more or less formally and without relation to the internal topology of the cluster.

3.4.3.2 Group Accessibility

Equal opportunity to react for any group irrespective of whether the group is a part of large or small molecule is a premise of the mean-field approach and is inherent to mass action law (proportionality to average concentration). Within the Eyring transition state model of chemical kinetics, it corresponds to the case when a pair of reacting groups forms a transition state complex which dissociates and reforms many times before a stable bond is formed. This equilibration before stable bond formation is characteristic for slow reaction. There is abundant experimental evidence that these conditions are met for many step growth polymerizations. If the reaction is fast, a stable bond is formed when the reactive groups meet (encounter control), and the reaction rate depends on their diffusivities. Large clusters experience interpenetration problems due to steric hindrances. Thermodynamic (osmotic) exclusion is the other factor operative especially in dilute system which makes the opportunity for two groups to meet unequal and cluster size and symmetry-dependent. Slow diffusion/partial exclusion of the critically large clusters was offered as explanation for non-mean-field value of critical exponents experimentally found even in the case of slow reactions.

The effects of such factors as volume exclusion can be simulated by applying the Smoluchowski equation with product k.

Kernel, where the exponent ω drops below the value 1 which is characteristic for mass action law control (Eq. 3.147).

$$K(x, y) = (xy)^{\omega} \tag{3.195}$$

This type of equations was investigated extensively by Leyvraz and Tschudi [277], and Leyvraz [278] who found that the value of ω is critical for gelation; for $\omega < {}^{1}\!/_{2}$, gelation does not occur. In the case of chemical reaction, one has to consider numbers of reactive groups l instead of of number of units x

$$k_{\text{inter}}(x, y) = k(l_x l_y)^{\omega} \tag{3.196}$$

Based on this proportionality, the effect of the value of the exponent was investigated on the development of weight-average degree of polymerization and shift of the gel point conversion to higher conversions [266] using Monte Carlo simulation of the Smoluchowski process. The exponent ω was related to the penetration depth of the clusters; $\omega = 2/3$ corresponded to the least penetrable spherical structures, i.e., only peripheral groups could form bonds. For acyclic step growth polyaddition of a three-functional monomer, the gel point conversion was shifted from $\alpha = 0.5$ to $\alpha = 0.636$. Increasing monomer functionality as well as formation of cycles further deteriorates the clusters interpenetrability and shifts the critical conversion to higher values.

3.4.3.3 Smoluchowski Equation and Postgel State

Some time ago, it was stated [29, 159] that the kinetics formulation of cluster evolution in gelling systems fails beyond the gel point. However, it was shown that solution can be extended beyond the gel point [176, 272] and that the result describes the distribution of the sol. The behavior of the Smoluchowski equation in the postgel stage was analyzed more in depth by Leyvraz [282]. The equations exclude the gel-gel reactions and the loss of sol by reaction with gel is not included in the balance as well. Therefore, that mass of the system (sum of mass of sol molecules) gives the sol fraction decreasing with increasing conversion. Dušek and Vojta [176] showed for random step growth reaction that the distribution was in 1–1 correspondence to that obtained statistically and that the dependence of extinction probability (used in statistical generation beyond the gel point) can be recovered from the results. This conclusion hints to the possibility that balances of reaction states of units in the sol obtained kinetically will give balances of units in the gel. Based on this information, substructures of the gel could be generated statistically.

3.4.3.4 Free-Radical Crosslinking Polymerizations

Chain growth systems, particularly the free-radical crosslinking polymerizations, occupy a broad area of research and their modeling; comparison with experiment has been a subject of many publications of the order of 10^3 in number. In this section, we will attempt to briefly review methods used for obtaining molecular weight distributions for free-radical and controlled free-radical polymerizations of multiusaturated (mainly bisunsaturated) monomers and their copolymerization. Compared to linear polymerization, pendant double bonds of the once reacted bisunsaturated monomer unit are involved in polymerization. The reactivities of double bonds of the crosslinker may differ or/and be dependent (substitution effect) which makes the sets of population balance equations quite complex. Moreover, in the course of (co)polymerization the kinetic chain length may change with reaction time (conversion). Kinetic chain length corresponds to the size of primary chains, i.e., linear molecules that would result from splitting connections between double bonds of the bisunsaturated monomer. It is evident that a statistical approach to crosslinking free-radical polymerizations is not suitable, unless the process is considered to consist of stages. The use of kinetics method as in the case of linear polymerizations is adequate and results in reaction time/conversion-dependent moments of distribution of degree of polymerization or molecular weights. Important advances have been done through decades mainly by groups of H. Tobita and M. R. P. F. N. Costa and R. P. F. Dias but also a large number of other researchers (cf., sample Refs. [283–292]).

Several problems had to be solved; it was first of all the large number of reaction types—each described by an infinite set of differential equations, and the divergence of the overall second and higher moments at the gel point which prevented to extend this kind of solution into the postgel region. Transformation of each set of differential equations into one differential equation for the generating function and manipulation with the finite set of differential equation for time change to get the moments of the distribution helped to rationalize the computation [240] very much. Perhaps even more interesting is the route of inversion of the generating functions for distribution into the distribution by Laplace transform (cf., e.g., [293, 294]).

Solution of the problem of gel point divergence was offered by Tobita and Hamielec [295] who extended the moment method based on population balances beyond the gel point by calculating the sol and gel fractions by the method of Flory [124].

In Flory's model, the sol fraction w_s is calculated from the assumption that, if a unit is part of sol, all adjoining units are parts of sol as well. In Flory's reasoning, w_s could serve as a weighting factor which was later in the cascade theory called "extinction probability v"; for cross-linking of monodisperse chains of degree of polymerization x) that $w_s = v$ because $w_s = (1 - \alpha + \alpha v)^x$; $v = (1 - \alpha + \alpha v)^{x-1}$.[4] Tobita and Hamielec [295] developed the concept of primary molecules (distribution of linear chains if all crosslinks are severed), and by probabilistic arguments (similar to those used in the recursive theories (cascade or Miller–Macosko) the probability was calculated that a given primary chain is a part of sol.

The "numerical" fractionation technique developed by Teymour and Campbell [296] to characterize the molecular weight distribution not only before but also beyond the gel point has been used in several studies. As the conventional kinetic approach, it considers the time transformations within active and inactive species. Instead of calculating moments of the overall populations, populations are defined that are simular in scale (degree of clustering), and the first generation includes linear chain plus chains added to them. If the first-generation chains keep adding branches and strands of linear polymer, they still are considered to belong to this generation. Transfer of the species to second generation is possible only if two first-generation species connet together, etc. When the reaction occurs between dissimilar generations the product belongs to the higher generation. The mathematical generation consist in calculating moments for each such generation. The generation moments then give the overall moments. Since the species in the generation are finite, also the moments are finite, and the divergence of the second and higher moments at the gel point characteristic for the moment method is overcome, but the gel point is clearly marked by a sharp maximum. Calculations can be then extended beyond the gel point because all considered generations are finite. From generation moments, the molecular weight distribution can be reconstructed. Similarly as in the conventional moment methods, the time evolution of the generation moments is obtained by numerical solution of system of differential equations. At high monomer conversions and high branching content, the reconstruction of the distribution requires a large number of generations in order to reduce approximation errors. Increasing model size and computational load are then necessary.

Of great help in characterization of distributions resulting from branching and crosslinking reaction is the application of software Predici® based on the Galerkin method developed by Wulkow [297, 298]. The polymer fractionation technique has been implemented to overcome the overall moment divergence at the gel point, so that distributions in the sol are available as well as the average branch point (crosslink) density in the gel. An example of application of the software is offered by Nikitin et al. [299].

The inversion method developed especially by Brandolin group is based on the transformation of the infinite population mass balances characterizing the polymer

---

[4] For crosslinking of primary chains of high molecular weight ($x >> 1$), the difference between w_s and v is negligible, and therefore, it is legitimate to use w_s  v. This approximation is not valid for smaller values of x or f.

process into a set of probability generating functions, obtaining a system of equations in which the dependent variable is the generating function transform of the distribution. Pgfs values are obtained by solving this system equations.

After these balances are transformed into the probability generating functions (for a very simple example Eqs. 3.153–3.157), the pgfs are inverted using numerical methods. Discrete points of the full molecular weight distribution result from this numerical inversion. Computation of each point is independent of any others, so that the accuracy of its prediction is unaffected by the number and location of the rest of the points. The number of points to be recovered must be set by the user. A large number will result in a smoother curve, but a larger equation system is needed. Also, the numerical method requires specification of the number of the terms of expansion of the polynomial. This number should be large enough to provide good accuracy in the calculation of MWD, but small enough to avoid producing excessive noise due to error propagation. Stefest's inversion method is often used. Details and application examples are described in refs. [29, 222, 223, 225, 294, 300]. For gelling systems, the method of numerical inverse Laplace transform is applicable as well, as already was shown by Dušek [177] on the example of crosslinking of primary chains. So far, only the evolutions in the pregel state were analyzed, but it seems feasible that the method could be extended beyond it because, generally, the solution of the Smoluchowski equations exists beyond the gel point.

Cyclization has been considered in some models of crosslinking polymerizations, their intensity being sometimes characterized by a single constant and limited only to the just growing primary chain. So far, the most comprehesive treatment of cyclization can be found in the work of Sarmoria et al. [240] who considered also secondary cyclization involving chains attached to the growing chains as well as statistical weightings according to the ring size. References to the previous treatments of cyclization in free-radical copolymerizations can be found in this source.

One of the other possible ways to simulate distribution in the pre- and postgel states is the Monte Carlo simulation working with finite systems which offers molecular weight distributions in the pregel and postgel stages where the largest molecule becomes by several orders of magnitude larger than the other molecules and represents the gel. At the present state of conventional hardware, there is no problem with the necessary memory requirements. As mentioned in the introduction, there are available several 3D computer simulation methods (like the bond fluctuation model) but their review is not subject of this chapter.

3.4.3.5 Special Features of Free-Radical Crosslinking Polymerizations

All models discussed above were based on the mean-field concept assuming that the rates were proportional to concentrations averaged over the whole reaction volume. In chain growth systems, this not always a good approximation. The free-radical crosslinking polymerization is a striking example. The is unequivocal experimental evidence of inhomogeneous state of the system when the concentration of crosslinker and the molecular weight of primary chains exceed certain limits (for orientation, very roughly about 5 mol.% crosslinker, and DP of primary chains about 50). The main evidence of this special reaction course is (1) much lower molecular weight averages of the pregel polymers than would correspond to ring-free case (factors 0.01–0.1); (2) a shift of the gel point conversion by a factor up to 10^2 compared to a ring-free case; (3) excessive amount of pendant double bonds remaining at high conversion; (4) loss of mobility of considerable part of units already before the gel point (NMR studies); (5) unexpected composition of the pregel polymers—at very low conversions, a large part of the bisunsaturated monomer units incorporated in the polymer are present in the form with both reacted double bonds; when the fraction of crosslinker in the feed increases, its fraction in the polymer reaches a certain upper limit, and all new units are incorporated as those with pendant double bond; (6) the intrinsic viscosity of (optically clean) pregel solutions reminds that of solutions of microgels. These conclusions were based on first experimental studies described mainly in Refs. [301–304] and supported by other studies such as Refs. [305–311] and reviewed in [312–315]. The NMR technique based in broadening of fine resolution signals has become a general technique in the case of microphase separation or vitrification.

Based on these findings, the mechanism of network formation by free-radical copolymerization at higher concentrations of the crosslinking monomer is shown in Figs. 3.53 and 3.54. The primary reason of inhomogeneous reaction course is cyclization. When polymerization starts, the macroradicals grow in dilute solution of monomers, and there are available for reaction very few pendant double bonds of other growing or dead chains, whereas the chance of reaction with pendant double bond on the same growing macroradical does not depend on dilution. At somewhat higher conversion, also ring closing involving pendant double bonds on attached chains (secondary cyclization) must be considered—some of these double bonds may be connected even by a shorter path of bonds than those pending from the growing chain (Fig. 3.53).

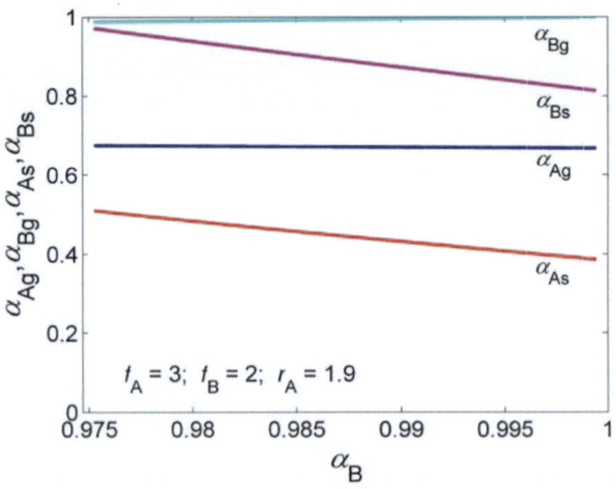

Fig. 3.53 Alternating copolymer A3 + B2; off-stoichiometric case, $r_A = [A]_0/[B]_0 = 1.9$. Conversions of groups A and B in sol and gel as a function of conversion: α_{As}, α_{Ag}, α_{Bs}, α_{Bg} conversions of groups A and B in sol and gel, respectively

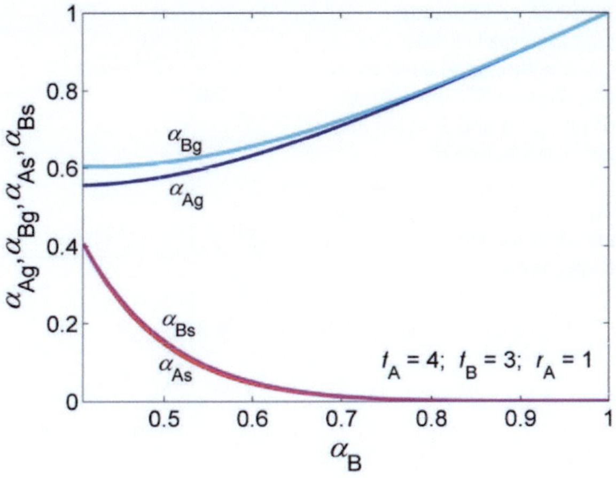

Fig. 3.54 Alternating copolymer A4 + B3; stoichiometric case, $r_A = [A]_0/[B]_0 = 1$. Conversions of groups A and B in sol and gel as a function of conversion: α_{As}, α_{Ag}, α_{Bs}, α_{Bg} conversions of groups A and B in sol and gel, respectively

Primary and secondary intramolecular crosslinking in early stages (still before the gel point is reached) give rise to formation of compact structures which sterically hinders contacts with double bonds of other such structures and thermodynamically repels them from each other. The units inside such compact molecules loose mobility and do not show up in the finer resolution NMR spectra. In this way, the relative intensity of intermolecular reactions further decreases, and the gel point is shifted more than would correspond to loss of a part of bonds in simple ring forming. Schematically, the situation is illustrated by Fig. 3.54.

It seems that the structural inhomogeneity reaches its maximum around the gel point where the sytems are most sensitive to thermodynamic phase separation. If phase separation does not occur and as conversion increases, the interticial space of microgel-like molecules is filled up with polymrizing monomers and the final network may seem homogenous. Figure 3.55 tries to visualize this situation somewhere on the midway between gel point and the final state.

Numerous equilibrium elasticity meaurements show that at conversions far from the gel point the experimentally determined concentration of EANCs increases linearly with increasing concentration of crosslinker in the feed, but the increases of modulus are smaller than the expected one (e.g., about 50% for polymethacrylates with small dimethacrylate crosslinkers).

No good theoretical model is available to cover this range of crosslinking polymerization where the mean-field element—proportionality to groups average concentrations—is an oversimplification. The group "effective reactivity" depends on molecule size and symmetry, and molecules overlap. This behavior can be somewhat simulated by the product Kernel of Smoluchowski equation with variable exponent (>1) discussed in Sect 3.4.3.3 but this is only a phenomenological, not structural approach.

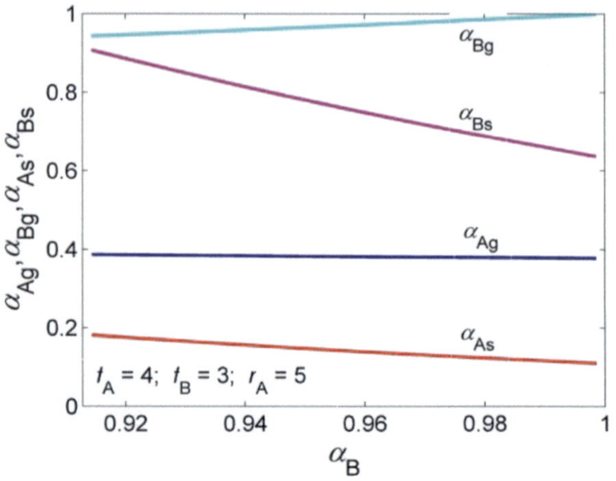

Fig. 3.55 Alternating copolymer A4 + B3; off-stoichiometric case, $r_A = [A]_0/[B]_0 = 5$. Conversions of groups A and B in sol and gel as a function of conversion: α_{As}, α_{Ag}, α_{Bs}, α_{Bg} conversions of groups A and B in sol and gel, respectively

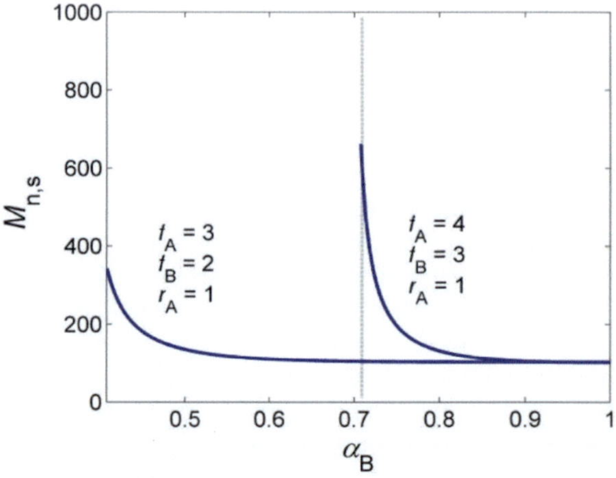

Fig. 3.56 Dependence of number-average molecular weight of the sol, $M_{n,s}$, on conversion for two stoichiometric systems of composition indicated; molecular weight of monomer units, $M_{0A} = 120$ and $M_{0B} = 90$ Dalton

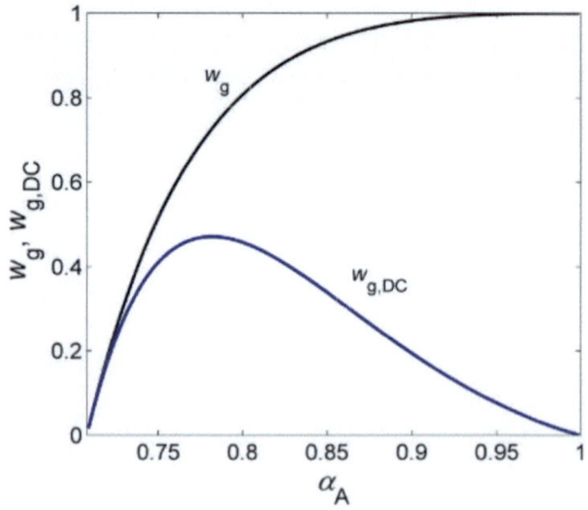

Fig. 3.57 Dependence of gel fraction w_g and weight fraction of dangling chains on conversion. System: A3 + B2, $r_A = 1$

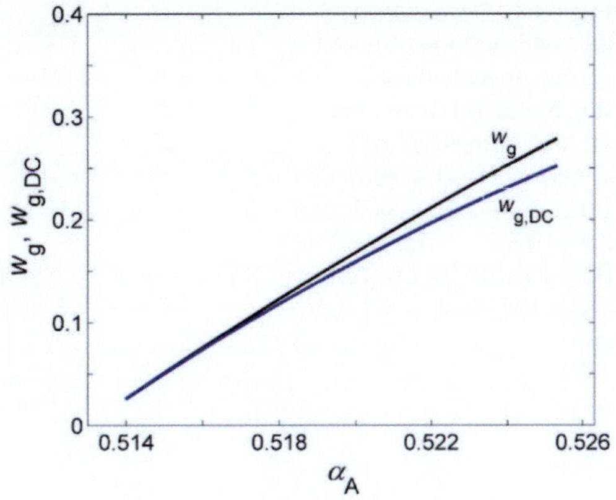

Fig. 3.58 Dependence of gel fraction w_g and weight fraction of dangling chains on conversion. System: A3 + B2, $r_A = 1.9$

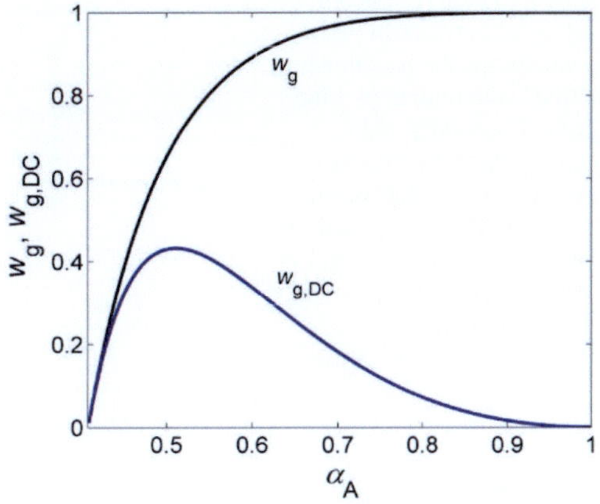

Fig. 3.59 Dependence of gel fraction w_g and weight fraction of dangling chains on conversion. System: A4 + B3, $r_A = 1$

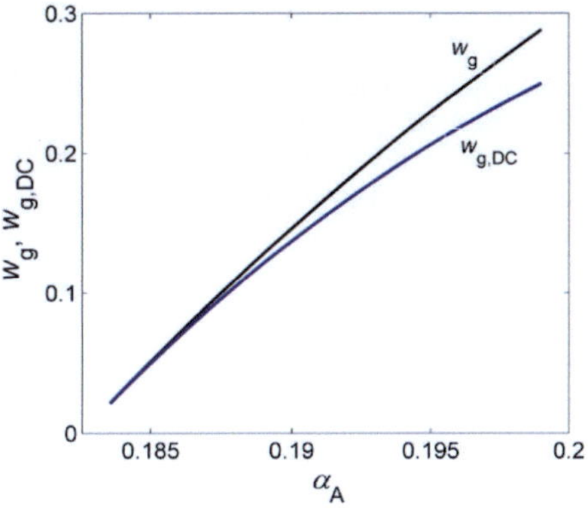

Fig. 3.60 Dependence of gel fraction w_g and weight fraction of dangling chains on conversion. System: A4 + B3, $r_A = 5$

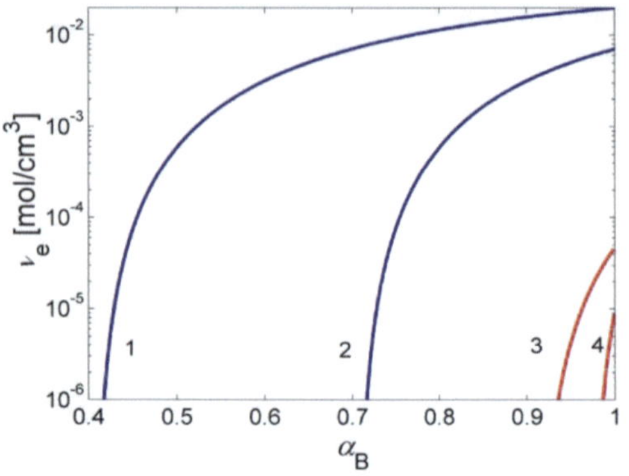

Fig. 3.61 Dependence of the concentration of EANCs, ν_e, on conversion for stoichiometric (blue) and off-stoichiometric (red) systems: (1) A4 + B3, $r_A = 1$, (2) A3 + B2, $r_A = 1$, (3) A4 + B3, $r_A = 5$, (4) A3 + B2, $r_A = 1.9$; $M_A = 120$, $M_B = 90$ g/mol, $\rho = 1.2$ g/cm^3

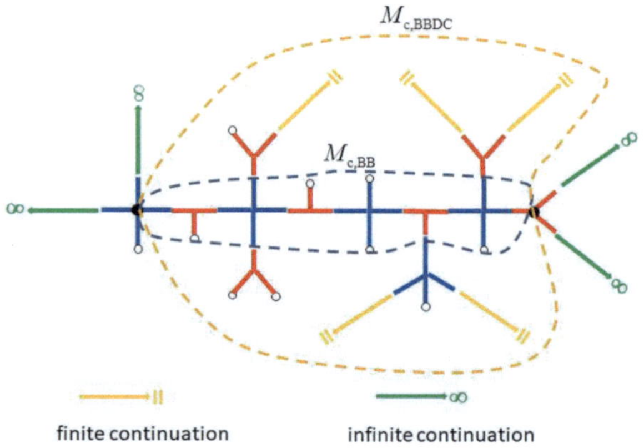

Fig. 3.62 Elastically active network chain (EANC) in a A4 + B3 network in two perceptions: the backbone EANC is delimited by blue dashed boundary and characterized by (number average) molecular weight $M_{c,BB}$; EANC composed by backbone units and dangling chains is delimited by orange dashed boundary and characterized by $M_{c,BBDC}$

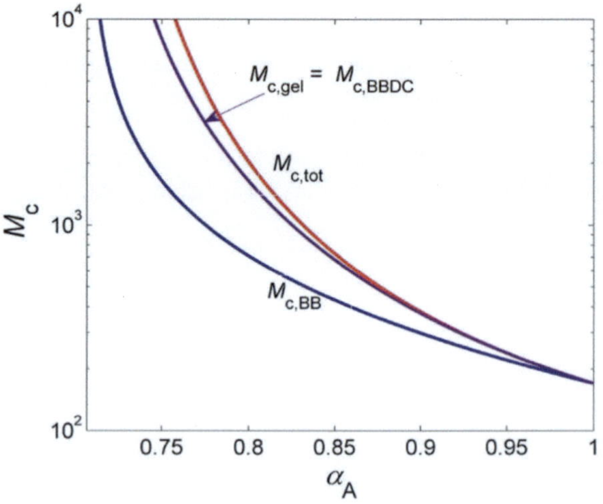

Fig. 3.63 Dependence of molecular weight of chains between crosslinks on conversion. $M_{c,tot}$—related to the whole system (Eq. 3.170); $M_{c,gel}$—related to the gel (Eq. 3.136), $M_{c,BB}$, related to the backbone units (Eq. 3.172), related to the backbone units including the dangling chains, $M_{c,BBDC}$ (Eq. 3.173). System A3 + B2, $r_A = 1$

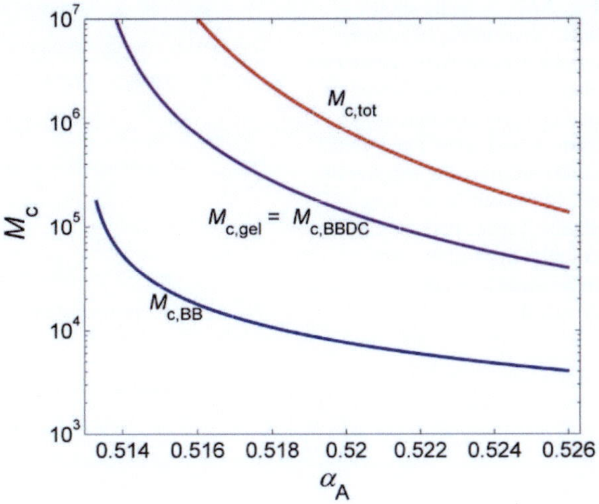

Fig. 3.64 Dependence of molecular weight of chains between crosslinks on conversion. $M_{c,tot}$—related to the whole system (Eq. 3.170); $M_{c,gel}$ related to the gel (Eq. 3.136), $M_{c,BB}$, related to the backbone units (Eq. 3.172), related to the backbone units including the dangling chains, $M_{c,BBDC}$ (Eq. 173). System A3 + B2, $r_A = 1.9$

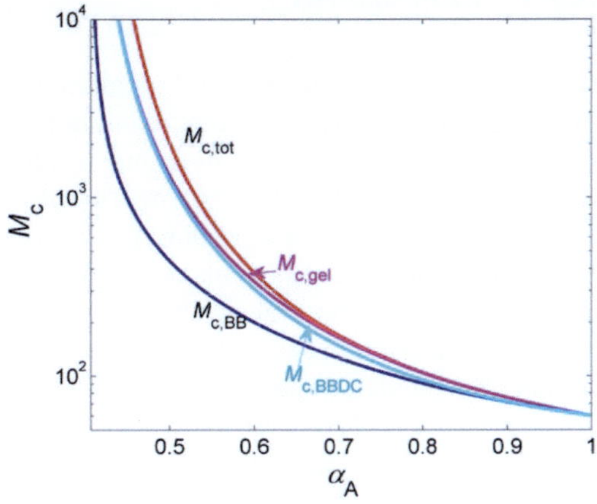

Fig. 3.65 Dependence of molecular weight of chains between crosslinks on conversion. $M_{c,tot}$—related to the whole system (Eq. 3.170); $M_{c,gel}$ related to the gel (Eq. 3.170), $M_{c,BB}$, related to the backbone units (Eq. 3.172), related to the backbone units including the dangling chains, $M_{c,BBDC}$ (Eq. 3.138). System A4 + B3, $r_A = 1$

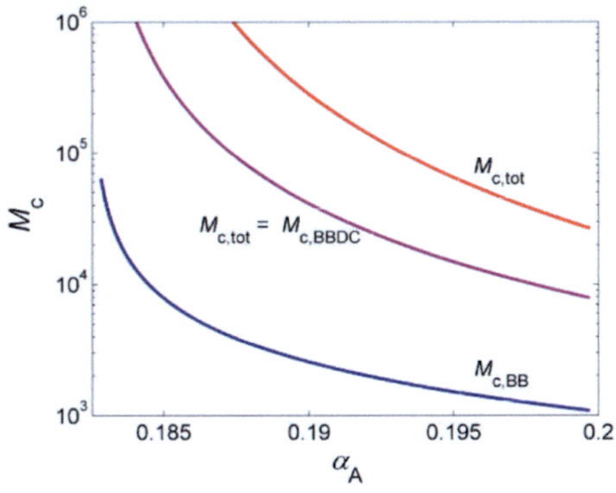

Fig. 3.66 Dependence of molecular weight of chains between crosslinks on conversion. $M_{c,tot}$—related to the whole system (Eq. 3.170); $M_{c,gel}$ related to the gel (Eq. 3.136), $M_{c,BB}$, related to the backbone units (Eq. 3.172), related to the backbone units including the dangling chains, $M_{c,BBDC}$ (Eq. 3.173). System A4 + B3, $r_A = 5$

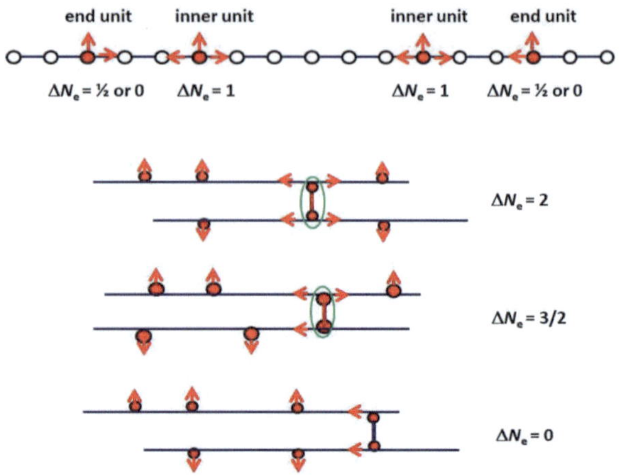

Fig. 3.67 Scheme shows how crosslinking of primary chains. Upper part: a primary chain showing crosslinked units having continuation to infinity (red arrow) and their contribution to the number of elastically active network chains. Lower part: possible combinations of end and inner units to form a crosslink

3.4.4 Conclusions

Several groups of modeling and simulation methods exist to describe branching, gelation, and polymer gel structure buildup. They differ in the depth of taking into account (a) chemical reactivities and reaction rules by which covalent bonds or their sequences are formed, (b) integrity of once formed structures for irreversible reactions or non-equilibrium states, (c) space filling (excluded volume) and physical interactions among components of reacting systems. The purpose of application and analysis of the theories can also be different. It can be applied in order to understand well general features of gelation processes and behavior in the critical region and to make long-range extrapolations possible. Aternatively, the theories are applied to particular systems in order to understand changes of physical and chemical properties and to make predictions of the effect of changes in composition and conditions of the formation process. In this case, the goal of application of the theory is optimization of technology and product properties wherever branched and crosslinked systems are applied. This applied use of theory is very important because the crosslinked material is very complex, and only a few physical methods are available for its characterization mainly. For leading industrial companies, for instance, in the protective organic coatings area, an input from the branching theory into research of new special binders is indispensable.

The statistical methods of assemblage of branched and crosslinked structures are relatively simple at least in its fsse (first-shell substitution effect version). The building units occurring in different reaction states are generated by chemical kinetics (or equilibrium) equations obeying the mass action (Guldberg–Waage) law and based on reaction mechanism. This approach is close to polymer chemist or technologist, or polymer physical chemist who can get direct answers in a language they understand. At present, the statistical methods offer the most detailed information on the gel in terms of the amount and size of various substructures like dangling chains, elastically

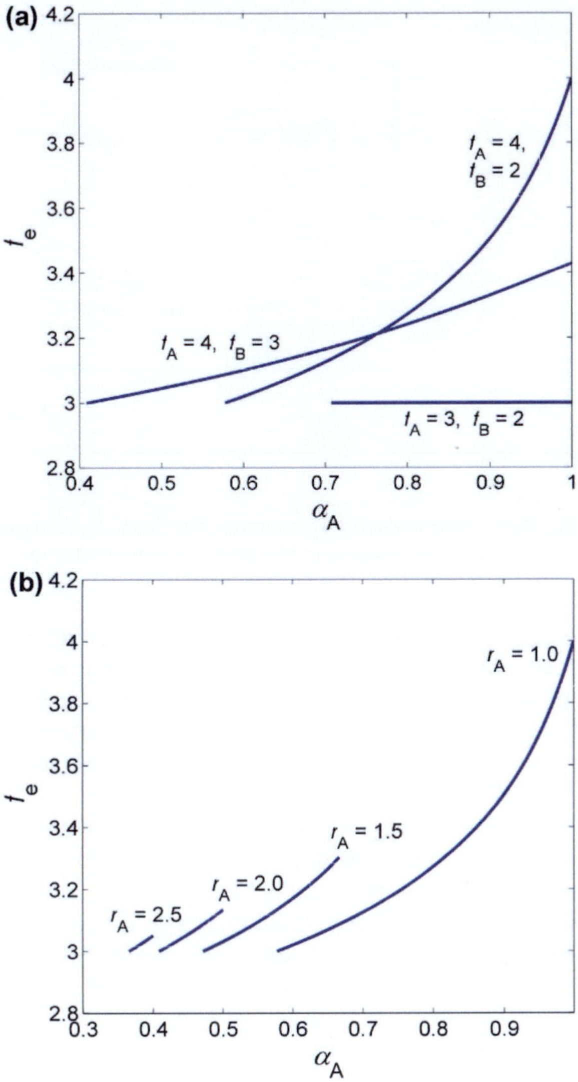

Fig. 3.68 **a** Dependence of effective functionality f_e of elastically active crosslinks in networks from $A_{fA} + B_{fB}$ monomers in stoichiometric ratio $r_A = 1$. **b** Dependence of effective functionality *fe* of elastically active crosslinks in networks from $A_{fA} + B_{fB}$ monomers, $f_A = 4, f_B = 2$; various molar ratios of functional groups r_A indicated

active network chains, or status of crosslinks. Also, statistical generation can deal with formation of chemically different clusters in a chemically different gel environment, or with networks formed in several stages by labeling the bonds of the same type by the stage number. Disadvantage of statistical assemblage is its disregard of long-range correlations in (reaction) time and space. Long-range time correlations involve the effect history of bond formation on formation of a new bond. As a long-range space interactions such problems like formation of cycles, or the existence of overcrowding in the gels should be considered.. For many systems of practical importance, the effects of such neglects are weak—within experimental error—and the statistical

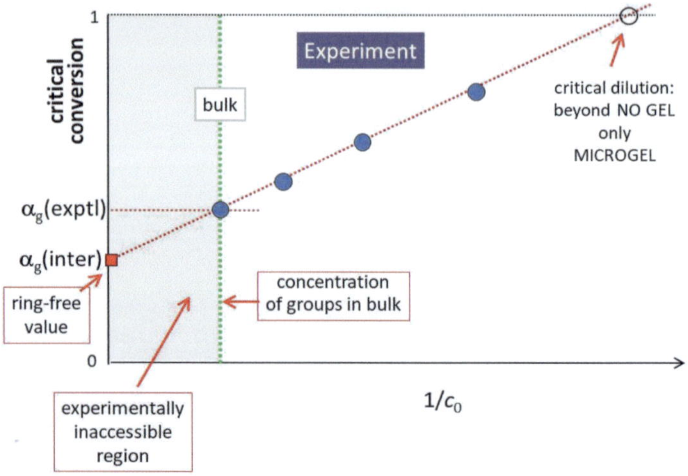

Fig. 3.69 Experimental determination of the fraction of bonds closing cycles from the dependence of the gel point conversion on reciprocal concentration of the system, $1/c_0$

methods can be used. Otherwise, perturbation methods can be used, like spanning-tree approximation for dealing with cyclization, or the combined method where the history sensitive processes are simulated kinetically and the history insensitive by statistical assemblage. An important contribution to modeling of branching in network buildup has been made recently by Piet Iedema's group in Amsterdam (cf., e.g., refs. [316, 317]).

There is a large group of systems where these long-range correlations and possibly combined with physical interaction of short range play important roles and where the statistical methods cannot be applied or are a poor approximation. It is first of all the chain growth buildup, where active species are generated by initiation and propagation, their length being modified by various chain transfer and termination mechanisms. The time correlations are related to changing of the content of crosslinker in primary chains and their degrees of polymerization as the reaction time (conversion) proceeds. The kinetics treatment of the chain growth systems is well elaborated and allows us to characterize the degree of polymerization / molecular weight distributions. Also, passage through the gel point, where the second and higher moments diverge, has been solved, and the degree of polymerization distribution of sol molecules has been characterized. The use of the generating function formalism in this area is increasing, and it has been of great help in treating more complex cases and obtaining of degree of polymerization by pgfs inversion. Ways of characterizing the gel component, when kinetics methods are used for its generation, are still to be developed, and cyclization should be made dependent on structure and properties of the polymerizing system. The long-range time correlations are also encountered in step growth system in the case of substitution effect but the exact results deviate from those obtained from the first-shell substitution effect (fsse) statistical treatment

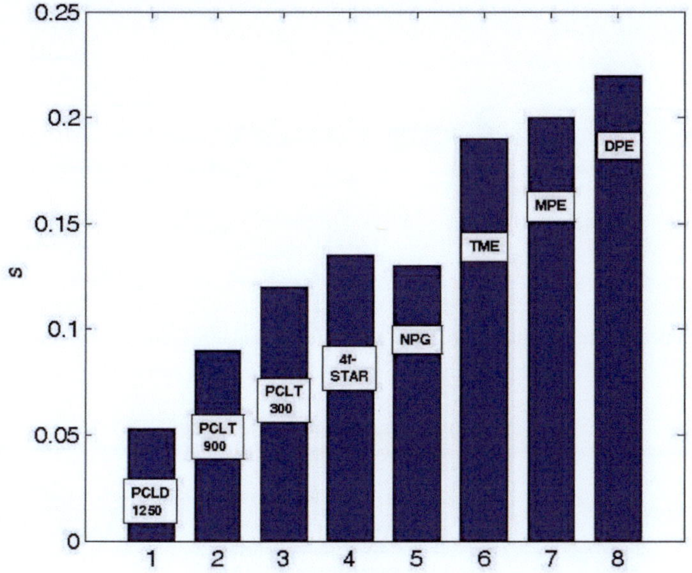

a)

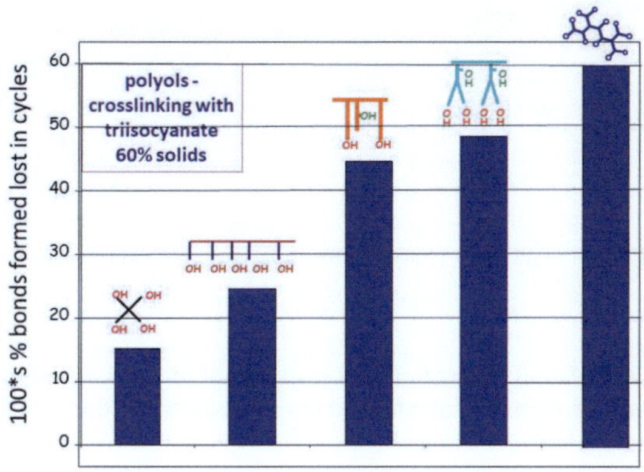

b)

Fig. 3.70 Fractions of bonds wasted in cycles at the gel point, *s*, (Eq. 3.139), determined by the dilution method. Figure 3.45a: reaction of triisocyanate (HDI trimer) with various polyols = polycaprolacton diols (CLD) and triols (CLT) of molecular weight indicated, tetrafunctional star (4f-STAR) based on tetrahydrophthalic anhydride—ethylene oxide chain extended pentaerythritol; data from [198, 255]; Fig. 3.45b: comparison of polyurethane formation from triisocyanate and polyols—tetrafunctional star, acrylic copolymer, acrylic copolymers of various architectures, and OH-functional hyperbranched polymer (poly(dimethylol propionic acid). Data not published before, systems partly described [199, 200, 257, 258]

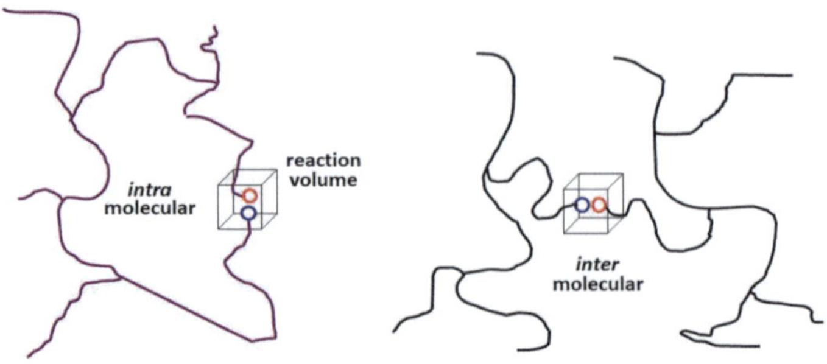

Fig. 3.71 Schematic representation of inter and intramolecular reaction for two functional groups A and B

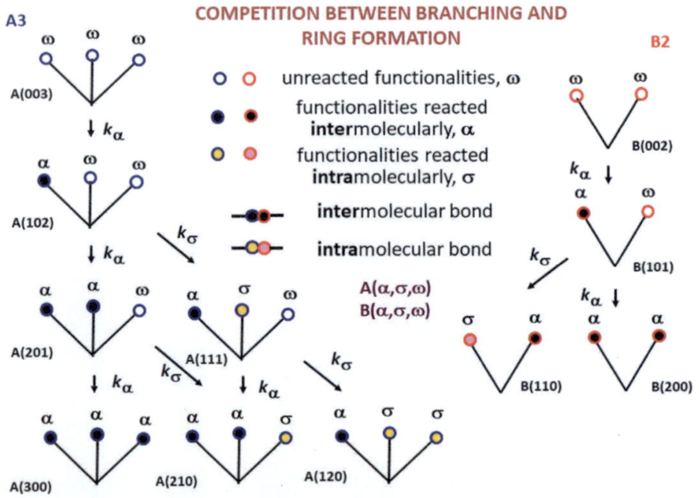

Fig. 3.72 Reaction scheme for evolution of distribution of building units having various numbers of functional groups participating in inter and intramolecular reactions and having not reacted. Case of alternating reaction between A3 and B2

only for some rather exotic combinations of rate constant. The kinetics methods are a special case of Smoluchowski aggregation equations. The aggregation rate is also proportional to average concentrations, but their Kernel (rate "constant") depends on specific properties of the components, for instance, their diffusivity or contact sites availability. For solution of processes as formulated by the Smoluchowski equation, Monte Carlo simulation in sufficiently large system is always at hand.

Computer simulation in 3D space is a non-mean-field method which can handle the long-range correlations as they exist in 3D space. It also can take into account physical interactions between segments and make predictions of some physical properties. The main problem is to find the most-probable state while respecting existing covalent and non-covalent interactions and structure connectivity. Sufficient mobility (time)

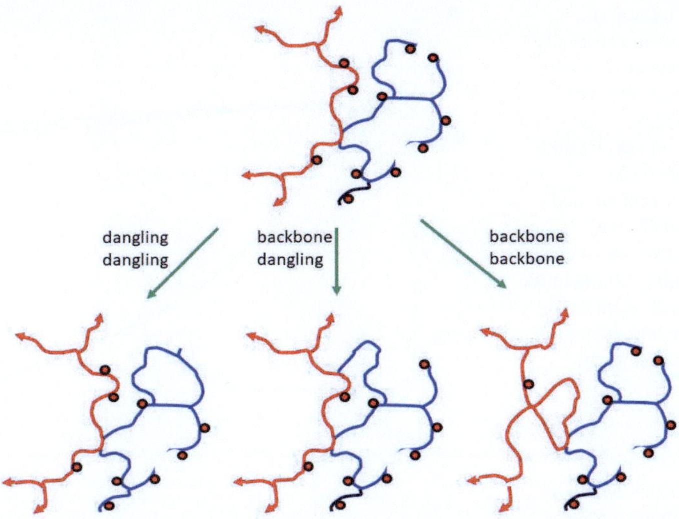

Fig. 3.73 Formation of loops by reaction of groups residing both on the same dangling chain, or groups one on a dangling chain and the other on its backbone EANC, or two groups both on the same backbone EANC; the backbone-dangling connection shown here corresponds to a double chain connection

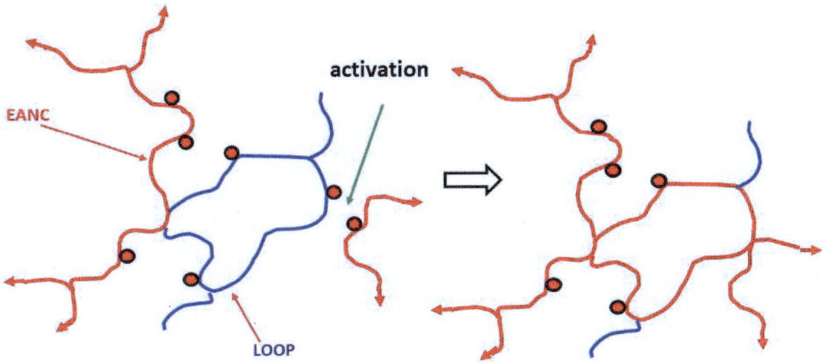

Fig. 3.74 Activation of loop by reaction of one of its functional groups with a functional group of an elastically active network chain

should be guaranteed for the rearrangement of the system to reach its minimum energy. This also means to check that the direction of rearrangements is correct (toward minimum), and the minimum is the deepest and not a local one. The structure and the minimum energy conditions change with formation of every new bond. Such simulation is a demanding task. The bond fluctuation model seems to be a good basis for such development. Multiscale molecular dynamics simulation methods mentioned in the introduction on which the network structure is mapped are another approach to simulation; however, mapping of graph-like structures onto a #D lattice

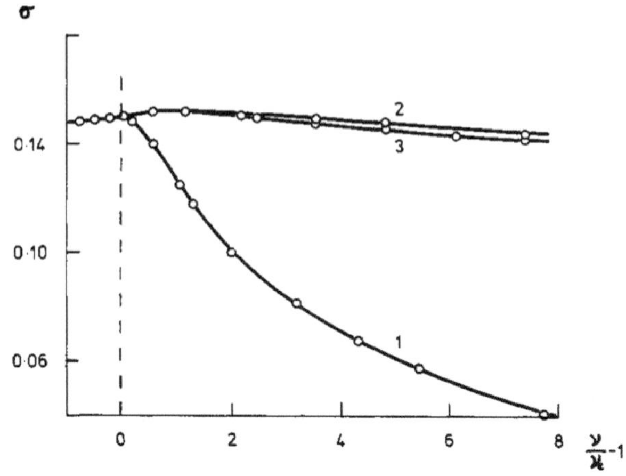

Fig. 3.75 Fraction of crosslinks lost in elastically inactive cycles in dependence on relative crosslink density, $(\nu - \nu_e)/\nu_e$. (1) cyclization only within sol; (2) cyclization within sol and gel; (3) as for (2) but including activation of cycle, bulk state, other details in the original source. Reprinted from [263] with permission of American Chemical Society

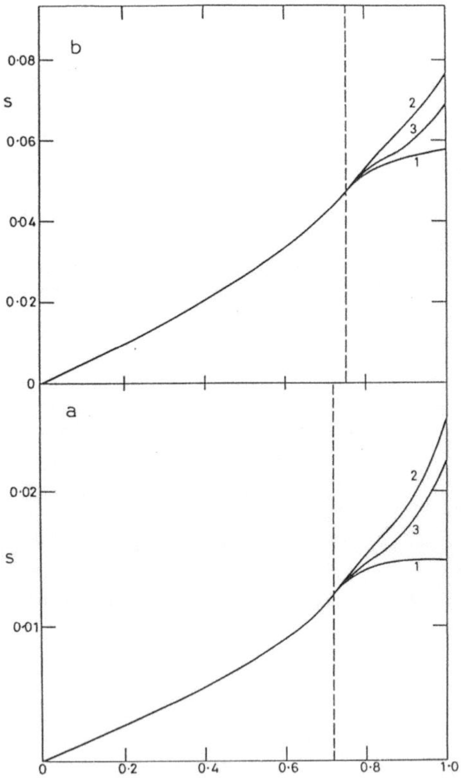

Fig. 3.76 Dependence of fraction of bonds lost in cycles (elastically inactive cycles beyond the gel point), s, on conversion for alternating addition of A3 + B2 monomers. The Λ parameter was calculated for $B =$ poly(oxypropylene) diol of $M_B = 1800$. (1) Contribution to cyclization involving only v-type bonds (bonds with finite continuation), no activation, (2) contribution to cyclization by v-type and $(1 - v)$-type bonds, (3) as (2) including consideration of the activation process. Figure 3.50a bulk system, Fig. 3.50b 25% solution. Reprinted from [175] with permission

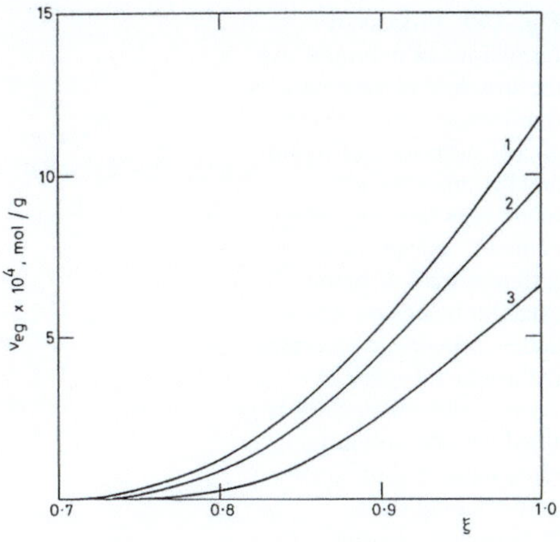

Fig. 3.77 Dependence of the concentration of EANCin the gel, ν_{eg}, on conversion, $\xi(=\alpha+\sigma)$, for alternating addition of $A3+B2$ monomers. (1) ring-free system, (2) including loop formation and activation, bulk system, (3) as (2) diluted system (to 25 vol.%). Reprinted from [175] with permission

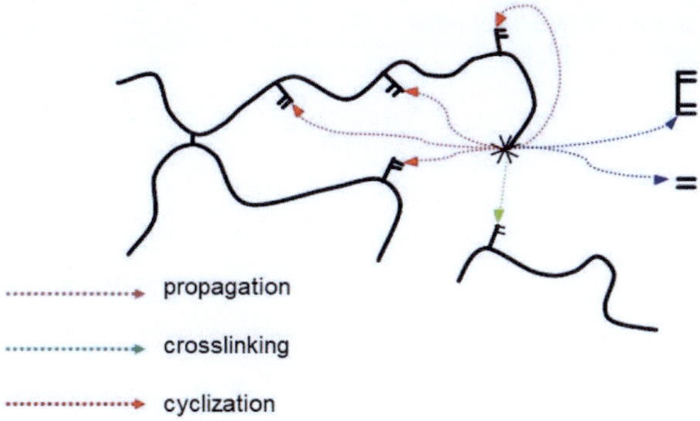

Fig. 3.78 Schematic presentation of propagation, crosslinking, and cyclization reactions for chain crosslinking copolymerization of a bisunsaturated and monounsaturated monomer

is somewhat artificial. Perhaps, in further development of branching and crosslinking theories, physical interactions affecting bond formation should be taken into account like in the lattice model of polymerization by Dudowicz et al. [317, 318, 318, 319, 321–330], and such approach will address polar systems governed by hydrogen bonding and hydrophobic associations. General comments to modeling of polymer systems can be helpful in this endeavor [320].

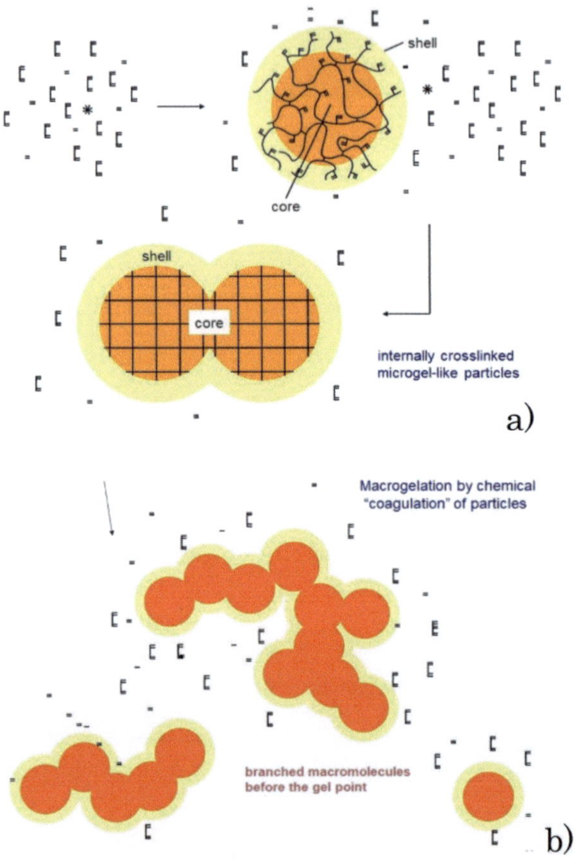

Fig. 3.79 Process of formations of microgel and the structure in the course of free-radical cross-linking copolymerization of mono- and bis-unsaturated monomers (in reality the compact structurers are of different size). If phase separation does not occur and as conversion increases, the interticial space of microgel-like molecules is filled up with polymrizing monomers, and the final network may seem homogenous. Figure 3.55 tries to visualize this situation somewhere on the midway between gel point and the final state.

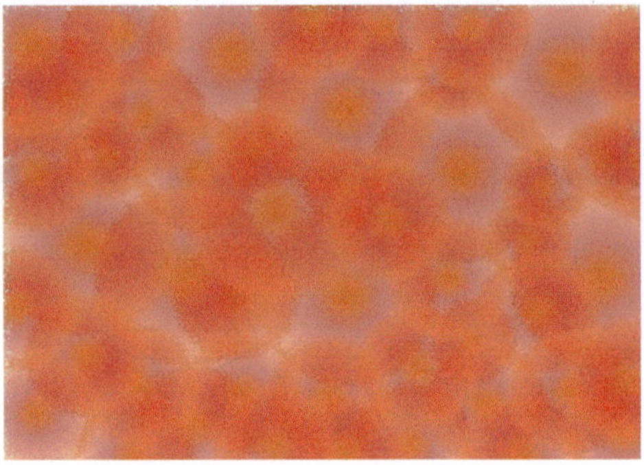

Fig. 3.80 Situation (overlap) at some distance from the gel point

References

1. IUPAC Recommendations 1996), Pure Appl. Chem. **68**(12), 2287–2311 (1996)
2. J.D. Ferry, *Viscoelastic Properties of Polymers*, 3rd edn. (Wiley, New York, 1980)
3. M. Doi, S.F. Edwards, *The Theory of Polymer Dynamics* (Clarenton, Oxford, 1986)
4. P.J. Flory, Disc. Faraday Soc. **57**, 7 (1974)
5. M. Gordon, Proc. Royal Soc. London A **268**, 240 (1962)
6. P.J. Flory, *Principles of Polymer Chemistry, Chapter 9 Ithaca* (Cornell University Press, New York, 1953)
7. P.G. de Gennes, *Scaling Concepts in Polymer Physics* (Cornell University Press, New York, 1979)
8. D. Stauffer, J. Chem. Soc. Faraday Trans. II **72**, 1354 (1976)
9. C.H. Bamford et al., *The Kinetics of Vinyl Polymerization by Radical Mechanisms* (Butterworths Scientific Publications Ltd., London, 1958)
10. c.f., ATRP involves organohalide and a metal halide. A metal depending on the different oxidation states abstracts an electron from the organohalide, creating a radical that then starts free radical polymerization. After initiation and propagation, the radical on the active chain end is reversibly terminated (with the halide) by reacting with the catalyst in its higher oxidation state. This is a redox process established a kind of an equilibrium between dormant (polymer-halide) and growing chains. The equilibrium is designed to heavily favor the dormant state, which effectively reduces the radical concentration to a sufficiently low level to limit bimolecular coupling. Polymerization is usually carried out at elevated temperatures (60–120 °C). Therefore, IUPAC recommends to use the term "reversible-deactivation radical polymerization" instead of "controlled" radical or "living" radical polymerization (IUPAC Recommendations 2010), Pure Appl. Chem. **82**(2), 483–491.)
11. N. Ide, T. Fukuda, Macromolecules **30**, 4268–4271 (1997)
12. K. Matyjaszewski, J. Xia, An excellent review written by the pioneer in the field. Atom Transf. Rad. Polym. Chem. Rev. **101**(9), 2921–2990 (2001)
13. N. Hu, G. Liu, J. Macromol. Sci. Pure Appl. Chem. **A32**, 949 (1995)
14. W.H. Stockmayer, J. Chem. Soc. **63**, 3083 (1943)
15. L.C. Case, J. Polym. Sci. **26**, 333 (1957)
16. H. Eyring, The activated complex in chemical reactions. J. Chem. Phys. **3**, 107–115 (1935)
17. E. Rabinowitch, Collision, co-ordination, diffusion and reaction velocity in condensed systems. Trans. Faraday Soc. **33**, 1225–1233 (1937)
18. K. Dušek, M. Dušková-Smrčková, Network structure formation during crosslinking of organic coating systems. Prog. Polym. Sci. **25**, 1215–1260 (2000)
19. M. Adam, Growth process of polymers near the gelation threshold. Macromol. Symposia **45**, 1–9 (1991)
20. K. Dušek, Diffusion control in the kinetics of cross-linking. Polym. Gels Networks **4**, 383–404 (1996). https://doi.org/10.1016/S0966-7822(97)89914-5
21. M.L. Williams, R.F. Landel, J.D. Ferry, The temperature dependence of relaxation mechanisms in amorphous polymers and other glass-forming liquids. J. Am. Chem. Soc. **77**, 3701–3706 (1955)
22. A.K. Doolittle, Studies in Newtonian flow. II. The dependence of the viscosity of liquids on free-space. J. Appl. Phys. **22**, 1471–1475 (1951). https://doi.org/10.1063/1.1699894
23. D. Turnbull, M.H. Cohen, Free-volume model of the amorphous phase: glass transition. J. Chem. Phys. **34**, 120 (1961). https://doi.org/10.1063/1.1731549
24. G. Adam, J.H. Gibbs, J. Chem. Phys. **43**, 139–148 (1965)
25. F.G.A. Huguenin, M.T. Klein, Intrinsic and transport-limited epoxyamine cure kinetics. Ind. Eng. Chem. Prod. Res. Dev. **24**, 166–171 (1985)
26. D.F. Rohr, M.T. Klein, Modeling diffusion and reaction in epoxy-amine linear polymerization kinetics. Ind. Eng. Chem. Prod. Res. Dev. **27**, 1361–1366 (1988)
27. D.F. Rohr, M.T. Klein, Ind. Eng. Chem. Prod. Res. Dev. **29**, 1210–1218 (1990)

28. I. Fraga, S. Montserrat, J.M. Hutchinson, Vitrification during the isothermal cure of thermosets: comparison of theoretical simulations with temperature-modulated DSC and dielectric analysis. Macromol. Chem. Phys. **209**, 2003–2011 (2008)
29. G. Van Assche, S. Swier, B. Van Mele, Modeling and experimental verification of the kinetics of reacting polymer systems. Thermochim. Acta **388**, 327–341 (2002)
30. S. Corezzi, D. Fioretto, G. Santucci, J.M. Kenny, Modeling diffusion-control in the cure kinetics of epoxy-amine thermoset resins: An approach based on configurational entropy. Polymer **51**, 5833–5845 (2010)
31. J.-P. Pascault, H. Sautereau, J. Verdu, R.J.J. Williams, *Thermosetting polymers* (M. Dekker, New York, 2002)
32. Y. Meng, S.L. Simon, Relation between mobility factor and diffusion factor for thermoset cure. Thermochim. Acta **437**, 179–189 (2005)
33. M. Javdanitehrana, D.C. Berg, E. Duemichen, G. Ziegmann, An iterative approach for isothermal curing kinetics modelling of an epoxy resin system. Thermochim. Acta **623**, 72–79 (2016)
34. S.L. Simon, J.K. Gillham, Cure kinetics of a thermosetting liquid dicyanate ester monomer/ high-Tg polycyanurate material. J. Appl. Polym. Sci. **47**, 461–485 (1993)
35. K. Dušek, I. Havlíček, Diffusion-controlled kinetics of crosslinking. Prog. Org. Coat. **22**, 145–159 (1993)
36. A.M. Rabea, S. Zhu, Modeling the influence of diffusion-controlled reactions and residual termination and deactivation on the rate and control of bulk atrp at high conversions. Polymers **7**, 819–835 (2015)
37. D.R. Dhooge, M.F. Reyniers, G.B. Marin, The crucial role of diffusional limitations in controlled radical polymerization. Macromol. React. Eng. **7**, 362–379 (2013)
38. A. Lee, G.B. McKenna, Effect of crosslink density on physical aging of epoxy networks. Polymer **29**, 1812–1817 (1988)
39. G.M. Odegard, A. Bandyopadhyay, Physical aging of epoxy polymers and their composites. J. Polym. Sci. B Polym. Phys. **49**, 1695–1716 (2011)
40. L.G. Lovell, K.A. Berchtold, J.E. Elliot, H. Lu, C.N. Bowman, Understanding the kinetics and network formation of dimethacrylate dental resin. Polym. Adv. Technol. **12**, 335–345 (2001)
41. G.A. O'Neil, M.B. Wisnudel, J.M. Torkelson, A critical experimental examination of the gel effect in free radical polymerization: Do entanglements cause autoacceleration? Macromolecules **29**, 7477–7490 (1996)
42. A. Fa1di, M. Tirrell, T.P. Lodge, Comparison between polymer diffusion and chain radical termination kinetics: the importance of polydispersity macromolecules **27**, 4176–4183 (1994)
43. C. Barner-Kowollik, G.T. Russell, Chain-length-dependent termination in radical polymerization: Subtle revolution in tackling a long-standing challenge. Prog. Polym. Sci. **34**, 1211–1259 (2009)
44. G.B. Smith, G.T. Russell, J.P.A. Heuts, Termination in dilute-solution free-radical polymerization: a composite model. Macromol. Theory Simul. **12**, 299–314 (2003)
45. G. Johnston-Hall, M.J. Monteiro, Bimolecular radical termination: new perspectives and insights. J. Polym. Sci. A Polym. Chem. **46**, 3155–3173 (2008)
46. J.C. Bevington, R.G.W. Norish, J. Chem. Soc. **771**, 11342 (1948)
47. J.C. Bevington, R.G.W. Norish, J. Chem. Soc. **11**, 482 (1949)
48. R. Kishi, M. Shishido, S. Tazuke, Macromolecules **23**, 3779 (1990)
49. Y. Osada, R. Kawamura, K. Sano, *Hydrogels of Cytoskeletal Proteins* (Springer, New York, 2016)
50. For example, V.A. Kabanov, A.B. Zezin, Macromol. Chem. **6**, 259 (1984)
51. A.B. Zezin, V.A. Kabanov, Macromol. Chem. Macromol. Symp. **26**, 249 (1989)
52. V.A. Kabanov, Polym. Sci. **36**, 143 (1994)
53. I.M. Papisov, V.A. Kabanov, Dokl. Akad. Nauk SSSR **199**, 1364 (1971)

54. Agar was discovered in Japan around 1658 by Japanese innkeeper, Mino Tarōzaemon. People said he discarded some seaweed soup and next day he found that it solidified later after a winter night's freezing. Agar gel is called in Japanese *kanten*, coined from the phrase *kan-zarashi tokoroten* or "cold-exposed agar" (M.J. Zimbro, D.A. Power, S.M. Miller, G.E. Wilson, J.A. Johnson, Difco and BBL Manual (PDF), 2nd edn. (Becton Dickinson and Company)
55. Agarose is the predominant component of agar and is a linear polymer consisting of the repeating monomeric unit of agarobiose (agarobiose is a disaccharide made of D-galactose and 3,6-anhydro-L-galactopyranose). Agaropectin is a mixture of alternating units of D-galactose and L-galactose which contain a sulfate and pyruvate groups as side groups
56. K. Dušek, M. Netopilík, P. Kratochvíl, Nonuniformities of distributions of molecular weights of grafted polymers. Macromolecules **45**, 3240–3246 (2012). https://doi.org/10.1021/ma3000187
57. K. Dušek, M. Dušková-Smrčková, Modeling of polymer network formation from preformed precursors. Macromol. React. Eng. **6**, 426–445 (2012). https://doi.org/10.1002/mren.201200028
58. H. Galina, J.B. Lechowicz, Mean-field kinetic modeling of polymerization: the Smoluchowski coagulation equation. Adv. Polym. Sci. **137**, 135–172 (1998). https://doi.org/10.1007/3-540-69685-7_4
59. F. Family, D.P. Landau, *Kinetics of aggregation and gelation* (Elsevier, Amsterdam, 1984)
60. C. Gauer, Z. Jia, H. Wu, M. Morbidelli, Aggregation kinetics of coalescing polymer colloids. Langmuir **25**, 9703–9713 (2009). https://doi.org/10.1021/la900963f
61. D. Stauffer, A. Coniglio, M. Adam, Gelation and critical phenomena, in *Polymer Networks*, ed. by K. Dušek (1982), pp. 103–158. https://doi.org/10.1007/3-540-11471-8_4
62. R. Bansil, H.J. Herrmann, D. Stauffer, Computer simulation of kinetics of gelation by addition polymerization in a solvent. Macromolecules **17**(5), 998–1004 (1984). https://doi.org/10.1021/ma00135a004
63. L.Y. Shy, B.E. Eichinger, Computer-simulations of radiation-cured networks. Macromolecules **19**, 2787–2793 (1986). https://doi.org/10.1021/ma00165a021
64. K.J. Lee, B.E. Eichinger, Computer-simulation of end-linked networks—telechelic poly(oxyethylene) cross-linked with plurifunctional isocyanates. Macromolecules **22**, 1441–1448 (1989). https://doi.org/10.1021/ma00193a073
65. V. Galiatsatos, B.E. Eichinger, Simulation of the formation of elastomers. Rubber Chem. Technol. **61**, 205–222 (1988). https://doi.org/10.5254/1.3536183
66. I. Carmesin, K. Kremer, The bond fluctuation method: a new effective algorithm for the dynamics of polymers in all spatial dimensions. Macromolecules **21**, 2819–2823 (1988)
67. H.P. Deutsch, K. Binder, Interdiffusion and self-diffusion in polymer mixtures: a Monte Carlo study. J. Chem. Phys. **94**, 2294 (1991). https://doi.org/10.1063/1.459901
68. P. Polanowski Polanowski, T. Pakula, Studies of polymer conformation and dynamics in two dimensions using simulations based on the Dynamic Lattice Liquid (DLL) model. J. Chem. Phys. **117**, 4022 (2002). https://doi.org/10.1063/1.1495836
69. M. Lang, Monomer fluctuations and the distribution of residual bond orientations in polymer networks. Macromolecules **46**, 9782–9797 (2013). https://doi.org/10.1021/ma402013b
70. P. Polanowski, J.K. Jeszka, K. Krysiak, K. Matyjaszewski, Influence of intramolecular crosslinking on gelation in living copolymerization of monomer and divinyl cross-linker. Monte Carlo simulation studies. Polymer **79**, 171–178 (2015). https://doi.org/10.1016/j.polymer.2015.10.018
71. C. Li, A. Strachan, Molecular simulations of crosslinking process of thermosetting polymers. Polymer **51**, 6058–6070 (2010). https://doi.org/10.1016/j.polymer.2010.10.033
72. A. Bandyopadhyay, P.K. Valavala, T.C. Clancy, K.E. Wise, G.M. Odegard, Molecular modeling of crosslinked epoxy polymers: The effect of crosslink density on thermomechanical properties. Polymer **52**, 2445–2452 (2011). https://doi.org/10.1016/j.polymer.2011.03.052
73. P.V. Komarov, C.Y. Tsung, C. Shih-Ming, P.G. Khalatur, P. Reineker, Highly cross-linked epoxy resins: an atomistic molecular dynamics simulation combined with a mapping/reverse mapping procedure. Macromolecules **40**, 8104–8113 (2007)

74. A.A. Gavrilov, P.V. Komarov, P.G. Khalatur, Thermal properties and topology of epoxy networks: a multiscale simulation methodology. Macromolecules **48**, 206–212 (2015). https://doi.org/10.1021/ma502220k
75. B. Demir, T.R. Walsh, A robust and reproducible procedure for cross-linking thermoset polymers using molecular simulation. Soft Matter **12**, 2453–2464 (2016). https://doi.org/10.1039/c5sm02788h
76. Q. Yang, X. Li, L. Shi, X. Yang, G. Sui, The thermal characteristics of epoxy resin: design and predict by using molecular simulation method. Polymer **54**, 6447–6454 (2013). https://doi.org/10.1016/j.polymer.2013.09.055
77. C. Li, A. Strachan, Evolution of net work topology of bifunctional epoxy thermosets during cure and its relationship to thermo-mechanical properties: a molecular dynamics study. Polymer **75**, 151–160 (2016). https://doi.org/10.1016/j.polymer.2015.08.037
78. D.M. Kroll, S.G. Croll, Influence of crosslinking functionality, temperature and conversion on heterogeneities in polymer networks. Polymer **2015**(79), 82–90 (2015). https://doi.org/10.1016/j.polymer.2015.10.020
79. D.M. Kroll, S.G. Croll, Heterogeneity in polymer networks formed by a single copolymerization reaction: I. Gelation and pre-gel structure. Polymer **115**, 154–163 (2017). https://doi.org/10.1016/j.polymer.2017.03.051
80. D.M. Kroll, S.G. Croll, Heterogeneity in polymer networks formed by a single copolymerization reaction: II. Post-gelation structure and pendants. Polymer **116**, 113–123 (2017). https://doi.org/10.1016/j.polymer.2017.03.059
81. G.R. Bowman, D.L. Ensign, V.S. Pande, Enhanced modeling via network theory: adaptive sampling of markov state models. J. Chem. Theory Comput. **6**, 787–794 (2010). https://doi.org/10.1021/ct900620b
82. A.N. Kolmogorov, N.A. Dmitriev, Branching random processes (Ветвящиеся случайные процессы). Dokl. Akad. Nauk SSSR **56**, 7–10 (1947)
83. T.E. Harris, *The Theory of Branching Proceses* (Springer, Berlin, 1963)
84. C.J. Mode, *Multitype Branching Processes* (American Elsevier Publishing Comp., New York, 1971)
85. M. Kimmel, D.E. Axelrod, in *Branching Processes in Biology* (Springer, New York, 2002). https://www.researchgate.net/publication/228792062_Branching_Processes_in_Biology
86. C.A. Macken, A.S. Perelson, Branching processes applied to cell surface aggregation phenomena, in *Book Series: Lecture Notes in Biomathematics*, vol. 58, (Springer, Berlin, 1985). https://doi.org/10.1007/978-3-642-52115-7
87. D. Ludwig, Stochastic population theories, in *Book Series: Lecture Notes in Biomathematics*, vol. 3, (Springer, Berlin, 1985). https://doi.org/10.1007/978-3-642-80883-8
88. F. Galton, Typical laws of heredity. Nature **15**, 512–514 (1877). https://doi.org/10.1038/015512b0
89. F. Galton, *Natural Inheritance* (Macmillan and Company, 1894)
90. F. Galton, *Educational Times* (1873), p. 17
91. H.W. Watson, *Solution to Problem 4001* (Educational Times, 1873), pp. 115–116
92. H.W. Watson, F. Galton, On the probability of the extinction of families. J. Royal Anthropol. Inst. **4**, 138–144 (1875)
93. I.J. Good, The number of individuals in a cascade process. Proc. Camb. Phil. Soc. **45**, 360–363 (1949). https://doi.org/10.1017/S030500410002497X
94. I.J. Good, The joint distribution for the sizes of the generations in a cascade process. Proc. Roy. Soc. A **268**, 256–259 (1952). https://doi.org/10.1098/rspa.1962.0137
95. I.J. Good, The joint distribution for the sizes of the generations in a cascade process. Proc. Camb. Phil. Soc. **51**, 240–242 (1955). https://doi.org/10.1017/S0305004100030115
96. I.J. Good, Generalizations to several variables of Lagrange's expansion, with applications to stochastic processes. Proc. Camb. Phil. Soc. **56**, 367–380 (1960). https://doi.org/10.1017/S0305004100034666
97. I.J. Good, Cascade theory and the molecular weight averages of the sol fraction. Proc. Roy. Soc. Lond. A **272**, 54–59 (1963). https://doi.org/10.1098/rspa.1963.0041

98. G. Allen, K. Dušek, J.W. Kennedy, R. Koningsveld, Professor manfred gordon: a general appreciation. Polym. Int. Br. Polym. J. **9**, 85–88 (1977). https://doi.org/10.1002/pi.4980090202
99. P. Whittle, Statistical processes of aggregation and polymerization. Math. Proc. Cambridge Phil. Soc. **61**, 475–495 (1965). https://doi.org/10.1017/S0305004100004047
100. P. Whittle, *Networks: Optimisation and Evolution* (Cambridge University Press, Cambridge, 2007)
101. M. Gordon, Good's theory of cascade processes applied to the statistics of polymer distributions. Proc. Roy. Soc. London A **268**, 240–256 (1962). https://doi.org/10.1098/rspa.1962.0136
102. K. Dušek, W. Prins, Structure and elasticity of non-crystalline polymer networks. Adv. Polym. Sci. **6**, 1–102 (1969). https://doi.org/10.1007/BFb0051042
103. D.R. Miller, C.W. Macosko, Molecular-weight relations for cross-linking of chains with length and site distribution. J. Polym. Sci. B Polym. Phys. **25**, 2441–2469 (1987). https://doi.org/10.1002/polb.1987.090251203
104. H. Staudinger, *Makromolekulare Chemie und Biologie* (Wepf & Co., Basle, 1947)
105. A. Charlesby, *Atomic Radiation and Polymers* (Pergamon Press, Oxford, 1960). *The Radiation Chemistry of Macromolecules*, ed. by M. Dole (Academic Press, New York, 1972)
106. M. Dole, *The Radiaion Chemistry of Mcromolecules* (Acamemic Press, New York, 1972)
107. O. Saito, On the effect of high energy radiation to polymers I. Cross-linking and degradation. J. Phys. Soc. Jpn. **13**, 198–206 (1958). https://doi.org/10.1143/JPSJ.13.198
108. O. Saito, Effects of high energy radiation on polymers II. End-linking and gel fraction. J. Phys. Soc. Jpn. **13**, 1451–1464 (1958). https://doi.org/10.1143/JPSJ.13.1451
109. O. Saito, Effects of high energy radiation on polymers III. Viscosity. J. Phys. Soc. Jpn. **13**, 1485–1496 (1958). https://doi.org/10.1143/JPSJ.13.1465
110. O. Saito, Effects of gamma rays on aqueous solution of polyvinyl alcohol. J. Phys. Soc. Jpn. **14**, 792–797 (1959). https://doi.org/10.1143/JPSJ.14.792
111. O. Saito, T. Takahashi, Production of polymer gel with least radiation dose. J. Phys. Soc. Jpn. **21**, 206 (1966). https://doi.org/10.1143/JPSJ.21.206-206
112. O. Saito, T. Takahashi, The number of intra-molecular linkages in cross-linked amorphous polymer. J. Phys. Soc. Jpn. **36**, 529–531 (1974). https://doi.org/10.1143/JPSJ.36.529
113. O. Saito, Statistical aspects of infinite network formations. Polym. Eng. Sci. **19**, 234–245 (1979). https://doi.org/10.1002/pen.760190403
114. A. Amemiya, M. Inokuti, Mathematical considerations in chemical kinetics of radiation-induced polymerization. J. Phys. Soc. Jpn. **16**, 949–960 (1961). https://doi.org/10.1143/JPSJ.16.949
115. A. Amemiya, Theory of crosslinking of polymers. J. Phys. Soc. Jpn. **20**, 2279–2288 (1965). https://doi.org/10.1143/JPSJ.20.2279
116. A. Amemiya, O. Saito, Mathematical theory of polymer gelation. J. Phys. Soc. Jpn. **26**, 1264–1276 (1966). https://doi.org/10.1143/JPSJ.26.1264
117. M. Inokuti, On the viscosity of irradiated polymers. J. Phys. Soc. Jpn. **14**, 79–85 (1959). https://doi.org/10.1143/JPSJ.14.965
118. M. Inokuti, K. Katsuura, On the viscosity of irradiated polymers. II. J. Phys. Soc. Jpn. **14**, 965–965 (1959). https://doi.org/10.1143/JPSJ.14.1379
119. M. Inokuti, Molecular size distribution and gelation of irradiated copolymers. J. Chem. Phys. **33**, 1607 (1960). https://doi.org/10.1063/1.1731470
120. M. Inokuti, Gel formation in polymers resulting from simultaneous crosslinking and scission. J. Chem. Phys. **38**, 2999–3005 (1963). https://doi.org/10.1063/1.1733633
121. M. Inokuti, M. Dole, Analysis of molecular weight changes and gel data of some irradiated vinyl polymers. J. Chem. Phys. **38**, 3006 (1963). https://doi.org/10.1063/1.1733634
122. M. Demjanenko, K. Dušek, Statistics of degradation and cross-linking of polymer chains with the use of the theory of branching processes. Macromolecules **13**, 571–579 (1980). https://doi.org/10.1016/0014-3057(80)90069-5

123. K. Dušek, M. Demjanenko, Application of the theory of branching processes (cascade theory) to polymer degradation and crosslinking: postgel stage. Int. J. Radiat. Appl. Instrum. C Radiat. Phys. Chem. **28**, 479–486 (1986). https://doi.org/10.1016/1359-0197(86)90173-6
124. P.J. Flory, *Principles of Polymer Chemistry* (Cornell Univ. Press, 1953)
125. P.J. Flory, Molecular size distribution in three dimensional polymers. III. Tetrafunctional branching units. J. Am. Chem. Soc. **63**, 3096–3100 (1941). https://doi.org/10.1021/ja01856a063
126. L.C. Case, Branching in polymers. I. Network defects. J. Polym. Sci. A Polym. Chem. **45**, 397–404 (1960). https://doi.org/10.1002/pol.1960.1204514609
127. G.R. Dobson, M. Gordon, Configurational statistics of highly branched polymer systems. J. Chem. Phys. **41**, 2389 (1964). https://doi.org/10.1063/1.1726276
128. M. Gordon, G.R. Scantlebury, Non-random polycondensation: statistical theory of the substitution effect. Trans. Faraday Soc. **60**, 604–621 (1964). https://doi.org/10.1039/TF9646000604
129. M. Gordon, G.N. Malcolm, D.S. Butler, Configurational statistics of copolymer systems. Proc. Roy. Soc. London A **295**, 29–54 (1966). https://doi.org/10.1098/rspa.1966.0224
130. M. Gordon, G.R. Scantlebury, Theory of ring-chain equilibria in branched non-random polycondensation systems, with applications to POCl3/P2O5. Proc. Roy. Soc. London A **41**, 380–402 (1966)
131. M. Gordon, G.R. Scantlebury, Theory of branching processes and kinetically controlled ring-chain competition processes. J. Polym. Sci. Pt. C Polym. Symp. **16**, 3933 (1968)
132. J.W. Aldersley, M. Gordon, Polyaddition and polycondensation. Substitution effects in polycondensation systems. J. Polym. Sci. Pt. C. Polym. Symp. **16**, 4567–4575 (1969)
133. M. Gordon, S. Kuchařík, T.C. Ward, The statistics of elastically active network chains and the efficiency of cross-linking in rubbers. Collect. Czech. Chem. Commun. **35**, 3252–3264 (1970). https://doi.org/10.1135/cccc19703252
134. M. Gordon, Network theory of the gel point and the incestuous polymerization of diallyl phthalate. J. Chem. Phys. **22**, 610–613 (1954). https://doi.org/10.1063/1.1740134
135. M. Gordon, S.B. Ross-Murphy, H. Suzuki, The graph-like state of matter. VI: combinatorial approach to the excluded-volume perturbation. Eur. Polym. J. **12**, 733–740 (1976). https://doi.org/10.1016/0014-3057(76)90085-9
136. M. Gordon, J.A. Torkington, S.B. Ross-Murphy, Graph-like state of matter. 9. Statistical thermodynamics of dilute polymer-solutions. Macromolecules **10**, 1090–1100 (1977)
137. W. Burchard, K. Kajiwara, M. Gordon, J. Kálal, J.W. Kennedy, Rayleigh scattering from solutions of critically branched polycondensates. Macromolecules **6**, 642–649 (1973). https://doi.org/10.1021/ma60034a033
138. W. Burchard, A. Thurn, E. Wachenfeld, Structure of branched polymers, in *Physics of Finely divided Matter. Proceedings in Physics 5*, eds. by N. Boccara, M. Daoud (Springer, Berlin, 1985), pp. 128–134. https://doi.org/10.1007/978-3-642-93301-1_16
139. K. Kajiwara, M. Gordon, Graphlike state of matter. V. Theory of Rayleigh scattering from randomly cross-linked chains of arbitrary primary distributions. J. Chem. Phys. **59**, 3623 (1973). https://doi.org/10.1063/1.1680529
140. I. Franken, W. Burchard, Application of the cascade theory to calculation of particle scattering factors of polydisperse systems of stiff chains. Macromolecules **6**, 848–855 (1973). https://doi.org/10.1021/ma60036a013
141. W. Burchard, Particle scattering factors of some branched polymers. Macromolecules **10**, 919–927 (1977). https://doi.org/10.1021/ma60059a008
142. W. Burchard, Application of cascade theory to calculation of quasielastic scattering functions. 1. Polydisperse, ideal, linear chains in dilute solutions. Macromolecules **11**, 455–459 (1978). https://doi.org/10.1021/ma60063a004
143. K. Kajiwara, W. Burchard, Hydrodynamic radius and dynamic scattering from randomly crosslinked chains. Polymer **22**, 1621–1628 (1981). https://doi.org/10.1016/0032-3861(81)90375-X

144. W. Burchard, K. Kajiwara, D. Nerger, Static and dynamic scattering behavior of regularly branched chains: a model of soft-sphere microgels. J. Polym. Sci. Pt. B Polym. Phys. **20**, 157–171 (1982). https://doi.org/10.1002/pol.1982.180200201
145. W. Burchard, Static and dynamic light scattering from branched polymers and biopolymers. Light Scatt. Polym. Adv. Polym. Sci. **48**, 1–124 (1983). https://doi.org/10.1007/3-540-12030-0_1
146. S. Bantle, W. Burchard, Branching in epoxy resins: molecular weight as a function of the extent of reaction and critical behavior. Polymer **27**, 728–734 (1986). https://doi.org/10.1016/0032-3861(86)90131-X
147. P. Lang, W. Burchard, Dynamic light scattering at the gel point. Macromolecules **24**, 814–815 (1991). https://doi.org/10.1021/ma00003a028
148. V. Trappe, W. Burchard, B. Steinmann, Anhydride-cured epoxies via chain reaction. 1. The phenyl glycidyl ether/phthalic acid anhydride system. Macromolecules **24**, 4738–4744 (1991). https://doi.org/10.1021/ma00017a002
149. K. Kajiwara, Statistics of randomly branched polycondensates. J. Chem. Phys. **54**, 296 (1971). https://doi.org/10.1063/1.1674607
150. K. Kajiwara, Statistics of randomly branched polycondensates: part 2. The application of Lagrange's expansion method to homodisperse fractions. Polymer **12**, 57–66 (1971). https://doi.org/10.1016/0032-3861(71)90012-7
151. K. Kajiwara, C.A.M. Ribeiro, Dilute-solution properties of randomly branched polymer systems. 1. Particle scattering factor. Macromolecules **7**, 121–128 (1974). https://doi.org/10.1021/ma60037a023
152. F. Tanaka, Thermodynamic theory of network-forming polymer-solutions. 1. Macromolecules **23**, 3784–3789 (1990). https://doi.org/10.1021/ma00218a012
153. F. Tanaka, Thermodynamic theory of network-forming polymer-solutions. 2. Equilibrium gelation by conterminous cross-linking. Macromolecules **23**, 3790–3795 (1990). https://doi.org/10.1021/ma00218a013
154. F. Tanaka, Gel formation with multiple interunit junctions. I. Molecules carrying different functional groups. J. Polym. Sci. Pt. B Polym. Phys. **41**, 2405–2412 (2003). https://doi.org/10.1002/polb.10586
155. F. Tanaka, W.H. Stockmayer, Thermoreversible gelation with junctions of variable multiplicity. Macromolecules **27**, 3943–3954 (1994). https://doi.org/10.1021/ma00092a039
156. F. Tanaka, M. Ishida, Elastically effective chains in transient gels with multiple junctions. Macromolecules **29**, 7571–7580 (1996). https://doi.org/10.1021/ma960604g
157. T. Nakao, F. Tanaka, S. Kohjiya, Cascade theory of substitution effects in nonequilibrium polycondensation systems. Macromolecules **35**, 5649–5656 (2002). https://doi.org/10.1021/ma011704b
158. T. Nakao, F. Tanaka, S. Kohjiya, New cascade theory of branched polymers and its application to size exclusion chromatography. Macromolecules **39**, 6643–6652 (2006). https://doi.org/10.1021/ma052686r
159. V.I. Irzhak, Statistical-theory of gelation—sol–fraction. Vysokomol. Soedin. Ser. A **17**, 535–545 (1975)
160. S.I. Kuchanov, E.S. Povolotskaya, Calculation of molecular mass-distribution of products of non-equilibrium polycondensation taking into account the substitution effect. Vysokomol. Soedin. Ser. A **24**, 2179–2189 (1982)
161. V.I. Irzhak, M.L. Tai, Applicability of statistical approach to description of processes of non-equilibrium polycondensation. Vysokomol. Soedin. Ser. A **25**, 2305–2311 (1983)
162. S.I. Kuchanov, T.V. Zharnikov, Stochastic branching process for description of non-random irreversible polycondensation. J. Stat. Phys. **111**, 1273 (2003). https://doi.org/10.1023/A:1023008401138
163. S.I. Kuchanov, H. Slot, A. Stroeks, Development of a quantitative theory of polycondensation. Prog. Polym. Sci. **29**, 563–633 (2004). https://doi.org/10.1016/j.progpolymsci.2004.02.001
164. S.I. Kuchanov, K.V. Tarasewich, T.V. Zharnikov, Configurational effects in statistical theory of branched non-random polycondensation. J. Stat. Phys. **122**, 875 (2006). https://doi.org/10.1007/s10955-005-9016-4

165. T.R. Deberdeev, R.M. Garipov, T.F. Irzhak, V.I. Irzhak, R.Y. Deberdeev, Critical conversion in nonrandom polycondensation. Dokl. Phys. Chem. **393**, 306–308 (2010). https://doi.org/10.1023/B:DOPC.0000003467.58638.14
166. T.F. Irzhak, V.I. Irzhak, Critical conversion during three-dimensional living copolymerization. Polym. Sci. Ser. B **53**, 427–430 (2011). https://doi.org/10.1134/S1560090411060054
167. V.I. Irzhak, Conformity of kinetic and statistical approaches in describing processes of formation of crosslinked polymers. Vysokomol. Soedin. Ser. B **20**(8), 606–609 (1978)
168. T.F. Irzhak, T.R. Deberdeev, V.I. Irzhak, Structural kinetics of the formation of hyperbranched polymers via condensation polymerization in the presence of the substitution effect. Polym. Sci. Ser. B **57**, 55–60 (2015). https://doi.org/10.1134/S1560090415010066
169. H. Galina, A. Szustalewicz, A kinetic approach to the network formation in an alternating stepwise copolymerization. Macromolecules **23**, 3833–3838 (1990). https://doi.org/10.1021/ma00197a039
170. H. Galina, J.B. Lechowicz, Monte–Carlo modeling of degradation of polymer networks. Polym. Gels Networks **6**, 103–111 (1998). https://doi.org/10.1016/S0966-7822(98)00004-5
171. H. Galina, J.B. Lechowicz, Monte–Carlo modeling of degradation of polymer networks: 3. Lattice networks. Polymer **41**, 615–619 (2000). https://doi.org/10.1016/S0032-3861(99)00221-9
172. H. Galina, J.B. Lechowicz, M. Walczak, Kinetic modeling of hyperbranched polymerization involving an AB2 monomer reacting with substitution effect. Macromolecules **35**, 3253–3260 (2002). https://doi.org/10.1021/ma011603d
173. K. Dušek, M. Ilavský, S. Luňák, Curing of epoxy resins. I. Statistics of curing of diepoxides with diamines. J. Polym. Sci. Polym. Symp. **53**, 29 (1975). https://doi.org/10.1007/BF02699642
174. K. Dušek, M. Ilavský, Cyclization in crosslinking polymerization. I. Chain polymerization of a bisunsaturated monomer (monodisperse case). J. Polym. Sci. Polym. Symp. **53**, 57s (1975)
175. S. Luňák, K. Dušek, Curing of epoxy resins. II. Curing of bisphenol a diglycidylether with diamines. J. Polym. Sci. Pt. C Polym. Symp. **53**, 45–55 (1975). https://doi.org/10.1002/polc.5070530108
176. K. Dušek, V. Vojta, Concentration of elastically active network chains and cyclization in networks obtained by alternating stepwise polyaddition. Polym. Int. Br. Polym. J. **9**, 164–171 (1977). https://doi.org/10.1002/pi.4980090214
177. K. Dušek, Correspondence between the theory of branching processes and the kinetic theory for random crosslinking in the post-gel stage. Polym. Bull. **1**, 523–528 (1979). https://doi.org/10.1007/BF00254479
178. V. Vojta, K. Dušek, Calculation of molecular weight distribution of crosslinked polymers using the theory of branching processes. Polym. Int. **12**, 1–4 (1980). https://doi.org/10.1002/pi.4980120102
179. J. Mikeš, K. Dušek, Simulation of the polymeric network formation by the Monte–Carlo method. Macromolecules **15**, 93–99 (1982). https://doi.org/10.1021/ma00229a018
180. M. Ilavský, K. Dušek, The structure and elasticity of polyurethane networks. I. Model networks from poly(oxypropylene) triols and diisocyanate. Polymer **24**, 981 (1983). https://doi.org/10.1016/0032-3861(83)90148-9
181. K. Dušek, The size of network chains. Macromolecules **17**, 716 (1984). https://doi.org/10.1021/ma00134a033
182. K. Dušek, Formation-structure relationships in polymer networks. Polym. Int. **17**, 185–189 (1985). https://doi.org/10.1002/pi.4980170218
183. K. Dušek, Formation of polymer networks: treatment of stochastic and spatial correlations using mean field approximation, in *Physics of Finely Divided Matter. Proceedings in Physics 5*, eds. by N. Boccara, M. Daoud (Springer, Berlin, 1985), pp. 197–212. https://doi.org/10.1007/978-3-642-93301-1_13
184. K. Dušek, J. Šomvársky, Build-up of polymer networks by initiated polyreactions. I. Comparison of kinetic and statistical approaches to living polymerization type build-up. Polym. Bull. **13**, 313–319 (1985). https://doi.org/10.1007/BF00262114

185. K. Dušek, Network formation in curing of epoxy resins. Adv. Polym. Sci. **78**, 1–59 (1986). https://doi.org/10.1007/BFb0035356
186. K. Dušek, B.J.R. Scholtens, G.P.J.M. Tiemersma-Thoone, Theoretical treatment of network formation by a multistage process. Polym. Bull. **17**, 239–245 (1987). https://doi.org/10.1007/BF00285356
187. Dušek, K., Networks from telechelic polymers: theory and application to polyurethanes, in *Telechelic Polymers: Synthesis and Applications*, ed. by E.J. Goethals (CRC Press, Boca Raton, 1989), pp 289–360
188. K. Dušek, M. Špírková, I. Havlíček, Network formation in polyurethanes due to side reactions. Macromolecules **23**, 1774–1781 (1990). https://doi.org/10.1021/ma00208a036
189. G.P.J.M. Tiemersma-Thoone, B.J.R. Scholtens, K. Dušek, M. Gordon, Theories for network formation in multi-stage processes. J. Polym. Sci. B Polym. Phys. **29**, 463–482 (1991). https://doi.org/10.1002/polb.1991.090290409
190. K. Dušek, J. Šomvársky, Chemical clusters in polymer networks. Faraday Discuss. Chem. Soc. **101**, 147–158 (1995). https://doi.org/10.1039/FD9950100147
191. K. Dušek, J. Šomvársky, Modelling of ring-free chain crosslinking copolymerization. Polym. Int. **44**, 225–236 (1997). https://doi.org/10.1002/(SICI)1097-0126(199711)44:3%3c225::AID-PI870%3e3.0.CO;2-C
192. L. Matějka, K. Dušek, P. Chabanne, J.-P. Pascault, Cationic polymerization of diglycidyl ether of bisphenol A. 2. Theory. J. Polym. Sci. Polym. Chem. Ed. **35**, 651–663 (1997)
193. K. Dušek, M. Dušková-Smrčková, Polymer networks from precursors of defined architecture. Activation of pre-existing branch points. Macromolecules **36**, 2915–2925 (2003). https://doi.org/10.1021/ma021666b
194. K. Dušek, M. Dušková-Smrčková, B. Voit, Highly-branched off-stoichiometric functional polymers as polymer networks precursors. Polymer **46**, 4265–4282 (2005). https://doi.org/10.1016/j.polymer.2005.02.037
195. K. Dušek, M. Dušková-Smrčková, L.A. Lewin, J. Huybrechts, R.J. Barsotti, Branching theories and thermodynamics used to help designing precursor architectures and binder systems. Surf. Coat. Int. Pt. B Coat. Trans. **82**, 121–131 (2006). https://doi.org/10.1007/BF02699642
196. K. Dušek, Applicability of statistical theories of network formation. Macromol. Symp. **256**, 18–27 (2007). https://doi.org/10.1002/masy.20075100
197. K. Dušek, M. Dušková-Smrčková, Polymer networks, in *Macromolecular Engineering. Precise Synthesis, Materials Properties, Applications, Structure-Property Correlation and Characterization Techniques*, eds. by K. Matyjaszewski, P. Gnanou, L. Leibler (Wiley, Weinheim, 2007), pp. 1687–1730
198. M. Dušková-Smrčková, H. Valentová, A. Ďuračková, K. Dušek, Effect of dilution on structure and properties of polyurethane networks: pregel and postgel cyclization and phase separation. Macromolecules **43**, 6450–6462 (2010). https://doi.org/10.1021/ma100626d
199. K. Dušek, M. Dušková-Smrčková, J. Huybrechts, A. Ďuračková, Polymer networks from preformed precursors having molecular weight and group reactivity distributions. Theory and application. Macromolecules **46**, 2767–2784 (2013). https://doi.org/10.1021/ma302396u
200. K. Dušek, M. Dušková-Smrčková, Polymer networks from nanosized functional precursors. Macromol. Symp. **372**, 14–27 (2017). https://doi.org/10.1002/masy.201600170
201. K. Dušek, M. Dušková-Smrčková, J. Huybrechts, Role of distributions in binders and curatives and their effect on network evolution and structure, in *Protective Coatings*, eds. by M. Wen, K. Dušek (Springer, 2017), pp 3–37. https://doi.org/10.1007/978-3-319-51627-1_1
202. J. Šomvársky, K. te Nijenhuis, M. Ilavský, Polyfunctional cross-linking of polymer chains. Macromolecules **33**, 3659–3670 (2000). https://doi.org/10.1021/ma98168m
203. C.W. Macosko, D.R. Miller, New derivation of average molecular-weights of nonlinear polymers. Macromolecules **9**, 199–206 (1976). https://doi.org/10.1021/ma60050a003
204. D.R. Miller, C.W. Macosko, New derivation of post gel properties of network polymers. Macromolecules **9**, 206–221 (1976). https://doi.org/10.1021/ma60050a004
205. D.R. Miller, C.W. Macosko, Network parameters for crosslinking of chains with length and site distribution. J. Polym. Sci. Pt. B Polym. Phys. **26**, 1–54 (1988). https://doi.org/10.1002/polb.1988.090260101

206. D.R. Miller, C.W. Macosko, Average property relations for non-linear polymerization with unequal reactivity. Macromolecules **11**, 656–662 (1978). https://doi.org/10.1021/ma60064a00
207. E.M. Valles, C.W. Macosko, Properties of networks formed by end linking of poly(dimethylsiloxane). Macromolecules **12**, 673–679 (1979). https://doi.org/10.1021/ma60070a025
208. D.R. Miller, C.W. Macosko, Substitution effects in property relations for stepwise polyfunctional polymerization. Macromolecules **13**, 1063–1069 (1980). https://doi.org/10.1021/ma60077a008
209. M. Gottlieb, C.W. Macosko, G.S. Benjamin, K.O. Meyers, E.W. Merrill, Equilibrium modulus of model poly(dimethylsiloxane) networks. Macromolecules **14**, 1039–1046 (1981). https://doi.org/10.1021/ma50005a028
210. A.M. Gupta, C.W. Macosko, Modeling strategy for systems with both stepwise and chainwise chemistry: amine-epoxy networks with etherification. J. Polym. Sci. Pt. B Polym. Phys. **28**, 2585–2606 (1990). https://doi.org/10.1002/polb.1990.090281309
211. G.L. Batch, C.W. Macosko, Kinetic-model for cross-linking free-radical polymerization including diffusion limitations. J. Appl. Polym. Sci. **44**, 1711–1729 (1992). https://doi.org/10.1002/app.1992.070441004
212. M. Lang, D. Göritz, S. Kreitmeier, Length of subchains and chain ends in cross-linked polymer networks. Macromolecules **36**, 4646–4658 (2003). https://doi.org/10.1021/ma034044e
213. M.A. Bibbo, E.M. Valles, Calculation of average properties of the pendant chains in a network. Macromolecules **15**, 1293–1300 (1982). https://doi.org/10.1021/ma00233a016
214. C. Sarmoria, E. Valles, D.R. Miller, Ring-chain competition kinetic-models for linear and nonlinear step-reaction copolymerizations. Macromol. Symp. **2**, 69–87 (1986). https://doi.org/10.1002/masy.19860020109
215. C. Sarmoria, E.M. Valles, D.R. Miller, Validity of some approximations used to model intramolecular reaction in irreversible polymerization. Macromolecules **23**, 580–589 (1990). https://doi.org/10.1021/ma00204a034
216. C. Sarmoria, D.R. Miller, Models for the first shell substitution effect in stepwise polymerization. Macromolecules **24**, 241833–2411845 (1991). https://doi.org/10.1021/ma00008a022
217. C. Sarmoria, M. Asteasuain, A. Brandolin, Prediction of molecular weight distributions in polymers using probability generating functions. Canad. J. Chem. Eng. **90**, 263–273 (2012). https://doi.org/10.1002/cjce.20699
218. C. Sarmoria, E. Valles, Copolymerization of a polydisperse oligomer with a monodisperse chain extender with intramolecular reaction allowed. Macromolecules **28**, 6244–6253 (1995). https://doi.org/10.1021/ma00122a035
219. D.R. Miller, C. Sarmoria, In-out" recursive probability modeling of branched step-growth polymerizations. Polym. Eng. Sci. **38**, 535–557 (1998). https://doi.org/10.1002/pen.10217
220. C. Sarmoria, D.R. Miller, Spanning-tree models for A(f) homopolymerizations with intramolecular reactions. Comput. Theor. Polym. Sci. **11**, 113–127 (2001). https://doi.org/10.1016/S1089-3156(99)00088-4
221. C. Sarmoria, E.M. Valles, Model for a scission-crosslinking process with both H and Y crosslinks. Polymer **45**, 5661–5669 (2004). https://doi.org/10.1016/j.polymer.2004.06.008
222. S. Pereda, A. Brandolin, E.M. Valles, C. Sarmoria, Copolymerization between A(3) and B-2 with ring formation and different intrinsic reactivity in one of the monomers. Macromolecules **34**, 4390–4400 (2001). https://doi.org/10.1021/ma001438m
223. A. Brandolin, M. Asteasuain, C. Sarmoria, A.R. López, K.S. Whiteley, B. del Amo Fernández, Numerical inversion techniques in the recovery of molecular weight distribution expressed in different transformed domains. Experimental validation. Polym. Eng. Sci. **41**, 1156–1170 (2001). https://doi.org/10.1002/pen.10817
224. A. Brandolin, M. Asteasuain, C. Sarmoria, A. Lopez-Rodriguez, K.S. Whiteley, B.D.A. Fernandez, Prediction of molecular weight distributions by probability generating functions: application to industrial autoclave reactors for high pressure polymerization of ethylene and ethylene-vinyl acetate. Polym. Eng. Sci. **41**, 1156–1170 (2001)

225. A.L.T. Brandão, J.B.P. Soares, J.C. Pinto, A.C. Alberton, When polymer reaction engineers play dice: applications of Monte Carlo models in PRE. Macromol. React. Eng. **9**, 141–185 (2015). https://doi.org/10.1002/mren.201500020
226. C. Fortunatti, C. Sarmoria, A. Brandolin, M. Asteasuain, Modeling of the bivariate molecular weight distribution-copolymer composition distribution in RAFT copolymerization using probability generating functions. Comput. Mater. Sci. **136**, 280–296 (2017). https://doi.org/10.1016/j.commatsci.2017.04.013
227. M.I. Aranguren, J. Borrajo, R.J.J. Williams, Statistics of novolacs. Ind. Eng. Chem. Prod. Res. Dev. **23**, 370–374 (1984). https://doi.org/10.1021/i300015a001-379
228. A. Vásquez, H.E. Addabo, R.J.J. Williams, Statistics of resols. Ind. Eng. Chem. Prod. Res. Dev. **23**, 375 (1984). https://doi.org/10.1021/i300015a001
229. C.C. Riccardi, R.J.J. Williams, Statistical structural model for the build-up of epoxy-amine networks with simultaneous etherification. Polymer **27**, 913–920 (1986). https://doi.org/10.1016/0032-3861(86)90304-6
230. R.J.J. Williams, C.C. Riccardi, K. Dušek, Build-up of polymer networks by initiated polyreactions. 3. Analysis of the fragment approach to the living polymerization type of build-up. Polym. Bull. **17**, 515–521 (1987). https://doi.org/10.1007/BF00283087
231. R.J.J. Williams, Statistics of free radical polymerizations revisited using a fragment approach. 1. Bifunctional monomers. Macromolecules **21**, 2568–2571 (1988). https://doi.org/10.1021/ma00186a042
232. R.J.J. Williams, C. Vallo, Statistics of free radical polymerizations revisited using a fragment approach. 2. Polyfunctional monomers. Macromolecules **21**, 2571–2575 (1988). https://doi.org/10.1021/ma00186a043
233. O. Georjon, J. Galy, J.P. Pascault, R.J.J. Williams, Statistical calculations of the network build-up in the postgel stage of A2-cyclotrimerization. Polym. Bull. **31**, 503–509 (1993). https://doi.org/10.1007/BF00329889
234. J. Huybrechts, K. Dušek, Star oligomers with different reactivity in low VOC polyurethane coatings: part II. Surf. Coat. Int. **82**, 172–177 (1998). https://doi.org/10.1007/BF02693857
235. B. Bruchmann, R. Koeniger, H. Renz, Tailor-made crosslinkers for high performance PUR coatings: hyperbranched polyisocyanates. Macromol. Symp. **187**, 271 (2002). https://doi.org/10.1002/1521-3900(200209)187:1%3c271::AID-MASY271%3e3.0.CO;2-L
236. B. Bruchmann, W. Schrepp, The AA*+B*B2 approach a simple and convenient synthetic strategy towards hyperbranched polyurea-urethanes. E-Polymers **014**, 191 (2003). https://doi.org/10.1515/epoly.2003.3.1.191
237. D. Tuerp, B. Bruchmann, Dendritic polyurea polymers. Macromol. Rapid Commun. **36**, 138–150 (2015). https://doi.org/10.1002/marc.201400510
238. J.L. Spouge, Equilibrium polymer size distributions. Macromolecules **16**, 121–127 (1983). https://doi.org/10.1021/ma00235a024
239. J.L. Spouge, Asymmetric bonding of identical units: a general AgRBf-g polymer model. Macromolecules **16**, 831–835 (1983). https://doi.org/10.1021/ma00239a021
240. C. Sarmoria, M. Asteasuain, A. Brandolin, Prediction of molecular weight distributions in polymers using probability generating functions. Can. J. Chem. Eng. **90**, 263–273 (2012). https://doi.org/10.1002/cjce.20699
241. R.C.S. Dias, M.R.P.F.N. Costa, Calculation of CLD using population balance equations of generating functions: Linear and non-linear ideal controlled radical polymerization. Macromol. Theory Simul. **19**, 323–341 (2010). https://doi.org/10.1002/mats.201000008
242. K. Kajiwara, W. Burchard, M. Gordon, Angular distribution of Rayleigh scattering from randomly branched polycondensates. Br. Polym. J. **2**, 110–115 (1970). https://doi.org/10.1002/pi.4980020203
243. R.M. Ziff, M.H. Ernst, E.M. Hendriks, Kinetics of gelation and universality. J. Phys. A Math. Gen. **16**, 2293–2320 (1983). https://doi.org/10.1088/0305-4470/16/10/026
244. B.E. Eichinger, Embeddable random networks: Method. Polymer **46**, 4258–4264 (2005). https://doi.org/10.1016/j.polymer.2005.03.018

245. N. Araújo, P. Grassberger, B. Kahng, K.J. Schrenk, R.M. Ziff, Recent advances and open challenges in percolation. Eur. Phys. J. Spec. Top. **223**, 2307–2321 (2014). https://doi.org/10.1140/epjst/e2014-02266-y
246. G.R. Dobson, M. Gordon, Theory of branching processes and statistics of rubber elasticity. J. Chem. Phys. **43**, 705 (1965). https://doi.org/10.1063/1.1696794
247. K. Dušek, Quasi-chemical equilibrium approach to crosslinked polymer solutions. J. Polym. Sci. Pt. C Polym. Symp. **39**, 83–106 (1972). https://doi.org/10.1002/polc.5070390108
248. J.A. Semlyen, *Cyclic Polymers* (Springer, Dordrecht, 2000)
249. R.F.T. Stepto, Non-linear polymerization, gelation and network formation, structure and properties, in *Polymer Networks. Principles of their Formation, Structure, and Properties*, ed. by R.F.T. Stepto (Blackie, London, 1998), pp. 14–63
250. R.F.T. Stepto, J.I. Cail, D.J.R. Taylor, I.M. Ward, R.A. Jones, Developments in understanding the formation, structure and properties of polymer networks. Macromol. Symp. **195**, 1–11 (2003). https://doi.org/10.1002/masy.200390109
251. J.I. Cail, R.F.T. Stepto, The gel point and network formation: theory and experiment. Polym. Bull. **58**, 15–25 (2007). https://doi.org/10.1007/s00289-006-0614-3
252. S. Dutton, H. Rolfes, R.F.T. Stepto, Comparison of Ahmad-Rolfes-Stepto theory, rate theory and Monte–Carlo modeling of gel point and network modulus. Polymer **35**, I4521-4526 (1994). https://doi.org/10.1016/0032-3861(94)90797-8
253. H. Rolfes, R.F.T. Stepto, Generalized statistical gelation theory. Makromol. Chem. Theory Simul. **1**, 245–260 (1992)
254. Z. Ahmad, R.F.T. Stepto, Approximate theories of gelation. Colloid Polym. Sci. **258**, 663–674 (1980). https://doi.org/10.1007/BF01384358
255. L. Matějka, K. Dušek, Formation of polyurethane networks studied by the gel point method. Polym. Bull. **3**, 489–495 (1980). https://doi.org/10.1007/BF00283825
256. A. Ďuračková, H. Valentová, M. Dušková-Smrčková, K. Dušek, Effect of diluent on the gel point and mechanical properties of polyurethane networks. Polym. Bull. **58**, 201–211 (2007)
257. M. Gordon, T.G. Parker, Graph-like state of matter. 1. Statistical effects of correlations due to substitution effects, including steric hindrance, on polymer distributions. Proc. Roy. Soc. Edin. Sect. A. **69**, 181–198 (1971). https://doi.org/10.1017/S0080454100008670
258. M. Dušková-Smrčková, H. Valentová, A. Ďuračková, K. Dušek, Diluent induced cyclization and phase separation in polymer networks. Macromol. Symp. **306–307**, 67–76 (2011). https://doi.org/10.1002/masy.201000133
259. H. Jacobson, W.H. Stockmayer, Intramolecular reaction in polycondensations. 1. The theory of linear system. J. Chem. Phys. **18**, 1600–1606 (1950). https://doi.org/10.1063/1.1747547
260. M. Gordon, W.B. Temple, The graph: like state of molecules. III. Ring-chain competition kinetics in linear polymerisation reactions. Makromol. Chem. **160**, 263–276 (1972). https://doi.org/10.1002/macp.1972.021600121
261. K. Dušek, J. Šomvársky, M. Smrčková, W.J. Simonsick, L. Wilczek, Role of cyclization in the degree-of-polymerization distribution of hyperbranched polymers: modelling and experiment. Polym. Bull. **42**, 489–496 (1999). https://doi.org/10.1007/s002890050493
262. E. Helfand, E. Al-Tonelli, Elastically ineffective polymer chains in rubbers. Macromolecules **7**, 832–834 (1974). https://doi.org/10.1021/ma60042a026
263. A.E. Tonelli, E. Helfand, Elastically ineffective cross-links in rubbers. Macromolecules **7**(1), 59–63 (1974). https://doi.org/10.1021/ma60037a012
264. K. Dušek, M. Gordon, S.B. Ross-Murphy, Graphlike state of matter. 10. Cyclization and concentration of elastically active network chains in polymer networks. Macromolecules **11**, 236–245 (1978). https://doi.org/10.1021/ma60061a044
265. M. Smoluchowski, Drei Vorträge über Diffusion, Brownsche Molekularbewegung und Koagulation von Kolloidteilchen". Physik. Z. **17**(557–571), 585–599 (1916)
266. M. Smoluchowski, Versuch einer mathematischen theorie der Koagulationskinetik kolloider Lösungen. Z. Phys. Chem. **92**, 129–168 (1917)
267. J. Šomvársky, K. Dušek, M. Smrčková, Kinetic modelling of network formation: size-dependent static effects. Comput. Theor. Polym. Sci. **8**, 201–208 (1998). https://doi.org/10.1016/S1089-3156(98)00036-1

268. T.A. Witten, L.M. Sander, Diffusion-limited aggregation, a kinetic critical phenomenon. Phys. Rev. Lett. **47**, 1400–1403 (1981). https://doi.org/10.1103/PhysRevLett.47.1400
269. R.J. Cohen, G.B. Benedek, Equilibrium and kinetic theory of polymerization and the sol–gel transition. J. Phys. Chem. **86**, 3696–3714 (1982)
270. P. Dongen, M.H. Ernst, Pre- and post-gel size distributions in (ir)reversible polymerization. J. Physics A: Math. Gen. **16**, L327–L332 (1983). https://doi.org/10.1088/0305-4470/16/10/003
271. E.M. Hendriks, M.H. Ernst, R.M. Ziff, Coagulation equations with gelation. J. Stat. Phys. **31**, 519–563 (1983). https://doi.org/10.1007/BF01019497
272. E.M. Hendriks, M.H. Ernst, Exactly soluble addition and condensation models in coagulation kinetics. J. Coll. Interf. Sci. **97**, 176–194 (1984). https://doi.org/10.1016/0021-9797(84)90285-6
273. R.M. Ziff, Kinetics of polymerization. J. Stat. Phys. **23**, 241–263 (1980). https://doi.org/10.1007/BF01012594
274. R.M. Ziff, G. Stell, Kinetics of polymer gelation. J. Chem. Phys. **73**, 3492–3499 (1980). https://doi.org/10.1063/1.440502
275. R.M. Ziff, E.D. McGrady, P. Meakin, On the validity of Smoluchowski's equation for cluster–cluster aggregation kinetics. J. Chem. Phys. **82**, 5269 (1985). https://doi.org/10.1063/1.448600
276. P. Sandkühler, M. Lattuada, H. Wu, J. Sefcik, M. Morbidelli, Further insight into the universality of colloidal aggregation. Adv. Colloid Interf. Sci. **113**, 65–83 (2005). https://doi.org/10.1016/j.cis.2004.12.001
277. F. Leyvraz, H.R. Tschudi, Singularities in the kinetics of coagulation processes. J. Phys. A **14**, 3389–3405 (1981)
278. F. Leyvraz, Scaling theory and exactly solved models in the kinetics of irreversible aggregation. Phys. Rep. **383**, 95–212 (2003)
279. E. Mastan, S. Zhu, Method of moments: a versatile tool for deterministic modeling of polymerization kinetics. Eur. Polym. J. **68**, 139–160 (2015). https://doi.org/10.1016/j.eurpolymj.2015.04.018
280. E. Mastan, X. Li, S. Zhu, Modeling and theoretical development in controlled radical polymerization. Prog. Polym. Sci. **45**, 71–191 (2015). https://doi.org/10.1016/progpolymsci.2014.12.003
281. H. Zhou, A. Szabo, Comparison between molecular dynamics simulations and the Smoluchowski theory of reactions in a hard-sphere liquid. J. Chem. Phys. **95**, 5948 (1991). https://doi.org/10.1063/1.461616
282. F. Leyvraz, Existence and properties of post-gel solutions for the kinetic equations of coagulation. J. Phys. A **16**, 2861–2873 (1983)
283. H. Tobita, N. Tani, T. Sakai, Postgel properties in the statistical crosslinking of heterochains II Free-radical crosslinking copolymerization. J. Polym. Sci. Pt. B Polym. Phys. **38**, 2342–2350 (2000)
284. H. Tobita, Molecular weight development in free-radical polymerization with polyfunctional chain-transfer agents. 2. Substitution effects. Macromol. Theory Simul. **10**, 581–586 (2001). https://doi.org/10.1002/1521-3919(20010701)10:6%3c581::AID-MATS581%3e3.0.CO;2-C
285. H. Tobita, Modeling controlled/living radical polymerization kinetics: bulk and miniemulsion. Macromol. React. Eng. **11–12**, 643–662 (2010). https://doi.org/10.1002/mren.201000029
286. H. Tobita, Free-radical polymerization with long-chain branching and scission: Markovian solution of the weight-average molecular weight. Macromol. Theory Simul. **23**, 477–489 (2014). https://doi.org/10.1002/mats.201400024
287. M.R.P.F.N. Costa, R.C.S. Dias, Prediction of sol fraction and average molecular weights after gelation for non-linear free radical polymerizations using a kinetic approach. Macromol. Theory Simul. **12**, 560–572 (2003). https://doi.org/10.1002/mats.200350011
288. L.G. Aguiar, M.A.D. Goncalves, V.D. Pinto, R.C.S. Dias, M.R.P.F.N. Costa, R. Giudici, Mathematical modeling of NMRP of styrene-divinylbenzene over the pre- and post-gelation periods including cyclization. Macromol. React. Eng. **8**, 295–313 (2014). https://doi.org/10.1002/mren.201300105

289. S. Lazzari, S. Hamzehlou, Y. Reyes, J.R. Leiza, M.R.P.F.N. Costa, R.C.S. Dias, G. Storti, Bulk crosslinking copolymerization: Comparison of different modeling approaches. Macromol. React. Eng. **8**, 678–695 (2014). https://doi.org/10.1002/mren.201400005
290. I. Kryven, J. Duivenvoorden, J. Hermans, P.D. Iedema, Random graph approach to multifunctional molecular networks. Macromol. Theory Simul. **25**, 449–465 (2016). https://doi.org/10.1002/mats.201600052
291. V. Schamboeck, I. Kryven, P.D. Iedema, Acrylic network formation by free-radical polymerization modeled using random graphs. Makromol. Theory Simul. **26**, 1700047 (2017)
292. A. Torres-Knoop, I. Kryven, V. Schamboeck, P.D. Iedema, Modeling the free-radical polymerization of hexanediol diacrylate (HDDA): a molecular dynamics and graph theory approach. Soft Matter **14**, 3404–3414 (2018). https://doi.org/10.1039/c8sm00451j
293. M. Asteasuain, A. Brandolin, Mathematical modeling of bivariate polymer property distributions using 2D probability generating functions. 1. Numerical inversion methods. Macromol. Theory Simul. **19**, 342–359 (2010). https://doi.org/10.1002/mats.200900096
294. A. Brandolin, M. Asteasuain, Mathematical modeling of bivariate distributions of polymer properties using 2D probability generating functions: part randformation of population mass balances of polymer processes. Macromol. Theory Simul. **22**, 273–308 (2013). https://doi.org/10.1002/mats.101200089
295. H. Tobita, A.E. Hamielec, Modeling of network formation in free radical polymerization. Macromolecules **22**, 3098–3105 (1989). https://doi.org/10.1021/ma00197a034
296. F. Teymour, J.D. Campbell, Analysis of the dynamics of gelation in polymerization reactors using the "numerical fractionation" technique. Macromolecules **27**, 2460–2469 (1994). https://doi.org/10.1021/ma00087a015
297. M. Wulkow, The simulation of molecular weight distributions in polyreaction kinetics by discrete Galerkin methods. Macromol. Theory Simul. **5**, 3393–3416 (1996). https://doi.org/10.1002/mats.1996.040050303
298. M. Wulkow, Computer aided modeling of polymer reaction engineering: the status of predici, 1-simulation. Macromol. React. Eng. **2**, 461–494 (2008). https://doi.org/10.1002/mren.200800024
299. A.N. Nikitin, M. Wulkow, C. Schűtte, Modeling of free-radical styrene/divinylbenzene copolymerization with the numerical fractionation technique. Macromol. Theory Simul. **22**, 473–479 (2013). https://doi.org/10.1002/mats.201300325
300. C. Fortunatti, C. Sarmoria, A. Brandolin, M. Asteasuain, Modeling of RAFT polymerization using probability generating functions: detailed prediction of full molecular weight distributions and sensitivity analysis. Macromol. React. Eng. **2**, 414–421 (2008). https://doi.org/10.1002/mren.201400020
301. K. Dušek, J. Spěváček, Cyclization in vinyl-divinyl copolymerization. Polymer **21**, 750 (1980). https://doi.org/10.1016/0032-3861(80)90290-6
302. K. Dušek, H. Galina, J. Mikeš, Features of network formation in the chain crosslinking (co)polymerization. Polym. Bull. **3**, 19–25 (1980). https://doi.org/10.1007/BF00263201
303. J. Spěváček, K. Dušek, Manifestation of microgel-like particles of styrene-ethylene dimethacrylate copolymers in 1H and 13C NMR spectra. J. Polym. Sci. Polym. Phys. Ed. **1980**, 18 (2027). https://doi.org/10.1002/pol.1980.180181003
304. H. Galina, K. Dušek, Z. Tuzar, M. Bohdanecký, J. Štokr, The structure of low conversion polymers of ethylene dimethacrylate. Eur. Polym. J. **16**, 1043–1046 (1980). https://doi.org/10.1016/0014-3057(80)90221-9
305. O. Okay, M. Kurz, K. Lutz, W. Funke, Cyclization and reduced pendant vinyl group reactivity during the free-radical cross-linking polymerization of 1,4-divinylbenzene. Macromolecules **28**, 2728–2737 (1995). https://doi.org/10.1021/ma00112a019
306. H. Hamamoto, K. Himei, S. Inoue, H. Aota, A. Matsumoto, Microgel-like network polymer precursor formation in free-radical cross-linking multi-allyl polymerization. Polym. J. **42**, 923–927 (2010). https://doi.org/10.1038/pj.2010.96
307. A. Matsumoto, Construction of crosslink-system-materials utilizing designed network-polymer-precursor modules. Polymer Symp. **306–307**, 89 (2011). https://doi.org/10.1002/masy.201000123

308. M. Wen, L.E. Scriven, A.V. McCormick, Kinetic gelation modeling: structural inhomogeneity during cross-linking polymerization. Macromolecules **36**, 4140–4150 (2003). https://doi.org/10.1021/ma010308q
309. Y.S. Gao, D.Z. Zhou, T.Y. Zhao, X. Wei, S. McMahon, J.O. Ahern, W. Wang, U. Greiser, B.J. Rodriguez, W.X. Wang, Intramolecular cyclization dominating homo-polymerization of multi-vinyl monomers toward single-chain cyclized/knotted polymeric nanoparticles. Macromolecules **48**, 6882–6889 (2015). https://doi.org/10.1021/acs.macromol.5b01549
310. J.G. Kloosterboer, G.M.M. Vandehei, H.M.J. Boots, Inhomogeneity during the photopolymerization of diacrylates: dsc experiments and percolation theory. Polym. Commun. **25**, 354–357 (1984)
311. J.G. Kloosterboer, Network formation by chain crosslinking photopolymerization and its applications in electronics. Adv. Polym. Sci. **84**, 1–61 (1988)
312. K. Dušek, Network formation by chain crosslinking (co)polymerization, in *Developments in Polymerization 3*, ed. by R. N. Haward (Applied Science Publ., Barking, 1982), pp. 143–206
313. K. Dušek, Special features of network formation by chain crosslinking copolymerization. Collect. Czech. Chem. Commun. **58**, 2245–2265 (1993). https://doi.org/10.1135/cccc19932245
314. O. Okay, Macroporous copolymer networks. Prog. Polym. Sci. **25**, 711–779 (2000). https://doi.org/10.1016/S0079-6700(00)00015-0
315. A. Matsumoto, Polymerization of multi-allyl monomers. Prog. Polym. Sci. **26**, 189–257 (2001). https://doi.org/10.1016/S0079-6700(00)00039-3
316. V. Schamboeck, P.D. Iedema, I. Kryven, Dynamic networks that drive the process of irreversible step-growth polymerization. Sci. Rep. **9**, 2276 (2019)
317. J. Dudowicz, K.F. Freed, J.F. Douglas, Lattice model of living polymerization. I. Basic thermodynamic properties. J. Chem. Phys. **111**, 7116 (1999). https://doi.org/10.1063/1.480004
318. J. Dudowicz, K.F. Freed, J.F. Douglas, Lattice model of living polymerization. II. Interplay between polymerization and phase stability. J. Chem. Phys. **112**, 1002 (2000)
319. J. Dudowicz, K.F. Freed, J.F. Douglas, Lattice model of living polymerization. III. Evidence for particle clustering from phase separation properties and "rounding" of the dynamical clustering transition. J. Chem. Phys. **113**, 434 (2000)
320. T.E. Gartner, A. Jayaraman, Modeling and simulations of polymers: a roadmap. Macromolecules **52**, 755–786 (2019)
321. T.F. Irzhak, V.I. Irzhak, Critical conversion in the processes of polymer formation. E-Polymers **118**, 154 (2010)
322. M. Ilavský, K. Dušek, The structure and elasticity of polyurethane networks. 5. The effect of diluent in the formation of model networks of poly(oxypropylene)triol and 4,4-diphenylmethane diisocyanate. Macromolecules **19**, 2139–2146 (1986). https://doi.org/10.1021/ma00162a006
323. D.R. Miller, E.M. Valles, C.W. Macosko, Calculation of molecular parameters for stepwise polyfunctional polymerization. Polym. Eng. Sci. **19**, 272–283 (1979). https://doi.org/10.1002/pen.760190409
324. S.E. Rankin, L.J. Kasehagen, A.V. McCormick, C.W. Macosko, Dynamic Monte Carlo simulation of gelation with extensive cyclization. Macromolecules **33**(20), 7639–7648 (2000). https://doi.org/10.1021/ma000132c
325. K. Fukui, T. Yamabe, Theory of gel formation with multifunctional interunit junctions. Bull. Chem. Soc. Japan **40**, 2052–2063 (1967). https://doi.org/10.1246/bcsj.40.2052
326. M. Asteasuain, A. Brandolin, C. Sarmoria, Recovery of molecular weight distributions from transformed domains. Part I: Application of numerical inversion methods. Polymer **43**, 2529–2541 (2002). https://doi.org/10.1016/S0032-3861(02)00035-6
327. L.A. Lewin, C.B. Douglas, K. Dušek, M. Dušková-Smrčková, P. Vlasák, Practical implications of solvent entrapment in ambient coatings. Eur. Coat. **81**, 21–39 (2005)
328. S.I. Kuchanov, *Methods of Kinetic Calculations in Polymer Chemistry* (Khimiya, Moscow, 1978)

329. J.B. McLeod, On an infinite set of non-linear differential equations. Quart. J. Math. **13**, 119–128 (1962). https://doi.org/10.1093/qmath/13.1.119
330. V. Orlova, I. Kryven, P.D. Iedema, Automated reaction generated for polymer networks. Comput. Chem. Eng. **112**, 37–47 (2018)

Chapter 4
Equilibrium Swelling of Gels

Karel Dušek

Abstract Swelling is an ubiquitous phenomenon of absorption of mostly liquid by solids resulting in an increase of volume of the solid. In this chapter, the swelling of cross-linked polymeric materials is concentrated and is focused mainly on modeling of swelling based on existing theories of polymer solutions and polymer chain elasticity. Phase separation in gels and during gel formation including volume phase transition is also explained.

Keywords Swelling · Gibbs energy · Flory–Huggins theory · Phase equilibrium · Gaussian network · Phase separation · Excluded volume · Volume phase transition

Swelling is an ubiquitous phenomenon of absorption of mostly liquid materials by solids resulting in an increase of volume of the solid. Here, we will disregard the medical aspects of swelling as well as swelling of mineral materials—mainly clays—as one of the most important causes of erosion. We will concentrate on the swelling of polymeric materials, especially those, that are permanently cross-linked. However, we will consider existence and effect of temporary bonds, which can be counted once as fully transient interactions that can dissociate and reform many times in the time window of experiment, and once as permanent within the time window of experiment when the degree of dissociation or reformation is below a limiting (negligible) value. We will focus mainly on modeling of swelling based on existing theories of polymer solutions and polymer chain elasticity. Also, we will mention the effect of constraints on swelling although a full account of the phenomenon will be given in another chapter of this book.

K. Dušek (✉)
Institute of Macromolecular Chemistry, Czech Academy of Sciences, Prague, Czech Republic

Y. Osada (ed.), *Science of Gel*, https://doi.org/10.1007/978-981-95-0301-8_4

4.1 What Is Swelling?

Why is the knowledge of the degree of swelling interesting? First of all, swelling brings the matter to its gel state—which is an intermediate state between liquid and solid states [references]. The gel materials have manifold applications in biosciences because the majority of organic materials in nature—animals or plants—are (mainly transient) gels. Various sophisticated associates formed by self-assembly serve as kinds of "cross-links" and prevent the hydrophilic constituents from dissolving. Many biomimetic assemblies are constructed nowadays for which the amount of absorption of biological fluids is important. Even for less sophisticated biomimetic constructs, such as implants, gel materials for controlled drug delivery or materials for tissue engineering, swelling in various fluids and control of its degree are important issues which are somewhat difficult to understand on molecular basis and to control them rationally. Of technological importance are various stimuli responsive "smart" gels, special types of hydrogels such as ion-exchange resins, moisture absorbing materials, concentrators of solutions, materials for chromatography. On the borderlines of the gel state domain, there are rheologically complex fluids (e.g., exhibiting thixotropy) on one side and on the other side, MOFs (Metal Organic Frameworks)—usually highly cross-linked materials absorbing large amount of gases such as hydrogen as serving for energy storage.

Polymer gels can be viewed as swollen networks in which the cross-links are covalent (permanent) or transient. In transient networks, the cross-links are a result of association of associative groups which can dissociate again as a result of changes state independent variables—temperature, pressure, and dilution. The change of the degree of association by changing external variables can be fast (weak uncorrelated interactions), so that equilibrium is established within the experimental window. Alternatively, the relaxation can be slow and equilibrium is not reached within the experimental window (strong bonds, correlated weak bonds) and the transient bonds behave as permanent ones. Networks formed by correlated weak bonds (sequences of weak bonds) that usually exhibit hysteresis loops depending on the direction the external variables are changed. Such a phenomenon appears when dissociation of sequences of weak bonds by zip mechanism exposes lyophilic groups buried inside the associate to contact with solvent.

Equilibrium swelling was treated in a number of monographs and textbook on polymer science starting from Flory [1] continuing by Dušek and Prins [2], Rubinstein and Colby [3] serving as examples. Among more recent reviews, those by Li et al. [4], Saunders et al. [5], and Quesada et al. [6] should be mentioned.

This chapter is organized as follows: After preceding Introduction, equilibrium swelling of non-ionized networks will be discussed, followed by swelling of ionized networks. The effect of constraints will be exposed briefly. In the last section, phase separation and phase transition phenomena will be briefly discussed. Equilibrium swelling is always considered as a result of interplay of the change of interactions between interacting sites of components in the system due to mixing and elastic forces generated by stretching/compression of network chains. The thermodynamics of mixing of polymers with solvents is quite elaborate as specific interactions are taken into account which lead to transient self-assembly. Water itself containing both H-bonding donors and acceptors can serve as an example. In the case of presence of charged groups, additionally the Donnan effect and electrostatic repulsion effect are taken into account. The models of swelling described in the literature differ in depth of refinement of interaction and elasticity effect and the interdependence between the parameters determining mixing interactions and network chain elasticity. Since swelling degrees generally extend over a wide scale ranging from few percent of liquid to volumes exceeding the dry volume several hundred times, it is appropriate to consider excluded volume effects at higher dilutions. This topic will be briefly addressed. The survey is concerned with mean field methods, techniques of molecular simulations (molecular dynamics, coarse-grained models) will not be discussed.

4.2 Determination of the Degree of Swelling and Ways of Its Expression

The degree of swelling can be determined either directly by measuring changes in weight, dimensions, or volume of the sample, or indirectly by determining related quantities such as refractive index, or polymer or solvent concentrations in the swollen sample. The change in weight (weight degree of swelling, W) is given by the ratio of weights (mass) of swollen m_{sw} and dry polymer m_{pol} sample

$$W = \frac{m_{sw}}{m_{pol}} \tag{4.1}$$

or expressed as solvent regain, R_w

$$R_W = \frac{m_{sw} - m_{pol}}{m_{pol}} = W - 1 \tag{4.2}$$

The volume degree of swelling, B, and volume solvent regain, R_B, are determined by changes in volume

$$B = \frac{V_{sw}}{V_{pol}} = \frac{L_x L_y L_z}{L_{xd} L_{yd} L_{zd}}, \qquad R_B = \frac{V_{sw} - V_{pol}}{V_{pol}} = B - 1 \tag{4.3}$$

where L_i are dimensions of the swollen sample and L_{id} dimensions of the dry sample in the directions x, y, z.

If the swollen sample does not contain (macro) pores, the weight and volume degrees of swelling or solvent regains can be converted to one another, as shown by Eq. 4.4:

$$B = \frac{m_{\text{dry}}/\rho_{\text{dry}} + m_{\text{solv}}/\rho_{\text{solv}}}{m_{\text{dry}}/\rho_{\text{dry}}} = 1 + (W-1)\frac{\rho_{\text{dry}}}{\rho_{\text{solv}}}$$

$$R_{\text{B}} = B - 1 = \frac{m_{\text{solv}}/\rho_{\text{solv}}}{m_{\text{dry}}/\rho_{\text{dry}}} = (W-1)\frac{\rho_{\text{dry}}}{\rho_{\text{solv}}} \tag{4.4}$$

where ρs are specific gravities.

If the sample contains pores, the situation is different: when pores are filled with water and the matrix still does not swell, the volume does not increase whereas the weight of the sample does. Generally, the swelling degree is contributed by both expansion of matrix and pores. If the increase of pore dimensions is affine with the expansion of macroscopic dimensions of the sample, the volume fraction of pores in the swollen state, $V_{\text{por,sw}}$ related to the total volume of the swollen gel, V_{sw}, can be calculated just from combined volume and weight degrees of swelling as

$$\phi_{\text{por}} = \frac{V_{\text{por,sw}}}{V_{\text{sw}}} = 1 - \frac{B}{1 + (W-1)\left(\rho_{\text{pol}}/\rho_{\text{solv}}\right)} \tag{4.5}$$

Equation 4.5 follows from the mass balance in combination with the assumption that $V_{\text{por,sw}} = BV_{\text{por,dry}}$. The volume and weight degrees of swelling, B and W, respectively, are now based on volumes containing pores, and empty pores are in the dry state and solvent filled pores are in the swollen state.

Alternatively, one can calculate the volume fraction of pores in the swollen state from the weight degree of swelling if the degree of swelling of the matrix is known. Such situation is met in the case when swelling of the matrix is isotropic and the swelling degree of the matrix is the same as swelling of the non-porous sample of the same composition—for instance, when porosity is generated by addition of an inert template. Also, for macroporous gels of poly(2-hydroxyethyl methacrylate) swollen in water, the swelling degree is the same in a range of low cross-link density (cf. Ref. [7]. Equation (A.4))[1]

$$\phi_{\text{por}} = 1 - \frac{(1/\phi_2)}{1 + (W-1)\left(\rho_{\text{pol}}/\rho_{\text{water}}\right)} \tag{4.5a}$$

where ϕ_2 is the volume fraction of polymer in the swollen polymer matrix. This suggests that

$$\phi_2 = 1/B \tag{4.6}$$

and, indeed, the increase of volume and B are the same for non-porous and porous gels. For many porous gels, the affinity condition is not fulfilled if dry and swollen states are compared because small pores gradually collapse on drying due to capillary condensation and are reformed when the xerogel is wetted (decrease of interfacial

[1] The symbol W in Ref. [7] was assigned to solvent regain, i.e., to the quantity $W - 1$, whereas W is used here for the weight degree of swelling.

tension). The expression of the welling degree through the volume fraction ϕ_2 is important because ϕ_2 is frequently used as concentration variable in thermodynamics of polymer solutions.

The above equations are based on the assumption of no volume change on mixing of polymer and solvent, i.e., on additivity of volumes of dry polymer and solvent. This is not fulfilled exactly because there can be a change of volume (usually contraction) by mixing, especially in the cases of strong polymer segment—solvent interaction or passage from glassy to rubbery states (devitrification volume). Rigorously, partial quantities—partial molar volumes, partial specific gravities, etc.—should be used. If the volume shrinkage is related to volume of polymer component, the shrinkage will typically be 2–5%. However, it can be higher if more free volume was frozen into the glassy state. The small error generated by the assumption volume additivities causes a small change in the interaction parameter χ (see below).

The weight degree of swelling W is determined experimentally by simple weighing of swollen and dry samples. The weight of the swollen sample refers to the extracted state, i.e., to a sample equilibrated with excess solvent (the solvent has been replaced several times to remove any soluble material that might be present). To obtain the swollen weight, traces of liquid solvent remaining on the surface of the object are removed, usually by gentle blotting with filtration paper or by gentle centrifuging, if the objects are too small. As the weight of the dry sample, one takes the weight of extracted sample, after the solvent has been removed by drying to constant weight. The volume degree of swelling is determined by directly measuring the dry and swollen volumes, or more often by measuring changes of dimensions using, for instance, a traveling microscope. If the sample swells anisometrically, changes of all dimensions (x, y, z) are measured.

4.3 Theoretical Approaches to Swelling: Equilibrium Swelling of Non-ionized Gels

4.3.1 Mixing and Chain Elasticity. Contributions and Their Additivity

Swelling of a cross-linked polymer is driven by the tendency of polymer segments to mix with the solvent molecules by which the network chains (precisely, the elastically active network chainsEANCs) expand. This expanding force is counterbalanced by contractive forces of stretched chains. In equilibrium, these two opposing forces are equal.

The thermodynamic state of a swollen cross-linked polymer in equilibrium with a liquid solvent or solvent vapor is determined by the Gibbs energy of the polymer network + solvent system. The Gibbs energy change accompanying swelling, ΔG_{sw}, is assumed to be a sum of changes due to mixing of the network with the solvent, ΔG_{mix}, and elastic term, associated with stretching of EANC in the network, ΔG_{el}:

$$\Delta G_{\text{sw}} = \Delta G_{\text{mix}} + \Delta G_{\text{el}} \tag{4.7}$$

The additivity (Eq. 4.7) was often questioned based on results of experiments where cross-linked gels and uncross-linked solutions were compared in a wide range of solvent content. The experiments included equilibrium solvent uptake from solvent vapors of varying fugacity or from solution in which solvent activity was decreased by a solute which did not penetrate the gel phase. In a series of papers by Eichinger et al. (cf., e. g., [8–11]) showed that a certain swelling function, according to Flory-Rehner theory should have been independent of the degree of swelling, exhibited a maximum. Likewise, Horkay et al. [12] analyzed the results of measurements of the swelling pressure and shear modulus of a set of poly(vinyl acetate) networks swollen to different degrees, when the solvent activity was decreased by dissolved polymer which did not penetrate into the gel phase. The difference between the cross-linked and uncross-linked states, when Flory-Rehner theory was applied, was in conflict with the assumption that the two mixing free energies are identical. Several other papers address this problem of additivity.

All these findings are related to tests of the validity of additivity assumption *only when the Flory-Rehner (FR) theory* is concerned, which is based on Flory–Huggins equation for a polymer of infinite molecular weight for the mixing term and Gaussian model for elasticity.

Indeed, in more sophisticated theories than the FR, ΔG_{mix} and ΔG_{el} are expected to be interdependent in the sense that ΔG_{mix} depends on parameters determining ΔG_{el} and vice versa for various reasons. We think, however, the additivity assumption (4.7), is a solid *ansatz* for viewing the swelling problem. Consideration of the effects of cross-links on mixing and interactions on elasticity is its integral part.

4.3.2 Mixing Contribution to Gibbs Energy—(Generalized) Flory—Huggins ((G)FH) Theory

We will start the analysis of the mixing contribution by analyzing the behavior using the (generalized) Flory–Huggins function applicable to concentrated polymer solutions keeping in mind that it has very weak predictive power in terms of polymer structure and type of interactions. The adjective *generalized* goes beyond the original FH model in that the interaction function is allowed to change with concentration and temperature more or less empirically fitting best the experimental dependences. Although of low predictive power, (G)FH can be used for analysis of concentration/temperature dependences of Gibbs energy/chemical potential with respect to phase stability and phase separation or transition.

Let us first examine ΔG_{mix} for concentrated solutions: Usually, the mixing contribution is represented by the Gibbs energy change corresponding to dissolution of a linear polymer of very high (infinite) molecular weight, independent of how many segments of network chains have been cross-linked to form f-functional junctions.

Although several more elaborate theories exist to describe binary mixtures involving a polymer, the Flory–Huggins form of ΔG_{mix} is used most frequently. ΔG_{mix} is expressed by the theory of regular solutions to which a Hildebrand—van Laar interaction term is added:

$$\Delta G_{\text{mix},N} = k_{\text{B}}T(N_1 \ln \phi_1 + N_2 \ln \phi_2 + g_N(\phi_2)N_1\phi_2) \tag{4.8}$$

where N_1 and N_2 are number of moles of component 1 and 2; k_{B} is the Boltzmann constant and $g_N(\phi_2)$ is the interaction function [13]; $\Delta G_{\text{mix},N}$ is expressed in its extensive form per N lattice sites. Alternatively, $\Delta G_{\text{mix,ls}}$ per lattice site

$$\Delta G_{\text{mix,ls}} = k_{\text{B}}T[(N_{1.\text{ls}}/m_1)\ln \phi_1 + \left(N_{2,\text{ls}}/m_2\right)\ln \phi_2 + g_{\text{ls}}(\phi_2)(N_1/m_1)\phi_2) \tag{4.9}$$

where $N_{1,\text{ls}}$ and $N_{2,\text{ls}}$ are the numbers of lattice sites occupied by molecules of component 1 and component 2, respectively, m_1 and m_2 are their numbers per molecule, and the interaction function g_{ls} refers to lattice site.

For polymers of very high degree of polymerization, $N_1 \gg N_2$ and, assumedly, for a polymer cross-linked into a network, $N_2 = 1$, so that the second term of Eq. 4.8 can be neglected

$$\Delta G_{\text{mix}} = kT(N_1 \ln \phi_1 + g_N(\phi_2)N_1\phi_2) \tag{4.10}$$

This form of mixing contribution implicitly suggests that the contribution changes when cross-links are introduced *before* the polymer molecule gets infinite, but it does not change any more when further cross-links are formed between segments of the infinite (branched) chain. Recent analysis has shown [14, 15] that the entropy continues to decrease when further cross-links are introduced between segments of the "infinite" molecule—the gel. This reduction of the degrees of freedom is proportional to the number of additional bonds; it is called effective *cycle rank*, Ξ_{e}.[2] This consideration leads to the expression for $\Delta G_{\text{mix},N}$ in the extensive form corresponding to Eq. 4.8 as

$$\Delta G_{\text{mix,n},N} = k_{\text{B}}T(N_1 \ln \phi_1 - \Xi_{\text{e}} \ln \phi_2 + g_N(\phi_2)N_1\phi_2) \tag{4.11}$$

where Ξ_{e} is related to the number of elastically active chains (EANC), N_{e}, as

$$\Xi_{\text{e}} = \frac{f_{\text{e}} - 2}{f_{\text{e}}} N_{\text{e}} \tag{4.12}$$

where f_{e} is the number-average functionality of junctions participating in the gel structure with three or more bonds with infinite continuation. Note that $\Delta G_{\text{mix,n}N}$

[2] The quantity cycle rank is defined in graph theory as the number of cuts by which the graph with cycles is transformed into a tree keeping all nodes connected. The adjective *effective* excludes from the rank of cycles small loops of the chains of which are not considered elastically active.

acquired the subscript n = network to distinguish it from polymer of finite molecular weight.

The reduced form of $\Delta G_{\text{mix,n}N}$ is usually expressed in molar quantities and reduced to lattice site[x] (the number of lattice sites is N):

$$\Delta G_{\text{mix,n,ls}} \equiv \Delta G_{\text{mix,n},N}/N = RT\left(\frac{N_{\text{ls},1}}{m_1}\ln\phi_1 - \Xi_{\text{e,ls}}\ln\phi_2 + g_{\text{ls}}(\phi_2)\big(N_{\text{ls},1}/m_1\big)\phi_2\right)$$

$$\Delta G_{\text{mix,ls}} = \frac{\Delta G_{\text{mix}}}{N_{\text{ls}}} = RT\left(\frac{\phi_1}{m_1}\ln\phi_1 - \frac{\xi_{\text{e}}}{m_2}\ln\phi_2 + g_{\text{ls}}(\phi_2)(\phi_1/m_1)\phi_2\right) \tag{4.13}$$

Alternatively, the volume of the lattice site is identified with volume of the solvent and the degree of polymerization is counted in terms of solvent molecules, so that $m_1 = m_2 = 1$

$$\Delta G_{\text{mix,n}} = RT(\phi_1\ln\phi_1 - \xi_{\text{e}}\ln\phi_2 + g(\phi_2)\phi_1\phi_2) \tag{4.14}$$

In Eq. 4.12, corresponding to Eq. 4.9, for simple solvents (diluents), $m_{1\text{s},1} > 1$ for polymeric diluents, $\Xi_{\text{e,ls}}$ is the effective cycle rank per lattice site, and $g_{\text{ls}}(\phi_2)$ is the polymer segment—solvent (diluent) segment interaction function per lattice site.

It is the common practice to identify the volume of the lattice site with the volume of the simple solvent molecule (component 1, $m_{1\text{s},1} \equiv m_1 = 1$). As molar volume of the solvent changes, it is better to keep constant reference volume for the intensity of interactions; $g(\phi_2)$ is related to mole of lattice sites, i.e., it should be independent of degree of polymerization of oligomeric or polymeric diluents. The reason why these seemingly subtle details of the mixing part of the Gibbs energy are discussed here at some length is important: when examining the effect of diluent size in phase separation studies, especially the reaction-induced phase separation, it is important to express the interactions per constant mass or volume of the component.

Passing from Eq. 4.12 to the change of chemical potential of components 1, one gets

$$\frac{\Delta\mu_{1,\text{mix,n}}}{RT} = \frac{1}{RT}\left[\frac{\partial\Delta G_{\text{mix},N}}{\partial N_{\text{ls},1}}\right]_{p,T} \tag{4.15}$$

Recalling that ϕ_2 is a function of $N_{\text{ls},1}$ and so that

$$\frac{\Xi_{\text{e,ls}}}{N_{\text{ls},1}} = \frac{\phi_2}{\phi_1}\zeta_{\text{e}}^* \tag{4.16}$$

where ζ_{e}^* is the dimensionless cycle rank related to the concentration (mol/volume) of EANCs, ν_{e}, as shown below.

$$\zeta_{\text{e}}^* = \zeta_{\text{e}}V_{0\text{ml}}, \quad \zeta_{\text{e}} = \big[(f_{\text{e}} - 2)/f_{\text{e}}\big]\nu_{\text{e}}, \quad \nu_{\text{e}} = \frac{N_{\text{e}}\rho_{\text{bu}}}{N_{\text{Avo}}M_{\text{bu}}} \tag{4.17}$$

where V_{0m1} is the molar volume of lattice site, N_{Avo} is the Avogadro number, N_e is the number of EANCs per building unit, and M_{bu} and ρ_{bu} are molecular weight and specific gravity of the building unit, respectively.

After substituting in Eqs. 4.16 and 4.17, Eq. 4.15 becomes

$$\frac{\Delta\mu_{1,\text{mix,n}}}{RT} = \left(\frac{1}{m_1}\ln\phi_1 + \frac{\phi_2}{m_1} + \frac{f_e - 2}{f_e}\nu_e V_{\text{mol},1}\phi_2 + \left(g(\phi_2) - \frac{\partial g(\phi_2)}{\partial\phi_2}\phi_1\right)\phi_2^2\right) \tag{4.18}$$

the subscript n means network.

Equation 4.18 can be used for determination of the equilibrium swelling of the network in contact with liquid solvent or its vapor.

$$\Delta\mu_{1,\text{mix}} = RT\ln a_1 = RT\ln\frac{f_1}{f_1^0} \approx RT\ln\frac{p_1}{p_1^0} \tag{4.19}$$

where a_1 is the activity of solvent in the swollen gel, f_1 is its fugacity (thermodynamically effective partial pressure), f_1^0 its value in standard state (saturated vapor), and p_1 and p_1^0 are partial pressures of solvent vapors over the solution and pure solvent, respectively. The solvent activity in solution is usually approximated by the ratio of vapor pressures p_1/p_1^0, although the difference of vapor pressure and fugacity for aqueous system may play a role and should be considered when hydrogels are equilibrated with air moisture.

The chemical potential of the polymer $\Delta\mu_2$ is obtained from Eq. 4.13 by differentiation with respect to the number of moles of lattice sites occupied by polymer segments.

$$\Delta\mu_{2,\text{n}}/RT = -\frac{\phi_1}{m_{1\text{s},1}} - \zeta_e V_{\text{mol,ls}}\phi_1 + \left(\frac{\partial g(\phi_2)}{\partial\phi_2}\phi_2 + g(\phi_2)\right)\phi_1^2 \tag{4.20}$$

4.3.2.1 The Interaction Function

The interaction functions $g(\phi_2)$ or $\chi(\phi_2)$ are temperature dependent and under real conditions also concentration dependent. As was pointed out above, the $g(\phi_2)$ function is usually used in conjunction with the Gibbs energy function, ΔG, whereas $\chi(\phi_2)$ is associated with the chemical potential function. The recalculation of $g(\phi_2)$ into concentration dependence of $\chi(\phi_2)$ is shown in Eq. (4.21)

$$\chi(\phi_2) = g(\phi_2) - \frac{\partial g(\phi_2)}{\partial\phi_2}\phi_1 \tag{4.21}$$

The temperature dependence of g can be factored out as shown by Eq. 4.22

$$g = \left[a + \frac{b\phi_2}{1 - c\phi_2}\right] B(T) \tag{4.22}$$

$$B(T) = b_0 + \frac{b_1 T_{\text{ref}}}{T} \tag{4.23}$$

where a, b, c, b_0, b_1, and T_{ref} are constants.

Several other forms of concentration and temperature dependences of $g(\phi_2)$ or $\chi(\phi_2)$ have been used, for instance, the power series form

$$g(\phi_2) = \sum_{i=0} g_i \phi_2^i \quad \chi(\phi_2) = \sum_{i=0} \chi_i \phi_2^i \tag{4.24}$$

or closed form expressions which can be expanded into a power series [13]

$$g(T, \phi_2) = a + \frac{b(T)}{1 - c\phi_2}, \quad b = b_0 + b_1/T + b_2 T \tag{4.25}$$

This form was satisfying the system polystyrene—cyclohexane. For more details, see Appendices 4 and 5 of Ref. [13]. An alternative form has been suggested

$$g = \left[a + \frac{b\phi_2}{1 - c\phi_2}\right] B(T); \quad B(T) = b_0 + \frac{b_1 T_{\text{ref}}}{T} \tag{4.25a}$$

As another example, for poly(ethylene glycol)—water system the following concentration dependence of $\chi(\phi_2)$ fitted well the experimental results [16]

$$\chi(T, \phi_2) = D(T)B(\phi_2); \quad D(T) = d_0 + d_1/T + d_2 \ln T; \quad B(\phi_2) = \frac{1}{1 - b\phi_2} \tag{4.26}$$

We see that, after transforming of Eq. 4.24 and using Eq. 4.22, the dependences of $\chi(\phi_2)$ on ϕ_2 are not the same, but the selected form is more a matter of closeness of fitting experimental data than a theoretical preference.

The interaction parameter is also dependent on pressure, p: it increasers with increasing pressure. This factor may be important for polymer blends (an also for network-polymer systems); release of pressure causes homogenization of blends, for instance, blend of poly(ethyl acrylate) and poly(vinylidene fluoride). The dependence of the interaction function on pressure has the following form [17].

$$g(T, \phi_2, p) = g_0 + g_1\phi_2$$
$$g_1 = g_{10} + g_{11}p(1 - p/p*), \quad g_0 = g_s + g_h/T, \quad g_h = g_{h0} + g_{h1}p + g_{h2}p^2 \tag{4.27}$$

4.3.2.2 Effect of Parameters on Concentration Dependence of Chemical Potential Using (G)FH Model

In the following paragraph, the effect of several variables on the change of chemical potential due to mixing of the solvent (without stretching of network chains) will be shown and discussed. The forms of Eq. 4.18 with concentration independent interaction function (interaction parameter) will be

$$g = \chi$$

and linearly dependent $g(\phi_2)$ and $\chi(\phi_2)$ will be used first

$$\begin{aligned} &g(\phi_2) = g_0 + g_1\phi_2, \quad \chi(\phi_2) = g_0 - g_1 + 2g_1\phi_2 \\ &\chi = \chi_0 + \chi_1 \cdot \phi_2, \quad \chi_0 = g_0 - g_1, \quad \chi_1 = 2g_1 \end{aligned} \tag{4.28}$$

One can see (Fig. 4.1) that the size of the solvent molecule has a large effect on miscibility of very large but *still uncross-linked polymer molecules*, a very well-known and widely documented phenomenon. The value of the interaction function $g = \chi = 0.45$ indicates that dissimilar "polymeric" solvents are very incompatible and their presence during cross-linking results in phase separation. If the interactions of polymer and solvent segments with each other are equal ($g = \chi = 0$, the systems are fully miscible if the polymer is not cross-linked. As soon as cross-links are introduced, the miscibility gap appears; for instance, for $m_1 = 10$, and the immiscibility gap extends to $\phi_2 \approx 0.6$ (Fig. 4.2). Also, the shape of curves is different in that inflexion point develops when the size is changed (Fig. 4.1), whereas the courses remain monotonous when only cross-link density increases.

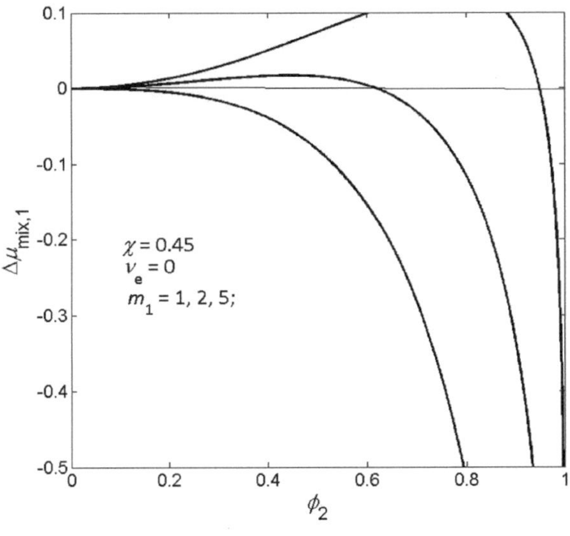

Fig. 4.1 Plot of $\Delta\mu_{1,mix} = RT \ln a_1$ versus volume fraction of polymer in the swollen gel in dependence on size (degree of polymerization) of the solvent molecule for concentration independent interaction function $g = \chi$. Values of other parameters indicated; m_1 upwards

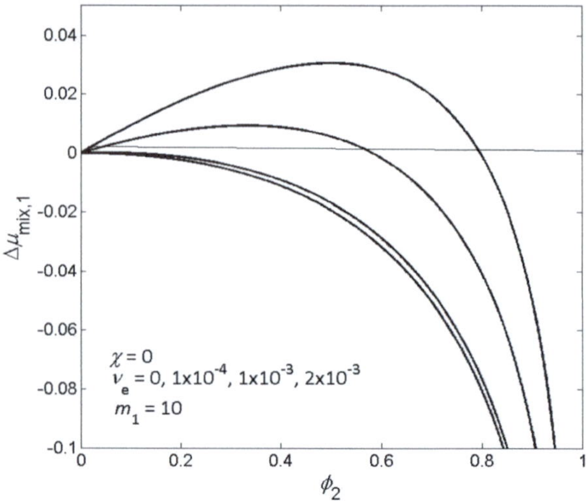

Fig. 4.2 Plot of $\Delta\mu_{1,\text{mix,n}} = RT \ln a_1$ versus volume fraction of polymer in the swollen gel in dependence on concentration of elastically active network chains (EANC), values of other parameters indicated; ν_e upwards

Figure 4.3 shows how the shape of the chemical potential dependence can change when the interaction function (interaction parameter) depends on polymer concentration. For simplicity, only linear dependence of $g(\phi_2)$ on ϕ_2 is examined. One can see that a dependence characterized by two extremes can develop even for uncross-linked systems when the constant g_1 of the power series dependence exceeds a certain limit. This characteristic shape with two extremes is called *van der Waals loop* because of its similarity with van der Waals equation describing condensation of water vapor.

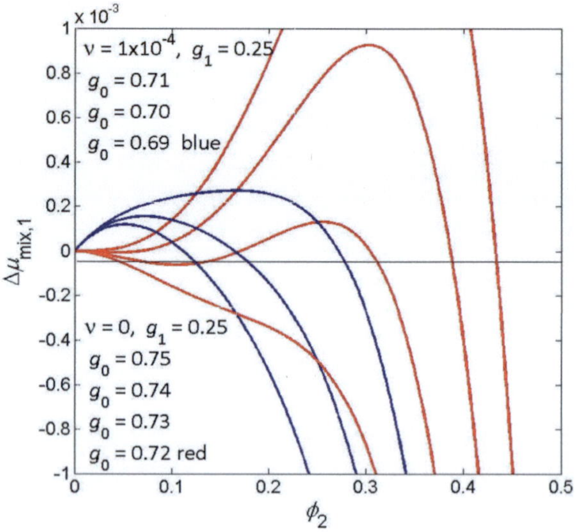

Fig. 4.3 Plot of $\Delta\mu_{1,\text{mix,n}} = RT \ln a_1$ versus volume fraction of polymer in the swollen gel in dependence on the concentration-dependent interaction function (linear dependence $g(\phi_2) = g_0 + g_1\phi_2$; $m_1 = 1$, values of other parameters indicated

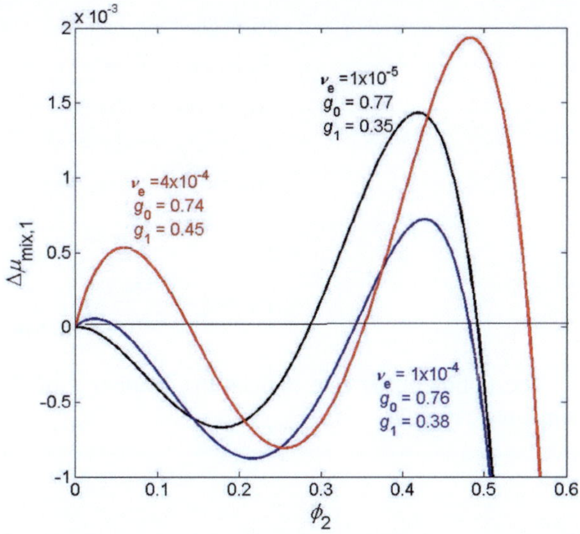

Fig. 4.4 Plot of $\Delta\mu_{1,\text{mix,n}} = RT \ln a_1$ versus volume fraction of polymer in the swollen gel as a function of the concentration-dependent interaction function (linear dependence $g(\phi_2) = g_0 + g_1\phi_2$); $m_1 = 1$, values of other parameters indicated

Figure 4.3 also shows that a large miscibility gap develops upon a subtle change of the constant g_0 (compare the family of red and blue curves) caused for instance by a temperature change (typical, for instance, for LCST system PNIPAM-water).

If the polymer segments are coupled with cross-links, the dependence in Fig. 4.3 with two extremes (with van der Waals loop) is transformed into a normal one (blue curves, without van der Waals loop). However, the sensitivity of swelling degree is extreme and related phase diagrams are of box-type, the term used by F. Tanaka for phase separation of systems with strong physical interactions at slightly different conditions (small change in the coefficients $g_0$and g_1).

The two extremes dependence is transformed into a three-extremes one (Fig. 4.4) pointing to the possibility of a three-phase equilibrium—pure solvent and two swollen gel phases of different degree of swelling—typical for the volume phase transition. The effect of the concentration dependence of the interaction function g on equilibrium swelling is not yet fully investigated especially in view of the possibility that two "off-zero" critical concentrations can exist for uncross-linked polymer of very high molecular weight.

4.3.2.3 Phase Equilibria

The general condition for thermodynamic equilibrium among phases $P = P(\text{a, b, c}, \ldots)$ in a system composed of components $C = C(i, j, k, \ldots)$ at constant temperature T and constant pressure p reads

$$\begin{aligned} p_{\mathrm{a}} &= p_{\mathrm{b}} = p_{\mathrm{c}} = \cdots \\ T_{\mathrm{a}} &= T_{\mathrm{b}} = T_{\mathrm{c}} = \cdots \\ \mu_{i\mathrm{a}} &= \mu_{i\mathrm{b}} = \mu_{i\mathrm{c}} = \cdots \\ \mu_{j\mathrm{a}} &= \mu_{j\mathrm{b}} = \mu_{j\mathrm{c}} = \cdots \\ \mu_{k\mathrm{a}} &= \mu_{k\mathrm{b}} = \mu_{k\mathrm{c}} = \cdots \\ &\vdots \\ \mu_i &= \left(\frac{\partial \Delta G_{\mathrm{mix}}}{\partial N_i}\right)_{T,p,N_j N_k,\ldots} \end{aligned} \tag{4.29}$$

The Gibbs phase rule defines the number f of *degrees of freedom* (called also *variance*); it reads

$$f = C + 2 - P \tag{4.30}$$

For *binary* polymer solution (C = 2), for example, the phase rule allows existence of a single liquid phase in equilibrium with the vapor phase at various temperatures and pressures; or two liquid phases can coexist at various temperatures *or* pressures ($P = 3, f = 1$). Three liquid phases in equilibrium with vapor phase can coexist but only at a fixed pressure and temperature (such system is non-variant, $f = 0$). The three-liquid phase systems have been found and are described in the monograph (Chap. 9.3 of Ref. [13]). Another classical example is represented by the three-phase equilibrium encountered in systems exhibiting phase volume transition [18].

Phase diagrams for strictly binary systems are characterized by *binodals* delimiting the single liquid phase region from stable two-phase region, *spinodals* delimiting the metastable region from the absolutely unstable region, and critical points showing the maximum (UCST) or minimum (LCST) temperature above or below which the system does not phase separate. These characteristics are given by the relations (Table 4.1).

Table 4.1 Thermodynamic functions determining phase equilibrium

Binodal	$\left(\frac{\partial \Delta G_{\mathrm{mix}}/N}{\partial \phi_1}\right)_{\text{phase } a} = \left(\frac{\partial \Delta G_{\mathrm{mix}}/N}{\partial \phi_1}\right)_{\text{phase b}}$ ...	$\Delta\mu_{1a} = \Delta\mu_{1b}$ $\Delta\mu_{2a} = \Delta\mu_{2b}$
Spinodal	$\left(\frac{\partial^2 (\Delta G_{\mathrm{mix}}/N)}{\partial \phi_2^2}\right)_{T,p} = 0$	$\left(\frac{\partial \Delta\mu_1}{\partial \phi_2}\right)_{T,p} = 0$
Critical point	$\left(\frac{\partial^3 (\Delta G_{\mathrm{mix}}/N)}{\partial \phi_2^3}\right)_{T,p} = 0$	$\left(\frac{\partial^2 \Delta\mu_1}{\partial \phi_2^2}\right)_{T,p} = 0$

For the (G)FH system (with the concentration-dependent interaction function), without the cross-linking term, the spinodal and critical point conditions read:

spinodal:

$$\frac{1}{m_1\phi_1} + \frac{1}{m_2\phi_2} - 2g + 2(\phi_1 - \phi_2)\frac{\partial g}{\partial \phi_2} + \phi_1\phi_2\frac{\partial^2 g}{\partial \phi_2^2} = 0 \tag{4.31}$$

critical point

$$\frac{1}{m_1\phi_1^2} + \frac{1}{m_2\phi_2^2} - \frac{6\partial g}{\partial \phi_2} + 3(1 - 2\phi_2)\frac{\partial^2 g}{\partial \phi_2^2} + \phi_2(1 - \phi_2)\frac{\partial^3 g}{\partial \phi_2^3} = 0 \tag{4.32}$$

4.3.2.4 Off-Zero Critical Concentration Systems

Such behavior is typical for systems showing the so-called *off-zero critical concentration*, the behavior mentioned first by Flory and Daoust [19] for poly(isobutylene) solutions, found later for solutions of poly(2-hydroxyethyl methacrylate) in aliphatic alcohols by Dušek [2], treated more generally by Erman and Floryv [20], and analyzed in depth by Šolc et al. [21] and eventually applied to volume phase transition in aqueous solutions of poly(isopropylacrylamide) (PNIPAM) by Moerkerke et al. [18] and poly(vinyl methyl ether) by Schaefer et al. [22].

What does *off-zero critical concentration* mean? Usually, the coexistence curve—binodal—delimiting one-phase system from a stable two-phase system is asymmetric for polymer–solvent systems and its maximum—the critical point—increases or decreases depending on whether the system is of UCST or LCST type and moves toward lower concentrations. It reaches concentration zero when its molecular weight is very large (see Fig. 4.5a for close to critical value of g_1). The ultimate temperature at $\phi_{2c} = 0$ is called the Θ-temperature; its value is characteristic for the given polymer–solvent pair. However, for a certain concentration dependences of the interaction function, for a polymer of very large molecular weight ($M \to \infty$) the critical point remains located at $\phi_2 > 0$ (Fig. 4.5b).

The analysis made by Šolc et al. in Ref. [21] shows that the (G)FH systems with quadratic concentration dependence of the interaction function $g(\phi_2) = g_0 + g_1\phi_2 + g_2\phi_2^2$ converge to the *off-zero critical concentration* behavior if $g_2 - g_1 > -1/6$ including the case $g_2 = 0$ ($g_1 > 1/6$). Thus the difference

$$\Delta g_2 \equiv g_2 - g_1$$

determines the number and type of limiting critical points. The analysis is based on equations for spinodal and critical point Eqs. 4.31 and 4.32 which, for g_i ($i > 2$) = 0, transforms to

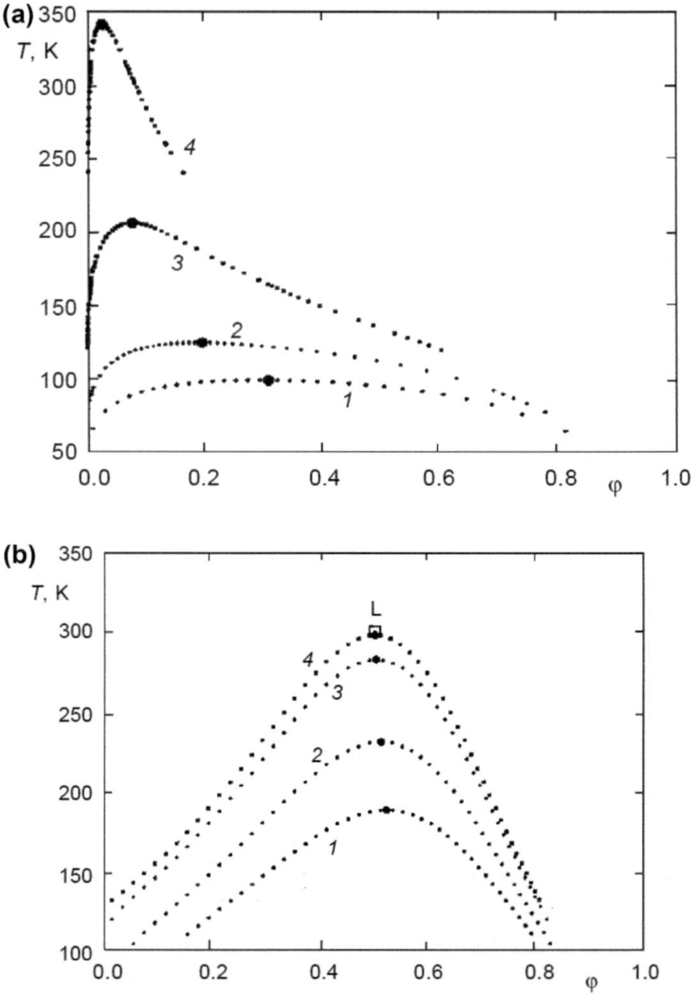

Fig. 4.5 **a** Phase diagrams for systems with $g_0 = 0.5 + 50/T$, $g_1 = 0.1$, $g_2 = 0.3$. Critical points computed for chain lengths (m_2): *1* 10, *2* 20, *3* 100, *4* 1 000; ❑ a limiting CP for $m \to \infty$. Reprinted from Ref. [18] with permission. **b** Phase diagrams for systems with $g_0 = 0.5 + 50/T$, $g_1 = 2/3$, $g_2 = 0.0$. Critical points (CP) computed for chain lengths (m_2): *1* 10, *2* 20, *3* 100, *4* 1 000; ❑, a limiting CP for $m \to \infty$. Reprinted from Ref. [21] with permission

$$\varphi_{2c}\left[1 + 2\Delta g_1 + \varphi_{2c}(1 + 6\Delta g_2) + \varphi_{2c}^2(1 + 12\Delta g_2) + \cdots\right] + m_{2w}^{-1} = 0 \qquad (4.33)$$

$$\varphi_{2c}^2\big[1 + 6\Delta g_2 + 2\varphi_{2c}(1 + 12\Delta g_2) + \cdots - m_{2z}m_{2w}^{-2} = 0$$
$$\Delta g_i = g_i - g_{i-1} \qquad (4.34)$$

Equations 4.31 and 4.32 further simplify for $m_2 \to \infty$.

The existence conditions for none, one or two "off-zero" critical concentrations are summarized in Table 4.2 [21].

The situation is illustrated graphically in Fig. 4.6.

Sometimes, the uncross-linked system shows up more complex behavior marked by occurrence of both zero and off-zero CPs. The phase diagram for such system is shown in Fig. 4.7.

The tie lines refer to equilibria between pure solvent and polymer solution (A), others to equilibria between two polymer solutions of different concentrations (B). In this case, the "new" sigmoidal binodal protrudes through the "regular" one, becoming partly stable. The two binodals intersect at $T3$ showing the existence of non-variant three-phase equilibrium: polymer solution of different concentrations in equilibrium with pure solvent (water). This is the case close to the PNIPAM-water system showing up volume phase transition. Cross-linking is not necessary. For application it is useful to keep the system in the gel state also in the region (at temperatures) when the chains are expanded and would fully dissolve in excess solvent when not kept together by covalent cross-links.

Table 4.2 Conditions for appearance of off-zero critical points

Δg_1	Δg_2		Number of "off-zero" critical points
$< -1/6$	Any value		None
		Criterion $G(g_2) \equiv 4g_2 - (12g_2)^{2/3}/2$	
$> -1/6$	$< 1/12$	$\Delta g_2 > G$	None
	$> 1/12$	$\Delta g_2 = G$	One = double CP
		$-1/6 < \Delta g_2 < G$	Two

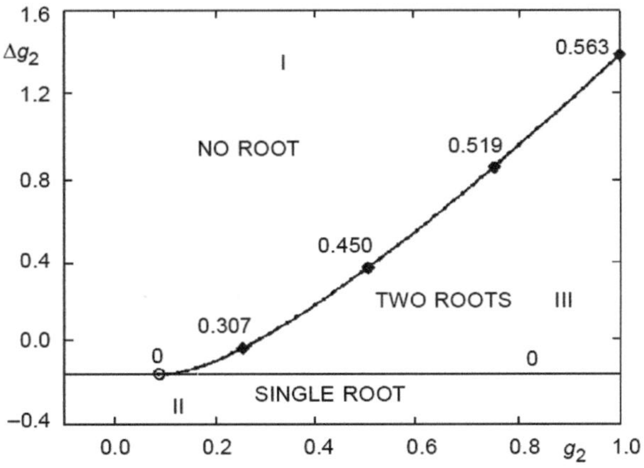

Fig. 4.6 Existential boundaries of various off-zero critical points as a function of interaction parameter coefficients. Quadratic expansion of interaction function. Numbers specify the off-zero critical concentrations. Roman numerals mark various regions. The zero CPs present for $\Delta g_2 > -1/6$ in regions I and III are not included. Reprinted from Ref. [21] with permission

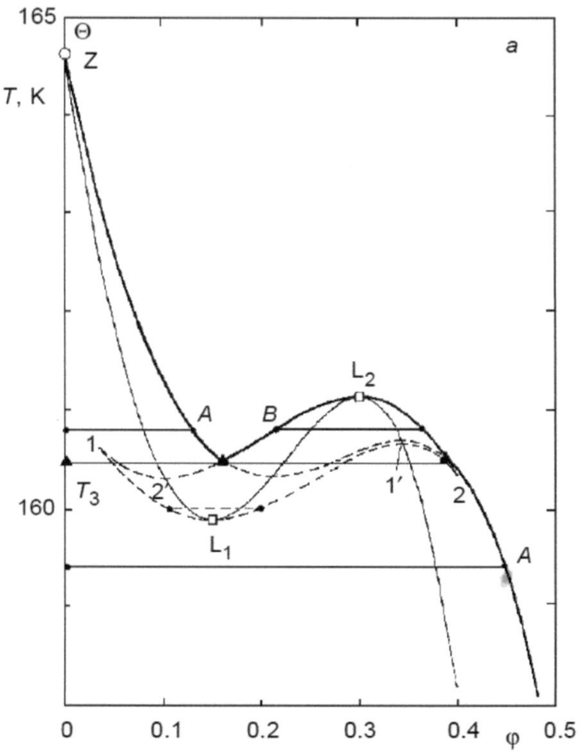

Fig. 4.7 Critical points Z (zero type) and L_1 and L_2 (off-zero type) for systems with interaction functions $g_0 = 0.5 + 50/T$, $g_1 = 8/27$, $g_2 = 16/81$ with $\varphi_{2c} = 0.15$ and 0.3, respectively, Some tie lines shown

4.3.2.5 Effect of Distributions

The degree of polymerization/molecular weight and composition distributions of the polymer component do not pose any important problem in the case of gels that do not contain any soluble fraction (sol). However, the distribution should be considered in the case of physical networks or combinations covalent/physical networks, when these networks are prepared from a polymer component bearing stickers or if aggregates are formed of lifetime of cross-links exceeding or commensurable with duration of experiment. Consideration of dispersity is of utmost importance when experimental results are analyzed to prove or disprove agreement of the theory with experiments. One should not forget that cloud-point curve is not a binodal and that the critical point is not located at maximum of the cloud-point curve. We will touch this problem only briefly.

In polymer thermodynamics, such systems are treated as *quasibinary*—the polymer component, or both polymer and solvent components are disperse in m but the mers of either component are of the same chemical composition. We will discuss quasibinary systems only briefly as a subcategory of multicomponent systems.

Let the polymer (component 2) disperse in degree of polymerization be characterized by sets of number fractions, m_k, and weight (second-moment) fractions, w_k, and their number- and weight-averages are m_n and m_w, respectively. We will distinguish distributions within component of type 1 solvent, m_{1k}, and polymer, m_{2l}. The mixing Gibbs energy within the so-called quasibinary description (for details, see Ref. [13], Chap. 10) is given by

$$\frac{\Delta G_\mathrm{mix}}{NRT} = \sum_k \left(\frac{\phi_{1k}}{m_{1k}}\right) \ln \phi_{1k} + \sum_l \left(\frac{\phi_{2l}}{m_{2l}}\right) \ln \phi_{2l} + g\phi_1\phi_2 \tag{4.35}$$

where $\phi_1 = \sum_k \phi_{1k}$ and $\phi_2 = \sum_l \phi_{2l}$

It is easier to deal first with the spinodal condition which is determined by the matrix of second derivatives of $\Delta G_\mathrm{mix}/N$, $\mathbf{G}^{(2)} = \left(G_{ij}^{(2)}\right)$ with respect to all independent concentration variables

$$G_{ij}^{(2)} = \left(\frac{\partial^2(\Delta G_\mathrm{mix}/N)}{\partial\phi_i\partial\phi_j}\right)_{T,p} \tag{4.36}$$

The spinodal condition reads

$$J_\mathrm{s} = \det\left|G_{ij}^{(2)}\right| = 0$$

The resulting *spinodal condition is independent of the type of distribution* and it is determined only by weight (second-moment) averages of degrees of polymerization

$$\frac{1}{m_\mathrm{w1}\phi_1} + \frac{1}{m_\mathrm{w2}\phi_2} - 2g + 2(\phi_1 - \phi_2)\frac{\partial g}{\partial\phi_2} + \phi_1\phi_2\frac{\partial^2 g}{\partial\phi_2^2} = 0 \tag{4.37}$$

For the critical point one gets (conditions in Table 4.1)

$$\frac{\xi_\mathrm{w1}}{m_\mathrm{w1}\phi_1^2} + \frac{\xi_\mathrm{w2}}{m_\mathrm{w2}\phi_2^2} - \frac{6\partial g}{\partial\phi_2} + 3(\phi_1 - \phi_2)\frac{\partial^2 g}{\partial\phi_2^2} + \phi_1\phi_2\frac{\partial^3 g}{\partial\phi_2^3} = 0 \tag{4.38}$$

In these equations and below

$$\xi_\mathrm{w1} = \frac{m_\mathrm{z1}}{m_\mathrm{w1}}, \quad \xi_\mathrm{n1} = \frac{m_\mathrm{w1}}{m_\mathrm{n1}}$$

$$m_\mathrm{w1} = \frac{\sum m_{1j}\phi_{1j}}{\sum \phi_{1j}}, \quad m_\mathrm{z1} = \frac{\sum m_{1j}^2\phi_{1j}}{\sum m_{1j}\phi_{1j}}, \quad m_\mathrm{n1} = \frac{\sum \phi_{1j}}{\sum \phi_{1j}/m_{1j}}$$

and analogous relations hold for component 2.

For the critical state of the zero concentration type, one can get explicit expressions for the critical concentration

$$\phi_{2,\text{crit}} = \frac{(m_{w1}\xi_{w1})^{1/2}}{(m_{w1}\xi_{w1})^{1/2} + (m_{w2}\xi_{w2})^{1/2}}$$
$$2\chi_{\text{crit}} = \left(m_{w1}^{-1/2} + m_{w2}^{-1/2}\right)^2 + \frac{\left(\xi_{w1}^{1/2} - \xi_{w2}^{1/2}\right)}{(m_{z1}m_{z2})^{1/2}} \tag{4.39}$$

for concentration independent interaction function $g = \chi$. For strictly binary systems, $m_n = m_w = m_z = m$; $\xi_w = 1$

4.3.2.6 Cloud Points and Coexistence Between Phases

When we enter the two-phase region by increasing $g(T)$, the onset of immiscibility is marked by appearance of turbidity. This point is called *cloud point*.

The dependence of the cloud-point temperature on polymer concentration may remind binodal characterizing the strictly binary system but it is very different from the binodal. An example is shown in Fig. 4.8 (Fig. 4.9).

This is because the condition of equality of chemical potentials in different phases is valid for each component (solvent—component 1, as well as each mer—component $2k$). The equilibria are determined by a set of equations

$$\frac{\Delta\mu_1/RT}{m_1} = \frac{\ln\phi_1}{m_1} + \frac{1}{m_1} - \frac{\phi_1}{m_1} - \frac{\phi_2}{m_{n2}} + \vartheta_1\phi_2^2$$
$$\frac{\Delta\mu_{2k}/RT}{m_{2k}} = \frac{\ln\phi_{2k}}{m_{2k}} + \frac{1}{m_{2k}} - \frac{\phi_1}{m_1} - \frac{\phi_2}{m_{n2}} + \vartheta_2\phi_1^2$$
$$\vartheta_1 = g + \phi_1\frac{\partial g}{\partial\phi_2}, \quad \vartheta_2 = g + \phi_2\frac{\partial g}{\partial\phi_2} \tag{4.40}$$

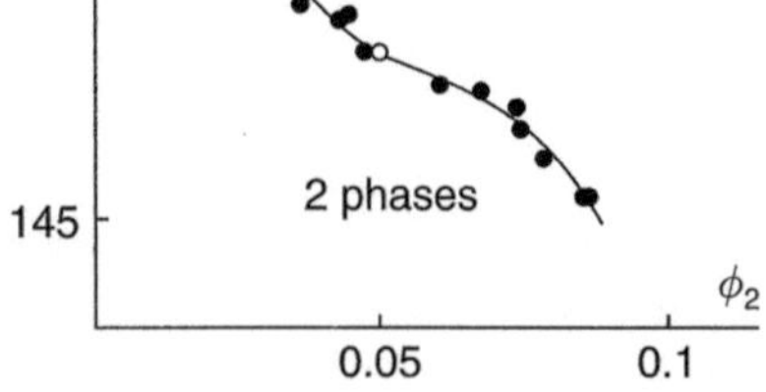

Fig. 4.8 Experimental cloud-point curve and critical point. Diphenyl ether—polyethylene (Mw = 140,000 g/mol; $M_w/M_n = 1.5$; $M_z/M_w = 1.9$); • cloud points, of critical point. Reprinted from [13] (Fig. 137c), with permission

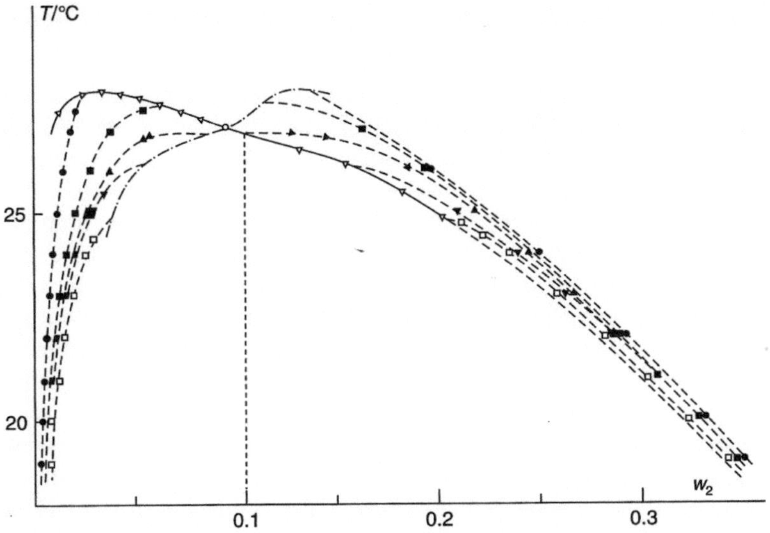

Fig. 4.9 Quasibinary phase diagram of solution of polystyrene in cyclohexane. Coexistence curves (dashed) at overall polymer concentrations 2 (●), 6 (■), 10 (▲), 15 (▼), and 20 (□) wt%. Critical point ○. Cloud-point curve (full), shadow curve (dash-dot), Mw = 210,000 g/mol, Mw/Mn = 1.7; Mz/Mw = 2.4. Reprinted from [13], Fig. 140, with permission

where k extends over the whole distribution of degrees of polymerization or its equivalent.

Note that the critical point is not located at the extremum of the cloud-point curve (CPC) but on its shoulder. Each point on CPC represents a bulk phase coexisting with a second phase which is called *shadow* phase (on its verge of appearing). The CPC and shadow curve intersect at the *critical point*. For the k-mer, there exist two equilibrium concentrations.

For more details as well as occurrence of three-phase equilibria determined by the molecular weight distribution, see, e.g., the monograph by Koningsveld et al. [13].

Concluding the reading of Sect. 4.3.2, one may ask—what is this whole exercise with phase stability of artificial systems good for? It has shown that a relatively simple concentration dependence of the interaction function (interaction parameter) can bring about a peculiar but interesting phase behavior of solutions of polymers of high molecular weight even without necessity of introducing covalent cross-links. At the beginning of this section, we showed that introduction of cross-links brings about a large and indisputable effect on mixing behavior. In the last part of the analysis made above, we did not discuss the cross-link effect. It could have been done easily, but it would have done this part perhaps too extensive. Nevertheless, addition of the cross-link term as in Eqs. (4.11)–(4.18) adds a new dimension to possible combinations.

4.3.3 Mixing Models—Alternatives of Flory–Huggins Theory

The classical theory of Flory and Huggins (FH) for mixing of polymers with solvent or polymer–polymer compatibility is based on a lattice model for polymer solutions wherein all lattice sites are occupied by segments of molecules, no vacant sites are allowed, and constant volume is maintained. Several simplifications and approximation were used to obtain the Helmholtz energy of mixing. The simplifications included the assumption of a large coordination number of the lattice, $z \rightarrow \infty$ reducing thus the interacting surface fraction to volume fraction and the interaction energy to a simple enthalpic contribution. Although the FH formulation has been the most widely used one for ΔG_{mix}, it was evident that it was an (unnecessary) oversimplification. For the derivation of combinatorial entropy, a lattice model is even not needed (cf., e.g., [13, 23]).

Since the original FH interaction parameter is proportional to $1/T$ which corresponds to the phase diagrams with Upper Critical Solution Temperature (UCST), the FH theory could explain neither the existence of the Lower Critical Solution Temperatures (LCST), nor closed loop and hour-glass immiscibility gaps, nor the peculiar behavior of aqueous solutions manifested by complicated concentration and temperature dependences of the interaction parameters (interaction function). These complicated and unpredictable dependences of the interaction parameters are due to the fact that the interaction function accumulates all entropic and enthalpic effects of mixing of chain-like molecules having a certain volume, compressibility of components and their mixtures, and specific interactions between chain-like molecules and solvents of various sizes. In other words, ΔG_{mix} should describe better the solution behavior of real uncross-linked system with weak as well as strong interactions like hydrogen bonding; and their dependence on thermodynamic properties of pure components should be taken into account. Also, it should not be forgotten that a part of the surface is excluded from interaction by introducing cross-links, and that the access to more distant units is limited.

The variety of types of phase diagrams is illustrated by Fig. 4.10.

Figure 4.10 shows that with increasing molecular weight of the polymer, the immiscibility region grows bigger and the coexistence curves get more asymmetric. When cross-links are introduced, the networks are insoluble, and the swollen phase is normally in equilibrium with pure solvent. Remember, however, that for special systems two-gel phases of different swelling degree may be in equilibrium. The cross-linked systems developed from the closed-loop systems by a minimum degree of swelling (maximum ϕ_2) and those developed from the hour-glass system are characterized by a maximum degree of swelling (minimum ϕ_2) at a certain temperature.

The analytical theories of thermodynamics of mixtures were first developed for mixtures of low-molecular-weight compounds and were modified to describe polymer solutions and blends. They can be grouped in three main categories: (a)

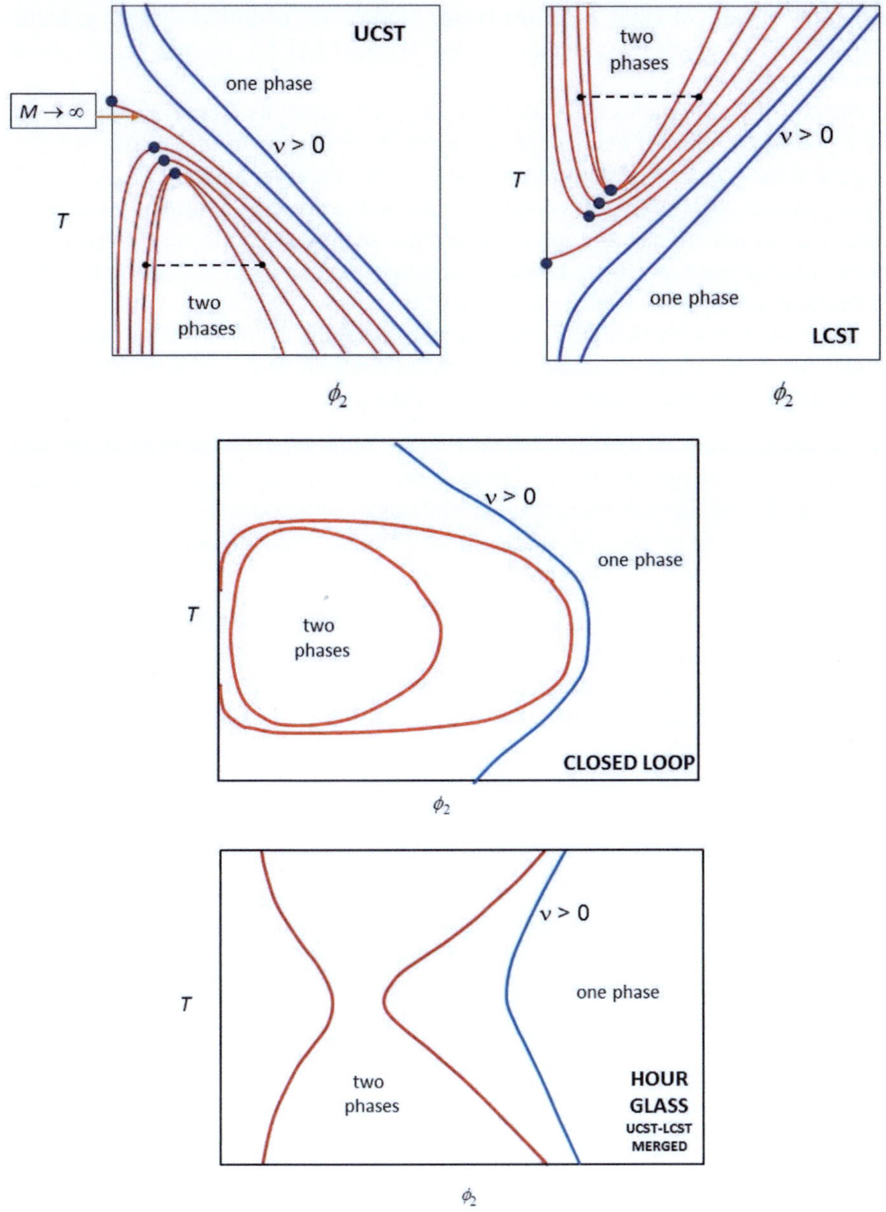

Fig. 4.10 Types of phase diagrams: systems with upper critical solution temperature (UCST), lower critical solution temperature (LCST), closed loop-shaped, and hour glass-shaped. Red curves—soluble polymers of finite (lower) molecular weight, blue curves—after cross-links were introduced

equations of state of cubic type; (b) lattice models; (c) perturbation theory, particularly SAFT (Statistical Associating Fluid Theory (SAFT)). Many hybrid variants were described.

Several review articles on non-dilute polymer solutions thermodynamics have been published, some of them emphasizing their application in equilibrium swelling of gels. Examples of the more recent ones can be found under Refs. [5, 6, 24, 25]. Highly swollen gel (hydrogel) systems correspond to the semi-dilute solutions where excluded volume effects are operative and the scaling relations are different from those valid for the concentrated solutions regime. A good overview can be found for the monograph by Rubinstein and Colby [23].

In this section, we will briefly discuss recent development of the lattice models and model based on equations of state constructed from properties of pure components and changes of enthalpy and entropy upon mixing.

(1) Consideration of compressibility (lattice fluid models: corresponding state theory, Prigogine-Patterson, lattice-fluid (LF) theory of Sanchez-Lacombe, Orwoll-Vrij-Flory, Koningsveld-Kleintjens)
(2) Consideration of specific interactions (different attracting and repulsing sites, hydrogen bonding, hydrophobic interactions

- universal functional-group activity coefficient (UNIFAC) (group contribution method)
- perturbed chain-statistical associating fluid theory (PC-SAFT)
- (modified) double lattice model (DL and MDL) 5, consideration of temporary network formation by strong interaction (association of "stickers", thermoreversible gels), sol–gel coexistence
- ΔG_{mix} through coefficients of virial expansion of osmotic pressure (Douglas).

In this section, we go neither into details of these approaches, nor offer final relations ready for applying to real systems. The purpose of this overview is to show how structural factors of components and strength and cooperativity of interactions determine the solution and swelling behavior of interactions of various strength. In fact, no comparison of theories has been made so far using selected experimental systems, so that preferences in selecting one of these methods is still to a certain degree a matter of taste.

4.3.3.1 Improvements of the Lattice Models

The simple lattice model of Flory and Huggins was challenged from the beginning by other models describing mixtures of low molecular weight compounds and adapted for high molecular weight components, and more molecular structure-based models.

A **Quasichemical equilibrium model**

One of the important models was the quasichemical equilibrium model developed by Guggenheim [26–28] and amended by Tompa [29, 30] and others. The solvent and polymer molecules are placed on lattice with finite coordination number, z, and the interaction contribution is made proportional to interacting surface (characterized by surface fraction ξ_2) rather than volume fraction.

$$\xi_1 = \frac{N_1}{N_1 + qN_2}, \quad \xi_2 = \frac{qN_2}{N_1 + qN_2}; \quad \phi_1 = \frac{N_1}{N_1 + mN_2}, \quad \phi_2 = \frac{mN_2}{N_1 + mN_2} \tag{4.41}$$

for binary system, where q is defined by

$$qz = m(z - 2) + 2 \tag{4.42}$$

so that for $m = 1$, $q = z$, but for $m > 1$, $q < z$.

The interaction contribution is determined by equilibrium constant characterizing the formation and dissociation of sites pairs (concentrations X_{12}, X_{11}, X_{22}, respectively)

$$\frac{(X_{12})^2}{4X_{11}X_{22}} = \eta^2 = e^{-2w/kT} \tag{4.43}$$

where the energy w has generally the character of free energy, f; η can vary from zero to a very large number corresponding to formation of a very strong (covalent) bond; consequently w can be positive or negative. The Guggenheim-Tompa model served as a basis of further development of thermodynamics of non-dilute polymer solutions.

For a binary system

$$\frac{\Delta G_{\text{mix}}}{RT} = \frac{\Delta G^*_{\text{mix}}}{RT} + \frac{1}{2} zN_1 \ln \frac{1 - \kappa\xi_2}{\xi_1} + \frac{1}{2} qzN_2 \ln \frac{1 - \kappa\xi_1}{\xi_2} \tag{4.44}$$

$$1 - \kappa = \kappa^2 \xi_2 \xi_2 \left(\eta^2 - 1\right) \tag{4.45}$$

where ΔG^*_{mix} corresponds to regular solution with $\kappa = \eta = 1$.

For the chemical potentials, one gets

$$\begin{aligned}\frac{\Delta\mu_1}{RT} &= \ln\phi_1 + \frac{z}{2}\left(\ln\frac{\xi_1}{\phi_1} + \ln\frac{1-\kappa\xi_2}{\xi_1}\right)\\ \frac{\Delta\mu_2}{RT} &= \ln\phi_2 + \frac{zq}{2}\left(\ln\frac{\xi_2}{\phi_2} + \ln\frac{1-\kappa\xi_1}{\xi_2}\right)\end{aligned} \tag{4.46}$$

Expansion of these expressions into power series of ϕ_2, the following equation is obtained

$$\frac{\Delta\mu_1}{RT} = \ln\phi_1 + \left(\frac{m-1}{m}\right)\phi_2 + \left[\frac{1}{z}\left(\frac{m-1}{m}\right)^2 + \frac{z}{2}(\eta^2-1)\left(\frac{q}{m}\right)^2\right]\phi_2^2 + \cdots \tag{4.47}$$

so that for $m \to \infty$, $\chi \approx 1/z + ((z-2)^2/2z)(\eta^2-1) + \cdots$.The interaction (free) energy w or f is usually related to monomer unit, so that in Eq. 4.43 w (or f) must decrease with z. Although originally w was considered as fully enthalpic, it reflects also entropic changes if they participate in the equilibrium constant.

Very useful was the extension of the quasichemical (QE) equilibrium approach to solutions of polymers containing *different kinds of segments* and especially *different kinds of contact surfaces* (Sects. 4.9 and 4.10 of Ref. [30], see also [29]). The latter case refers to systems with specific and oriented interactions and relates to hydrophilic polymer systems interacting with H-bond-forming solvents and correlated hydrophobic interactions. The interacting sites are distinguished in their interaction type and strength (α, β, γ, ...) and distributed in the i-th component.

$$\sum_\alpha q_{\alpha i} = q_i \tag{4.48}$$

The quasichemical equilibrium Eq. (4.43) was replaced by a set of equations

$$\frac{4X_{\alpha\alpha}X_{\beta\beta}}{X_{\alpha\beta}^2} = \eta_{\alpha\beta}^2 \tag{4.49}$$

which eventually leads to chemical potentials in the form

$$\frac{\Delta\mu_i}{RT} = \frac{\Delta\mu_i^*}{RT} + \frac{z}{2}\sum_\alpha q_{\alpha i}\ln\frac{X_{\alpha\alpha}/X_{\alpha\alpha}^*}{X_{\alpha\alpha}^{i0}/X_{\alpha\alpha}^{i0*}} \tag{4.50}$$

where the asterisk refers to $\eta_{\alpha\beta} = 1$ and the superscript 0 to pure compounds. One has to solve simultaneously the sets of equilibrium equations for concentrations X, knowing all values of $\eta_{\alpha\beta}$ but certain simplifications are possible (see below).

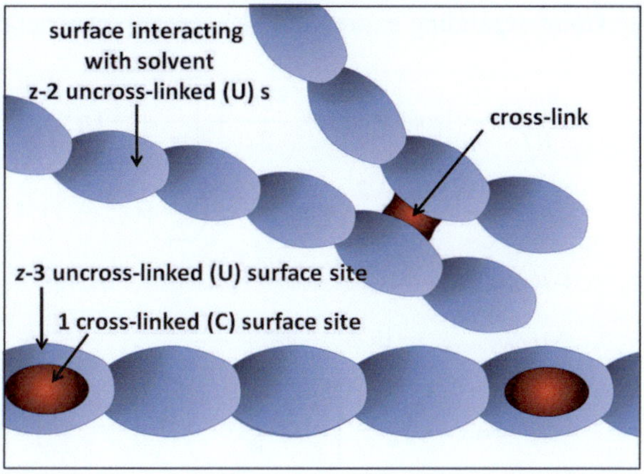

Fig. 4.11 Schematic representation showing of a part of cross-linked polymer; cross-linked surface sites are exposed from which a cross-link is formed

B Quasichemical equilibrium model and effect of cross-links on contribution by mixing

The latter approach was used by Dušek [31, 32] to model the effect of cross-links (cross-linked units) on the interaction of polymer chains with solvent. In the simplest case, the cross-linked surface sites (marked c) form a very strong bonds, so that $X_{cc} = X_c^0/2$; the remaining sites are uncross-linked (marked u) and available to interaction with other uncross-linked sites. The fraction of monomer bearing cross-linked surface sites is denoted by v and the fraction of the uncross-linked units is equal to $1 - v$. The corresponding fractions are denoted as u and c, respectively.

$$zq_u = (1 - v)[m(z - 2) + 2] + v[m(z - 3) + 2], \quad zq_c = mv \tag{4.51}$$

We consider X_{ss}, X_{uu}, X_{us} while $X_{cs} = X_{cu} = 0$, $X_{cc} = X_c^0/2$.

A part of chain bearing cross-linked and uncross-linked surface sites is schematically shown in Fig. 4.11.

After manipulation similar to that used for linear chains, one gets for the chemical potential of the solvent and polymer the equations

$$\begin{aligned} \frac{\Delta\mu_{1,n}}{RT} &= \ln\phi_1 + \frac{z}{2}\left(\ln\frac{\xi_1}{\phi_1} + \ln\frac{1 - \kappa u \xi_2}{\xi_1}\right) \\ \frac{\Delta\mu_{2,n}}{RT} &= \ln\phi_2 + \frac{zq}{2}\left(\ln\frac{\xi_2}{\phi_2} + u\ln\frac{1 - \kappa\xi_1}{\xi_2} - c\ln\xi_2\right) \end{aligned} \tag{4.52}$$

Note the difference when Eq. 4.52 is compared with equations describing solutions of linear chains. This difference is marked by subscript n, $\Delta\mu_{i,n}/RT$. The non-random factor κ is determined by the relation

$$1 - \kappa(1 - c\xi_2) = \kappa^2\xi_1\xi_2 u\left(\eta^2 - 1\right) \tag{4.53}$$

Upon expanding expression (4.71) into power series of ϕ_2, one gets

$$\begin{aligned}
\frac{\Delta\mu_{1,\mathrm{n}}}{RT} &= \ln\phi_1 + \phi_2 + \frac{\nu\phi_2}{2} + \left[\frac{1}{z} + \frac{z}{2}(\eta^2 - 1)\left(\frac{z-2}{z}\right)^2\right]\phi_2^2 \\
&+ \nu\left[\frac{1}{z} - \frac{z-2}{z}(\eta^2 - 1) + \left(\frac{\nu}{2z}\right)(\eta^2 - 1) + \frac{\nu}{2z}\right]\phi_2^2 + \cdots \\
\Delta\mu_{1\mathrm{mix},,\mathrm{n}} &= \Delta\mu_{1\mathrm{mix,comb}} + \left(\Delta\mu_{1\mathrm{mix\,,excess}}\right)_{\mathrm{u}} + \left(\Delta\mu_{1\mathrm{mix,excess}}\right)_{\mathrm{c}} \\
\Delta\mu_{1\mathrm{mix,comb}} &= \ln\phi_1 + \phi_2 + \frac{\nu\phi_2}{2} \\
\left(\Delta\mu_{1\mathrm{mix,excess}}\right)_{\mathrm{u}} &\approx \left[\frac{1}{z} + \frac{z}{2}(\eta^2 - 1)\left(\frac{z-2}{z}\right)^2\right]\phi_2^2 \\
\left(\Delta\mu_{1\mathrm{mix,excess}}\right)_{\mathrm{c}} &\approx \nu\left[\frac{1}{z} - \frac{z-2}{z}(\eta^2 - 1) + \left(\frac{\nu}{2z}\right)(\eta^2 - 1) + \frac{\nu}{2z}\right]
\end{aligned} \tag{4.54}$$

One can see that, in addition to the combinatorial term for linear chains, there is another combinatorial term due to cross-links, $\nu\phi_2/2$. The pairwise interactions term also depends on the fraction of surface sites engaged in cross-links. When $z \to \infty$, $\nu = 0$, the FH equation for solutions of linear polymer of very high degree of polymerization is recovered.

The model was extended to systems where the networks that are formed stepwise with cross-links having functionality different from 4 [32] and more than one kind of interacting sites exist—functional groups, formed groupings, backbone units. The interacting sites and cross-linked sites distribution was modeled by the statistical branching theory and conversion and composition dependences were taken into account. Also, the composition differences between sol and gel were taken into account to describe correctly swelling of an extracted gel.

C Other versions of the quasiequilibrium model

Prange, Hooper, and Prausnitz [33] developed a quasichemical equilibrium model for segments containing different interacting surface sites which is suitable to deal with systems governed by strong interactions like hydrogen bonding—an approach similar to that discussed above. The non-random factors, called, Γ, similar to the variable X in Eq. 4.49 were determined by the quasichemical equilibrium relations of the form

$$\frac{\left(\Gamma_{ij}^{\rho\sigma}\right)^2}{\Gamma_{ij}^{\rho\rho}\Gamma_{ij}^{\sigma\sigma}} = e^{-2w_{ij}^{\rho\sigma}/kT} \tag{4.55}$$

where i and j designate components and ρ and σ types of surface sites (for instance H-donating sites(α), H-accepting sites β, and sites interacting by weak dispersion forces (D). The interchange (free) energies w are related to (free) energies of pure components (subscripts ii, jj)

$$w_{ij}^{\rho\sigma} = u_{ij}^{\rho\sigma} - u_{ii}^{\rho\rho}/2 - u_{jj}^{\sigma\sigma}/2 \tag{4.56}$$

A binary system ($i = 1, j = 2$) with three types of interacting sites (α, β, D) is described by 21 non-random factors Γ; their values can be calculated numerically if the interchange energies w are known. The chemical potential of component 1 (solvent, e.g., for $m = 1, q = 1$) for a solution of linear or branched polymer of high molecular weight reads

$$\begin{aligned}\frac{\Delta\mu_{1,m}}{RT} &= \ln\phi_1 + \phi_2 - \sum_{k=1}^{m_1}\frac{1}{2}q_1(\rho_1|k)(z_1^{\alpha}|k)\ln\frac{\Gamma_{11}^{\alpha\alpha}|_{\mathrm{kk_p}}}{\Gamma_{11}^{\alpha\alpha}|_{\mathrm{kk_m}}}\\ &\quad - \sum_{k=1}^{m_1}\frac{1}{2}q_1(\rho_1|k)(z_1^{\beta}|k)\ln\frac{\Gamma_{11}^{\beta\beta}|_{\mathrm{kk_p}}}{\Gamma_{11}^{\beta\beta}|_{\mathrm{kk_m}}}\\ &\quad - \sum_{k=1}^{m_1}\frac{1}{2}q_1(\rho_1|k)(z_1^{D}|k)\ln\frac{\Gamma_{11}^{\mathrm{DD}}|_{\mathrm{kk_p}}}{\Gamma_{11}^{\mathrm{DD}}|_{\mathrm{kk_m}}}\end{aligned} \tag{4.57}$$

where

$$\frac{\Gamma_{ij}^{S_iS_j}|_{kl}}{\left(\Gamma_{ii}^{S_iS_j}|_{kk}\right)\left(\Gamma_{ii}^{S_iS_j}|_{ll}\right)} = \exp\left(-\frac{2w^{S_iS_j}|_{kl}}{kT}\right)$$

and $\rho_1|k$ is the number fraction of a k-type segment present in the solvent. Subscripts p and m refer to pure and mixture states, respectively. A k-th segment of a given molecule can interact through $z_1^{\alpha}|k$ hydrogen-bond donating sites, $z_1^{\beta}|k$ hydrogen-bond accepting sites, and $z_1^{\mathrm{D}}|k$ sites interacting through dispersion forces; z_1^{α}, z_1^{β}, z_1^{D}, $z_1^{\alpha} + z_1^{\beta} + z_1^{D} = z$ are numbers of interacting sites in component 1 (solvent) of the respective type. In Eq. 4.55, the authors [33, 34] used the FH form for the combinatorial part of $\Delta\mu_1$ (i.e., $z \to \infty$), although it was possible to use the quasichemical equilibrium form of Eqs. 4.46 or 4.47.

To get the values of the interchange energies may be a problem but certain simplifications can be made. Input from experiments (vapor pressures of pure components and mixtures), results from molecular dynamics simulation can help to solve the problems as well. Example of application of this approach was demonstrated in the works of Freitas group [34, 35].

The polymer mixing theory based on the Guggenheim-Tompa approach has been further developed by Yang and Hu, et al. [36–39]. Equation (4.58) illustrates their way of approach to the problem (for $m_1 = 1$)

$$\begin{aligned}\frac{\Delta G_{\mathrm{mix}}}{kT} &= \phi_1\ln\phi_1 + \frac{\phi_2}{m_2}\ln\phi_2 + \frac{z}{2}\left[\phi_1\ln\frac{\xi_1}{\phi_1} + \phi_2\frac{q_2}{m_2}\ln\frac{\xi_2}{\phi_2}\right]\\ &\quad + \frac{z}{2T^*}\phi_1\phi_2 - \frac{z}{4T^{*2}}\phi_1^2\phi_2^2 - \frac{z}{12T^{*3}}\phi_1^2\phi_2^2(\phi_1^2 + \phi_2^2)\end{aligned}$$

$$-\frac{m_2-1+\lambda}{m_2}\phi_2\ln\left(\frac{1+\phi_1(\exp(1/T^*)-1)}{1+\phi_1\phi_2(\exp(1/T^*)-1)}\right) \quad (4.58)$$

where $T^* = kT/\Delta\varepsilon$ is inverse dimensionless interaction energy. Using this equation, the authors were able to generate closed-loop miscibility diagrams.

Lattice models similar to that of Prange [33] were developed from the chemical engineering models applied to mixtures of low-molecular-weight compounds like UNIQUAC (UNIversal QUAsiChemical) theory [40], or NRTL (Non-Random Two Liquids) theory [41, 42].

The model of Lipson et al. [43–45] also belong to the Guggenheim-generic category; it describes systems with simple type of interacting surfaces for each component and also takes into account the volume change on mixing (see next section). The binary incompressible [43] and compressible [44, 45] lattice mixtures are treated using the Born–Green–Yvon (BGY) integral equation approach with the Kirkwood superposition approximation. The polymer and solvent molecules are placed on lattice with coordination number $z = 6$; the solvent occupies m_1 lattice site and the polymer m_2 sites; the interactions between segments of the same type are designated by ε_{ii} and lattice site volume by v. The system is made compressible by leaving some lattice sites vacant; ϕ_1 and ξ_1 are volume and interacting surface fractions, respectively, and ϕ_h and ξ_h are quantities related to fractions of vacant sites (holes). The relation for the chemical potential of component 1 due to mixing for the compressible case reads

$$\begin{aligned}\frac{\Delta\mu_1}{RT} &= \ln\phi_1 + \frac{z}{2}\ln\frac{\xi_1}{\phi_1} - m_1\left(\xi_h + \frac{z}{2}\ln\frac{\xi_h}{\phi_h}\right) + \left(1 - m_1 - \frac{z}{2}(q_1 - m_1)\right)\\ &- \frac{zq_1}{2}\ln(\xi_1\exp(-\varepsilon_{11}/kT) + \xi_2\exp(-\varepsilon_{12}/kT) + \xi_h)\\ &+ \frac{z}{2}(q_1 - m_1)(1 - \xi_h)\\ &- \frac{zq_1}{2}\left(\frac{\xi_1\exp(-\varepsilon_{11}/kT) - m_1/q_1}{\xi_1\exp(-\varepsilon_{11}/kT) + \xi_2\exp(-\varepsilon_{12}/kT) + \xi_h}\right)\\ &- \frac{zq_1}{2}\left(\frac{\xi_2\exp(-\varepsilon_{12}/kT) - m_1/q_1}{\xi_1\exp(-\varepsilon_{11}/kT) + \xi_2\exp(-\varepsilon_{12}/kT) + \xi_h}\right)\end{aligned} \quad (4.59)$$

where ξ_h is the surface fraction of holes; the other symbols have their usual meaning. The model describes well phase equilibria for polyisobutylene-benzene systems and some polymer blends.

Many other papers using off-FH and improved Guggenheim-like theoretical variants have been published and they can serve at least as inspiration for how to get values of system parameters from experiment.

In the models presented above, the formation of cross-links by complete association of pairs of cross-linked sites discussed above and the resulting in loss of entropy and depletion of the number of physically interacting sites were not taken into account. However, it can be accounted for easily as shown in this section. It should be included into the combinatorial part of the mixing energy, either in the FH–form, or in the quasichemical equilibrium form depending on how the combinatorial part is formulated. The number of c–c contacts is a constant which does not depend on volume.

The quasichemical equilibrium model taking into account specific (oriented) interactions between different kind of interacting surface sites can be further refined by making the equilibria dependent on the "states" of component segments, i.e., in how many bonds and of what kind the surface sites are engaged. Alternatively, the dependence can be extended even on the states of next-neighbor segments. In chemical kinetics, such dependences are called, respectively, *first-shell* and *second-shell substitution effects*, *fsse* and *ssse*. They correspond to the first- and second-order Markovian statistics. The ssse can generate growth of large sequences of bonds. For handling of *fsse* and *ssse*, the statistical branching theories are well-adapted.

D **Lattice cluster model**

- Freed and collaborators, in their *lattice cluster* (LC) theory [46–53] made the FH lattice model exact by placing rigorously polymer chains on lattice avoiding multiple occupancy of lattice sites and considering the polymer and solvent components to be composed of several kinds of segments. The authors also considered possible asymmetry of segments arrangement. In this model, the entropic part of the interaction parameter arises not only from considering interacting surfaces but also from the packing of monomers with different sizes and shapes (linear, branched). Also asymmetry of shapes of monomer units could be taken into account. Non-random mixing was considered as well as lattice compressibility when it was important. The ability to include the effects of different chain and solvent morphologies on the free energy of mixing is one of the principal strengths of the lattice cluster theory. These molecular shape and packing details determine the entropic part of the interaction parameter and they also affect its temperature dependence. This quite general treatment resulted in rather complicated expressions for the free energy. Similar to Mayer's theory for non-ideal gases, they developed a double expansion in power series with respect to the reciprocals of the coordination number, $1/z$ and the temperature, ε/kT. The power series was truncated. Under this provision, the free energy of mixing for the non-compressible case could be formulated in a simple equation

$$\frac{\Delta F_{\mathrm{mix}}}{NkT} = \frac{\phi_1}{m_1}\ln\phi_1 + \frac{\phi_2}{m_1}\ln\phi_2 + \phi_1\phi_2\sum_{n=0}^{\infty} g_n(\varepsilon/kT)^n \tag{4.60}$$

where g_n are the coefficients of a reciprocal temperature expansion of an effective χ parameter. Calculation of the values of coefficients g_n is feasible but not a simple task. It is interesting to note Madden's et al. [54] finding that the athermal free energy of mixing for the lattice cluster theory is nearly identical with that of the athermal limit of the Guggenheim theory but distinctly different from the results based on traditional Flory–Huggins theory.

In a later series of papers, Dudowicz and Freed [55, 56] considered the compressibility effect in a polymer–solvent multicomponent system. The excess (liquid-like) free volume was modeled through the presence of voids (holes)—unoccupied lattice sites that are smaller than the extended monomer structures. The algebraic complexity of treating the corrections to Flory–Huggins theory for the athermal limit of packing entropy and particularly for the internal energy was considerably diminished [56]. The mixing free energy expression is similar to Eq. (4.60)—it contains additional combinatorial term for the fraction of holes and the coefficients of the interaction term reflect the dispersity of components as well as the presence of holes.

$$\frac{\Delta F_{\text{mix}}}{NkT} = \frac{\phi_1}{m_1}\ln\phi_1 + \frac{\phi_2}{m_2}\ln\phi_2 + \phi_{\text{h}}\ln\phi_{\text{h}} + \sum_{i=1}^{\infty}\sum_{j,l=1,2}^{\infty} a_i^{jl}\left(\varepsilon_{jj}/kT + \varepsilon_{ll}/kT + \varepsilon_{jl}/kT\right)^i \tag{4.61}$$

where ϕ_{h} is the fraction of holes.

The lattice cluster theory was extended to describe the influence of monomer structure and local correlations on the free energy of strongly interacting and self-assembling polymer systems [57]. This extension combined a systematic high dimension ($1/d$) and high temperature expansion (that is appropriate for weakly interacting systems) with a direct treatment of strong interactions. The theory was illustrated for a binary polymer blend whose two components contain “sticky” donor and acceptor groups, respectively. The free energy was determined as an explicit function of the donor–acceptor contact probabilities; these probabilities depend on the local structure and both the strong and weak interactions.

Effects of strong interactions of different strength were modeled by Xu and Freed [58]. The backbone and side groups had different interaction strengths: three energy parameters were present: backbone-backbone, side group-side group, and backbone-side group interaction energies. Because of the great algebraic complexity of this extension, the authors tried to retain maximal simplicity by further specializing this initial study to models of polymer melts. Complexity could be simplified for specific cases. A similar system was treated by Dudowicz et al. [59] who considered solutions of polymers B with strongly interacting solvent A as a collection of associates A_iB dissolved in A as a quasiternary system (A—component 1, B—component 2, A_iB ($i \leq 1$)—component 3; interaction energies ε_{AA}, ε_{BB}, ε_{AB}) placed on lattice with coordination number z. The relation to the FH interaction parameter is a relatively complex function of composition.

It can be concluded that the lattice cluster theory is the most elaborated approach to polymer solutions thermodynamics capable to take into account fine details of component structure. Mathematical complexity is its disadvantage. Hu et al. [60–64] offered a revised Freed’s model and used it in the double lattice (DL) model, the latter was designed to account for the orientation effect of hydrogen bonding. These models provide better description of the experimental phase behavior including those with a lower critical solution temperature, loop and hour-glass type liquid–liquid coexistence curves.

4.3.3.2 Consideration of Free Volume

In the Flory–Huggins theory, no volume change on mixing was considered, $\Delta V_{\text{mix}} = 0$. The case of $\Delta V_{\text{mix}} \neq 0$ was considered first by Prigogine et al. [65] initially called Prigogine-Flory theory but it was developed by Patterson and Delmas in a practical version (Patterson's theory or corresponding states theory) [66, 67]. The dependence on pressure was accounted for by introducing vacancies in the lattice description, as for example in the *lattice- fluid theory* (LF) of Sanchez and Lacombe [68, 69] models of Flory-Orwoll-Vrij-Eichinger [70–79], and Flory and others (cf., e.g., [80, 81]), by Kleintjens and Koningsveld [82] and others [83, 84].

This dependence on volume changes can be accounted for by introducing vacancies either in the Flory–Huggins or GuggenheimTompa, or more advanced lattice descriptions (see Sect. 3.3.1.) as for example in the lattice-fluid theories of Sanchez and Lacombe (Lattice-Fluid model) (LF) or of KleintjensKoningsveld, or Flory-Orwoll-Vrij-Eichinger equations of state. Consideration of compressibility can explain not only the dependence of miscibility on pressure, but also the occurrence of LCST—mainly, by the dissimilarity of free volume of the dense polymer liquids and expanded solvents. Two liquid components have generally different free volumes or degrees of thermal expansion. A liquid polymer has extremely low free volume because of chain connectivity. Then, mixing can be viewed as condensation of a gas into the dense polymer. Associated with such process is an exothermic process and negative contribution to entropy. As temperature rises, the dissimilarity between polymer and solvent increases and so does the contribution to the χ parameter. According to this theory, two liquids of differing free volume will experience a net contraction on mixing and, as a result of this, negative contributions appear in both ΔH_{mix} and ΔS_{mix}. The net contribution to the free energy of mixing, ΔG_{mix}, is predicted to be positive, i.e., unfavorable to mixing. Compressibility of the mixtures is especially important when the solvent approach their boiling and critical points.

The models are based on properties (equations of state) of pure compounds; for binary systems, proper mixing rules are introduced. All models consider variation of volume by changes of pressure and temperature associated with enthalpy and (possibly large) entropy changes. The free volume is dependent on temperature and pressure; it is modeled as vacancies existing in lattice sites or as empty lattice sites that mix with lattice sites occupied by segments.

The *Prigogine-Patterson* corresponding states theory works with reduced quantities: temperature ($\tilde{T}$), volume ($\tilde{V}$), and pressure ($\tilde{P}$) defined as

$$\tilde{T} = T/T^*, \quad \tilde{V} = V/V^* \approx v/v^*, \quad \tilde{P} = P/P^* \tag{4.62}$$

where the starred quantities are constant reduction parameters and V^*, v^* have the significance of being the molar and specific volumes at 0 K, i.e., the "core volumes".

At ordinary temperature, $\tilde{P} = 0$, and $\tilde{V}$ and $\tilde{T}$ are linked by equation of state as shown below. The reduced volume is a function of the expansion coefficient α and the reduced temperature is a function of reduced volume

$$\tilde{V}^{1/3} - 1 = \frac{\alpha T}{3(1 + \alpha T)} \tag{4.63}$$

$$\tilde{T} = \frac{\tilde{V}^{1/3} - 1}{\tilde{V}^{4/3}} \tag{4.64}$$

For $\alpha = 0$, $\tilde{V}^{1/3} = 1(0\text{ K})$, at $\alpha \to \infty$, $\tilde{V}^{1/3} = 4/3$ (vapor–liquid critical point). The free volume difference between component 1 and component 2 is characterized by

$$\tau = 1 - \frac{T_1^*}{T_2^*} \approx 0.6\frac{\Delta\alpha}{\alpha} \tag{4.65}$$

The equation of state is

$$\frac{\tilde{P}\tilde{V}}{\tilde{T}} = \frac{\tilde{V}^{1/3}}{\tilde{V}^{1/3} - 1} - \frac{1}{\tilde{V}\tilde{T}} \tag{4.66}$$

The Patterson's theory applied to polymer mixtures shows that the FH interaction parameter has its enthalpic and entropic parts and that the concentration dependence of interaction parameter is determined by changes of free volume [66].

Patterson showed how the interaction parameter can vary with temperature (temperature affects free volume) and the contribution by free volume change and temperature change on a simple example depicted on Figs. 4.12 and 4.13 parameter for a mixture of polymers interacting with dispersion forces with $\chi_{12} = 10$ J cm^{-3} against $V_1^{1/3}$ and temperature t, with different values of τ. Polystyrene is the reference polymer; further details in Ref. [67]. Reprinted from Ref. [67] with permission.

The Flory-Orwoll-Vrij-Eichinger theory is somewhat similar to the Prigogine-Patterson approach in that it works with reduced quantities and considers changes of "free volume" generated by mixing. The volume unit hosting a segment is larger than the core volume; at 0 K, free volume is equal to zero. It is assumed that the intermolecular energy depends only on the volume, and a hard sphere repulsive potential is adopted for segments of the chain. The segments are endowed with hard sphere type repulsive potential and soft attraction of unspecified type. The equation of state for a single component has the same form as that given before (Eq. 4.66).

For pure components

$$\tilde{T} = \left(\tilde{v}^{1/3} - 1\right)/\tilde{v}^{1/3} \qquad \tilde{v}^{1/3} - 1 = (\alpha T/3)/(1 + \alpha T) \qquad p^* = \gamma T \tilde{v}^2$$

where $\tilde{v}$ is reduced volume, α is thermal expansion coefficient, and γ thermal pressure coefficient at $p = 0$.

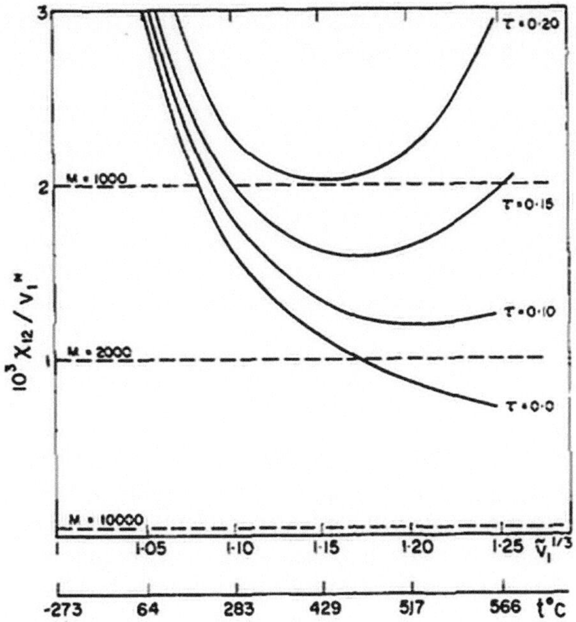

Fig. 4.12 Polymer–polymer system with $\chi_{12} > 0$. The χ_{12}/v_1^*

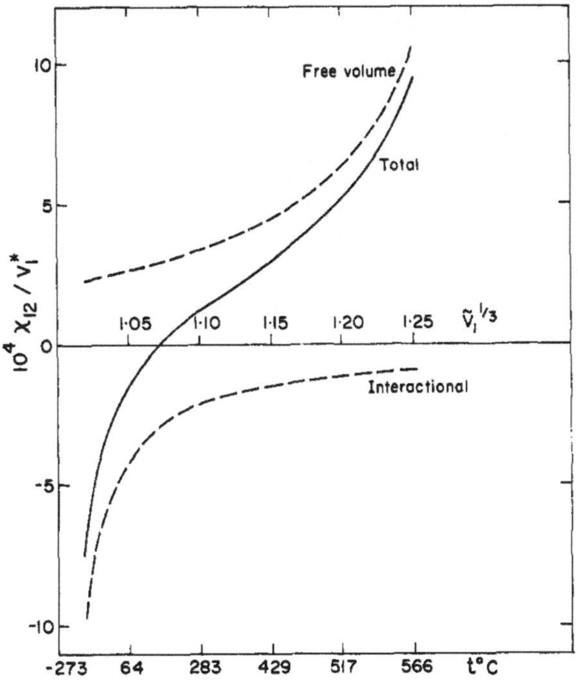

Fig. 4.13 Polymer–polymer system with $\chi_{12} < 0$. For mixtures of compatible polymers with $\chi_{12} = -1.3\ \mathrm{J\ cm^{-3}}$ plotted against $V_1^{1/3}$ and temperature t showing the favorable interaction contribution. Polystyrene is the reference polymer. Reprinted from Ref. [67] with permission

For binary mixtures, this model offers for the chemical potential of the solvent the following equation [72]

$$\frac{\Delta\mu_1}{kT} = RT\left(\ln\phi_1 + \left(1 - \frac{v_1^*}{v_2^*}\right)\right) + p_1^* v_{\mathrm{sp},1}^* \left\{3\tilde{T}_1 \ln\frac{\tilde{v}_1^{1/3} - 1}{\tilde{v}^{1/3} - 1} + (\tilde{v}_1^{-1} - \tilde{v}^{-1})\right\}$$
$$+ (v_{\mathrm{sp},1}^* X_{12}/\tilde{v})\xi_2^2 \qquad (4.67)$$

The interaction function X_{12} reflects entropy and enthalpy changes due to pairwise interactions between segments and volume changes. It can be presented in the form

$$X_{12} = \frac{s_1 \Delta\eta}{2(v^*)^2}; \qquad \Delta\eta = \eta_{11} + \eta_{22} - 2\eta_{12}; \qquad (4.68)$$

In these equations, the characteristic quantities $v_{\mathrm{sp},1}^*$ and $\tilde{T}$ can be evaluated from the specific volume and thermal expansion coefficient; the reduced quantities without subscript relate to the mixture, otherwise the kind of component is indicated; s is the number of interacting surface contacts per unit, η_{ij} is the energy of contact $i-j$.

In the *Sanchez-Lacombe* theory, holes are introduced to account for variations in compressibility and density, i.e., the mixture density may vary by increasing the fraction of holes. PVT properties of pure components are obtained which can interact with intermolecular potential. Again, the state of the system is described by reduced variables including reduced density $\tilde{\rho} = 1/\tilde{v} = V^*/V$.

Later, the compressibility issues addressed in the lattice-fluid theory were combined with the effect of strong or specific interactions between components in the Sanchez-Balazs [69] approach. In either case, a thermodynamic analysis indicated that the phase separation is entropically driven. In their original formulations, the models did not take into account specific interactions and the associated entropy effects. Strong or specific interactions should induce some local ordering or degree of non-randomness in a mixture. The LF model has been generalized for specific interactions while its original simplicity was preserved. Instead of Guggenheim's quasichemical equilibrium approach, the simple Brinke and Karasz incompressible model of binary mixtures with specific interactions was used. This modification resulted in a generalized LF model addressing both the compressibility and strong interaction issues. For the chemical potential of component 1, one gets a relatively simple equation (Eq. 4.35 of Ref. [69])

$$\frac{\Delta\mu_1}{RT} = \ln\phi_1 + \frac{z}{2}\ln\frac{\xi_1}{\phi_1} - m_1\left(\xi_\mathrm{h} + \frac{z}{2}\ln\frac{\xi_\mathrm{h}}{\phi_\mathrm{h}}\right) + \left(1 - m_1 - \frac{z}{2}(q_1 - m_1)\right)$$
$$- \frac{zq_1}{2}\ln\left(\xi_1\exp(-\varepsilon_{11}/kT) + \xi_2\exp(-\varepsilon_{12}/kT) + \xi_\mathrm{h}\right) + \frac{z}{2}(q_1 - m_1)(1 - \xi_\mathrm{h})$$
$$- \frac{zq_1}{2}\left(\frac{\xi_1\exp(-\varepsilon_{11}/kT) - m_1/q_1}{\xi_1\exp(-\varepsilon_{11}/kT) + \xi_2\exp(-\varepsilon_{12}/kT) + \xi_\mathrm{h}}\right)$$

$$- \frac{zq_1}{2}\left(\frac{\xi_2 \exp(-\varepsilon_{12}/kT) - m_1/q_1}{\xi_1 \exp(-\varepsilon_{11}/kT) + \xi_2 \exp(-\varepsilon_{12}/kT) + \xi_h}\right) \quad (4.69)$$

where $\tilde{\rho}$ is reduced density, $1 - \tilde{\rho}$ is the fraction of vacancies, and ϕ_1 and ϕ_2 are volume fractions at $\tilde{\rho} = 1$, m are numbers of segments, ε_{ij}^* are interaction energies of the segments i and j, $\delta\varepsilon^*$ is a free energy parameter and accounts for specific interactions; subscripts relate to components. The Sanchez-Balasz concept was used by Wu et al. to treat strong interactions within the NRTL (Non-Random Two Liquids) equation [85].

4.3.3.3 Double Lattice (DL) and Modified Double Lattice (MDL) Models

Strong interactions, like hydrogen bonding or other types of donor–acceptor bonding, strong dipole–dipole interactions, sequences templating, etc. dominate the mixing behavior when present. Such strong bonds are formed not only by association of individual groups, but also by cooperative arrangement of individually weaker bonds into blocks of higher strength. This is the case when bond formation is assisted by the state of structural or spatial neighbors. Within the Guggenheim-Freed approach, strong bonds between individual associating groups can be modeled by the method "segments with different kind of contact points" (Sect. 3.3.1.), meaning that the units of participating components (polymer and solvent molecules in binary or quasibinary systems) bear some surface contacts binding strongly to specified other (or same) contacts on units of the polymer and solvent molecules, in addition of some other that interact by weak van der Waals forces. Poly(vinyl alcohol) (PVA) solution in water is a typical example: each unit of PVA bears one H-bond acceptor and one donor and unspecified number of "weak" contact, each molecule of water has two donors and two acceptors. Obviously, the strength of existing H-bonds is not the same. Application of the "different contact point" method to some simple or special cases was already discussed in Sect. 3.3.1—existence of simple strong bonds [33, 34] and effect of cross-links [31, 86]. Competition of contacts in formation of various strong bonds differing in strengths and effect of states of other contacts in the same unit or nearest neighbor units has not been considered as yet, but the experience gained in solution of the first-shell and the second-shell substitution effects would be helpful.

One of the methods developed in recent decades to take into account the strong bonds effect is the double lattice model. Hu et al. [60–63] revised the Freed's lattice cluster theory and developed a simple double lattice (DL) model by taking into account specific interactions such as hydrogen bonding. Conventional weak physical interactions in polymer solutions were described by the primary lattice, while a secondary lattice was introduced as a perturbation to account for oriented interactions. This secondary lattice required an additional parameter related to the energy of the oriented interaction and one empirical parameter. The secondary lattice was a simple

Ising lattice where the size of each component was of size equal to one ($m_1 = m_2 = 1$) and occupies one site.

The contribution by strong (oriented) bonds reads

$$\frac{\Delta F_{ij}}{N_{ij}kT} = \sum_i \frac{\phi_i}{m_i} \ln \phi_i + \frac{b_2}{2} \sum_i \sum_j \phi_i \phi_j (m_i^{-1} + m_j^{-1})^2 + \sum_i \sum_j a_{ij} \phi_i \phi_j - \left(\sum_i \sum_j a_{ij}^* \phi_i \phi_j \right)^2 \tag{4.70}$$

where η is the fraction of surface contacts that can participate in oriented interactions and the fraction $1 - \eta$ cannot participate. The choice of the value for η depends on the choice of the interaction energy, $\delta\tilde{\varepsilon}_{ij}$. The values of the constants c are obtained from simulation.

An effective chain-insertion probability was used in the derivation of the entropy term and the energy term was expanded into power series. The unknown coefficients in the model are determined by a few computer-simulation data and by referring to the rigorous Freed's theory. The combined Helmholtz energy of mixing as formulated for multicomponent system (cf. also Ref. [87]) is given by the equation

$$\frac{\Delta F_{\text{sec},ij}}{kTN_{ij}} = \frac{2}{z}\Big[\eta_{ij} \ln \eta_{ij} + (1 - \eta_{ij}) \ln(1 - \eta_{ij}) + 3\delta\tilde{\varepsilon}_{ij}\eta_{ij}(1 - \eta_{ij}) - 1.5(\delta\tilde{\varepsilon}_{ij})^2 \eta_{ij}^2 (1 - \eta_{ij})^2 - c_3 (\delta\tilde{\varepsilon}_{ij})^3 \eta_{ij}^3 (1 - \eta)^3 - 3\delta\tilde{\varepsilon}_{ij}\eta_{ij}\Big] \tag{4.71}$$

The coefficients b_2, a_{ij}, and a_{ij}^* were calculated from data in the literature based on rigorous solution obtained for Ising lattice obtained by Scesney [88], combined with simulations of Madden et al. [54, 89]

Hu et al. [63] later redefined the original derivation of $\Delta F_{\text{sec},ij}$ [60] and obtained for contribution by the secondary lattice expression

$$\frac{\Delta F_{\text{sec},ij}}{kTN_{ij}} = \frac{2}{z}\Big[\eta_{ij} \ln \eta_{ij} + (1 - \eta_{ij}) \ln(1 - \eta_{ij}) + 3\delta\tilde{\varepsilon}_{ij}\eta_{ij}(1 - \eta_{ij}) - 1.5(\delta\tilde{\varepsilon}_{ij})^2 \eta_{ij}^2 (1 - \eta_{ij})^2 - c_3 (\delta\tilde{\varepsilon}_{ij})^3 \eta_{ij}^3 (1 - \eta)^3 - 3\delta\tilde{\varepsilon}_{ij}\eta_{ij}\Big] \tag{4.72}$$

The primary contribution (conventional mixing of the Freed type) for binary system can be simplified as

$$\frac{\Delta F_{\text{prim}}}{N_m kT} = \frac{\phi_1}{m_1} \ln \phi_1 + \frac{\phi_2}{m_2} \ln \phi_2 + \sum_m \sum_n a_{mn} \phi_1^m \phi_2^n \tag{4.73}$$

With this formulation of the additional Helmholtz energy for strong bonds, the authors could successfully describe various UCST, LCST and closed-loop phase diagrams.

Oh and Bae [90] modified the double lattice model of Hu [60] by introducing a new interaction parameter and simplifying the expression. This modified theory was called Modified Double Lattice (MDL) model. The secondary lattice was modified by an additional temperature-dependent energy contribution. To improve the mathematical approximation defect and to reduce the number of parameters, a new Helmholtz energy of mixing in the fractional form was defined. The new expression reads

$$\frac{\Delta F_{\mathrm{sec},ij}}{kTN_{ij}} = \frac{2}{z}\left[\eta \ln \eta + (1-\eta)\ln(1-\eta) + \frac{zC_\alpha \delta\tilde{\varepsilon}_{ij}\eta(1-\eta)}{1+\eta(1-\eta)}\right] \tag{4.74}$$

where the last term is a convoluted series of the free energy contribution due to strong bonds. For aqueous solutions, the best fit value of C_α was 0.488.

The expression for the chemical potential of solvent can be presented in a simple form where the factors C_α, C_β, and C_χ express entropy and energy (ϕ_2-dependent) contributions by both lattices

$$\frac{\Delta\mu_1}{kT} = \ln(1-\phi_2) - m_1\left(\frac{1}{m_2}-\frac{1}{m_1}\right)\phi_2 + m_1\left[C_\beta\left(\frac{1}{m_2}-\frac{1}{m_1}\right)^2 + \left(\left(\frac{1}{m_2}-\frac{1}{m_1}\right)+C_\gamma\tilde{\varepsilon}\right)\tilde{\varepsilon} + \left(2+\frac{1}{m_2}\right)\tilde{\varepsilon}\right]\phi_2^2$$
$$-2m_1\left[\left(\left(\frac{1}{m_2}-\frac{1}{m_1}\right)+C_\gamma\tilde{\varepsilon}\right)\tilde{\varepsilon} + C_\gamma\tilde{\varepsilon}^2\right]\phi_2^2 + 3m_1C_\gamma\tilde{\varepsilon}^3\phi_2^4 \tag{4.75}$$

The MDL model was used to simulate various phase equilibria and phase transitions in linear and cross-linked polymer dissolved in water and other polar solvents (alcohols) and solvent mixtures (cf., e.g., Refs. [90–96]). For instance, it was found [95] that hydrogen bonds between PEO and water dissociate with increasing temperature more rapidly than the water—water ones. In the temperature range of existence phase separated system both mixing energy and radius of gyration showed lower values than those of the homogenous phase, which corresponds well to the closed-loop type miscibility behavior. For multicomponent systems, a version of MDL model was used by Ryu and Bae [87]. They proposed a thermodynamic model for multicomponent systems that takes the chain length dependence of the polymer and oriented interactions into account.

An example of phase equilibrium in poly(ethylene glycol)—water system that the agreement of theory with experiment is almost perfect.

Somewhat related to DL and MDL models is the Yang's model. Yang et al. [36, 38] developed a lattice model for thermodynamics of polymer mixtures based on a three-step process for construction of ΔF_{mix}:

(1) dissociate the pure chains to form pure monomers;
(2) mix solvents and monomers to form a mixture;
(3) associate the monomers into chain molecules

$$\Delta F_{\mathrm{mix}} = -T\Delta S_{\mathrm{mix0}} + \Delta F_1 + \Delta F_2 + \Delta F_3 \tag{4.76}$$

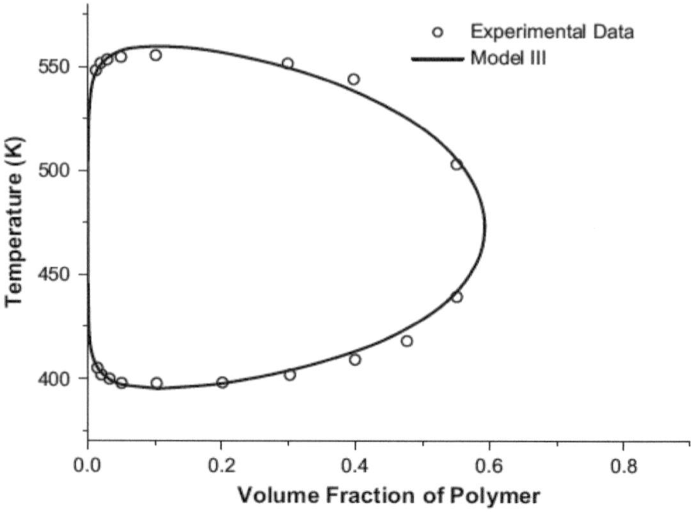

Fig. 4.14 Coexistence curves for PEG/water system. The open circles are experimental data for PEG molecular weights of 8000 full curve calculated by the DL theory. Reprinted from Ref. [91] (Fig. 4.14) with permission

The entropy ΔS_0 is the athermal entropy of mixing based on Guggenheim's model

$$\frac{\Delta S_{\text{mix0}}}{kN_{\text{latt}}} = \phi_1 \ln \phi_1 + \frac{\phi_2}{m_2} \ln \phi_2 + \frac{z}{2}\left[\phi_1 \ln \frac{\xi_1}{\phi_1} + \phi_2 \frac{q_2}{m_2} \ln \frac{\xi_2}{\phi_2}\right] = \frac{\Delta S_{\text{Gugg,ath}}}{kN_{\text{latt}}} \tag{4.77}$$

For the step (1) and step (3), the formalism of chemical association through the cavity correlation function based on the sticky-point model of Zhou and Stell [97–99] was employed in conjunction with the sticky-chain model based on particle correlation function g and energy ε by Hu et al. [63]

$$\Delta F_1 = N_2 \varepsilon^{(m_2)} \tag{4.78}$$

where $\varepsilon^{(m_2)}$ is the attractive energy of a chain. ΔF_2 is the residual Helmholtz energy of mixing of an Ising lattice. Yan et al. [100] expressed this contribution by a truncated polynomial

$$\Delta F_2 = \Delta F_{2-\text{Ising,mix}} = N_{\text{latt}} kT\left[\frac{z}{2T*}\phi_1\phi_2 - \frac{z}{4T*^2}\phi_1^2\phi_2^2 - \frac{z}{12T^{*3}}\phi_1^2\phi_2^2(\phi_1^2 + \phi_2^2)\right]$$
$$T^* = kT/\varepsilon \tag{4.79}$$

where ε is exchange energy between solvent-segment pairs.

For the association energy, one can write

$$\Delta F_3 = -kTN_2 \ln g^{(m_2)} - N_2 \varepsilon^{(m_2)} \tag{4.8}$$

where $g^{(m_2)}$ is the m-particle correlation function which is approximated by

$$g^{(m_2)} = \left(g^{(2)}\right)^{m_2 - 1 + \lambda} \tag{4.81}$$

In this expression

$$g^{(2)} = \frac{1 + \phi_1(\exp(1/T^*) - 1)}{1 + \phi_1\phi_2(\exp(1/T^*) - 1)} \quad \text{while} \quad \lambda = \frac{(m_2 - 1)(m_2 - 2)}{m_2^2}(am_2 + b) \tag{4.82}$$

The parameters a and b are estimated from computer simulation ($a = 0.132$ and $b = 0.592$ and $m_2 = 4$–200).

The total change of Helmholtz energy on mixing is equal to

$$\begin{aligned}
\Delta F_{\text{mix}} &= -T\Delta S_{\text{Gugg,ath}} + \Delta F_{2,\text{Ising,mix}} - kTN_2 \ln g^{(m_2)} \\
&= \frac{\Delta F_{\text{mix}}}{kN_{\text{latt}}} = \phi_1 \ln \phi_1 + \frac{\phi_2}{m_2} \ln \phi_2 + \frac{z}{2}\left[\phi_1 \ln \frac{\xi_1}{\phi_1} + \phi_2 \frac{q_2}{m_2} \ln \frac{\xi_2}{\phi_2}\right] \\
&+ \left[\frac{z}{2T*}\phi_1\phi_2 - \frac{z}{4T*^2}\phi_1^2\phi_2^2 - \frac{z}{12T*^3}\phi_1^2\phi_2^2(\phi_1^2 + \phi_2^2)\right] \\
&- \frac{m_2 - 1 + \lambda}{m}\phi_2 \ln \frac{1 + \phi_1(\exp(1/T^*) - 1)}{1 + \phi_1\phi_2(\exp(1/T^*) - 1)}
\end{aligned} \tag{4.83}$$

The simulated coexistence curves are much better than using FH theory but they also are better than those obtained by Freed's theory. This model is extended [37] to cover strong interactions: The double lattice concept is introduced by making the exchange energy composed of several contributions $i = 1, 2, 3, \ldots$ positive or negative.

$$\frac{\varepsilon}{kT} = \sum_i \frac{\varepsilon_i}{(kT/1000)^i} \tag{4.84}$$

The same approach and equation were used for the mixing part of the Helmholtz/Gibbs energy to describe equilibrium swelling. For uncharged hydrogels of the PNIPAM type, the reduced exchange energy was used in the form [39]

$$\frac{1}{T*} = \frac{\varepsilon}{kT} = \frac{\varepsilon_1}{kT} + \frac{\varepsilon_2}{(kT)^2} + \frac{\varepsilon_3}{(kT)^3} \tag{4.85}$$

with $\varepsilon_1/k = 9.0$, $\varepsilon_2/k^2 = -8.0$, $\varepsilon_3/k^3 = -60$ to fit PNIPAM—water systems. A similar model was used for a three-component system (PNIPAM-water–ethanol) [101] considering binary interactions between the components. For charged groups, the Donnan and possibly the terms covering electrostatic repulsion along the chains were added and Gaussian chain or Kovac's version of finite extensibility model network chains was considered [102].

4.3.3.4 Perturbed Chain Statistical Associating Fluid Theory (PC-SAFT)

The PC-SAFT method and its hybrids offer equations of state which are able to describe and predict phase behavior of polymer systems as a function of pressure, temperature, polymer concentration, polymer molecular weight, and polydispersity. The SAFT equation of state, developed in the late eighties, is a non-cubic, segment-based equation of state which was designed specifically to deal with systems containing polymers and associating fluids. Two reviews on SAFT applications are available [103, 104]. A number of modifications of SAFT have also appeared in the last two decades, including SAFT-VR and PC-SAFT.

This theory developed originally by Chapman et al. [105, 106] and extended to polymeric systems by Huang et al. [107] and further developed and applied to polymers and gels, recently mainly by Sadowski's group in Dortmund [108–113].

The method is a perturbation theory based on pure-component parameters, which are physically meaningful. The interactions of molecules are divided into a repulsive part and a contribution due to the attractive part of the potential. The attractive interactions are further separated into dispersive interactions and a contribution due to association. The theory is able to explicitly account for the effects of various molecular properties such as the non-spherical shape, the ability to form hydrogen bonds or segment, solvent polarity, etc. In the framework of PC-SAFT, molecules are assumed to be chains of freely jointed spherical segments exhibiting attractive forces among each other.

The basic idea is that excess Helmholtz (ΔF_{exc}) or Gibbs (ΔG_{exc}) energies, can be split into that describing the deviations form of the reference system (ΔF_{ref}) and that due to perturbation (ΔF_{pert}) parts

$$\Delta F = \Delta F_{\text{id}} + \Delta F_{\text{exc}} = \Delta F_{\text{id}} + \Delta F_{\text{ref}} + \Delta F_{\text{pert}} \qquad (\Delta F_{\text{id}} = 1) \tag{4.86}$$

Deviations from the reference system occur due to attractive interactions (dispersion forces), formation of hydrogen bonds (association forces), or deviations from spherical shape of the molecules as, for instance, arrangement of monomer units into chains of monomer units. A chain of hard units is taken as reference state: it is a sum of hard sphere and chain formation contributions

$$\Delta F_{\text{exc}} = \Delta F_{\text{hc}} + \Delta F_{\text{disp}}(m) + \Delta F_{\text{assoc}} \tag{4.87}$$

The hard-chain contribution, ΔF_{hc}, is derived from the hard sphere contribution, ΔF_{hs}. Modeling with PC-SAFT needs three pure-component parameters for a non-associating molecule: segment diameter σ (corresponds to van der Waals parameter b), energy-related parameter u (van der Waals parameter a) and number of segments, m. The asymmetry parameter k_{ij} is introduced to compensate for the deviation of the interaction energy parameters from their geometric means. For associating molecules, the models need five pure-component parameters. In the case of solvents, these parameters can be obtained from their vapor pressures and liquid densities. For the polymer component, these parameters are obtained from properties of binary systems. Alternatively, pure-component parameters can be determined by simultaneously fitting liquid density data and binary phase equilibrium data—an approach of pragmatic nature. For copolymer systems, a mean value of segment diameter $\sigma_{ij} = (\sigma_{ii} + \sigma_{jj})/2$ and energy $u_{ij} = \sqrt{u_i u_j}(1 - k_{ij})$ can be obtained; the asymmetry parameter k_{ij} can be calculated by the group contribution method.

Explicitly, the excess hard sphere Helmholtz energy for a binary system reads

$$\Delta F_{\mathrm{hc}} = m\Delta F_{\mathrm{hs}} - NkT \sum_i x_i(m_i - 1)\lg g_{ii}\left(d_{ii}^{+}\right) \tag{4.88}$$

where the subscript i relates to degree of polymerization distribution and x_i is the molar fraction of i-mers. In the expression for ΔF_{hc}

$$\begin{aligned}
\Delta F_{\mathrm{hs}} &= \frac{1}{\zeta_0}\left[\frac{3\zeta_1\zeta_2}{1-\zeta_3} + \frac{3\zeta_2^3}{\zeta_3(1-\zeta_3)^2} + \left(\frac{\zeta_2^3}{\zeta_3^2} - \zeta_0\right)\ln(1-\zeta_3)\right] \\
m &= \sum_i x_i m_i \\
\zeta_k &= \frac{\pi}{6}\rho \sum_i x_i m_i d_i^k, \qquad \zeta_3 = \eta, \qquad k \in (0, 1, 2, 3)
\end{aligned} \tag{4.89}$$

The quantity $g_{ij}\left(d_{ij}^{+}\right)$ is the value of radial distribution function at contact—a function of the distance between segments i and j

$$\ln g_{ij}\left(d_{ij}^{+}\right) = \frac{1}{1-\zeta_3} + \left(\frac{d_i d_j}{d_i + d_j}\right)\frac{3\zeta_2}{(1-\zeta_3)^2} + \left(\frac{d_i d_j}{d_i + d_j}\right)^2 \frac{2\zeta_2^2}{(1-\zeta_3)^3} \tag{4.90}$$

The quantity $g_{ii}\left(d_{ii}^{+}\right)$ is identical with the segment diameter; the segment diameter is temperature-dependent

$$d_i = \sigma_i\left[1 - 0.12\exp\left(\frac{-3\varepsilon_i}{kT}\right)\right] \tag{4.91}$$

Five material and physically meaningful parameters are needed for each component of the associative systems. In addition to the number of segments m_{i}, segment diameter σ_{i}, and dispersion energy u_{i}, the association energy $\varepsilon^{\mathrm{A_iB_i}}$ and associative volume parameter $\sigma^{\mathrm{A_iB_i}}$ are needed. The association contribution is then calculated according to Eq. 4.92 where x_{i} is the molar fraction of component i and for interacting sites A and B, $X^{\mathrm{A_i}}$ and $X^{\mathrm{B_i}}$ are the fractions of non-bonded interacting sites;

$\Delta^{A_iB_j}$ characterizes the association strength (for more details, see Ref. [114])

$$\Delta F_{\text{assoc}} = NkT \sum_i x_i \sum_{A_i} \left(\ln X^{A_i} - X^{A_i}/2 + 1/2\right) \tag{4.92}$$

The quantity X^{A_i} can be iteratively derived from the following Eq. 4.93

$$X^{A_i} = \left(1 + \rho \sum_{v} x_j \sum_{B_j} X^{B_j} \Delta^{A_iB_j}\right) \tag{4.93}$$

The bonding strength between each of two associating sites $\Delta^{A_iB_j}$ depends on the radial distribution function. Each molecule may have two different association sites, for instance, one donor and one acceptor of hydrogen bond. Huang et al. analyzed results obtained for a great number of binary and ternary systems in order to parametrize the equation of state and accumulated data on interaction strength [107, 114].

When applied to PNIPAM with 1% cross-linker [112], it was necessary to consider a small elastic contribution generated by deformation of PNIPAM chains. A typical flat "binodals" characteristic for the "off-zero critical concentration" [18, 21] or the "box" shape] were obtained (Fig. 4.15). This was achieved by adding an elastic term based on Gent finite extensibility function (see [14, 115]) proportional to the fraction of PNIPAM segments in the system.

$$\Delta F_{\text{exc}} = \Delta F_{\text{hc}} + \Delta F_{\text{disp}}(m) + \Delta F_{\text{assoc}} + \Delta F_{\text{elast}} \tag{4.94}$$

The maximum extensibility corresponded to a selected molar fraction of cross-linker ($y \leq 0.01$).

In the PC-SAFT treatment [112] of PNIPAM solutions and gels it is assumed that the main interaction is complexation of water molecules with proton donating and proton accepting groups of PNIPAM as shown in Fig. 4.16.

The SAFT-VR (statistical associating fluid theory for potentials of variable range) was developed as a variant of the SAFT method [116] which was applied for PEG-water system [117]. It takes into account the delicate balance between the water—water, water—PEG, and PEG—PEG hydrogen-bonding interactions as well as the usual repulsive and attractive/dispersive contributions. SAFT-VR allows for accounting the effect of pressure in a rigorous fashion. It is well-suited to the study of water-soluble polymers as both the polymer–solvent size asymmetry and hydrogen bonding can be accounted for explicitly. The closed-loop phase separation equilibria typical for poly(ethylene glycol) were studied in detail. The strong interactions treatment was based on the model of components bearing interacting surfaces shown in Fig. 4.17.

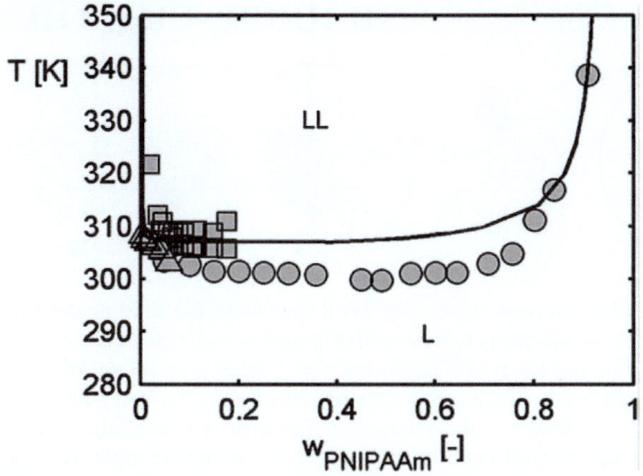

Fig. 4.15 Liquid–liquid phase equilibria of PNIPAM in water, the line represents modeled values with a molecular weight corresponding to a theoretical fraction of cross-linker $y = 0.01$, symbols are experimental data taken from the literature with different molecular weights (circles, 53,000 g mol^{-1}; squares, 615,000 g mol^{-1}; triangles, 2,100,000 g mol^{-1}). Reprinted from Ref. [112] with permivssion of American Chemical Society

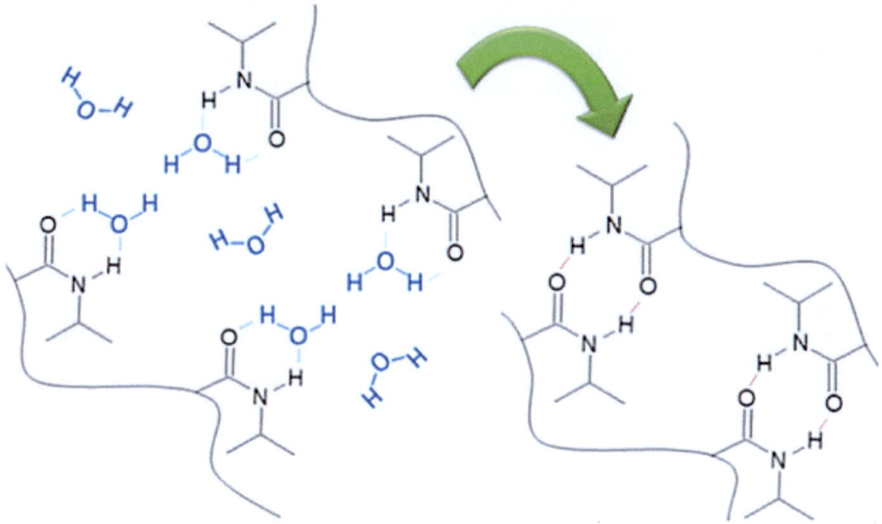

Fig. 4.16 Interactions in system PNIPAM-water before and after the collapse transition; reprinted from Ref. [112] with permission of American Chemical Society

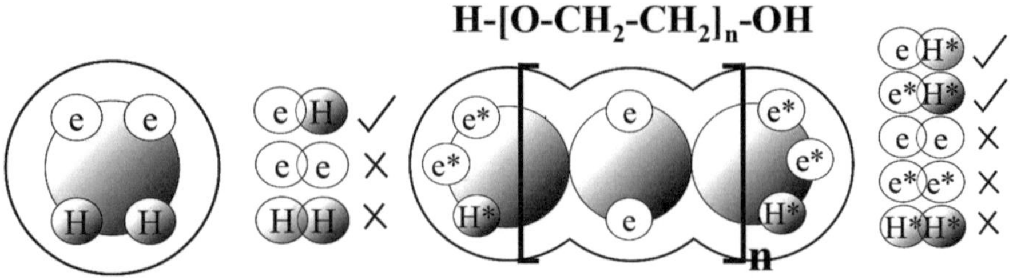

Fig. 4.17 Model of water used in the SAFT-VR studies of the fluid phase equilibria of aqueous solutions of PEG. The molecule is represented by a square-well core of diameter σ_{11} with dispersive interactions characterized by a well depth ε_{11} and a range λ_{11}; model of PEG: the molecule is represented by a number m of square-well segments of diameter σ_{22} with dispersive interactions characterized by a well depth ε_{22} and a range λ_{22}. The hydrogen-bonding interactions are mediated through off-center square-well bonding sites of types e (ethylene oxide (EO) repeat unit oxygen electron lone pairs), e* (hydroxyl OH terminal oxygen electron lone pairs), and H* (hydroxyl OH terminal hydrogen atoms) located halfway between the center and surface of the segment core. Reprinted from Ref. [117] (Figs. 4.1, 4.2) with permission of American Chemical Society

SAFT-VR was further applied to multicomponent systems—mixed solvents and disperse polymer systems. Figure 4.18 show SAFT-VR simulated dependences of the degree of swelling in mixtures of alcohols with water.

Although the theory is based on experimentally available parameters, a crucial aspect for the success of modeling with SAFT and PC-SAFT or SAFT-VR equations is the correct selection of the parameters of the pure components.

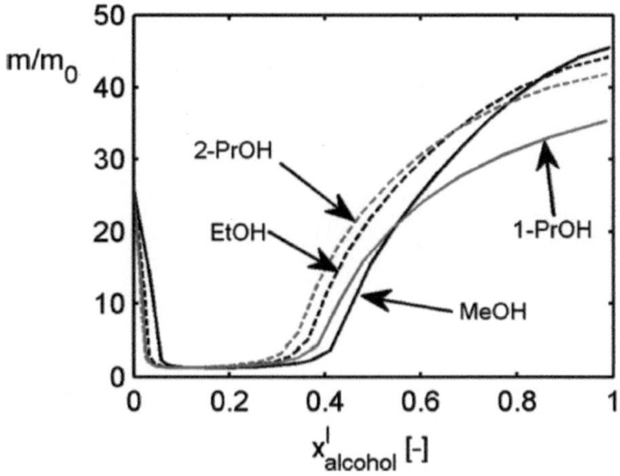

Fig. 4.18 Modeled mass-swelling of PNIPAM in aqueous alcohol solutions at 25 °C depending on the molar alcohol fraction in the solution. Reprinted from Ref. [117] with permission of American Chemical Society Phase diagram of thermoreversible gels

4.3.3.5 Strong Interactions Viewed as Reversible Clusters and Networks

When the components, including the solvent, bear associative groups (contact sites, stickers), clusters of varying sizes and possibly a (reversible) network can be formed which should be taken into account in a thermodynamic description. Increase or decrease of association is manifested by changes of viscosity or gelation (cf., (thermo) reversible gels). Likewise, sequences of weakly interacting surface sites in a medium of different polarity can form associates much stronger than a bond formed by coupling of two single sites. These temporary bonds have different strengths measured by their energy; they also dissociate and reassociate faster or slower and have different lifetimes. When the associationdissociation rates are high (lifetimes are short) and much higher than the relaxation of the whole structure (clusters, network), equilibrium is established at any time. The structure (distribution of cluster sizes, sol–gel distribution, and number of bonds per unit in the network) is reestablished corresponding to the new conditions. If however the dissociation is slow, slower than the relaxation of the whole structure, the once-formed structure remains unchanged and the temporary bonds behave as permanent. Ionic associates in apolar medium represent such strong bonds. The association or dissociation are assisted by changes of external variables—temperature, pressure, and also mechanical stress.

What is the practical implication of these considerations for swelling? A mixture of oligomers or polymers with solvent can be in the gel state. The gel may change into solution when more solvent is added. Upon dilution the system can phase separate into a gel phase and a solution phase, or two solution phases. Alternatively with strong bonds, after more solvent is added, the gel may take up some more solvent corresponding to the new equilibrium and this equilibrium remains unchanged in the time window of our observation. The case when a volume change is observed but a steady value cannot be reached within our observation time span corresponds to the intermediate situation. Such strong bonds are almost equivalent to labile chemical bonds, the dissociation and association of which is the basis of self-healing polymers. For such gels, elasticity contribution to the free energy must be considered. Therefore, a part of reversible gels can be counted as belonging to the solution class, and another part to rubbery solid, depending on the temperature range of investigation.

The theoretical description of systems with strong bonds and (thermo) reversible gels has been a research subject of several groups. Both equilibrium states and the dynamics of formation and reformation were considered. Semenov and Rubinstein, [3] considered association of polymer chains bearing many equidistant stickers in a solvent under the conditions of bond reversibility and considering two- and three-body interactions (interaction parameter linearly dependent on concentration). The fraction of associated stickers into bonds, p, determines the distribution of chains according to the number of associated stickers. In that sense, the treatment is similar to that considering solutions of cross-linked by permanent bonds, constructed from polymers bearing cross-linked and uncross-linked units and coupling of the cross-linked ones [31] discussed in section. The distribution of degrees of polymerization and gelation conditions were obtained by implementing cross-linking statistics.

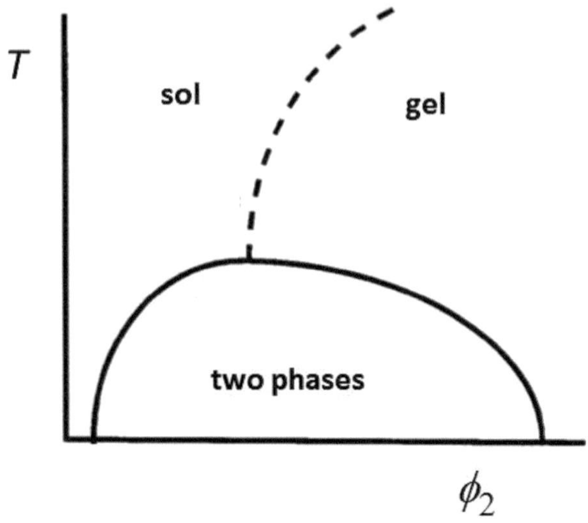

Fig. 4.19 Phase diagram of thermoreversible gels

Coupling of stickers can lead to phase separation and redistribution of chains differing in the number of coupled stickers.

A typical UCST phase diagram is schematically shown in Fig. 4.19.

The diagram in Fig. 4.19 is valid for polymer concentrations above the concentration c^*. For more dilute systems, micelles are formed.

In the dynamics studies [3], the authors used Rouse model and found that the dynamics was governed by two parameters: the effective energy of bond and the activation energy for bond formation. The dynamics is not affected by dissociation of stickers well below the gel point. Beyond the gel point, the viscosity starts to increase. The lifetime of a cluster is defined by a time during which it breaks into two parts of "comparable" sizes; comparable size means cluster backbone size. Further aspect and detail discussed by this group can be found in Refs.

A similar approach to the thermodynamics of reversible gels formed by association of associative groups and composed of distribution of solvent, finite clusters of chains, and gel was used by F. Tanaka and collaborators. The distribution of tree-like finite clusters in terms of their degrees of polymerization before the gel point and in the sol beyond the gel point as well as the number of formed bonds in the gel were calculated using the off-lattice branching theory as described by Flory; this whole distribution was placed on lattice and solvent (water) molecules were added.

Tanaka et al. considered the change of the Helmholtz energy of mixing the collection of clusters formed from linear chains with stickers and the network as

$$\Delta F_{\text{mix}} = \Delta F_{\text{rea}} + \Delta F_{\text{FHm}} \tag{4.95}$$

where ΔF_{rea} is the free energy required to bring the system from the reference state into an intermediate state in which the polymer chains are combined into clusters whose size distribution remains exactly the same as that after mixing with solvent.

ΔF_{FHm} is the FH free energy of solution composed of distribution of m-clusters and the network.

$$\Delta F_{\text{rea}}/kT = \sum_m \Delta_m N_m + \delta_\infty N^{\text{G}} \tag{4.96}$$

where Δ_m is the free energy to form an m-cluster in the intermediate state from m primary polymer chains; $\delta_\infty N^{\text{G}}$ is the number of chains in the gel, and δ_∞ is the contribution by one chain to a very large number of chains. This quantity depends on the network model; for random cross-linking it would be proportional to the cycle rank per chain.

If water is considered as solvent, a fraction of water molecules are bound to H-bond donors and H-bond acceptors of the polymer, i.e., water molecules are distinguished as free or bound. The cooperativity effect makes the change of swelling very abrupt. The effect of cooperativity of water molecules binding was observed experimentally. For PNIPAM, if a water molecule succeeds in forming a H-bond with an amide group of polymer chain, a second water molecule can form a H-bond with the chain more easily than the first one because the first molecule causes some displacement of the isopropyl group, thus creating more access space for the next water molecule. The cooperativity effect is introduced by formulating a set of equilibrium constants for formation and dissociating of H-bonds depending on the state of nearest neighbor unit—hydrated or not hydrated. Such cooperative hydration leads to "miscibility square behavior" of aqueous polymer solutions which, in fact, is the type found for the case "off-zero critical concentration" discussed before. A part of a large immiscibility gap for PNIPAM-water system and the fraction of bound water are shown in Fig. 4.20.

F. Tanaka and collaborators focused on phase equilibria involving PNIPAM in single solvents as well as in solvent mixtures (cosolvency and anti-cosolvency). It is assumed that for single PNIPAM molecules in water, the hydrophobic parts of several neighboring units collapse and the units between collapsed parts are cooperatively hydrated.

Related to F. Tanaka's approach is the model by Pouchlý's et al. They consider an off-lattice model of polymer chains and solvent molecules with stickers which form associates of variable multiplicity of stickers (2 and more stickers associate). They used Veitsman's formulation of the partition function for distribution of states. The distribution of clusters of polymer and solvent molecules is controlled by association constants. The resulting Helmholtz energy of mixing has the form

$$\frac{\Delta F_{\text{mix}}}{RT} = \phi_1 \ln \phi_1 + \left(\frac{\phi_2}{r}\right) \ln \phi_2 + \Delta F_{\text{assoc}}/RT + \chi \phi_1 \phi_2 \tag{4.97}$$

where $\chi = a_0 + a_1/T$ is concentration independent interaction parameter which relates to interaction by weak dispersion forces and ΔF_{assoc} is the additional change of Helmholtz energy due to association of stickers. The association contribution has the following form

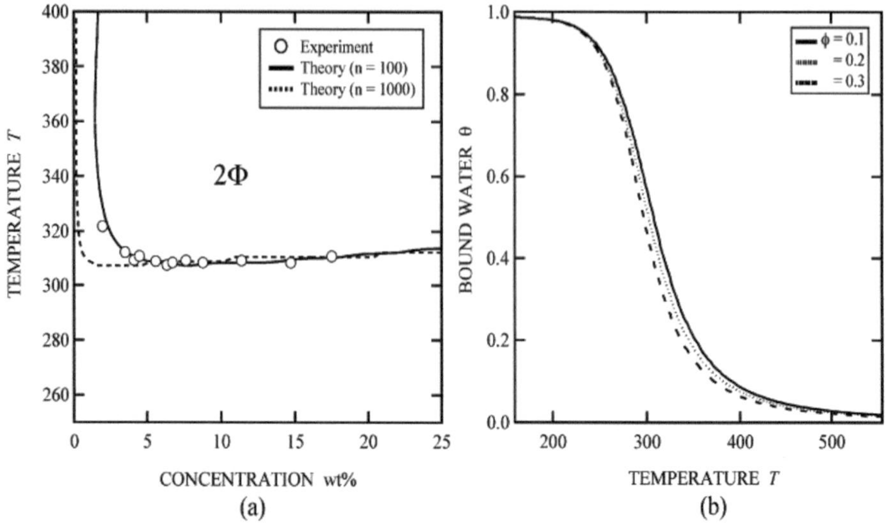

Fig. 4.20 **a** Phase diagrams of aqueous PNIPAM solution. Experimental data (o) of the spinodal curve for is compared with theoretical calculation. The DP of the polymer is $m = 100$ (solid line) and $m = 1000$ (broken line). Theoretical parameters used is $\Theta_0 = 555$ and $\lambda_0 = 0.002$ for $m = 100$, and $\Theta_0 = 565$ and $\lambda_0 = 0.003$ for $m = 1000$. Other parameters are fixed at $\gamma = 3.5$, $\sigma = 0.3$; **b** content of bound water plotted against temperature for three polymer volume fractions for m = 100. Reprinted from Ref. [Okada2005] with permission of American Chemical Society

$$\Delta F_{\text{assoc}}/RT = \left(\frac{f\phi_2}{r}\right)\ln\left(\frac{y_2}{y_2\phi_2}\right) + \phi_1\ln\left(\frac{y_1}{\phi_1}\right) + b + (1 - 1/m)\left(X - \phi_2 X_2^0\right) \tag{4.98}$$

with

$$X = k^{m-1} y_2^m y_1^{bm}$$

and mass-balance relation

$$f\phi_2/r = y_2 + X$$
$$\phi_1 = y_1 + bX$$

In these relations f is the number of stickers per chain, m is junction multiplicity (number of stickers forming a junction), bm is number of solvent molecules in one junction, X, y_1, y_2 are number densities of associated stickers, unassociated solvent molecules, and unassociated stickers, respectively, k is association constant. The superscript 0 refers to standard state (pure component). The distribution of clusters, the gel point, and post-gel state (gel, distribution of sol molecules) were generated by the theory of branching processes of tree-like clusters employing the extinction probability to extinguish bonds with finite and infinite continuation (cf. [118, 119]).

The phase behavior was characterized by analyzing the spinodal behavior given by $\partial^2(\Delta F_{\text{mix}}/RT)/\partial\phi_2^2 = 0$. Depending on k, and m, the spinodal curve can have very different shapes and, in some cases, two extremes on the dependences are seen at different concentrations. It may correspond to cases when two off-zero critical concentrations are found (cf., Sect. 3.2.3).

Douglas, Freed, Dudowicz, and collaborators view the problem of structure growth in solution with strongly interacting (associating) contact points as supermolecular self-organization. They consider polymer chain formation in the presence of solvent on a lattice; analyze approach to ceiling temperature and thermodynamic phase stability or instability. The Helmholtz energy change considers the dispersity of the polymer changers

$$\frac{\Delta F_{\text{mix}}}{kT} = \phi_1 \ln \phi_1 + \phi_{\text{m}} \ln \phi_{\text{m}} + \sum_{i=1}^{\infty} \left(\frac{\phi_i}{i}\right) \ln \phi_i + \chi \phi_1 \phi_{\text{m}} + \chi \phi_1 \sum_{i=2}^{\infty} \phi_i + \sum_{i=2}^{\infty} \phi_i f_i \tag{4.99}$$

where ϕ_{m} is the volume fraction of monomer which is thermodynamically the same as solvent (1), and f_i is dimensionless specific energy of an i-mer. This specific energy is not only a function of the bond energy but also of entropy. For fully flexible chain

$$\begin{aligned} f_2 &= \frac{1}{2} \ln\left[\frac{2}{2z}\right] + 1 - \frac{1}{2} + \frac{1}{2}\frac{\Delta f_p}{kT} \\ f_i &= \frac{1}{i} \ln\left[\frac{2(z-1)^2}{iz}\right] + \frac{i-1}{i} - \ln(z-1) + \frac{i-1}{i}\frac{\Delta f_p}{kT} \quad \text{for } i > 3 \end{aligned} \tag{4.100}$$

The bonding energies are also available for semi-flexible and stiff chains.

The dynamics of systems containing physical clusters and transient networks is determined by structure relaxation coupled with associating-dissociation kinetics. The equilibration process after a temperature jump from high-temperature unassembled state was measured [120] as the fraction of particles in the assembled state at time t. This process of association-dissociation for linear chains of two-patchy particles was described by Smoluchowski equation

$$\begin{aligned} \frac{\Delta N_l}{kT} &= -k_{\text{breaking}}(l-1)N_l + 2k_{\text{breaking}} \sum_{j=l+1}^{\infty} N_j \\ &\quad + \frac{1}{2}\frac{k_{\text{bonding}}}{V} \sum_{j=1}^{i-1} N_j N_{l-j} - N_l \frac{k_{\text{bonding}}}{V} \sum_{j=1}^{\infty} N_j \end{aligned} \tag{4.101}$$

The bond formation reactions depend on volume because they are of second order. Smoluchowski coagulation equations can be used for any number of patches and the kernel can be modified to allow for various kinds of hindrances making the rate "constants" dependent on cluster size.

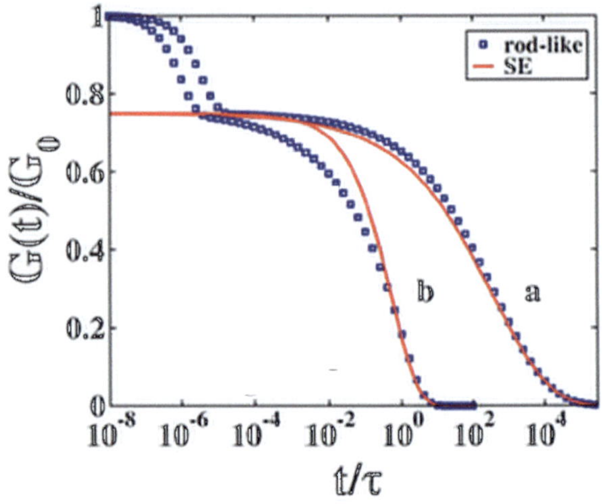

Fig. 4.21 Multistep stress relaxation function for polydisperse mixtures of **a** unbreakable rod-like clusters, **b** living rods. Reprinted from Ref. [121] (modified Fig. 4.2) with permission

Stukalin et al. [121] concluded that relaxation of the self-assembling clusters proceeded by motions associated either with cluster rotations, diffusive internal chain dynamics, or interchain entanglement interactions. All models yield a multistep relaxation for low polymer scission rates of "persistent polymers". The short time relaxation is nearly exponential and is dominated by the monomeric species and solvent; the long time relaxation is approximately a stretched exponential function, $\exp\left(-(t/\tau)^{\beta}\right)$, a behavior that arises from averaging over the equilibrium chain length distribution and internal relaxation modes of the assembled structures. The macroscopic viscoelastic and dielectric properties of these fluids appear to exhibit a common pattern of relaxation characterized by emergence of fast and slow relaxation processes that are separated by an intermediate plateau in the stress relaxation (Fig. 4.21).

The short time relaxation is nearly exponential and is dominated by the monomeric species and solvent, and the long time relaxation is approximately a stretched exponential. The apparent activation energy for the long time relaxation becomes temperature dependent, but the fast monomeric relaxation process remains Arrhenius.

4.3.3.6 Phase Separation in Transiently Cross-linked System

The strength of the physical bonds as well as the distribution of their numbers per unit determine whether the system is liquid-like or gel-like and whether the gel is stable or yield the imposed stress within our observation window. Kumar and Douglas [122] studied such systems. There are two extreme types of gels formed by physically associating polymer solutions. When the stickers pairing energy (ε_{ass}) is comparable or smaller than kT, the stickers pair and unpair rapidly and the system is liquid-like despite of surpassing its geometrical percolation. When the sticker bond lifetimes are very large, effectively infinite, the paired stickers or their associates behave as covalent bonds. In the intermediate range, after changing the external variables, full relaxation caused by rearrangement can be slow but equilibrium can be eventually reached within the available window of observation. Similarity has been found [122] between the dynamics of associating polymer solutions and vitrification of glass-forming liquids. Kumar et al. have found that the vitrification of typical liquids is remarkably similar to the thermo-reversible gelation of associating polymers (see also [123]). The results of simulation are shown in Fig. 4.22.

Audus et al. [124] developed a "patchy" model for thermally reversible gelling systems controlled by interplay between isotropic and highly directional and strong interactions that lead to a mutual coupling of phase separation and self-assembly. Patchy models have been used extensively to describe small globular proteins. Patch is a part of surface of a particle or monomer unit that is distinct with respect to its interaction type and strength from other parts of the particle or unit surface. Systems were considered having up to five attractive patches on one particle. Results obtained by renormalized mean field theory based on SAFT-VR model, percolation on lattice allowing cyclization, and tree-like, acyclic model of Stockmayer and Flory were compared with Monte Carlo simulation of the cluster size distribution. The predictions by the analytic mean field get more accurate when the strong interactions increase. Surprisingly, the Stockmayer-Flory tree-like model was found to be a good approximation as compared with simulation when the cluster size increased.

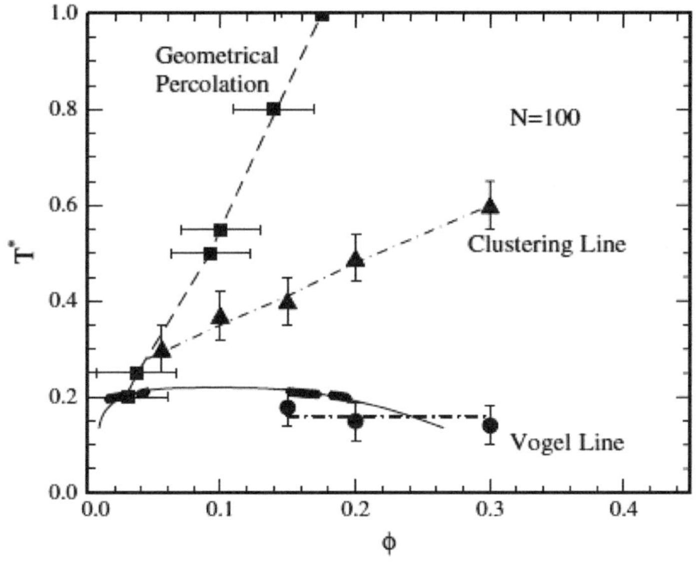

Fig. 4.22 Reduced temperature—concentration behavior of chains of length $N = 100$ with sticker of different binding strength. The broken lines delimit non-percolating system from percolating gel (geometrical percolation), reversible gel from permanent-like gel (clustering line), and rubber elastic state from glassy state (Vogel line). Full line corresponds to phase separation using an Ising model scaling function. Reprinted from [122] (Fig. 4.1) with permission

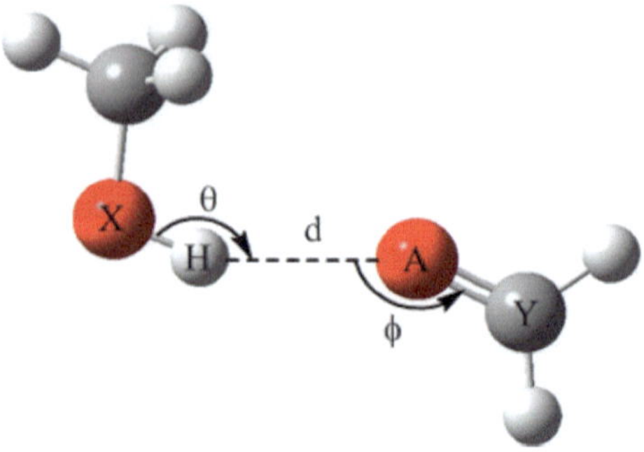

Fig. 4.23 Schematic illustration of hydrogen bonding

4.3.3.7 Hydrogen Bonding—Water as Solvent

Among strong physical bonds, the hydrogen bond has a dominant place in hydrophilic as well as hydrophobic systems. Its strength combined with its mobility and ability to reorganize and helps to increase and control the toughness of engineering polymeric materials as well as hydrophilic gels. It plays an important role in self-healing systems.

The hydrogen bond is formed between proton donors, i.e., compounds having labile hydrogen (e.g. groups, like OH, NH, FH, BH) and proton accepting electron pairs on O, N, NH, NH_2, etc. (see the scheme of Fig. 4.23).

Zheng et al. [125] have proposed to characterize hydrogen-bonding strength by so-called "hydrogen bonding index" for pairs of H-bond donors with H-bond acceptors. They found that the hydrogen-bonding strength and its corresponding hydrogen-bond distance for simple mono-hydrogen bond systems have an exponential relationship. The hydrogen-bonding index is also suggested as an additional criterion for virtual screening besides the widely accepted Lipinski's rule of five.[3] Also, a "Hydrogen-Bond Index Estimator (HBIE)" module was implemented in the so-called "Visual Force Field Derivation Toolkit (VFFDT)" program to estimate the hydrogen-bonding indices of any small molecules in batch and screen possible stronger donors or acceptors from the small-molecule database. Such predictions are very useful for organic polymer system, e.g., for protective coatings since regulation of mechanical properties through transient bonds in addition to covalent bond is very important.

Below, we show comparison of strengths and bond lengths for combination of various donors and acceptors (Fig. 4.23). In table the abbreviations of groupings are explained and Table 4.3 lists corresponding energies and bond lengths (Fig. 4.24).

The bonds characteristics are shown in the following Table 4.4.

The more negative is E_{HB}, the stronger is the H-bond. There exists correlation between and hydrogen-bonding distance in the hydrogen-bonding index I_{donor} or I_{acceptor} which was expressed through constants k_1 and b: $I = -\exp(k_1 d_{\mathrm{HB}} + b)$. The

[3] Lipinski [Lipinski2004]] formulated condition for a compound to be a successful drug in terms of the number of hydrogen-bond donors and acceptors and other properties of the molecule.

Table 4.3 Hydrogen-bonding strength (E_{HB}, kcal/mol) and optimal hydrogen-bonding distance (d_{HB}, Å) computed by the MP2 method; horizontal line: donors, left column: acceptors [125]

	wat		meo		eto		phy		eta	
	E_{HB}	d_{HB}	E_{HB}	d_{HB}	E_{HB}	d_{HB}	E_{HB}	d_{HB}	E_{HB}	d_{HB}
wat	– 4.26	2.1	– 4.51	2.0	– 4.44	2.0	– 5.67	2.0	– 2.32	2.3
met	– 4.83	2.0	– 5.38	1.9	– 5.35	1.9	– 7.64	1.8	– 2.92	2.2
fad	– 3.27	2.2	– 3.28	2.1	– 3.24	2.1	– 4.68	2.1	– 1.78	2.4
act	– 4.16	2.1	– 4.21	2.1	– 4.18	2.1	– 6.07	2.0	– 2.32	2.3
ace	– 5.20	2.0	– 5.29	2.0	– 5.24	2.0	– 7.63	1.9	– 2.72	2.3
frn	– 2.65	2.2	– 3.14	2.1	– 3.13	2.1	– 4.64	2.0	– 1.82	2.3
cn	– 4.40	2.2	– 4.48	2.1	– 4.43	2.1	– 6.39	2.1	– 2.26	2.4
py	– 5.91	2.0	– 6.58	2.0	– 6.56	2.0	– 9.62	1.8	– 3.32	2.3
dms	– 1.69	2.8	– 2.08	2.7	– 2.07	2.7	– 3.21	2.6	– 1.30	2.9
prt	– 0.84	3.1	– 0.90	2.9	– 0.91	3.0	– 1.69	2.9	– 0.77	3.1
tph	– 0.56	3.0	– 0.85	2.9	– 0.87	2.9	– 1.21	2.9	– 0.71	3.1
	mea		pri		aln		ett		bnt	
	E_{HB}	d_{HB}	E_{HB}	d_{HB}	E_{HB}	d_{HB}	E_{HB}	d_{HB}	E_{HB}	d_{HB}
wat	– 2.66	2.3	– 2.10	2.4	– 3.91	2.2	– 2.16	2.3	– 3.36	2.3
met	– 3.11	2.2	– 2.98	2.2	– 4.82	2.1	– 2.77	2.2	– 4.30	2.1
fad	– 2.13	2.3	– 1.88	2.4	– 3.05	2.2	– 1.64	2.4	– 2.67	2.4
act	– 2.77	2.3	– 2.43	2.3	– 3.94	2.2	– 2.16	2.3	– 3.54	2.3
ace	– 3.54	2.2	– 3.19	2.3	– 4.79	2.1	– 2.71	2.3	– 4.40	2.2
frn	– 1.74	2.3	– 1.88	2.4	– 3.04	2.2	– 1.64	2.4	– 2.71	2.3
cn	– 2.83	2.4	– 2.42	2.5	– 3.94	2.3	– 2.12	2.5	– 3.41	2.4
py	– 3.67	2.3	– 3.42	2.3	– 5.63	2.1	– 3.30	2.3	– 5.15	2.2
dms	– 1.27	2.9	– 1.20	3.0	– 2.29	2.8	– 1.25	2.9	– 2.26	2.9
prt	– 0.78	3.1	– 0.89	3.1	– 1.31	2.9	– 0.77	3.1	– 1.50	3.0
tph	– 0.53	3.1	– 0.66	3.2	– 1.17	2.9	– 0.68	3.1	– 1.25	3.1

relation between and E_{HB} and acceptor–donor distance passes through a minimum (Figs. 4.25 and 4.26).

4.3.3.8 Water as Solvent

Water is extremely important; it is the exclusive medium for life on Earth as well as solvent and cosolvent in various technologies and applications (including hydrogels). The most important property in this respect is its hydrogen-bonding ability. Because the hydrogen bonds between water molecules are much stronger than the bond S–H–S hydrogen bonds, water is a liquid while the heavier dihydrogen sulfide, H_2S, is a gas under normal conditions. Also, water molecules self-associate through hydrogen

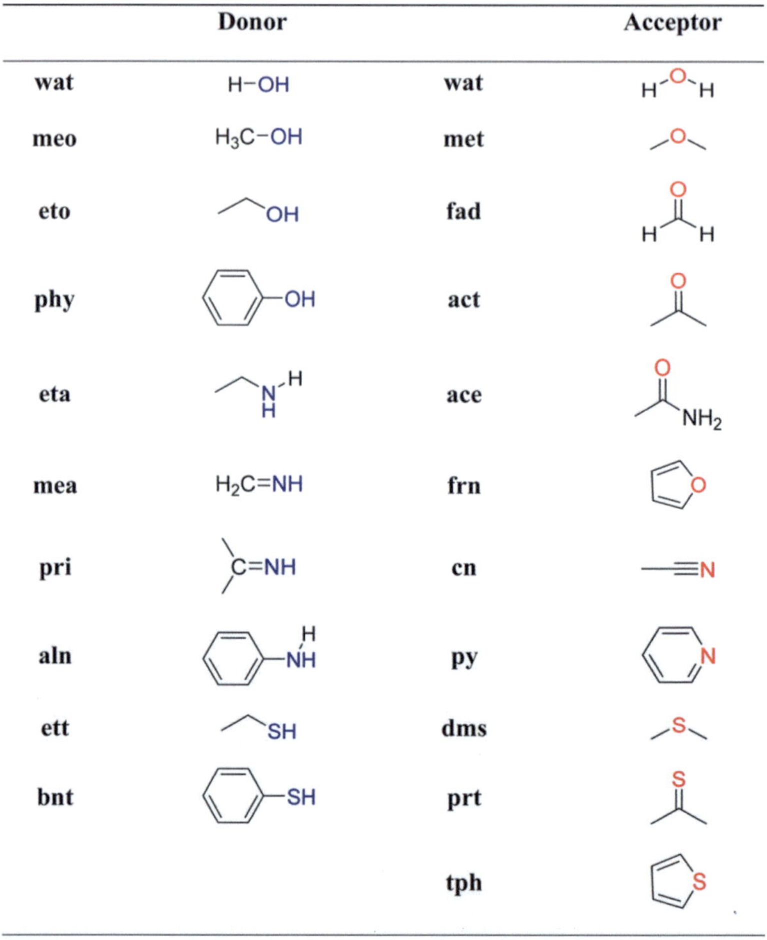

Fig. 4.24 Model molecules for hydrogen-bond donors and acceptors [125]

bonding with other water molecules and in the liquid state exist they exist in the form of distribution of molecular clusters. This self-association must be taken into account in considering water solution properties. The effect of hydrophobic parts of organic molecules on degree of association of water molecules is the basis of understanding aqueous solutions; this type of these interaction is called "hydrophobic interactions"; the underlying theory was formulated by Neméthy and Scheraga [126–128] and developed by Scheraga group [129–133] and Ben Naim [134, 135] and others [136, 137].

Prominent is the Wertheim's model [138–141] which was used as basis of several other thermodynamics theories of water solutions (e.g., SAFT-VR). Mixing of water with cosolvent or cosolutes modifies water structure in that heteroclusters are formed and the solvent power increases or sometimes decreases.

Table 4.4 Forms of ΔF_{el} for various models of networks with finite extensibility of network chains

Gauss	$\Delta F_{\mathrm{el}}/kT = \frac{N_{\mathrm{e}}}{2}\left(\lambda_{\mathrm{x}}^2 + \lambda_{\mathrm{y}}^2 + \lambda_{\mathrm{z}}^2 - 3\right)$
3-chain	$\frac{\Delta F_{\mathrm{el}}}{kT} = \frac{Nn^{1/2}}{3}\left[\lambda_x\beta_x + n^{1/2}\ln\left(\frac{\beta_x}{\sinh\beta_x}\right) + \lambda_y\beta_y + n^{1/2}\ln\left(\frac{\beta_y}{\sinh\beta_y}\right) + \lambda_z\beta_z + n^{1/2}\ln\left(\frac{\beta_z}{\sinh\beta_z}\right)\right]$ $\ell(\beta) = \frac{\lambda}{n^{1/2}} = \coth\beta - \frac{1}{\beta};\quad \beta = \ell^{-1}\left(\frac{\lambda}{n^{1/2}}\right);\quad \beta = \ell^{-1}\left(\frac{\lambda}{n^{1/2}}\right) \approx \frac{\lambda n^{-1/2}(3 - \lambda^2 n^{-1})}{1 - \lambda^2 n^{-1}}$
8-chain	$\Delta F_{\mathrm{el}}/kT = N_{\mathrm{e}}n^{1/2}\left[\lambda_{\mathrm{chain}}\beta_{\mathrm{chain}} + n^{1/2}\ln\left(\frac{\beta_{\mathrm{chain}}}{\sinh\beta_{\mathrm{chain}}}\right)\right]$ $\beta_{\mathrm{chain}} = \ell^{-1}\left(\frac{\lambda_{\mathrm{chain}}}{n^{1/2}}\right);\quad \lambda_{\mathrm{chain}} = \left(\frac{1}{3}\left(\lambda_x^2 + \lambda_y^2 + \lambda_z^2\right)\right)^{1/2}$
3-chain amended	$\frac{\Delta F_{\mathrm{el}}}{kT} = \frac{N_{\mathrm{e}}n^{1/2}}{3}\left[\lambda_x\beta_x + n^{1/2}\ln\left(\frac{\beta_x}{\sinh\beta_x}\right) + \lambda_y\beta_y + n^{1/2}\ln\left(\frac{\beta_y}{\sinh\beta_y}\right) + \lambda_z\beta_z + n^{1/2}\ln\left(\frac{\beta_z}{\sinh\beta_z}\right)\right]$ $-\frac{1}{3}N_{\mathrm{e}}\ln\frac{\beta_x\beta_y\beta_z}{\lambda_x\lambda_y\lambda_z}$
8-chain amended	$\Delta F_{\mathrm{el}}/kT = N_{\mathrm{e}}n^{1/2}\left[\lambda_{\mathrm{chain}}\beta_{\mathrm{chain}} + n^{1/2}\ln\left(\frac{\beta_{\mathrm{chain}}}{\sinh\beta_{\mathrm{chain}}}\right)\right] - N_{\mathrm{e}}\ln\left(\frac{\beta_{\mathrm{chain}}}{\lambda_{\mathrm{chain}}n^{-1/2}}\right)$
Gent	$\Delta F_{\mathrm{el}}/kT = -\frac{EJ_M}{6}\ln\left[1 - \frac{J_1}{J_{\mathrm{M}}}\right];\quad J_1 = I_1 - 3;\quad I_1 = \lambda_x^2 + \lambda_y^2 + \lambda_z^2$ $J_{\mathrm{M}} = I_{1\mathrm{M}} - 3;\quad I_{1\mathrm{M}} = I_1\left(\max\lambda_i = n^{1/2}\right)$
Kovac	$\Delta F_{\mathrm{el}}/kT = -\frac{N_{\mathrm{e}}n}{2}\left[\ln\left(1 - \lambda_x^2\frac{1}{n}\right) + \ln\left(1 - \lambda_y^2\frac{1}{n}\right) + \ln\left(1 - \lambda_z^2\frac{1}{n}\right) + \mathrm{const}\right]$

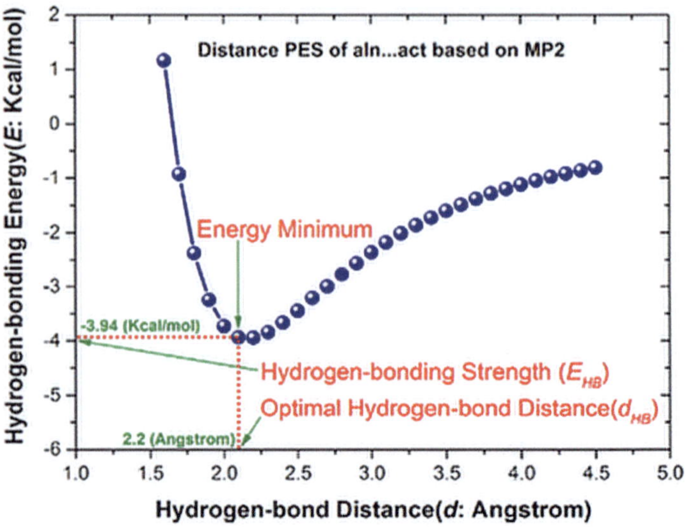

Fig. 4.25 Profile of the hydrogen-bonding energy and hydrogen-bond distance for the hydrogen-bond complex; reprinted from Ref. [125] Fig. 4.2 with permission of American Chemical Society

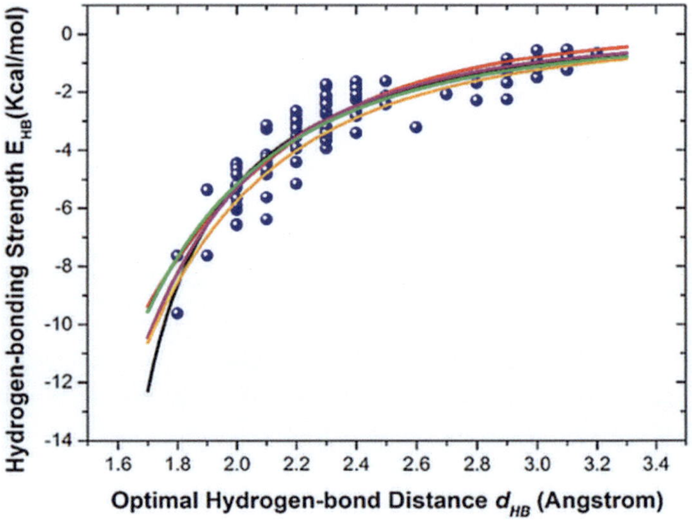

Fig. 4.26 Plot of hydrogen-bonding strength (E_{HB}) and optimal hydrogen-bond distance (d_{HB}) obtained by MP2 computation. The red, black, pink, and green curves fit to various correlation equations. Reprinted from Ref. [125] Fig. 4.2 with permission of American Chemical Society

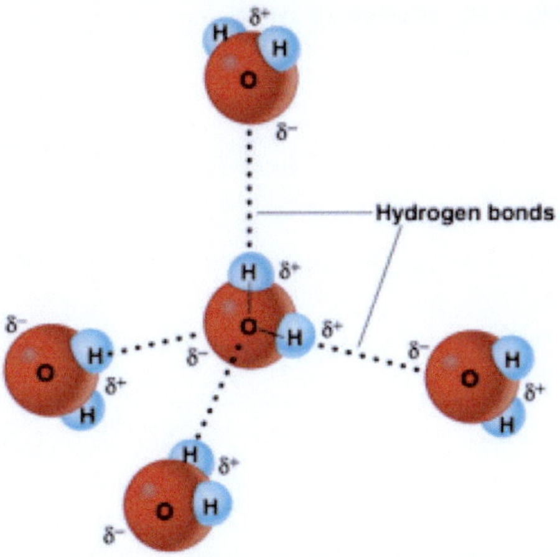

Fig. 4.27 Hydrogen-bonding sites of a water molecule

Among molecular clusters, water clusters are of paramount importance due to their unusual properties. The loci on a water molecule can be visualized as 2 donating and 2 accepting patches on the surface of the four arms of a tetrahedron centered on the oxygen atom—two H-bond donors and two H-bond acceptors (Fig. 4.27).

Liquid water structure in terms of populations of various aggregates has been a subject of extensive studies over recent decades (cf., Ref. [142]). For the dimer, the open form was found to be more stable over the cyclic configuration, whereas for trimers through hexamers, the cyclic forms were more favorable than their open analogues. Among the small-sized water clusters, the hexamer is of substantial interest since it represents the transition from monocyclic two-dimensional to multicyclic three-dimensional structures. The studies have been extended to large-sized water clusters. The overall stability of the clusters was also found to depend on the relative orientation of the non-H-bonded hydrogen atoms. This may be due to the dispersion interactions originating from the hydrogen pairs not participating in HB formation in the water clusters. In liquid water, regions of unassociated water are rare (Figs. 4.28 and 4.29).

The water associates were characterized by force spectroscopy. For instance, the cyclic trimer is a quite abundant species and has been well-characterized [143].

In a certain P, T range, a percolating network of hydrogen bonds is assumed to exist.

However, there is at present consensus that liquid water exists as a continuously rearranging H-bonded network and that inclusion of many body forces into the model potential is necessary. In liquid water, amid these flickering clusters, there are only few non-bonded water molecules. The lifetime of the water clusters is estimated to be in the range of picoseconds (single H-bond) to nanoseconds. The rearrangement dynamics involves collective motion of water molecules and energy fluctuations. It was shown that individual water molecules exhibit enormously large energy fluctuation (picoseconds) sometimes reaching up to 20 kcal/mol, which is associated with these collective motions. The power spectrum of the potential energy

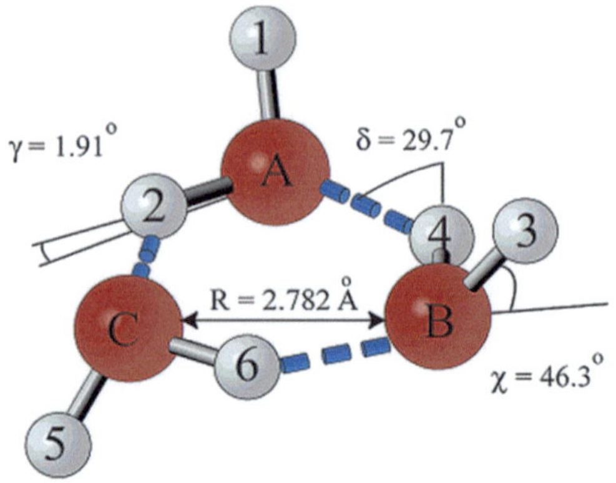

Fig. 4.28 Structure of cyclic water trimer

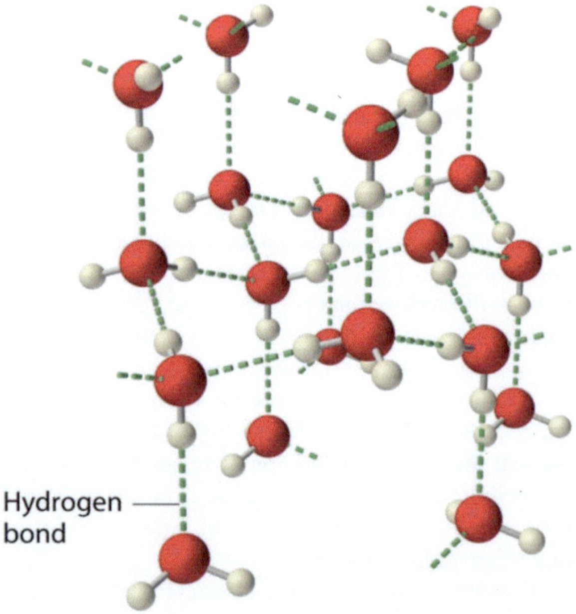

Fig. 4.29 Hydrogen-bonded network of water

fluctuation in liquid water showed frequency dependence indicating existence of multiple relaxation processes of the water hydrogen-bond network [142]. This spectrum is in strong contrast to that of simple atomic liquids, such as liquid argon, which yields nearly flat "white-noise" spectra. An average lifetime of individual hydrogen bonds is a few picoseconds, but global hydrogen-bond network structures are seen to persist for longer time. The formation of H-bonds between water molecules is highly cooperative [144] and it should also be accounted for in aqueous solution.

In the Wertheim's model [138– 141] of water solutions, the water molecule is modeled as a single spherical square-well segment of hard-core diameter σ_{11}; the square well is characterized by a depth ε_{11} and a range λ_{11}.

Four additional off-center square-well associating sites are used to mediate the hydrogen bonding (cf. Fig. 4.17). Two e sites represent the electronegative oxygen atom (the two lone pairs of electrons), and two H sites represent the hydrogen atoms. Only e–H bonding is permitted, allowing for up to two hydrogen bonds per water molecule, since a donor hydrogen or acceptor lone pair of electrons contributes half of a hydrogen bond.

Not only water molecules associate, they also become ordered along the organic structure. When Watson and Crick first proposed the double helix structure in 1953 [145], they already predicted the delicate relationship between DNA structure and aqueous environments. In later studies, a new structure was uncovered buried just at the surface of DNA: a chiral, helical superstructure of water molecules. Water molecules that follow the contours of double helix DNA likely participate in energy dissipation, playing a role in shielding DNA from excess heat and UV photodamage [146, 147].

With alcohols (both H-bond acceptors and donors) especially with methanol, water molecules form heteroclusters. The methanol−water heterodimer was found to be the most stable isomer. The addition of two water molecule breaks an H-bond between two glycol molecules, indicating the dominance of the ethylene glycol−water interactions over the ethylene glycol−ethylene glycol and water−water ones. The existence of associates in methanol was taken into account to fit reasonably swelling data for polyurethanes by considering methanol as trimer [148].

For aqueous solutions, the hydrophobic interactions are equally important as hydrogen bonds. They cause non-polar substance or their non-polar parts to minimize their contact with water and together with H-bonds they are the main driving force for polypeptides to fold. Hydrogen bonds are very specific and oriented while the interaction between hydrophobic is much less specific; the strength of hydrophobic bonds is in their multiplicity—large interacting surfaces. To review the theoretical treatments of interactions of solutions in water and of the hydrogen bond and hydrophobic group containing compounds is beyond the scope of this chapter.

4.3.3.9 Ternary and Multicomponent Systems

Ternary and multicomponent systems are complicated by the necessity to consider a number of interactions and because consideration of pairwise interaction may not be sufficient and ternary interactions have to be considered. One can distinguish between systems.

(1) single solvent—distribution of polymer molecules
(2) single polymer—mixture of two or more solvents
(3) two (or more) solvents—distribution of polymer molecules.

All types have practical significance. System (a) is the most usual case—a solution of polymer disperse molecular weights. There is no problem in formulating ΔG_{mix} and $\Delta\mu_1$, $\Delta\mu_{2i}$ dependence on polymer concentration, and to use them for one-phase system, but when the temperature changes and the system phase separates fractionation of the polymer occurs. The polymer–solvent interaction parameters are the same except for the shortest ones if they are present. This case was already briefly discussed in Sect. 3.2.4.

Systems two (or more) solvents—one polymer is very important. For instance, almost all solvent-based protective coatings are formulated as solvent blends. The classical Gibbs energy for an ideal ternary system reads

$$\frac{\Delta G_{\mathrm{mix}}}{NRT} = \frac{\phi_1}{m_1}\ln\phi_1 + \frac{\phi_2}{m_2}\ln\phi_1 + \frac{\phi_3}{m_3}\ln\phi_3 + g_{12}\phi_1\phi_2 + g_{13}\phi_1\phi_3 + g_{23}\phi_2\phi_3 \tag{4.102}$$

Formally, one can get each of the interaction constants g_{12}, g_{13}, g_{23} by independent measurements of solutions of binary systems: solution of polymer 3 in solvent 1, polymer 3 in solvent 2, and g_{12} for the two solvents from liquid–vapor equilibria. If the solvents are similar (similar size and similar polarity), the dependence of g_{12} on composition is more or less reasonable, however, for dissimilar solvents, the dependence of g_{12} on composition can be quite "wild" pointing to the fact that the mixtures are not regular solutions. "Single liquid approximation"—weighted mean of the interaction parameters—makes the thermodynamics very simple but their applicability in phase equilibria is very limited because it ignores differences in solvent composition between phases.

Ternary and multicomponent systems can be modeled as well by other theoretical approaches listed above (cf., e.g., publications of the Bae's group using the modified double lattice method [87], or PC-SAFT modeling [111]. Either the respective equations for the Gibbs energy and chemical potentials are available explicitly or the binary version can be extended.

Very interesting is the effect of cosolvency or co-non-solvency. It often happens that a mixture of solvent is better than each of the pure ones—especially the polar and hydrogen–bonding solvent mixtures form often cosolvent mixtures—for instance, low molecular weight alcohols may break the water clusters; the resulting smaller clusters are more miscible with the organic solutes. As an example, alcohol-water mixtures are better solvents for poly(2-hydroxyethyl methacrylate) (PHEMA) than water and monomer alone (Fig. 4.30) shows that the dependence of the equilibrium swelling in 2-hydroxyethyl methacrylate mixture passes through a maximum. This fact is of great importance for the formation of macroporous morphology when water is taken as a diluent in HEMA polymerization [149] (Fig. 4.31).

On contrary, water-alcohol mixtures exhibit a co-non-solvency for poly(N-hydroxypropylacrylamide) [95]. The degree of swelling of gels swollen in 2-propanol mixture shows minimum degree of swelling and the immiscibily region is maximum approximately for the 1:1 blend.

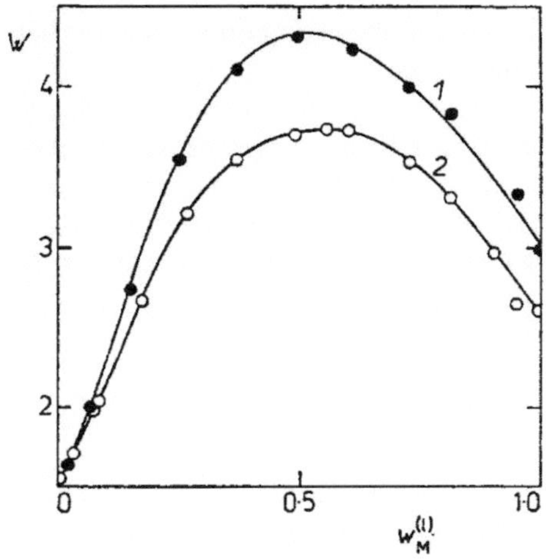

Fig. 4.30 Weight degree of swelling of poly(2-hydroxyethyl methacrylate) gels in mixtures 2-hydroxyethyl methacrylate-water (50 °C); w_M is the weight fraction of 2-hydroxyethyl methacrylate in the liquid phase. Concentration of ethylene dimethacrylate in the monomer in wt%: 1 0.08; 2 0.2

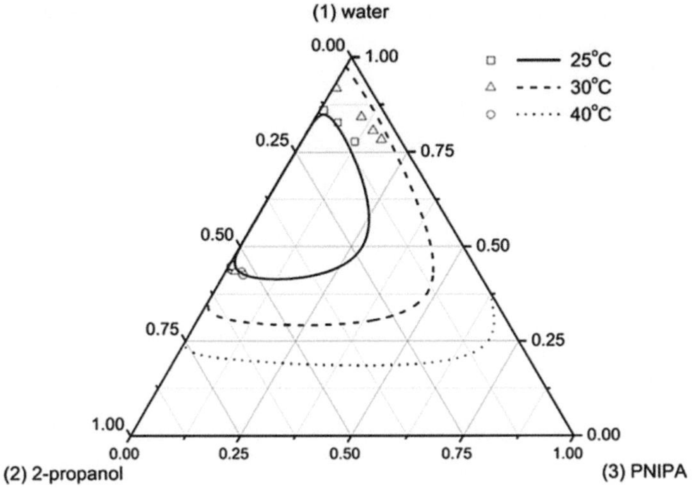

Fig. 4.31 Phase behavior of water/2-propanol/PNIPA solutions; transition temperatures at a fixed PNIPAM concentration of 1.0 wt%. PNIPAM molecular weight was 20 kDa. Reprinted from Ref. [95] Fig. 4.3 with permission

Other examples of cosolvency and co-non-solvency can be found in Refs. [112, 150–152].

In this section, the author has tried to show that mixing of polymer with the solvent alone can be very intricate and bring about effects sometimes attributed to the network elasticity. The mixing effects play a decisive role in systems where the strong physical bonds are involved and transient physical network can be formed. Such systems can show up interesting phenomena like volume phase transition just because of the mixing effects and the network structure may play only a secondary, fixation role.

The underlying theoretical studies are very helpful in understanding interactions and their reflection in properties of solutions and gels. At the moment, it is not easy to say which is "better" and which is "worse" because they accent different aspects of the phenomenon. The understanding of the interaction mechanisms and intensities is very inspiring for development of new gels and understanding how nature lives.

4.3.4 Elasticity Contribution

The osmotic pressure generated by mixing and discussed in the preceding section is opposed by force generated by stretching of network chains. If cross-links are not present, the stretching of chains is only temporary due to entanglements or strong interactions but in time the chains relax and adopt conformations of minimum energy. However, we have seen in Sect. 4.3.3 that certain temporary networks can behave as permanent network, i.e., their relaxation time is much larger than the observation time. In this section, we will discuss elastic contribution by the network, $\Delta F_{\mathrm{el,n}}$, to the free energy change on swelling. Adding it to ΔG_{mix} will give us ΔG_{sw} and conditions for equilibrium in a liquid solvent or their mixtures, or in solvent vapors.

There are several rubber elasticity theories available in the literature. We do not try to scrutinize them but just to comment on the most frequently used ones. Most theories are based on conformational changes of single network chains. We should distinguish the contribution to entropy of the system by a network chain embedded in volume V_0 where the chain ends fluctuate more or less freely around the mean positions and the swollen volume when the volume is increased by $\Delta V = V - V_0$ (V_0 is the reference volume) after adding solvent. The probability of finding an unperturbed chain or its chain end in volume V relative to volume V_0 is proportional to V_0/V and entropy change to $\ln(V/V_0)$: as regards the entropy contribution by deformation relative to the reference state, the ΔF_{el} contribution (no volume change) is distinguished from $\Delta F_{\mathrm{el,n}}$(with volume change by swelling).

$$\Delta G_{\mathrm{el,n}} \approx \Delta F_{\mathrm{el,n}} = \Delta F_{\mathrm{el}} - RTX_{\mathrm{e}} \ln\left(\frac{V}{V_0}\right) \tag{4.103}$$

where X_{e} is the number of elastically active elements, for instance N_{e} for Gaussian affine, or Ξ_{e} for phantom networks.

Next, we will review the most frequently used forms of ΔF_{el} for Gaussian networks, finite chain extensibility networks, and entangled networks. More on various elasticity models can be found in Ref. [153].

4.3.4.1 Gaussian Networks

The network is built-up from freely jointed chains which obey the Gaussian distribution of distances between chain ends. The chains ends are connected in f-functional junctions. In a perfect network, all chains between junctions are elastically active and their number, N_{chain} is equal to the product $(f/2)N_{\mathrm{junction}}$. In real networks, not all chains are elastically active (number of elastically active chains N_{e}); due to network defects, sometime introduced intentionally, some chains between junctions are not active. In the model, the junctions, also called cross-links, fluctuate more or less freely and the chains and junctions have no volume and the chains can freely cross one another. The most prominent model for such network is the junction-fluctuation model of Flory and Erman [153–156]. Two limiting situations are considered: phantom networks and affine networks. They are distinguished by fluctuation of cross-links. In phantom networks, the cross-links fluctuate freely around their mean positions which are displaced affinely with external strain. In affine networks, the chains do not fluctuate and the cross-links are displaced affinely with external strain. The suppression of fluctuation is caused by unspecified interactions between network chains. The corresponding change of Helmholtz energy caused by external deformation

$$\Delta F_{\mathrm{el}}/kT = \frac{\Xi_{\mathrm{e}}}{2}\left(\lambda_{\mathrm{x}}^2 + \lambda_{\mathrm{y}}^2 + \lambda_{\mathrm{z}}^2 - 3\right) \quad \text{for phantom network} \tag{4.104}$$

and

$$\Delta F_{\mathrm{el}}/kT = \frac{N_{\mathrm{e}}}{2}\left(\lambda_{\mathrm{x}}^2 + \lambda_{\mathrm{y}}^2 + \lambda_{\mathrm{z}}^2 - 3\right) \quad \text{for affine network} \tag{4.105}$$

where N_{e} and Ξ_{e} are number of elastically active and elastically active cycle rank, and λ_x, λ_y, and λ_z are the deformation ratios of the sample along the x, y, z coordinates with respect to reference state ($\lambda_x = L_x/L_{x0}$, etc.); the reference state is isotropic $L_{x0} = L_{y0} = L_{z0} = L_0$ The relation between N_{e} and Ξ_{e} is given by

$$\Xi_{\mathrm{e}} = \frac{f_{\mathrm{e}} - 2}{f_{\mathrm{e}}} N_{\mathrm{e}} \quad \text{or } \xi_{\mathrm{e}} = \frac{f_{\mathrm{e}} - 2}{f_{\mathrm{e}}} \nu_{\mathrm{e}} \tag{4.106}$$

where ν_{e} and ξ_{e} are the concentration versions of N_{e} and Ξ_{e}, and f_{e} is the elastically effective functionality of the cross-link defined as

$$f_{\mathrm{e}} = \frac{\sum_{i=3}^{f_{\mathrm{chem}}} i t_i}{\sum_{i=3}^{f_{\mathrm{chem}}} t_i} \tag{4.107}$$

where t_i is the fraction of cross-links participating in paths of bonds having infinite continuation (a cross-link counted as elastically active if $i \geq 3$); this quantity can be obtained from cross-linking statistics [32] (see also [119]); f_{e}increases from 3 near the gel point and may reach chemical functionality far from the gel point but not always; in off-stoichiometric network it may be lower.

Erman and Flory [156, 157] introduced a bridging function for transition from the phantom limit to the affine limit, considered additive to $(\Delta F_{el})_{phantom}$

$$\Delta F_{el} = (\Delta F_{el})_{phantom} + \frac{1}{2}\xi_e kT \frac{\mu}{\xi} \sum_{i=x,y,z} [B_i + D_i - \ln(B_i + 1) - \ln(D_i + 1)] \tag{4.108}$$

where μ is the number concentration of junctions, and

$$D_i = \lambda_i^2 \kappa^{-1} B_i; \quad B_i = \kappa^2 (\lambda_i^2 - 1)(\lambda_i^2 + \kappa)^{-2} \tag{4.109}$$

μ is related to ξ as $\mu = (2/(f-2))\xi = 2\nu/f$ for perfect network or to ξ as $\mu_e = (2/(f_e - 2))\xi_e = 2\nu_e/f_e$ for a network with defects.

Thus, the additive contribution to $(\Delta F_{el})_{phantom}$ is only a function of κ which is a measure of concentration of entanglements or of other interactions proportional to junction-junction contacts; for a given system, it has to be obtained by fitting of experiments.

There is ample experimental evidence that the elastic moduli of swollen polymer networks decrease with increasing degree of swelling unless the finiteness of the real-chain extensibility becomes noticeable. Phenomenologically, the constant C2 in the Mooney-Rivlin equation between stress and tensile strain decreases.

$$f_{extension}/a = (C_1 + C_2/\Lambda_x)(\Lambda_x - \Lambda_x^{-2}) \tag{4.110}$$

where Λ_x is the extension ratio with respect to the isotropic non-deformed sample. This fact can be explained by weakening of transient polymer–polymer interactions. When the networks are prepared in the presence of diluents, the number of permanent chain-chain interactions (trapped entanglements) decreases but one must be aware of the fact that too much diluent may induce formation of inhomogeneities or even lead to phase separation.

Over decades, a great effort was spent to resolve the phantom-affine problem, i.e., whether real networks adhere more to that or the other class, which resulted in hundreds of publications. Networks of low cross-link density and prepared in the presence of diluent are believed to fulfill the conditions of the phantom model. Clear resolving of the problem was hampered by uncertainties about the network perfectness. At the content of about 1% of cross-linkable units, at such low concentration not all of functional groups may have reacted and most spectroscopic methods have resolution limits at 0.1–1% content. Such networks may have much less elastically active network chains (EANC) that can be calculated from chemical composition assuming full conversion of cross-linkable groups. Also, formation of cycles that do not contribute to the number of elastically active cross-links lowers the concentration of EANCs and there is hardly any experimental method to make their determination possible. The results are scattered between these limits and extend far above that limit. Despite these uncertainties, we believe that for some systems the phantom behavior can be reached. A recent publication suggests that such system has been found [158]. Similarly as for compressible solutions, the lattice gas concept was used for swelling in such solvents concept was for swelling of cross-linked polymers [159, 160].

4.3.4.2 Non-gaussian Networks

In systems having equilibrium moduli clearly above the affine limit, evidently, trapped entanglements contribute. However, constraints by trapped entanglements and similar interactions are different from covalent cross-links. Several theoretical solutions of the problem are available. The constrained-chain model of Erman and Monnerie, the slip-link model, or the diffused constraints model can be named as examples (cf., e.g., [153]). The functions additive to $(\Delta F_{el})_{phantom}$ have been formulated. Edwards and Vilgis [161] elaborated a model of a network with sliding entanglements and finite extensibility of network chains. Since the finite extensibility issue will be discussed in the next paragraph, it will be discussed there.

Finite Extensibility of Network Chains

When the EANCs are strained by swelling and possibly by additional mechanical stresses, they may reach the degree of strain at which the finite extensibility starts affecting the Gaussian distribution dependence. This problem is especially important for highly swellable hydrogels. It is a special feature of chains described by the Gauss function that they can be extended far beyond its contour length because the Gauss function of squared variables is defined and remains continuous in the interval from zero to infinity. The most frequently used finite extensibility models were recently briefly summarized in ref. [119] and reviews by Boyce et al. and Horgan et al. [162–166]. That the end-to-end distance cannot exceed the contour length of the chain, nl, (n is number of segments and l is the segment length) is the necessary condition for the model. The exact Rayleigh distribution of population of end-to-end distances of a chain composed of n statistical segments was transformed by Kuhn and Grün [167] into a functional form containing the inverse Langevin function of the deformation ratio taken relative to the maximum extension nl, λ_{rel}. The chain deformation λ is taken relative to the unperturbed state $(r/n^{1/2}l)$, so that $\lambda_{rel} = \lambda n^{-1/2}$. The Kuhn–Grün function has been used in various models such as the three-chain model and eight-chain model [168– 173]. The semi-empirical function developed by Gent et al. [115, 174, 175] is based on assumption of proportionality of the deformation energy to the logarithm of the first invariant of the Cauchy-Green strain vector, $J_1 = \lambda_x^2 + \lambda_y^2 + \lambda_z^2 - 3$, as a way of generalization of inclusion of higher powers of the first strain invariant. The invariant $J_1 = I_1 - 3$ is taken relative to its maximum value J_M. The first invariant describes ΔF_{el} of the Gaussian network and the E or G equilibrium moduli are interpreted in terms of the concentration of EANCs and state of network chains in the same way as for the Gaussian model. It is also favored because it correctly converges to the Gaussian limit for large values of n and low degrees of swelling. Horgan and Saccomandi [174] have shown that the functional form of the Gent's function is close to that of the three-chain model if the inverse Langevin function is expressed by the Paddé approximant. Kovac [176] proposed a modified Gaussian model based on the end-to-end distance distribution function for a chain subjected to equilibrium force which can extend the chain up to its maximum length equal to nl. The Kovac's and Gent's models are functionally similar.

Modified tube models represent the third group of theories; however, more input parameters are required. The application of these models was mainly focused on stress–strain behavior of dry and swollen networks; not much to equilibrium swelling. For the current state of molecular tube models, the reader is referred to reviews by Hoei [177] and Klüppel [178, 179].

In this table, N_e is the number of elastically active network chains (EANC), n number of statistical segments per EANC, $\beta = \ell^{-1}\left(\lambda/n^{1/2}\right)$ inverse Langevin function of the Langevin function $\ell(\beta) = \coth\beta - 1/\beta$; $\beta = \ell^{-1}\left(\lambda/n^{1/2}\right)$ was approximated by the Paddé approximant $\beta \approx \lambda n^{-1/2}\left(3 - \lambda^2 n^{-1}\right)/\left(1 - \lambda^2 n^{-1}\right)$.

The amendment of the three-chain and eight-chain models was made by Jernigan and Flory [180, 181] in order that the function gave the correct limit for $\phi_2 = 1$.

The expressions for contribution of the chemical potential of solvent was derived from Helmholtz energy of Table 4.5 for isotropic swelling and the corresponding expressions are summarized in Table 4.6.

The dependences of $\left(\Delta_{\mu_1}\right)_{el}$ on ϕ_2 for various models listed in Table 4.6 are shown in Fig. 4.32a, b.

All listed theories show an upturn before the network chains reach their contours lengths which depends on n (number of statistical segments of an EANC, taken as number average). There exists a relation between n and ν_e

$$\nu_e = \rho_2/nM_s \tag{4.111}$$

Table 4.5 Forms of $(\Delta\mu_1)_{el}$ for various models of networks with finite extensibility of network chains

Gauss	$(\Delta\mu_1)_{el}/RT = V_{1m}\nu_e\phi_2^{1/3}\left(\phi_2^0\right)^{2/3}$
3-chain	$(\Delta\mu_1)_{el}/RT = \frac{1}{3}V_{1m}\nu_e n^{1/2}\left[(\beta + \beta\prime\lambda) + n^{1/2}\frac{\beta\prime}{\beta}(1 - \beta\coth\beta)\right]\phi_2^{2/3}\left(\phi_2^0\right)^{1/3}$
8-chain	The same as 3-chain, because $\lambda_{chain} = \lambda$
3-chain amended	The same as 3-chain model plus term $\frac{1}{3}V_{1m}\nu_e\left[\frac{\beta\prime}{\beta} - \frac{1}{\lambda}\right]\phi_2^{2/3}\left(\phi_2^0\right)^{1/3}$
8-chain amended	The same as 3-chain model amended
Gent	$\left(\Delta_{\mu_1}\right)_{el}/RT = \nu_e V_{1m}\frac{\left(\phi_2^0\right)^{2/3}\phi_2^{1/3}}{1-\left[\left(\phi_2^0/\phi_2\right)^{2/3}-1\right](n-1)}$
Kovac	$(\Delta\mu_1)_{el}/RT = \nu_e V_{1m}\frac{\phi_2^{1/3}\left(\phi_2^0\right)^{2/3}}{1-\left(\phi_2^0/\phi_2\right)^{2/3}/n}$

Table 4.6 Values used to generate data plotted in Fig. 4.30a,b and identification of curves

$\chi(\phi_2) = a_0/(1 - b\phi_2); f_e = 3$, $V_{mol,ls} = 100\ cm^3/mol$; $\rho_2 = 1.2\ g/cm^3$ $M_s = 200\ g/mol$	a_0	b	Color used in figure
	0.10	0	Green
	0.20	0	Black
	0.40	0.20	Red
	0.50	0.40	Blue

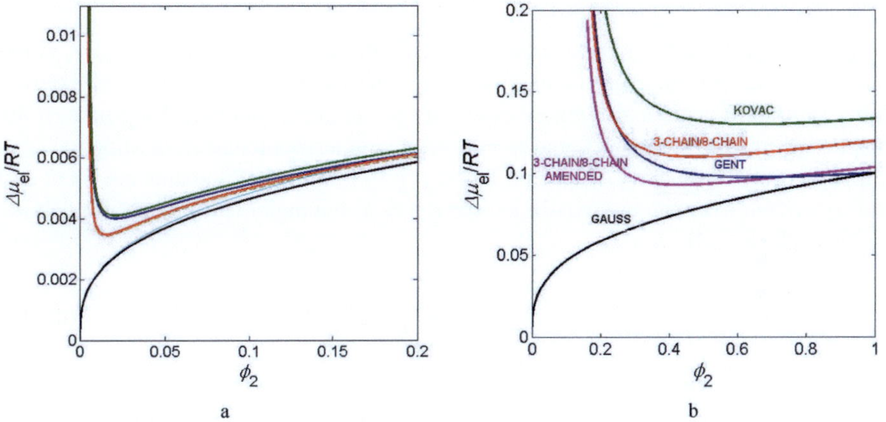

Fig. 4.32 **a**, **b** Plot of elastic contributions to the change of chemical potential of the solvent $(\Delta\mu_1)_{el}/RT$ by free swelling listed in Table 4.6 versus volume fraction of the polymer ϕ_2. Concentration of EANCs, ν_e, and average number of statistical segments per EANC, n: a) $\nu_e = 10^{-4}$ mol/cm^3, $n = 40$; b) $\nu_e = 10^{-3}$ mol/cm^3, $n = 4$; values of other parameters: $\phi_2^0 = 1$, molar volume of solvent (=lattice site) $V_{1m} = 100$ cm^3/mol, molecular mass of a statistical segment $M_s = 300$ g/mol. Identification of the curves (see Table 4.6): black—Gauss, red—3-chain and 8-chain, magenta—3-chain and 8-chain amended, blue—Gent, green—Kovac

where M_s is the molecular weight of statistical segment. Divergence of the chemical potential function is reached when the nl limit is reached. For high cross-link densities, the differences are noticeable but for moderate values of ν_e, they are small. All functions should converge to the same value at $\phi_2^0 = 1$ and they do after correction (amendment)—also that of Kovac if corrected. On contrary to divergence of the chemical potential function at the high swelling limit, the Gauss function converges to zero. The same form for 3-chain and 8-chain models is valid only for the triaxial symmetry, for asymmetric deformation modes, they are different (Fig. 4.31).

The theory by Edwards and Vilgis [161] considers both trapped entanglement (slip-link model) and finite extensibility (based on tube model). For $(\Delta F)_{el}$ the model is described by the following function

$$\Delta F_{\mathrm{el}} = \frac{1}{2}N_{\mathrm{e}}kT\left[\frac{(1-\alpha^2)\left(\lambda_x^2+\lambda_y^2+\lambda_z^2\right)}{1-\alpha^2\left(\lambda_x^2+\lambda_y^2+\lambda_z^2\right)} + \ln\left(1-\alpha^2\left(\lambda_x^2+\lambda_y^2+\lambda_z^2\right)\right)\right]$$
$$+\frac{1}{2}N_{\mathrm{s}}kT\left[\frac{(1-\alpha^2)(1+\eta)}{1-\alpha^2\left(\lambda_x^2+\lambda_y^2+\lambda_z^2\right)}\left(\frac{\lambda_x^2}{1+\eta\lambda_x^2}+\frac{\lambda_y^2}{1+\eta\lambda_y^2}+\frac{\lambda_z^2}{1+\eta\lambda_z^2}\right)\right.$$
$$\left.+\ln\left(1+\eta\lambda_x^2\right)\left(1+\eta\lambda_y^2\right)\left(1+\eta\lambda_z^2\right)\right) + \ln\left(1-\alpha^2\left(\lambda_x^2+\lambda_y^2+\lambda_z^2\right)\right)\Big] \tag{4.112}$$

In this equation, N_{e} is the number of elastically active network chains, N_{s} is the number of trapped entanglements, $\alpha = l/a$ is the measure of limitation of extensibility (length of statistical segment of chain l per length of statistical segment of the respective primitive paths a), and $\eta = \varepsilon^2/L$ characterizes measure of slippage (ε is slippage probability and L the chain length); $\eta = 0$, all entanglements are fixed and are equivalent to covalent cross-links; for $\eta >> 1$, entanglements do not contribute.

Thus, for $\alpha = 0 \quad \eta = 0$

$$\Delta F_{\mathrm{el}} = \frac{1}{2}N_{\mathrm{e}}kT\left(\lambda_x^2+\lambda_y^2+\lambda_z^2\right) + \frac{1}{2}N_{\mathrm{s}}kT\left(\lambda_x^2+\lambda_y^2+\lambda_z^2\right) \tag{4.113}$$

This theory seems to be quite universal but its application is limited by the necessity to estimate the values of N_{s}, η and α. For the given system, estimates can be made from dynamic and static thermomechanical properties.

In conclusion of Sect. 3.4: the Gaussian theory in and between its phantom and affine limits is being widely used for swollen systems. However, for certain types of gels (e.g., highly swollen (hydro)gels and deformed hydrogels) the finite extensibility effects should be considered. The finite extensibility effect should also be considered in some double network types of IPNs. In this section, only some non-Gaussian theories have been characterized. The reader can find more in the reviews listed above.

4.3.5 Equilibrium Swelling of Non-ionized Networks

Once the mixing and elasticity contribution have been formulated in Sects. 4.3.3 and 4.3.4, respectively, the condition for swelling equilibrium is determined by the additivity condition (Eq. 4.7)

$$\Delta \mathrm{G}_{\mathrm{sw}} = \Delta \mathrm{G}_{\mathrm{mix,n}} + \Delta \mathrm{G}_{\mathrm{el,n}} \tag{4.114}$$

where $\Delta G_{\mathrm{mix,n}}$ has been formulated so far for the Flory–Huggins (Eqs. 4.11 and 4.18) and Guggenheim models (Eqs. 4.52 and 4.54) but could be formulated for other theories. The relation between $\Delta G_{\mathrm{el,n}}$ and ΔG_{el} is given by Eq. 4.103 and respects the increase of volume when solvent is added. In terms of chemical potential of the solvent and FH model of the mixing part

$$\frac{\Delta\mu_{1,\mathrm{mix,n}}}{RT} = \left(\frac{1}{m_1}\ln\phi_1 + \frac{\phi_2}{m_1} + \frac{f_\mathrm{e}-2}{f_\mathrm{e}}\nu_\mathrm{e}V_{\mathrm{mol,1}}\phi_2 + \left(g(\phi_2) - \frac{\partial g(\phi_2)}{\partial\phi_2}\phi_1\right)\phi_2^2\right) \tag{4.115}$$

and Gaussian model of elasticity in its affine limit

$$\frac{\Delta\mu_{1,\mathrm{el,n}}}{RT} = \nu_\mathrm{e}V_\mathrm{ls}\left(\phi_2^{1/3}\left(\phi_2^0\right)^{2/3} - \phi_2\right) \tag{4.116}$$

The condition for equilibrium with pure solvent or solvent of reduced activity, or solvent vapors is the given

$$\frac{1}{m_1}\ln\phi_1 + \frac{\phi_2}{m_1} + \frac{f_\mathrm{e}-2}{f_\mathrm{e}}\nu_\mathrm{e}V_{\mathrm{mol,ls}}\phi_2 + \left(g(\phi_2) - \frac{\partial g(\phi_2)}{\partial\phi_2}\phi_1\right)\phi_2^2 + \nu_\mathrm{e}V_{\mathrm{mol,ls}}\left(\phi_2^{1/3}\left(\phi_2^0\right)^{2/3} - \phi_2\right) = \ln a_1 \tag{4.117}$$

We can discuss qualitatively the effect of variable parameters on the swelling degree:

- m_1 volume of solvent molecule with respect to the volume of lattice site: increasing causes decrease of swelling degree;
- f_e effective functionality can vary between 3 and chemical functionality f_{chem}; the factor $(f_\mathrm{e}-2)/f_\mathrm{e}$ varies between 1/3 and 1; far from the gel point it is equal to $(f_{\mathrm{chem}}-2)/f_{\mathrm{chem}}$;
- ν_e concentration of elastically active network chains; positively contribute, swelling degree decreases;
- molar volume of lattice site: a constant, e.g., 100 cm^3, mol or is made equal to molar volume of monomer unit of the polymer; its magnitude must correspond to the magnitude of the interaction function g or χ, whatever is used;
- $g(\phi_2)$ the interaction function and its concentration dependence: increasing $g(\phi_2)$ decreases the degree of swelling, the concentration dependence can bring about interesting effect, for instance in phase separation (see Sect. 4.7)
- $\phi_2^0 = 1 - \phi_1^0$, ϕ_1^0 the volume fraction of diluent during network formation: increasing ϕ_1^0 increases degree of swelling; too high ϕ_1^0 may cause reaction-induced phase separation during formation of the network.

For some of other mixing theories (papers cited here), excess functions for uncross-linked polymers have been formulated as additional to the FH, but cross-linked structures have not been addressed.

For system where finite chain extensibility was addressed, one of the elasticity contributions listed in Table 4.6 can be used. For instance, using the Gent's function together interaction function formulated as $\chi(\phi_2)$, the swelling equilibrium condition reads

$$\frac{1}{m_1}\ln\phi_1+\frac{\phi_2}{m_1}+\frac{f_e-2}{f_e}\nu_e V_{\mathrm{mol,ls}}\phi_2+\chi(\phi_2)\phi_2^2$$
$$+\nu_e V_{\mathrm{mol,ls}}\left(\frac{\left(\phi_2^0\right)^{2/3}\phi_2^{1/3}}{1-\left[\left(\phi_2^0/\phi_2\right)^{2/3}-1\right](n-1)}-\phi_2\right)$$
$$=\ln a_1;\qquad \chi(\phi_2)=\frac{a_0}{1-b\phi_2} \tag{4.118}$$

Equilibrium values of ϕ_2 in dependence on the concentration of EANCs for various values of the interaction parameter $\chi(\phi_2)$ specified in Table 4.6 are shown in Fig. 4.33a, b

Figure 4.29a, b shows that in the range of moderate cross-link densities there is marked difference in cross-link density depending on whether experimental swelling

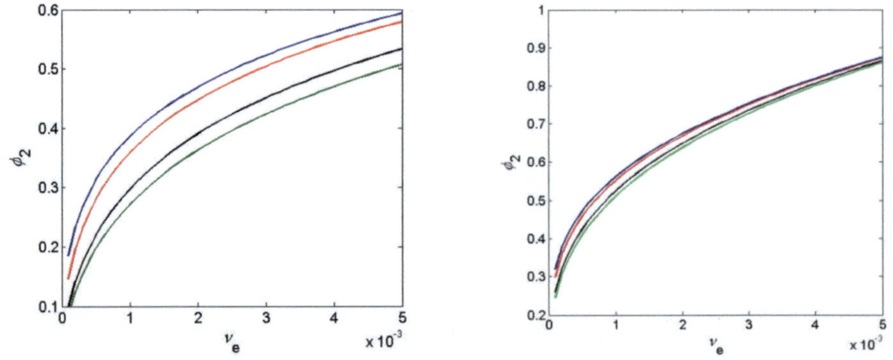

a b

Fig. 4.33 **a, b** Volume fraction of polymer in the swollen gel, ϕ_2, as a function of concentration of EANCs, ν_e(mol/cm^3), for (28a) the model of Gaussian network, affine limit (Eq. (4.117) and (28b) Gent's model of network with finite extensibility of network chains (Eq. 4.118); different values of the interaction parameter specified in Table 4.7. Modified from Ref. [Dusek2017] with permission

Table 4.7 Scaling relations for root-mean-square end-to-end distance of polymer chains for various dilution ranges

Regime	rms ETE $\langle R^2\rangle^{1/2}$
θ-solvents + conc. solution	$b^2N^{1/2}$
Semi-dilute athermal	$bN^{1/2}(\phi_2)^{-\frac{\upsilon-1/2}{3\upsilon-1}}$
Semi-dilute good	$bN^{1/2}\left(\phi_2 b^3/\mathrm{v}\right)^{\frac{2\upsilon-1}{2(1-3\upsilon)}}$

data are processed disregarding or taken into account the finite extensibility of network chains. In the region of high cross-link densities the finite extensibility limit predominates, in the region of small concentrations of EANCs, its effect is weaker but still noticeable (Table 4.7).

The effect of constraints on equilibrium swelling will be discussed in Sect. 4.6.

4.4 Swelling of Gels with Charged Groups

Polymer networks and gels bearing ionizable groups and engaged in ionic interactions are of great interest because of their important applications like ion exchangers, moisture absorbers, components of medical materials—implants, cell cultivation scaffolds, drug delivery materials, or cross-linkers of rubbery materials. Even more important is the fact that charged groups and ionic interactions play an important role in living nature being engaged in responsivity and coding. The presence of ionized groups in polymer gels affects their degree of swelling in several ways: ionized groups are more hydrated to maintain electroneutrality, they attract low molecular weight counter ions which are also hydrated. Also, equally charged groups placed along the chain may repulse each other. Addition of low-molecular-weight salts screens off the effect of charges. Opposite charges attract each other and the ionic bonds are stronger physical bonds than are hydrogen bonds. Ionic bonds are usually not thermo-reversible but medium reversible (pH, ionic liquids. certain salts). In non-polar media however, the charges groups are ion-paired with their counter ions and these ion pairs associate into clusters (ionic vulcanization).

4.4.1 Modifications of FH Lattice Models

In this section, we will briefly deal with models describing equilibrium swelling of cross-linked polyelectrolyte gels in aqueous media in a similar way as we did with electrically uncharged polymers. We will focus only on the main features of systems with equal sign of charges, leaving amphiphilic gels aside. He effect of ionic interactions on the free energy (Helmholtz/Gibbs) is considered additive to the effects of mixing and elasticity. The charge contribution is composed of the osmotic contribution of ions (Donnan effect), ΔG_{ion}, and charge repulsion (electrostatic effect) ΔG_{elst}:

$$\Delta G_{\text{sw}} = \Delta G_{\text{mix,n}} + \Delta G_{\text{el,n}} + \Delta G_{\text{ion}} + \Delta G_{\text{elst}} \tag{4.119}$$

Sometimes this additivity scheme is formulated in terms of solvent chemical potential contributions

$$\Delta\mu_{1,\text{sw}} = \Delta\mu_{1,\text{mix,n}} + \Delta\mu_{1,\text{el,n}} + \Delta\mu_{1,\text{ion}} + \Delta\mu_{1,\text{elst}} \tag{4.120}$$

or osmotic pressure contributions

$$\Pi_{\mathrm{sw}} = -\Delta\mu_{1,\mathrm{sw}}/V_{1,\mathrm{m}} = \Pi_{\mathrm{mix,n}} + \Pi_{\mathrm{el,n}} + \Pi_{\mathrm{ion}} + \Pi_{\mathrm{elst}} \tag{4.121}$$

The contributions by charged groups to polyelectrolyte solutions and gels were first formulated by Katchalsky and Michaeli [182, 183] and extended by Hasa et al. [184, 185] to cover large swelling degrees and equilibrium mechanical properties. Several variants of this approach were offered since then (cf., e.g., Refs. [102, 186–191]). The topic has been discussed in several reviews, such as [5, 6, 192, 193] and in chapters of two volumes of Advances in Polymer Science on volume phase transition [14, 194], particularly in chapters by Shibayama and Tanaka [195], Onuki [196], Ilavsky [197], Khokhlov [198], Siegel [199], and Gehrke [200]. The mixing and elasticity contributions were expressed usually by the mixing contribution for the infinitely long, uncross-linked chain and, respectively, the Flory-Erman junction-fluctuation model or its liming forms for phantom or affine models; in some papers, a finite extensibility models have been applied [185–189].

The contribution of ions to osmotic pressure can be expressed by

$$\Pi_{\mathrm{ion}} = RT\sum_j \left(z_j c_j^{\mathrm{gel}} - z_j c_j^{\mathrm{ext}}\right) \tag{4.122}$$

where z_j is the ion charge and c_j^{gel}, c_j^{ext} are concentrations of ions.

For polyelectrolyte gel with ionizable groups of degree of dissociation i and added low molecular weight salt composed of counter ions and coions (ions of the same sign of charge as the fixed ions in the gel) Π_{ion} (Donnan equilibrium) has the following form [185, 197]

$$\Pi_{\mathrm{ion}} = \frac{RT}{M_0}\left\{\alpha_{\mathrm{i}}\rho_2\phi_2 - 2f_{\mathrm{co}}c_{\mathrm{co}}M_0\left[\left(1 + \frac{\alpha_{\mathrm{i}}\rho_2\phi_2}{f_{\mathrm{co}}c_{\mathrm{co}}M_0}\right)^{1/2} - 1\right]\right\} \tag{4.123}$$

where M_0 is the molecular weight of the monomer unit, α_{i} i is degree of ionization, ρ_2 is density of dry gel and c_{co} and f_{co} are, respectively; concentration and activity coefficient of coions. If no salts are added, $c_{\mathrm{co}} = 0$, and the second term in Eq. 4.123 vanishes, and

$$\Pi_{\mathrm{ion}} = RT\frac{\alpha_{\mathrm{i}}\rho_2\phi_2}{M_0} \tag{4.124}$$

The electrostatic contribution corresponds to interaction of fixed charges surrounded by their ionic atmospheres along the chains and can be negative or positive [184, 197, 201, 202]

$$\Pi_{\mathrm{elst}} = \left[\frac{\nu_{\mathrm{e}} N_{\mathrm{A}} z \alpha_{\mathrm{i}}^2 e^2 \phi_2^{4/3}}{3D\vec{h}_0^2 \left(\phi_2^0\right)^{1/3}} \left(\frac{5A/2}{1+A} - \ln(1+A) \right) \right]^{1/2}$$
$$A = \frac{9\left(\phi_2^0\right)^{2/3} \phi_2^{-2/3} DkTM_0}{\pi N_{\mathrm{A}} e^2 \vec{h}_0^2 (2M_0 c_{\mathrm{co}} + \alpha_i \rho_2)} \tag{4.125}$$

where $\vec{h}_0^2$ is the mean-square end-to-end distance of network chain, N_{A} is Avogadro number, z is number of charges per monomer unit, D is dielectric constant of the medium; the other quantities have been defined before. Π_{elst} becomes important in very dilute systems when the charges do not interact with charges of other chains. In more concentrated systems $\Pi_{\mathrm{elst}} << \Pi_{\mathrm{ion}}$ can be neglected.

4.4.2 Counter Ion Condensation into Ion Pairs

However, a limitation holds for the degree of ionization in that it cannot exceed a certain limit because ion pairs are formed. This means that the degree of ionization starts exceeding the degree of neutralization above a certain charge density. Manning [203, 204] developed a theory for aqueous solutions of polyelectrolytes that is based on the concept of what he called "counter ion condensation". It is caused by strong Coulombic interactions between a highly charged polymer backbone and its counter ions, i.e., a polyelectrolyte is not completely dissociated even when dissolved in a large surplus of water. This ion pairs formation between fixed ions and mobile counter ions occurs as a so-called "territorial binding", i.e., caused by purely long-range electrostatic interactions when counter ions tend to be non-specifically bound to the poly ion and are delocalized. Selective binding of the counter ions to the poly ions is attributed to short-range interactions. Manning estimated the remaining charge of the poly ion using the quasichemical equilibrium method.

Based on a lattice model, Manning and Ray [205] (see also Ref. [206]) derived a contribution to the Gibbs energy due to ion pair formation including the effect of added salt. According to Manning, at equilibrium the number of ion pairs is determined by two factors: the energy gain of electrostatic attraction of ions $\sim z^2e^2/Db$ and the entropy loss due to reduction of translational motion of ions involved in ion pairs. The electrostatic potential of counter ion condensation is called *Manning parameter* ξ

$$\xi = \frac{z^2 e^2}{4\pi DkTb} \tag{4.126}$$

where b is the average distance between two neighboring charged segments along the poly ion chain. The quantity b is given by

$$b = \frac{\sigma_p m_p}{m_{cp}}, \quad m_{cp} = \alpha_i M_p / M_m, \quad M_m = \alpha_i M_m(\text{neutralized})$$
$$+ (1 - \alpha_i) M_m(\text{non-neutralized}) \qquad (4.127)$$

where σ_p, m_p, and m_{cp} are the segment diameter, number of segments of the poly ion, and number of charged segments of poly ion, respectively; M_p and M_m are molecular weights of the polyelectrolyte chain and average molecular weight of the monomer segment, respectively. The Bjerrum length l_B is defined by

$$l_B = \frac{e^2}{DkT} \qquad (4.128)$$

The interpretation of ξ is such that values higher than one are unfavorable and lead to a condensation of counter ions with fixed ions. This neutralizes a certain number of charges along the poly ion chain increasing the charge distance b and thus reducing ξ until it equals one. The process of counter ion condensation can be considered as an equilibrium

$$A^- + C^+ \rightleftarrows A^- C^+$$

and the dimensionless association constant K_a relates to ξ as

$$\ln K_a = \xi \qquad (4.129)$$

The fraction β of the charged poly ion segments that are obtained by neutralization but form ion pairs is a function of K_a

$$K_a = \frac{1}{\Phi_c} \frac{\beta}{(1 - \beta)^2} \quad \Phi_c = \frac{x_p m_{cp} \sigma_p^3}{\sum x_i m_i \sigma_i^3} \qquad (4.130)$$

where m_{cp} is defined by Eqs. 4.128, x_p is the molar fraction of charged unit in a polyelectrolyte chain, σ_p is segment diameter. Then, for $\xi = 1$, the value of β can be calculated. It gives the fraction of poly ion segments that are effectively charged

$$x_{c,eff} = (1 - \beta) \frac{m_{cp}}{m_p} \qquad (4.131)$$

Following the effect of limited ionization upon neutralization of acid or basic group, it means that the quantity α_i occurring in Eqs. 4.124 and 4.125 can reach only a limiting value $\alpha_{i,lim} < 1$ depending on the Bjerrum number l_B and b, the distance between neighboring charged groups on the polymer chain.

Experimental evidence of Manning effect is still needed, but strong polyelectrolytes like polystyrene sulfonic acid and the corresponding ion-exchange resins can be titrated fully, i.e., their exchange capacity corresponds to their content of –SO_3H groups.

4.4.3 Other Models for Charged Gels

Similarly as the Flory–Huggins approach, Hooper et al. [207] modified the quasi-chemical equilibrium (QCE) model by adding the Donnan term in the form of Eq. 4.123. Applied to cationic quaternary ammonium gels, they got good agreement of the dependences of the degree of swelling on the degree of ionization and low molecular weight salt, but the effect of dilution did not agree with the theory. Also Zhi et al. [102] used a QCE model with coordination number $z = 6$ amended by Donnan contribution and electrostatic interaction contribution similar to those used by Hasa et al. [185] and got a good agreement for the dependence of the degree of swelling on pH.

Lee and Bae [208] in their non-random contact (NRC) model treated hydrogen-bonding and polyelectrolyte systems with added salts interacting with specific contact surface areas on the polymer chain. Similarly as in the modified double lattice model (MDL) discussed in Sect. 3.3.3, they combined the Guggenheim QCE model with Ising lattice that considers the occupation of salt molecules on the nearest lattice sites. Salt-dependent interchange energy ($\tilde{\varepsilon}_{salt}$) is introduced to the mixing contribution term to overcome a major limitation of classical Donnan-type approaches of poor applicability of conventional ionic models to non-charged gel systems and to thermosensitive gels. Various interaction parameters serving as input information are considered: $\delta\varepsilon_1$ hydrophobic interaction parameter between water and the gel, $\delta\varepsilon_{1,salt}$ salt-dependent hydrophobic interaction parameter between water and the gel. $\delta\varepsilon_2$ hydrophilic interaction parameter between water and the gel, $\delta\varepsilon_{2,salt}$ salt-dependent hydrophilic interaction parameter between water and the gel, $\delta\varepsilon_{23}$ hydrophilic interaction parameter between water and the gel $\delta\varepsilon_{23,salt}$ salt-dependent hydrophilic interaction parameter between water and the gel (MDL and MFH models), $\delta\varepsilon_{ij}$ interaction energy parameter between $i - j$ pairs contributed from the oriented interactions in the secondary lattice, $\delta\tilde{\varepsilon}_{ij}$ reduced interaction energy parameter contributed from the oriented interactions, ε_{ij} interaction energy parameter between $i - j$ pairs contributed from the ordinary interactions, $\tilde{\varepsilon}_{ij}$ reduced interchange energy, $\tilde{\varepsilon}_{ij,salt}$ salt-dependent interchange energy. When the polymer contained ionized group, the conventional ionization (Donnan) term was added. The NRC method is closely related to Xin's et al. mixing model for copolymers [209], in which the Helmholtz energy of mixing is based on a multicomponent Ising lattice where the interactions between segments is accounted for. Segments of different kind on the polymer are involved in a number of specific interactions and the bond-forming probability is affected by the nearest neighbors. These methods perfectly formalize the variety of interactions but from practical view point it may be questionable how the values of the various interaction parameters can be estimated.

It has been proposed to determine the interaction parameters from calculations for non-charged hydrogel systems: (1) Before applying the NRC model, the swelling equilibrium at a salt-free system is calculated. (2) The volume fraction is then substituted at the volume phase transition temperature of the given system into the degree of changed contacts. (3) After converting the conventional interaction parameters into the salt-dependent interaction parameters, the NR factor θ_{salt} (the ratio of the reduced coordination number of salt molecule by excluding the NRC between salts and the hydrophilic elements of a gel molecule) is determined by fitting to the experimental data at salt conditions. To account rigorously for salt effects is a very difficult task because the degree of swelling is ion, especially anion, specific and the dependence on salt concentration has various shapes (increasing, decreasing, curved upward or downward (cf. Ref. [210]).

Also, Perturbed Chain—Statistical Associating Fluid Theory (PC SAFT) has been adapted to cover electrically charged systems (ePC-SAFT) [211–214]. The ePC-SAFT model developed by Cameretti et al. [215] as a combination of the PC-SAFT EOS by Gross and Sadowski [108] and the Debye–Hückel (DH) contribution which accounts for the Coulombic interactions. It considers the ionic species independent of the salt they are part of. Only two parameters are used to characterize each ion. The first one is the ionic diameter of hydrated ion σ. As the DH contribution only accounts for Coulombic forces among the ions, the interactions between ion and water (hydration) is described by means of dispersive energy u_j/k, which reflects the strength of ionic hydration.

The ePC-SAFT equation of state is composed by contributions to the Helmholtz energy

$$\Delta F^{\text{res}}/N = \Delta f^{\text{res}} = \Delta f^{\text{hc}} + \Delta f^{\text{disp}} + \Delta f^{\text{assoc}} + \Delta f^{\text{ion}} \tag{4.132}$$

where N is the total number of molecules, Δf^{hc} represents the hard-chain repulsion of the reference system; Δf^{disp}, Δf^{assoc}, and Δf^{ion} are contributions due to dispersive, associative, and Coulomb interactions, respectively. The expressions for Δf^{disp} and Δf^{assoc} are used as in the original PC-SAFT model of Gross and Sadowski [108], and Cameretti et al. [215] used a Debye–Hückel term to describe the Helmholtz energy contribution Δf^{ion} that is caused by charging the species j

$$\frac{\Delta f^{\text{ion}}}{kT} = -\frac{\kappa}{12\pi kT\varepsilon}\sum_j x_j q_j^2 \chi_j \tag{4.133}$$

where x_j is mole fraction and q_j charge of ion j, κ is inverse Debye screening length, and ε is dielectric constant. The ions can approach each other to the distance a_j which is equivalent to the ion diameter σ_j; χ_j is given by the equation

$$\chi_j = \frac{3}{\left(\kappa a_j\right)^3}\left(\frac{3}{2} + \ln\left(1 + \kappa a_j\right) - 2\left(1 + \kappa a_j\right) + \frac{1}{2}\left(1 + \kappa a_j\right)^2\right) \tag{4.134}$$

and

$$\kappa = \left[\frac{N_{\text{A}}}{kT\varepsilon}\sum_j q_j^2 c_j\right]^{1/2} = \left[\frac{\rho_{\text{N}} e^2}{kT\varepsilon}\sum_j z_j^2 x_j\right]^{1/2} \tag{4.135}$$

where c_j is molar concentration, ρ_N number density of the system, and N_A is Avogadro number.

To describe salt solutions, the conventional Berthelot–Lorenz combining rules are used:

$$\sigma_{ij} = \frac{1}{2}(\sigma_i + \sigma_j)$$
$$u_{ij} = \sqrt{u_i u_j} \tag{4.136}$$

The approach is used for weak electrolytes [212] by implementing an association/dissociation mechanism using the acid dissociation constant K_a in combination with the classical ePC-SAFT model. To account for the association of ions (ion pairing), the approach of an association/dissociation equilibrium based on mass action law is used where an ion pair is in equilibrium with its respective free ions. The Manning limiting law discussed in the preceding section was applied. To determine the Helmholtz energy contribution Δf^{ion} for a polyelectrolyte solution, the Debye-Hückel term was applied to the free counter ions as well as to the charged segments of the polymer chain and Δf^{ion} was transformed and used in the ePC-SAFT on a segmental basis, and Manning criterion was applied.

The effect of charged monomers within the polyelectrolyte chain is explicitly taken into account in the reference term by replacing the hard-chain contribution of the PC-SAFT model by a charged hard-chain contribution. Moreover, counter ion condensation is accounted for. The Helmholtz energy contribution for the charged hard-chain is now be written similarly as by Eq. 4.132

$$\Delta f^{\text{res}} = \Delta f^{\text{chc}} + \Delta f^{\text{disp}} + \Delta f^{\text{assoc}} + \Delta f^{\text{ion}} \tag{4.137}$$

where however Δf^{chc} accounts for charged hard chain. Considering the Manning's limits for charge density, it is assumed that the fraction of charged–charged segmental interactions cc is identical with the fraction of poly ion segments that are effectively charged and can be calculated for any polyelectrolyte

$$cc = (1 - \beta)\frac{m_{\text{cp}}}{m_{\text{p}}} \tag{4.138}$$

where β is the extent of counter ions condensation.

The contribution for non-ionized hard chain

$$\Delta f^{\text{hc}} = \widetilde{m}\Delta f^{\text{hs}} - kT\sum_i x_i(m_i - 1)\ln y_{ii}(d_{ii}), \quad \widetilde{m} = \sum_i x_i m_i$$

Where x_i and m_i are mole fraction and number of segments, respectively; $y_{ii}(d_{ii})$ is the cavity correlation function (CCF) of segments of two different molecules of same type, is transformed into contribution for charged hard chain Δf^{hcc}

$$\Delta f^{\text{hcc}} = \tilde{m}\Delta f^{\text{hs}} - kT\sum_i x_i(m_i - 1)\left[(1 - cc)\ln y_{00}(d_{ii}) + cc\ln y_{ii}(d_{ii})\right] \quad (4.139)$$

where $y_{00}(d_{ii})$ is the CCF for two non-charged segments.

For modeling, the pure-component parameters of the polyelectrolyte as well as the binary interaction parameter between water and the poly ion were used as determined for the non-charged polymers. Using only the information of the polyelectrolyte degree of neutralization and without fitting additional parameters, the model is able to predict the vapor–liquid equilibria of polyelectrolyte systems in a satisfactory agreement with experimental data. The model was also used to calculate phase equilibria in ternary systems polyelectrolyte-water-salt [213].

The ion pairing becomes very important in less polar system of lower dielectric constant. The Bjerrum length increases with increasing dielectric constant and the majority of counter ions become ion paired or associated in multiplets of ion pairs. Philippova et al. [216] proposed a simple theoretical approach based on the usual additive scheme in which

$$\Delta F_{\text{total}} = \Delta F_{\text{elast}} + \Delta F_{\text{int}} + \Delta F_{\text{charge}} + \Delta F_{\text{aggreg}} \quad (4.140)$$

where the first three terms have the usual form (Gaussian for ΔF_{elast}, FH for ΔF_{int}, Donnan for ΔF_{charge}). The ΔF_{aggreg} term cares for association into ion pairs and their multiplets.

$$\Delta F_{\text{aggreg}} = Nf_{\text{i}}\left[p + \frac{3}{2}q + 2\ln(1 - p - q)\right] \quad (4.141)$$

where Nf_{i} is the number of ionized groups in a chain of N units, p is the fraction of ion pairs and q the fraction of ion pairs dimers. The quantities$p$ and q are related to concentration (volume fraction) of polymer through a mass action equilibrium

$$\frac{p}{(1 - p - q)^2} = k_{\text{p}}f_{\text{i}}\phi_2, \quad \frac{q}{p^2} = 2k_{\text{m}}\phi_2$$
$$k_{\text{p}} = \exp\left(\frac{u_0}{(1 - \delta D\phi_2)b/a}\right) \quad (4.142)$$

where δD is the relative dielectric constant difference between polymer and solvent and b/a is the relative size of ion pair being of order of unity. Under certain conditions, the degree of swelling can show up as a function of f_{i} a discontinuous transition from collapsed to expanded states, sometimes followed by re-transition to the collapsed state again caused by ion pairing.

It can be concluded the charged polymer solutions can be considered as belonging to the most complicated polymer systems, especially when the ionic interactions are not predominating and compete with other non-ionic interactions, strong and weak in which the ionized groups also participate.

4.5 Semi-dilute Systems: Excluded Volume and Scaling

The Flory–Huggins and other theoretical treatments of mixing and elasticity are based on the mean field assumption and characterize the situation in the concentrated regime where segments of the network chains do not recognize whether the neighbors belong to the same chain or the other chains. Such situation is found either under θ-conditions or in at sufficiently high polymer concentrations where the segment density fluctuations are caused exclusively by thermal fluctuations. The thermal blob size and the correlation length are the same. However, in reality many swollen networks may be classified with respect to their segment concentration, ϕ_2, as occurring in the semi-dilute region where the correlation length exceeds the thermal blob size. At short distances, the chain sees mainly its own segments and their excluded volume affects chain conformation, whereas at larger distances the segments are not distinguishable from segments of other chains. So, in the semi-dilute region, the excluded volume is expected to affect the modulus of elasticity and degree of swelling.

The effect of excluded volume on swelling was dealt with by the scaling method first by de Gennes [217] (c^*-theorem, which was found not to be generally applicable); the present status is explained in the monograph by Rubinstein and Colby [23]. The analysis of swelling of tetra-PEG gels can serve as a recent example of application of the scaling method [218].

The effect of dilution at preparation conditions, (ϕ_2^0), also depends on the regime—concentrated versus semi-dilute region. The spatial distribution of chain conformations is not affected by interactions with the solvent in the concentrated regime, $\phi_2 > \phi_2^{**}$. Below the concentration ϕ_2^{**}, excluded volume becomes operative, and the semi-dilute ($\phi_2^{**} - \phi_2^{*}$) and dilute ($\phi_2 < \phi_2^{*}$) regimes are distinguished $\phi_2 > \phi_2^{**}$. The differences are characterized by the scaling relations between root-mean-square end-to-end distance (rms ETE), the number of chain segments N, and excluded volume parameter υ, $\upsilon \approx b^3$, where b^3 is the volume of the monomer unit (Table 4.8).

v is excluded volume, positive in good solvent and negative in poor solvent. The c^* concentration depends itself on excluded volume as

Table 4.8 Scaling relations for equilibrium shear modulus and degree of swelling

Regime	Concentration range	G_e	ϕ_2
θ-solvents + conc. solution	$\phi_2 > \phi_2^{**}$, $\phi_2^0 > \phi_2^{**}$	$\propto (\phi_2^0)^{2/3} \phi_2^{1/3}$	$\propto (\phi_2^0)^{1/4} \nu_e^{3/8}$
Semi-dilute athermal	$\phi_2^{*} < \phi_2 < \phi_2^{**}$	$\propto (\phi_2^0)^{\frac{1}{3(3\upsilon-1)}} \phi_2^{\frac{9\upsilon-4}{3(3\upsilon-1)}}$	$\propto (\phi_2^0)^{\frac{1}{4}} \nu_e^{\frac{3(3\upsilon-1)}{4}}$
Semi-dilute good	$\phi_2 < \phi_2^0 < \phi_2^{**}$	$\propto (\phi_2^0)^{-\frac{1}{3(3\upsilon-1)}} \phi_2^{\frac{1}{3}}$	$\propto (\phi_2^0)^{\frac{1}{4}} \nu_e^{\frac{3(3\upsilon-1)}{4}}$
	$\phi_2 < \phi_2^{**} < \phi_2^0$	$\propto (\phi_2^0)^{2/3} \phi_2^{\frac{9\upsilon-4}{3(3\upsilon-1)}}$	$\propto (\phi_2^0)^{\frac{3\upsilon-1}{2}} \nu_e^{\frac{3(3\upsilon-1)}{4}}$

Table 4.9 Values of parameters used in simulation of dependences of reduced chemical potential of solvent

Figure	m_1	ν_e mol/cm^3	g_0	g_1	α_i
4.60a	0.9–1.1	3×10^{-4}	0.755	0.60	0
4.60b	1.0	3×10^{-4}	0.730–0.780	0.60	0
4.60c	1.0	$(2–4) \times 10^{-4}$	0.755	0.60	0
4.60d	1.0	4×10^{-4}	0.755	0.60	0–0.03
4.60e	1.0	1×10^{-3}	1.0	0.40	0.15–0.17

$f_e = 4$; $V_{ls} = 100$, $\phi_2^0 = 0.5$

$$\phi_2^* \approx \left(b^3/\mathrm{v}\right)^{6\upsilon-3} N^{1-3\upsilon} \tag{4.143}$$

The exponents for the dependences of the equilibrium shear modulus and swelling degree expressed through ϕ_2 are summarized in Table 4.9.

The excluded volume effect can be quite important for highly swollen polyelectrolyte gels [3]. To get full scale quantitative relations for the degree of swelling and fluctuations, one has to resort to Monte Carlo simulations.

4.6 Swelling Under Constraints

Mechanical constraint and deformations affect the equilibrium degree of swelling. Also, external constraints, isotropic or anisotropic, change the amount of absorbed solvent. The geometric constraints, characteristic for gel constructs, often give rise to gradients in swelling degree as, for instance, in shells of core–shell particles or in the swollen cross-linked matrix of particulate composites. In other cases, swelling can be only unidirectional like in the case of adhering cross-linked coating films but such systems are prone to development of mechanical instabilities manifested in formation of surface pattern like buckling or wrinkling [219–223]. Corrugated surfaces may be sometimes of lower free energy than planar surface films.

4.6.1 Swelling of Deformed Gels

Calculation of the degree of swelling as a function of deformation (cf., Sect. 3.4),

$$\Delta \mathrm{G}_{\mathrm{sw}} = \Delta \mathrm{G}_{\mathrm{mix,n}} + \Delta \mathrm{G}_{\mathrm{el,n}}\left(\lambda_x, \lambda_y, \lambda_z\right) \tag{4.144}$$

where $\Delta G_{\mathrm{mix,n}}$ includes the combinatorial part of the effect of cross-links on solubility and $\Delta G_{\mathrm{el,n}}(\lambda_x, \lambda_y, \lambda_z)$ covers the elastic response of the network immersed in theta solvent and is dependent on macroscopic deformation given by the deformation ratios along the coordinate axes λ_x, λ_y, λ_z. At equilibrium, the chemical potentials of the solvent are equal. For simplicity reasons, we assume that no soluble components are present, i.e., the dilute phase is composed of pure liquid solvent or solvent vapor:

$$\Delta\mu_{1,\mathrm{mix,n}} + \Delta\mu_{1,\mathrm{el,n}}(\lambda_x, \lambda_y, \lambda_z) = \mathrm{RT}\ln a_1 = \mathrm{RT}\ln f_1 \approx \mathrm{RT}\ln(p_1/p_1^0) \quad (4.145)$$

The deformation ratios λ_j are related to the reference state, but practical deformation ratios used, e.g., in experiments, are related to the isotropic non-deformed state, Λ_j. The deformation ratios are correlated (assuming the chains are in references state at network formation) as

$$\lambda_j = \frac{L_j}{L_i}\frac{L_i}{L_0} \approx \frac{L_j}{L_i}\frac{L_i}{L_{\mathrm{dry}}}\frac{L_{\mathrm{dry}}}{L_{\mathrm{nf}}} = \Lambda_j\phi_{2,i}^{-1/3}\left(\phi_2^0\right)^{1/3} = \frac{L_j}{L_{\mathrm{dry}}}\frac{L_{\mathrm{dry}}}{L_{\mathrm{nf}}} = \lambda_{j,\mathrm{d}}\left(\phi_2^0\right)^{1/3} \quad (4.146)$$

where $\phi_{2,i}$ is the volume fraction of polymer in swollen *isotropic* network; $\lambda_{j,\mathrm{d}}$ is the macroscopic deformation ratio related to the dry state which has the advantage that it is, similarly as λ_j, independent of the degree of swelling.

For particular types of deformation, the following bounds always hold

$$\lambda_x\lambda_y\lambda_z = V/V_0, \quad \lambda_{x,\mathrm{d}}\lambda_{y,\mathrm{d}}\lambda_{z,\mathrm{d}} = V/V_{\mathrm{dry}}, \quad \Lambda_x\Lambda_y\Lambda_z = V/V_i \quad (4.147)$$

For a Gaussian network in uniaxial extension or compression

$$\Delta G_{\mathrm{el}}(\lambda_x, \lambda_y, \lambda_z) = RT\frac{N_e}{2N_A}\left(\phi_2^0\right)^{2/3}\left(\lambda_{x,\mathrm{d}}^2 + 2\lambda_{x,\mathrm{d}}^{-1}\frac{V}{V_\mathrm{d}} - 3\right) \quad (4.148)$$

and

$$\Delta\mu_{1,\mathrm{el}} = \left(\frac{\partial\Delta G_{\mathrm{el}}}{\partial N_1}\right)_{p,T,\lambda_{x,\mathrm{d}}} = RT\nu_e V_{\mathrm{m,ls}}\left(\phi_2^0\right)^{2/3}\phi_{2,i}^{1/3}\Lambda_x^{-1} \quad (4.149)$$

After adding the mixing term, the swelling equation of uniaxially deformed gel conforming the generalized FH model reads

$$\begin{aligned}(\Delta\mu_1/RT) &= \ln a_1 = \ln\phi_1 + \phi_2 + \xi_e V_{\mathrm{m,ls}}\phi_2 + \chi(\phi_2)\phi_2^2 \\ &+ XV_{\mathrm{m,ls}}\left(\phi_{2,i}^{1/3}\left(\phi_2^0\right)^{2/3}/\Lambda_x - \phi_2\right) \\ \xi_e &= \frac{f_e - 2}{f_e}\nu_e; \quad X_{\mathrm{affine}} = \nu_e; \quad X_{\mathrm{phantom}} = \xi_e\end{aligned} \quad (4.150)$$

where $\phi_{2,i}$ is the volume fraction of polymer in the undeformed isotropic state while ϕ_2 characterizes the swelling in the deformed (anisotropic) sample.

One can see that for $\Lambda_x > 1$ (extension) the swelling degree increases but for $\Lambda_x < 1$, it decreases with increasing deformation relative to free swelling conditions. Note that the volume fraction of solvent in the elastic contribution refers to equilibrium swelling of the isotropic, non-deformed swollen sample, whereas the other quantities are related to the deformed sample.

Proceeding in the same way, one gets for equibiaxial planar extension

$$\ln a_1 = \ln \phi_1 + \phi_2 + \xi_e V_{1m}\phi_2 + \chi(\phi_2)\phi_2^2 + XV_{1mol}\left(\phi_2^{-1}\phi_{2,i}^{4/3}\left(\phi_2^0\right)^{2/3}/\Lambda_{x,y}^4 - \phi_2\right) \tag{4.151}$$

where $\Lambda_{x,y} = \Lambda_x = \Lambda_y$. The biaxial deformation has a stronger effect on the elastic term than the uniaxial one.

In a similar way, one can use combinations of other theoretical forms of $\Delta G_{mix,c}$ and $\Delta G_{el,n}(\lambda_x, \lambda_y, \lambda_z)$ establishing relations between $\lambda_x, \lambda_y, \lambda_z$ corresponding to the chosen geometry. For the Gent's model of finite extensibility of network chains (cf. Table 4.5)

$$(\Delta\mu_1)_{el}/kT = \left(\frac{\partial \Delta G_{el}/kT}{\partial N_1}\right)_{P,T,\lambda} = \frac{N_e}{2}\left(\frac{J_M}{J_M - J_1}\right)\left(\frac{\partial J_1}{\partial N_1}\right)_{\lambda_x} \tag{4.152}$$

Introducing the conditions for uniaxial extension/compression $\Lambda_y^2 = \Lambda_z^2 = \Lambda_x^{-1}(V/V_i)$

$$\begin{aligned} &(\Delta\mu_1)_{el}/RT = \nu_e V_{1m}\frac{J_M}{J_M - \left(\Lambda_x^2 P_i^{2/3} + 2\Lambda_x^{-1}P^{2/3}\phi_2^{-1/3}\right)}\left(\Lambda_x^{-1}\phi_{2i}^{1/3}\left(\phi_2^0\right)^{2/3}\right) \\ &J_M = J_{Mx} = n + 2n^{-1/2}P - 3 \quad \text{(extension)} \\ &J_M = J_{Myz} = 2n + n^{-2}P^2 - 3 \quad \text{(compression)} \\ &P = \phi_2/\phi_2^0, \quad P_i = \phi_{2i}/\phi_2^0 \end{aligned} \tag{4.153}$$

Other examples of constrained network chains even in the absence of external forces are chains of the first network in the sequential homogeneous IPNs (double networks) in which the first network is swollen in monomer of the second network which is then polymerized. In the relaxed state, the chains of the first network in the IPN are isotropically extended over their reference dimensions. Dimensions of network chains in networks cross-linked in the state of strain after relaxation of the network exceed the reference dimensions if the deformation is of expansive type. Swelling of gels subjected to deformation was studied among others by Drozdov [224, 225].

4.6.2 Swelling Pressure

If a gel is subjected to pressure in three dimensions while kept in contact with solvent at atmospheric pressure, it swells less than it would in its free state, i.e., after the pressure has been released. Alternatively, the dry or partially swollen gel can be confined in a (isometric or anisometric) cavity and pressure will develop when the gel is brought into contact with solvent, for instance, through porous walls of the container. The pressure acting on walls, Π_{sw}, is directly related to the chemical potential of the solvent, and since the chemical potential of the solvent is generally dependent on deformation ratios, λ_x, λ_y, λ_z,through its elastic component also he swelling pressure has three components

$$\Pi_{sw} = -\frac{\Delta\mu_{1,sw}}{V_{1m}}, \quad \Pi_{sw}(x, y, z) = -\frac{\Delta\mu_{1,sw}(\lambda_x, \lambda_y, \lambda_z)}{V_{1m}} \tag{4.154}$$

The swelling pressure can be quite high. The ancient peasants used swelling pressure to split tree stumps (see the cartoon in Fig. 4.34). They bored a hole, pressed peas into it poured water on peas; after some time, large cracks developed and the stump was split.

Swelling pressure offers important information on gel structure and has been studied by several groups (cf., e.g. Refs. [226–231]). While Horkay and Zrinyi

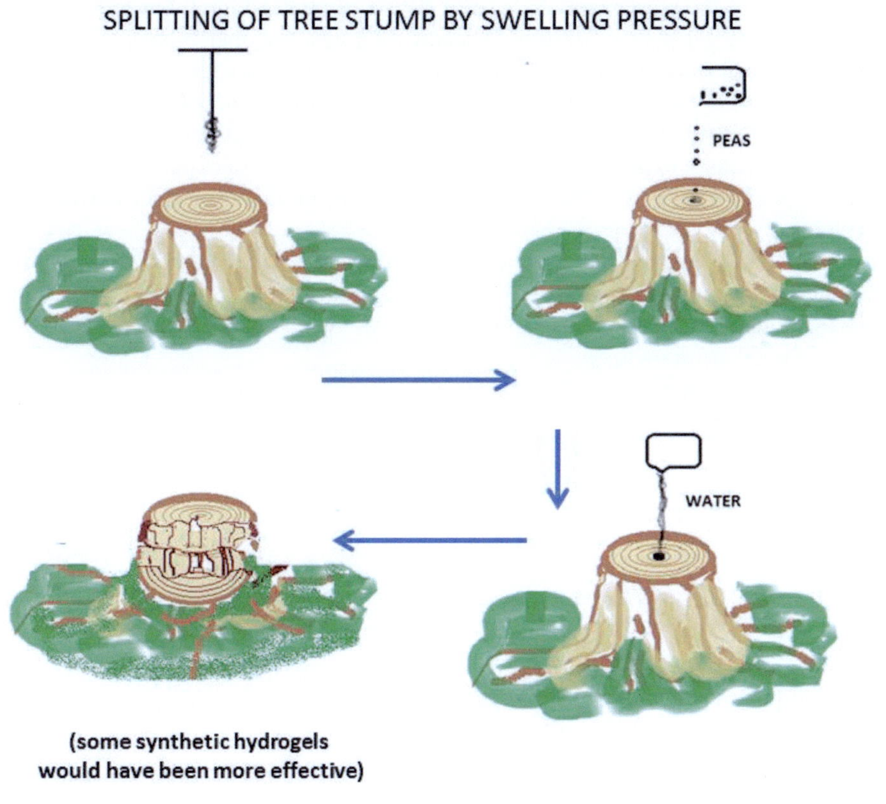

Fig. 4.34 Splitting of tree stumps by swelling pressure of swelling peas

measure swelling pressure as osmotic pressure by wrapping the gel in a semipermeable membrane and immersing it in solution of a polymer of lowered solvent activity, in other studies the pressure was measured directly with pressure gauges [232, 233] or by using a balance. Very interesting is the method that uses an analytical ultracentrifuge. The advantage of this method is that in contrast to the known swelling pressure instruments, swelling pressures are obtained from only a single equilibrium experiment. The Schlieren optical system of the ultracentrifuge allows the observation of the concentration gradient during the process of deswelling and swelling. The gel concentration is detected at every point. The osmotic deswelling method and direct pressure determination should give comparable results for homogeneous sample but different results when the gels are inhomogeneous or porous [231, 234, 235].

4.6.3 Swelling of Gel Constructs

A gel construct is an object composed of at least two phases swelling to different degrees; one of the phases may be a solid not swelling itself at all but affecting the swelling of the gel phase. Adhering organic coatings, polymer bilayers, layer-by-layer films or objects, or cross-linked core–shell particles may serve as examples.

Swelling of a cross-linked film of protective coating adhering well to a rigid substrate is a classical example of geometric constraint. The adhering film in contact with solvent cannot expand in plane (x, y) and the only way of expanding is vertical (z) (see upper part of Fig. 4.35), unless mechanical instability by surface corrugation develops.

Therefore, $\lambda_y = \lambda_z = 1$; $\lambda_x = \lambda = \phi_2{}^{-1}$; the changes of the Helmholtz energy for such case respecting the finite extensibility have been summarized in Ref. [14]; here, only the equations for the Gaussian affine model, Gent's model and Arruda-Boyce 8-chain model are shown.

Gauss constrained

$$(\Delta\mu_1)_{\text{el}}/RT = V_{\text{mol,ls}}\nu_{\text{e}}\left(\phi_2^0\right)^2\phi_2^{-1} \tag{4.155}$$

Gent constrained

$$(\Delta\mu_1)_{\text{el}}/RT = V_{\text{mol,ls}}\nu_{\text{e}}\frac{\left(\phi_2^0\right)^2\phi_2^{-1}}{1-\left(\left(\phi_2^0/\phi_2\right)^2-1\right)/(n-1)} \tag{4.156}$$

8-chain model amended by Jernigan and Flory

$$(\Delta\mu_1)_{\text{el}}/RT$$
$$=\frac{1}{3}V_{\text{mol,ls}}\nu_{\text{e}}n^{1/2}\left[\left(\beta_{\text{chain}}+\beta'_{\text{chain}}\lambda_{\text{chain}}\right)+n^{1/2}\frac{\beta'_{\text{chain}}}{\beta_{\text{chain}}}\left(1-\beta_{\text{chain}}\coth\beta_{\text{chain}}\right)\right]\phi_2^0$$

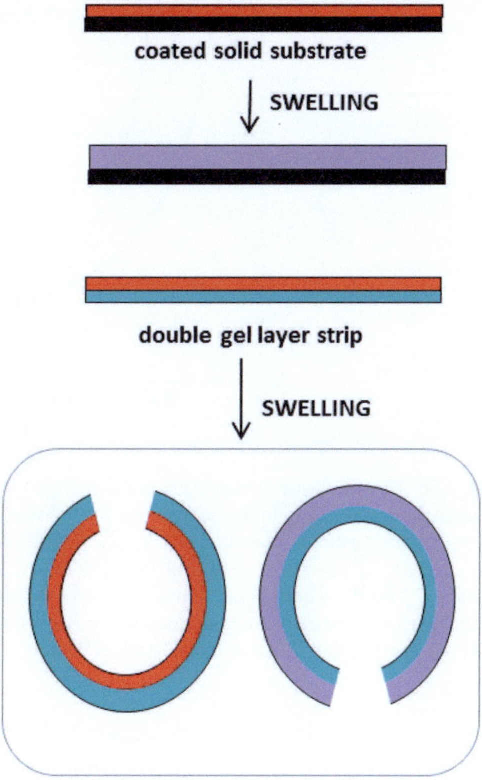

Fig. 4.35 Swelling of an adhering coating film and a bilayer film

$$+\frac{1}{3}V_{\mathrm{mol,ls}}\nu_{\mathrm{e}}\left[\frac{\beta'_{\mathrm{chain}}}{\beta_{\mathrm{chain}}}-\frac{1}{\lambda_{\mathrm{chain}}}\right]\phi_2^0 \tag{4.157}$$

with $\lambda_{\mathrm{chain}}=\left(\frac{1}{3}\left(\lambda_x^2+2\right)\right)^{1/2}$ and the inverse β_{chain} of the Langevin function $\ell(\beta_{\mathrm{chain}}) = \coth\beta_{\mathrm{chain}} - 1/\beta_{\mathrm{chain}}$; which is approximated by $\beta_{\mathrm{chain}} \approx \frac{\lambda_{\mathrm{chain}}n^{-1/2}\left(3-\lambda_{\mathrm{chain}}^2n^{-1}\right)}{1-\lambda_{\mathrm{chain}}^2n^{-1}}$;

β'_{chain}' is derivative of β_{chain} with respect to λ_{chain}.

To characterize the model, only the dependences for the highest concentration of elastically active network chains are shown (Fig. 4.36).

It can be seen that the strongest constraints are developed by applying the 3-chain, Kovac's, and Gent's models. Swelling of adhering films is often used for characterizing the cross-link density and the effect of constraint is not taken into account. The following figures characterize the error produced by neglect of the constraint.

It can be seen that the neglect of adhesion is always serious, especially at high swelling degrees. The effect of finite extensibility is small at high degrees of swelling but becomes serious when the network chains are short. When in a bilayer film the

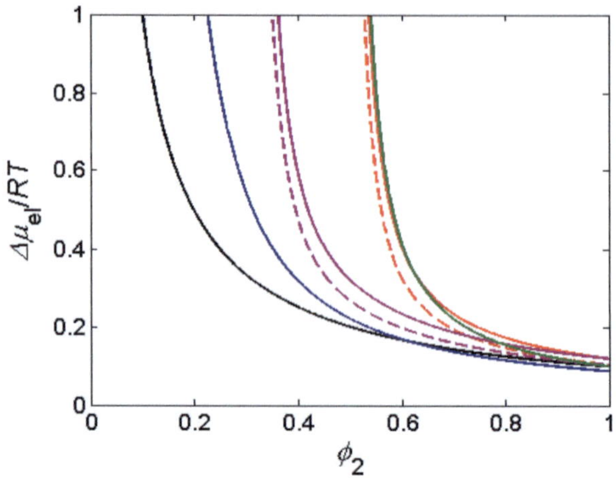

Fig. 4.36 Plot of elastic contribution to the change of chemical potential of the solvent $\Delta\mu_{1,\mathrm{el}}/RT$ by constrained swelling in dependence on the volume fraction of polymer in the swollen network. $\phi_2^0 = 1$, molar volume of lattice site $V_{\mathrm{mol,ls}} = 100$ cm^3, specific gravity of polymer $\rho_2 = 1.2$ g/cm^3, molecular mass of a statistical segment $M_s = 300$ g/mol; identification of the curves (see Table 4.5): black—Gauss, blue Smith [14], red—3-chain, red dashed—3-chain amended, violet—8-chain, violet dashed—8-chain amended, green—Gent, magenta—Kovac. Reprinted from Ref. [14] with permission of American Chemical Society

layers swell differently, bending occurs as shown in Fig. 4.37. This effect is utilized in characterizing stresses developed during drying of coatings films. One of the layers is a thin steel membrane. It is bent by contractile forces (shrinkage) produced by the drying film; either the deflection is measure or the force necessary to prevent bending. The application of degree of swelling for characterization of curing and structure of final coating films and pitfalls of the method are discussed in Ref. [236].

Core–shell particles are more complicated constructs. Core–shell structures have been a subject of thousands of articles in the few recent decades; yet, their structure, especially in swollen state is not well-understood. If one considers swelling of polymeric core–shell particles, swelling of core affects the swelling of shell and vice versa. If core and shell are homogeneous phases, then after swelling the core remains homogeneous, but in the shell a gradient of the degree of swelling. A special feature of the core–shell structure is the special role of the core to serve as a test tube where a negative (expansive) pressure can be generated which offers a rare possibility of investigating polymer–solvent systems while weakening the cohesive forces. In view of the wide application of core–shell particles, modeling of their swelling is important, especially of the gradients in shells. Although some simple cases can be solved numerically by variational calculus (cf., e.g., Ref. [237]), the problem of swelling of such complex constructs is too demanding and a more capable method had to be used. Such method is finite element method (FEM) which was applied recently to a number of problems involving swelling of gels (cf., e.g., Refs. [238–243]).

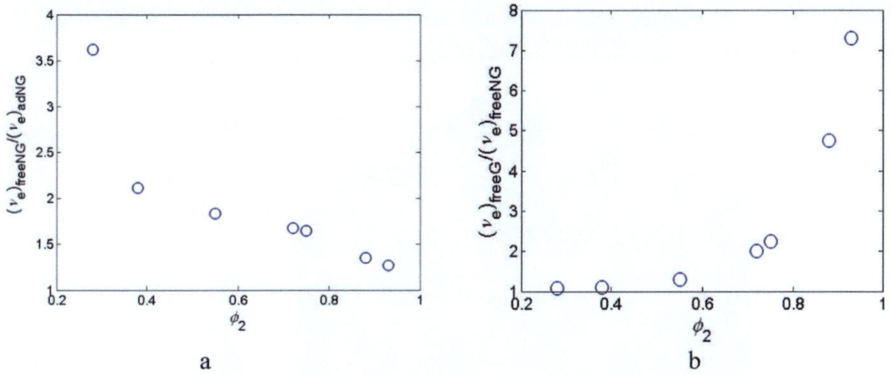

Fig. 4.37 Effect of adhesion constraints on the concentration of EANCs calculated from swelling equation in Ref. [14] on plasma cross-linked poly(ethylene glycol) films. a: ratio of values of ν_e calculated using the free-swelling model and adhesion constraint model, both using the finite extensibility model of Gent; b: ratio of values of ν_e for free swelling using the Gaussian affine model and the Gent finite extensibility model. Reprinted from Ref. [14] with permission of American Chemical Society

Recently, the FEM method was applied also to core–shell spherical particles, composed of differently swelling cores and shells [15, 244]. Figure shows an example when the shell swells more than the shell (a)n and the opposite case when it is the core that swells the shell. The colors on the right scale show the volume degree of swelling. The thin black circles show the size of the dry particle. The figure shows that in the case of more swelling core, the gradient in swelling degree is well-developed, whereas in the opposite case, the gradient in the compresses less swelling core is hardly seen (Fig. 4.38).

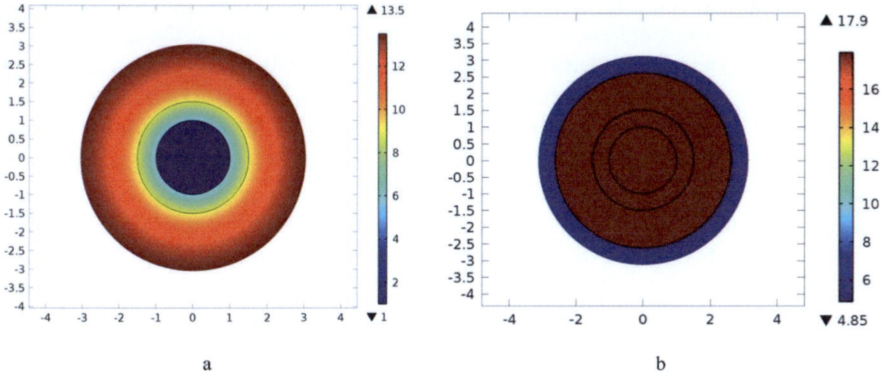

Fig. 4.38 Core–shell particle with cross-linked core and shell. Volume degree of swelling on the right scales. **a** Shell swells more than the core; **b** core swells more than the shell. Results of FEM simulation

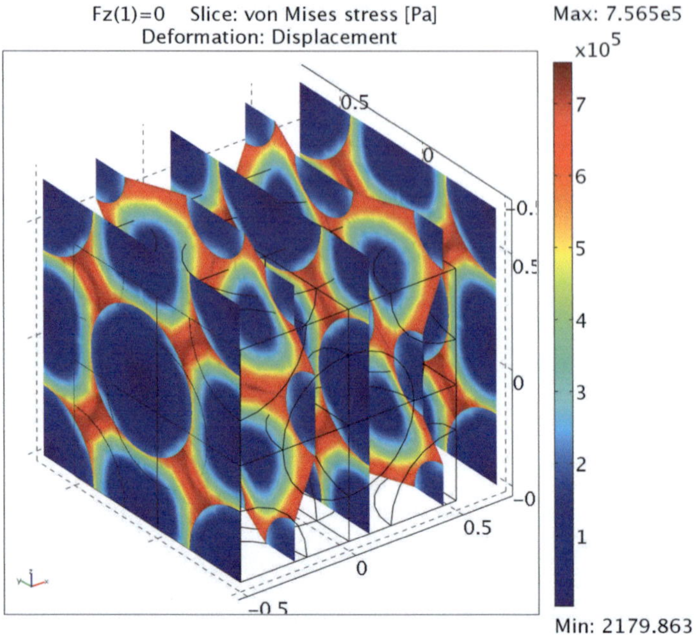

Fig. 4.39 Distribution of von Mises stresses in a swollen particulate composite as the degree of filling increases. Generated by J. Šomvársky

Another, more complicated case is represented by swelling of a particulate composite with cross-linked matrix. In the model experiments on swelling, spherical filler particles are regularly distributed on a cubic lattice [15, 244].

The stress distribution in the swollen composite is shown on Fig. 4.39. As expected, the largest stress develops in parts of matrix where the distances between filler surface are the smallest, i.e., the matrix confinement there is the largest.

Figure 4.40 shows distribution of the degrees of swelling (vol-vol) as it changes when the filling degree, v_{fill}, (volume fraction of filler) increases. At the lowest degree of filling, the swelling degree is hardly affected, except of the closest vicinity of the particles. With increasing v_{fill}, the degree of swelling is more and more variable and near the filler surface loci start appearing of the highest degree of swelling. In the last system of the highest degree of filling, $v_{\mathrm{fill}} = 0.69$, the filler spheres are bridged by very extended "bundles" of chains of the highest degree of swelling. This finding may seem unexpected and is at variance with the intuitive view that a composite is a collection of fused filler particles (Voronoi polyhedra) coated with the matrix polymer [245, 246].

Since the degree of swelling of coated particles decreases toward the filler–polymer interface, the locus of closest distance should be a locus of lowest degree of swelling. However, the opposite result of FEM simulation can be understood, because it is the swelling force of the matrix part at the farthest distance from the

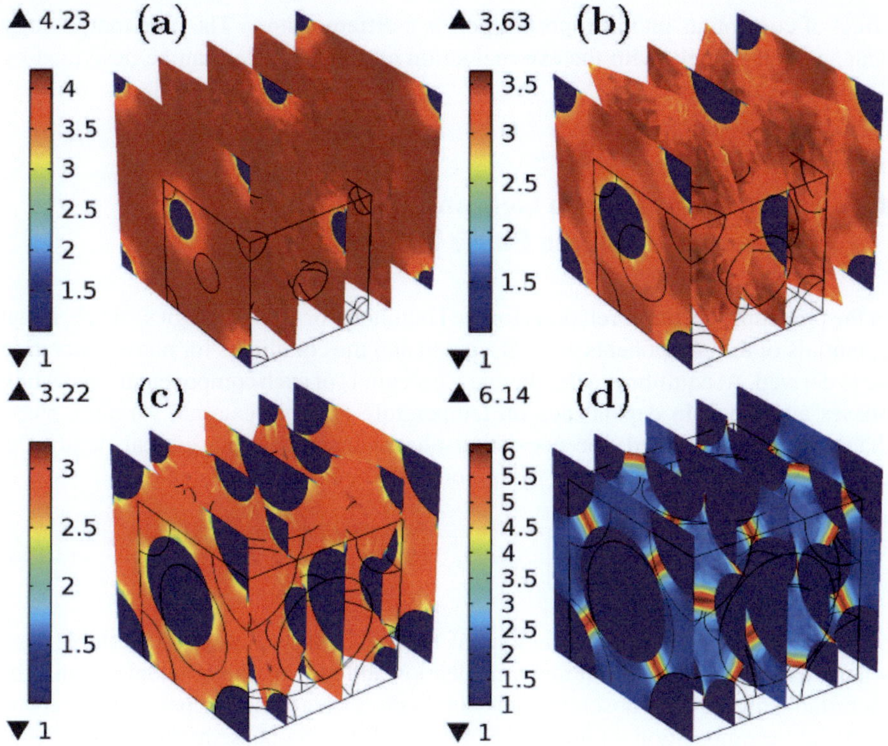

Fig. 4.40 Profiles of the degree of swelling in a particulate composite—increasing degree of filling: ν_{fill}, = 0.05 (**a**), 0.25 (**b**), 0.50 (**c**), 0.69 (**d**); generated by J. Šomvársky

interface which pushes filler sphere apart. These forces concentrate in the loci of the shortest distances (Fig. 4.40).

In order for FEM to be successful for simulation studies, the software should make possible to implement proper polymer physics—relations derived theoretically to describe real behavior. Not all commercial software is enough user-friendly to make it possible.

Understanding of the effects of constraints imposed on gels is very important because (1) gels are subjected to deformations during their applications (implants, tissue engineering, composites), (2) they are mutually constrained in gel-gel constructs or gel-non-swelling solids, (3) the degree of swelling changes with deformation during mechanical tests. Modeling of the constraint effect can be done by changing the elastic part of Helmholtz energies which is dependent of the deformation ratios along the Cartesian exes. This way becomes difficult for application to constructs of different geometric morphologies. The Finite Element Method (FEM) is very suitable for such cases, but the corresponding relation describing the behavior of the involved material must be implemented. In addition to equilibrium properties, the dynamics of the processes leading to equilibrium can be modeled as well. The effect of constraints on macroporous gels is a different story. The pore morphology changes non-affinely with the external strain and the changes can be described by the poroelastic theory.

4.7 Phase Separation in Gels and During Gel Formation—Volume Phase Transition

In the preceding sections, relations for the Helmholtz or Gibbs energies and chemical potentials of the components were discussed and the conditions for phase separation were derived. At equilibrium the chemical potentials of each component in coexisting phases are equal. In dependence on temperature or other external variable, phase diagrams were obtained; the coexisting phases merged at the critical point. The phase state is also characterized by spinodal defined by

$$\left[\frac{\partial^2 \Delta G_{\mathrm{sw}}}{\partial \phi_2^2}\right]_{T,p} = 0, \quad \text{or} \left[\frac{\partial \Delta \mu_1}{\Delta \phi_2}\right]_{T,p} = 0 \tag{4.158}$$

While the binodal delimits regions of full miscibility from the metastable state where the second phase develops by nucleation and growth, the spinodal delimits the metastable region from absolutely unstable region where phase separation occurs by spinodal decomposition without development of distinct phase boundaries.

For strictly binary FH polymer–solvent systems, the spinodal curve is given by the equation valid for $\phi_2 > 0$

$$\frac{1}{m_1\phi_1} + \frac{1}{m_2\phi_2} - 2g + 2(\phi_1 - \phi_2)\frac{\partial g}{\partial \phi_2} + \phi_1\phi_2\frac{\partial^2 g}{\partial \phi_2^2} = 0 \tag{4.159}$$

and for a quasibinary system (disperse in m_2)

$$\frac{1}{m_{1\mathrm{w}}\phi_1} + \frac{1}{m_{2\mathrm{w}}\phi_2} - 2g + 2(\phi_1 - \phi_2)\frac{\partial g}{\partial \phi_2} + \phi_1\phi_2\frac{\partial^2 g}{\partial \phi_2^2} = 0 \tag{4.160}$$

Unlike for phase equilibria, one gets a single curve for a system disperse in molecular weight. This is important for description of thermoreversible gels before the gel point or beyond the gel point.

Ternary systems are also important in phase separation studies (polymer—solvent—non-solvent). There, the spinodal is described by the equation [13]

$$Z_{22}Z_{33} - Z_{23}^2 = 0, \quad Z_{l,k} = Z_{k,l} = \left(\frac{\partial^2(\Delta G_{\mathrm{sw}})}{\partial \phi_k \partial \phi_l}\right)_{p,T} \tag{4.161}$$

For an extracted cross-linked gel, the spinodal for a FH binary system ($\phi_2 > 0$) and affine model of network elasticity is given by

$$\frac{1}{m_1\phi_1} - \frac{V_{1\mathrm{m}}\xi_\mathrm{e}}{\phi_2} - \frac{V_{1\mathrm{m}}\nu_\mathrm{e}}{\phi_2}\left[\frac{1}{3}\left(\frac{\phi_2^0}{\phi_2}\right)^{2/3} - 1\right] - 2g + 2(\phi_1 - \phi_2)\frac{\partial g}{\partial \phi_2} + \phi_1\phi_2\frac{\partial^2 g}{\partial \phi_2^2} = 0 \tag{4.162}$$

Binodals and spinodals for cross-linked gels are compared with uncross-linked polymer of infinite molecular weight in Fig. 4.41 using Eq. 4.162 for concentration independent interaction function $g = \chi$.

It is interesting to note that the metastable region becomes wide with increasing degree of cross-linking which points to increasing difficulty to pass quickly to the absolutely unstable region located left of the spinodals. This shows that cross-linked gels in contrast to the uncross-linked ones are less prone to phase separation by spinodal decomposition. For systems other than FH, relation 145 can be used substituting the respective equations for ΔG_{sw} or $\Delta\mu_1$. The spinodal decomposition mechanism of phase separation will be briefly discussed in the next section as to whether phase separation by this mechanism is likely for gels.

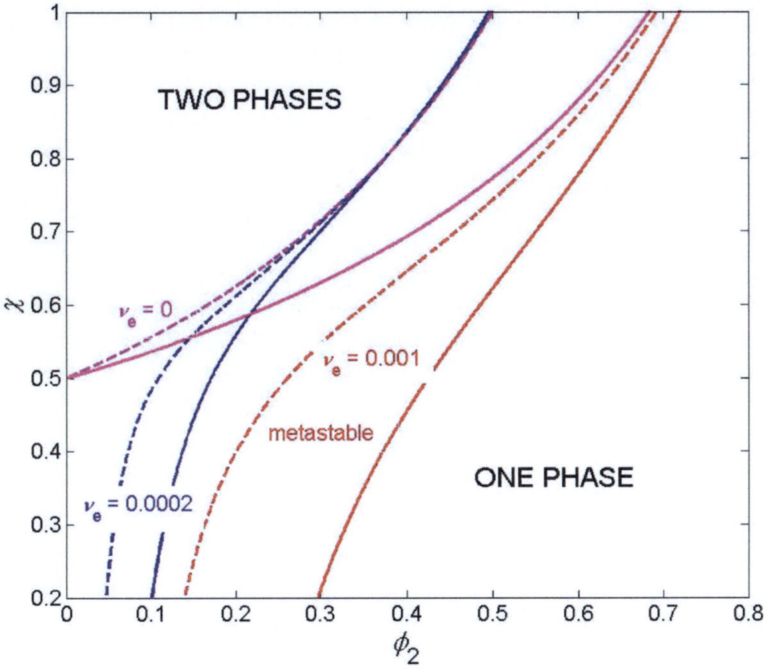

Fig. 4.41 Binodals (full lines) and spinodals (dashed lines) for two levels of cross-linking $\nu_\mathrm{e} = 0.0002$ mol/cm^3(blue) and $\nu_\mathrm{e} = 0.001$ mol/cm^3(red), compared with uncross-linked polymer of very large molecular weight (magenta) with critical point at Θ-state $\chi_. = 1/2$

Yet, there are more kinds of two-phase gels; for instance, such that were formed by coalescence of particles of two-phase systems (suspension). These include condensed dispersions (latexes), or suspensions of colloidal aggregates called "equilibrium gels of limited valence colloids" [247]. The phase structure already existed and the gel state was formed afterward.

4.7.1 Macro and Microphase Separation—Macro and Microsyneresis—In Gels

The phase diagrams as calculated from Gibbs or Helmholtz energies correspond to equilibrium between two or more bulk phases where the interphase energy contribution is negligible just because the amount of material affected is small. Separation into bulk phases (layers) is typical for low molecular weight systems and for well cross-linked gels if phase separation starts when the network structure is well-developed. Nevertheless, some time for reaching the bulk phase separated state is needed; it is preceded by transient formation of interdispersed phases manifested by appearance of turbidity. In more concentrated and viscous systems this transient, non-equilibrium state may have a very long duration. The kinetics of phase separation processes determine not only whether or not the thermodynamically favored transition will occur, but also to what extent the transition will take place and in what form (nucleation and growth or spinodal decomposition). If the composition of the system is located at polymer concentration higher than that of the critical point, the solution demixes by nucleation and growth of droplets embedded in the polymer-rich phase. The droplets can then grow and the interface decreases.

Solutions quenched into the area enclosed by the spinodal are unstable and fluctuations in composition decrease the free energy. If the fraction of the minor phase is sufficiently high, spinodal decomposition proceeds via formation of bicontinuous structures. As time passes, the gradients get larger and the structure coarsens and eventually two fully separated phases develop. Thus, in the later stages, it is difficult to distinguish between systems starting to separate by spinodal decomposition or nucleation and growth. In fact, the concentrations of phases are now such that they have moved out of the absolutely unstable region. Mechanisms most frequently considered for coarsening are Ostwald ripening (also known as evaporation–condensation) or Smoluchowski coagulation (coalescence).

Typical morphology of phase structure obtained by spinodal decomposition is shown in Fig. 4.42.

The main argument for control by spinodal decomposition is the bicontinuous morphology and occurrence of a predominant spacing between polymer-rich regions which would not be expected from phase separation based on nucleation and growth occurring at random points in the sample. Nevertheless, whatever the time scales of microphase separation mechanisms are, the existence of bicontinuous structures and diffuse interphases are very challenging because of excellent properties of the

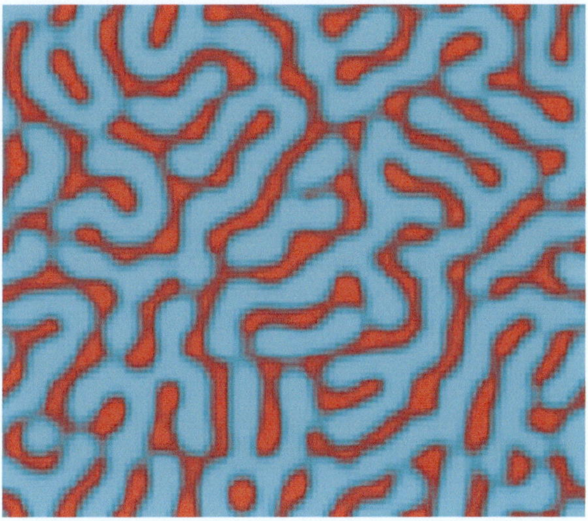

Fig. 4.42 Typical phase morphology of a system formed by spinodal decomposition

materials such as resistance to mechanical impact. Also important are the possibilities of stabilizing such morphological structures by transfer one of the components into the glassy state or by photochemical cross-linking. Therefore, in the following few paragraphs, the spinodal decomposition mechanism will be addressed.

The spinodal decomposition studies in polymer systems, especially the isothermal development of structure was reviewed by Hashimoto [248] and Nose [249] and in the van de Witte's et al. paper [250]. There and elsewhere (Refs. [251, 252]), it was shown that the mechanism of observed structure growth depended on the depth of quench and its speed, shallow quenches and slower temperature change were more favorable for nucleation growth. Tsai's et al. study illustrated the importance of choosing a system viscous enough to avoid phase separation by nucleation and growth during quenching to the unstable region. Su et al. [253] prepared porous membranes from polysulfone/N-methyl-pyrrolidinone solutions by exposing the layers to humid air and found that near the surface the phase separated structure was typical for spinodal decomposition, whereas in the interior structure resembled that characteristic for nucleation and growth. The differences in structure could be explained by the differences in the speed of structure evolution.

The theoretical framework of spinodal decomposition was formulated by Cahn and Hilliard in 1958 [254] who included into the Helmholtz energy the free energy gradient; under diffusional constraints, it is energetically more favorable to proceed through the formation of gradient in composition of certain periodicity and deepen the gradient while preserving the periodicity rather than forming an interface. To explain the solution of the problem, the procedure used by Caneba et al. [255] is partly outlined below.

According to Cahn [254], the free energy of a phase-separated system depends not only on the concentration but also on the concentration gradient. This leads to an expression for a phenomenological, coarse-grained free energy, $\Delta G(c)$, given by

$$\Delta G(c) = \int_V \left[\Delta f(c) + \kappa(\nabla c)^2\right] d\mathbf{r} \tag{4.163}$$

where $d\mathbf{r}$ corresponds to an infinitesimal volume, $\Delta f(c)$ is the local free energy of mixing as a function of polymer concentration, c, κ is the so-called gradient-energy coefficient, and ∇c the gradient; the second term to free energy in Eq. 4.163 expresses the diffuse interphase contribution. The concentration and temperature-dependent mobility, $M(c)$, is defined by the flux $\mathbf{J}$ which is driven by the difference of chemical potentials of polymer and solvent

$$\mathbf{J} = -M(c)\nabla(\mu_2 - \mu_1) \tag{4.164}$$

The difference between the chemical potential of the polymer and solvent is obtained by minimizing the integral of Eq. 4.163

$$\int_V (c - c_0) d\mathbf{r} = 0 \tag{4.165}$$

where c_0 is the average concentration over the whole system. Thus,

$$\mu_2 - \mu_1 = \frac{\partial \Delta f(c)}{\partial c} - 2\kappa \nabla^2 c \tag{4.166}$$

A differential mass balance of polymer segments within an infinitesimal control volume is involved. Considering only diffusional contributions to the flux of material, as the system is stagnant

$$\frac{\partial c}{\partial t} - -\nabla \mathbf{J} \tag{4.167}$$

After substitution, one gets

$$\frac{\partial c}{\partial t} = \nabla \cdot \left[M(c)\nabla\left[\frac{\partial \Delta f}{\partial c} - 2\kappa\nabla^2 c\right]\right] \tag{4.168}$$

To introduce periodic perturbation, the sine function can be applied. For early stages, this equation can be linearized along the average composition

$$\frac{\partial c}{\partial t} = M_0\left[\left(\frac{\partial^2 \Delta f}{\partial c^2}\right)_0 \nabla^2 c - 2\kappa\nabla^4 c\right] \tag{4.169}$$

The quantities $\left(\frac{\partial^2 \Delta f}{\partial c^2}\right)_0$ and M_0 are evaluated for average composition. This linearized problem has a closed-form solution

$$c - c_0 = \sum_{\mathbf{R}} \exp[r(\mathbf{K})t][A(\mathbf{K})\cos(\mathbf{Kr}) + B(\mathbf{K})\sin(\mathbf{Kr})] \tag{4.170}$$

where $r(\mathbf{K})$ is the amplification factor. This equation allows us to calculate the resonant wavelength of the most rapidly growing peak, λ_m, in the frequency spectrum during spinodal decomposition in the initial stage

$$\lambda_m = 2\left(2^{1/2}\right)\pi\left[-\frac{\left(\partial\Delta f/\partial c^2\right)_0}{2\kappa}\right]^{-1/2} \tag{4.171}$$

An example of evolution of concentration profiles resulting from the calculation is shown in Fig. 4.40 (Fig. 4.43).

The solution of the Cahn–Hilliard problem and its variations as well as the solution methods have met with great attention in the literature [151] [255–259] and will be helpful to anyone interested in practical applications. Very good description of the

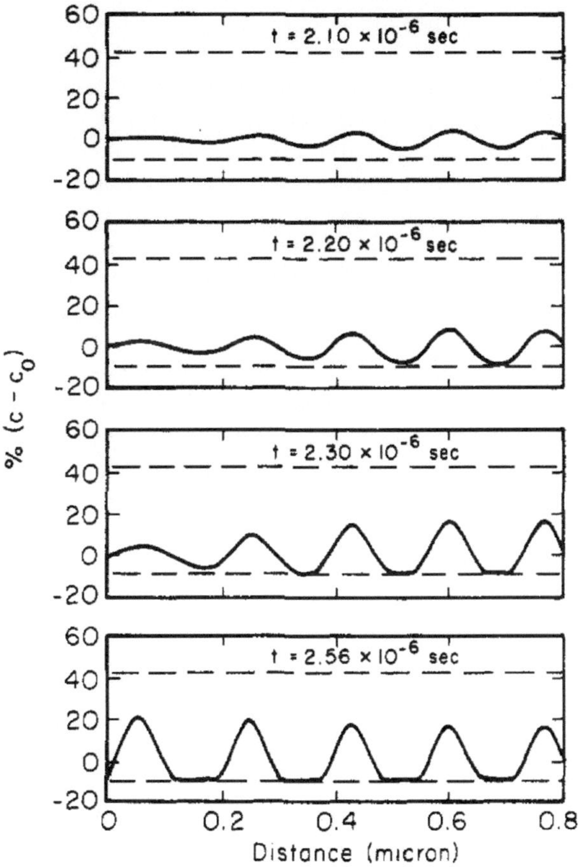

Fig. 4.43 Polymer segment profiles as functions of time during the evolution of structure for a polymer–solvent system specified in Ref. [255]. Dashed lines are binodal compositions. Reprinted from Ref. [255] (Fig. 4.14) with permission of American Chemical Society

phase diagram with the properties of interfaces and the nucleation barrier that needs to be overcome to form a droplet (or bubble) can be found in a review by Kremer et al. [260].

An illustration of structure evolution by spinodal decomposition is made available on Internet by NIST Gaithersburg, VA [video].

4.7.2 Phase Separation in Gels

The equivalent for macro and microphase separation in polymer blends and solutions are the terms *macrosyneresis* and *microsyneresis* coined in late sixtieths [2]. Indeed, *syneresis* derived from *Greek synairein*—to contract—means expulsion of liquid from the gel while the gel contracts. This is what happens as a result of macrosyneresis: upon change of external conditions (temperature, solvent exchange, pressure), the gel deswells. Deswelling takes place while the gel remains clear, and the final, equilibrium state is governed by the swelling equation as discussed in Sect. 4.3. When phase separation by microsyneresis is effective, the outer dimensions of the gel do not change but the gel gets turbid by excess solvent separated in microdomains. The formation of microdomains as the final state can occur by the mechanism of nucleation and growth or spinodal decomposition. Here, we are talking about permanently cross-linked gels not containing sol fraction—sol fraction is either extracted or negligible when the cross-linking degree is sufficiently high. When sol is present, fractionation of the sol occurs between the gel and liquid phases; the separation equilibria depends on the degree of cross-linking. When microsyneresis proceeds via spinodal decomposition, polymer-rich (less swollen) and polymer-poor (more swollen) domains are separated by a gradient in the degree of swelling. The existence of a permanent network is not favorable for the spinodal decomposition mechanism because (a) the metastable region is much wider than for uncross-linked systems, and (b) deformation of network chains in the gradient positively contribute to the free energy and makes such arrangement less stable. The situation may be different for temporary network because phase separation may lead to bond rearrangement.

Almost 50 years ago, the following observations have been made [261, 262]: Series of clear poly(2-hydroxyethyl methacrylate) gels was swollen at elevated temperature in n-butanol, and the temperature was lowered. When the cross-link density was low, turbidity developed; it slowly decreased with time while keeping the temperature constant. Eventually, the gel was clear again but only after several months. Similar behavior was observed with PHEMA gels swollen in water at room temperature when the temperature was raised from 20 to 50 °C [263], or as a result of solvent exchange [264]. The turbidity changed in time and was measured at different wavelengths; the sizes of scatters were characterized by the turbidity ratio method based on Mie's theory of light scattering [265, 266]. The calculated size of particles responsible for appearance of turbidity varied in the range 150–200 nm, being somewhat smaller for aqueous systems than for butanol. However, during

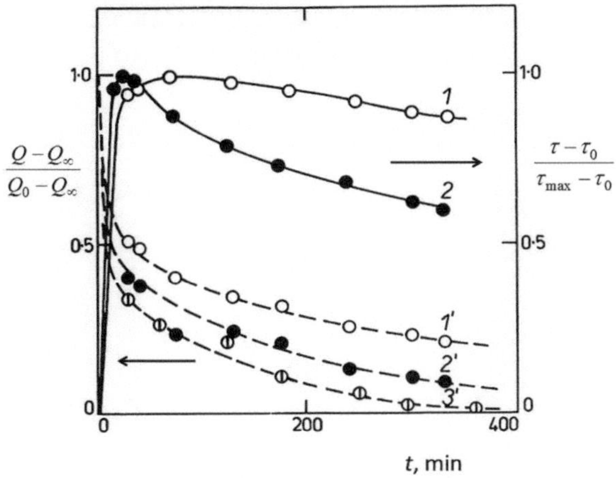

Fig. 4.44 Relative decrease of the volume degree of swelling (Q) and relative turbidity (τ) for poly(2-hydroxyethyl methacrylate) gel swollen in n-butanol when the temperature was decreased from 80 to 35 °C (turbidity starts developing at 50–60 °C). Concentration of ethylene dimethacrylate cross-linker: 1 0.2, 2 1.0, 3 5.0 wt%. The subscript 0 means initial value, ∞ after a long time, max maximum. Reprinted and modified from [263] with permission (Fig. 4.6)

the turbidity decay with time, the particle size remained constant while their number (concentration) decreased slowly until after several months the turbidity disappeared.

For gels of higher cross-link density, the decrease of temperature caused only volume shrinkage without development of turbidity. Such changes are illustrated by Fig. 4.41. PHEMA gels of cross-link density corresponding to 5 wt% cross-linker and higher only deswell without turbidity developing (Fig. 4.44).

Schematically, the droplet evolution and disappearance is shown in Fig. 4.45.

The analysis revealed a very interesting finding that, on a wide time scale of turbidity decay at constant temperature, the size of the separated particles remained constant while their number (concentration) decreased. These features were interpreted by formation of small droplets of separated solvent, their size being determined by forces necessary to form a void in the gel increasing with their size and interphase energy increasing by increasing the interface area per unit volume (decreasing droplet size). Using the experimentally determined interphase energy and a model of elasticity of a network respecting finite extensibility of network chains, a shallow minimum was found for diameters of droplets between 100 and 300 nm. These results were amended by other studies on the same system [267, 269].

It seems that the nucleation mechanism is operative; ripening manifested by domains size increase is not observed, the appearance of turbidity and restoration of transparence is relatively rapid. Thermal fluctuations in concentration and static inhomogeneities can serve as nucleation sites. Static inhomogeneities in gels are a big issue. We are not going to discuss this issue here—their formation can be induced by the gel formation process (especially by chain cross-linking (co)polymerization) or by hydrophobic-hydrophilic association of various kinds. The inhomogeneities have

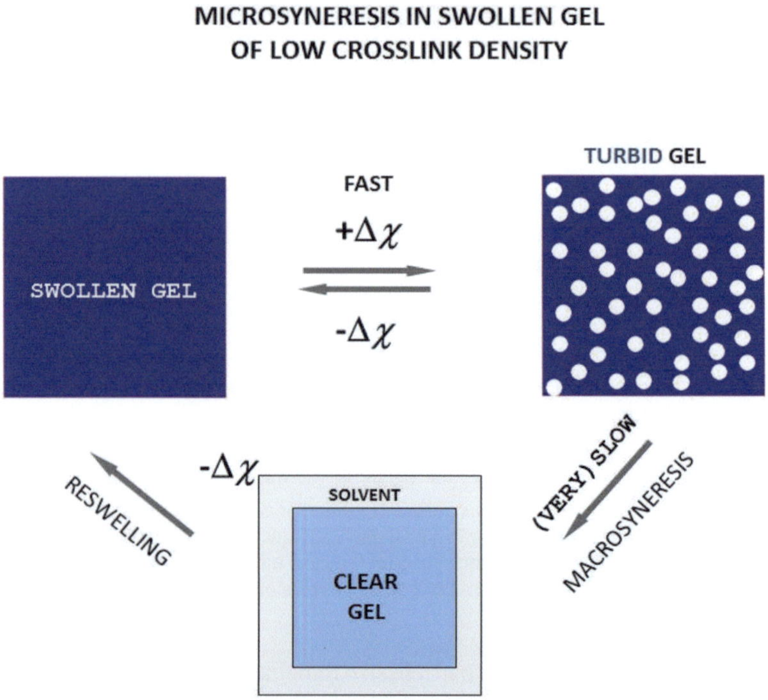

Fig. 4.45 Droplet model of microphase separation in a gel

been seen by radiation scattering methods—of the vast literature on this subject, see, e.g., the work of Shibayama's group [270–273] or recent review by Seiffert [274]. For Gaussian type of network, large expansion of a seed inhomogeneity on the level of mesh size is allowed, but limits are imposed if one considers finite extensibility of network chains. Then, in order to get an inflated cavity of diameter of 100 nm, one would need the seed to be not smaller than 10 nm; for 1% cross-linker the mesh size of stretched chains for uniform network would not exceed 3–4 nm. Therefore, the hypothesis of droplet formation needs further clarification.

However, all these observations cannot exclude spinodal decomposition as being operative at least in the very initial phases of nucleation. The phase boundaries evolve from concentration gradients. This is why phase separation in gels is often considered as being controlled by spinodal decomposition mechanism. In the literature, one finds a few studies using mainly time-resolved light scattering. In-depth studies were published by Bansil et al. [275, 276] aimed at elucidation of structural changes occurring during turbidity development resulting from temperature quenches or solvent exchange in gelatin gel and chemically cross-linked poly(N-isopropylacrylamide) (PNIPAM) gel. In both cases, the early stages of turbidity development could be well-described the linear Cahn–Hilliard-Cook (CHC) [277] theory. The resulting structure was quite different in the later stages being dependent on the nature of

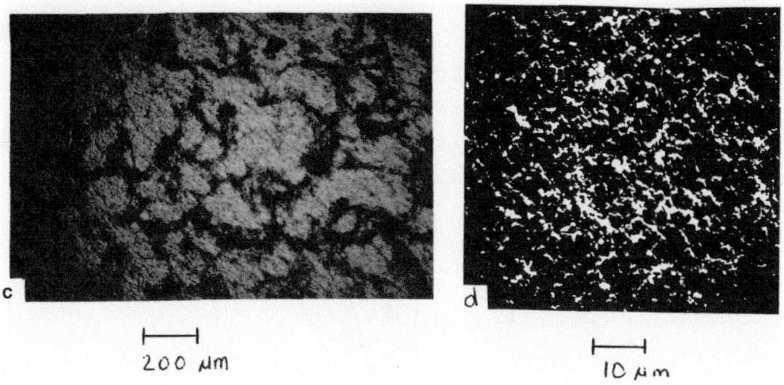

Fig. 4.46 Kinetics of spinodal decomposition in a 7% solution of gelatin in water–methanol mixture quenched to different temperatures below the spinodal; **c** optical micrograph showing the morphology of the phase-separated domains 24 h after the sample was quenched to 22 °C. The connected morphology characteristic of spinodal decomposition is preserved due to gelation, leaving the sample in a microphase-separated state; **d** scanning electron micrograph of an immersion precipitate membrane made from gelatin showing a highly connected porous structure with gelatin fibers and micron-sized holes. Reprinted from Ref. [276] with permission (Fig. 4.47)

the cross-link, i.e., whether it was a permanent covalent cross-link or a temperature reversible weak bond.

The optical microscopy and SEM records show the enormously large-scaled inhomogeneities in gelatin pointing to a large-scale reorganization of physical cross-links (Fig. 4.46).

Microphase separation in agarose gels was also studied by Morita et al. [278]. The morphology developed by quenching or very slow cooling of gelatin solutions in water or in mixtures of water and methanol, which is a poor solvent system for agarose. The phase diagram is shown below (Fig. 4.47).

The spinodal curve appears below the cloud-point curve and both curves are entirely buried below the sol–gel transition curve in the aqueous agarose system. It is concluded that the first event that occurs during the cooling process of the aqueous agarose solution is the transition from the sol state to the gel state. For fast quench, the characteristic spinodal ring from scattered light was observed and the dependences with distinct maxima in small-angle X-ray developed (Fig. 4.48), the maxima were moving to smaller angles as quench temperature increased.

The structure of agarose gel was observed by the confocal laser scanning microscope (CLSM) imaging technique that revealed that the density fluctuations were grown up to micrometer scale. The real space image of the quenched agarose gel indicated that the gel consisted of the porous network. The authors believe that dangling chains and other molecular defects can be responsible for the phase separation and the coarsening of the dense region. However, from the point of view of dimensions it does not seem too likely.

For turbidity development in a covalently cross-linked PNIPAM gel, it was found by SALS that early after a temperature quench, the linear Cahn–Hilliard-Cook (CHC)

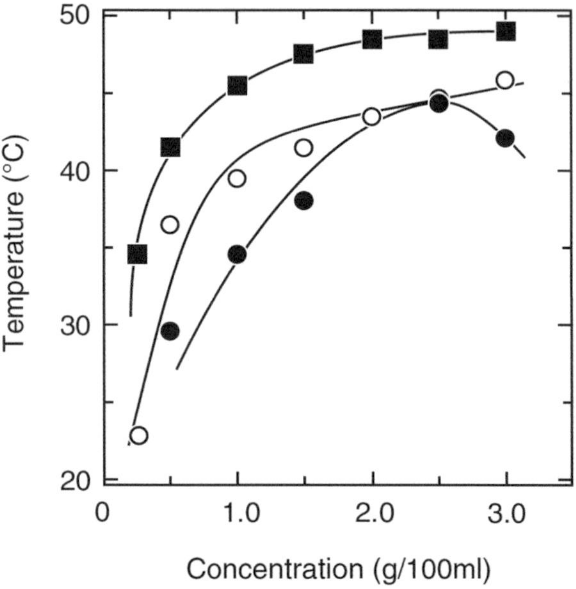

Fig. 4.47 Phase diagram of agarose solution. The closed squares, the open circles, and the closed circles represent, the sol–gel transition point, the cloud point, and the spinodal point, respectively. Reprinted from [278] with permission (Fig. 4.1)

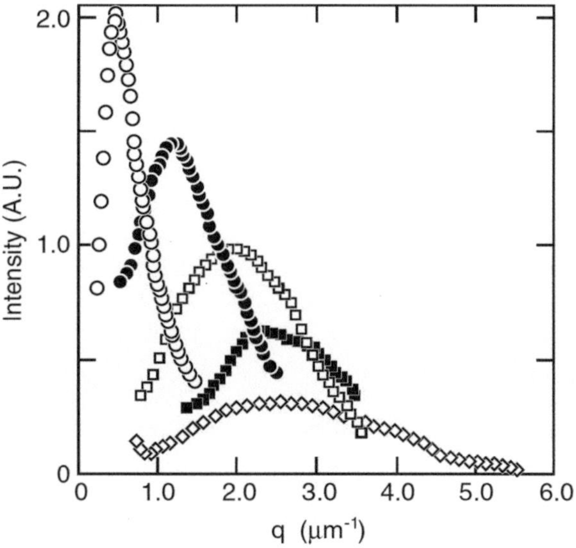

Fig. 4.48 Small-angle light scattering profiles from the quenched agarose gel. The concentration of agarose is 2.5 g/100 ml. The quenched temperatures are 43.0 °C, open circles; 35.0 °C, closed circles; 25.0 °C, open squares; 15.0 °C, closed squares, and 5.0 °C, open diamonds. Reprinted from [278] with permission (Fig. 4.2)

theory could be used to describe the time evolution of the scattered intensity and turbidity. Surprisingly, the initial length scale on which fluctuations grow was found to be much larger than the concentration correlation length. Also, the diffusion constant characteristic of the phase-separating gel was much lower than the collective diffusion constant. It appears that the initial process of phase separation in these gels is driven by fluctuations on micron–sized regions, perhaps representing large-scale variations in the degree of swelling. When considering the phase structure changes, one should not forget that PNIPAM is a system undergoing volume phase transition, i.e., phase separation does not form only liquid and gel phases, but two phases of different swelling degree, possibly coexisting with pure liquid. In that sense, the system is different and the results cannot be generalized as valid for simple hydrogels.

In conclusion from the relatively scarce data on microphase separation in cross-linked gels, one can say that most probably the very early stages are governed by spinodal decomposition which soon goes over to growth by other mechanisms like that of nucleation and growth, the grows being arrested by existence of the network meshes. It is to be made clear what the composition of the separated phase is, whether it is just excess solvent or a gel of (much) larger degree of swelling. For the void formation model, the observed size of inhomogeneities is too large to leave the network structure undamaged. More experiments are needed on gel where preparation-induced inhomogeneities (e.g., by free radical polymerization) can be eliminated. Based on this approach, also the volume shrinkage accompanying or following microphase separation involving a rapid (liquid) component can be addressed.

Some time ago the "viscoelastic phase separation" theory of microphase separation was put forward which stressed the important role of viscoelasticity and viscoelastic asymmetry of the components in the dynamics of phase-separated systems and pattern evolution [279]. The domain shape is affected by mechanical balance of elastic forces and interfacial forces. For polymer—small molecule solvents (dynamically asymmetric mixture), a long-lived interaction network (transient network) of slow component (polymer) is assumed to be formed if concentration is sufficiently high and interactions are sufficiently slow. Then, domains of the fast components are nucleated in transient gel even if the slow component phase is a minority phase [280].

4.7.3 Phase Separation During Network Formation

Phase separation can occur during formation of the network as a consequence of structural changes resulting from the polymerization and cross-linking processes. Usually, a non-reactive diluent is added, but phase separation can also occur when the monomer itself is a poor solvent for the formed polymer. Network formation proceeds usually at constant temperature, but not necessarily. As network formation proceeds, the molecular weights increase before the gel point and number of cross-links and gel fraction increase after the gel point is passed. These structural changes decrease the miscibility and contribute to phase separation. Also, the polymer–polymer and polymer–solvent interactions change, partly because the formed cross-links deplete a part of the interacting surface, but also because reactive groups are transformed into bonding groups and their interactions may be quite different—for instance from hydrogen donation as well as hydrogen-accepting OH and COOH groups only hydrogen-accepting –OCO– ester groups are formed. Then, the compatibility of an apolar solvent with the formed polymer can get better and act against phase separation.

Reaction-induced phase separation has become an important technology for production of porous polymeric materials, especially macroporous separation materials—membranes, separation media for various kinds of chromatography, ion-exchange resins, absorbents, as well as a part of micro and mesoporous storage materials—metal–organic frameworks (MOF). Very important role in biomedical field is played by macroporous hydrogels which are mainly prepared by reaction-induced phase separation (tissue engineering, implants, controlled drug delivery). Several reviews exist on this subject—some of them can be found under Refs. [281].

Similarly as in a gel induced by a change of interactions, phase separation caused by conversion shift can occur on macro and microscale. On macroscale, the gel shrinks, deswells, and excess liquid diluent is separated as a bulk phase. On microscale, a dispersion of phases is formed without changing outer dimension, or both micro and macroforms of phase separation are combined—the phases are partly interdispersed or part of the separated liquid is expelled outside of the gel. The morphology of the phase separated can vary from dispersion of droplets of liquid, to various co-continuous forms up to chemically sintered spherical particles. The morphology depends on several factors, such as poorness of the diluent, its volume with respect to monomer unit, depth of structure development, interfacial energy, or reaction *vs.* diffusion rates. Typical transformations are shown in Fig. 4.49.

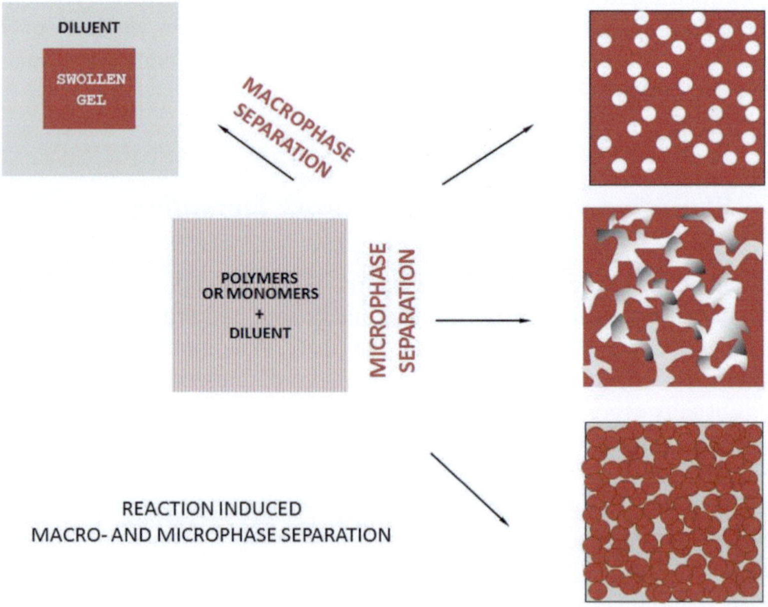

Fig. 4.49 Reaction-induced macro and microphase separation

It is characteristic for reaction-induced phase separation that the reaction rate can be faster than realization of equilibrium, which means that non-equilibrium structures are fixed by introduction of cross-links. However, the driving forces of morphology development are based on thermodynamics and are driven by gradient of chemical potential.

The control of macroporous structure development by thermodynamics of mixtures of cross-linked polymer with diluent was made clear for the first time in the studies of copolymerization of styrene and divinylbenzene in the presence of low and high molecular weight additives [281] based on formulation of conditions for incipience of phase separation [282] that were indentified with crossing the binodal from thermodynamically stable into the metastable region.

Let us look what happens when a solution of polymer is being cross-linked. We will consider the simple case when the system has passed the gel point and the sol fraction is already small. The system is considered as binary system, the diluent being component 1. Making the usual assumption (as the first approximation) that network chains are in reference (unstressed) state when they are formed, the contribution arising from stretching of chains relatively to their reference state is zero, i.e., $\Delta\mu_{1,\mathrm{el}} = 0$, and the thermodynamics is controlled purely by the mixing contribution (cf. Section 4.4)

$$\frac{\Delta\mu_{1,\mathrm{mix,n}}}{RT} = \left(\frac{1}{m_1}\ln\phi_1 + \frac{\phi_2}{m_1} + \frac{f_\mathrm{e}-2}{f_\mathrm{e}}\nu_\mathrm{e}V_{\mathrm{mol,ls}}\phi_2 + \left(g(\phi_2) - \frac{\partial g(\phi_2)}{\partial\phi_2}\phi_1\right)\phi_2^2\right)$$
$$= \ln a_1 = \ln\frac{f_1}{f_1^0} \approx \ln\frac{p_1}{p_1^0} \qquad (4.172)$$

where $V_{\mathrm{mol,ls}}$ is the molar volume of lattice site, the diluent molecule occupies$m_1V_{\mathrm{mol,ls}}$ lattice sites;a_1 is the diluent activity. The volume fraction of diluent is identical with the volume fraction of solvent $\phi_2 = \phi_2^0$;f_1 and$f_1^0(= 1)$ are fugacities of diluent in the vapor phase and in its standard state, and approximately equal to the ratio of vapor pressure of the diluent over the gelp_1 and vapor pressure of pure diluent,p_1^0. When network formation starts, the solvent activit $a_1 < 1$ and the system is thermodynamically stable and in equilibrium with solvent vapor. By introducing cross-links,a_1 increases until it reaches the value of 1 which means equilibrium with pure solvent is established. At that point and the maximum degree of swelling is given by $\phi_2 = \phi_2^0$. In equilibrium,a_1 cannot grow any further and to maintain the value 1, ϕ_2 must increase and the corresponding amount of diluent must leave the gel phase. Schematically, the situation is shown in Fig. 4.50.

When the gel shrinks, the network chains are compressed relative to their reference state and $\Delta\mu_{1,\mathrm{el}} \neq 0$. The shrinkage is continuous which means that network chains formed in the earlier stages are more compressed (they were subjected to larger volume change) than chains formed in the later stages. For expressing this contribution to elasticity, the two-network hypothesis of Andrews et al. [283] is applied. According to this hypothesis, the contributions to elasticity by network A and network B formed at different states of strain are additive

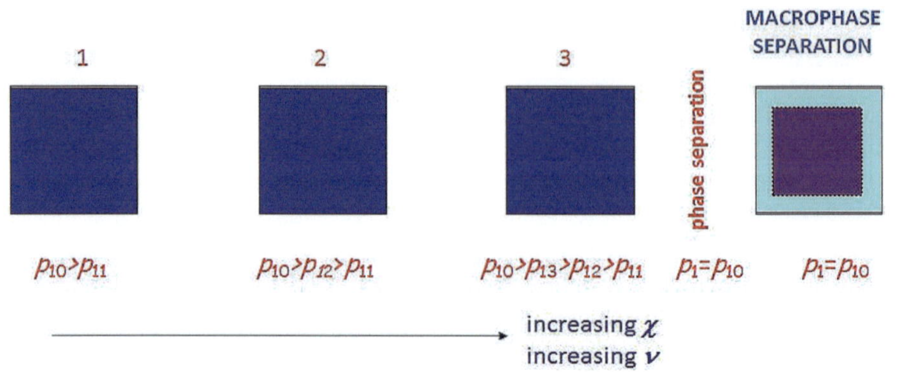

Fig. 4.50 Equilibria with diluent vapors and liquid phase during cross-linking at constant temperature

$$\frac{\Delta F_{\mathrm{el}}}{kT} = \frac{N_{\mathrm{eA}}}{2}\left(\lambda_{\mathrm{A}x}^2 + \lambda_{\mathrm{A}y}^2 + \lambda_{\mathrm{A}z}^2 - 3\right) + \frac{N_{\mathrm{eB}}}{2}\left(\lambda_{\mathrm{B}x}^2 + \lambda_{\mathrm{B}y}^2 + \lambda_{\mathrm{B}z}^2 - 3\right) \tag{4.173}$$

which for initially isotropic system differing only in reference volumes of network A and B, Eq. 4.173 transforms into the swelling Eq. 4.174

$$\begin{aligned}\frac{\Delta\mu_1}{RT} &= \frac{1}{m_1}\ln\phi_1 + \frac{\phi_2}{m_1} + \frac{f_{\mathrm{e}}-2}{f_{\mathrm{e}}}\nu_{\mathrm{e}}V_{\mathrm{mol,ls}}\phi_2 + \chi\phi_2^2 \\ &+ V_{\mathrm{mol,ls}}\left(\nu_{\mathrm{eA}}\left(\phi_{2\mathrm{A}}^0\right)^{2/3} + \nu_{\mathrm{eB}}\left(\phi_{2\mathrm{B}}^0\right)^{2/3}\right)\phi_2^{1/3} - \nu_{\mathrm{e}}V_{\mathrm{mol,ls}}\phi_2 = 0 \\ \nu_{\mathrm{eA}} &+ \nu_{\mathrm{eB}} = \nu_{\mathrm{e}}\end{aligned} \tag{4.174}$$

In the general context, the factors $\phi_{2\mathrm{A}}^0$ and $\phi_{2\mathrm{B}}^0$ lose their volume fraction meaning since they may acquire values > 1, if the chain are prestretched like in sequential IPNs (network A is prestretched by swelling in monomer B).

For the cross-linking reaction-induced macrophase separation, the memory factor ϕ_2^0 is constant until phase separation starts and the maximum degree of swelling is characterized by $\phi_2 < \phi_2^0$. When the binodal is reached, and phase separation starts (in the case of very slow reaction), the reference volume for the newly formed chains starts decreasing and the factor ϕ_2^0 increasing; for the increment of concentration of EANCs $\delta\nu_{\mathrm{e}}$ and the contraction $\delta\phi_2 = \delta\phi_2^0$. This leads to a cross-linking reaction-dependent memory factor (conversion parameter α)

$$\begin{aligned}\phi_2^0(\alpha)\nu_{\mathrm{e}}(\alpha) &= (\nu_{\mathrm{e}})_{\mathrm{crit}}\phi_2^0 + \int_{(\nu_{\mathrm{e}})_{\mathrm{crit}}}^{\nu_{\mathrm{e}}(\alpha)} \phi_2(\alpha)d\nu_{\mathrm{e}} = (\nu_{\mathrm{e}})_{\mathrm{crit}}\phi_2^0 + \int_{\alpha_{\mathrm{crit}}}^{\alpha} \phi_2(\alpha)\frac{d\nu_{\mathrm{e}}(\alpha)}{d\alpha}d\alpha \\ \phi_2^0(\alpha) &= \frac{(\nu_{\mathrm{e}})_{\mathrm{crit}}\phi_2^0}{\nu_{\mathrm{e}}(\alpha)} + \frac{1}{\nu_{\mathrm{e}}(\alpha)}\int_{\alpha_{\mathrm{crit}}}^{\alpha} \phi_2(\alpha)\frac{d\nu_{\mathrm{e}}(\alpha)}{d\alpha}d\alpha\end{aligned} \tag{4.175}$$

where $(\nu_e)_{crit}$ and α_{crit} are the concentration of elastically active chains and conversion of the cross-linking reaction at incipient phase separation, respectively, ϕ_2^0 and $\phi_2^0(\alpha)$ are the constant memory terms for $\alpha \leq \alpha_{crit}$ and conversion dependent for $\alpha > \alpha_{crit}$, respectively. The values of $\nu_e(\alpha)$ are obtained from cross-linking theory and $\phi_2(\alpha)$ is obtained beyond the point of incipience of phase separation by solving integral Eq. 4.176. Since the ϕ_2^0 *vs.* α dependence is not a continuous function, it is split into two parts

for $\alpha \leq \alpha_{crit}$

$$\frac{1}{m_1}\ln\phi_1 + \frac{\phi_2}{m_1} + \frac{f_e - 2}{f_e}\nu_e V_{mol,ls}\phi_2 + \chi\phi_2^2 + V_{mol,ls}\nu_e\left(\phi_2^0\right)^{2/3}\phi_2^{1/3} - \nu_e V_{mol,ls}\phi_2 = 0 \tag{4.176a}$$

and for $\alpha > \alpha_{crit}$

$$\frac{1}{m_1}\ln\phi_1 + \frac{\phi_2}{m_1} + \frac{f_e - 2}{f_e}\nu_e V_{mol,ls}\phi_2 + \chi\phi_2^2 + V_{mol,ls}\phi_2^{1/3}\left[\frac{(\nu_e)_{crit}\phi_2^0}{\nu_e} + \frac{1}{\nu_e}\int_{\alpha_{crit}}^{\alpha}\phi_2\frac{d\nu_e}{d\alpha}d\alpha\right]^{2/3} - \nu_e V_{mol,ls}\phi_2 = 0 \tag{4.176b}$$

There is no turbidity development seen while the gel is shrinking and the shrinkage rate can be simply described by viscoelasticity of the network and diffusion of excess solvent out of the gel. Not many studies have been done on macrosyneresis; macrophase separation induced by cross-linking in stepwise polyurethane network formation was addressed along these lines in Refs. [118] [284, 285]. Macrophase separation was monitored during formation of polyurethane network in the presence of solvent and compared with a pseudoternary model consisting of solvent (component 1), sol (component 2) and cross-linked gel (component 3). The interaction parameters χ_{12}, χ_{13}, χ_{23}, and their dependence on composition as a function of conversion of the cross-linking reaction were obtained by combination of swelling experiments with the branching theory (evolution of molecular weights and composition and amount of sol and gel). Using this dataset, solvent activities at constant solvent concentrations were constructed [118] and plotted as a function of conversion. The solvent activity increased with conversion. When the system was closed (no evaporation), macrophase separation started at a certain conversion of isocyanate groups when the solvent activity reached the value 1. This was in agreement with experiment; the conversion at which the solvent started sweating out of the gel was monitored by ATR FTIR spectroscopy.

When the solution is deposited as film on a substrate and evaporation of solvent allowed (weight loss determined by automatic weighing in an environmental chamber), macrophase separation does not take place because the concentration increase matches or overrides the activity increase due to worsening of polymer solvent interaction and formation of cross-links. The situation is shown in (Figs. 4.51 and 4.52).

While macrophase separation is relatively easy to understand and to simulate on the basis of a thermodynamic model, the microphase separation can result in a variety of morphological forms which greatly depend on the type of polymerization and on the polymerization rate. Although microphase separation is sometimes an unwanted phenomenon, in the majority of cases it is intentional and is used for preparation of porous products.

Microphase form of phase separation (microsyneresis) occurs in the case the degree of cross-linking is still low, so that the retractive force to local deformation—formation of a void—is weak, i.e., close to the gel point or even before the gel point. Phase separation depends very much on whether network formation proceeds stepwise or chainwise. In a stepwise process, there is a continuous distribution of oligomers of which a gradually increasing part transforms into gel. Much more frequently used are the chainwise processes, like free-radical cross-linking copolymerization, for preparation of macroporous networks and hydrogels. In this process,

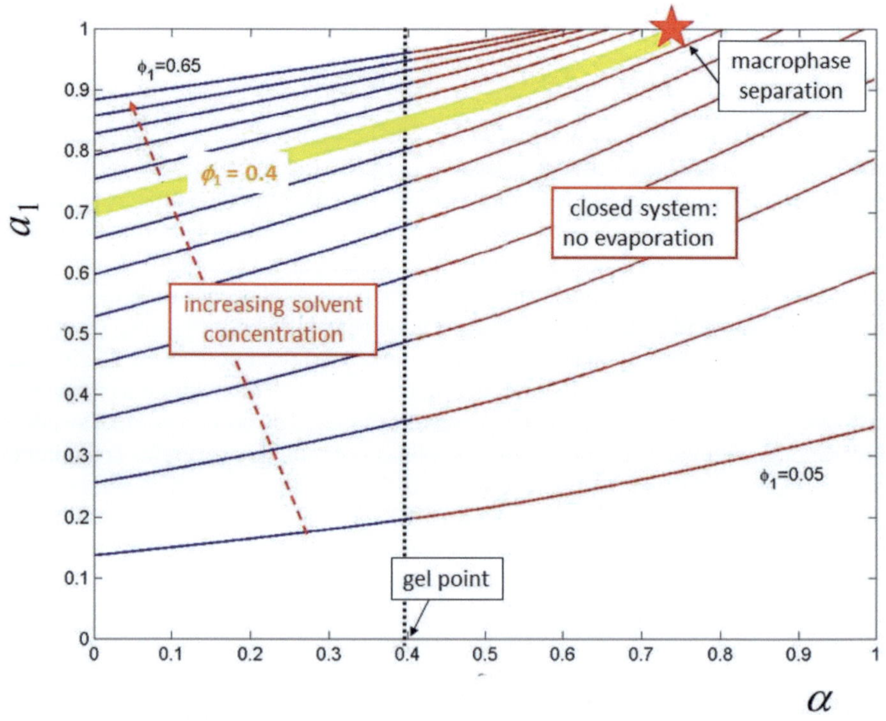

Fig. 4.51 Activity of solvent (methyl amyl ketone) as a function of conversion for reaction of a star shape polyester tetrol with triisocyanate in closed system at different initial concentrations of solvent at 25 °C. Red star marks the onset of macrophase separation; yellow curve is the activity of solvent at $\phi_1 = 0.4$. Details on the system in Ref. [118]

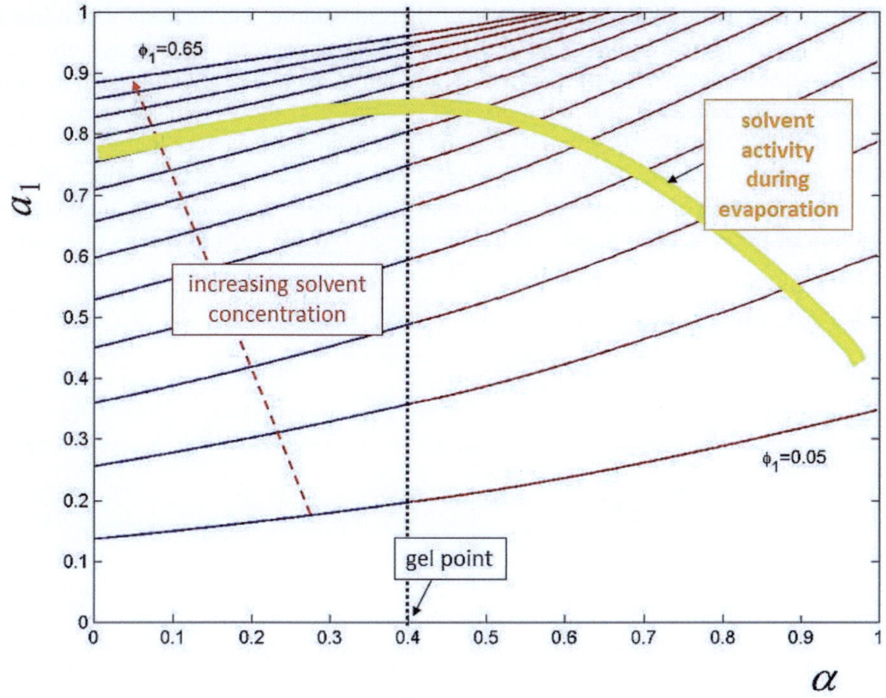

Fig. 4.52 Solvent activity changes for system of Fig. 4.51 when solvent evaporates from the reacting film. The rate of solvent loss can be read off from course of the yellow curve. In that case, the solvent activity never reaches the value 1

high molecular polymers are formed from the beginning and gelation occurs at low conversion of double bonds. Because of that, the low molecular weight monomer serves as a solvent and its thermodynamic interaction with the polymer must be considered. To get porous network, usually one has to add a diluent and the system is to be considered at least as a pseudoternary one. This means that the molar ratios of monomer and diluent are different in the gel and in the liquid phase and, as polymerization proceeds, monomer is consumed in the gel phase and the liquid phase by different rates and mass transfer between microphases occurs. Moreover, the cross-linking free-radical polymerization at higher concentration of cross-linker is non-ideal in that due to enhanced cyclization, clusters (even microgel-like particles) are formed with a part of pendant double bonds being buried inside these clusters. These special features have several consequences liked delayed gel point discussed extensively in the literature—see, e.g., Refs. [118, 194] 286–290]. The consequence relevant for phase separation is an increased inhomogeneity close to the gel point which facilitates the microform of phase separation.

The first studies on formation of macroporous network were published by Millar et al. [291, 292] and a systematic research has been completed by Seidl et al. [281] they could be fully understood in terms of polymer thermodynamics [281, 282] as phase separation induced by a change of thermodynamic interactions and diluent size (χ-syneresis) and by increase of the cross-linking degree (ν-syneresis), often as combination of both. All these factors $-\chi$, m_1, and ν_e increase the chemical potential of the solvent. Indeed, this is what original results [281] show (Fig. 4.53a–d).

Certain fraction of poor solvent 2,2,4-trimethylpentane is necessary before a new stable phase is formed (about 15–20 vol%, Fig. 4.53a), at higher fractions of the diluent at the end of the reaction, practically all diluent separated, but the volume fraction of pores is even higher than that amount (50–60 vol%). This is due partly

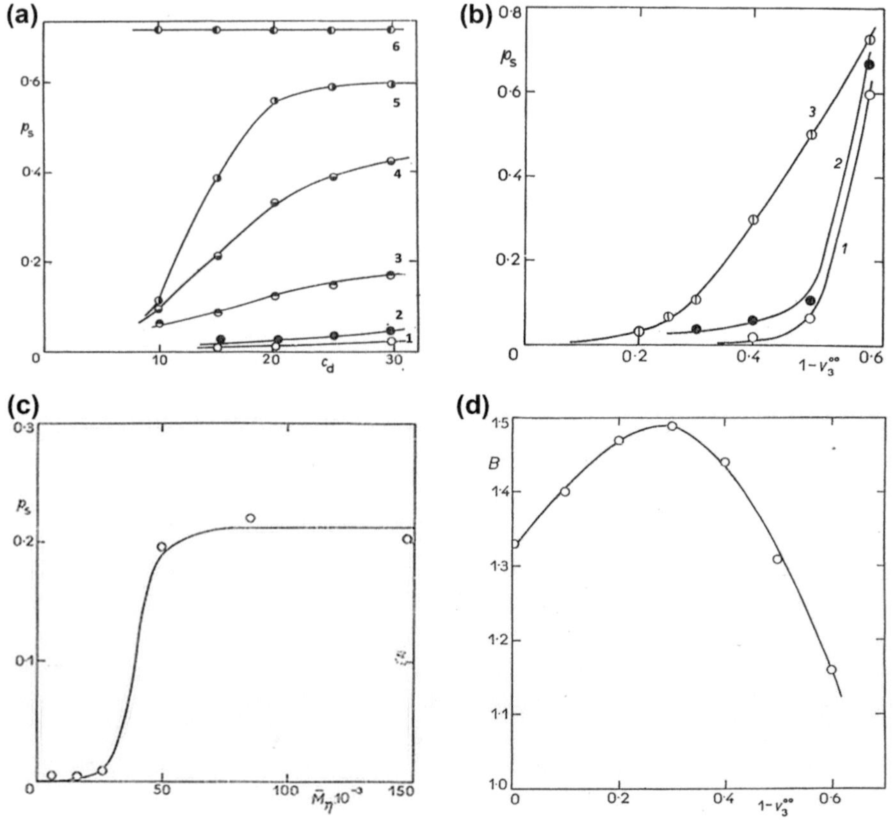

Fig. 4.53 **a** Porosity of the dry ST-DVB copolymer prepared in the presence of 2,2,4-trimethylpentane as a function of the degree of cross-linking. The volume fraction of 2,2,4-trimethylpentane in the starting mixture; volume fraction of diluent: *1* 0.10, *2* 0.20, *3* 0.30, ’*4* 0.40, *5* 0.50, *6* 0.60; p_s is the volume fraction of pores in the dry copolymer. **b** Porosity of the dry ST-DVB copolymer plotted versus the dilution of the monomers by aliphatic hydrocarbons $(1 - v_3^{00})$; concentration of DVB = 8%; hydrocarbons: *1* n-hexane; *2* n-octane; *3* linear alkanes C12–C21•. **c** Porosity of the dry copolymer as a function of the viscosity average molecular weight of the added polystyrene; 20% (w/w) of polystyrene, concentration of DVB, $c_d = 10$ wt%. **d** Volume swelling degree, *B*, as a function of dilution of the monomers by 2,2,4-trimethylpentane ($c_d = 20$ wt%). Reprinted from [281] with permission

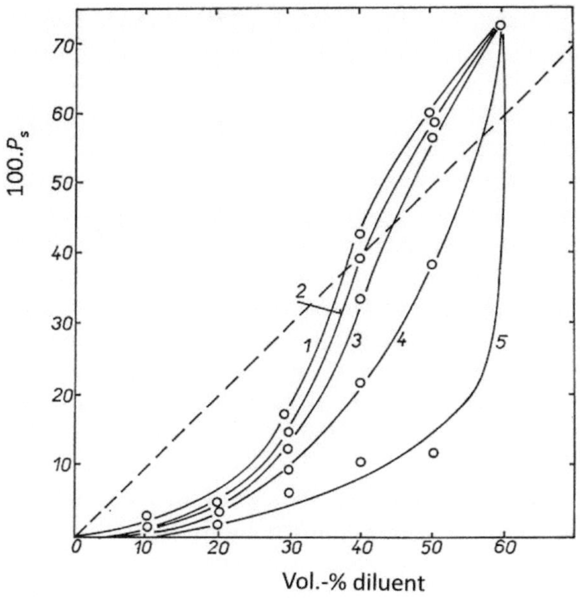

Fig. 4.54 Volume fraction of pores (in vol%) in dry ST-DVB networks in dependence on dilution with 2,2,4-trimethylpentane for different contents (in wt%) of cross-linker (DVB); *1* 30, *2* 25, *3* 15, *4* 10, *5* 5. Reprinted from [281] with permission (Fig. 4.16)

due to the presence of non-polymerizable compounds in commercial DVB (about 30 wt%), but at > 60 wt%- DVB, the pore volume is even higher than the sum of all non-polymerizable compounds (cf. Fig. 4.53a), the total volume is larger than the initial volume, and the glass ampules used as polymerization containers broke by swelling pressure. Surprising in Fig. 4.54 is a steep increase by which the final value was approached. These samples are less coherent because phase separation occurred before the gel point and the separated domains were chemically sintered in at the later stages.

According to the model, such cases are possible. After the pore structure has been well-developed, the non-polymerized monomers are transferred from the liquid phase into the gel phase and transformed there into the polymer, the volume increases and the network chains formed earlier are stretched. The network is then stiffer than that prepared in a diluent that is a better solvent for the polymer.

Figure 4.52b, c shows the important role of size of diluent molecules: aliphatic hydrocarbons (48b) and polystyrene as polymeric diluent (Fig. 4.53c). The sharp transition into macroporous morphology with increasing molecular weight of polystyrene is remarkable; it is the value of m_1 which causes the difference[4] (see also Sect. 4.3.1). Figure 4.53d shows that initially a large part of the diluent is contained in the gel phase and acts as a normal diluent increasing equilibrium.

[4] In the previous papers, the effect of diluent size was expressed through molar volume of the solvent or diluent, rather than their "polymerization degree" in the number of lattice sites they occupy; the volume of lattice site is usually chosen as equal to the volume of polymer segment. Both ways are transferable to each other, however, the first (molar volume) way leads to a change of the interaction parameter with diluent size ($\chi \rightarrow m_1\chi$) which is not practical.

Results of several other experiments were published by Okay (e.g., [293–295]) and the state-of-the-art by the year 2000 was reviewed in his review. Newer results can be found in the references placed in the introductory part of Sect. 7.2.

Modeling of phase separation was attempted by formulating the Gibbs or Helmholtz energy changes as a function of cross-linking degree which depends on the fraction of cross-linked units or polymerization conversion. In these simulations, deviation from ideality (cyclization, group reactivity obstructions, polydispersity, etc.) were not taken into account, as well as the effect of interface which is unimportant in the case of macrosyneresis but becomes important especially for development of morphology of the microseparated system. Also, the transition over the whole range of semi-dilute region and excluded volume effect was not taken into account. Concerning the mechanism of microphase separation, spinodal decomposition with fast ripening cannot be excluded, especially for very fast reaction, but otherwise, the systems are quite fast (phase separation occurs at low conversion, i.e., at relatively high dilution).

It is to be noted that the size of the pores varies from micron size to nanometer size. For characterization of pores sizes, gas (nitrogen, argon) adsorption methods and mercury porosimetry are used. For styrene–divinylbenzene networks, the pore sizes vary from 10^2 nm if polymeric additives (polystyrene) are used; for poor solvents (precipitants), the typical pore sizes are in the range of tens of nanometers, and for good solvents the pores are small, below 1 nm [281] For these small-pore solids, the specific surface obtained by gas adsorption vary between several hundreds to over thousand m^2/g. Now, these systems are classified as MOFs (metal organic frameworks).

Macroporous hydrogels are important materials for medical applications, especially in tissue engineering [296– 298]. Macroporous hydrogels are prepared by cross-linking polymerization in the presence of additives that are not fully compatible with the polymer. As in the case of hydrophobic system, the choice of additives varies between less compatible low molecular weight compounds, such as water, alcohols, some ketones, oligomeric hydrophilic polymers, solution of salting-out compounds. Very helpful is the control of pore morphology by surface active additives which are important modifiers of pore size and morphology. The cross-linker content also affects the onset of microphase separation and pore volume and size but, generally, for soft hydrogels the cross-linker concentration is kept low, usually not exceeding 3–5%.This means that the type of microphase separation is the χ-syneresis. Interesting are diluents in which the weak segment-segment interactions are combined with strong physical interactions, like hydrogen bonding and positive–negative ion pairing.

The role of thermodynamic interactions will be demonstrated by microsyneresis in PHEMA-water systems. Because the PHEMA gel swells in water only moderately, the effect of cross-linker in the range of 0–2% on equilibrium swelling is negligible, while in better solvents (e.g., dimethylsulfoxide or aqueous solutions of salting-in salts like magnesium perchlorate solution) the effect of cross-linking degree is seen. The critical amount of water above which phase separation starts for gels with concentration of cross-linker about or below 1 mol/%, is about 41 wt%. When this critical amount is surpassed, the gel becomes hazy when swollen but is clear when dry. For 50–60 wt% of water, the gels are opaque, but permanent porosity in dry state is found in gels prepared in the presence 70 wt% and more of water. When the affinity of diluent to the gel is impaired by adding sodium chloride, the critical amount of the solution and the borderlines between hazy, opaque, and permanently porous gels shifts to lower concentration of diluent. If, on contrary, water solutions of salting-in salts [22] is used, these borderlines move to higher concentrations of diluent, and eventually, all gels are clear. More details on the morphology can be found in papers by Karpushkin et al. [149, 299]. Typical morphologies are shown in Fig. 4.55.

Macroscopic appearance of all these samples in their swollen state is white and non-transparent. Gels were polymerized in aqueous NaCl solutions of varying concentrations and then reswollen in water while salt was thoroughly washed away. The gel prepared in the presence of just water is not homogeneous (white in the swollen state) but the pores are not well-developed. If the diluent is just water, the appearance of the macroporous structure is delayed to higher concentrations of water (80 wt%) the morphology is clearly fused-spheres type (Fig. 4.56).

The mechanism of morphologies build-up and their final states for PHEMA-water systems can be understood by analysis of polymerization of HEMA in the presence of water. The examples shown here belong clearly to χ-type microsyneresis. The

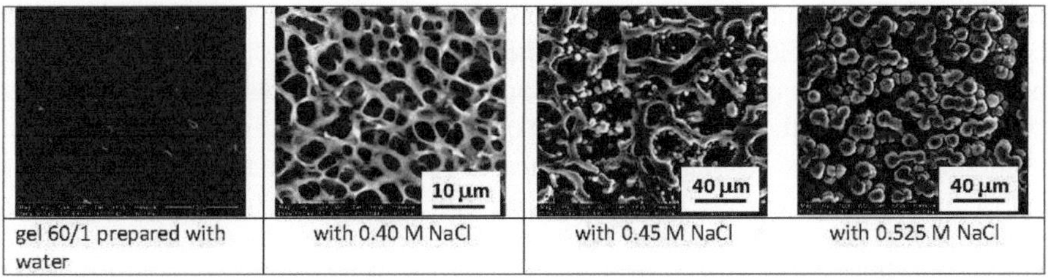

Fig. 4.55 Effect of diluent quality at gel preparation on hydrogel morphology as revealed by ESEM of swollen samples. PHEMA gel with 1 wt% cross-linker in the presence of 60 vol% diluent. The diluent composition is indicated

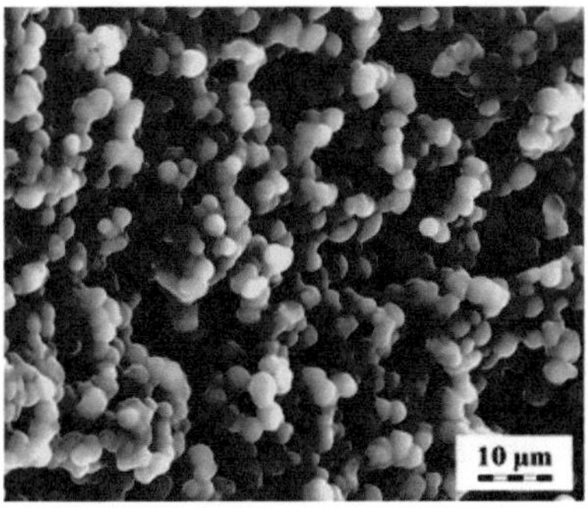

Fig. 4.56 ESEM pictures (swollen state, 2 °C, 507–535 Pa) showing morphology of hydrogel prepared with 80 wt% of water as diluent, observed at equilibrium swollen state in water

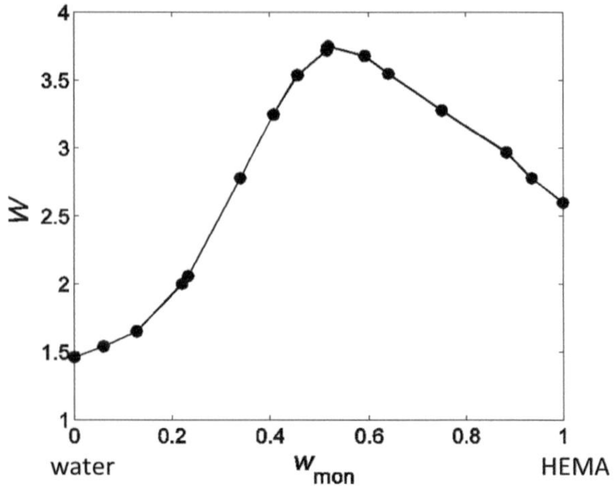

Fig. 4.57 Weight degree of swelling in of cross-linked PHEMA gel (1% cross-linker) in dependence on the weight fraction water in water-HEMA mixture in the gel phase. Reprinted from Ref. [301] with permission

mixture of HEMA monomer with water functions as diluent in the polymerization. The mixtures HEMA-water show up cosolvency which is manifested a maximum degree of swelling of a homogeneous PHEMA gel [300] (Fig. 4.57).

After phase separation starts and the system is being depleted of the HEMA monomer the solvation power, the solvation power of the liquid mixture deteriorates rapidly and distinct phase separated structure is developed. Figure 4.58 shows the difference between the water content in the whole system (dashed dependence) and the swelling capacity of the gel as a function of conversion. Clearly, for $w_w = 0.4$, the swelling capacity of the gel over the whole conversion range is higher than the amount of diluent (red full vs. red dashed lines) and full line touches dashed one only at the end of the reaction. For $w_w = 0.6$, polymerization proceeds first in homogeneous system and phase separation starts only later and the size of separated domains is limited by the formed network. The system of $w_w = 0.8$ is phase separated from the beginning very likely before or in close vicinity of the point.

Water soluble oligomers and oligomers of poly(oxyethylene) diols or copolymers of ethylene glycol with propylene glycol are also used for preparation of macroporous gels but systematic studies are not available. Kumacheva et al. [302] showed the way of combining compatibility and interfacial tension effects to change the gel morphology.

In principle, phase separation during network formation could occur by the form of volume phase transition between two gel phases. Suitable conditions for such type of phase separation exist as soon as the gel point is surpassed and the coexistence curve and the spinodal shoot to infinity (+ ∞ or $-\infty$, depending on whether the systems is UCST or LCST type). If gelation occurs at very low conversions, phase separation would occur almost immediately after passing the gel point; otherwise the mechanism would depend on whether the van der Waals loop could develop, in other words, on the dependences of $\Delta\mu_{1,\text{el,n}}$ and, $\Delta\mu_{1,\text{el,n}}$(i.e., $g(\phi_2(\alpha))$ and ν_e)

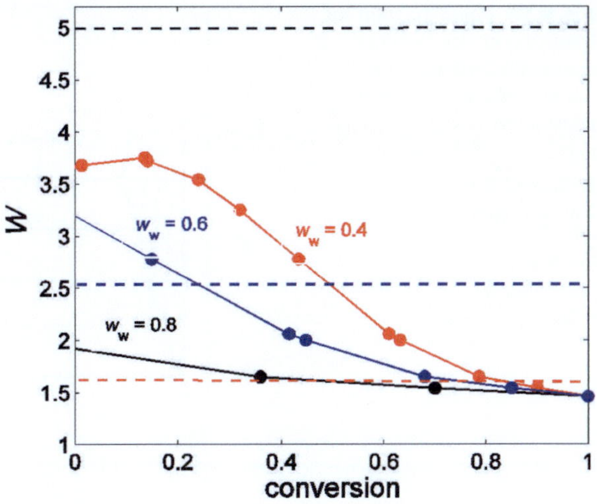

Fig. 4.58 Dependence of the maximum degree of swelling in water-HEMA mixtures as a function of conversion of double bonds for three levels of water content indicated by dashed lines. The degree of swelling (full curves) and the corresponding water content (dashed curves) are compared. Reprinted from Ref. [301] with permission

on conversion. There is a little hope that these two types could be distinguished experimentally.

4.7.4 Conditions for Volume Phase Transition

Phase separation in gels proceeds not only by macro and microsyneresis, but also by volume phase transition by which the gel of a certain degree of swelling is abruptly transformed into a gel of different degree of swelling, and theoretically, these two gel phases co-exists with the pure solvent phases. Such three-phase equilibrium in a binary system is invariant of temperature, i.e., for a given system, it can exist only at a certain temperature, if the system fulfills certain conditions. Research on systems exhibiting volume phase transition (or collapse) and their application is nowadays very extensive and very interesting. Thousands of papers and many reviews in journal, book, and monograph forms exist on this subject. The goal of this short section is just to remind to relation of the phase volume transition to polymer thermodynamics.

This type of phase separation was first theoretically predicted by Dušek and Patterson in 1968 by analyzing chemical potential dependences on polymer volume fraction by varying the polymer–solvent interaction parameter, cross-link density, and dilution of the system during network formation. For certain combinations, the characteristic van der Waals loop appeared in the solvent chemical potential dependence on ϕ_2 which indicated a possibility of three-phase equilibrium. By solving the conditions for phase equilibrium, $\Delta\mu_1' = \Delta\mu_1'' = 0$ and $\Delta\mu_2' = \Delta\mu_2''$ the phase diagram was obtained (Fig. 4.59).

It was found that it would be difficult to prepare homogeneous gels at such dilutions and sufficiently high cross-link density and that very probably the reaction-induced phase separation would interfere and prevent preparation of a clear gel. However, at

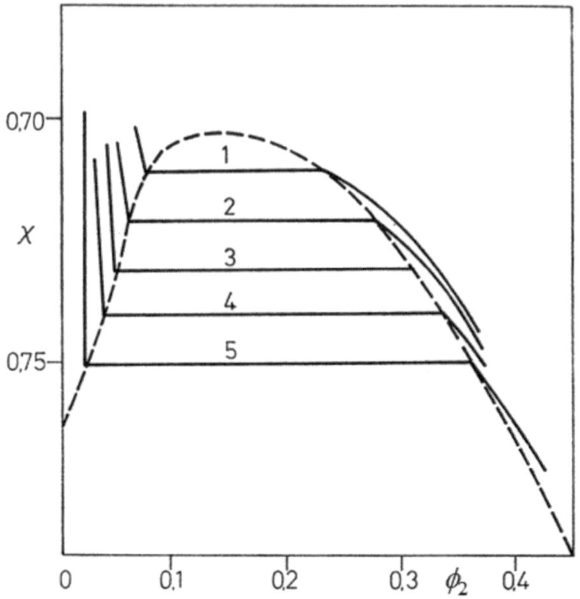

Fig. 4.59 Phase diagram for transition in free swelling caused the change of the interaction parameter χ at constant values of $\nu_e V_{1,ls} = 0.07$, $(\phi_2^0)^{2/3} = 0.05$. Reprinted from Fig. 4.2, with permission

that time only simple UCST systems were considered and in combination with the FH mixing contribution and χ-parameter independent of concentration. However, it was found that mechanical extension modes facilitate phase transition, as shown in Fig. 4.60.

Also, it was found that concentration dependence of χ facilitated the transition in softening the limits of high ν_e combination with low ϕ_2^0; however this part was later removed on request to make the manuscript shorter [bookpref]. Almost 10 years later, this transition experimentally was discovered and interpreted independently as the volume phase transition by Tanaka et al. [303, 304]. He was able to see the transition because he was working with a charged system and water was the solvent—i.e., a combination of swelling increasing ionic Donnan term with hydrophilic-hydrophobic interactions characteristic for aqueous solutions. This breakthrough discovery started a large-scale research on responsive gels in various countries. The wide stream of ionic systems was then amended by aqueous systems of poly(N-isopropylacrylamide) (PNIPAM) fixed with a small amount of cross-links to prevent its dissolution when PNIPAM gel is in its expanded state.

Breaks in the degree of swelling as well as in related properties like modulus of elasticity, or dielectric losses were found around a certain temperature, pH, solvent-non-solvent compositions, and other stimuli. Examples of recent reviews and comments can be found in Refs, [305, 306, [5, 193, 307].

As experimental examples of volume phase transition the classical results by Shibayama and Tanaka [195] of temperature induced transition of PNIPAM with variable content of ionized carboxyl groups are shown in Fig. 4.61.

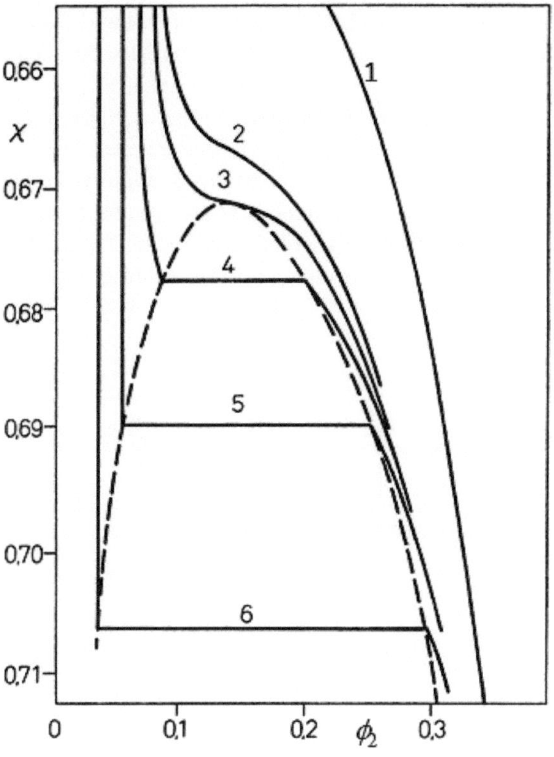

Fig. 4.60 Phase diagram calculated for volume phase separating system by the change of the interaction parameter χ at different uniaxial extensions, $\nu_e V_{1,ls} = 0.07; (\phi_2^0)^{2/3} \Lambda^{-1}$: *1* 0.050, *2* 0.025, *3* 0.0235; *4* 0.0211, *5* 0.0175, *6* 0.0133. Reprinted from Fig. 4.3, with permission

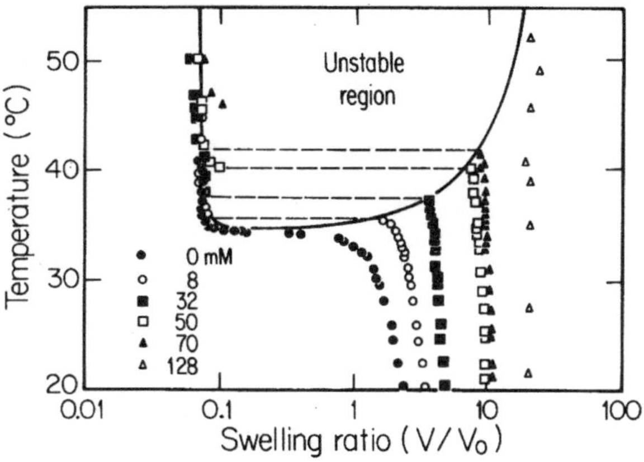

Fig. 4.61 Volume swelling relative to the volume at preparation (V/V_0) of ionized poly(N-isopropylacrylamide) in dependence on the degree of ionization (data inserted in the figure) Reprinted from [195] Fig. 4.7 with permission

Another way of inducing the phase volume transition of hydrogels is changing the solvent quality, for instance to use solvent-non-solvent mixtures. Figure 4.62 shows the variation of swelling degree and shear modulus for cross-linked acrylamide/sodium methacrylate copolymer gels as a function of composition of mixtures acetone/water (acetone is a poor solvent and water is a good solvent). It can be seen that not only the degree of swelling but also the shear modulus show up a discontinuity as the acetone concentration increases [197].

According to the thermodynamic, a phase transition of the first order should be associated with a discontinuity in the dependence of swelling degree on the external variable, however, when changing gradually the external variable, the discontinuity is not seen. On the other hand, some systems theoretically not fulfilling the conditions for phase transitions are characterized by a quite sharp but continuous volume change with the continuous change of the external variable. The problem was analyzed by

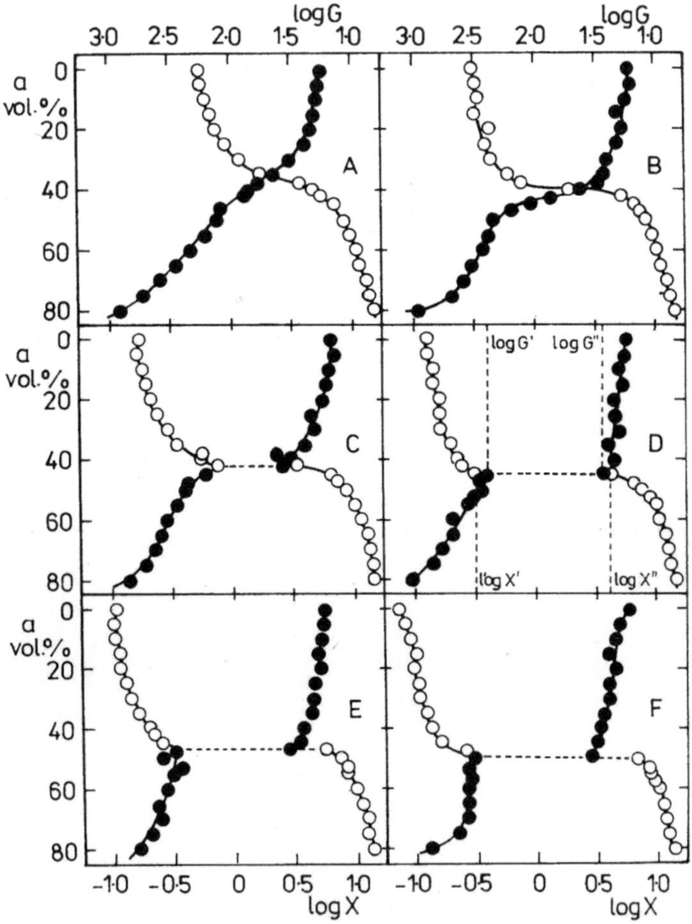

Fig. 4.62 Changes of relative degree of swelling $X = \phi_2^0/\phi_2$ and equilibrium shear modulus G on the composition of the swelling liquid (vol% of acetone) for the cross-linked acrylamide/sodium methacrylate copolymers with various molar fraction of sodium methacrylate; content of sodium methacrylate in molar %: A 0, B 0.004, C 0.008, D 0.012, E 0.016, F 0.024. Reprinted from [197] Fig. 4.6 with permission

Fig. 4.63 Appearance of a sample of polyacrylamide/sodium methacrylate gel at critical composition of acetone–water mixture. Reprinted from [309] (Fig. 4.3) with permission of American Chemical Society

Wu and Zhou [308] who came to a conclusion that the continuity in contrast to expected discontinuity is caused by distribution of lengths of network chains which resembles distribution of subnetworks of different cross-link density. This is possible especially at high degrees of swelling, but then the mean field model working with average segment densities is not adequate. However, the explanation may be even more prosaic. Changing the temperature means heat transfer which takes a certain time and the volume change spans over a certain time. However, it is likely that the transition proceeds through the stage of microphase separation, or, as illustrated by Fig. 4.63 through the stage of evolution of mechanical instabilities (corrugations, etc.) and also the transition between energies of expanded and collapsed equilibrium proceeds through intermediate states.

What are the parameters in the swelling equation that control development of the van der Waals loop on the $\Delta\mu_{1,\mathrm{sw}}$ *versus* ϕ_2 dependence, i.e., possibility of a volume phase transition? For a crude analysis, we will use the simple FH model amended by the simple Donnan term for salt-free system. As follows from Sect. 4.4, the relevant equation reads

$$\frac{\Delta\mu_{1,\mathrm{mix,n}}}{RT} = \left(\frac{1}{m_1}\ln\phi_1 + \frac{\phi_2}{m_1} + \frac{f_\mathrm{e}-2}{f_\mathrm{e}}\nu_\mathrm{e} V_{1\mathrm{s}}\phi_2 + \left(g(\phi_2) - \frac{\partial g(\phi_2)}{\partial\phi_2}\phi_1\right)\phi_2^2\right) \tag{4.177}$$

$$\frac{\Delta\mu_{1,\mathrm{el,n}}}{RT} = \nu_\mathrm{e} V_{1\mathrm{s}}\left(\phi_2^{1/3}\left(\phi_2^0\right)^{2/3} - \phi_2\right) \quad \text{(for affine network)} \tag{4.178}$$

$$\frac{\Delta\mu_{1.\mathrm{ion}}}{RT} = \frac{V_{1\mathrm{s}}\alpha_\mathrm{i}\rho_2\phi_2}{M_0} \to \alpha_\mathrm{i}\phi_2 \quad (\text{if } M_0/\rho_2 = V_{1\mathrm{s}}) \tag{4.179}$$

$$\frac{\Delta\mu_{1.\mathrm{sw}}}{RT} = \frac{\Delta\mu_{1.\mathrm{mix,n}}}{RT} + \frac{\Delta\mu_{1.\mathrm{el,n}}}{RT} + \frac{\Delta\mu_{1.\mathrm{ion}}}{RT} = \ln a_1 \tag{4.180}$$

For equilibrium with pure solvent $a_1 = 1$ and the phase transition occurs at a certain external conditions; when $a_1 < 1$(solvent vapors, activity of solvent decrease by non-permeable component), the transition can occur at different conditions and can have different width.

The independently controllable parameters are:

- m_1 size of the solvent molecule (increasing size means not only higher molecular weight, but proportional increase of the respective interacting surfaces),
- ν_e concentration of elastically active network chains,
- $g_0, g_1, g_2,\ldots$ coefficients of the power expansion of the interaction function $g(\phi_2)$
- $\phi_2^0 = 1 - \phi_1^0$; $\phi_2^0 = 1 - \phi_1^0$ the volume fraction of diluent at network formation,
- α_i fraction of charged monomer units.

To get information on the system as to whether it can show up volume phase transition, the Maxwell solution and examination of the corresponding plot are very instructive. The Maxwell condition for phase equilibrium between two polymer phases and a solvent or its vapor can be obtained from the Gibbs–Duhem equation. It is determined by the condition

$$\int_{\phi_2^\alpha}^{\phi_2\beta} \frac{1}{\phi_2^2}\Delta\mu_{1.\mathrm{sw}} = 0 \tag{4.181}$$

The roots ϕ_2^α and ϕ_2^β indicating the composition of coexisting gel phases, are shown in Fig. 4.64 where one can also see the equality of red and blue striped areas.

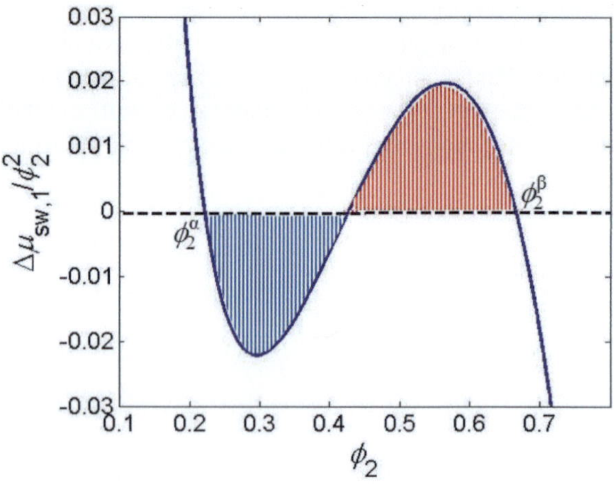

Fig. 4.64 Maxwell plot and calculation of composition of coexisting phases

Examination of development of the vd W loop is a very convenient way of estimation of system inclination to exhibit volume phase transitiona hint to what to do in order to develop the van der Waals loop and to shift to $a_1 = 1$ region, or possibly to a fixed $a_1 < 1$ level (application, for instance, to a sensor indicating reaching a certain humidity level). The following series of Fig. (4.65a–e) indicates the effect of changing external variables.

The above examples show that to design a gel showing up phase volume transition manifested by an abrupt and large change in the degree of swelling one has to consider a very intricate interplay of various factors. The effect of various variables was demonstrated here using the most simple model of polymer gel. However, one can use more structure and interaction-based models for various parts of $\Delta\mu_{1,sw}$ some of which were discussed in the preceding sections.

4.8 Conclusions

Swelling remains one the most important properties of living nature and important property of many man-made materials. It is the most characteristic property of hydrogels. Three quarters century ago, we were fascinated by the discovery of rational relationship of swelling to molecular structure of man-made materials, today we realize how little we know about swelling when we move toward natural materials and their synthetic analogues. However, since three quarters century ago we continue perceiving swelling as competition between forces trying to expand opposed by forces trying to hold together—mixing and elasticity. In polymer science these two aspects went often their own ways—polymer solutions and rubbery materials. Today they seem to come again closer to each other. Despite of that in many contemporary research reports the mixing part of swelling is reduced to the simplest model of linear polymer of very large molecular weight in an almost ideal solvent. That is why in this short review more space than usually was allotted to mixing effects. It seems to be helpful to view swollen systems as interacting two networks—one of covalent bonds if they are present and the second of temporary bonds governed by physical interactions whose lifetime varies from picoseconds to months or years where it becomes indistinguishable from the covalent network.

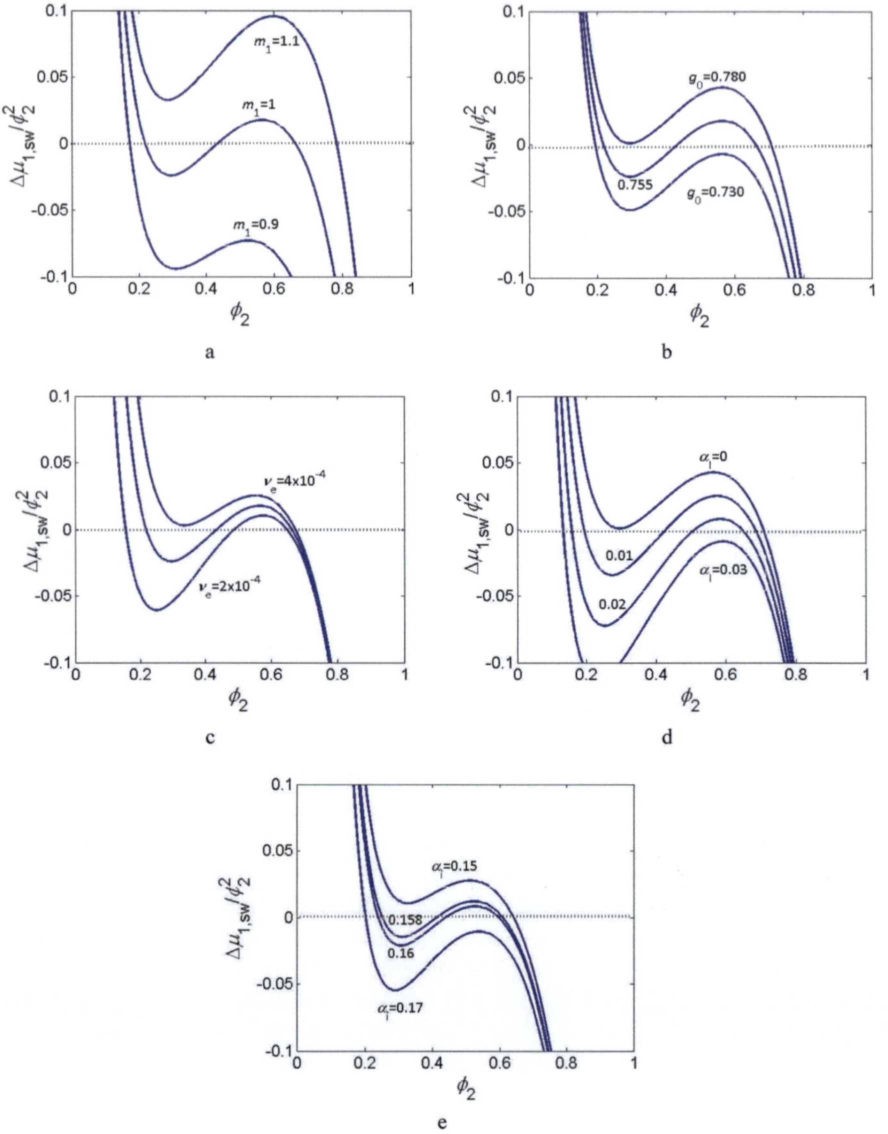

Fig. 4.65 **a–e** Effect of various factors on shape and position of the reduced chemical potential of the solvent: **a** size of the solvent molecule; **b** parameter g_0 of the interaction function; **c** concentration of elastically active network chains; **d** fraction of ionized groups as modifiers; **e** fraction of ionized groups in a polyelectrolyte gel; values of parameters used in simulation are given in Table 4.9

References

1. P.J. Flory *Principles of Polymer Chemistry* (1953)
2. K. Dušek, W. Prins. Structure and elasticity of non-crystalline polymer networks, in *Fortschritte der Hochpolymeren-Forschung* (Springer Berlin Heidelberg, Berlin, Heidelberg 1969)
3. M. Rubinstein et al., Elastic modulus and equilibrium swelling of polyelectrolyte gels. Macromolecules **29**, 398–406 (1996)
4. H. Li, *Smart Hydrogel Modelling* (2009)
5. J.R. Saunders et al., Modeling theories of intelligent hydrogel polymers. J. Comput. Theor. Nanosci. **5**, 1942–1960 (2008)
6. M. Quesada-Pérez et al., Gel swelling theories: the classical formalism and recent approaches. Soft Matter **7**, 10536–10547 (2011)
7. M. Přádný et al., Macroporous 2-hydroxyethyl methacrylate hydrogels of dual porosity for cell cultivation: morphology, swelling, permeability, and mechanical behavior. J. Polym. Res. **21**, 1–12 (2014)
8. R.W. Brotzman, B.E. Eichinger, Volume dependence of the elastic equation of state. 2. Solution-cured poly(dimethylsiloxane). Macromolecules, **14**(5), 1445–1448 (1981)
9. R.W. Brotzman, B.E. Eichinger, Volume dependence of the elastic equation of state. 3. Bulk-cured poly(dimethylsiloxane). Macromolecules **15**(2), 531–535 (1982)
10. R.W. Brotzman Jr., B.E. Eichinger, Swelling of model poly(dimethylsiloxane) networks. Macromolecules **16**(7), 1131–1136 (1983)
11. N.A. Neuburger, B.E. Eichinger, Critical experimental test of the Flory-Rehner theory of swelling. Macromolecules **21**, 3060–3070 (1988)
12. F. Horkay, A.M. Hecht, E. Geissler, The effects of cross-linking on the equation of state of a polymer solution. J. Chem. Phys. **91**, 2706–2711 (1989)
13. R. Koningsveld, W.H. Stockmayer, E. Nies. *Polymer Phase Diagrams: A Textbook* (2001)
14. K. Dusek, et al., *Responsive Gels: Volume Transitions 1* (Springer Berlin Heidelberg, 2014)
15. K. Dušek, M. Dušková-Smrčková, J. Šomvársky, Effect of constraints on swelling of polymer networks. Macromol. Symp. **358**(1), 120–127 (2015)
16. L.V. Mohite, V.A. Juvekar, Quantification of thermodynamics of aqueous solutions of poly(ethylene glycols): role of calorimetry. Fluid Phase Equilib. **278**, 41–53 (2009)
17. Y. Suzuki et al., Pressure effects on phase diagrams in mixtures of poly(ethyl acrylate) and poly(vinylidene fluoride). J. Polym. Sci.: Polym. Lett. Ed. **20**, 563–568 (1982)
18. R. Moerkerke et al., Phase transitions in swollen networks. Macromolecules **28**(4), 1103–1107 (1995)
19. P.J. Flory, H. Daoust, Osmotic pressures of moderately concentrated polymer solutions. J. Polym. Sci. **25**, 429–440 (1957)
20. B. Erman, P.J. Flory, Critical phenomena and transitions in swollen polymer networks and in linear macromolecules. Macromolecules **19**(9), 2342–2353 (1986)
21. K. Šolc et al., "Zero" and "Off-Zero" critical concentrations in solutions of polydisperse polymers with very high molar masses. Collect. Czech. Chem. Commun. **60**, 1661–1688 (1995)
22. H. Schäfer-Soenen, et al., Zero and off-zero critical concentrations in systems containing polydisperse polymers † with very high molar masses. 2. The system water-poly(vinyl methyl ether). Macromolecules **30**, 410–416 (1997)
23. M. Rubinstein, R.H. Colby, *Polymer Physics* (Oxford University Press, 2003)
24. P. Knychala et al., 50th Anniversary perspective: phase behavior of polymer solutions and blends. Macromolecules **50**, 3051–3065 (2017)
25. Y. Guerrieri, et al., *A Survey of Equations of State for Polymers* (2012)
26. E.A. Guggenheim, Statistical thermodynamics of mixtures with zero energies of mixing. Proc. R. Soc. London. Ser. A. Math. Phys. Sci. **183**, 203–212 (1944)
27. E.A. Guggenheim, Statistical thermodynamics of mixtures with non-zero energies of mixing. Proc. R. Soc. London. Ser. A. Math. Phys. Sci. **183**, 213–227 (1944)

28. E.A. Guggenheim, *Mixtures: The Theory of the Equilibrium Properties of Some Simple Classes of Mixtures Solutions and Alloys* (1952)
29. H. Tompa, Quasi-chemical treatment of mixtures of oriented molecules. J. Chem. Phys. **21**, 250–258 (1953)
30. H. Tompa, *Polymer Solutions*, vol. xiv (Butterworths Scientific Publications, London, United Kingdom, 1956), 325 p
31. K. Dušek, Quasichemical equilibrium approach to crosslinked polymer solutions. J. Polym. Sci. Part C: Polym. Symp. **39**(1), 83–106 (1972)
32. K. Dušek, Parameters of the swelling equation and network structure. Faraday Discuss. Chem. Soc. **57**, 101–109 (1974)
33. M.M. Prange, H.H. Hooper, J.M. Prausnitz, Thermodynamics of aqueous systems containing hydrophilic polymers or gels. AIChE J. **35**(5), 803–813 (1989)
34. É.D.D. Oliveira, A.S. Silva, R.F.S. Freitas, Contributions to the thermodynamics of polymer hydrogel systems. Polymer **45**, 1287–1293 (2004)
35. R.O.R. Costa, R.F.S. Freitas, Phase behavior of poly(N-isopropylacrylamide) in binary aqueous solutions. Polymer **43**, 5879–5885 (2002)
36. J. Yang et al., A generic molecular thermodynamic model for linear and branched polymer solutions in a lattice. Fluid Phase Equilib. **244**, 188–192 (2006)
37. J. Yang et al., Liquid–liquid equilibria of polymer solutions with oriented interactions. Fluid Phase Equilib. **249**, 192–197 (2006)
38. J. Yang et al., A molecular thermodynamic model for binary lattice polymer solutions. Polymer **47**, 5187–5195 (2006)
39. Y. Huang et al., A molecular thermodynamic model for the swelling of thermo-sensitive hydrogels. Fluid Phase Equilib. **263**, 96–101 (2008)
40. K. Poschlad, S. Enders, Thermodynamics of aqueous solutions containing poly (N-isopropylacrylamide). J. Chem. Thermodyn. **43**, 262–269 (2011)
41. M.B. Yu, H. Nishiumi, J.D.S. Arons, Thermodynamics of phase separation in aqueous solutions of polymers. Fluid Phase Equilibr. **83**, 357–364 (1993)
42. W. Li, Z. Zhu, M. Li, Measurement and calculation of liquid-liquid equilibria of binary aqueous polymer solutions. Chem. Eng. J. **78**, 179–185 (2000)
43. J.E.G. Lipson, A Born–Green–Yvon integral equation treatment of incompressible lattice mixtures. J. Chem. Phys. **96**, 1418–1425 (1992)
44. J.E.G. Lipson, P.K. Brazhnik. *A BornGreenYvon Integral Equation Treatment of Compressible Lattice Mixtures* (1993)
45. E.A. Clark, J.E.G. Lipson, LCST and UCST behavior in polymer solutions and blends. Polymer **53**(2), 536–545 (2012)
46. K.F. Freed, New lattice model for interacting, avoiding polymers with controlled length distribution. J. Phys. A **18**, 871–877 (1985)
47. K.F. Freed, J. Dudowicz, Lattice cluster theory for pedestrians: the incompressible limit and the miscibility of polyolefin blends. Macromolecules **31**, 6681–6690 (1998)
48. M.G. Bawendi, K.F. Freed, Statistical mechanics of the packing of rods on a lattice: cluster expansion for systematic corrections to mean field. J. Chem. Phys. **85**(5), 3007–3022 (1986)
49. M.G. Bawendi, K.F. Freed, U. Mohanty, A lattice model for self-avoiding polymers with controlled length distributions. II. Corrections to flory–huggins mean field. J. Chem. Phys. **84**, 7036–7047 (1986)
50. M.G. Bawendi, K.F. Freed, U. Mohanty, A lattice field theory for polymer systems with nearest-neighbor interaction energies. J. Chem. Phys. **87**(9), 5534–5540 (1987)
51. A.M. Nemirovsky, M.G. Bawendi, K.F. Freed, Lattice models of polymer solutions: Monomers occupying several lattice sites. J. Chem. Phys. **87**, 7272–7284 (1987)
52. A.I. Pesci, K.F. Freed, Lattice models of polymer fluids: Monomers occupying several lattice sites. II. Interaction energies. J. Chem. Phys. **90**, 2003–2016 (1989)
53. K.F. Freed, J. Dudowicz, Lattice cluster theory for pedestrians: models for random copolymer blends. Macromol. Symp. **149**, 11–16 (2000)

54. W.G. Madden, A.I. Pesci, K.F. Freed, Phase equilibria of lattice polymer and solvent: tests of theories against simulations. Macromolecules **23**(4), 1181–1191 (1990)
55. J. Dudowicz, K.F. Freed, Effect of monomer structure and compressibility on the properties of multicomponent polymer blends and solutions: 1. Lattice cluster theory of compressible systems. Macromolecules **24**(18), 5076–5095 (1991)
56. J. Dudowicz, M.S. Freed, K.F. Freed, Effect of monomer structure and compressibility on the properties of multicomponent polymer blends and solutions. 2. Application to binary blends. Macromolecules **24**(18), 5096–5111 (1991)
57. K.F. Freed, Extension of lattice cluster theory to strongly interacting, self-assembling polymeric systems. J. Chem. Phys. **130**(6), 061103 (2009)
58. W.-S. Xu, K.F. Freed, Lattice cluster theory for polymer melts with specific interactions. J. Chem. Phys. **141**(4), 044909 (2014)
59. J. Dudowicz, K.F. Freed, J.F. Douglas, Solvation of polymers as mutual association. I. General theory. J. Chem. Phys. **138**(16), (2013)
60. Y. Hu et al., Double-lattice model for binary polymer solutions. Macromolecules **24**(15), 4356–4363 (1991)
61. Y. Hu et al., Binary liquid-liquid equilibria from a double-lattice model. Fluid Phase Equilib. **67**, 65–86 (1991)
62. Y. Hu et al., Molecular thermodynamics of polymer solutions. Fluid Phase Equilib. **83**, 289–300 (1993)
63. Y. Hu, H. Liu, J.M. Prausnitz, Equation of state for fluids containing chainlike molecules. J. Chem. Phys. **104**, 396–404 (1996)
64. Y. Hu, H. Liu, Y.-Y. Shi, Molecular thermodynamic theory for polymer systems. I. A close-packed lattice model. Fluid Phase Equilib. **117**, 100–106 (1996)
65. I. Prigogine, N.J. Trappeniers, V. Mathot, Statistical thermodynamics of r-MERS and r-MER solutions. Discuss. Faraday Soc. **15**, 93–107 (1953)
66. D. Patterson, Free volume and polymer solubility. A qualitative view. Macromolecules **2**(6), 672–677 (1969)
67. D. Patterson, A. Robard, Thermodynamics of polymer compatibility. Macromolecules **11**(4), 690–695 (1978)
68. I.C. Sanchez, R.H. Lacombe, Statistical thermodynamics of polymer solutions. Macromolecules **11**, 1145–1156 (1978)
69. I.C. Sanchez, A.C. Balazs, Generalization of the lattice-fluid model for specific interactions. Macromolecules **22**, 2325–2331 (1989)
70. P.J. Flory, R.A. Orwoll, A.A. Vrij, Statistical thermodynamics of chain molecule liquids. I. An equation of state for normal paraffin hydrocarbons. J.Am. Chem. Soc. **86**, 3507–3514 (1964)
71. P.J. Flory, R.A. Orwoll, A.A. Vrij, Statistical thermodynamics of chain molecule liquids. II. Liquid mixtures of normal paraffin hydrocarbons. J. Am. Chem. Soc. **86**, 3515–3520 (1964)
72. P.J. Flory, Statistical thermodynamics of liquid mixtures. J. Am. Chem. Soc. **87**, 1833–1838 (1965)
73. P.J. Flory, B.E. Eichinger, R.A. Orwoll, Thermodynamics of mixing polymethylene and polyisobutylene. Macromolecules **1**(3), 287–288 (1968)
74. R.A. Orwoll, P.J. Flory, Equation- of-state parameters for normal alkanes. Correlation with chain length. J. Am. Chem. Soc. **89**(26), 6814–6822 (1967)
75. R.A. Orwoll, P.J. Flory, Thermodynamic properties of binary mixtures of n-alkanes. J. Am. Chem. Soc. **89**(26), 6822–6829 (1967)
76. B.E. Eichinger, P.J. Flory, Thermodynamics of polymer solutions. Part 1.—Natural rubber and benzene. Trans. Faraday Soc. **64**, 2035–2052 (1968)
77. B.E. Eichinger, P.J. Flory, Thermodynamics of polymer solutions. Part 2.—Polyisobutylene and benzene. Trans. Faraday Soc. **64**, 2053–2060 (1968)
78. B.E. Eichinger, P.J. Flory, Thermodynamics of polymer solutions. Part 3.—Polyisobutylene and cyclohexane. Trans. Faraday Soc. **64,** 2061–2065 (1968)
79. B.E. Eichinger, P.J. Flory, Thermodynamics of polymer solutions. Part 4.—Polyisobutylene and n-pentane. Trans. Faraday Soc. **64**, 2066–2072 (1968)

80. H. Höcker, P.J. Flory, Thermodynamics of polystyrene solutions. Part 2.—Polystyrene and ethylbenzene. Trans. Faraday Soc. **67**, 2270–2274 (1971)
81. A. Abe, P.J. Flory, The thermodynamic properties of mixtures of small, nonpolar molecules. J. Am. Chem. Soc. **87**(9), 1838–1846 (1965)
82. L.A. Kleintjens, R. Koningsveld, Liquid-liquid phase separation in multicomponent polymer systems. Colloid Polym. Sci. **258**(6), 711–718 (1980)
83. K. Gauter, R. Heidemann, A proposal for parametrizing the Sanchez-Lacombe equation of state. Ind. Eng. Chem. Res. **39**, 1115–1117 (2000)
84. A.M. Kisselev, E. Manias, Phase behavior of temperature-responsive polymers with tunable LCST: An equation-of-state approach. Fluid Phase Equilib. **261**, 69–78 (2007)
85. Y.T. Wu et al., A modified NRTL equation for the calculation of phase equilibrium of polymer solutions. Fluid Phase Equilib. **121**, 125–139 (1996)
86. K. Dušek, Quasichemical approach to crosslinked polymer solutions and the swelling equation for polycondensation networks. J. Polym. Sci. Polym. Phys. Ed. **12**(6), 1089–1107 (1974)
87. S.K. Ryu, Y.C. Bae, Understanding phase behaviors of multicomponent polymer mixtures based on a molecular thermodynamic framework combined with molecular simulation. Polymer **53**, 1339–1346 (2012)
88. P.E. Scesney, Mobile-electron ising ferromagnet. Phys. Rev. B **1**, 2274–2288 (1970)
89. J. Dudowicz, K.F. Freed, W.G. Madden, Role of molecular structure on the thermodynamic properties of melts, blends, and concentrated polymer solutions: comparison of Monte Carlo simulations with the cluster theory for the lattice model. Macromolecules **23**(22), 4803–4819 (1990)
90. K.S. Oh et al., Effect of cross-linking density on swelling behavior of NIPA Gel particles. Macromolecules **31**, 7328–7335 (1998)
91. S.Y. Oh, Y.C. Bae, Closed miscibility loop phase behavior of polymer solutions. Polymer **49**, 4469–4474 (2008)
92. S.Y. Oh, Y.C. Bae, Molecular thermodynamics approach for polymer–polymer miscibility. Eur. Polymer J. **46**, 1328–1333 (2010)
93. S.Y. Oh, Y.C. Bae, Understanding liquid mixture phase miscibility via pair energy parameter behaviors with respect to temperatures determined from molecular simulations. J. Phys. Chem. B **115**(19), 6051–6060 (2011)
94. S.Y. Oh, Y.C. Bae, Role of intermolecular interactions for upper and lower critical solution temperature behaviors in polymer solutions: molecular simulations and thermodynamic modeling. Polymer **53**, 3772–3779 (2012)
95. S.Y. Oh, H.J. Kim, Y.C. Bae, Molecular thermodynamic analysis for phase transitions of linear and cross-linked poly(N-isopropylacrylamide) in water/2-propanol mixtures. Polymer **54**, 6776–6784 (2013)
96. S.Y. Oh, H.E. Yang, Y.C. Bae, Molecular simulations and thermodynamic modeling for closed-loop phase miscibility of aqueous PEO solutions. Macromol. Res. **21**, 921–930 (2013)
97. G. Stell, Y. Zhou, Chemical association in simple models of molecular and ionic fluids. J. Chem. Phys. **91**, 3618–3623 (1989)
98. Y. Zhou, G. Stell, Chemical association in simple models of molecular and ionic fluids. II. Thermodynamic properties. J. Chem. Phys. **96**, 1504–1506 (1992)
99. Y. Zhou, G. Stell, Chemical association in simple models of molecular and ionic fluids. III: the cavity function. J. Chem. Phys. **96**, 1507–1515 (1992)
100. Q. Yan, H. Liu, Y. Hu, Analytical expressions of Helmholtz function of mixing for Ising model. Fluid Phase Equilib. **218**, 157–161 (2004)
101. D. Zhi et al., A molecular thermodynamic model for temperature- and solvent-sensitive hydrogels, application to the swelling behavior of PNIPAm hydrogels in ethanol/water mixtures. Chem. Eng. Sci. **65**, 3223–3230 (2010)
102. D. Zhi et al., Molecular thermodynamic model for swelling behavior and volume phase transition of multi-responsive hydrogels. Fluid Phase Equilib. **312**, 106–115 (2011)
103. E.A. Müller, K.E. Gubbins, Molecular-based equations of state for associating fluids: a review of SAFT and related approaches. Ind. Eng. Chem. Res. **40**, 2193–2211 (2001)

104. I.G. Economou, Statistical associating fluid theory: a successful model for the calculation of thermodynamic and phase equilibrium properties of complex fluid mixtures. Ind. Eng. Chem. Res. **41**, 953–962 (2002)
105. W.G. Chapman et al., SAFT: equation-of-state solution model for associating fluids. Fluid Phase Equilib. **52**, 31–38 (1989)
106. W.G. Chapman et al., New reference equation of state for associating liquids. Ind. Eng. Chem. Res. **29**(8), 1709–1721 (1990)
107. S.H. Huang, M. Radosz, Equation of state for small, large, polydisperse, and associating molecules: extension to fluid mixtures. Ind. Eng. Chem. Res. **30**, 1994–2005 (1991)
108. J. Gross, G. Sadowski, Perturbed-chain SAFT: an equation of state based on a perturbation theory for chain molecules. Ind. Eng. Chem. Res. **40**, 1244–1260 (2001)
109. G. Sadowski, Thermodynamics of polymer systems. Macromol. Symp. **206**(1), 333–346 (2004)
110. K. Albers, G. Sadowski, Minimal experimental data set required for estimating PCP-SAFT parameters. Ind. Eng. Chem. Res. **50**, 11746–11754 (2011)
111. G. Sadowski, Modeling of polymer phase equilibria using equations of state, in *Polymer Thermodynamics: Liquid Polymer-Containing Mixtures*, ed. by B.A. Wolf, S. Enders (Springer Berlin Heidelberg, Berlin, Heidelberg, 2011), pp. 389–418
112. M.C. Arndt, G. Sadowski, Modeling poly(N-isopropylacrylamide) hydrogels in water/alcohol mixtures with PC-SAFT. Macromolecules **45**(16), 6686–6696 (2012)
113. S.-J. Chen, I.G. Economou, M. Radosz, Phase behavior of LCST and UCST solutions of branchy copolymers: experiment and SAFT modelling. Fluid Phase Equilib. **83**, 391–398 (1993)
114. S.H. Huang, M. Radosz, Equation of state for small, large, polydisperse, and associating molecules. Ind. Eng. Chem. Res. **29**, 2284–2294 (1990)
115. A.N. Gent, A new constitutive relation for rubber. Rubber Chem. Technol. **69**, 59–61 (1996)
116. G.N.I. Clark et al., Developing optimal Wertheim-like models of water for use in statistical associating fluid theory (SAFT) and related approaches. Mol. Phys. **104**(22–24), 3561–3581 (2006)
117. G.N.I. Clark et al., Modeling and understanding closed-loop liquid−liquid immiscibility in aqueous solutions of poly(ethylene glycol) using the SAFT-VR approach with transferable parameters. Macromolecules **41**(17), 6582–6595 (2008)
118. K. Dušek, M. Dušková-Smrčková, Network structure formation during crosslinking of organic coating systems. Prog. Polym. Sci. **25**(9), 1215–1260 (2000)
119. K. Dušek, M. Dušková-Smrčková, Modeling of polymer network formation from preformed precursors. Macromol. React. Eng. **6**(11), 426–445 (2012)
120. F. Sciortino, C. De Michele, J.F. Douglas, Growth of equilibrium polymers under non-equilibrium conditions. J. Phys. Condens. Matter **20**, 155101 (2008)
121. E.B. Stukalin, J.F. Douglas, K.F. Freed, Multistep relaxation in equilibrium polymer solutions: a minimal model of relaxation in "complex" fluids. J. Chem. Phys. **129**(9), 094901 (2008)
122. S.K. Kumar, J.F. Douglas, Gelation in physically associating polymer solutions. Phys. Rev. Lett. **87**, 188301 (2001)
123. J. Dudowicz, K.F. Freed, J.F. Douglas, Lattice model of living polymerization. I. Basic thermodynamic properties. J. Chem. Phys. **111**(15), 7116–7130 (1999)
124. D.J. Audus, F.W. Starr, J.F. Douglas, Coupling of isotropic and directional interactions and its effect on phase separation and self-assembly. J. Chem. Phys. **144**(7), 074901 (2016)
125. S. Zheng et al., Proposed hydrogen-bonding index of donor or acceptor reflecting its intrinsic contribution to hydrogen-bonding strength. J. Chem. Inf. Model. **57**(7), 1535–1547 (2017)
126. G. Némethy, H.A. Scheraga, Structure of water and hydrophobic bonding in proteins. I. A model for the thermodynamic properties of liquid water. J. Chem. Phys. **36**(12), 3382–3400 (1962)
127. G. Némethy, H.A. Scheraga, Structure of water and hydrophobic bonding in proteins. II. Model for the thermodynamic properties of aqueous solutions of hydrocarbons. J. Chem. Phys. **36**, 3401–3417 (1962)

128. G. Némethy, H.A. Scheraga, the structure of water and hydrophobic bonding in proteins. III. The thermodynamic properties of hydrophobic bonds in proteins1,2. J. Phys. Chem. **66**, 1773–1789 (1962)
129. A.T. Hagler, H.A. Scheraga, G. Némethy, Structure of liquid water. Statistical thermodynamic theory. J. Phys. Chem. **76**, 3229–3243 (1972)
130. B.R. Lentz, A.T. Hagler, H.A. Scheraga, Structure of liquid water. II. Improved statistical thermodynamic treatment and implications of a cluster model. J. Phys. Chem. **78**, 1531–1550 (1974)
131. J.C. Owicki, H.A. Scheraga, Monte Carlo calculations in the isothermal-isobaric ensemble. 1. Liquid water. J. Am. Chem. Soc. **99**(23), 7403–7412 (1977)
132. J.C. Owicki, H.A. Scheraga, Monte Carlo calculations in the isothermal-isobaric ensemble. 2. Dilute aqueous solution of methane. J. Am. Chem. Soc. **99**(23), 7413–7418 (1977)
133. J.H. Griffith, H.A. Scheraga, Statistical thermodynamics of aqueous solutions. I. Water structure, solutions with non-polar solutes, and hydrophobic interactions. J. Mol. Struct.-Theochem **682**, 97–113 (2004)
134. A. Ben-Naim, *Hydrophobic Interactions,* vol. xiii (Plenum Press, 1980), 311 p.
135. A. Ben-Naim, *Molecular Theory of Water and Aqueous Solutions* (World Scientific. v., 2009)
136. D. Chandler, Interfaces and the driving force of hydrophobic assembly. Nature **437**(7059), 640–647 (2005)
137. K.D. Schwenke, Charles Tanford: the hydrophobic effect: formation of micelles and biological membranes. 200 Seiten, zahlreiche Abb. und Tab., John Wiley & Sons, New York, London, Sydney, Toronto 1973. Preis: 7,90 £. Nahrung-Food **20**, 449–449 (1976)
138. M.S. Wertheim, Fluids with highly directional attractive forces. I. Statistical thermodynamics. J. Stat. Phys. **35**, 19–34 (1984)
139. M.S. Wertheim, Fluids with highly directional attractive forces. II. Thermodynamic perturbation theory and integral equations. J. Stat. Phys. **35**, 35–47 (1984)
140. M.S. Wertheim, Fluids with highly directional attractive forces. III. Multiple attraction sites. J. Stat. Phys. **42**, 459–476 (1986)
141. M.S. Wertheim, Fluids with highly directional attractive forces. IV. Equilibrium polymerization. J. Stat. Phys. **42**, 477–492 (1986)
142. S.R. Gadre, S.D. Yeole, N. Sahu, Quantum chemical investigations on molecular clusters. Chem. Rev. **114**(24), 12132–12173 (2014)
143. F.N. Keutsch, J.D. Cruzan, R.J. Saykally, The water trimer. Chem. Rev. **103**(7), 2533–2578 (2003)
144. B.D. Marshall, A second order thermodynamic perturbation theory for hydrogen bond cooperativity in water. J. Chem. Phys. **146**(17), 174104 (2017)
145. J.D. Watson, F.H.C. Crick, Molecular structure of nucleic acids: a structure for deoxyribose nucleic acid. Nature **171**, 737–738 (1953)
146. E.A. Perets, E.C.Y. Yan, The H_2O Helix: the chiral water superstructure surrounding DNA. ACS Cent. Sci. **3**(7), 683–685 (2017)
147. M.L. McDermott et al., DNA's chiral spine of hydration. ACS Cent. Sci. **3**, 708–714 (2017)
148. Z.S. Petrović et al., Swelling of model networks. Macromolecules **20**, 1088–1096 (1987)
149. E.A. Karpushkin et al., Rheology and porosity control of poly(2-hydroxyethyl methacrylate) hydrogels. Polymer **54**, 661–672 (2013)
150. S.C. Jung, S.Y. Oh, Y.C. Bae, Reentrant swelling behavior of thermosensitive N-isopropylacrylamide nano-sized gel particles. Polymer **50**, 3370–3377 (2009)
151. S. Lee, Y.C. Bae, Enhanced solvation effect of re-collapsing behavior for cross-linked PMMA particle gel in aqueous alcohol solutions. Polymer **55**, 4684–4692 (2014)
152. Y.G. Kim, Y.C. Bae, Upper critical solution temperature-type thermosensitive hydrogel phase equilibrium of poly(2-hydroxyethyl methacrylate)/water/n-alkanol mixtures. Polymer **55**, 3987–3994 (2014)
153. B. Erman, J.E. Mark, *Structures and Properties of Rubberlike Networks* (1997)
154. P.J. Flory, Molecular theory of rubber elasticity. Polym. J. **17**, 1–12 (1979)

155. P.J. Flory, Statistical thermodynamics of random networks. Proc. R. Soc. London. A. Math. Phys. Sci. **351**, 351–380 (1976)
156. P.J. Flory, B. Erman, Theory of elasticity of polymer networks. 3. Macromolecules **15**(3), 800–806 (1982)
157. B. Erman, P.J. Flory, Relationships between stress, strain, and molecular constitution of polymer networks. Comparison of theory with experiments. Macromolecules **15**(3), 806–811 (1982)
158. F. Horkay, K. Nishi, M. Shibayama, Decisive test of the ideal behavior of tetra-PEG gels. J. Chem. Phys. **146**(16), 164905 (2017)
159. M. Marchetti, S. Prager, E.L. Cussler, Thermodynamic predictions of volume changes in temperature-sensitive gels. 1. Theory. Macromolecules **23**, 1760–1765 (1990)
160. M. Marchetti, S. Prager, E.L. Cussler, Thermodynamic predictions of volume changes in temperature-sensitive gels. 2. Experiments. Macromolecules **23**, 3445–3450 (1990)
161. S.F. Edwards, T. Vilgis, The effect of entanglements in rubber elasticity. Polymer **27**(4), 483–492 (1986)
162. M.C. Boyce, E.M. Arruda, Constitutive models of rubber elasticity: a review. Rubber Chem. Technol. **73**(3), 504–523 (2000)
163. C.O. Horgan, G. Saccomandi, Phenomenological hyperelastic strain-stiffening constitutive models for rubber. Rubber Chem. Technol. **79**, 152–169 (2006)
164. W.H. Han, F. Horka, G.B. McKenna, Mechanical and swelling behaviors of rubber: a comparison of some molecular models with experiment. Math. Mech. Solids **4**, 139–167 (1999)
165. M.C. Boyce, E.M. Arruda, Swelling and mechanical stretching of elastomeric materials. Math. Mech. Solids **6**(6), 641–659 (2001)
166. C.O. Horgan, G. Saccomandi, Constitutive models for compressible nonlinearly elastic materials with limiting chain extensibility. J. Elast. **77**, 123–138 (2004)
167. W. Kuhn, F. Grün, Beziehungen zwischen elastischen Konstanten und Dehnungsdoppelbrechung hochelastischer Stoffe. Kolloid-Zeitschrift **101**, 248–271 (1942)
168. M.C. Wang, E. Guth, Statistical theory of networks of non-gaussian flexible chains. J. Chem. Phys. **20**, 1144–1157 (1952)
169. E.M. Arruda, M.C. Boyce, A three-dimensional constitutive model for the large stretch behavior of rubber elastic materials. J. Mech. Phys. Solids **41**(2), 389–412 (1993)
170. J.E. Bischoff, E.M. Arruda, K. Grosh, A new constitutive model for the compressibility of elastomers at finite deformations. Rubber Chem. Technol. **74**(4), 541–559 (2001)
171. A.S. Johal, D.J. Dunstan, Energy functions for rubber from microscopic potentials. J. Appl. Phys. **101**, 084917 (2007)
172. J.G. Curro, J.E. Mark, *A non-Gaussian theory of rubberlike elasticity based on rotational isomeric state simulations of network chain configurations. II. Bimodal poly(dimethylsiloxane) networks*. J. Chem. Phys. **80**(9), 4521–4525 (1984)
173. A. Kloczkowski, B. Erman, J.E. Mark, Effect of non-Gaussian chains on fluctuations of junctions in bimodal networks. Polymer **43**, 2569–2574 (2002)
174. C.O. Horgan, G. Saccomandi, A molecular-statistical basis for the gent constitutive model of rubber elasticity. J. Elast. **68**, 167–176 (2002)
175. C.O. Horgan, The remarkable gent constitutive model for hyperelastic materials. Int. J. Non-Linear Mech. **68**, 9–16 (2015)
176. J. Kovac, Modified Gaussian model for rubber elasticity. Macromolecules **11**, 362–365 (1978)
177. Y. Hoei, Analysis of anomalous swelling by a rubber elasticity model for entangled networks. J. Macromol. Sci., Part B **52**, 190–198 (2013)
178. M. Klueppel, Finite chain extensibility and topological constraints in swollen networks. Macromolecules **27**(24), 7179–7184 (1994)
179. M. Klüppel, R.H. Schuster, G. Heinrich, Structure and properties of reinforcing fractal filler networks in elastomers. Rubber Chem. Technol. **70**, 243–255 (1997)
180. R.L. Jernigan, P.J. Flory, Distribution functions for chain molecules. J. Chem. Phys. **50**(10), 4185–4200 (1969)

181. A. Elías-Zúñiga, M.F. Beatty, Constitutive equations for amended non-Gaussian network models of rubber elasticity. Int. J. Eng. Sci. **40**, 2265–2294 (2002)
182. A.K. Katchalsky, I. Michaeli, Polyelectrolyte gels in salt solutions. J. Polym. Sci. **15**, 69–86 (1955)
183. I. Michaeli, A. Katchalsky, Potentiometric titration of polyelectrolyte gels. J. Polym. Sci. **23**(104), 683–696 (1957)
184. J. Hasa, M. Ilavský, Deformational, swelling, and potentiometric behavior of ionized poly(methacrylic acid) gels. II. Experimental results. J. Polym. Sci.: Polym. Phys. Ed. **13**(2), 263–274 (1975)
185. J. Hasa, M. Ilavský, K. Dušek, Deformational, swelling, and potentiometric behavior of ionized poly(methacrylic acid) gels. I. Theory. J. Polym. Sci.: Polym. Phys. Ed. **13**(2), 253–262 (1975)
186. W. Oppermann, Swelling behavior and elastic properties of ionic hydrogels, in *Polyelectrolyte Gels* (American Chemical Society, 1992), pp. 159–170
187. A. Fernández-Nieves et al., Charge controlled swelling of microgel particles. Macromolecules **33**(6), 2114–2118 (2000)
188. K. Ogawa, Y. Ogawa, E. Kokufuta, Effect of charge inhomogeneity of polyelectrolyte gels on their swelling behavior. Colloids Surf. A **209**, 267–279 (2002)
189. P. Košovan, T. Richter, C. Holm, Modeling of polyelectrolyte gels in equilibrium with salt solutions. Macromolecules **48**, 7698–7708 (2015)
190. S. Morozova, M. Muthukumar, Elasticity at swelling equilibrium of ultrasoft polyelectrolyte gels: comparisons of theory and experiments. Macromolecules **50**, 2456–2466 (2017)
191. D. Melekaslana, O. Okaya, *Swelling of Strong Polyelectrolyte Hydrogels in Polymer Solutions: Effect of Ion Pair Formation on the Polymer Collapse* (2000)
192. F.A. Escobedo, J.J.D. Pablo, Molecular simulation of polymeric networks and gels: phase behavior and swelling. Phys. Rep. **318**, 85–112 (1999)
193. E.Y. Kramarenko, O.E. Philippova, A.R. Khokhlov, Polyelectrolyte networks as highly sensitive polymers. Polym. Sci., Ser. C **48**(1), 1–20 (2006)
194. K. Dušek, *Responsive Gels: Volume Transitions II* (1993)
195. M. Shibayama, T. Tanaka, Volume phase transition and related phenomena of polymer gels, in *Responsive Gels: Volume Transitions I*, ed. by K. Dušek (Springer Berlin Heidelberg, Berlin, Heidelberg, 1993), pp. 1–62
196. A. Onuki, *Theory of Phase Transition in Polymer Gels* (1993)
197. M. Ilavský, Effect of phase transition on swelling and mechanical behavior of synthetic hydrogels, in *Responsive Gels: Volume Transitions I*, ed. by K. Dušek (Springer Berlin Heidelberg, Berlin, Heidelberg, 1993), pp 173–206
198. A.R. Khokhlov, S.G. Starodubtzev, V.V. Vasilevskaya, Conformational transitions in polymer gels: theory and experiment, in *Responsive Gels: Volume Transitions I*, ed. by K. Dušek (Springer Berlin Heidelberg, Berlin, Heidelberg, 1993) p. 123–171
199. R.A. Siegel, Hydrophobic weak polyelectrolyte gels: Studies of swelling equilibria and kinetics. Adv. Polym. Sci. **109**, 233–267 (1993)
200. S.H. Gehrke, *Synthesis, Equilibrium Swelling, Kinetics, Permeability and Applications of Environmentally Responsive Gels* (1993)
201. A.K. Katchalsky, S. Lifson, J. Mazur, The electrostatic free energy of polyelectrolyte solutions. I. Randomly kinked macromolecules. J. Polym. Sci. **11**, 409–423 (1953)
202. F, O., *Polyelectrolytes, Chap. 5.* (1971)
203. G.S. Manning, Limiting laws and counterion condensation in polyelectrolyte solutions II. Self-diffusion of the small ions. J. Chem. Phys. **51**, 934–938 (1969)
204. G.S. Manning, Limiting laws and counterion condensation in polyelectrolyte solutions I. Colligative properties. J. Chem. Phys. **51**, 924–933 (1969)
205. G.S. Manning, J. Ray, Counterion condensation revisited. J. Biomol. Struct. Dyn. **16**(2), 461–476 (1998)
206. S. Naeem, G. Sadowski, pePC-SAFT: modeling of polyelectrolyte systems: 1. Vapor–liquid equilibria. Fluid Phase Equilib. **299**(1), 84–93 (2010)

207. H.H. Hooper et al., Swelling equilibria for positively ionized polyacrylamide hydrogels. Macromolecules **23**(4), 1096–1104 (1990)
208. C.H. Lee, Y.C. Bae, Effect of salt on swelling behaviors of thermosensitive hydrogels: applicability of the nonrandom contact model. Macromolecules **48**, 4063–4072 (2015)
209. Q. Xin et al., A molecular thermodynamic model for random copolymer solutions. Fluid Phase Equilib. **267**(2), 163–171 (2008)
210. K. Dušek, M. Bohdanecký, V. Vošický, Solubilization of poly(2-hydroxyethyl methacrylate) with aqueous salt solutions; swelling of gels. Collect. Czech. Chem. Commun. **42**, 1599–1614 (1977)
211. C. Held, L.F. Cameretti, G. Sadowski, Modeling aqueous electrolyte solutions: Part 1. Fully dissociated electrolytes. Fluid Phase Equilib. **270**, 87–96 (2008)
212. C. Held, G. Sadowski, Modeling aqueous electrolyte solutions. Part 2. Weak electrolytes. Fluid Phase Equilibria **279**, 141–148 (2009)
213. S. Naeem, G. Sadowski, PePC-SAFT: modeling of polyelectrolyte systems 2. Aqueous two-phase systems. Fluid Phase Equilib. **306**(1), 67–75 (2011)
214. M.C. Arndt, G. Sadowski, Thermodynamic model for polyelectrolyte hydrogels. J. Phys. Chem. B **118**(35), 10534–10542 (2014)
215. L.F. Cameretti, G. Sadowski, J.M. Mollerup, Modeling of aqueous electrolyte solutions with perturbed-chain statistical associated fluid theory. Ind. Eng. Chem. Res. **44**(9), 3355–3362 (2005)
216. O.E. Philippova et al., New type of swelling behavior upon gel ionization: theory vs experiment. Macromolecules **46**, 9359–9367 (2013)
217. P.-G.D. Gennes, *Scaling Concepts in Polymer Physics*. (Cornell University Press, 1979), 324 p
218. T. Katashima, U.-I. Chung, T. Sakai, Effect of swelling and deswelling on mechanical properties of polymer gels. Macromol. Symp. **358**, 128–139 (2015)
219. O. Ortiz et al., Surface instabilities in ultrathin, cross-linked poly(N-isopropylacrylamide) coatings. Langmuir **26**(22), 17489–17494 (2010)
220. J. Dervaux, M.B. Amar, Mechanical instabilities of gels. Annu. Rev. Condens. Matter Phys. **3**, 311–332 (2012)
221. B. Li et al., Mechanics of morphological instabilities and surface wrinkling in soft materials: a review. Soft Matter **8**, 5728–5745 (2012)
222. S. Singamaneni, V.V. Tsukruk, Buckling instabilities in periodic composite polymeric materials. Soft Matter **6**, 5681–5692 (2010)
223. M.K. Kang, R. Huang, Swelling-induced instability of substrate-attached hydrogel lines. Int. J. Appl. Mech. **03**, 219–233 (2011)
224. A.D. Drozdov, J. de Claville Christiansen, Mechanical response and equilibrium swelling of temperature-responsive gels. Eur. Polym. J. **94**, 56–67 (2017)
225. A.D. Drozdov, J.D.C. Christiansen, Volume changes in hydrogels subjected to finite deformations. Mech. Res. Commun. **50**, 33–38 (2013)
226. F. Horkay, M. Zrínyi, Studies on the mechanical and swelling behavior of polymer networks based on the scaling concept. 4. Extension of the scaling approach to gels swollen to equilibrium in a diluent of arbitrary activity. Macromolecules **15**, 1306–1310 (1982)
227. E. Geissler, F. Horkay, A.M. Hecht, Osmotic and scattering properties of chemically crosslinked poly(vinyl alcohol) hydrogels. Macromolecules **24**(22), 6006–6011 (1991)
228. G.B. McKenna, F. Horkay, Effect of crosslinks on the thermodynamics of poly(vinyl alcohol) hydrogels. Polymer **35**, 5737–5742 (1994)
229. B.G. Stubbe et al., Tailoring the swelling pressure of degrading dextran hydroxyethyl methacrylate hydrogels. Biomacromol **4**(3), 691–695 (2003)
230. W. Borchard, Theory of swelling of a crosslinked substance in equilibrium with a pure solvent in various phases. Part II: The variation of the pressure. Colloid Polym. Sci. **271**(9), 843–851 (1993)
231. H. Hinsken, W. Borchard, Continuous swelling pressure equilibria of the systemκ-carrageenan/water. Colloid Polym. Sci. **273**, 913–925 (1995)

232. E.J. van de Kraats et al., Polymer network characterization by means of swelling pressure and unilateral compression data. Recl. Trav. Chim. Pays-Bas **88**(4), 449–464 (1969)
233. I.S. Han et al., Constant-volume hydrogel osmometer: a new device concept for miniature biosensors. Biomacromol **3**(6), 1271–1275 (2002)
234. P. Höhne, K. Tauer, How much weighs the swelling pressure. Colloid Polym. Sci. **292**, 2983–2992 (2014)
235. G. Holtus, W. Borchard, Swelling pressure equilibrium of physical networks in the field of an analytical ultracentrifuge. Colloid Polym. Sci. **267**, 1133–1138 (1989)
236. K. Dušek, M. Dušková-Smrčková, C. Brent Douglas, Swelling of coating films, in *Protective Coatings: Film Formation and Properties*, ed. by M. Wen, K. Dušek (Springer International Publishing: Cham, 2017), pp. 271–291
237. S.S. Sternstein, Inhomogeneous swelling in filled elastomers. J. Macromol. Sci., Part B **6**, 243–262 (1972)
238. J. Dolbow, E. Fried, H. Ji, Chemically induced swelling of hydrogels. J. Mech. Phys. Solids **52**(1), 51–84 (2004)
239. H. Ji et al., Kinetics of thermally induced swelling of hydrogels. Int. J. Solids Struct. **43**(7), 1878–1907 (2006)
240. Q.X. Wang, H. Li, K.Y. Lam, Meshless simulation of equilibrium swelling/deswelling of pH-sensitive hydrogels. J Polym Sci B **44**, 326–337 (2006)
241. R.A. Paxton, A.M. Al-Jumaily, Swelling of polyelectrolyte hydrogels using a finite element model. Polymer **47**(16), 5997–6003 (2006)
242. T. Wallmersperger, et al. *Modeling and Simulation of Hydrogels for the Application as Bending Actuators* (2013)
243. S. Baek, T.J. Pence, Inhomogeneous deformation of elastomer gels in equilibrium under saturated and unsaturated conditions. J. Mech. Phys. Solids **59**(3), 561–582 (2011)
244. J. Šomvársky, K. Dušek, M. Dušková-Smrčková, Constraints effects in swollen particulate composites with hyperelastic polymer matrix of finite extensibility modeled by FEM. J. Phys. Conf. Ser. **490**, 012207 (2014)
245. J. Berriot et al., Reinforcement of model filled elastomers: experimental and theoretical approach of swelling properties. Polymer **43**(23), 6131–6138 (2002)
246. H. Montes et al., Particles in model filled rubber: dispersion and mechanical properties. Eur. Phys. J. E **31**, 263–268 (2010)
247. F. Sciortino, E. Zaccarelli, Equilibrium gels of limited valence colloids. Curr. Opin. Colloid Interface Sci. **30**, 90–96 (2017)
248. T. Hashimoto, Dynamics in spinodal decomposition of polymer mixtures. Phase Transit. **12**, 47–119 (1988)
249. T. Nose, Kinetics of phase separation in polymer mixtures. Phase Transit. **8**, 245–260 (1987)
250. P.V.D. Witte, et al., Phase-separation processes in polymer-solutions in relation to membrane formation. J. Membr. Sci. 117, 1–31 (1996)
251. J. Tao et al., Phase separation induced by a two-step temperature jump in polystyrene/poly(2-chlorostyrene) blends: dependence on duration and temperature of the first step. Polymer **36**, 3909–3917 (1995)
252. M. Hatanaka, H. Saito, In-Situ investigation of liquid—liquid phase separation in polycarbonate/carbon dioxide system. Macromolecules **37**, 7358–7363 (2004)
253. Y.S. Su et al., Interplay of mass transfer, phase separation, and membrane morphology in vapor-induced phase separation. J. Membr. Sci. **338**, 17–28 (2009)
254. J.W. Cahnand, J.E. Hilliard, Free energy of a nonuniform system. I. Interfacial free energy. J. Chem. Phys. **28**(2), 258–267 (1958)
255. G.T. Caneba, D.S. Soong, Polymer membrane formation through the thermal-inversion process. 2. Mathematical modeling of membrane structure formation. Macromolecules **18**(12), 2545–2555 (1985)
256. K.V. Kurada, S. De, Modeling of solution thermodynamics: a method for tuning the properties of blend polymeric membranes. J. Membr. Sci. **540**, 485–495 (2017)

257. X. Sun, G. Sun, X. Wang, Morphology modeling for polymer monolith obtained by non-solvent-induced phase separation. Polymer **108**, 432–441 (2017)
258. D. Jeong, J. Kim, A practical numerical scheme for the ternary Cahn-Hilliard system with a logarithmic free energy. Physica A-statistical Mechanics and Its Applications **442**, 510–522 (2016)
259. A. Novick-Cohen, Chapter 4 The Cahn–Hilliard equation, in *Handbook of Differential Equations: Evolutionary Equations*, ed. by C.M. Dafermos, M. Pokorny (North-Holland, 2018). pp. 201–228
260. K. Binder, et al., Polymer + solvent systems: phase diagrams, interface free energies, and nucleation, in *Advanced Computer Simulation: Approaches for Soft Matter Sciences I*, ed. by C. Holm, K. Kremer (Springer Berlin Heidelberg, Berlin, Heidelberg, 2005), pp. 1–110
261. K. Dušek, B. Sedláček, Structure and properties of hydrophilic polymers and their gels. XI. Microsyneresis in swollen poly(ethylene glycol methacrylate) gels induced by changes of temperature. Collect. Czechoslov. Chem. Commun. **34**, 136–157 (1969)
262. K. Dušek, B. Sedláček, Effect of diluents on the microsyneresis in poly(2-hydroxyethyl methacrylate) gels induced by temperature changes. Collect. Czech. Chem. Commun. **36**, 1569–1577 (1971)
263. K. Dusek, Inhomogeneities induced by crosslinking in the course of crosslinking copolymerization, in *Polymer Networks: Structure and Mechanical Properties*, ed. by A.J. Chompff, S. Newman (Springer US, Boston, MA, 1971), pp. 245–260
264. K. Dušek, B. Sedláček, Structural changes in swollen polymer gels induced by solvent exchange. Collect. Czech. Chem. Commun. **38**, 3434–3439 (1973)
265. W. Heller, W.J. Pangonis, Theoretical investigations on the light scattering of colloidal spheres. I. The specific turbidity. J. Chem. Phys. **26**, 498–506 (1957)
266. G. Mie, Beiträge zur optik trüber medien, speziell kolloidaler metallösungen. Ann. Phys. **330**(3), 377–445 (1908)
267. M. Ilavsky, W. Prins, Rheo-optics of poly(2-hydroxyethyl methacrylate) gels. I Effect of nature and amount of diluent. Macromolecules **3**(4), 415–425 (1970)
268. M. Ilavsky, W. Prins, Rheo-optics of poly(2-hydroxyethyl methacrylate) gels. II. Effect of cross-linking density and stage of dilution during network formation. Macromolecules **3**(4), 425–433 (1970)
269. M.B. Huglin, D.C.F. Yip, Microsyneresis region in poly(2-hydroxyethyl methacrylate) hydrogels. Macromolecules **25**(4), 1333–1337 (1992)
270. M. Shibayama, Small-angle neutron scattering on polymer gels: phase behavior, inhomogeneities and deformation mechanisms. Polym. J. **43**, 18–34 (2011)
271. M. Shibayama, T. Norisuye, Gel formation analyses by dynamic light scattering. Bull. Chem. Soc. Jpn **75**, 641–659 (2002)
272. F. Ikkai, M. Shibayama, Inhomogeneity control in polymer gels. J Polym Sci B **43**, 617–628 (2005)
273. M. Shibayama, Universality and specificity of polymer gels viewed by scattering methods. Bull. Chem. Soc. Jpn **79**, 1799–1819 (2006)
274. S. Seiffert, Origin of nanostructural inhomogeneity in polymer-network gels. Polym. Chem. **8**, 4472–4487 (2017)
275. R. Bansil, G. Liao, P. Falus, Kinetics of spinodal decomposition in chemically crosslinked gels. Physica A **231**(1), 346–358 (1996)
276. R. Bansil, Phase separation in polymer solutions and gels. Le Journal de Physique IV **03**(C1), C1-225–C1-235 (1993)
277. H.E. Cook, Brownian motion in spinodal decomposition. Acta Metall. **18**(3), 297–306 (1970)
278. T. Morita et al., Phase behaviors of agarose gel. AIP Adv. **3**, 042128 (2013)
279. H. Tanaka, Viscoelastic phase separation. J. Phys.: Condens. Matter **12** (2000)
280. H. Tanaka, T. Koyama, T. Araki, Network formation in viscoelastic phase separation. J. Phys.: Condens. Matter **15** (2003)
281. K. Dušek, J.H. Seidl, J. Malinský, Phase separation in the copolymerization of styrene and divinylbenzene in the presence of diluents: comparison of experiment with theory. Collect. Czech. Chem. Commun. **32**, 2766–2778 (1967)

282. K. Dušek, Phase separation during the formation of three-dimensional polymers. J. Polym. Sci. Part C: Polym. Symp. **16**(3), 1289–1299 (1967)
283. R.D. Andrews, A.V. Tobolsky, E.E. Hanson, The theory of permanent set at elevated temperatures in natural and synthetic rubber vulcanizates. J. Appl. Phys. **17**(5), 352–361 (1946)
284. M. Dušková-Smrčková, et al., Effect of dilution on structure and properties of polyurethane networks. Pregel and postgel cyclization and phase separation. Macromolecules **43**(15), 6450–6462 (2010)
285. M. Dušková-Smrčková et al., Diluent induced cyclization and phase separation in polymer networks. Macromol. Symp. **306–307**(1), 67–76 (2011)
286. K. Dušek, J.I. Spěváček, Cyclization in vinyl-divinyl copolymerization. Polymer **21**, 750–756 (1980)
287. J.I. Spěváček, K. Dušek, Manifestation of microgel-like particles of styrene-ethylene dimethacrylate copolymers in solution in 1H and 13C NMR spectra. J. Polym. Sci. Part B. **18**, 2027–2035 (1980)
288. H. Galina et al., The structure of low conversion polymers of ethylene dimethacrylate. Eur. Polymer J. **16**, 1043–1046 (1980)
289. K. Dušek, H. Galina, J. Mikeš, Features of network formation in the chain crosslinking (co)polymerization. Polym. Bull. **3**, 19–25 (1980)
290. L. Matějka, S. Pokomý, K. Dušek, Network formation involving epoxide and carboxyl groups. Polym. Bull. **7**(2), 123–128 (1982)
291. J.R. Millar, et al., 33. Solvent-modified polymer networks. Part I. The preparation and characterisation of expanded-network and macroporous styrene–divinylbenzene copolymers and their sulphonates. J. Chem. Soc. (Resumed) 218–225 (1963)
292. J.R.A. Millar, D.G. Smith, T.R.E. Kressman, 45. Solvent-modified polymer networks. Part IV. Styrene–divinyl-benzene copolymers made in the presence of non-solvating diluents. J. Chem. Soc. (Resumed) 304–310 (1965)
293. O. Okay, Phase separation in free-radical crosslinking copolymerization: formation of heterogeneous polymer networks. Polymer **40**, 4117–4129 (1999)
294. S. Durmaz, O. Okay, Phase separation during the formation of poly(acrylamide) hydrogels. Polymer **41**(15), 5729–5735 (2000)
295. O. Okay, Formation of macroporous styrene-divinylbenzene copolymer networks: theory vs experiments. J. Appl. Polym. Sci. **74**, 2181–2195 (1999)
296. Y. Osada, Polymer gels as artificial soft tissue. Polym. Sci., Ser. C **59**(1), 3–10 (2017)
297. V. Raeisdasteh Hokmabad et al., Design and fabrication of porous biodegradable scaffolds: a strategy for tissue engineering. J. Biomater. Sci. Polym. Ed. **28**, 1797–1825 (2017)
298. G. Chen, T. Ushida, T. Tateishi, Scaffold design for tissue engineering. Macromol. Biosci. **2**(2), 67–77 (2002)
299. E. Karpushkin et al., Rheological properties of homogeneous and heterogeneous poly(2-hydroxyethyl methacrylate) hydrogels. Polym. Int. **61**(2), 328–336 (2012)
300. K. Dušek, B. Sedláček, Phase separation in poly(2-hydroxyethyl methacrylate) gels in the presence of water. Eur. Polymer J. **7**(9), 1275–1285 (1971)
301. M. Dušková-Smrčková et al., The manifold varieties of poly(2-hydroxyethyl methacrylate) hydrogels−IPNs. Macromol. Symp. **372**, 28–42 (2017)
302. S. Dubinsky et al., Toward controlling the surface morphology of macroporous copolymer particles. Macromolecules **42**(6), 1990–1994 (2009)
303. T. Tanaka, Collapse of gels and the critical endpoint. Phys. Rev. Lett. **40**, 820–823 (1978)
304. T. Tanaka et al., Phase transitions in ionic gels. Ferroelectrics **30**, 97–97 (1980)
305. H.G. Schild, Poly(N-isopropylacrylamide): experiment, theory and application. Prog. Polym. Sci. **17**, 163–249 (1992)
306. W. Richtering, B.R. Saunders, Gel architectures and their complexity. Soft Matter **10**(21), 3695–3702 (2014)
307. S. Wu et al., Modeling investigation of hydrogel volume transition. Macromol. Theory Simul. **13**, 13–29 (2004)

308. C. Wu, S. Zhou, Volume phase transition of swollen gels: discontinuous or continuous? Macromolecules **30**, 574–576 (1997)
309. M. Ilavsky, Phase transition in swollen gels. 2. Effect of charge concentration on the collapse and mechanical behavior of polyacrylamide networks. Macromolecules **15**(3), 782–788 (1982)

Chapter 5
Swelling and Shrinking Kinetics

Mitsuhiro Shibayama

Abstract Polymer gels are sensitive to their environment, such as temperature, solvent, and pH, and swell or shrink with slight changes in these environmental parameters. Since the phenomena of gel swelling, the degree of swelling, and the thermodynamic picture of swelling and shrinking are described in detail in Chap. 4 (Thermodynamics of Gels), this chapter describes the dynamic picture of swelling and shrinking of polymer gels. Starting from the continuum theory of polymer gels, theoretical equations for swelling and shrinking of polymer gels are derived. Then, examples of the application of the theory to various gels are given to enhance the reader's understanding.

Keywords Tanaka-Fillmore theory · Swelling equation · Cooperative diffusion · Poly(*N*-isopropylacrylamide) (PNIPAM) gel · Critical slowing down · Ion exchange resin

Polymer gels are sensitive to their environment, such as temperature, solvent, and pH, and swell or shrink with slight changes in these environmental parameters. Since the phenomena of gel swelling, the degree of swelling, and the thermodynamic picture of swelling and shrinking are described in detail in Chap. 4 (Thermodynamics of Gels), this chapter describes the dynamic picture of swelling and shrinking of polymer gels. Starting from the continuum theory of polymer gels, theoretical equations for swelling and shrinking of polymer gels are derived. Then, examples of the application of the theory to various gels are given to enhance the reader's understanding.

M. Shibayama (✉)
Neutron Science and Technology Center, Comprehensive Research Organization for Science and Society (CROSS), Ibaraki, Japan
e-mail: m_shibayama@cross.or.jp

Y. Osada (ed.), *Science of Gel*, https://doi.org/10.1007/978-981-95-0301-8_5

5.1 Phenomenology of Gel Swelling

You have ever probably seen a toy of gold fish or turtle made of gel sold at night stalls and other places. If you put it in water, it gradually swells during overnight. Next morning, you find a swollen fish or turtle with the same shape but in a large size. This phenomenon seems so natural that no one pays attention to why it occurs. Koya-tofu (dried tofu), on the other hand, a traditional Japanese food, is a dried gel in a desiccated state. Since it is cooked after being soaked in water for a while, the swelling of koya-tofu in water can be observed at home. From a scientific point of view, these phenomena can be divided into water penetration, gel softening, and gel swelling. Similar phenomena can be observed in other polymer materials, such as polymer water absorbents and ion exchange resins. In addition, wet gels swell or shrink by changing the environment, such as pH, temperature, and salinity. In general, the rate of swelling and shrinking of wet gels is much slower than that of dry gels. This is mainly because dry gels are porous and the surface area of the dry gel is much larger than that of the wet gel. The kinetics of gel swelling is based on the theory of water diffusion into the gel. In 1965, the theory of water diffusion in ion exchange resins was proposed by Dusek [1]. In the 1970s, a comprehensive theory of cooperative diffusion (collective diffusion) in polymer gels [2], followed by the theory of gel swelling [3] proposed by Tanaka et al.

Swelling and shrinking of gels are essentially different phenomena from the diffusion of liquids and gases. Figure 5.1 schematically illustrates (a) the diffusion of an ink (pigment) stain on a filter paper and (b) the swelling of a cylindrical gel. When a few drops of water are dropped on the filter paper, the ink stain begins to diffuse as indicated by the thin arrow on the right. As a result, the stain blurs as shown in the left bottom of Fig. 5.1a. On the other hand, when water is added to a cylindrical gel, it swells while maintaining its shape. Schematically, the dotted rectangle (before swelling) changes to a solid rectangle (after swelling) as shown on the right bottom of Fig. 5.1b. The obvious difference between (a) the ink diffusion on the filter paper and (b) the swelling of the gel is the shape after diffusion or swelling. In the former, the four corners are rounded, while in the latter, the four corners are sharp. Why do these differences occur? The answer is as follows. On the filter paper, the pigment particles of the ink diffuse independently from their original position. As a result, the envelope of the most distantly diffused ink pigment particles becomes smooth. In the gel, on the other hand, they cannot diffuse (move) independently because of the interconnected chains. The chains must move in such a way that the strain energy of the entire gel is minimized, which is satisfied by maintaining the same shape. The above example suggests that gels must be treated as a continuum when discussing the phenomenon of gel swelling.

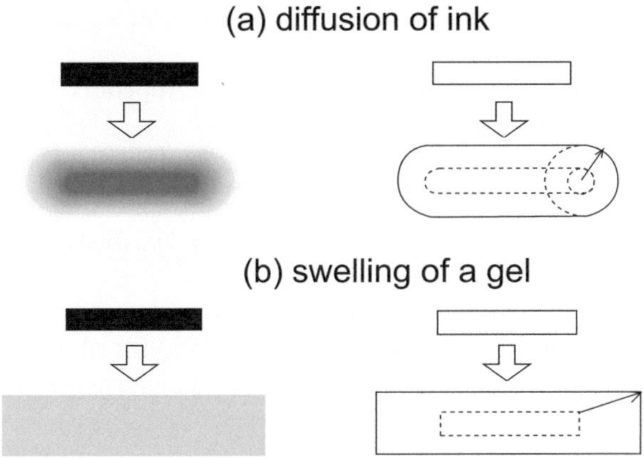

Fig. 5.1 Comparison of **a** diffusion of ink and **b** swelling of a gel

5.2 Cooperative Diffusion Theory of Polymer Networks

A full-fledged theory of gel swelling was proposed by Tankaka et al. as the cooperative diffusion theory of polymer networks to explain the spectrum of scattered light originating from thermally induced chain fluctuations in gels [2]. This theory states that the dynamics of polymer networks is cooperative diffusion (or collective diffusion), i.e., each diffusing element interacts with each other via chain linkages. Cooperative diffusion is essentially different from self-diffusion, interdiffusion, probe diffusion, etc., as described in many physical chemistry textbooks. In order to understand the cooperative diffusion theory, it is necessary to have a basic understanding of elasticity theory [4].

5.2.1 Strain Tensor and Stress Tensor

In this section, the theory of elasticity is explained using tensors. The basics of tensors are summarized in Appendix 1 at the end of the chapter.

A. **Strain tensor**

Consider the deformation of the continuum shown in Fig. 5.2, where two points $\mathbf{r}_1$ and $\mathbf{r}_2$ have moved to $\mathbf{r}_1'$ and $\mathbf{r}_2'$, respectively, due to deformation (swelling/shrinking, elongation/compression, rotation, etc.).

$$\mathbf{u}_1 = \mathbf{r}_1' - \mathbf{r}_1, \quad \mathbf{u}_2 = \mathbf{r}_2' - \mathbf{r}_2 \tag{5.1}$$

where $\mathbf{u}_i$ is called the displacement vector. If the change in the displacement vector due to deformation is $\delta\mathbf{u}$, the vector $\delta\mathbf{r}'$ connecting $\mathbf{r}_1'$ and $\mathbf{r}_2'$ after deformation is given by

$$\delta\mathbf{r}' = \delta\mathbf{r} + \delta\mathbf{u} \tag{5.2}$$

The square of $\delta\mathbf{r}'$ is also given by (Appendix 3.1)

$$\begin{aligned}(\delta\mathbf{r}')^2 &\cong (\mathrm{d}x_i^2) + \frac{\partial u_i}{\partial x_j}\mathrm{d}x_i\mathrm{d}x_j + \frac{\partial u_i}{\partial x_k}\mathrm{d}x_i\mathrm{d}x_k \\ &= (\mathrm{d}x_i^2) + 2u_{ik}\mathrm{d}x_i\mathrm{d}x_k\end{aligned} \tag{5.3}$$

From this point on, the summation convention is used (i.e., Repeated indices are implicitly summed over). u_{ik} in Eq. (5.3) is called the strain tensor, and is defined as the strain in the direction i acting on a plane perpendicular to k, as shown in Fig. 5.3 for the case of $i = x$ or z and $k = x$,

$$u_{ik} \equiv \frac{1}{2}\left(\frac{\partial u_i}{\partial x_k} + \frac{\partial u_k}{\partial x_i}\right) \tag{5.4}$$

$u_{ik.}$ is the symmetry tensor. Its diagonal sum (trace)u_{ii} is given by

$$\begin{aligned}u_{ii} &\equiv \sum_i (u_{xx} + u_{yy} + u_{zz}) \\ &= \left(\frac{\partial u_x}{\partial x} + \frac{\partial u_y}{\partial y} + \frac{\partial u_z}{\partial z}\right) = \frac{\partial u_i}{\partial x_i} = \nabla \cdot \mathbf{u}\end{aligned} \tag{5.5}$$

This expression implies that u_{ii} is nothing but the divergence $\nabla\mathbf{u}$ of the displacement vector $\mathbf{u}$. In other words, u_{ii} represents pure dilation or compression.

The strain tensor can be decomposed into a pure shear component and a hydrostatic compression component and is expressed as

$$u_{ik} = \left(u_{ik} - \frac{1}{3}\delta_{ik}u_{ll}\right) + \frac{1}{3}\delta_{ik}u_{ll} \tag{5.6}$$

Note that $\delta_{ik} = 3$. Thus, the change in free energy due to deformation, F, is given by

$$F = \mu\left(u_{ik} - \frac{1}{3}\delta_{ik}u_{ll}\right)^2 + \frac{1}{2}Ku_{ll}^2 \tag{5.7}$$

where μ and K are the shear modulus of elasticity and bulk modulus, respectively, and represent the resistance to deformation against shear and compression, respectively.

B. **Stress tensor**

The stress tensor σ_{ik} is defined, as in the case of the strain tensor, as a stress in the direction i acting in a plane perpendicular to k. The stress is a second-order tensor and is given in terms of free energy as

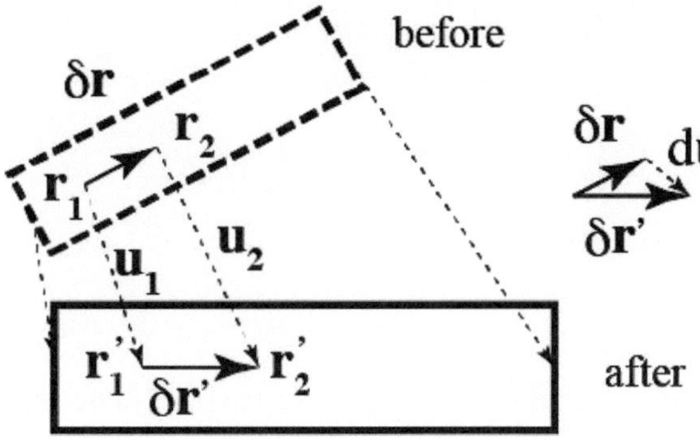

Fig. 5.2 Deformation and the displacement vector, **u**

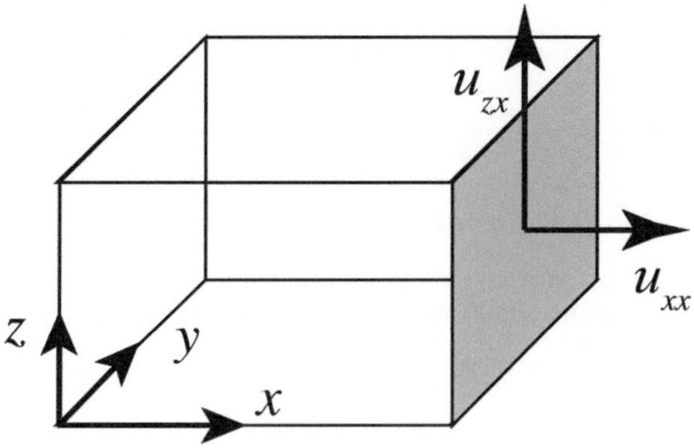

Fig. 5.3 Strain tensor

$$dF = -SdT + \sigma_{ik} du_{ik} \tag{5.8}$$

Therefore, σ_{ik} is given by

$$\sigma_{ik} = \left(\frac{\partial F}{\partial u_{ik}} \right)_T \tag{5.9}$$

Substituting Eq. (5.7) into Eq. (5.9) and differentiating with respect to u_{ik}, we obtain

$$\begin{aligned} \sigma_{ik} &= \frac{\partial}{\partial u_{ik}} \left[\mu \left(u_{ik} - \frac{1}{3} \delta_{ik} u_{ll} \right)^2 + \frac{1}{2} K u_{ll}^2 \right] \\ &= 2\mu \left(u_{ik} - \frac{1}{3} \delta_{ik} u_{ll} \right) + K u_{ll} \delta_{ik} \end{aligned} \tag{5.10}$$

Here, the following relational equation was used

$$\frac{\partial u_{ll}^2}{\partial u_{ik}} = 2u_{jj}\frac{\partial u_{ll}}{\partial u_{ik}}$$
$$= 2\left(u_{xx} + u_{yy} + u_{zz}\right)\frac{\partial\left(u_{xx} + u_{yy} + u_{zz}\right)}{\partial u_{ik}} = 2u_{ll}\delta_{ik} \tag{5.11}$$

Using Eq. (5.5), we obtain

$$u_{ll}\delta u_{ik} = \nabla \cdot \mathbf{u}\delta_{ik} \tag{5.12}$$

Thus, the stress tensor is given as the sum of the shear and hydrostatic compression components as

$$\sigma_{ik} = 2\mu\left(u_{ik} - \frac{1}{3}\nabla \cdot \mathbf{u}\delta_{ik}\right) + K\nabla \cdot \mathbf{u}\delta_{ik} \tag{5.13}$$

5.2.2 Cooperative Diffusion Equation for Gels

A. **Equations of motion of a continuum**

The equation of motion for a small volume element in a continuum is given by

$$\text{force} = \nabla \cdot \tilde{\sigma} - f\frac{\partial u}{\partial t} \tag{5.14}$$

where $\tilde{u}$ is the stress tensor ($\tilde{u} \equiv \sigma_{ik}$) and f is the friction coefficient. The force is given by using the acceleration and mass density ρ of the microvolume element as

$$\text{force} = \rho\frac{\partial^2 \mathbf{u}}{\partial t^2} \tag{5.15}$$

By assuming that Eqs. (5.14) and (5.15) are equal and substituting Eq. (5.13), the equation of motion of the polymer gel is obtained.

$$\rho\frac{\partial^2 \mathbf{u}}{\partial t^2} = \frac{\partial}{\partial x_k}\left[2\mu\left(u_{ik} - \frac{1}{3}\nabla \cdot \mathbf{u}\delta_{ik}\right) + K\nabla \cdot \mathbf{u}\delta_{ik}\right] - f\frac{\partial u_i}{\partial t} \tag{5.16}$$

By performing the divergence operation, we obtain the following equation, which will ultimately serve as the starting point when discussing the motion of the gel [2].

$$\rho\frac{\partial^2 \mathbf{u}}{\partial t^2} = \mu\nabla^2\mathbf{u} + \left(K + \frac{1}{3}\mu\right)\nabla(\nabla \cdot \mathbf{u}) - f\frac{\partial \mathbf{u}}{\partial t} \tag{5.17}$$

The derivation process of Eq. (5.17) is presented in Appendix 3.2.

B. **Cooperative diffusion equation**[3]

In polymer gels, the polymer chains in a network are not free to move, and their motion is constrained by their interactions with other polymer chains. Therefore, the diffusion of polymer chains is not self-diffusion but cooperative diffusion. The diffusion rate of cooperative diffusion of polymer chains is much slower than that of solvent self-diffusion. When considering the motion of the polymer network in a gel, the left-hand side of Eq. (5.17), i.e., the term of inertia force, is expected to be much smaller than the other terms and can be neglected. Therefore, Eq. (5.17) can be rewritten as

$$f\frac{\partial \mathbf{u}}{\partial t} = \mu\nabla^2\mathbf{u} + \left(K + \frac{1}{3}\mu\right)\nabla(\nabla\cdot\mathbf{u}) \tag{5.18}$$

Now, take the displacement vector parallel to the z-axis and define $\mathbf{u}$ as $\mathbf{u} \equiv (0,\ 0,\ u_z(z))$. This simplifies Eq. (5.18) to the following equation

$$\frac{\partial u_z}{\partial t} = \frac{1}{f}\left(K + \frac{4}{3}\mu\right)\frac{\partial^2 u_z}{\partial z^2} \tag{5.19}$$

The coefficient on the right-hand side is the diffusion coefficient defined as

$$D \equiv \frac{1}{f}\left(K + \frac{4}{3}\mu\right) \tag{5.20}$$

The well-known one-dimensional (z-axis) diffusion equation can be obtained by redefining it as

$$\frac{\partial u_z}{\partial t} = D\frac{\partial^2 u_z}{\partial z^2} \tag{5.21}$$

where D is the diffusion coefficient. Equations (5.19) or (5.21) is the starting equation when discussing gel swelling/contraction and dynamic light scattering [2].

5.3 Swelling and Shrinking

When the environment of a gel in swelling equilibrium with the surrounding solvent is changed, the gel swells or shrinks. This process is quite slow for gels of a few centimeters in size, which are often used in experiments, and takes from a few minutes to several days to relax. Therefore, it is very easy to measure the kinetics of gels by observing the size change of the gel visually or with an optical microscope. In this section, we first discuss the kinetics of swelling and shrinking of spherically symmetric gels, followed by the kinetics of swelling and shrinking of non-spherical gels, and then discuss their applications.

5.3.1 Swelling Equation in Spherical Coordinates

Since the displacement vector **u** is spherically symmetric, we can write

$$\mathbf{u}(\mathbf{r}, t) = u_r(r, t)\frac{\mathbf{r}}{r} \tag{5.22}$$

The radial stress tensor is given by

$$\sigma_{rr} = \left(K + \frac{4}{3}\mu\right)\frac{\mathrm{d}u}{\mathrm{d}r} + 2\left(K - \frac{2}{3}\mu\right)\frac{u}{r} \tag{5.23}$$

For gel swelling, the acceleration term in Eq. (5.17) can be neglected. In other words, the acceleration term can be written as follows

$$\frac{\partial^2 \mathbf{u}}{\partial t^2} = 0 \tag{5.24}$$

Thus, Eq. (5.17) can be rewritten as

$$f\frac{\partial \mathbf{u}}{\partial t} = \left(K + \frac{1}{3}\mu\right)\nabla(\nabla \cdot \mathbf{u}) + \mu\Delta\mathbf{u} \tag{5.25}$$

Here, **u** is written as follows in the spherical coordinate

$$u = (u_r, u_\theta = 0, u_\phi = 0) \tag{5.26}$$

(Change u by **u** (vector)

Then, the following equation is obtained (Appendix 3.3) [3].

$$\frac{\partial u_r}{\partial t} = D\frac{\partial}{\partial r}\left\{\frac{1}{r^2}\left[\frac{\partial}{\partial r}(r^2 u_r)\right]\right\} \tag{5.27}$$

where D is the cooperataive diffusion coefficient defined in Eq. (5.20). Equation (5.27) is called the swelling equation.

5.3.2 Solution of the Swelling Equation [3]

Suppose a spherical gel of radius a_0 begins to swell at time $t = 0$ and reaches swelling equilibrium at $t = \infty$ with radius a (Figs. 5.4 and 5.5). This section describes how the gel changes from the initial state to the point at swelling equilibrium. We will also derive equations for the local stress and strain in the radial direction over time.

To obtain these relations, we begin with the swelling Eq. (5.27). Before solving the equations, let us review the initial and boundary conditions.

Initial conditions

At $t = 0$,, the only stress on the gel is osmotic pressure (Fig. 5.4 left). Hence,

$$\sigma_{ik} = \Pi_0 \delta_{ik} \tag{5.28}$$

Since the gel is spherically symmetric, the strain vector has only a radial component.

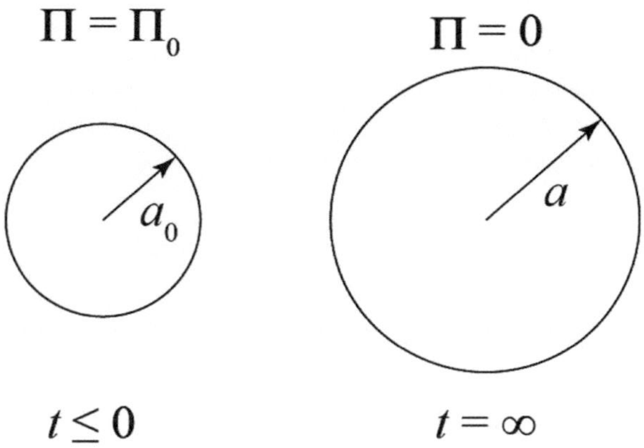

Fig. 5.4 Swelling of a spherical gel (left) before immersion (right), at swelling equilibrium after immersion

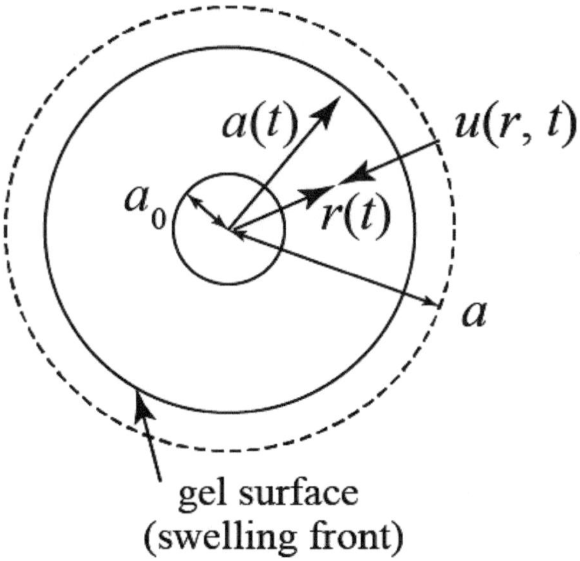

Fig. 5.5 The strain $u'(r, t)$ on a spherical gel from the initial state $r = a_0$ at $t = 0$ to the final state r $= a$ at $t = \infty$

$$\mathbf{u} = (u_r, 0, 0) \equiv (u, 0, 0) \tag{5.29}$$

In addition, the initial and final conditions are given by (See, Fig. 5.5),

$$u(0, t) = 0, \quad \Delta a_0 \equiv a - a_0 \tag{5.30}$$

and

$$u(r, t) = \frac{r}{a}[a - a(t)] \tag{5.31}$$

Boundary conditions

Before immersion in water, the gel has a positive osmotic pressure Π_0 and a positive stress. However, when the gel is immersed in water, the gel surface comes into contact with water and $\sigma_{\mathrm{normal}} = \sigma_{rr} = 0$ at the gel surface, i.e., the stress at the gel surface is zero. In other words,

$$\sigma_{rr} = \left(K + \frac{4}{3}\mu\right)\left(\frac{\partial u}{\partial r}\right) + 2\left(K - \frac{2}{3}\mu\right)\left(\frac{u}{r}\right) = 0 \quad (r = a) \tag{5.32}$$

Assuming $K \gg \mu$ for simplicity of computation, Eq. (5.32) becomes

$$\sigma_{rr} = K\left[\left(\frac{\partial u}{\partial r}\right) + 2\left(\frac{u}{r}\right)\right] = \frac{K}{r^2}\left[\frac{\partial}{\partial r}(r^2 u)\right] = 0 \quad (r = a) \tag{5.33}$$

In a typical polyacrylamide gel (2.5 wt%), $K \approx 8 \times 10^2$ Pa and $\mu \approx 2 \times 10$ Pa [3]. Therefore, in most cases, the assumption that $K \gg \mu$ (compressive modulus is much greater than shear modulus) is reasonable. From Eq. (5.33), we obtain

$$\frac{K}{r^2}\left[\frac{\partial}{\partial r}(r^2 u)\right] = 0 \tag{5.34}$$

and solving this equation in the spherical coordinate yields

$$\frac{\partial u_r}{\partial t} = D\frac{\partial}{\partial r}\left\{\frac{1}{r^2}\left[\frac{\partial}{\partial r}(r^2 u_r)\right]\right\} \tag{5.35}$$

The strain is obtained from Eq. (5.33) by taking account of the above initial and boundary conditions as follows

$$u(r, t) = \frac{6}{\pi}\Delta a_0 \sum_{n=1}^{\infty} \exp\left(-\frac{t}{\tau_n}\right)\frac{(-1)^{n+1}}{n}\left[\frac{\sin(\beta_n r) - (\beta_n r)\cos(\beta_n r)}{(\beta_n r)^2}\right] \tag{5.36}$$

In addition, the stresses are as follows

$$\sigma_{rr} = 2\Pi_0 \sum_{n=1}^{\infty} \exp\left(-\frac{t}{\tau_n}\right)(-1)^{n+1}\frac{\sin(\beta_n r)}{\beta_n r} \tag{5.37}$$

β_n and τ_n are given as follows

$$\beta_n \equiv \frac{n\pi}{a} \tag{5.38}$$

$$\tau_n = \frac{a^2}{n^2 D\pi^2} \tag{5.39}$$

where

$$\Pi_0 = 3\frac{\Delta a_0}{a}K \tag{5.40}$$

See Appendix 3.4 for the derivation of Eqs. (5.36) and (5.37).

It is more practical to express the swelling behavior using the degree of swelling Q/Q_0 instead of $u(r, t)$, where Q and Q_0 are the volume or weight after and before swelling. The degree of swelling is given by

$$\begin{aligned}\left(\frac{Q}{Q_0}\right)^{1/3} &\equiv \frac{1 - u(a,t)/a}{1 - \Delta a_0/a}\\ &= \left(1 - \frac{\Pi_0}{3K}\right)^{-1}\left\{1 - \frac{2\Pi_0}{\Pi^2 K}\sum_{n=1}^{\infty}\frac{1}{n^2}\exp(-\beta_n^2 Dt)\right\}\end{aligned} \tag{5.41}$$

The swelling rate is given by

$$\tau_n^{-1} \equiv D\beta_n^2 = D\left(\frac{n\pi}{a}\right)^2 \tag{5.42}$$

However, since diffusion is mainly governed by the longest relaxation time τ_1, we obtain

$$\tau_1 \cong \tau = \frac{a^2}{\pi^2 D} \tag{5.43}$$

Furthermore, for $t > \tau_1$, the change in radius is approximated by the following equation

$$\frac{a - a(t)}{a - a_0} = \frac{a - a(t)}{\Delta a_0} \cong \frac{6}{\pi^2}\exp\left(-\frac{t}{\tau}\right) \tag{5.44}$$

5.3.3 Strain, Stress, and Swelling Rate of Spherical Gels

In this section, the swelling equation is discussed in more detail. Figure 5.6 shows the relationship between the reduced radius r/a and the normalized strain $u(r, t)/\Delta a_0$ (Eq. (5.36)) at various reduced times $t/\tau \quad (\tau \equiv \tau_1)$: at $t/\tau = 0$ the strain is a linear function of r. The strain is zero at the center of the gel, *i.e.* $r/a = 0$, and maximum at $r/a = 1$. After immersing the gel in water, the gel begins to swell from the gel surface $r/a = 1$, as indicated by the decrease in $u(r = a, \ t)$. Thereafter, the envelope of the swollen gel surface gradually moves into the interior of the gel. At $t = \infty$ ($t/\tau = 100$ in the figure), $u(r, \ t)$ is zero over the entire gel.

Figure 5.7 shows the relationship between the reduced radius r/a and the ratio of the gel stress to the osmotic pressure difference $\sigma_{\rm rr}(r, t)/\Delta\Pi_0$ at various reduced times t/τ. The time shown here is normalized by the relaxation time $\tau = a^2/D\pi^2$ (Eq. (5.43)). Inside the gel, the reduced stress $\sigma_{rr}(r, t)/\Delta\Pi_0$ is uniformly 1 at $t/\tau = 0$. However, when immersed in water, a finite stress is applied inside the gel, but it quickly becomes zero at the gel surface. As can be seen from the curves at various times t/τ, the stress away from the surface decreases with time; at $t = \infty$ (where $t/\tau = 100$), and the stress on the entire gel vanishes.

Figure 5.8 shows the change in strain (ratio of radius change) of a spherical gel over time. The strain has a maximum at $t/\tau = 0$, rapidly changes immediately up on the start of swelling ($0 \ll t/\tau < 0.5$), and then decreases exponentially (the gel swells). Note that the change in swelling of the spherical gel over time is universal with respect to the reduced relaxation time $\tau = a^2/D\pi^2$. The relaxation time of swelling is proportional to the square of the gel size. As shown in Fig. 5.9, for a

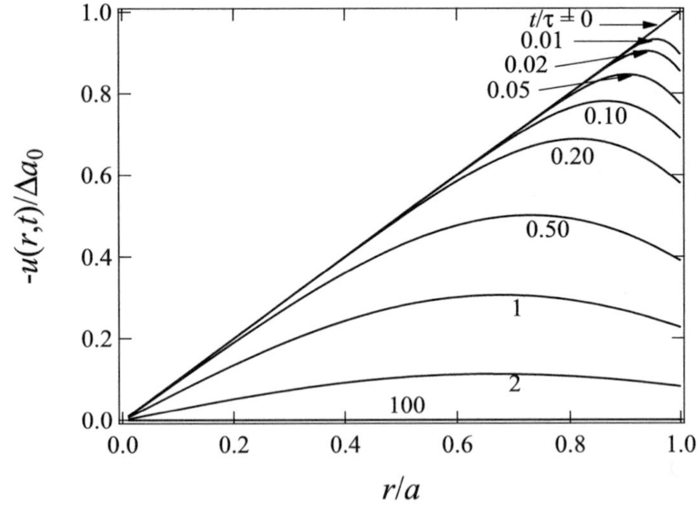

Fig. 5.6 Relative position dependence of the reduced strainr/a of spherical gels for various values of t/τ

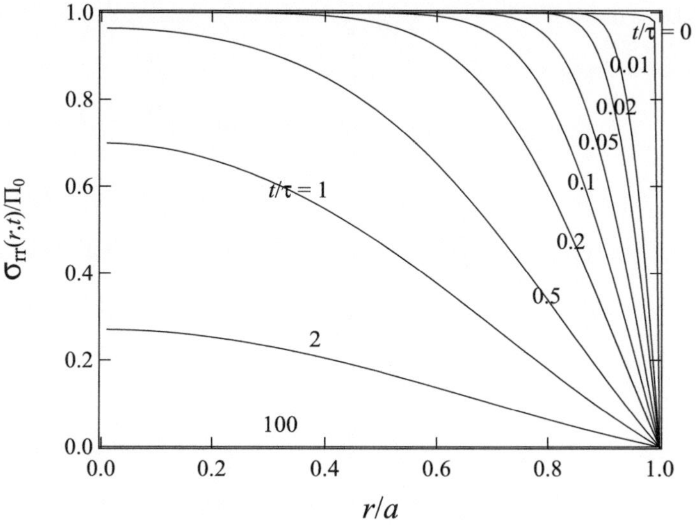

Fig. 5.7 Reduced radius r/a dependence of the ratio of stress $\sigma_{rr}(r, t)$ to the osmotic pressure difference $\Delta\Pi_0$ for various values of t/τ

spherical gel with a diffusion coefficient $D = 10^{-7}$ cm^2/s, the relaxation time of swelling at $a = 0.001$ cm is about 1 second, while at $a = 0.01$ cm it takes 100 seconds. $a = 0.1$ cm takes 10^4 seconds (about 3 hours) and $a = 1$ cm takes 10^6 seconds (about 11.5 days).

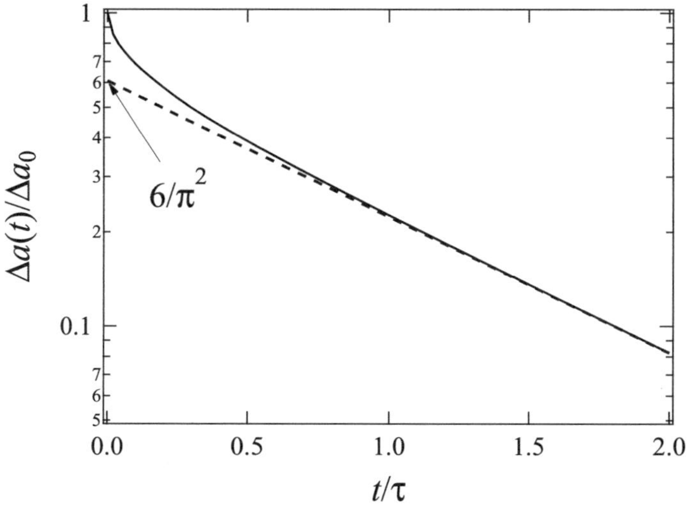

Fig. 5.8 Relationship between the strain (ratio of radius change) and the reduced time t/τ for spherical gels

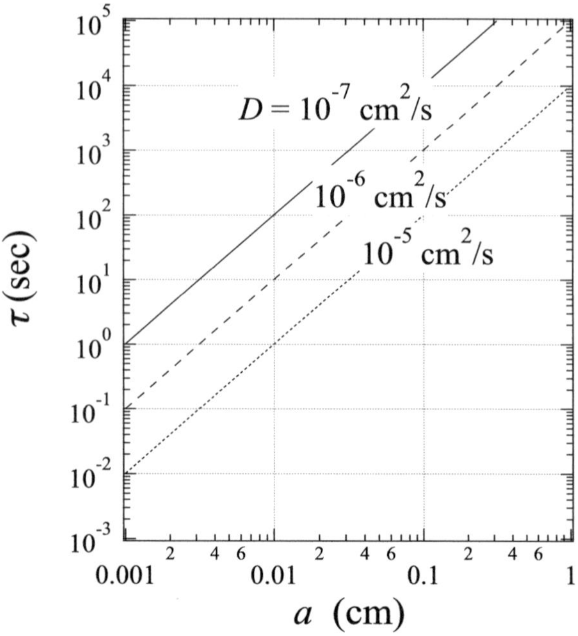

Fig. 5.9 Relation between final gel radius a and relaxation time τ for various diffusion coefficients

5.3.4 Swelling of Non-spherical Gels

For non-spherical gels, such as long cylindrical gels and thin disk gels, the strain energy accumulates during swelling because the strain is anisotropic. The gel spontaneously tries to minimize this strain energy, which determines its shape during and after swelling. This is why the gel maintains its shape during swelling (see Fig. 5.1). The shear energy is given by

$$F_{\mathrm{sh}} = \mu \int_V \left[\left(u_{xx} - \frac{T}{3} \right)^2 + \left(u_{yy} - \frac{T}{3} \right)^2 + \left(u_{zz} - \frac{T}{3} \right)^2 \right] \mathrm{d}V \tag{5.45}$$

where $T(= u_{ll})$ is the trace of the strain tensor u_{ik} (Eq. (5.5)) and the integration is over the entire gel. The conditions for minimizing the strain energy are

$$\delta F_{\mathrm{sh}} = 0 \tag{5.46}$$

For spherical gels, this condition is automatically satisfied, but for cylindrical gels or thin disk gels with large aspect ratios, the swelling equation under the constraints of Eq. (5.46) must be solved.

For long cylindrical gels, the following relationship was derived by Li and Tanaka [5]

$$\frac{u_z(z,t)}{z} = \frac{u_r(a,t)}{a} \tag{5.47}$$

where z is the axial coordinate of the cylinder, r is the radial coordinate, and a is the radius of the gel. Equation (5.47) shows that for a cylindrical gel, the axial and radial strains are the same. That is, the swelling deformation is similar to the deformation schematically shown in Fig. 5.1. Similarly, for a thin disk-shaped gel of thickness $2a$, we obtain the following equation

$$\frac{u_r(r,t)}{r} = \frac{u_z(a,t)}{a} \tag{5.48}$$

Hence, the swelling deformation of the thin disk-shaped gel is similarly isotropic.

Once the swelling equations are derived for the radial direction ($u_r(a, t)$ for cylindrical gels and $u_z(a, t)$ for disk-shaped gels), $u_z(z, t)$ and $u_r(r, t)$ are obtained from Eqs. (5.47) and (5.48), respectively. The strain of a long cylindrical gel is obtained, after a rather tedious calculation, as follows

$$\begin{aligned} u_r(r,t) &= \Delta a \sum_n A_n J_1(q_n r) \exp\left(-D_n q_n^2 t\right) \\ u_z(z,t) &= \frac{z}{a} \Delta a \sum_n A_n J_1(\alpha_n) \exp\left[-D_n(a,t) q_n^2 t\right] \end{aligned} \tag{5.49}$$

where A_n, α_n, and q_n are constants and eigenvalues ($\alpha_n = q_n a$) determined by the initial and boundary conditions, respectively. The exact form is not given here because these are complex functions and are beyond the scope of this document. For more details, please refer to the original paper [5].

Let us consider the case where the shear modulus is not negligible. We define R as the ratio of the shear modulus to the longitudinal shear modulus.

$$R = \frac{\mu}{K + (4/3)\mu} \tag{5.50}$$

Note that R varies from 0 ($\mu = 0$) to 3/4 ($\mu/K = \infty$). We now simplify Eq. (5.49) as follows

$$\begin{aligned} u_r(r,t) &\approx \Delta a \sum_n B_n \exp(-t/\tau) \\ u_z(z,t) &\approx \frac{z}{a} \Delta a \sum_n B_n \exp(-t/\tau) \end{aligned} \tag{5.51}$$

However,

$$B_n \equiv A_n J_1(\alpha_n) \tag{5.52}$$

The relationship between the first-order eigenvalue α_1 and the coefficient B_1 and R is shown in Fig. 5.10. Equation (5.52) is further simplified to,

$$u_r(r,t) \approx \Delta a \frac{r}{a} \exp(-t/\tau)$$
$$u_z(z,t) \approx \Delta a \frac{z}{a} \exp(-t/\tau) \tag{5.53}$$

The diffusion coefficient is a function of gel position and time, but is constant at $r = a$,

$$D_e = \frac{2}{3} D_0 \tag{5.54}$$

Similarly to the long cylindrical gel, the following equation is obtained for the thin disk-shaped gel

$$u_r(r,t) \approx \Delta a \frac{r}{a} \sum_n B_n \exp(-t/\tau_n)$$
$$u_z(z,t) \approx \Delta a \frac{z}{a} \sum_n B_n \exp(-t/\tau_n) \tag{5.55}$$

$$u_r(r,t) \approx \Delta a \frac{r}{a} \sum_n B_n \exp(-t/\tau)$$
$$u_z(z,t) \approx \Delta a \frac{z}{a} \sum_n B_n \exp(-t/\tau) \tag{5.56}$$

and

$$D_e = \frac{1}{3} D_0 \tag{5.57}$$

B_1 and τ are obtained from a plot of $\ln[u_r(r,t)]$ or $\ln[u_z(z,t)]$ versus t in a swelling experiment; the intercept and slope at $t = 0$ give B_1 and τ, respectively. Figure 5.11 shows the relationship between the diffusion coefficient $D_e(R)$ normalized by the diffusion coefficient D_0 of the spherical gel and the distance from the gel center. As shown here, $D_e(R)$ is a function of R, but at the gel surface it is 2/3 and 1/3 of the spherical gel regardless of R.

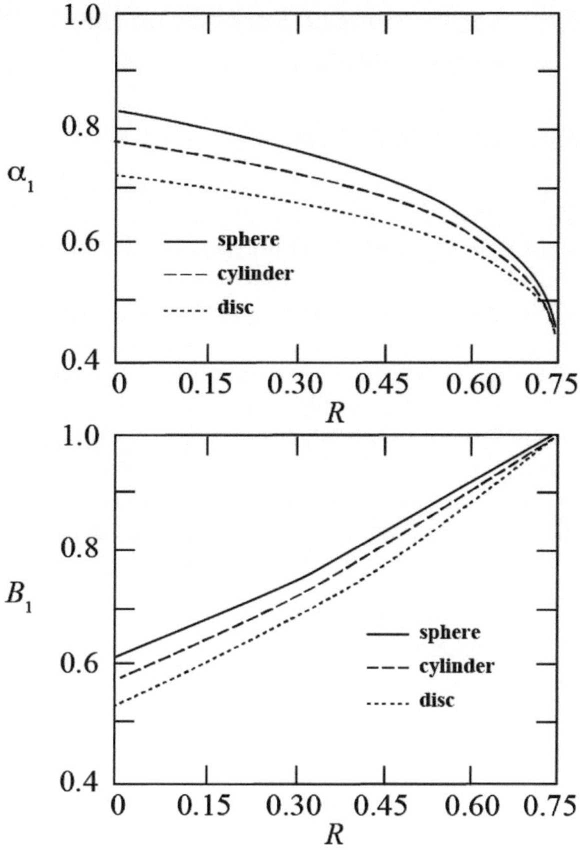

Fig. 5.10 Ratio of the shear modulus to the longitudinal shear modulus, R, and the first-order eigenvalue α_1 and the first-order coefficient B_1 (Li and Tanaka [5]; reproduced with permission, © American Institute of Physics)

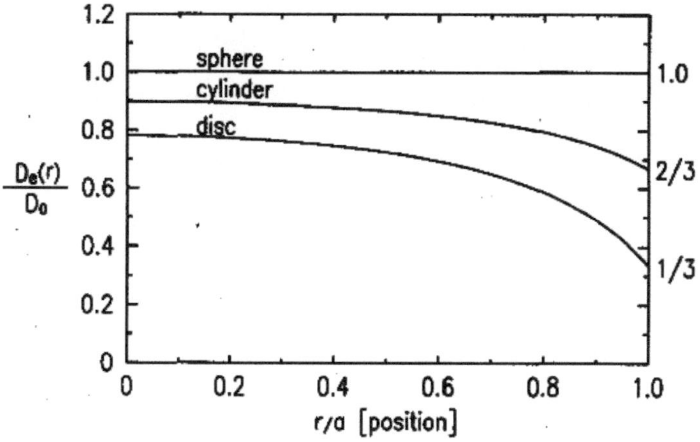

Fig. 5.11 Relation between the effective cooperative diffusion coefficient $D_e(R)$ normalized by the diffusion coefficient of spherical gel $D_0(= (K + 4\mu/3)/f)$ and the reduced distance r/a from the gel center (Li and Tanaka [5]; reproduced with permission, © American Institute of Physics)

5.4 Applications of Theories on the Kinetics of Swelling and Shrinking

5.4.1 *Spherical Gels*

The validity of the kinetics of gel swelling was experimentally confirmed by Tanaka and Fillmore for spherical polyacrylamide gels [3]. 5 g of acrylamide and 0.133 g of *N,N'*-methylenebisacrylamide were added to 100 mL of water and 5 mL of the solution was used to prepare the gel. 5 mL of the solution was used for gel preparation. The gelation reaction was initiated by the addition of *N,N,N',N'*-tetraethylethylenediamine in the presence of ammonium persulfate. Spherical gels were prepared by pipetting the gel precursor solution into paraffin oil to obtain spherical gels with several diameters a ranging from 1.0 mm to 2.8 mm. The diffusion coefficient obtained by swelling measurement was $D = 3.2 \times 10^{-7}$ cm^2/s, which was in good agreement with the value obtained by dynamic light scattering (DLS) measurement ($D_{DLS} = 3(\pm 0.5) \times 10^{-7}$ cm^2/s) [3].

5.4.2 *Non-spherical Gels*

The theory of swelling and shrinking of non-spherical gels was developed by Li and Tanaka [5] and experimentally verified. Polyacrylamide gels prepared by free radical polymerization were used as samples (same preparation method as for spherical gels). Three types of gels were prepared: (1) short cylindrical gels of equal height and diameter (control for spherical gels), (2) long cylindrical gels of 1.33 mm diameter ($2a = 1.33$ mm), and (3) thin disk-shaped gels of 1.33 mm thickness ($2a = 1.33$ mm). The aspect ratios of the long cylindrical gel (length to diameter ratio) and the thin disk-shaped gel (diameter to thickness ratio) are 30 and 20, respectively. Figure 5.12 shows the time-dependent results of the relative strain for (a) cylindrical gels of equal diameter and length, (b) elongated cylindrical gels, and (c) thin disk gels in the swelling experiments. The results show that (1) the relaxation times of cylindrical gels with equal diameter and length are the same both in the diameter and length, and (2) the relaxation times of the diameter for long cylindrical gels (b) and of the thickness for thin disk-shaped gels are much larger than those for the equisized cylindrical gels (a). The fact (1) means that the gel swells by maintaining its shape, i.e., affine swelling, and the fact (2) means that gels with large aspect ratios swell more slowly than spherical gels.

The swelling of spherical gels has three degrees of freedom, while that of cylindrical gels has two degrees of freedom (*x*- and *y*-directions) and that of thin disk gels is only one degree of freedom (*z*-direction). Therefore, the effective diffusion coefficients of long cylindrical gels and thin disk gels are 2/3 and 1/3 of those of spherical gels, respectively. For example, the degree of swelling of a thin disk-shaped gel is determined by its thickness ($2a$). This means that the swelling velocity in the

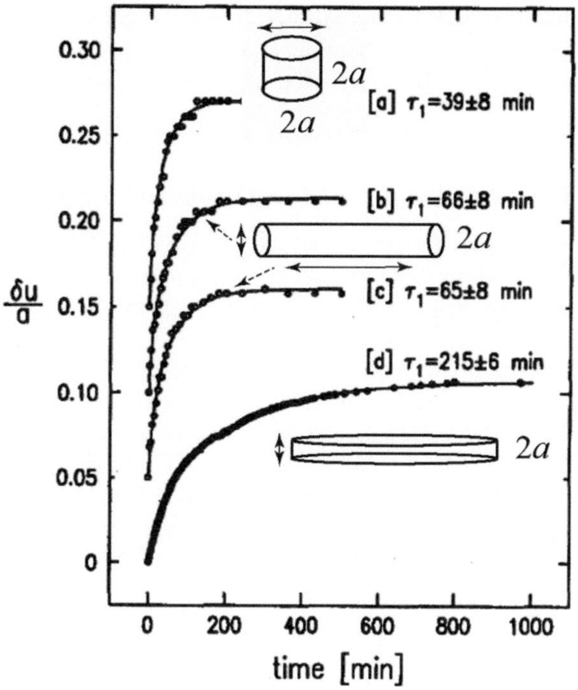

Fig. 5.12 Position dependence of the effective cooperative diffusion coefficient normalized by the diffusion coefficient $D_0 = ((K + 4\mu/3)/f)$ for spherical gels. The relative strain over time is shown for the diameter for cylindrical gels with equal diameters and lengths (**a**), and diameter (**b**) and length (**c**) of a long cylindrical gel and thickness (**d**) of a thin disk-shaped gel (Li and Tanaka [5]; reproduced with permission, © American Institute of Physics)

z-direction is somewhat smaller, whereas in the radial direction it is much larger than that of the spherical gel.

5.4.3 *Rapid Swelling*

When environmentally sensitive polymer gels are applied to actuators, sensors, and drug delivery systems, it is desirable for the gels to contract and swell rapidly. However, the rate of swelling and shrinking is in principle governed by cooperative diffusion. This means that the characteristic time τ of swelling/shrinking depends on the gel size R and the co-diffusion coefficient D, i.e.

$$\tau \approx \frac{R^2}{D} \tag{5.58}$$

In order to accelerate the rate of gel shrinking, special gels have been designed and prepared, such as (i) gels with hydrophilic chains or water molecules pathways through pores, (ii) gels with strong shrinking tendency due to the presence of dangling chains [6, 7], or (iii) gels with micropores created by gamma irradiation [8]. In the case of (ii), the dangling chains in the gel are free on one side, so the gel can be easily disintegrated by external stimuli, and it is considered important to add heterogeneous structures or conformations to the gel to speed up the response time. These findings have opened the door for the application of gels to drug delivery systems and sensors.

5.4.4 Critical Slowing Down and Phase Separation

The swelling equation can also be applied to the shrinking kinetics. However, as the gel approaches the critical miscibility point, the driving force for shrinking becomes negligible, resulting in a rapid deceleration of shrinking [9]. Figure 5.13 shows the temperature dependence of (a) the degree of swelling, (b) the relaxation rate, and (c) the thermal expansion coefficient of poly(*N*-isopropylacrylamide) (PNIPAM) gel. At low temperatures, the gel is in a swollen state. As the temperature is increased, the gel gradually shrinks, and at 34 °C, the gel dramatically shrinks to a collapsed state. The change in relaxation rate is shown in Fig. 5.13b. The change in relaxation rate becomes almost zero at the volume phase transition temperature. The thermal expansion coefficient, which represents the degree of change in gel volume with increasing temperature, becomes infinite at the volume phase transition temperature, as shown in Fig. 5.13c.

5.4.5 Ion Exchange Resin

Ion exchange resins are widely used in various applications such as water purification, dealkalization of water, recovery of useful substances, purification of sugars, catalysts, and dehydrating agents because of their excellent exchange capacity for a wide variety of ions. Although the precursors of ion exchange resins, such as styrene-divinylbenzene copolymers prepared by bulk or suspension polymerization, have been well studied, the microstructure of ion exchange resins in the gel state has not been well investigated.

Ikkai et al. investigated the swelling characteristics of cation exchange resins and their precursors with different crosslink densities (CD) in toluene in aqueous solutions of pH and salt concentration C_{salt} [10]. The microstructures of ion exchange resins (deuterated water) and their precursors (deuterated toluene solvent) were analyzed by small-angle neutron scattering (SANS), and the relationship between the microstructures and macroscopic swelling behavior of ion exchange resins and precursors was discussed based on the effects of crosslink density on swelling kinetics, equilibrium swelling degree and structural parameters. Ion exchange resins were prepared by sulfonation of the precursor and the degree of sulfonation estimated by titration was 40 wt%. Several precursors and cation exchange resins with different crosslink densities were prepared in particulate form. Measurements yielded diffusion coefficients $D = 4.40 \times 10^{-6}$ cm²/s for the precursor and $D = 3.18 \times 10^{-5}$ cm²/s for the ion exchange resin, indicating that the D of the ion exchange resin is about

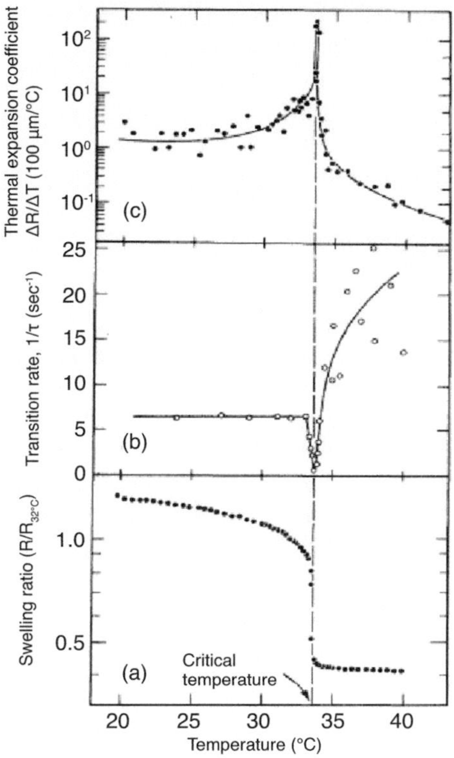

Fig. 5.13 Critical slowing down of poly(N-isopropylacrylamide) (PNIPAM) gels near the volume phase transition temperature. Temperature dependence of **a** the degree of swelling, **b** the relaxation rate, and **c** the thermal expansion coefficient [6] (Tanaka et al. [9]; reproduced with permission, © American Physical Society)

10 times greater than that of the precursor. The rapid swelling behavior of the ion exchange resin is attributed to electrostatic repulsive interactions and the Donnan potential generated by ion penetration. It was further discussed based on SANS data that porosity may affect the rapid swelling of ion exchange resins.

5.4.6 Polyvinyl Alcohol (PVA)-Ion Complex Gel (Slime)

Polyvinyl alcohol (PVA) undergoes gelation in aqueous solutions containing boric acid ions. This gel is called slime and is a type of physical gel (a gel in which cross-linking is in dynamic equilibrium). Gelation is caused by the formation of complexes (ion complexes) between PVA and boric acid ions, which have been industrially applied in the production of high-strength PVA fiber [11]. The mechanism and kinetics of complex formation in PVA gel have been extensively studied.

PVA can be chemically cross-linked with glutaraldehyde. Chemically cross-linked PVA gels have interesting borate concentration dependence. Shibayama et al. investigated the effect of borate ion concentration *b* on the swelling and shrinking behavior of chemically cross-linked PVA gels complexed with borate ions and their change over time [12]. Figure 5.14 shows the change in diameter d of a cylindrical PVA gel as a function of borate ion concentration b. The PVA gel shrinks at low borate ion concentrations, and the gel is opaque. However, at high boric acid concentrations ($b \geq 0.04$ mol/L), the gel swells and becomes transparent. Figure 5.15 shows the change over time of the diameter of PVA gels immersed in aqueous solutions of various borate concentrations. The swelling and shrinking behavior was analyzed using the Li-Tanaka theory of cylindrical gels, and the following important results were obtained.

(1) The cooperative diffusion coefficient D in the swelling of chemically cross-linked PVA gel is larger than that of free swelling without borate ions ($b = 0$). On the other hand, the shrinking rate is about 1/10 that of free swelling. These are explained by the fact that the ionization of PVA chains promotes swelling and the decrease in the miscibility (intramolecular cross-linking) slows down the shrinking.
(2) The macroscopic diffusion coefficient obtained by swelling measurements was about 4.0×10^{-7} cm^2/s, which is in good agreement with the value obtained by dynamic light scattering measurements.

This result also supports the validity and applicability of the kinetics of gel swelling and shrinking.

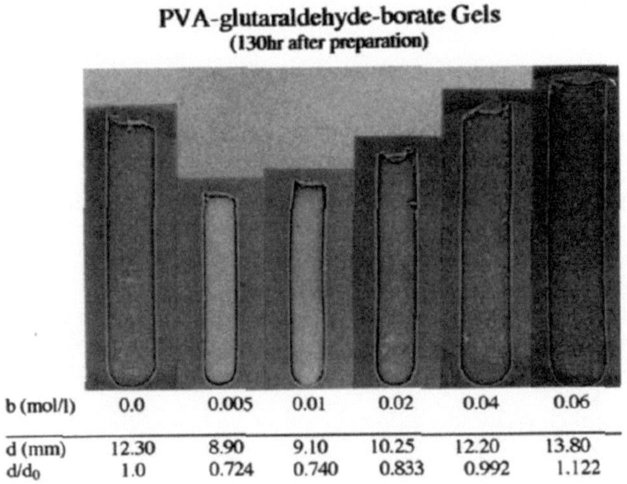

Fig. 5.14 Series of photographs showing the change over time of chemically cross-linked PVA gels soaked in a borate solution. The borate concentration b, the gel diameter dependence d, and the degree of swelling d/d_0 d_0 are also shown in the figure (Shibayama et al. [12]; reproduced with permission, © American Chemical Society)

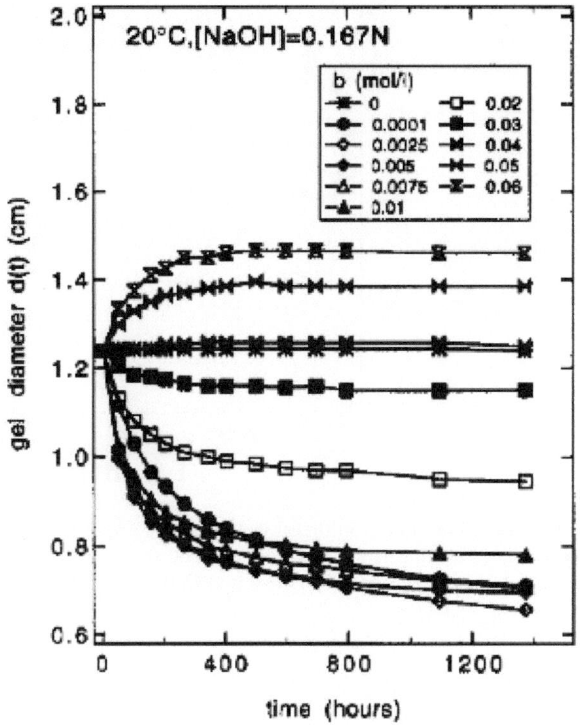

Fig. 5.15 Change over time in diameter $d(t)$ during swelling or shrinking of chemically cross-linked PVA gels after immersion in aqueous solutions containing borate ions at various concentrations b (Shibayama et al. [12]; reproduced with permission, © American Chemical Society)

5.5 Swelling and Shrinking Around the Volume Phase Transition Temperature

5.5.1 Weakly Charged Gels

In Sect. 5.4.4, we showed that the rate of shrinking decelerates rapidly as the volume phase transition temperature is approached (critical slowing down). In this section, we present our experimental findings on the kinetics of gel shrinking and swelling after a "temperature jump (T-jump)" above the volume phase transition temperature.

The kinetics of shrinking and swelling of poly(*N*-isopropylacrylamide/acrylic acid) (P(NIPAM/AAc)) weakly charged gels after a T-jump has been studied by Hirose and Shibayama [13]. Figure 5.16 shows the change over time of linear swelling d/d_0 during quasi-static heating from 20 °C (below T_c) to 40 °C (above T_c). The shrinking process is characterized by three steps. First, a sharp decrease in d/d_0 is observed (stage 1), followed by a plateau region (stage 2). Subsequently, d/d_0 again decreased linearly with time (Stage 3). In the plateau region, two coexisting phases were observed, as shown schematically in (c), indicating that the shrinking transition occurs at 40 °C in the case of T-jumping, while a higher temperature (i.e.,

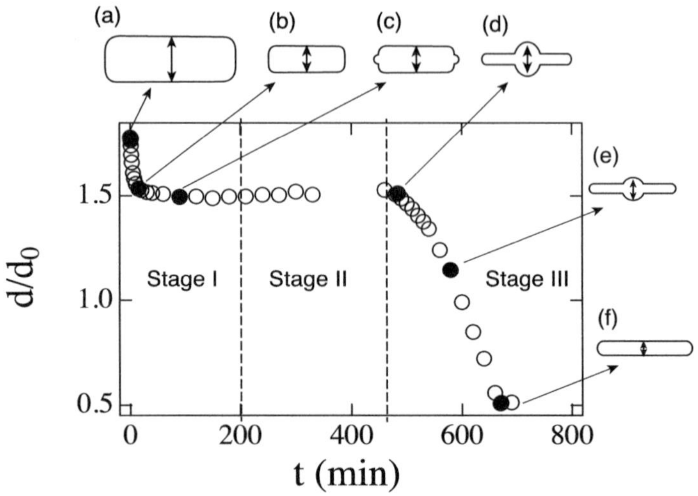

Fig. 5.16 Time course of linear swelling ratio d/d_0 of weakly charged gel P(NIPAM/AAc) during quasi-static heating from 20 to 40 °C. The shapes of the gel during the shrinking process are schematically shown in (**a**–**f**) (Hirose and Shibayama [13]; reproduced with permission, © American Chemical Society)

43 °C) is required for the gel to shrink in the case of quasi-static heating. This marked difference in behavior during the shrinking process can be explained as follows.

In the case of quasi-static heating, it always takes time for the gel to adjust to the new equilibrium state as the temperature changes. Above the Θ temperature of the PNIPAM gel (34 °C), hydrophilic regions are forms around the charged AAc segment. The hydrophilic regions remain swollen until the attractive interactions (hydrophobic segregation) exceed the repulsive interactions (electrostatic interactions and Donnan potential). This is why the gel does not shrink up to 43 °C and maintains a large d/d_0 value. On the other hand, at the T-jump, the gel does not form such a hydrophilic region, resulting in a low swelling state. This leads to an unstable state in the gel, with slow shrinking starting at temperatures lower than 40 °C (i.e., about 34 °C).

In contrast, the swelling process is well explained by the Tanaka-Fillmore theory. Figure 5.17 shows the change in d/d_0 during the swelling process as the temperature was varied from initial temperatures of 60, 50, and 40 °C ($\geq T_c$) to the destination temperature of 20 °C ($< T_c$). The observed diffusion coefficient was evaluated to be $\sim 2.2 \times 10^{-7}$ cm^2/s according to Eq. (5.60) (independent of temperature from 40 to 60 °C). This value is close enough to the literature value (3.7×10^{-7} cm^2/s) [3].

$$d(t) \approx d(\infty) + [d(0) - d(\infty)]\frac{6}{\pi^2}\exp(-t/\tau) \qquad (t > \tau) \tag{5.59}$$

$$D = \left(\frac{d(\infty)}{2\pi}\right)^2 \frac{1}{\tau} \tag{5.60}$$

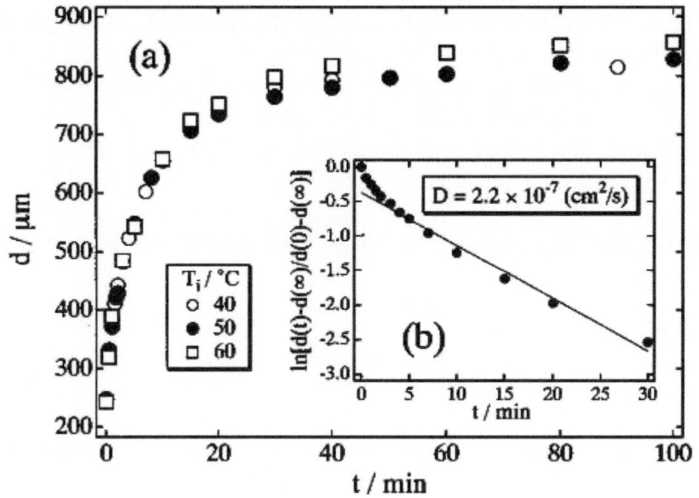

Fig. 5.17 Time course of the gel diameter d of P(NIPAM/AAc) gels swollen by T-jumps from 60, 50, and 40 to 20 °C. **a** Linear plot, **b** semi-logarithmic plot (Hirose and Shibayama [13]; reproduced with permission, © American Chemical Society)

Note that Eq. (5.59) is a practical formulation of Eq. (5.44). There is a marked difference in the shrinking behavior of weakly charged and uncharged gels [13]. In the case of P(NIPAM/AAc) gel, a three-stage shrinking process was observed. Stage 1 is the shrinking process from the swollen state to the swollen state, characterized by an exponential decrease in the gel diameter. In stage 2, the gel is shrunken from both sides of the cylindrical gel, but the diameter of the gel remains constant as in the swollen state. In Stage 3, the cylindrical gel is completely shrunken from both sides and the two sides come into contact. The diameter of the gel decreases rapidly and linearly with the passage of time.

In the case of neutral gels consisting only of PNIPAM, the shrinking process is greatly slowed down by macrophase separation. A more systematic study was performed by Shibayama and Nagai [14].

5.5.2 *Uncharged Gels: Kinetics of Shrinking and Phase Separation*

Shrinking and swelling of uncharged PNIPAM hydrogels upon T-jump have also been studied with respect to the effects of monomer concentration C_{NIPAM}, destination temperature T_d, and cross-linker concentration C_{BIS} [14]. Figure 5.18 shows the change over time of the morphology of PNIPAM gels when T-jumped from $T = 20$ °C to various destination temperatures T_d. When T-jumped to $T_d = 35$ °C, the surface roughness of the gel first became rough, then the gel became opaque and shrunk. The shrinking took about 20 min. On the other hand, at $T_d = 45$ °C, the gel first became opaque (appears black in the picture) and remained unchanged for more than a few hours. More interestingly, at $T_d = 55$ °C, vigorous bubble formation occurred first, followed by rapid shrinking 2 min later.

To understand this complex phenomenon, the following analysis is useful. Figure 5.19 shows the T_d dependence of the time $t_{1/2}$ required for the gel to shrink to half its size and the size of the gel d_{opaque} at the time the gel undergoes phase separation. Here, the characteristic time $t_{1/2}$ of gel shrinking is defined to satisfy the following relationship

$$\frac{d(t_{1/2}) - d(\infty)}{d(0) - d(\infty)} = \frac{1}{2} \tag{5.61}$$

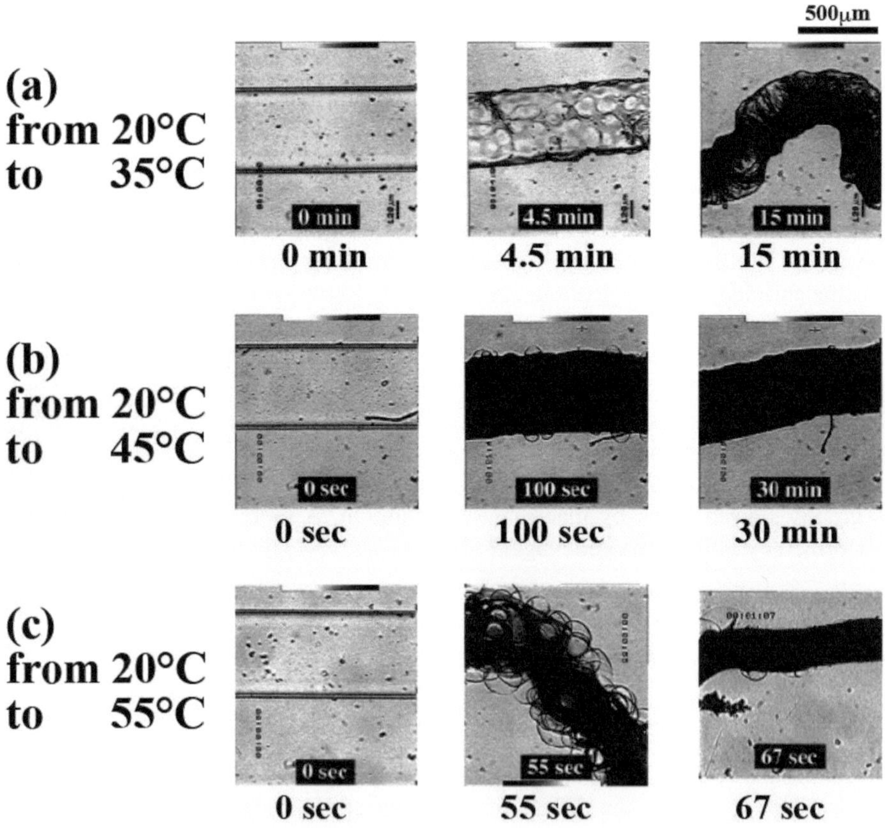

Fig. 5.18 Optical micrographs showing the morphological changes of PNIPA gel ($C_{\text{NIPAM}} = 690$ mM, $C_{\text{BIS}} = 8.62$ mM) after T-jumping from 20 °C to the destination temperatures. Destination temperatures T_d are **a** 35 °C, **b** 45 °C, and **c** 55 °C (Shibayama and Nagai [14]; reproduced with permission, © American Chemical Society)

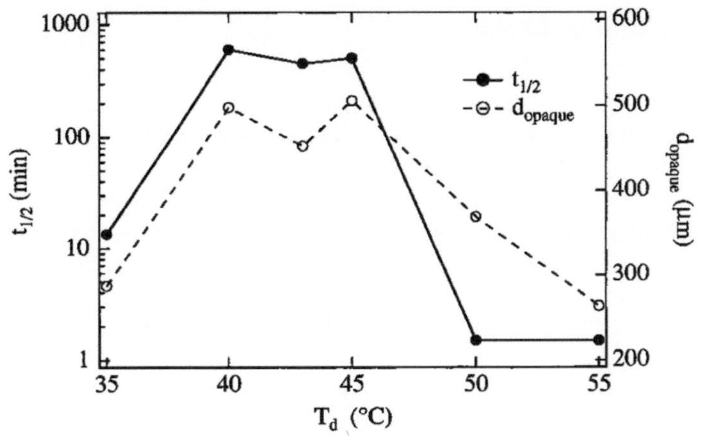

Fig. 5.19 T_d dependence of $t_{1/2}$ (the gel to shrink to half its size) and the d_{opaque} (the size of the gel d_{opaque} at phase separation) (Shibayama and Nagai [14]; reproduced with permission, © American Chemical Society)

Figure 5.19 suggests that the rate of gel shrinking (i.e., $t_{1/2}$) is closely related to phase separation (gel turbidity). Figure 5.18b also shows that the gel shrinking freezes when phase separation occurs during the T-jump. However, if the gel shrinks faster than the phase separation, as shown in Fig. 5.19c, the gel shrinks rapidly.

The shrinking behavior is strongly dependent on the gel parameters (i.e., C_{NIPAM} and C_{BIS}) as well as the environmental parameters (i.e., T_d). When C_{NIPAM} drops from 690 mM (reference concentration) to 300 mM, the gel shrinks rapidly, forming coarse bubbles. On the other hand, when $C_{\text{NIPAM}} = 1000$ mM, the contraction is frozen as in Fig. 5.19b.

Figures 5.18 and 5.19 strongly suggest that the kinetics of gel shrinking is governed by many factors. It is influenced not only by environmental parameters such as initial and destination temperatures, but also by the composition of the gel, such as monomer and cross-linker concentrations.

Unlike shrinking behavior, the swelling behavior does not seem to be affected by these parameters. In fact, the swelling behavior is similar to that of weakly charged gels discussed in Sect. 5.5.2, and the kinetics of swelling is well explained by the Tanaka-Fillmore theory. It should also be noted that the diffusion coefficients obtained by swelling experiments are in good agreement with those obtained by dynamic light scattering measurements [14].

5.6 Concluding Remarks

In this chapter, the kinetics of gel swelling/shrinking is discussed from the viewpoints of theory, experiments, and applications. The characteristic features of the kinetics of polymer gels are the following: (1) Gel swelling is a similarity deformation (isotropic swelling). (2) Swelling phenomena of gels are similar to diffusion and the relaxation time for swelling is proportional to the square of the gel size. (3) For gels with large aspect ratios, the rate of swelling/shrinking is determined by the smallest axis, e.g., the diameter of long cylindrical gels and thickness of thin-disk gels. (4) The effective diffusion coefficient is dependent on the dimension of swelling, and is 1/3 and 2/3 for thin-disk gel and long cylindrical gel, with respect to that of a spherical gel, respectively. (5) The effective diffusion coefficient of gel swelling can be evaluated by swelling experiment, which is in good agreement with that obtained by dynamic light scattering.

The theory of kinetics of polymer gels has been developed by Tanaka-Fillmore by assuming that (1) the shear modulus is negligible in comparison with the bulk modulus and (2) the effective diffusion is constant during swelling. This theory significantly contributed to the understanding of gels. However, it is pointed out by Candau et al. [15] and by Onuki [16] and also in the Li-Tanaka theory [5] that the shear modulus is not negligibly small, particularly in good solvent systems. Then, Li and Tanaka pointed out that the effective diffusion coefficient is not constant but changes depending on the degree of swelling and on the position in the gel, which are of particular importance when discussing gel drying process and volume phase transitions.

More elaborate theories have been developed which take into account of these issues. However, it is beyond of the scope of this chapter, and it should be clear enough and should not go beyond the Tanaka-Fillmore and Li-Tanaka theories for understanding the basic physics behind the swelling/shrinking phenomena of polymer gels.

Appendix 1: Tensors

Tensors are used to relate physical quantities such as currents and electric fields. They are also used for coordinate transformations: an nth-rank tensor in an m-dimensional space is a mathematical object that has n indices and m^n components and obeys certain transformation rules. Each index of the tensor ranges over the number of dimensions of space. According to this rule, a scalar is a tensor of rank 0, *i.e.*, $(n, m) = (0, 1)$, and a vector (in the three dimension) $\mathrm{r} = (x,\ y,\ z)$ is a tensor of the first rank, *i.e.*, $(n, m) = (1, 3)$. In this section, we summarize the properties and rules of tensors.

Example: Conductivity Tensor

The Ohm's law for anisotropic solids is given by

$$\mathbf{J} = \sigma \mathbf{E} \tag{5.62}$$

Here, **J** is the current density and **E** is the electric field. In general, it reads

$$J_i = \sum_k \sigma_{ik} E_k \tag{5.63}$$

In ordinary three-dimensional space, the scalar conductivity σ is given by a set of nine elements σ_{ik}. This array of nine elements forms a tensor,

$$\begin{pmatrix} \sigma_{11} & \sigma_{12} & \sigma_{13} \\ \sigma_{21} & \sigma_{22} & \sigma_{23} \\ \sigma_{31} & \sigma_{32} & \sigma_{33} \end{pmatrix} \tag{5.64}$$

Example: Coordinate Rotation

If the coordinates are rotated by an angle θ from O-xy to O-$x'y'$, then (x,' y') is obtained with (x, y)

$$\begin{pmatrix} x' \\ y' \end{pmatrix} = \begin{pmatrix} \cos\theta & -\sin\theta \\ \sin\theta & \cos\theta \end{pmatrix} = \begin{pmatrix} x \\ y \end{pmatrix} \tag{5.65}$$

Note that the elements cannot be chosen arbitrarily and must satisfy the following relationship.

Contravariant Vector

The vector **r**' = $\{x_i'\}$ is written with **r** = $\{x_i\}$ by partial differentiation.

$$x_i' = \sum_j \frac{\partial x_i'}{\partial x_j} x_j \tag{5.66}$$

By learning this relation, if one can write,

$$A'_i = \sum_j \frac{\partial x'_i}{\partial x_j} A_j \tag{5.67}$$

A in Eq. (5.67) is called the contravariant vector.

Covariant Vector

On the other hand, other vector transformations are known. The gradient of scalar $\nabla\varphi$ defined by

$$\nabla\varphi = \mathbf{i}\frac{\partial\varphi}{\partial x_1} + \mathbf{j}\frac{\partial\varphi}{\partial x_2} + \mathbf{k}\frac{\partial\varphi}{\partial x_3} \tag{5.68}$$

can be transformed as follows

$$\frac{\partial\varphi'}{\partial x'_i} = \sum_j \frac{\partial\varphi}{\partial x_j}\frac{\partial x_j}{\partial x'_i} \tag{5.69}$$

where ϕ is a scalar function and $\phi \equiv \phi(x, y, z) = \phi(x', y', z') = \phi'$. Here, $\nabla\phi$ in Eq. (5.69) is the definition of covariant vector. In the following, as long as we are distinguishing contravariant and covariant, let us use a superscript for contravariant and a subscript for covariant. There are three types of tensor of rank 2;

(i) Contravariant tensor (coordinate transformation type: Eq. (5.68))

$$A'^{ij} = \sum_{kl} \frac{\partial x'_i}{\partial x_k}\frac{\partial x'_j}{\partial x_l} A^{kl} \tag{5.70}$$

(b) Mixed tensor

$$B'^{i}_{j} = \sum_{kl} \frac{\partial x'_i}{\partial x_k}\frac{\partial x_l}{\partial x'_j} B^k_l \tag{5.71}$$

(iii) Covariant tensor (gradient type: Eq. (5.69))

$$C'_{ij} = \sum_{kl} \frac{\partial x_k}{\partial x'_i}\frac{\partial x_l}{\partial x'_j} C_{kl} \tag{5.72}$$

In Cartesian coordinate systems, there is no difference between covariant and contrariant transformations. Hence, we simply write as

$$\frac{\partial x_j}{\partial x'_i} = \frac{\partial x'_i}{\partial x_j} = a_{ij} \tag{5.73}$$

In other coordinate systems, covariant and contrariant transformations are different.

Rules

(i) Addition:

If

$$\mathbf{A} + \mathbf{B} = \mathbf{C} \tag{5.74}$$

$$A^{ij} + B^{ij} = C^{ij} \tag{5.75}$$

(b) Symmetric tensor:

If

$$A^{mn} = A^{nm} \tag{5.76}$$

A is called a symmetric tensor. On the other hand,

(d) Antisymmetric tensor:

If

$$A^{mn} = -A^{nm} \tag{5.77}$$

it is called antisymmetric tensor.

(v) Decomposition into symmetric and antisymmetric tensors:

Clearly each tensor (of rank 2) can be decomposed to symmetric and antisymmetric

tensors. That is,

$$A^{mn} = \frac{1}{2}\left(A^{mn} + A^{nm}\right) + \frac{1}{2}\left(A^{mn} - A^{nm}\right) \tag{5.78}$$

Here, the first term of the RHS is a symmetric tensor and the latter is an antisymmetric tensor. Note that this decomposition is important in the theory of elasticity.

(f) Inner product:

$$\mathbf{A} \cdot \mathbf{B} = \sum_i A_i B_i \tag{5.79}$$

By contracting,

$$A_i B_i \tag{5.80}$$

This is an inner product of the vector.

(vii) Direct product:

By multiplying the components of the covariant vector (first-order tensor) a_i and the components of the contravariant vector (first-order tensor) b^j component by component, we obtain the general term $a_i b^j$

$$a'_i b'^j = \frac{\partial x_k}{\partial x'_i} a_k \frac{\partial x'_j}{\partial x_l} b^l \tag{5.81}$$

This is a second-rank tensor. By contracting

$$a'_i b'^j = a_k b^k \tag{5.82}$$

This is called direct product. A typical example is $\nabla \mathbf{E}$ in electromagnetism.

Note

Summation convention: When a subscript (letter, not number!) appears twice on one side of an equation, summation with respect to that subscript is implied.

Contraction: Contraction consists of setting two unlike indices (subscripts) equal to each other and then summing as implied by the summation convention.

Appendix 2: Bessel Functions

Bessel function:

$$\begin{aligned} J_{1/2}(z) &= \sqrt{\frac{2}{\pi z}} \sin z, \quad J_{-1/2}(z) = \sqrt{\frac{2}{\pi z}} \cos z \\ J_{3/2}(z) &= \sqrt{\frac{2}{\pi z}} \left(\frac{\sin z}{z} - \cos z \right), \quad J_{-3/2}(z) = \sqrt{\frac{2}{\pi z}} \left(\sin z + \frac{\cos z}{z} \right) \end{aligned} \tag{5.83}$$

Differentiation:

$$\begin{aligned} J'_\nu(z) \equiv \frac{\mathrm{d}J_\nu(z)}{\mathrm{d}z} &= \nu z^{-1} J_\nu(z) - J_{\nu+1}(z) \\ &= J_{\nu-1}(z) - \nu z^{-1} J_\nu(z) \\ &= \frac{1}{2} \left[J_{\nu-1}(z) - J_{\nu+1}(z) \right] \end{aligned} \tag{5.84}$$

Bessel's differential equation:

$$\frac{\mathrm{d}^2u}{\mathrm{d}z^2}+\frac{1}{z}\frac{\mathrm{d}u}{\mathrm{d}z}+\left(1-\frac{\nu^2}{z^2}\right)u=0 \tag{5.85}$$

The solution of the Bessel differential equation is

$$u(z)=AJ_{\nu}(z)+BJ_{-\nu}(z) \tag{5.86}$$

Appendix 3: Derivation of Equations

Derivation of Eq. (5.3)

$$\begin{aligned}\left|\delta\mathbf{r}'\right|^2&=(\mathrm{d}x_i+du_i)^2=\left(\mathrm{d}x_i+\frac{\partial u_i}{\partial x_k}\mathrm{d}x_k\right)^2\\&\cong \mathrm{d}x_i^2+2\frac{\partial u_i}{\partial x_k}\mathrm{d}x_i\mathrm{d}x_k\\&=|\delta\mathbf{r}|^2+2u_{ik}\mathrm{d}x_i\mathrm{d}x_k\end{aligned} \tag{5.87}$$

Derivation of Eq. (5.17) from Eq. (5.16)

Let us note the component of divergence, $\partial/\partial\mathrm{k}\equiv\nabla\mathrm{k}$. From Eq. (5.4), we see that

$$\frac{\partial u_{ik}}{\partial x_k}=\frac{1}{2}\left(\frac{\partial^2u_i}{\partial x_k^2}+\frac{\partial^2u_k}{\partial x_i\partial x_k}\right)=\frac{1}{2}\left(\nabla^2u_i+\frac{\partial}{\partial x_i}\nabla\cdot\mathbf{u}\right) \tag{5.88}$$

On the other hand, the

$$\frac{\partial}{\partial x_k}(\nabla\cdot\mathbf{u})\delta_{ik}=\frac{\partial}{\partial x_i}\nabla\cdot\mathbf{u} \tag{5.89}$$

Substituting Eqs. (5.88) and (5.89) into the first term in the right-hand parentheses of Eq. (5.16), we obtain

$$\frac{\partial}{\partial x_k}\left(u_{ik}-\frac{1}{3}\nabla\cdot\mathbf{u}\delta_{ik}\right)=\frac{1}{2}\left(\nabla^2u_i+\frac{\partial}{\partial x_i}\nabla\cdot\mathbf{u}\right)-\frac{1}{3}\frac{\partial}{\partial x_i}\nabla\cdot\mathbf{u}$$

$$= \frac{1}{6}\frac{\partial}{\partial x_i}\nabla \cdot \mathbf{u} + \frac{1}{2}\nabla^2 u_i \tag{5.90}$$

Therefore, the

$$\frac{\partial}{\partial x_k}\left\{2\mu\left(u_{ik} - \frac{1}{3}\nabla \cdot \mathbf{u}\delta_{ik}\right) + K\nabla \cdot \mathbf{u}\delta_{ik}\right\} = \mu\nabla^2 u_i + \frac{1}{3}\mu\frac{\partial}{\partial x_i}\nabla \cdot \mathbf{u} + K\frac{\partial}{\partial x_i}\nabla \cdot \mathbf{u} \tag{5.91}$$

$$\rho\frac{\partial^2 \mathbf{u}}{\partial t^2} = \mu\nabla^2\mathbf{u} + \left(K + \frac{1}{3}\mu\right)\nabla(\nabla \cdot \mathbf{u}) - f\frac{\partial \mathbf{u}}{\partial t} \tag{5.92}$$

Derivation of Eq. (5.27) from Eq. (5.25)

In spherical coordinates

$$\nabla \cdot \mathbf{u} = \frac{1}{r^2}\frac{\partial}{\partial r}\left(r^2 u_r\right) + \frac{1}{r\sin\theta}\frac{\partial}{\partial\theta}(\sin\theta u_\theta) + \frac{1}{r\sin\theta}\frac{\partial}{\partial\phi}u_\phi \tag{5.93}$$

Thus, we have

$$\begin{aligned}\nabla(\nabla \cdot \mathbf{u}) = \Bigg(&\frac{\partial}{\partial r}\left[\frac{1}{r^2}\frac{\partial}{\partial r}\left(r^2 u_r\right) + \frac{1}{r\sin\theta}\frac{\partial}{\partial\theta}(\sin\theta u_\theta)\right.\\ &\left.+\frac{1}{r\sin\theta}\frac{\partial}{\partial\phi}u_\phi\right], \ldots, \ldots\Bigg)\end{aligned} \tag{5.94}$$

Substituting $\mathbf{u} = \left(u_r, u_\theta = 0, u_\phi = 0\right)$, we obtain

$$[\nabla(\nabla \cdot \mathbf{u})]_r = \frac{\partial}{\partial r}\left[\frac{1}{r^2}\frac{\partial}{\partial r}\left(r^2 u_r\right)\right] \tag{5.95}$$

This is because $u_\theta = 0$ and $u_\phi = 0$. By manipulating equation

$$\begin{aligned}[\nabla(\nabla \cdot \mathbf{u})]_r &= \frac{\partial}{\partial r}\left[\frac{1}{r^2}\frac{\partial}{\partial r}\left(r^2 u_r\right)\right] = \frac{\partial}{\partial r}\left[\frac{1}{r^2}\left(2ru_r + r^2\frac{\partial u_r}{\partial r}\right)\right]\\ &= \frac{\partial}{\partial r}\left(\frac{2u_r}{r} + \frac{\partial u_r}{\partial r}\right)\\ &= \frac{2}{r}\frac{\partial u_r}{\partial r} + \frac{\partial^2 u_r}{\partial r^2} - \frac{2u_r}{r^2}\end{aligned} \tag{5.96}$$

On the other hand, the

$$\Delta \mathbf{u} = \nabla^2 \mathbf{u} = \nabla(\nabla \cdot \mathbf{u}) \tag{5.97}$$

Thus, $[\nabla(\nabla \cdot \mathrm{u})]_r$ and $(\Delta \mathbf{u})_r$ are equivalent. Therefore, we have

$$\frac{\partial u_r}{\partial t} = D\frac{\partial}{\partial r}\left[\frac{1}{r^2}\frac{\partial}{\partial r}(r^2 u_r)\right] \tag{5.98}$$

As a result, we obtain Eq. (5.27).

Derivation of Eqs. (5.36) and (5.37) from Eq. (5.35)

Equation (5.99) is

$$\frac{\partial u}{\partial t} = D\frac{\partial}{\partial r}\left[\frac{1}{r^2}\frac{\partial}{\partial r}(r^2 u)\right] \tag{5.99}$$

By letting

$$c(r,t) \equiv r^2 u(r,t) \tag{5.100}$$

we obtain

$$\frac{\partial c}{\partial t} = Dr^2\frac{\partial}{\partial r}\left(\frac{1}{r^2}\frac{\partial c}{\partial r}\right) = D\left(\frac{\partial^2 c}{\partial r^2} - \frac{2}{r}\frac{\partial c}{\partial r}\right) \tag{5.101}$$

Furthermore, by letting

$$c(r,t) \equiv f(r)e^{-kt} \tag{5.102}$$

we obtain

$$f'' - \frac{2}{r}f' + \frac{k}{D}f = 0 \tag{5.103}$$

Furthermore, by letting

$$f(r) \equiv r^{3/2} g(r) \tag{5.104}$$

the following equation is obtained

$$g'' + \frac{1}{r}g' + \left(\beta^2 - \frac{9/4}{r^2}\right)g = 0 \tag{5.105}$$

$$\beta \equiv \sqrt{\frac{k}{D}} \tag{5.106}$$

Since Eq. (5.105) is a Bessel differential equation (Eq. (5.84)), the solution is

$$g(r) = AJ_{3/2}(\beta r) + BJ_{-3/2}(\beta r) \tag{5.107}$$

Thus, we have

$$f(r) = r^{3/2} g(r) = r^{3/2}\left[AJ_{3/2}(\beta r) + BJ_{-3/2}(\beta r)\right] \tag{5.108}$$

$$c(r) = e^{-kt} f(r) = e^{-kt} r^{3/2}\left[AJ_{3/2}(\beta r) + BJ_{-3/2}(\beta r)\right] \tag{5.109}$$

Here, the boundary conditions are given for gel swelling

Boundary conditions:

In Eq. (5.28), the only stress on the gel at time $t = 0$ is the osmotic pressure Π_0. Thus,

$$\sigma_{rr} = \frac{K}{r^2}\frac{\partial}{\partial r}\left(r^2 u\right) = \Pi_0,\ (t = 0) \tag{5.110}$$

Also, the stress at the gel surface ($r = a$) during equilibrium swelling is zero,

$$\sigma_{rr} = \frac{K}{r^2}\frac{\partial}{\partial r}\left(r^2 u\right) = 0,\ (r = a) \tag{5.111}$$

Now, we want to solve $\frac{\partial}{\partial r}\left(r^2 u\right) = \frac{\partial c}{\partial r}$. However, we do not need the time-dependent part, we solve $\frac{\partial f}{\partial r}$.

From Eq. (5.107), we have

$$f'(r) = \beta r^{3/2}\left[AJ_{1/2}(\beta r) - BJ_{-1/2}(\beta r)\right] \tag{5.112}$$

Here, we used the formula (5.84) for the derivative of the Bessel function. Thus, we have

$$\sigma_{rr} = K\beta e^{-kt} r^{-1/2}\left[AJ_{1/2}(\beta r) - BJ_{-1/2}(\beta r)\right] \tag{5.113}$$

where, when $t = 0$, we have

$$\sigma_{rr}|_{t=0} = K\beta r^{-1/2}\left[AJ_{1/2}(\beta r) - BJ_{-1/2}(\beta r)\right] \tag{5.114}$$

From Eq. (5.109), B must be 0, i.e., $B = 0$, for $\sigma_{rr}|_{t=0}$ at $r \to 0$. Also, from Eq. (5.110), we have

$$J_{1/2}(\beta a) = 0 \tag{5.115}$$

Thus, we have

$$\beta a = n\pi \tag{5.116}$$

where n is an integer. Then, by putting

$$\beta_n = \frac{n\pi}{a} \tag{5.117}$$

Then, we obtain

$$\sigma_{rr}(r,t) = Kr^{-1/2}\sum_{n=1}^{\infty} e^{-k_n t}\beta_n A_n J_{1/2}(\beta_n r) \tag{5.118}$$

Here,

$$k_n = \beta_n^2 D = \frac{n^2\pi^2}{a^2}D \tag{5.119}$$

Similarly,

$$u(r,t) = r^{-1/2}\sum_{n=1}^{\infty} e^{-k_n t} A_n J_{3/2}(\beta_n r) \tag{5.120}$$

Next, let us obtain the coefficient A_n. First, it is written,

$$u(r,t=0) = r^{-1/2}\sum_{n=1}^{\infty} A_n J_{3/2}(\beta_n r) \tag{5.121}$$

And, from

$$u(r=a,t=0) = \Delta a_0 \tag{5.122}$$

we have

$$\Delta a_0 = a^{-1/2}\sum_{n=1}^{\infty} A_n J_{3/2}(\beta_n a) \tag{5.123}$$

Furthermore, we have

$$a^{1/2}\Delta a_0 = \sum_{n=1} A_n J_{3/2}(\beta_n a)$$

$$
\begin{aligned}
&= \sum_{n=1}^{\infty} A_n J_{3/2}(n\pi) \\
&= \sum_{n=1}^{\infty} A_n \sqrt{\frac{2}{\pi^2 n}} \left(\frac{\sin(n\pi)}{n\pi} - \cos(n\pi) \right) \\
&= \sum_{n=1}^{\infty} A_n \sqrt{\frac{2}{\pi^2 n}} (-1)^{n+1} \qquad (5.124)
\end{aligned}
$$

Then,

$$
\frac{\pi}{\sqrt{2}} a^{1/2} \Delta a_0 = \sum_{n=1}^{\infty} A_n \frac{(-1)^{n+1}}{n^{1/2}} \qquad (5.125)
$$

Here, by putting

$$
A_n \equiv \frac{(-1)^{n+1}}{n^{3/2}} \overline{A} \qquad (5.126)
$$

we have

$$
\frac{\pi}{\sqrt{2}} a^{1/2} \Delta a_0 = \sum_{n=1}^{\infty} \frac{1}{n^2} \overline{A} = \frac{\pi^2}{6} \overline{A} \qquad (5.127)
$$

Using these equations, we obtain

$$
\begin{aligned}
u(r, t) &= 3\sqrt{2\pi} \frac{\Delta a_0}{a} r^{-1/2} \sum_{n=1}^{\infty} e^{-k_n t} \frac{(-1)^{n+1}}{\beta_n^{3/2}} J_{3/2}(\beta_n r) \\
&= \frac{6}{\pi} \Delta a_0 \sum_{n=1}^{\infty} \exp\left(-\frac{t}{\tau_n} \right) \frac{(-1)^{n+1}}{n} \left[\frac{\sin(\beta_n r) - (\beta_n r) \cos(\beta_n r)}{(\beta_n r)^2} \right] \qquad (5.128)
\end{aligned}
$$

Thus, we obtain Eq. (5.36). Here,

$$
\tau_n = \frac{a^2}{n^2 \pi^2 D} \qquad (5.129)
$$

Similarly, we obtain

$$
\sigma_{rr}(r, t) = 6K \frac{\Delta a_0}{a} \sum_{\infty}^{n=1} e^{-k_n t} (-1)^{n+1} \frac{\sin(\beta_n r)}{\beta_n r} \qquad (5.130)
$$

Furthermore, by using

$$\Delta a_0 = \frac{\Pi_0 a}{3K} \tag{5.131}$$

We obtain

$$\sigma_{rr}(r, t) = 2\Pi_0 \sum_{\infty}^{n=1} e^{-k_n t}(-1)^{n+1} \frac{\sin(\beta_n r)}{\beta_n r} \tag{5.132}$$

Thus, we obtain Eq. (5.37).

References

1. K. Dusek, Collection Czechoslov. Chem. Commun. **30**, 3804 (1965)
2. T. Tanaka, L.O. Hocker, G.B. Benedek, Spectrum of light scattered from a viscoelastic gel. J. Chem. Phys. **59**, 5151–5159 (1973)
3. T. Tanaka, D.J. Fillmore, Kinetics of swelling of gels. J. Chem. Phys. **70**, 1214 (1979)
4. L. D. Landau, E.M. Lifshitz, Nauka, SSSR (1965)
5. Y. Li, T. Tanaka, Kinetics of swelling and shrinking of gels. J. Chem. Phys. **92**, 1365 (1990)
6. R. Yoshida, K. Uchida, Y. Kaneko, K. Sakai, A. Kikuchi, Y. Sakurai, T. Okano, Comb-type grafted hydrogels with rapid deswelling response to temperature changes. Nature **374**(6519), 240–242 (1995)
7. Y. Kaneko, K. Sakai, A. Kikuchi, R. Yoshida, Y. Sakurai, T. Okano, Influence of freely mobile grafted chain length on dynamic properties of comb-type grafted poly (N-isopropylacrylamide) hydrogels. Macromolecules **28**(23), 7717–7723 (1995)
8. R. Kishi, O. Hirasa, H. Ichijo, Fast responsive poly (N-sopropylacrylamide) hydrogels prepared by γ-ray irradiation. Polym. Gels Netw. **5**(2), 145–151 (1997)
9. T. Tanaka, E. Sato, Y. Hirokawa, S. Hirotsu, J. Peetermans, Critical kinetics of volume phase transition of gels. Phys. Rev. Lett. **55**(22), 2455 (1985)
10. F. Ikkai, M. Shibayama, Microstructure and swelling behavior of ion-exchange resin. J. Polym. Sci. Part B Polym. Phys. **34**(9), 1637–1645 (1996)
11. H. Fujiwara, M. Shibayama, J.H. Chen, S. Nomura, Preparation of high-strength poly (vinyl alcohol) fibers by crosslinking wet spinning. J. Appl. Polym. Sci. **37**(5), 1403–1414 (1989)
12. M. Shibayama, T. Takeuchi, S. Nomura, Swelling/shrinking and dynamic light scattering studies on chemically cross-linked poly (vinyl alcohol) gels in the presence of borate ions. Macromolecules **27**(19), 5350–5358 (1994)
13. H. Hirose, M. Shibayama, Kinetics of volume phase transition in poly (N-isopropylacrylamide-co-acrylic acid) gels. Macromolecules **31**(16), 5336–5342 (1998)
14. M. Shibayama, K. Nagai, Shrinking kinetics of poly (N-isopropylacrylamide) gels T-jumped across their volume phase transition temperatures. Macromolecules **32**(22), 7461–7468 (1999)
15. A. Peters, S.J. Candau, Kinetics of swelling of polyacrylamide gels. Macromolecules **19**(7), 1952–1955 (1986)
16. A. Onuki, Adv. Polym. Sci. 109, 63–121 (1993)

Chapter 6
Mechanics and Rheology of Gels (Urayama)

Kenji Urayama

Abstract This chapter presents a comprehensive overview of the mechanical and rheological properties of polymer gels, emphasizing the distinctive features that set gels apart from conventional solids. Section 6.1 introduces the fundamentals of gel elasticity—such as low elastic moduli, high deformability, and reversible shape changes—arising from the entropic elasticity of their polymer networks. Theoretical developments are traced from the affine and phantom network models to more advanced frameworks that incorporate finite extensibility, network defects, and deformation-mode dependence. Section 6.2 explores *diffusio-mechanical coupling*, where strain-induced solvent migration leads to time-dependent stress relaxation and swelling/shrinking phenomena. Various deformation scenarios are analyzed, including constant strain, strain rate, oscillatory strain, and compression, along with centrifugal and shear-induced effects. Section 6.3 addresses the sol-gel transition, highlighting its critical behavior through the lens of percolation and fractal theories. The chapter outlines rheological criteria for identifying the gel point—such as power-law scaling of the dynamic modulus—and discuss both universal and system-specific aspects of the critical exponents. Experimental data are examined to link viscoelastic properties with the underlying network structure and dynamics, offering insights into gelation process. Together, the sections illustrate how molecular structure, solvent transport, and mechanical deformation collectively govern the emergent properties of gels.

Keywords Gel elasticity · Rheology · Diffusio-mechanical coupling · Viscoelasticity · Critical gel · Strain-induced swelling · Sol-gel transition · Critical exponents · Gelation kinetics

Gels are exceptionally soft yet do not exhibit macroscopic flow. Even when placed in a container for a long time, a gel retains its original shape rather than conforming to the container's form. According to the principles of rheology, this behavior classifies

K. Urayama (✉)
Department of Material Chemistry, Kyoto University, Kyoto, Japan
e-mail: urayama.kenji.2s@kyoto-u.ac.jp

Y. Osada (ed.), *Science of Gel*, https://doi.org/10.1007/978-981-95-0301-8_6

gels as solids. What distinguishes gels from conventional solids is their extremely low elastic modulus and their remarkable ability to undergo reversible, high deformability—properties that are rarely found in other solid materials. These characteristics stem from rubber-like entropic elasticity, as described in Chap. 2. Due to their unique mechanical behavior, gels are heretical among solids—unconventional material with properties that make it irreplaceable in various industrial applications. For example, when considering materials that mimic biological tissues, gels are among the closest analogs, as much of the human body consists of gel-like substances.

At the microscopic level, the polymer networks in a gel exhibit high motility, with network strands undergoing micro-Brownian motion and the single bonds in the main chain rotating freely. Compared to rubber, gels have a lower network density, which results in a significantly lower elastic modulus. Despite this, gels share many mechanical characteristics with rubber. For instance, while metals and glasses undergo finite volume changes when deformed, rubber and gels exhibit minimal volume variation with a Poisson's ratio close to 0.5, indicating near-incompressibility. This behavior arises because polymer chains in rubber possess liquid-like high mobility.

Gels are thermodynamically semi-open systems composed of a polymer network and a solvent. They expel or absorb solvent in response to external stimuli, leading to substantial changes in volume. As discussed in Chaps. 2 and 4, gels exhibit strong sensitivity to environmental factors such as solvent composition, temperature, pH, and salinity, which influence their swelling behavior. Additionally, when mechanical deformation is applied to a gel, it can thermodynamically induce solvent migration, causing volume changes that significantly affect its mechanical response. This strain-induced volume change results in mechanical relaxation governed by the kinetics of swelling and shrinking, meaning that the relaxation time depends primarily on the gel's size rather than on conventional viscoelastic relaxation mechanisms. This phenomenon, known as diffusio-mechanical coupling, is a defining feature of soft matter systems [1].

The influence of strain-induced volume changes on gel mechanics depends on timescale and sample size. Because the diffusion constant of gels is relatively low $10^{-10}\text{m}^2/\text{s}$, a gel with a characteristic size of approximately 10 mm requires over a day to exhibit noticeable volume changes. Consequently, in standard mechanical tests—where deformation occurs over much shorter timescales—volume change effects are negligible, and gels behave similarly to rubber. However, for long-duration deformations, or in small-scale and fibrous gel systems, the impact of strain-induced volume change becomes significant even at conventional strain rates.

The elasticity of gels originates from their three-dimensional polymer network structure. This infinite network, formed through the linking of monomers or precursor polymers, has been a classical topic in polymer science, originally studied by Flory and others [2, 3]. More recently, the formation of polymer networks has been explored from the viewpoints of phase transitions, percolation theory, and critical phenomena. A key development in this field has been the ability to experimentally quantify the critical point—known as the gel point—at which an infinite polymer cluster forms. Rheological techniques allow researchers to trace the sol-to-gel transition by monitoring changes in mechanical behavior. At the gel point, the system exhibits a fractal structure with no characteristic length or time scale, a feature that can be detected through rheological measurements. Quantifying the gel point provides valuable insights into how key properties such as cluster size, viscosity, and elastic modulus evolve near the transition.

This chapter is structured as follows:

Sect. 6.1 covers the fundamentals of rubber elasticity, providing a foundation for understanding the mechanical behavior of gels.

Sect. 6.2 discusses strain-induced volume changes and the resulting mechanical relaxation due to diffusio-mechanical coupling.

Sect. 6.3 introduces the sol-gel transition, explains rheological methods for quantifying the gel point, and presents experimental insights into the gel formation process.

6.1 Elasticity of Gels

6.1.1 Entropy Elasticity

In this and subsequent chapters, we explore how the elasticity of gel and rubbers arises from a reduction in the conformational entropy of their constituent polymer chains under deformation. We also explain how the elastic modulus and stress-strain behavior correlate with the structural characteristics of the polymer network. For simplicity, we consider the mechanics of gels under conditions where volume remains constant during deformation. Additionally, we neglect the influence of polymer-solvent interactions on mechanical behavior. Under these assumptions, the fundamental physics governing the elasticity of gels is essentially the same as that of rubber elasticity. Unless otherwise stated, the primary difference between gels and rubbers lies in their network concentration.

To understand the origin of the force generated by elastic, isovolumetric deformation, we begin with a thermodynamic approach. Consider a solid of length L, which is stretched by an infinitesimal amount $\mathrm{d}L$ under an applied tension f under a constant temperature T. According to the first law of thermodynamics, the mechanical work $f\mathrm{d}L$ corresponds to a change in the Helmholtz energy A (where $A = U - TS$; U is internal energy, and S is entropy): $\mathrm{d}A = f\mathrm{d}L$. The force f is then given by the partial derivative of A with respect to L:

$$f = \left(\frac{\partial A}{\partial L}\right)_{T,V} \tag{6.1}$$

This expression can be rewritten in terms of U and S:

$$f = \left(\frac{\partial U}{\partial L}\right)_{T,V} - T\left(\frac{\partial S}{\partial L}\right)_{T,V} = f_U + f_S \tag{6.2}$$

This equation shows that the total force f consists of two components: (i) f_U, the energetic contribution arising from changes in internal energy U, and (ii) f_S, the entropic contribution due to changes in entropy S from deformation. In materials such as metals, glasses, and ceramics, deformation alters interatomic distances, and elasticity is governed primarily by interatomic potential energy (U). This type of elasticity is known as energetic elasticity, typically resulting in high elastic moduli.

In contrast, for gels and rubbers, the internal energy change upon deformation is negligible ($\mathrm{d}U \approx 0$). This is because the micro-Brownian motion of network chains allows them to adopt new conformations in response to macroscopic deformation without significantly altering interatomic distances or bond angles. When a network chain is stretched from its relaxed state, its end-to-end distance increases, reducing the number of accessible conformations Ω (Fig. 6.1). According to Boltzman's entropy formula:

$$S = k_\mathrm{B} \log \Omega \tag{6.3}$$

where k_B is the Boltzmann constant. The force f in gels and rubbers thus originates from the entropy decrease caused by stretching $[(\partial S / \partial L)_{T,V} < 0]$, leading to the concept of entropic elasticity:

$$f \approx f_S = -T\left(\frac{\partial S}{\partial L}\right)_{T,V} = -kT\left(\frac{\partial \log \Omega}{\partial L}\right)_{T,V} \tag{6.4}$$

As evident from this expression, f depends on how Ω decreases with increasing L. If we determine the functional form of $S(L)$ or Ω (L), we can derive the force-extension relationship $f(L)$ for the gel. To build intuition, we draw an analogy with the pressure p of an ideal gas, which is also entropic in origin (Fig. 6.1b). Under isothermal conditions, and assuming $[(\partial U/\partial V)_\mathrm{T} = 0]$, the pressure p is given by:

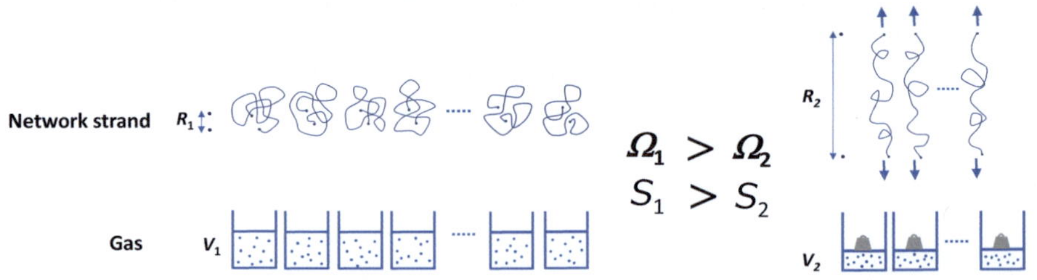

Fig. 6.1 Entropy elasticity of **a** single polymer chain and **b** gas. Stress of rubber and gel originates from a reduction in Ω (i.e., S) caused by the deformation ($R_1 \rightarrow R_2$) of the constituent chain, while pressure of gas results from a reduction in Ω (i.e., S) caused by the compression ($V_1 \rightarrow V_2$)

$$p = -\left(\frac{\partial A}{\partial V}\right)_T = T\left(\frac{\partial S}{\partial V}\right)_T = kT\left(\frac{\partial \log \Omega}{\partial V}\right)_T \tag{6.5}$$

Thus, p results from entropy decrease upon compression ($\mathrm{d}V < 0$). If the functional form of $S(V)$ is known, one can derive the pressure-volume relationship $p(V)$. For an ideal gas, the number of accessible states $\Omega(V)$ for N molecules in a volume V is proportional to $V^N/N!$:

$$\Omega(V) = aV^N/N! \tag{6.6}$$

where a is a proportionality constant. Using Eq. (6.3), S is given by:

$$S(V) = Nk_{\mathrm{B}} \log V + B \tag{6.7}$$

where B is a V-independent constant. From these expressions, the well-known Boyle-Charles law follows:

$$p(V) = \frac{NkT}{V} \tag{6.8}$$

Similarly, to derive $f(L)$ for a gel, we must determine Ω (L), the counterpart of Ω (V) in a gas. However, unlike the ideal gas case, Ω (L) for a real polymer network is challenging due to its complex structure. This challenge remains one of the classic problems in polymer physics. Over the years, many theoretical models have been proposed to account for the structure and chain interactions within real polymer networks. Just as the ideal gas law was refined (e.g., by van der Waals state equation) to include intermolecular interactions and molecular volume, theories of rubber elasticity have evolved to better describe real systems. In the 1940s, Kuhn and others introduced a statistical theory of rubber elasticity based on an ideal network [4, 5]. Since then, the theory has been progressively improved, incorporating more complex structural and molecular considerations. In the following chapters, we will examine the historical development and modern understanding of rubber elasticity, beginning with idealized models and progressing toward descriptions of real networks.

6.1.2 Elasticity of Single Polymer Chain

The elasticity of rubber originates from the entropy elasticity of its constituent flexible chains within the polymer network. This section examines the elasticity of a single flexible polymer chain. Following the approach outlined in the previous section, we first consider the conformational entropy of a polymer chain with an end-to-end distance R. For this purpose, we use the freely jointed chain model, which is the simplest model for a polymer chain. This model assumes no restrictions on the angles between adjacent bonds. When one end of a freely jointed chain, consisting of N segments of length b, is fixed at the origin, the probability of finding the other end at position $\mathbf{R} = (R_x, R_y, R_z)$ within a small volume element $dR_x dR_y dR_z$ is given by $p(\mathbf{R})dR_x dR_y dR_z$. When the chain is sufficiently long ($N >> 1$) and R is much smaller than the fully extended length ($R << R_{max} = Nb$), the probability density function $p(\mathbf{R})$ follows a Gaussian distribution function:

$$p(\mathbf{R})dR_x dR_y dR_z = D\exp\left(-\frac{3R^2}{2Nb^2}\right)dR_x dR_y dR_z \tag{6.9}$$

where D is an R-independent constant given by $D = [3/(2Nb^2)]^{3/2}$. A polymer chain governed by this distribution is referred to as a Gaussian chain. The mean-square end-to-end distance of a Gaussian chain is given by $\langle R_0{}^2\rangle = Nb^2$, where $\langle\ldots\rangle$ denotes averaging over all accessible conformations. From the ratio $\langle R_0{}^2\rangle/R_{max}{}^2 = N^{-1}$, it is evident that most of conformations are significantly shrunken. For instance, when $N = 10^4$, the relative size of the typical end-to-end distance compared to the fully extended length is merely $\langle R_0{}^2\rangle^{1/2}/R_{max} = 0.01$, meaning that the chain is largely coiled. Since the number of conformations of a chain with an end-to-end vector $\mathbf{R}$, denoted as $\Omega(\mathbf{R})$, is related to the total number of conformations W through $p(\mathbf{R}) \sim \Omega(\mathbf{R})/W$, the conformational entropy S_{chain} of a Gaussian chain is given by:

$$S_{chain}(\mathbf{R}) = k\log p(\mathbf{R}) + C = -\frac{3k}{2R_0^2}R^2 + C' \tag{6.10}$$

where C and C' are constants independent of $\mathbf{R}$. This equation indicates that S_{chain} decreases in proportionally to R^2 (as shown by the dashed line in Fig. 6.2a).

When one end of a Gaussian chain is fixed at the origin, the force required to isothermally displace the other end by $d\mathbf{R}$ can be derived using the thermodynamic relation:

$$\mathbf{f}_{chain} = -T\left(\frac{\partial S_{chain}}{\partial \mathbf{R}}\right)_T = \frac{3kT}{R_0^2}\mathbf{R} \tag{6.11}$$

Thus, the Gaussian chain behaves as a Hookean spring with an effective spring constant of $3kT/R_0{}^2$. The reduction in S as R increases leads to an increase in the restoring force f_{chain}. However, this model predicts that f_{chain} continues to increase linearly with R, allowing the chain to extend beyond its theoretical maximum length R_{max}, as illustrated by the dashed line in Fig. 6.2b. This unphysical prediction arises because the probability p remains finite even at $R = R_{max}$. A physically realistic model requires $p = S = 0$ at $R = R_{max}$, which is not the case for the Gaussian chain. Consequently, the Gaussian chain model fails to accurately describe mechanical behavior in highly stretched states (Fig. 6.2b).

To address this limitation, the inverse Langevin chain model provides a more rigorous description of $p(\mathbf{R})$. The corresponding probability density distribution for a freely jointed chain is given by:

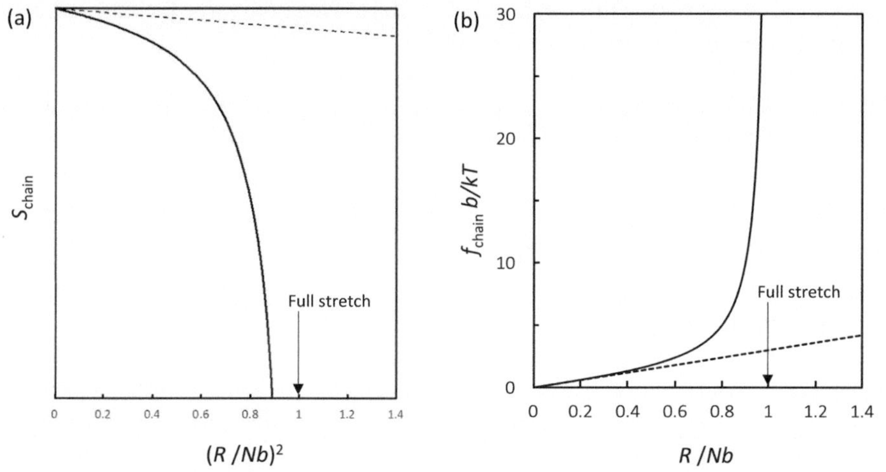

Fig. 6.2 **a** Configurational entropy as a function of $(R/Nb)^2$ for a single Gaussian chain (dashed line) and a single inverse-Langevin chain (solid line). **b** Force-stretch curves of a single Gaussian chain (dashed line) and a single inverse-Langevin chain (solid line) of $Nb = 10$

$$\ln p(R) = -N\left(\frac{R}{Nb}L^{-1}\left(\frac{R}{Nb}\right) + \ln\frac{L^{-1}\left(\frac{R}{Nb}\right)}{\sinh L^{-1}\left(\frac{R}{Nb}\right)}\right) + C'' \tag{6.12a}$$

where the Langevin function $L(b)$ is defined as:

$$L(\beta) = \frac{R}{\text{Nb}} = \coth\beta - \frac{1}{\beta} \tag{6.12b}$$

The conformational entropy S_{chain} and the entropic force f_{chain} for an inverse Langevin chain are then given by:

$$S_{chain}(\mathbf{R}) = -kN\left(\frac{R}{Nb}L^{-1}\left(\frac{R}{Nb}\right) + \frac{L^{-1}(R/Nb)}{\sinh(R/Nb)}\right) + C''' \tag{6.13}$$

$$\begin{aligned}\mathbf{f}_{chain} &= \frac{kT}{b}L^{-1}\left(\frac{R}{Nb}\right)\\ &= \frac{kT}{b}\left\{3\left(\frac{R}{Nb}\right) + \frac{9}{5}\left(\frac{R}{Nb}\right)^3\right.\\ &\quad\left.+\frac{297}{175}\left(\frac{R}{Nb}\right)^5 + \frac{1539}{875}\left(\frac{R}{Nb}\right)^7 + \cdots\right\}\end{aligned} \tag{6.14}$$

In the inverse Langevin chain model, the fully stretched state ($R_{max} = Nb$) corresponds to $p = S = 0$ and an infinite force $f_{chain} = \infty$ (solid line in Fig. 6.2b). This behavior accounts for the finite extensibility effect, where the force increases steeply as the chain approaches its fully extension.

For small deformations ($R << R_{max}$), Eq. (6.14) reduces to Eq. (6.11) for the Gaussian chain, since the higher-order corrections become negligible. This means that for moderate elongations, both models yield nearly identical force predictions.

6.1.3 *Affine Network Model*

6.1.3.1 Elastic Energy

Kuhn et al. [4, 5] developed a statistical mechanics model of rubber elasticity, assuming an ideal polymer network composed of Gaussian chains without structural defects, as described in the previous section. This model, known as the affine network model, is based on the following assumptions:

1. The dimension of a network strand (a chain between adjacent junctions) in the relaxed state is identical to that of the corresponding strand when isolated from the network.
2. The volume of the network remains unchanged during deformation.
3. The total entropy of the network is the sum of the entropies of the individual chains.
4. The end-to-end vector of a network strand scales proportionally with the macroscopic distortion applied to the network, a behavior known as affine displacement.
5. The junctions are embedded in the rubber matrix and do not exhibit thermal fluctuations.

A modified version of this model, which accounts for thermal fluctuations of the junctions by removing assumptions (4) and (5), is called the phantom network model. This alternative model will be discussed later. Collectively, the affine and phantom network models are referred to as classical rubber elasticity theories.

To derive the entropy of the affine network model, consider a cubic rubber sample with initial unit lengths, deformed such that its dimensions in each principal direction become λ_x, λ_y or λ_z (Fig. 6.3). By assumption (2), the volume remains constant:

$$\lambda_x \lambda_y \lambda_z = 1 \tag{6.15}$$

Under deformation, the end-to-end vector $\mathbf{R_0} = (R_{x0}, R_{y0}, R_{z0})$ in the relaxed state $\left(\lambda_x = \lambda_y = \lambda_z = 1\right)$ transforms into $\mathbf{R} = (R_\mathrm{x}, R_\mathrm{y}, R_\mathrm{z})$. From assumption (4), we obtain:

$$R_x = \lambda_x R_{x0}, R_y = \lambda_y R_{y0}, R_z = \lambda_z R_{z0}, \tag{6.16}$$

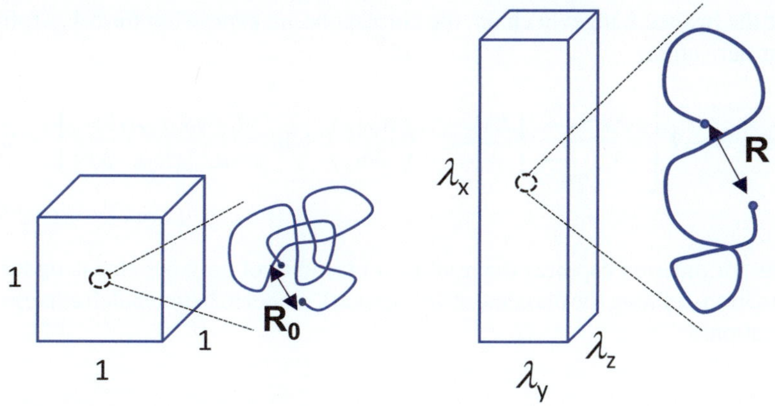

Fig. 6.3 Displacement of a network strand

The entropy change of a single Gaussian chain due to deformation, based on Eq. (6.10), is given by:

$$S_{\text{chain}}(\boldsymbol{R}) - S_{\text{chain}}(\boldsymbol{R}_0) = -\frac{3k}{2Nb^2}\left[\left(\lambda_x^2 - 1\right)R_{x0}^2 + \left(\lambda_y^2 - 1\right)R_{y0}^2 + \left(\lambda_z^2 - 1\right)R_{z0}^2\right] \tag{6.17}$$

The entropy of the network composed of n strands is given by summing S_{chain} (assumption (iii)) as

$$\Delta S_{\text{net}} = -\frac{3k}{2Nb^2}\left[\left(\lambda_{\text{x}}^2 - 1\right)\sum^n R_{\text{x0}}^2 + \left(\lambda_{\text{y}}^2 - 1\right)\sum^n R_{\text{y0}}^2 + \left(\lambda_{\text{z}}^2 - 1\right)\sum^n R_{\text{z0}}^2\right] \tag{6.18}$$

Since network strands are isotropically distributed in the relaxed state, we use the relationship $\Sigma R_{i0}{}^2 = \Sigma R_0{}^2/3 = nR_0{}^2/3$ (for $i = x,y,z$). Using this relationship and the assumption (i), we obtain

$$\Delta S_{\text{net}} = -\frac{nk}{2}\left(\lambda_x^2 + \lambda_y^2 + \lambda_z^2 - 3\right) \tag{6.19}$$

When no internal energy change occurs during deformation, the change in Helmholtz energy ΔA_{net} is given by:

$$\Delta A_{\text{net}} = -T\Delta S_{\text{net}} = \frac{nkT}{2}\left(\lambda_x^2 + \lambda_y^2 + \lambda_z^2 - 3\right) \tag{6.20}$$

This expression represents the stored elastic energy, which corresponds to the work done by the deformation.

For the inverse Langevin chain, the corresponding expression for ΔA_{net} follows a similar derivation:

$$\Delta A_{\text{net}} = -T\left[\frac{nNk}{3}\sum_{i=x,y,z}\left\{\frac{R_0\lambda_i}{Nb}L^{-1}\left(\frac{R_0\lambda_i}{Nb}\right)+\ln\frac{L^{-1}(R_0\lambda_i/Nb)}{sinhL^{-1}(R_0\lambda_i/Nb)}\right\}-S_0\right] \tag{6.21}$$

This formulation captures the nonlinear elasticity of polymer chains under large deformation, refining the classical affine network model to better match experimental observations.

6.1.3.2 Stress-Strain Relationships in Various Deformation Modes

The stress-strain behavior for any type of deformation can be derived from ΔA_{net}. Here, we obtain the stress-strain relationships for several typical deformations from Eqs. (6.20) and (6.21).

A. **Uniaxial Extension**

In uniaxial extension, a load is applied in a single direction (the x-direction). From Eq. (6.15), the lateral stretches are given by $\lambda_y = \lambda_z = \lambda_x^{-1/2}$. Using this condition in Eq. (6.20) yields the following expression for the elastic energy in uniaxial extension:

$$\Delta A_{\text{net}} = \frac{nkT}{2}\left(\lambda_x^2 + \frac{2}{\lambda_x} - 3\right)(\text{Uniaxial extension}) \tag{6.22}$$

The force f_x is calculated from Eq. (6.1) as:

$$f_x = \left(\frac{\partial A_{\text{net}}}{\partial L_x}\right)_T = \frac{1}{L_{x0}}\frac{\partial A_{\text{net}}}{\partial \lambda_x} = \frac{nkT}{L_{x0}}\left(\lambda_x - \frac{1}{\lambda_x^2}\right) \tag{6.23}$$

Before deformation, the cross-sectional area subjected to force is $L_{y0}L_{z0}$, while in the deformed state it is L_yL_z. For simplicity, we assume $L_{x0} = L_{y0} = L_{z0} = 1$ (Fig. 6.4). The engineering (or nominal) stress σ_x, defined as force per unit cross-sectional area in the unreformed state, is expressed as:

$$\sigma_x = \frac{f_x}{L_{y0}L_{z0}} = \frac{nkT}{V_0}\left(\lambda_x - \frac{1}{\lambda_x^2}\right) \tag{6.24}$$

where $V_0 = L_{x0}L_{y0}L_{z0}$ represents the initial volume. The true stress s_x, which considers the deformed cross-sectional area, is expressed as:

$$s_x = \frac{f_x}{L_yL_z} = \frac{nkT}{V_0}\left(\lambda_x^2 - \frac{1}{\lambda_x}\right) = \sigma_x\lambda_x \tag{6.25}$$

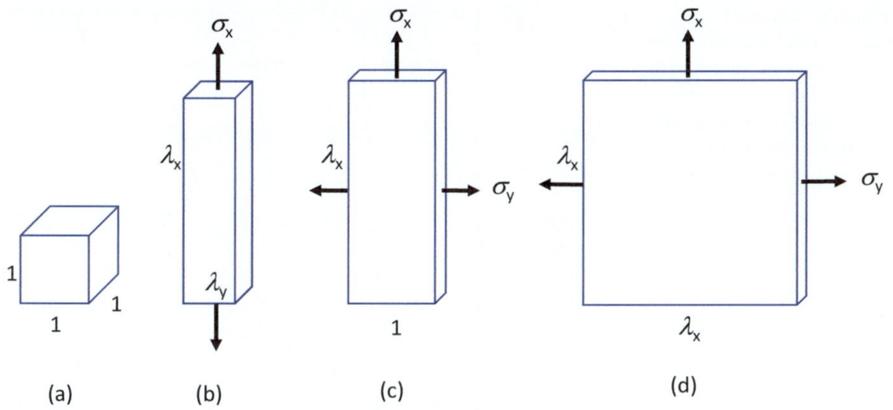

Fig. 6.4 **a** Undeformed state, **b** uniaxial, **c** planar and **d** equi-biaxial extension

For a stretching deformation ($\lambda_x > 1$), s_x is larger than σ_x by a factor of λ_x, reflecting the decrease in cross-sectional area due to incompressibility. While the force of a single Gaussian chain is linearly proportional to displacement (Eq. 6.11), the force of a network composed of Gaussian chains increases nonlinearly with displacement, as illustrated by the broken line in Fig. 6.5. Unlike a single one-dimensional polymer chain, a three-dimensional network exhibits lateral contraction (proportional to $1/\lambda_x^2$) due to incompressibility. The Gaussian chain network can stretch infinitely without experiencing a finite extensibility effect, as seen in the behavior of single chains (Fig. 6.2).

For small deformation ($\lambda_x \approx 1$), we can approximate the stretch as $\lambda_x = 1 + \varepsilon_x$ (with $\varepsilon_x \approx 0$), leading to the linearized relation:

$$\sigma_x = \frac{3nkT}{V_0}\varepsilon_x \tag{6.26}$$

where ε_x is the nominal strain defined by $\varepsilon_x = (L_x - L_{x0})/L_{x0}$. This equation confirms that the stress-strain relationship is linear for small deformations (Hooke's law). It also defines the Young's (elongation) modulus E_{af} and the shear modulus G_{af} for the affine network model as:

$$E_{\mathrm{af}} = \frac{3nkT}{V_0} = 3\nu kT \tag{6.27a}$$

$$G_{\mathrm{af}} = \frac{nkT}{V_0} = \nu kT \tag{6.27b}$$

where ν is the number of network strands per unit volume. Here, the standard linear elasticity relation $E = 3G$ for incompressible materials is employed. Equation (6.27) indicates that the modulus of rubber and gel increases proportionally with the network

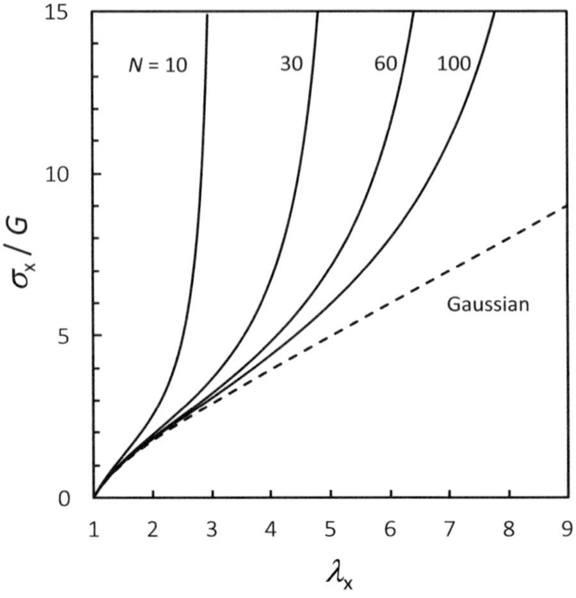

Fig. 6.5 Reduced stress-stretch curves of the affine network model with inverse-Langevin chains (solid lines) and Gaussian chains (dashed line) in uniaxial stretching

strand density ν. Since gels contain a significant amount of solvent, their strand density ν is lower than that of rubbers, resulting in a lower modulus. Equations (6.24) and (6.27) highlight key characteristics of entropic elasticity:

1. Force and modulus are directly proportional to temperature T.
2. They are also proportional to number strand density ν.

These characteristics mirror ideal gas behavior, where pressure p (in Pascals, the same unit as stress and modulus) is proportional to T and the molecular number density N/V.

The stress σ_x for a network composed of inverse Langevin chains can be derived similarly from Eq. (6.21):

$$\sigma_x = \frac{GN^{1/2}}{3}\left\{L^{-1}\left(\frac{\lambda_x}{N^{1/2}}\right) - \lambda_x^{-3/2}L^{-1}\left(\frac{1}{\lambda_x^{1/2}N^{1/2}}\right)\right\} \tag{6.28}$$

This equation predicts an upturn in stress at high elongations, diverging to infinity at $\lambda_{max}\left(= N^{1/2}\right)$, where the network strand reaches full extension (Fig. 6.5). This behavior reflects the mechanics of an individual inverse Langevin chain (Fig. 6.2b). The upturn occurs at smaller elongations for networks with shorter strands (i.e., smaller N). Additionally, for small deformations ($\lambda \ll \lambda_{max}$), Eq. (6.28) reduces to Eq. (6.24), recovering the stress response of a network composed of Gaussian chains.

B. **Biaxial Extension**

In biaxial extension, loading is applied simultaneously in two orthogonal directions (x- and y-axes), as illustrated in Fig. 6.4c and d. From Eq. (6.15), the deformation along the z-direction is given by $\lambda_z = (\lambda_x \lambda_y)^{-1}$. Substituting this into Eq. (6.20), we obtain the net energy change ΔA_{net} under biaxial extension:

$$\Delta A_{\text{net}} = \frac{nkT}{2}\left(\lambda_x^2 + \lambda_y^2 + \frac{1}{\lambda_x^2 \lambda_y^2} - 3\right) \text{(biaxial extension)} \tag{6.29}$$

Following the same procedure as in the uniaxial stretching case (discussed in previous section), we derive the stress components σ_x and σ_y:

$$\sigma_x = G\left(\lambda_x - \frac{1}{\lambda_x^3 \lambda_y^2}\right) \tag{6.30a}$$

$$\sigma_y = G\left(\lambda_y - \frac{1}{\lambda_x^2 \lambda_y^3}\right) \tag{6.30b}$$

Figure 6.6a shows the $\sigma - \lambda_x$ curves for two characteristic cases of biaxial stretching: equi-biaxial extension ($\lambda_x = \lambda_y$; Fig. 6.4d) and planar extension, where stretching occurs in the x-direction while the y-dimension remains fixed ($\lambda_y = 1$; Fig. 6.4c). At moderate λ_x, a noticeable difference in σ_x appears between these two deformation modes due to lateral constraint. However, at sufficiently high λ_x, this difference vanishes. This asymptotic behavior occurs because $\sigma_x \approx G\lambda_x$ becomes independent of λ_y at large λ_x.

The biaxial stress-strain relationships for networks composed of inverse Langevin chains can be derived analogously:

$$\sigma_x = \frac{GN^{1/2}}{3}\left\{L^{-1}\left(\frac{\lambda_x}{N^{1/2}}\right) - \lambda_x^{-2}\lambda_y^{-1}L^{-1}\left(\frac{1}{\lambda_x \lambda_y N^{1/2}}\right)\right\} \tag{6.31a}$$

$$\sigma_y = \frac{GN^{1/2}}{3}\left\{L^{-1}\left(\frac{\lambda_y}{N^{1/2}}\right) - \lambda_x^{-2}\lambda_y^{-1}L^{-1}\left(\frac{1}{\lambda_x \lambda_y N^{1/2}}\right)\right\} \tag{6.31b}$$

In Fig. 6.6b, the finite extensibility effect produces an upturn in σ_x for each deformation mode. However, as in the cause of Gaussian chain network model, the difference in σ_x between different biaxial stretching modes diminishes at large λ_x. As will be discussed later, real gels and elastomers often retain a finite difference in σ_x across different stretching modes due to coupling between the x- and y-directional stretches. This behavior cannot be captured within the framework of classical rubber elasticity theory.

C. Deformation of Swollen Gels

Gel and rubber undergo isotropic swelling or shrinking when immersed in a solvent. Degree of swelling or shrinking is governed by thermodynamics, as discussed in Chap. 4. In this section, we describe the mechanical behavior of swollen and shrunken gels using the affine network model. Let us consider uniaxial extension of a gel

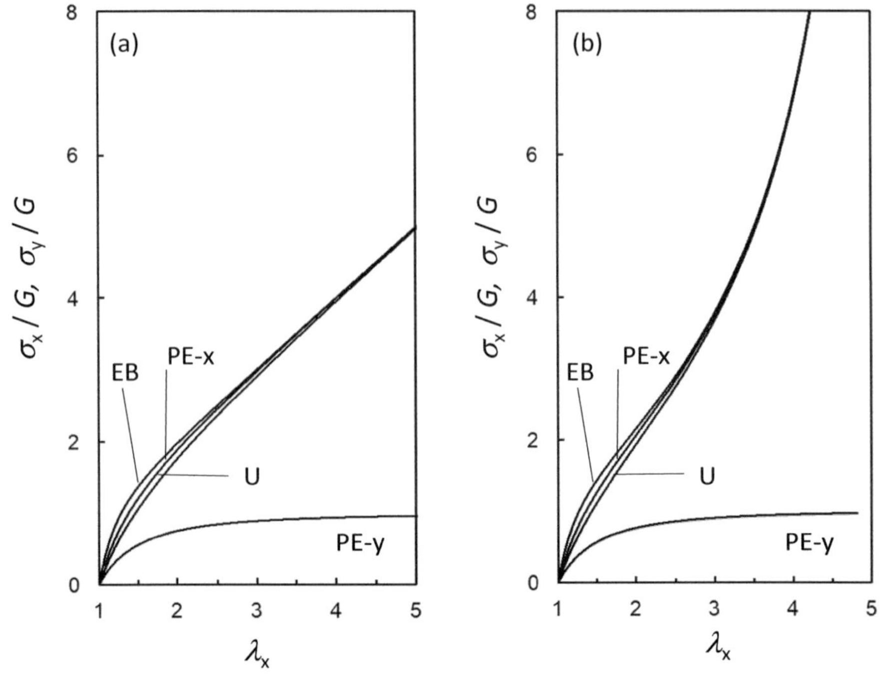

Fig. 6.6 Reduced stresses-stretch curves of the affine network models with **a** Gaussian chains and **b** inverse-Langevin chains of $N = 30$ in equi-biaxial (EB), planar (PE) and uniaxial extension (U)

that has undergone swelling (or shrinking) after preparation. The total deformation from the preparation state λ_i ($i = x, y, z$) can be expressed as the product of two contributions: isotropic swelling (or shrinking), and deformation imposed on the swollen (or shrunken) gel (Fig. 6.7: for simplicity, the initial dimensions are set to unity in the figure). Denoting the total principal stretches by $\lambda_i (i = x, y, z)$, the deformation can be separated as

$$\lambda_i = \frac{L_i}{L_{i0}} = \frac{L_i L_s}{L_s L_{i0}} = \left(\frac{V}{V_0}\right)^{1/3} \alpha_i = Q^{1/3}\alpha_i (i = x, y, z) \tag{6.32}$$

where $V = L_x{}^s L_y{}^s L_z{}^s$ is the volume in the swollen (or shrunken) state, $Q = V/V_0$ is the volume ratio, and α_i is the deformation imposed on the swollen (or shrunken) state. We assume that this imposed deformation is incompressible: $\alpha_x\alpha_y\alpha_z = 1$. Substituting into Eq. (6.20), we obtain the elastic energy change for a swollen gel under uniaxial extension:

$$\Delta A_{\text{net}} = \frac{nkT}{2} Q^{2/3}\left(\alpha_x^2 + \frac{2}{\alpha_x} - 3\right) \text{(Swollen state; Uniaxial extension)} \tag{6.33}$$

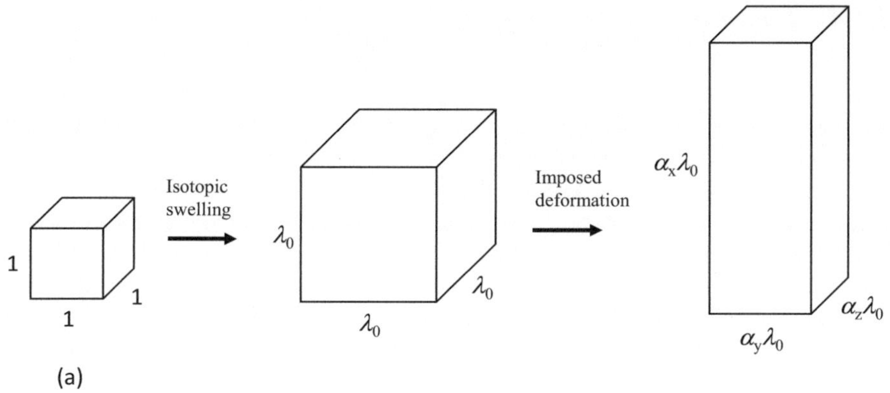

Fig. 6.7 Deformation of a swollen gel

The nominal stress, calculated using the cross-sectional area in the swollen state ($L_y{}^s L_z{}^s$), is obtained by taking the partial derivative of A with respect to α as

$$\sigma_\mathrm{i} = \frac{1}{V}\left(\frac{\partial A}{\partial \alpha_\mathrm{i}}\right)_{T,V} \quad (i = x, y, z) \tag{6.34}$$

and

$$\sigma_x = \frac{f_x}{L_y^s L_z^s} = \frac{nkT}{V_0}\frac{1}{Q^{1/3}}\left(\alpha_x - \frac{1}{\alpha_x^2}\right) = G_s\left(\alpha_x - \frac{1}{\alpha_x^2}\right) \tag{6.35}$$

where

$$G_\mathrm{s} = GQ^{-1/3} \tag{6.36}$$

This result shows that shear modulus in the swollen state, G_s, is reduced by a factor of $Q^{-1/3}$ compared to the modulus in the preparation state, G. Importantly, the strain-dependent term $\left(\alpha_x - \alpha_x^{-2}\right)$ in the stress remains identical to that in the preparation state expression $\left(\lambda_x - \lambda_x^{-2}\right)$. The reduction in modulus is not simply proportional to Q^{-1}, as might be naively anticipated from straightforward dilution of network chains. Instead, it is $Q^{-1/3}$, which arises from two combined effects: (1) dilution of network chains reduces strand density by a factor of Q^{-1}, and (2) swelling increases the dimensions of the network by a factor of $Q^{2/3}$, enhancing the chain length scale. Together, these yields the effective modulus scaling as $Q^{-1/3}$ For biaxial extension of swollen gels, the nominal stress can be similarly obtained by replacing G and λ with G_s and α, respectively, in the original stress expressions.

6.1.4 Phantom Network Model

Since network strands undergo micro-Brownian motion, it is natural to consider that the position of cross-links also fluctuate thermally over time. The affine network model described in the previous section assumes that cross-links are fixed within a matrix and deform affinely with macroscopic deformation. The phantom network model, proposed by James and Guth, incorporates the thermal fluctuations of cross-links into the Gaussian network framework [6]. This model assumes that all cross-links are free to fluctuate thermally, excepting for those located at the outer wall of the gel, which are constrained by the surface as boundary conditions. A key conclusion of the phantom network model is that, although the instantaneous positions of cross-links deviate from affine displacements due to thermal motion, their time-averaged positions still follow affine displacement. As a result, the thermal fluctuations of cross-links effectively reduce the elastic energy of the network, A_{net}. The expression for ΔA_{net} in the phantom network model is given as

$$\Delta A_{net} = \frac{(n-m)kT}{2}\left(\lambda_x^2 + \lambda_y^2 + \lambda_z^2 - 3\right) \tag{6.37}$$

where m is the total number of cross-links. As evident from comparing Eqs. (6.20) and (6.37), the effect of thermal fluctuations emerges in the modulus, not in the strain dependence of the stress. The shear modulus in the phantom network model (G_{ph}) is expressed as:

$$G_{ph} = \frac{(n-m)kT}{V_0} = (\nu - \mu)kT \tag{6.38}$$

where ν and μ is the number densities of network strands and cross-links, respectively. The difference $(\nu - \mu)$, known as the cycle rank, corresponds to the number density of independent loops in the network. In an ideal, defect-free network, μ is related to ν by 2 ν/f where f is the functionality of each cross-link (i.e., the number of strands connected to a cross-link). For a network with $f = 4$, this relation yields $G_{ph} = G_{affine}/2$, indicating that thermal fluctuations reduce the modulus to half the value predicted by the affine network model (Eq. 6.27b).

The term "phantom chain" is commonly employed in polymer physics to refer to an idealized chain with no excluded volume that can pass through other chains without entangling. Although the name of "phantom network model" may seem misleading, it is important to note that both the affine and phantom network models assume a network composed of "phantom chains" and neglect interactions such as chain entanglement and excluded volume effects.

6.1.5 Relationship Between Modulus and Network Structure in Real Gel and Rubber

Equations (6.27) and (6.38) relate the modulus of a polymer network with its structural parameters, such as the number densities of strands (ν) and cross-links (μ). The validity of these equations can be assessed by comparing theoretical predictions with experimentally measured moduli for samples with known structural parameters. However, such comparison is not straightforward, since real gels and rubbers inevitably contain structural defects, and direct evaluation of ν and μ by current analytical methods is challenging. Real networks exhibit imperfections such as dangling chains (with one free end), looped chains, trapped entanglements-entanglements that cannot be disentangled due to topological constraints imposed by cross-links (Fig. 6.8). Although dangling chains and loops are physically part of the network, they cannot sustain long-term mechanical stress and therefore do not contribute to the equilibrium modulus (G), which is measured over long times. In contrast, trapped entanglements behave differently: unlike the temporary entanglements in uncross-linked melts (which relax with time), trapped entanglements have effectively infinite relaxation times, thereby contributing to the equilibrium modulus. In uncross-linked polymer melts and concentrated solutions, the rubbery plateau region—a time domain where the modulus remains nearly constant—is attributed to transient entanglement networks [7]. The width of this plateau increases with molecular weight due to slower disentanglement dynamics. The tube model successfully explains this behavior [8]. By analogy, it is reasonable to expect that trapped entanglements contribute to the equilibrium modulus in cross-linked networks as well.

To rigorously evaluate the respective contributions of cross-links and trapped entanglements to modulus, model networks with well-defined structures are needed. These can be synthesized via end-linking reactions (Fig. 6.9), such as between end-functionalized polymers of known molecular weight M_p and multi-functional cross-linkers [9], or between star-shaped end-functionalized precursors [10]. Ideally, these reactions yield uniform networks with strand length M_p, but in practice, incomplete reactions lead to unavoidable defects such as dangling chains. Miller and Macosko developed a theoretical model for end-linked networks using random branching theory, which allows estimation of ν and μ as a function of conversion p [11]. This model assumes no loop formation and uniform reactivity of functional groups. For a system of linear precursors and tetra-functional cross-linkers, ν and μ are given by:

$$\nu = \frac{n_c}{(1 - w_{sol})}\left[6\alpha(1-\alpha)^3 + 2(1-\alpha)^4\right] \tag{6.39a}$$

$$\mu = \frac{n_c}{(1 - w_{sol})}\left[4\alpha(1-\alpha)^3 + (1-\alpha)^4\right] \tag{6.39b}$$

where n_c denotes the number of cross-linkers, and α, β are the probabilities that a strand is not connected to the infinite network via the cross-linker or precursor,

Fig. 6.8 Schematic of real network with structural defects and trapped entanglements

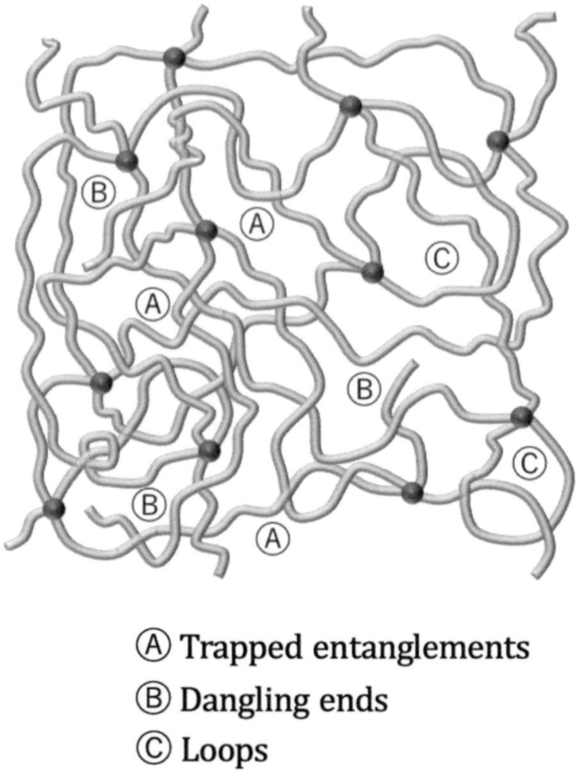

respectively. The w_{sol} stands for the weight fraction of soluble (non-networked) components. The values of α and β depend on the functional group ratio r (with $r = 1$ indicating stoichiometry) and conversion p:

$$\alpha = \left(\frac{4 - 3rp^2}{4rp^2}\right)^{1/2} - \frac{1}{2} \tag{6.40a}$$

$$\beta = rp\left[\left(\frac{4 - 3rp^2}{4rp^2}\right)^{1/2} - \frac{1}{2}\right]^3 + 1 - rp \tag{6.40b}$$

The relationship between w_{sol} and p is given by:

$$w_{\mathrm{sol}} = w_{\mathrm{p}}\beta^2 + w_{\mathrm{c}}\alpha^4 \tag{6.41}$$

where w_{p} and w_{c} are the weight fractions of precursor and cross-linker, respectively. The conversion p can be evaluated by chemical analysis or estimated using Eq. (6.41) based on extraction measurements of w_{sol}. If p is known, Eqs. (6.39) and (6.40) can be

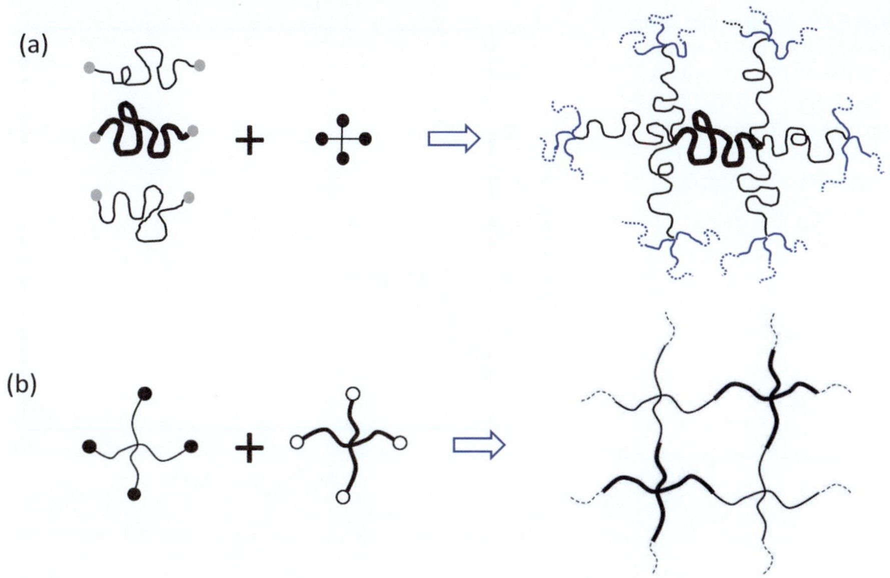

Fig. 6.9 **a** Uniform ideal networks via end-linking bi-functional linear precursors with tetra-functional cross-linker. **b** Uniform ideal networks via cross-linking the two types of end-reactive tetra-arm star precursors

used to calculate ν and μ, allowing comparison between theoretical and experimental modulus.

Figure 6.10 shows G as a function of ν for the end-linked polydimethylsiloxane (PDMS) networks with various precursor molecular weights M_p [12]. The soluble fraction w_{sol} is low (< 1 wt%), indicating minimal structural defects. The modulus G increases linearly with ν, consistent with the affine network model. However, a finite modulus remains as $\nu \to 0$ (i.e., $M_p \to \infty$), suggesting that G consists of two additive contributions:

$$G(\nu) = G_c(\nu) + G_e = \nu kT + G_e \tag{6.42}$$

where G_c denotes the contribution from covalent cross-links, and G_e represents the contribution from trapped entanglements, with a magnitude comparable to the rubbery plateau modulus $G_N{}^0$ of uncross-linked PDMS (where $G_N{}^0/RT \approx$ 80 mol m^{-3}). This suggests that G_c follows affine behavior, and the phantom-like reduction due to thermal fluctuation is suppressed in networks formed in the melt. Similar findings were reported in a simulation work [13].

In the networks containing solvent with a network volume fraction at preparation ϕ_0, both G_c and G_e decrease upon dilution. In end-linked gels, we expect $G_c \sim \phi_0^1$ and $G_e \sim \phi_0^2$, since G_c is governed by strand density and G_e by entanglement coupling, which is a two-bodu interaction. This leads to the expression:

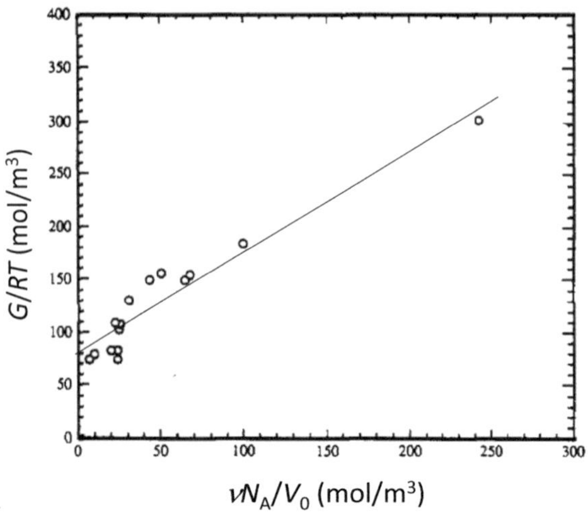

Fig. 6.10 Equilibrium shear modulus as a function of the number density of network strands for end-linked PDMS elastomers. Reprinted with permission from Ref. [12]. Copyright 1992 American Chemical Society

$$G(\phi_0) = G_e(\phi_0) + G_e(\phi_0) = \phi_0 G_e(1) + \phi_0^2 G_e \tag{6.43}$$

This relation holds for end-linked PDMS gels prepared in the concentrated regime ($0.4 < \phi_0 < 1$) (Fig. 6.11) [14].

In the semi-dilute regime near $\phi_0 \approx \phi*$ ($\phi*$: overlapping concentration of precursor chains), G_e becomes negligible and $G \approx G_c$. Figure 6.12 illustrates the ϕ_0 dependence of G for Tetra-PEG gels formed from four-arm star precursors in the semi-dilute regime [15]. In the figure, the measured modulus G is reduced by the affine network modulus (G_{af}) calculated using the ν value of each network. The G values near $\phi*$ ($1 < \phi_0/\phi* < 3$) are close to the phantom network moduli (G_{ph}). It appears that G approaches G_{af} as the overlapping of precursor chains at preparation ($\phi_0\phi*$) increases. Using a parameter h ($0 \leq h \leq 1$) to quantify thermal fluctuation of cross-links, G can be expressed as:

$$G = (\nu - h\mu)kT \tag{6.44}$$

where $h = 0$ for full fluctuation (phantom behavior; $G = G_{ph}$), and $h = 1$ for no fluctuation (affine behavior; $G = G_{af}$). This equation captures the observed ϕ_0-dependence of G, with h increasing as precursor overlap increases.

Overall, the relationships between G and network structure for end-linked gels over a broad ϕ_0 range—covering both unentangled and entangled regimes—can be summarized by [16, 17]:

$$G(\phi_0) = [\nu(\phi_0) - h(\phi_0)\mu(\phi_0)]k_B T + \phi_0^2 G_e \tag{6.45}$$

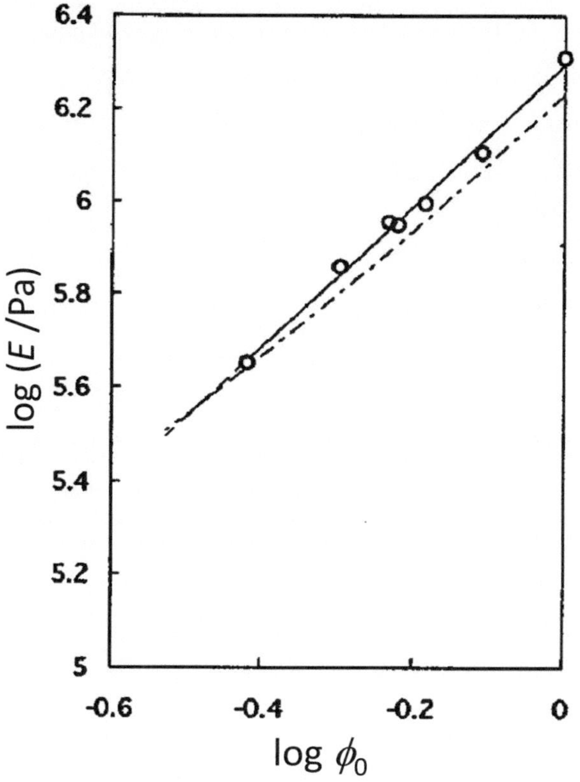

Fig. 6.11 Young's modulus as a function of volume fraction of network in preparation for end-linked PDMS gels. The dash-dotted line represents Eq. (6.43). The solid line is the fitted line to the data. Reprinted from Ref. [14] with the permission of AIP Publishing

This expression shows that modulus alone does not uniquely determine network structure, as ν, μ, h and G_e are interconnected. For example, assuming an affine ($h = 1$) or phantom ($h = 0$) model can lead to vastly different estimates of ν. Furthermore, in highly concentrated gels, the contributions of trapped entanglements are non-negligible and embedded in the observed modulus. Therefore, when interpreting experimental G using theoretical models, one must be cautious of such inherent ambiguities. Efforts to formulate more accurately the relationship between network structure and elasticity in real gels—such as by considering higher-order loop structures—are ongoing [18].

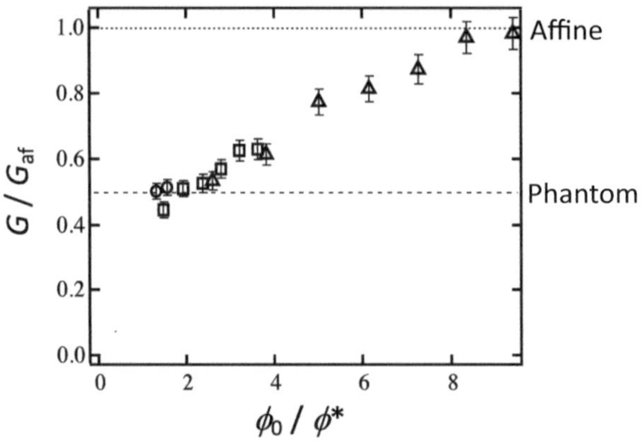

Fig. 6.12 Shear modulus (G) as a function of volume fraction of network in preparation (ϕ_0) for Tetra-PEG gels with various molecular weights of precursor (10,000, circle; 20,000, square; 40,000, triangle). The values of G and ϕ_0 of each specimen are reduced by the corresponding affine network modulus (G_{af}) and the value of ϕ for the onset of overlapping of precursor chain ($\phi*$). Reprinted with permission from Ref. [15]. Copyright 2013 American Chemical Society

6.1.6 *Finite Deformation Behavior of Real Gel and Rubber, and Extended Models of Rubber Elasticity*

In the previous section, we discuss how the linear modulus at infinitesimal strain is related to network structure and can be using classical rubber elasticity theories such as the affine and phantom network models. This section extends the discussion to finite deformation, comparing the behavior of real gels and rubbers with predictions from these classical models. As noted in Sect. 6.1.4, the affine and phantom network models differ only in the predicted modulus; they share the same stress-strain functional form.

Figure 6.13 presents the σ-λcurves for a polyacrylamide gel under equi-biaxial, planar and uniaxial extension [19]. At large stretches, the differences in σ_x between these extension types become prominent, contrary to predictions from classical models (Fig. 6.6). For the same stretch ratio λ_x, σ_x follows the order: equibiaxial > planar > uniaxial extension. This indicates a strong mutual influence between strains and stresses in orthogonal directions—for example, strain in the y-direction significantly affects stress in the x-direction, and vice versa—a phenomenon referred to as the strain cross-effect.

Classical models assumes that network strands are aligned along the principal axes (Fig. 6.14a) and attribute deformation coupling solely to the incompressibility condition $\left(\lambda_x\lambda_y\lambda_z = 1\right)$. This is reflected in the absence of explicit cross-terms like $\lambda_i\lambda_j(i, j = x, y, z)$ in the expression for ΔA_{net} (Eqs. (6.20) and (6.37)). Consequently, these models cannot capture the strain cross-effect observed in experiments. Accurate modeling of finite deformation therefore requires incorporating such cross-effect explicitly. Broadly, two primary strategies have been developed to account for the

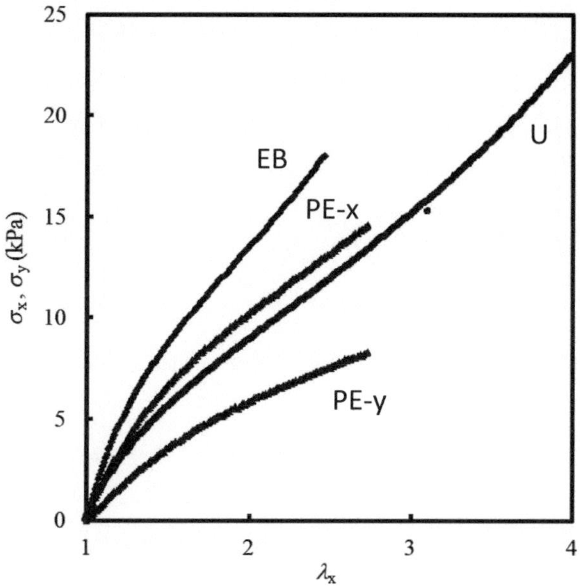

Fig. 6.13 Nominal stresses as a function of λ_x in equi-biaxial (EB), planar extension (PE) and uniaxial extension for a polyacrylamide hydrogel. Reprinted with permission from Ref. [19]. Copyright 2012 the Royal Society of Chemistry

strain cross-effect: (1) geometric consideration of network strand orientation; (2) introduction of second strain invariant into the strain energy function. Under approach (1), instead of assuming that strands align with principal axes, some models introduce strand orientations at finite angles. In such configurations, strain applied in one direction influences responses in others. Notable examples include the four-chain [20] and eight-chain [21] models (Fig. 6.14b and c). Classical models with strands along three orthogonal axes are often referred to as three-chain models. In the eight-chain model, each undeformed strand has a length of $3^{1/2}/2$ (assuming a unit cell of length 1). Under deformation, the strand length scales as: $\lambda_{\text{chain}} = \left[\left(\lambda_x^2 + \lambda_y^2 + \lambda_z^2\right)/3\right]^{1/2}$. This introduces a direct cross-effect into the deformation response. When accounting for finite extensibility, the elastic energy becomes:

$$\Delta A_{\text{net}} = \nu kT\sqrt{N}\left[\lambda_{\text{chain}} L^{-1}\left(\frac{\lambda_{\text{chain}}}{\sqrt{N}}\right) + \ln\frac{L^{-1}\left(\lambda_{\text{chain}}/\sqrt{N}\right)}{\sinh L^{-1}\left(\lambda_{\text{chain}}/\sqrt{N}\right)}\right] \quad (6.46)$$

These models help visualizing the source of strain cross-effects intuitively, but the strength of the cross-effect is fixed by geometry, often limiting agreement with experimental data.

In approach (2), the strain energy density function W, corresponding to ΔA_{net} per unit volume, is expressed strain invariants I_1, I_2 and I_3 of the left Cauchy-Green strain tensor:

$$W = \Delta A_{\text{net}}/V_0 = W(l_1, I_2, I_3) \quad (6.47)$$

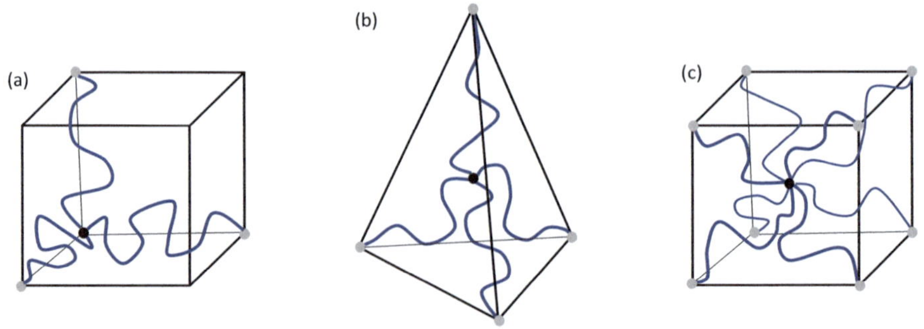

Fig. 6.14 Schematics for **a** three-chain model, **b** four-chain model and **c** eight-chain model

$$I_1 = \lambda_x^2 + \lambda_y^2 + \lambda_z^2 \tag{6.48}$$

$$I_2 = \lambda_x^2\lambda_y^2 + \lambda_y^2\lambda_z^2 + \lambda_z^2\lambda_x^2 \tag{6.48b}$$

$$I_3 = \lambda_x^2\lambda_y^2\lambda_z^2 \tag{6.48c}$$

For incompressible materials, $I_3 = 1$. Therefore, W for rubber and gel which are nearly incompressible is generally written as a function of I_1 and I_2. The invariant I_1 reflects principal stretch magnitudes, while I_2 accounts for surface-area changes and embodies strain cross-effects via $\lambda_i^2\lambda_j^2 (i \neq j)$. The nominal stress under uniaxial and biaxial deformations becomes:

$$\sigma_i = \frac{\partial W}{\partial \lambda_i} = \frac{2}{\lambda_i}\left(\lambda_i^2 - \frac{1}{\lambda_i^2\lambda_j^2}\right)\left(\frac{\partial W}{\partial I_1} + \lambda_i^2\frac{\partial W}{\partial I_2}\right) \quad (i, j = x, y) \tag{6.49}$$

In classical models, W is linear in I_1, and this yields the neo-Hookean model:

$$W(I_1) = (G/2)(I_1 - 3) \tag{6.50}$$

A simple extension is the Mooney-Rivlin model, which adds a linear I_2 term [22, 23]:

$$W(I_1, I_2) = C_{10}(I_1 - 3) + C_{01}(I_2 - 3) \quad (C_{10}, C_{01} : \text{constants}) \tag{6.51}$$

This model improves fitting for uniaxial data but often fails to predict stress under other deformation modes (e.g., biaxial). Figure 6.15 shows a clear mismatch when Eq. (6.51) is applied to biaxial extension [24]. In such cases, C_{01} lacks physical significance. This highlights a key point: relying solely on uniaxial data can lead to misleading conclusions, since uniaxial extension is just one case among many valid deformation modes [25].

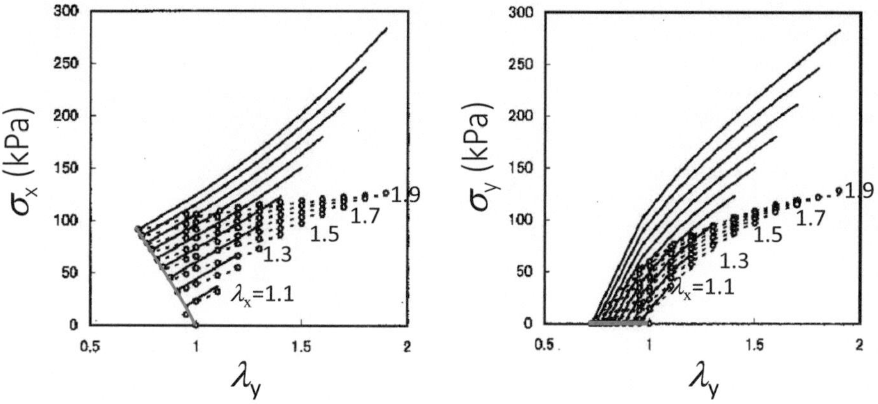

Fig. 6.15 Comparison of the biaxial stress-elongation data (open symbols) for a PDMS gel with the prediction of Eq. (51) (solid lines) with the coefficients C_{10} and C_{01} evaluated from the uniaxial data (gray lines). The dotted lines are guides for eyes for the data at same λ_x Reprinted with permission from Ref. [24]. Copyright 2001 American Chemical Society

To enhance fitting flexibility, W is often expanded as a polynominal in I_1 and I_2 [23]:

$$W(I_1, I_2) = \sum C_{ij}(I_1-3)^i(I_2-3)^j \tag{6.52}$$

For instance, fitting to biaxial data in Fig. 6.15 requires a second-order expansion:

$$\begin{aligned} W(I_1, I_2) &= C_{10}(I_1-3) + C_{01}(I_2-3) + C_{11}(I_1-3)(I_2-3) \\ &\quad + C_{20}(I_1-3)^2 + C_{02}(I_2-3)^2 \end{aligned} \tag{6.53}$$

Figure 6.16 shows that Eq. (6.53) successfully describes the biaxial stress–strain data of an end-linked PDMS gel ($\phi = 0.7$) [24]. Although the physical interpretation of each term is complex, the relative magnitudes $|C_{ij}/C_{10}|$ diminish as ϕ decreases. [84] This trend suggests that W of dilute gels approaches the neo-Hookean form (Eq. 6.50), while higher-order terms mainly reflect entanglement effects.

Beyond phenomenological approaches, many molecular entanglement models have been developed to describe the large deformation behavior [26–31]. These models typically express ΔA_{net} as the sum of two contributions: The chemical network formed by covalent cross-links (usually modeled by affine or phantom models, incorporating finite extensibility), and the entanglement contribution, which varies depending on the specific model. To validate these molecular models, it is crucial to connect fitting model parameters to physical structural quantities. As shown in Fig. 6.15, fitting only uniaxial data is insufficient because many models can adequately describe such behavior [32]. Validation requires the stress-strain data from multiple deformation modes, including biaxial extension with varying strain ratios

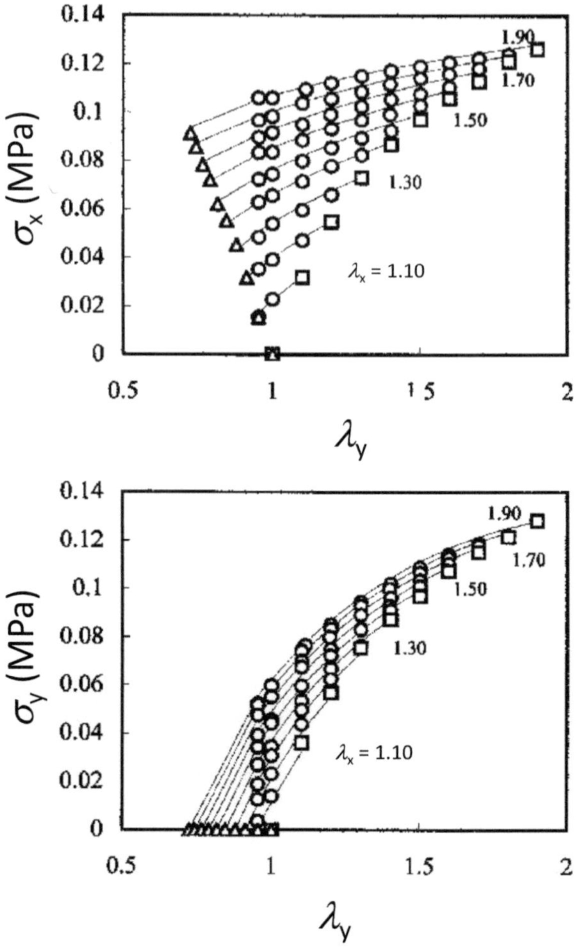

Fig. 6.16 Comparison of the biaxial stress-elongation data for an end-linked PDMS gel with the prediction of Eq. (6.53) (solid lines). Reprinted with permission from Ref. [24]. Copyright 2001 American Chemical Society

and simultaneous measurements of shear and normal stresses in shear deformation [33].

The slip-link model [28] and extended tube model [29] are two representative molecular models. In the slip-link model, entanglements are treated as slip-links—mobile constraints that can slide along network strands. The corresponding expression for ΔA_{net} is given by

$$\begin{aligned}\frac{\Delta A_{\mathrm{net}}}{RT} &= \frac{1}{2}N_c\left[\frac{\left(1-\alpha^2\right)\sum\lambda_i^2}{1-\alpha^2\sum\lambda_i^2}+\ln\left(1-\alpha^2\sum\lambda_i^2\right)\right]\\ &+\frac{1}{2}N_s\left[\sum\left\{\frac{\lambda_i^2(1+\eta)\left(1-\alpha^2\right)}{\left(1+\eta\lambda_i^2\right)\left(1-\alpha^2\sum\lambda_i^2\right)}+\ln\left(1+\eta\lambda_i^2\right)\right\}\right.\end{aligned}$$

$$+\ln\left(1-\alpha^2\sum\lambda_i^2\right)\Big] \tag{6.54}$$

where N_c and N_s are the numbers of chemical network strands and slip-links, respectively, η is the mobility factor of the slip-link, and α accounts for the finite extensibility of network strands. These parameters correlate with the network structure as $\eta = M_e/M_c$ (M_e: entanglement molecular weight; M_c: molecular weight between adjacent cross-links) and $\alpha = 1/\lambda_{max} = N_k{}^{-1/2}$ (λ_{max}: the ultimate stretch of network strand; N_k: the number of Kuhn segments per strand). The slip-link model successfully describes the biaxial stress–strain data for entangled, end-linked PDMS gels (Fig. 6.17), and the fitted values of η and α satisfactorily agree with independently estimated structural parameters [34, 35]. This demonstrates that the same data can be interpreted using both the phenomenological approach (Eq. 6.53) and the molecular approach (Eq. 6.54).

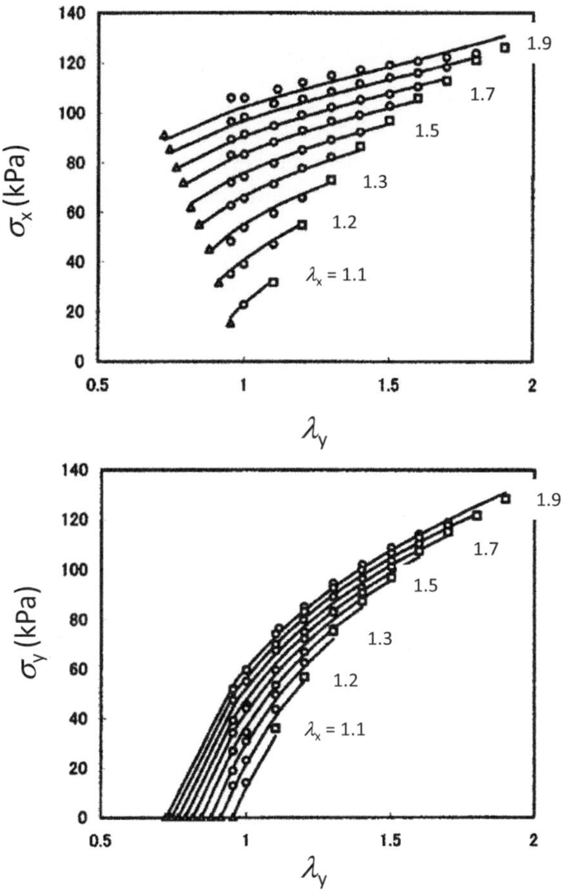

Fig. 6.17 Comparison of the biaxial stress-strain data of an end-linked PDMS gel with the fitted curves of the slip-link model (solid lines). Reprinted with permission from Ref. [34]. Copyright 2001 American Chemical Society

The extended tube model refines the classical tube concept by allowing the tube diameter—which reflects entanglement constraints—to change with imposed deformation. The corresponding W is expressed by:

$$W = \frac{G_c}{2}\left[\frac{(1-\delta^2)(I_1-3)}{1-\delta^2(I_1-3)} + \ln\left(1-\delta^2(I_1-3)\right)\right] + \frac{2G_e}{\beta^2}\sum_{i=1}^{3}\left(\lambda_i^{-\beta}-1\right) \quad (6.55)$$

where β represents the strength by entanglement constraints, and δ captures the finite extensibility of network strands. This model reproduce stress-strain data under various deformation modes for several rubbers [29].

The strain cross-effect—i.e., how deformation in one direction affects stress in another direction—depends strongly on network structure and cross-linking type. A practical way to assess this effect is via the stress ratio $\left(\sigma_x/\sigma_y\right)$ in planar extension. Assuming $\partial W/\partial I_2 = 0$ (i.e., neglecting the I_2 term), Eq. (6.49) yields [36]:

$$\frac{\sigma_y}{\sigma_x} = \frac{\lambda_x}{\lambda_x^2+1} (\text{Planar extension; } \partial W/\partial I_2 = 0) \quad (6.56)$$

Notably, this relation holds for any model where W depends solely on I_1, regardless of its functional form. Figure 6.18 compares experimental data with Eq. (6.56).[36] Data for randomly cross-linked polyacrylamide gels with heterogeneous structures (Fig. 6.13) and the entangled end-linked PDMS gels (Fig. 6.16) lie well above the theoretical line, indicating $\partial W/\partial I_2 > 0$ and a substantial contribution from I_2. Even in Tetra-PEG gels prepared near $\phi*$, the data deviate from the line, though less markedly. While the small-strain modulus in these gels is not affected by entanglements (Fig. 6.12), finite contribution from I_2 appear at large strains [37]. The W for Tetra-PEG gels can be modeled as a combination of I_1- and I_2-dependent terms (Fig. 6.19) [37, 38]:

$$W\left(I_1, I_2\right) = \frac{G}{2}\left\{-(1-r)(I_m-3)\ln\left(1-\frac{I_1-3}{I_m-3}\right) + r(I_2-3)\right\} \quad (6.57)$$

Here, r quantifies the relative strength of the I_2-term ($0 < r < 1$). The I_1-term corresponds to the Gent model, which extends the classical model by introducing the finite extensibility effect [39]:

$$W(I_1) = -\frac{G}{2}(I_m-3)\ln\left(1-\frac{I_1-3}{I_m-3}\right) \quad (6.58)$$

where I_m is the limiting value of I_1 value where the strand reaches its extensibility limit and stress diverges. In the limit $I_m \to \infty$, Eq. (6.58) reduces to the neo-Hookean model (Eq. (6.50)). In Tetra-PEG gels, r decreases with decreasing ϕ, and the I_2-term vanishes as $\phi \to 0$ [37, 38]. This implies that even networks with nearly uniform

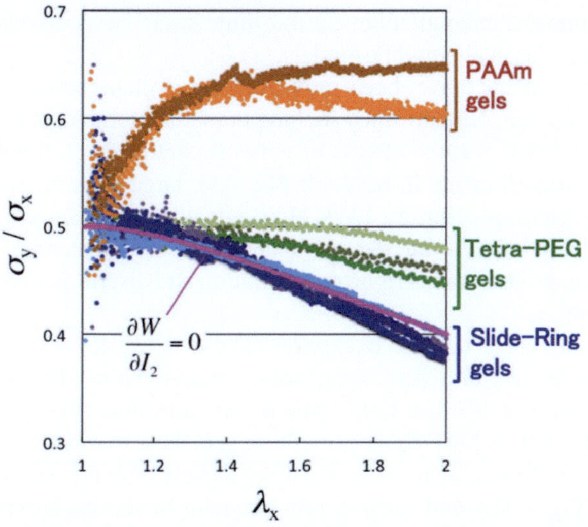

Fig. 6.18 Stress ratio (σ_y/σ_x) as a function of stretching (λ_x) in planar extension for PAAm gels, Tetra-PEG gels, and slide-ring gels. The directions x and y are the axes of stretching and constraint, respectively. Data for specimens prepared with various concentrations of cross-linker and precursor are included. The pink solid line depicts the case $\partial W/\partial I_2 = 0$ (Eq. (6.56)). Reprinted from Ref. [36] with the permission of AIP Publishing

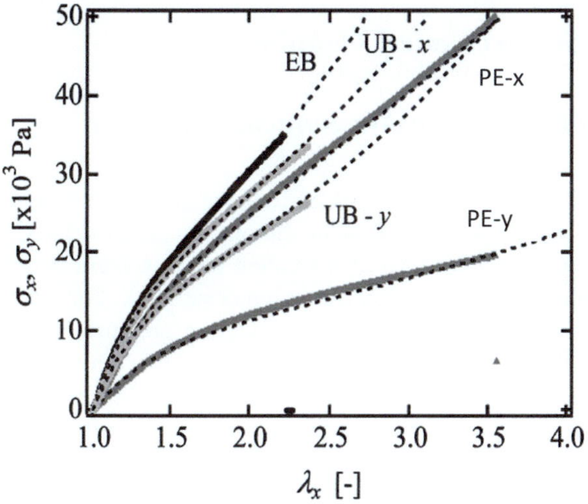

Fig. 6.19 Comparison of the biaxial stress-strain data for a Tetra-PEG gel with the predictions of Eq. (6.57) (dotted lines): EB and PE denote the equi-biaxial and planar extension, respectively, and UB represents the unequal biaxial stretching with a relation of $\lambda_y = (\lambda_x + 1)/2$. Reprinted with permission from Ref. [37]. Copyright 2012 the Royal Society of Chemistry

structure and minimal entanglement exhibit finite strain cross-effects, contradicting predictions from classical network models.

Several mechanisms have been proposed to explain strain cross-effects in systems without entanglement. They include nematic interactions between network strands [40], excluded volume effects of network strands [41], topological interactions between ring elements in network [42, 43], large-scale thermal fluctuations constrained by incompressibility [31]. However, these models are often assessed using only uniaxial data, and the associated interaction parameters tend to be phenomenological. This makes rigorous validation and discrimination between these theories difficult.

Slide-ring gels are a notable exception. These networks are formed by cross-linking ring molecules threaded along polyrotaxane chains (Fig. 6.20a). The key feature is the mobility of cross-links, which can slide along the polymer strands in response to deformation [44]. As a result, these gels show minimal strain cross-effect. Experimental data from planar extension align closely with the theoretical line given by Eq. (6.56) (Fig. 6.18), and stress–strain behavior in various biaxial deformations is well described by the Gent model (Eq. 6.58) [36], which depends only on I_1 (Fig. 6.20b and c). This indicates that the strain cross-effect in slide-ring gels arises solely from incompressibility, and internal freedom to rearrange cross-link positions effectively suppresses it.

6.2 Diffusion-Mechanical Coupling

6.2.1 What Is Diffusion-Mechanical Coupling?

A gel, composed of a polymer network and a solvent, is a thermodynamic semi-open system that allows the solvent to flow in and out. As described in Chap. 4, the equilibrium volume (or composition) of a swollen gel in solvent is determined by the balance of various forces acting on the polymer network. For a neutral, uncharged gel, this balance is primarily between the swelling pressure (which drives solvent absorption) and the elastic restoring force of the polymer network. When a mechanical stress or strain is applied to such an equilibrium swollen gel in solvent, the force balance shifts, leading to a new equilibrium state. This often results in a change in gel volume. Much like environmental changes—such as variations in solvent composition, temperature, or pH—mechanical deformation can induce swelling or deswelling. This strain-induced volume change significantly influences the stress response of gels. In contrast to ideal rubber, which contains no solvent and behaves purely elastically (maintaining constant stress under a fixed strain), a gel exposed to constant strain exhibits a gradual decrease in stress. This stress reduction arises from solvent inflow or outflow accompanying volume change (Fig. 6.21). The relaxation rate is governed by the kinetics of this volume change and characterized by a time constant that scales with the square of the gel's characteristic size. Although

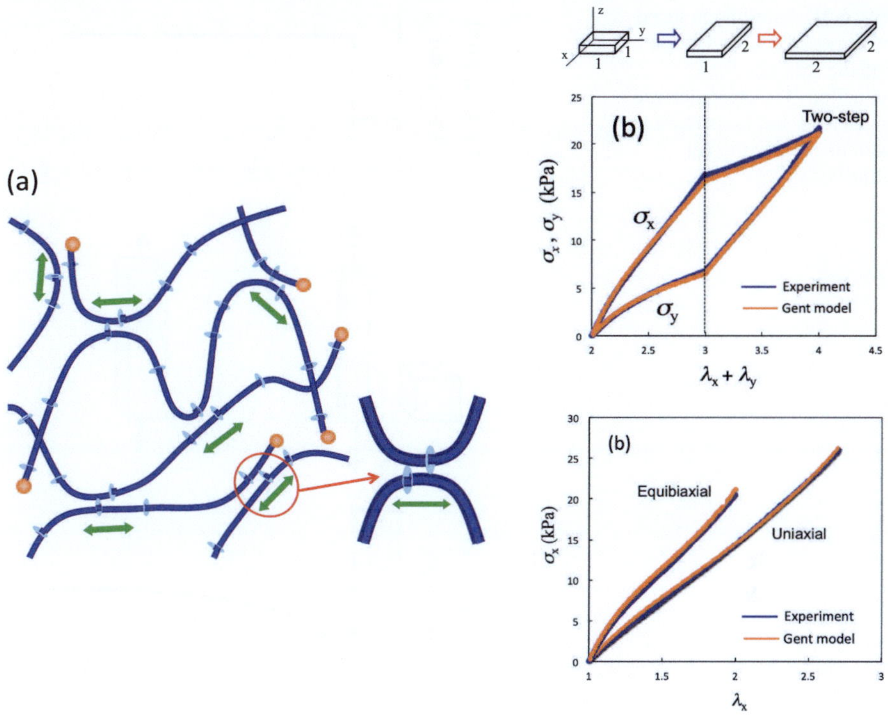

Fig. 6.20 **a** Schematics of slide-ring gel. Nominal stress-stretch curves for a slide-ring gel in **b** two-step planar extension and **c** uniaxial and equi-biaxial extension with the fitting curves of Gent model. Reprinted from Ref. [36] with the permission of AIP Publishing

this behavior resembles viscoelastic stress relaxation, the underlying mechanism is fundamentally different. In viscoelastic materials, relaxation is due to internal molecular rearrangements, whereas in gels it stems from mass transport (solvent diffusion). This phenomenon—where deformation-induced mass transfer influences mechanical behavior—is referred to as "diffusio-mechanical coupling," a defining characteristic of gels as semi-open thermodynamic systems [1].

To assess when diffusio-mechanical coupling becomes significant during gel stretching, consider the characteristic time for volume change, denoted τ_{sw}. This can be estimated by the ratio of the square of the gel's thickness to the solvent diffusion constant D (~ 10^{-9} m^2 /s). For example, a gel with length $L = 100$ mm and thickness $d = 10$ mm yields $\tau_{\mathrm{sw}} = \mathrm{d}^2/\mathrm{D} \sim 10^5\mathrm{s}$. The corresponding characteristic strain rate, $\dot{\varepsilon}_{sw} = \tau_{sw}^{-1}$, is 10^{-5} s^{-1}, giving a critical extension rate $v_c = L \times \dot{\varepsilon}_{sw} \sim 1\mu\mathrm{m/s}$. Since conventional tensile tests are typically performed at much higher extension rates than this v_c value, strain-induced swelling can usually be neglected. However, for a gel fiber with both length and diameter on the order of 100 μm, the corresponding v_c becomes ~ 10 mm/s—comparable to or exceeding conventional stretching speeds. In such cases, strain-induced swelling significantly affects the mechanical response.

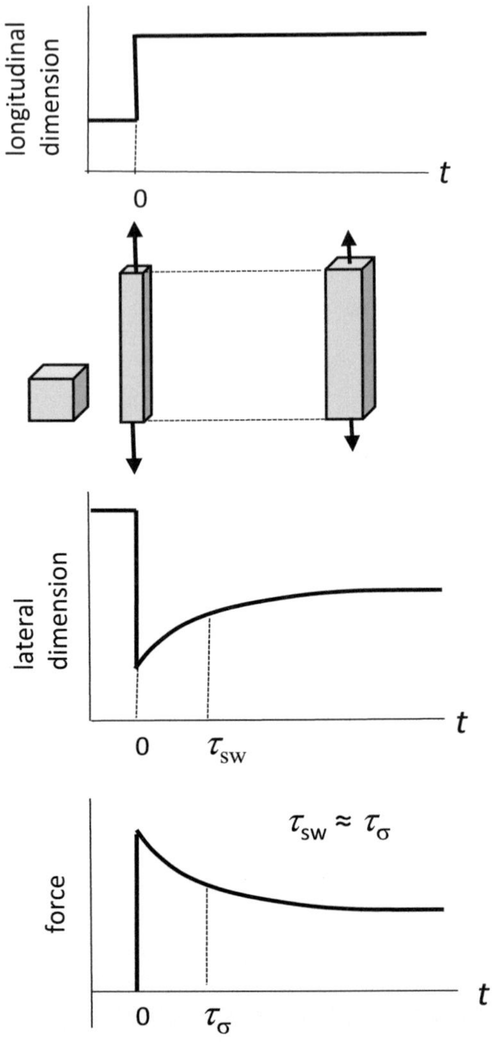

Fig. 6.21 Swelling induced by imposed constant strain and the resultant force reduction. The characteristic times $\tau_{\max}$ and τ_s increase with increasing the gel dimensions

Thus, diffusio-mechanical coupling must be considered when evaluating the mechanical behavior of gels with fine structures, such as biological tissues, fast-actuating gel-based devices, and miniaturized gels used in micro electro mechanical system (MEMS). Accurate interpretation and optimal material design in these systems require accounting for this coupling effect.

The next section discusses how diffusio-mechanical coupling manifests under different types of external deformation and loading conditions.

6.2.2 Thermodynamics of Swelling Under Deformation

This section derives how imposed deformation alters gel volume through solvent inflow or outflow, and subsequently changes the stress, based on the thermodynamic framework introduced in Chap. 4 [45, 46]. Consider a gel with an initial network volume fraction of ϕ_{00} and volume V_{00}, which swells freely in a solvent until equilibrium is attained. A constant deformation α_x (where $\alpha_x > 1$ denotes extension, and $\alpha_x < 1$ denotes compression) is then applied and held along the x-direction (Fig. 6.22). According to Flory-Rehner theory, the Gibbs energy change (ΔA) of a neutral gel is the sum of two contributions: the elastic energy (ΔA_{el}) due to deformation, and the mixing energy (ΔA_{mix}) between the polymer network and solvent. Adopting the classical network model for ΔA_{el}, we obtain:

$$\begin{aligned}\Delta A/k_BT &= (\Delta A_{mix} + \Delta A_{el})/k_BT \\ &= N_s[\ln(1-\phi) + \chi\phi] + N_c\left(\lambda_x^2 + \lambda_y^2 + \lambda_z^2 - 3\right)/2 \end{aligned} \tag{6.59}$$

where N_s is the number of solvent molecules, N_c is the number of elastically effective network strands, ϕ is the current network volume fraction, and χ is the Flory-Huggins interaction parameter. The total deformation $\lambda_i = \lambda_0\,\alpha_i$ (for $i = x, y, z$) is decomposed into the free-swelling deformation $\lambda_0 = (\phi_{00}/\phi_0)^{1/3}$ and the imposed deformation α_i. From ΔA, the true stress in direction i, s_i (force per unit area in the deformed state), is given by:

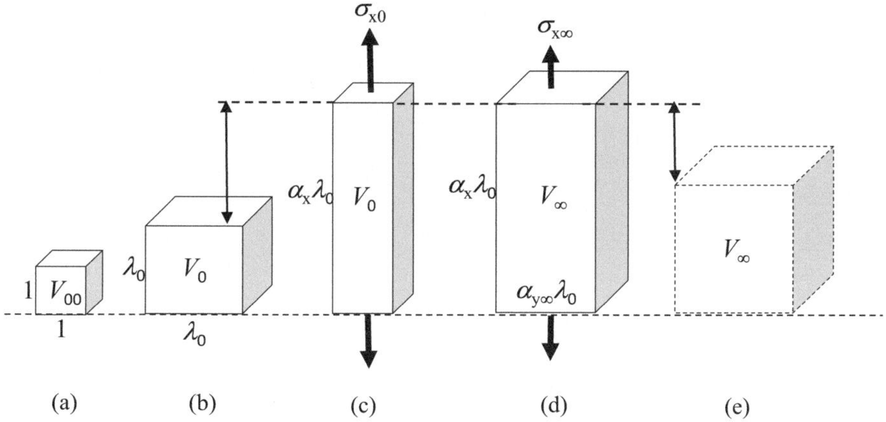

Fig. 6.22 Gels under various states: **a** reference state; **b** freely swollen state; **c** immediately after the imposition of a stretch of α_x on (**b**) (isovolumetric deformation); **d** equilibrium swollen state under a constant stretch of α_x; **e** immediately after the release of the imposed stretch from (**d**) (isovolumetric deformation). Note that the effective strain acting on tensile force, which is proportional to the length of the vertical arrows in (**b**) and (**e**), is reduced by the stretch-induced further swelling, resulting in a reduction in tensile force ($\sigma_{x0} > \sigma_{x\infty}$)

$$s_i = \frac{1}{l_j l_k}\left(\frac{\partial \Delta A}{\partial l_i}\right) = \frac{1}{V_{00}\lambda_j\lambda_k}\left(\frac{\partial \Delta A}{\partial \lambda_l}\right) \quad (6.60)$$

At equilibrium, the stresses s_y and s_z are zero, giving:

$$s_y/k_BT = s_z/k_BT = \frac{N_s}{(1-\phi)V}\left[\ln(1-\phi)+\phi+\chi\phi^2\right] + \frac{N_c\phi_0^{1/3}}{V_{00}\alpha_x} = 0 \quad (6.61)$$

where $n_s = N_s/[(1-\phi)V_\infty]$ is the solvent number density and $n_c = N_c/V_{00}$ is the strand density. This condition also holds for the free-swollen state (α_x=1, ϕ=ϕ_0):

$$\frac{N_s}{(1-\phi_0)V}\left[\ln(1-\phi_0)+\phi_0+\chi\phi_0^2\right] + \frac{N_c\phi_0^{1/3}}{V_{00}} = 0 \quad (6.62)$$

For highly swollen gels ($\phi << 1$ and $\phi_0 << 1$), combining Eqs. (6.61) and (6.62) yields the volume change ratio due to imposed strain:

$$V_\infty/V_0 = \alpha_\mathrm{x}^{1/2} \quad (6.63)$$

This result shows that elongation ($\alpha_x > 1$) leads to volume increase (solvent inflow), while compression ($\alpha_x < 1$) leads to volume decrease (solvent outflow). The sign of the pressure just after deformation reflects this:

$$s_y^0/k_BT \approx \frac{N_c\phi_0^{1/3}}{V_{00}}\left(\frac{1}{\alpha_x} - 1\right) \quad (6.64)$$

Thus, elongation causes negative pressure ($s_y{}^0 < 0$; driving solvent in), while compression induces positive pressure ($s_y{}^0 > 0$; driving solvent out). These effects were confirmed experimentally [47, 48], closely match the prediction of Eq. (6.63) (Fig. 6.23). From Eq. (6.63), the relationship between transverse and longitudinal stretches is:

$$\alpha_\mathrm{y} = \alpha_\mathrm{x}^{-1/4} \quad (6.65)$$

where $\phi_{00}/\phi = \lambda_\mathrm{x}\lambda_\mathrm{y}\lambda_\mathrm{z}$ and $\phi_0/\phi = \alpha_x\alpha_y\alpha_z$. The finite deformation Poisson's ratio (μ) is defined as:

$$\mu = -\frac{\varepsilon_y}{\varepsilon_y} = -\log\alpha_y/\log\alpha_x \quad (6.66)$$

If no volume change occurs before and after imposed deformation, $\mu = 1/2$. Under equilibrium, however, Poisson's ratio μ_∞ becomes 1/4, independent of deformation magnitude. This value was confirmed experimentally over a wide strain range (inset of Fig. 6.23).

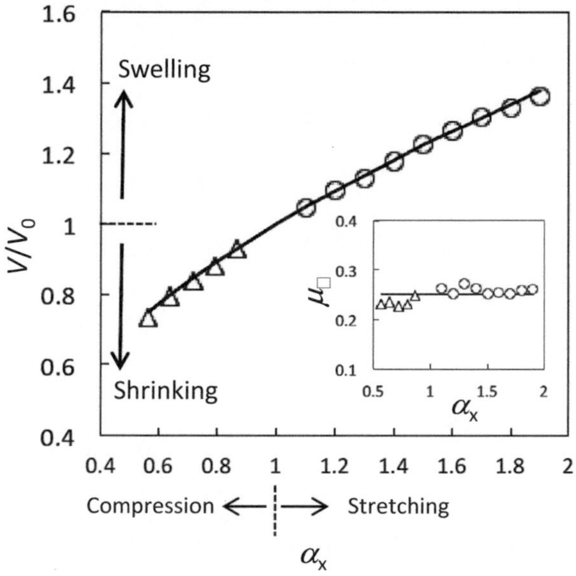

Fig. 6.23 Strain-induced swelling of a poly(acrylamide) hydrogel in water under constant elongation ($\alpha_x > 1$)[47], and strain-induced shrinking of a poly(vinyl acetate) gel under constant compression ($\alpha_x < 1$) [48]. The inset shows the α_x dependence of the equilibrium Poisson's ratio μ_∞. The solid line corresponds to Eq. (6.63). Reprinted with permission from Ref. [45]. Copyright 2012 the Royal Society of Chemistry

The true stress at equilibrium, $s_{x\infty}$, corresponds to a nominal stress $\sigma_{x\infty}$ (referenced to the undeformed area):

$$\begin{aligned}\alpha_{x\infty} = s_x\alpha_y\alpha_z = s_x\alpha_x^{-2\mu} &= \frac{N_c k_B T \phi_0^{1/3}}{V_{00}}\left(\alpha_x - \alpha_x^{-1-2\mu}\right) \\ &= G_0\left(\alpha_x - \alpha_x^{-1-2\mu}\right) = G_0\left(\alpha_x - \alpha_x^{-3/2}\right)\end{aligned} \tag{6.67}$$

where $G_0 = N_c kT\,\phi_0{}^{1/3}/V_{00}$ is the shear modulus in the freely swollen state ($\phi = \phi_0$). Immediately after applying strain (before swelling), assuming $\mu = 1/2$, the nominal stress, σ_{x0}, is:

$$\sigma_{x0} = G_0\left(\alpha_x - \alpha_x^{-2}\right) \tag{6.68}$$

Thus, under constant, the stress relaxes from σ_{x0} to $\sigma_{x\infty}$ due to volume change (corresponding to the change of μ from 1/2 to 1/4). This occurs because the effective strain decreases as the gel volume changes (Fig. 6.22). The relative stress reduction, $\Delta\,\sigma\,/\sigma_0 = (\sigma_0 - \sigma_\infty)/\sigma_0$, is approximately 17% for small strains ($\alpha_x \approx 1$). The α_x-dependence of this stress reduction ratio matches experimental data for poly(acrylamide) (PAAm) hydrogels (Fig. 6.24b).

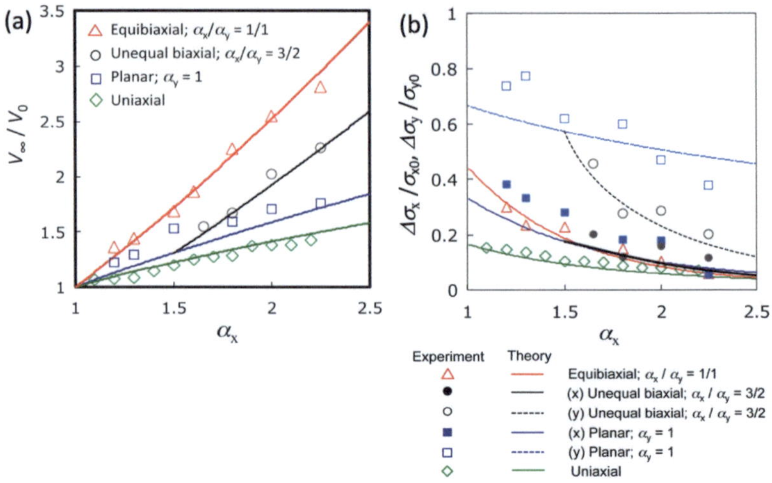

Fig. 6.24 **a** Degree of strain-induced volume change as a function of α_x in various types of deformation for a polyacrylamide hydrogel. The solid lines represent the theoretical predictions calculated using Eq. (6.69). Reproduction from Ref. [49]. **b** Degree of total reduction of nominal stress in each direction as a function of α_x for a polyacrylamide in water. $\Delta\sigma_i$ is reduced by the initial value at each deformation. The lines represent the theoretical predictions (Eq. 6.70). Reprinted with permission from Ref. [49]. Copyright 2015 American Chemical Society

Strain-induced swelling and associated stress reduction strongly depend on deformation type. For biaxal stretching with imposed deformations α_x and α_y, the results are [49]:

$$V_\infty/V_0 = \left(\alpha_x\alpha_y\right)^{2/3} \tag{6.69}$$

$$\sigma_{x\infty} = s_{x\infty}\alpha_y\alpha_z = G_0\left(\alpha_x - \frac{1}{\alpha_x^{5/3}\alpha_y^{2/3}}\right) \tag{6.70a}$$

$$\sigma_{y\infty} = s_{y\infty}\alpha_z\alpha_x = G_0\left(\alpha_y - \frac{1}{\alpha_y^{5/3}\alpha_x^{2/3}}\right) \tag{6.70b}$$

Figure 6.24 shows how volume change and stress reduction vary with α_x under different deformation modes [49]. For $\alpha_x = 2$, uniaxial stretching leads to $V_\infty/V_0 \approx 1.4$, while equi-biaxial stretching yields $V_\infty/V_0 \approx 2.5$. The corresponding stress reduction are ~ 10% and ~ 50%, respectively. Notably, stress reduction becomes anisotropic when the biaxial strains are unequal. For example, in planar extension, $\Delta\sigma_{i\infty}/\sigma_{i0} \approx 20\%$ in the stretching direction, but ~ 70% in the y-direction (constraint direction) (Fig. 6.25). This anisotropy is a hallmark of swelling-induced stress relaxation and contrasts with viscoelastic relaxation, which is isotropic. Since strain-induced swelling or shrinking is isotropic, the imposed anisotropic strain results in

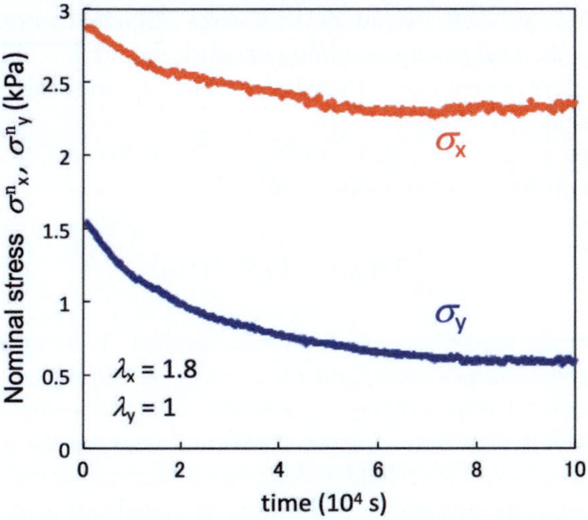

Fig. 6.25 Nominal stresses in the two directions as a function of time after the imposition of planar extension $(\lambda_x = 1.8; \lambda_y = 1)$ for a polyacrylamide hydrogel in water. The degrees of the total stress-reduction are different in the two directions: ca. 20 and 70% in x and y directions, respectively. Reprinted with permission from Ref. [49]. Copyright 2015 American Chemical Society

direction-dependent effective strain reductions, leading to anisotropic stress relaxation. Theoretical predictions (Eqs. (6.69) and (6.70)) agree well with experimental results.

This analysis adopts the classical network model for ΔA_{el}. While it does not fully capture nonlinear stress–strain behavior, it yields reliable estimates of relative changes (e.g., V_∞/V_0, σ_∞/σ_0), which are the focus here. More accurate modeling of absolute stress and volume would require more sophisticated elasticity theories. However, the current formulation prioritizes mathematical simplicity and physical transparency.

This discussion assumes highly swollen gels ($\phi << 1$, $\phi_0 << 1$), typically found in good solvents. In low-swelling (near-θ) solvents, μ_∞ increases to ~ 0.33, as observed experimentally [50, 51]. This shift can be captured numerically within the same framework by setting $\chi = 0.5$. Scaling theory, which considers excluded volume effects, predicts $\mu_\infty = 0.258$ in good solvents and $\mu_\infty = 1/3$ in θ conditions [50, 52].

6.2.3 *Kinetics of Volume Change Under Imposed Strain*

This section discusses the time-dependent behavior of strain-induced volume change in gels and the associated stress reduction. To rigorously model the kinetics of volume change, it is essential to consider the relative motion between the polymer network and the solvent[85, 86] Here, we adopt the Tanaka-Fillmore theory (Chap. 5) [53], which simplifies the analysis by assuming network motion in a quiescent solvent. This approach offers mathematical tractability while capturing the essential physics of the phenomenon.

For a rectangular gel subjected to uniaxial deformation, the kinetics of volume change follow a diffusion-type equation [46]:

$$\frac{\partial}{\partial t}\mathrm{Tr}(u_{\mathrm{ij}}) = D_{\mathrm{L}}\nabla^2\mathrm{Tr}(u_{\mathrm{ij}}) \tag{6.71}$$

where u is the strain tensor of a small volume element, Tr denotes the trace (i.e., the sum of diagonal components), and $D_L = (K_{\mathrm{OS}} + 4G/3)/f$ is the longitudinal diffusion coefficient (defined in Chap. 5), with K_{OS} being the osmotic bulk modulus, G the shear modulus, and f the friction coefficient between the polymer and the solvent. In a long rectangular gel held under constant uniaxial strain $\varepsilon_x = \varepsilon_{x0}$, volume change primarily occurs via changes in the cross-sectional area in the yz-plane. Thus, Eq. (6.71) reduces to a two-dimensional diffusion problem. Solving this with the initial condition $\mathrm{Tr}(u_{ij}) = \varepsilon_{y0} + \varepsilon_{z0} = -2m_0\varepsilon_{x0}$ at $t = 0$, and the boundary condition $\mathrm{Tr}(u_{ij}) = \varepsilon_{y\infty} + \varepsilon_{z\infty} = -2m_\infty\varepsilon_{x0}$, yields:

$$\mathrm{Tr}(u_{ij}) = -2\mu(y, z, t)\varepsilon_{x0} \tag{6.72a}$$

$$\mu(y, z, t) = \mu_\infty + \sum_{m,n}^{\mathrm{odd}} \frac{16(\mu_0 - \mu_\infty)}{mn\pi^2} \sin\frac{m\pi y}{a_r} \sin\frac{n\pi z}{a_r} \exp\left[-\frac{(m^2 + n^2)t}{2\tau_L}\right] \tag{6.72b}$$

where $\varepsilon_{\mathrm{i}0}$ and $\varepsilon_{\mathrm{i}\infty}$ (i = y, z) are the initial and equilibrium transverse strains, respectively, and μ_0 and μ_∞ are the corresponding initial and equilibrium Poisson's ratios. The quantity a_{r} denotes the gel dimension in the y- and z-directions, and $\tau_{\mathrm{sw}} = a_{\mathrm{r}}^2/(2\pi^2 D)$ is the longest characteristic time associated with swelling. The macroscopic strain in the transverse direction is obtained by integrating the local strain solution over the yz-plane:

$$\varepsilon_y(t) = \varepsilon_z(t) = \frac{1}{2a_r^2}\int\int_0^{a_r} \mathrm{d}y\mathrm{d}z\mathrm{Tr}(u_{ij}) \tag{6.73}$$

This expression describes the time evolution of the gel dimensions perpendicular to the imposed strain.

Figure 6.26a shows the time-dependent y-direction dimension for a PAAm hydrogel in water under constant uniaxial strain. The solid lines represent calculations based on Eq. (6.73) using $\mu_0 = 1/2$, $\mu_\infty = 1/4$, and the experimentally determined value of τ_{sw}. The agreement with experimental data is excellent [47]. To evaluate stress reduction, the time-dependent Poisson's ratio $\mu(t) = -\varepsilon_y(t)/\varepsilon_x$ is substituted into Eq. (6.67). The resulting stress evolution reproduces the observed decrease in nominal stress over time (Fig. 6.26b). The diffusion coefficient $D = 3.3 \times 10^{-10}$ m^{2}s^{-1}, evaluated from the measured τ_{sw}, closely matches values obtained from free-swelling experiments. This consistency confirms that the kinetics of this phenomenon are governed by the diffusion of polymer network.

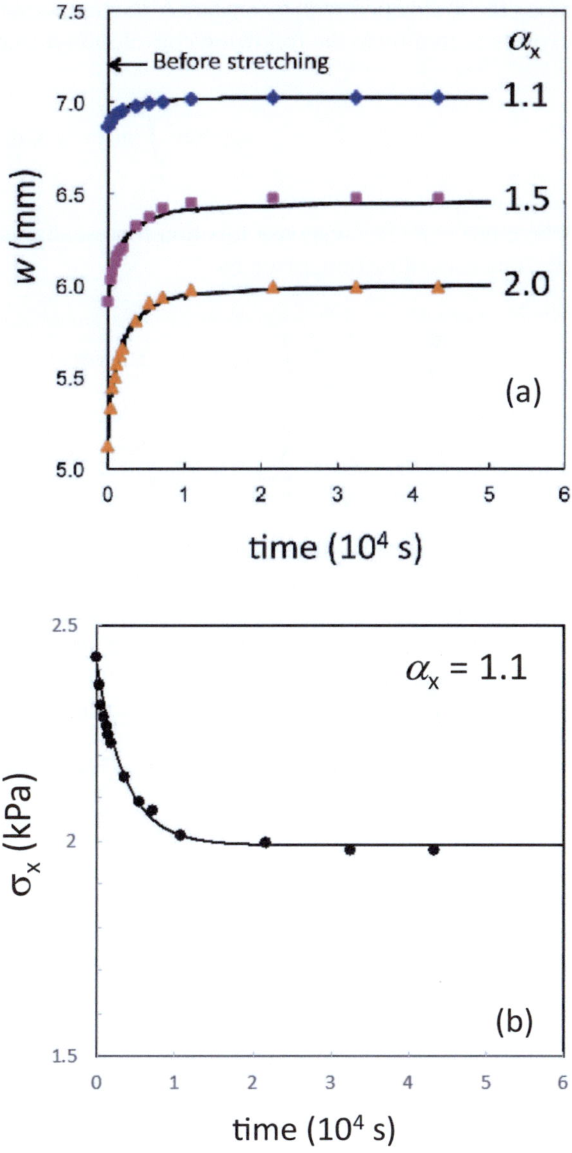

Fig. 6.26 **a** Lateral dimension (w) and nominal stress (σ_x) as a function of the time that has elapsed since a constant stretch (α_x) was imposed for a polyacrylamide hydrogel in water. The solid lines are the theoretical curves obtained from Eqs. 6.67 and 6.73 with the experimental values of τ_{sw}. Reprinted with permission from Ref. [47]. Copyright 2011 American Chemical Society

6.2.4 Gel Subjected to Stretching with Constant Strain Rate

In this section, we examine the mechanical response of a gel immersed in solvent when subjected to stretching at a constant strain rate. The resulting stress-strain behavior depends strongly on the relationship between the applied strain rate ($\dot{\varepsilon}$) and the characteristic rate of volume change, given by τ_{SW}^{-1}. When a gel is stretched along the x-direction with a constant $\dot{\varepsilon}$, the transverse strain in the y-direction, $\varepsilon_y(t)$, evolves according to the following convolution integral [47, 54]:

$$\varepsilon_{\mathrm{y}}(t) = \int_0^t m(t - t')\dot{\varepsilon}\mathrm{d}t' \tag{6.74a}$$

where $m(t - t')$ is a response function representing time-dependent deformation in the transverse direction, given by:

$$\begin{aligned} m(t - t') &= -\frac{1}{a_r^2}\int\int_0^{a_r} \mu(y, z, t - t\prime)\mathrm{d}y\mathrm{d}z \\ &= -\mu_\infty - \sum_{m,n}^{\mathrm{odd}} \frac{64(\mu_0 - \mu_\infty)}{m^2 n^2 \pi^4} \exp\left[-\frac{(m^2 + n^2)(t - t')}{2\tau_L}\right] \end{aligned} \tag{6.74b}$$

Figure 6.27a shows the experimental data for the lateral true strain ε_y as a function of the longitudinal true strain ε_x for a PAAm hydrogel (initial length $l_{x0} = 60$ mm, thickness $d = 1.2$ mm) stretched in water at various stretching speeds (v).[47] When the gel is stretched at a constant speed, the strain rate $\dot{\varepsilon}$ decreases over time due to sample elongationg:$\dot{\varepsilon} = v/(l_{\mathrm{x0}} + vt) = \dot{\varepsilon}_0/(1 + \dot{\varepsilon}_0 t)$ where $\dot{\varepsilon}_0$ $(= v/l_{\mathrm{x0}})$ is the initial strain rate. The solid curves in Fig. 6.27a, calculated using Eq. (6.74), agree well with the experimental data.

The gradient of each curve at a given time corresponds to the instantaneous Poisson's ratio, as defined in Eq. (6.66): $\mu(t) = -\mathrm{d}\varepsilon_{\mathrm{y}}/\mathrm{d}\varepsilon_{\mathrm{x}}$. The behavior of $\mu(t)$ depends on the product $\dot{\varepsilon}_0\tau_{sw}$, which compares the deformation rate to the rate of volume change: Specifically, for fast deformation of $\dot{\varepsilon}_0\tau_{sw} > 60$, the gel deforms nearly at constant volume, with $\mu = \mu_0 = 1/2$. For intermediate rates of $10 < \dot{\varepsilon}_0\tau_{sw} < 60$, strain-induced swelling occurs to a limited extent, and $\mu < 1/2$. For slow deformation of $\dot{\varepsilon}_0\tau_{sw} < 0.1$, the gel deforms under near-equilibrium conditions, and $\mu \to \mu_\infty = 1/4$.

Figure 6.27b displays the evolution of μ as a function of normalized time t/τ_{sw}. Remarkably, the results from stretching at various strain rates collapse onto a single master curve when plotted against $t/t/\tau_{sw}$, coinciding with the $\mu(t)$ curve obtained under constant strain (as discussed in Sect. 6.2.3). This alignment indicates that strain-induced swelling is primarily governed by the reduced time $t/t/\tau_{sw}$, regardless of whether strain is held constant or continuously increased.

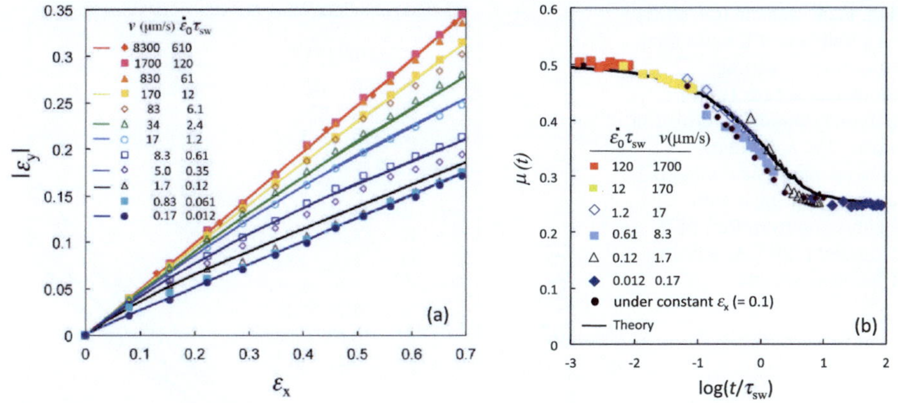

Fig. 6.27 **a** Transverse true strain as a function of longitudinal true strain at various crosshead speeds (v) for a poly(acrylamide) hydrogel in water. The solid lines correspond to the theoretical curves. **b** Poisson's ratio (μ) as a function of elapsed time (t) for the tensile stretching measurements carried out with varying v. The solid line represents the theoretical prediction. The t dependence of μ obtained from the measurement under a constant stretch of $\alpha_x = 1.1$ (Fig. 6.26) is also shown. Reprinted with permission from Ref. [47]. Copyright 2011 American Chemical Society

Figure 6.28 presents the true stress-true strain relationships calculated using the time-dependent Poisson's ratio $\mu(t)$ and the constitutive relation in Eq. (6.67), along with the corresponding experimental data [47]. When compared at the same level of ε_x, the stress at intermediate $\dot{\varepsilon}$ decreases with decreasing $\dot{\varepsilon}$. This reduction reflects the effect of time-dependent solvent inflow, which reduces the effective strain and thereby lowers the stress level.

6.2.5 Gel Subjected to Sinusoidal Strain

Consider a gel immersed in solvent and subjected to sinusoidal strain with angular frequency ω. Based on the preceding discussions, strain-induced swelling is expected to occur when the frequency is sufficiently low—specifically, when $\omega < \omega_{sw}$, where $\omega_{sw} = 1/\tau_{sw}$ is the characteristic angular frequency associated with swelling kinetics. For relatively thick gels (e.g., 10 mm), ω_{sw} is extremely low—on the order of 10^{-6}s^{-1} making it difficult to observe strain-induced swelling in standard rheological tests. However, in thinner systems such as gel fibers or films with characteristic dimensions around 100 μm, ω_{sw} increases to approximately 10^{-2}s^{-1}, a range accessible via conventional oscillatory testing.

Figure 6.29a shows the ω dependence of Poisson's ratio (μ) for a PAAm gel fiber with a thickness of approximately 400 μm [55, 56]. In this test, a sinusoidal tensile load is applied along the axial (||) direction, and the resulting dimensional changes are measured in both the axial (||) and transverse ($\perp$) directions. The strains in each direction can be expressed as:

Fig. 6.28 Tensile true stress as a function of longitudinal true strain at various crosshead speeds (v) for a poly(acrylamide) hydrogel in water. The solid lines correspond to the theoretical curves. Reprinted with permission from Ref. [47]. Copyright 2011 American Chemical Society

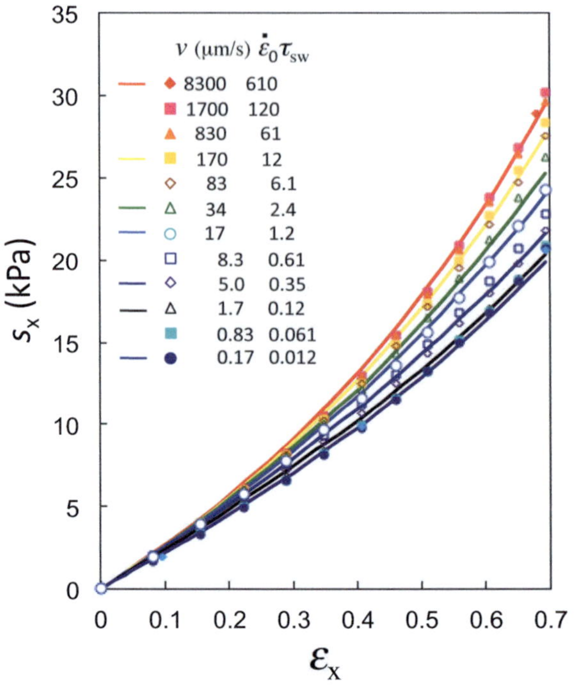

$$\varepsilon_{\parallel}^{*}(t) = \varepsilon_{\parallel} \exp\left[i\left(\omega t + \delta_{\parallel}\right)\right] \tag{6.75a}$$

$$\varepsilon_{\perp}^{*}(t) = \varepsilon_{\perp} \exp[i(\omega t + \delta_{\perp})] \tag{6.75b}$$

Here, $\varepsilon_{||}$ and $\varepsilon_{\perp}$ are the strain amplitudes in the axial and transverse directions, respectively, and $\delta_{||}$, $\delta_{\perp}$ are the corresponding phase angles. The dynamic Poisson's ratio, defined as $|\mu^*| = -|\varepsilon_{\perp}/\varepsilon_{||}|$, characterizes the frequency-dependent deformation behavior. In silicone oil, a non-solvent for the gel, μ^* remains constant at approximately 0.5 across all frequencies, indicating that deformation without volume change. In contrast, for the same gel in water, the $|\mu^*(\omega)| - \log\omega$ curve closely resembles the μ(t)-log$t$ curve obtained under constant strain (Fig. 6.27b). At high frequencies ($\omega >> 1/\tau_{sw}$) volume change cannot keep pace with the oscillating strain, and $|\mu^*| \approx 0.5$. At low frequencies ($\omega << 1/\tau_{sw}$), the gel approaches equilibrium swelling during each cycle, resulting in $|\mu^*| \approx 0.25$. The phase angles $\delta_{||}$ and $\delta_{\perp}$ exhibit extrema near $\omega_{sw} \approx 0.01\text{s}^{-1}$, as shown in Fig. 6.29b. In the intermediate frequency range ($\omega_{sw}/10 < \omega < 10\omega_{sw}$), $|\mu^*|$ decreases with decreasing ω, reflecting a transition from elastic-like to swelling-mediated deformation. Notably, the characteristic time inferred from ω_{sw}^{-1}($\approx$ 100 s) is consistent with τ_{sw}($\approx$ 80 s), as independently obtained from the time evolution of μ(t) under constant strain for the same gel (inset of Fig. 6.29a). This agreement confirms that strain-induced swelling

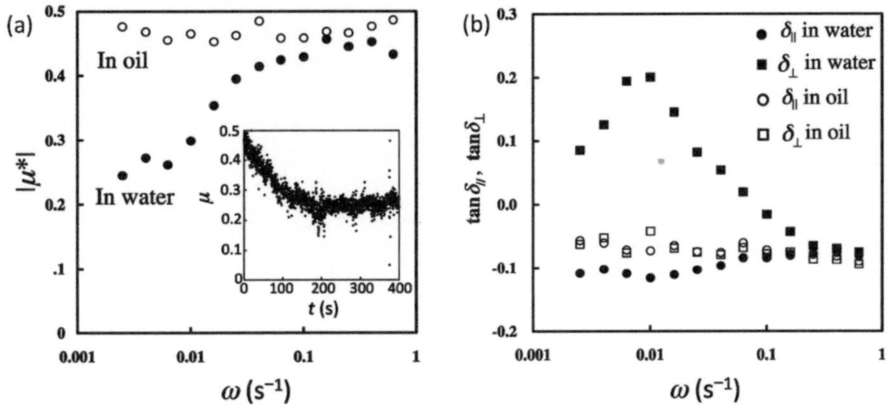

Fig. 6.29 **a** Absolute value of complex Poisson's ratio ($|\mu^*|$) as a function of angular frequency (ω) for a poly(acrylamide) hydrogel fiber with a diameter of 400 μm in water and oil. The inset shows μ for the same gel fiber in water as a function of elapsed time since a constant force is imposed. **b** Tangent of the phase lag in the direction parallel or normal to the imposed force as a function of ω in water and oil. Reprinted with permission from Ref. [56] with the permission of Springer Nature

under sinusoidal deformation is governed by the same kinetics as observed under step-strain conditions.

6.2.6 Toughening of Gel by Ultra-Slow Compression

Consider a gel subjected to compression with a constant speed. As discussed in earlier sections, mechanical compression thermodynamically drives shrinking through solvent outflow, just as stretching induces swelling. The underlying mechanism is the same as that for stretching-induced swelling, with the direction of mass transport reversed. However, unlike stretching, which dilutes the network by absorbing solvent, compression concentrates the polymer network as water is expelled. This increase in network concentration can significantly alter the fracture behavior, in some cases resulting in pronounced toughening. This is a distinctive feature of compression-induced shrinking, in contrast to the softening effect by stretching-induced swelling.

Figure 6.30a shows the compression behavior of a cylindrical hydrogel (diameter: 11.5 mm, height: 10 mm, polymer concentration: 1.33 wt%) made of gellan, saccharide-based hydrogel, tested across a wide range of compression speeds (v) [57]. At high compression speeds ($v > 1$ mm/min), the gel undergoes brittle fracture at moderate strains (tens of percent), as shown in Fig. 6.30b. Under these conditions, volume change is negligible because the compression rate far exceeds the rate of solvent outflow. Macroscopic cracks form at approximately 45° to the compression axis—consistent with failure along planes of maximum shear stress, a hallmark of

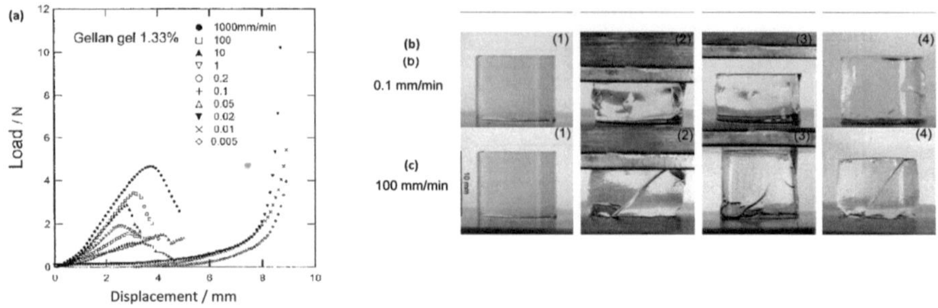

Fig. 6.30 **a** Compressive load-displacement curves at various crosshead speeds for a gellan hydrogel with a diameter of 11.5 mm and a height of 10 mm. Photographs of the compressed gellan gels at a crosshead speed of **b** 100 mm/min and **c** 0.005 mm/min: (1) undeformed state, (2) compressed state, (3) unloaded state and (4) reswollen state of the unloaded gels (3) in water for 2 days. Reprinted from Ref. [57] with the permission of Elsevier publishing

brittle behavior. In contrast, at low compression speeds ($v < 0.2$ mm/min), the gel behaves dramatically differently. It can be compressed into a thin disk—down to just 10% of its original height—without fracturing (Fig. 6.30c). During this process, a large amount of water is expelled, and the stress rises steeply at high strains (> 80%). Remarkably, when the compressed gel is re-immersed in water, it re-swells to nearly its original dimensions, indicating that no structural damage has occurred in the network (Fig. 6.30c). This toughening under ultra-slow compression arises from network densification enabled by solvent outflow. The expelled water increases the local concentration of polymer strands, enhancing the gel's resistance to crack propagation. In contrast to brittle failure at high speeds, the system becomes tougher and more compliant at low rates due to diffusion-mechanical coupling. Boundary conditions also play a crucial role in this behavior [58]. In the experiments shown in Fig. 6.30c, the gel is compressed while maintaining its diameter. As a result, the final polymer concentration increases roughly tenfold—from 1.33 wt% to 13.3 wt%—reflecting the 90% reduction in height. The characteristic compression timescale, $t_c = l_0 / v_c$, where l_0 is the initial height and v_c is the compression speed, closely matches the characteristic time τ_{sw} for solvent outflow under fixed compressive strain. This agreement confirms that the high compressibility and toughening observed at low compression rates originate from diffusio-mechanical coupling.

6.2.7 *Shrinking of Gel Under Ultracentrifugal Field*

Consider a gel placed in a centrifuge, as shown in Fig. 6.31. If a density difference exists between the polymer network and the solvent, the network experiences relative motion with respect to the solvent. In steady state, the network reaches a position where the centrifugal force is balanced by the elastic restoring force. Analyzing this process allows for evaluation of key material parameters such as the friction coefficient f between the network and solvent, and the osmotic bulk modulus L [59, 60]. As discussed earlier in this chapter, rigorous treatment of swelling kinetics requires considering the relative motion between the polymer network and the solvent[85,86]. However, for mathematical simplicity, the model used in previous sections assumed motion of the network in a quiescent solvent. In the case of centrifugal deformation, this simplification is fully justified, since the governing equations for relative motion are equivalent under both models.

Let us now analyze the behavior of a gel under centrifugation (Fig. 6.31). Define r_1 and r_2 as the positions of the network/solvent interface and the bottom of the container, respectively. Initially, at $t = 0$, the interface at $r = r_1$ corresponds to the air/gel interface. Let $u(r,t)$ denote the radial displacement of the network at time t and position r. The governing equation for the displacement is given by [59]:

$$\frac{\partial u(r,t)}{\partial t} = D\frac{\partial^2 u(r,t)}{\partial r^2} + \frac{\phi \Delta\rho\omega^2}{f} r \tag{6.76}$$

where inertia is neglected, assuming that the shrinking process is sufficiently slow. In this equation, D is the diffusion coefficient given by $D = (K_{OS} + 4/3G)/f = L/f$, $\Delta\rho$ is the density difference between network and solvent, and ω is the angular velocity of the centrifuge. When the centrifugal field is applied at $t = 0$ and $\Delta\rho > 0$, the solution of Eq. (6.76) under suitable initial and boundary conditions is:

$$u(r,t) = \sum_{n=1}^{\infty} A_n \exp\left(-\frac{(2n-1)^2 t}{\tau_L}\right) + u(r,\infty)$$

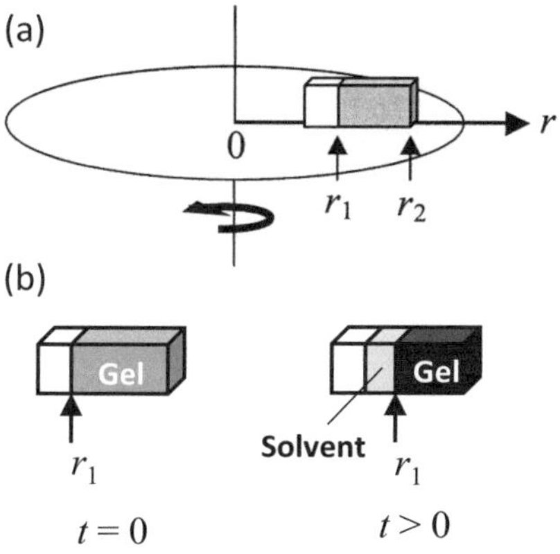

Fig. 6.31 **a** Schematic for a gel under centrifugal field. The gray part indicates a gel confined in a cell. Each of r_1 and r_2 corresponds to the position of the moving boundary and immobile bottom wall of the cell. **b** Schematic for a gel network compressed by a centrifugal field at $t > 0$. The density of network is assumed to be higher than that of solvent

$$A_n(r) = \frac{16a^2\phi\Delta\rho\omega^2}{\pi^4 L}\left\{\frac{(-1)^n\pi r_2}{(2n-1)^3} + \frac{2a}{(2n-1)^4}\right\} \times \cos\left\{\frac{(2n-1)\pi(r-r_1)}{2a}\right\}$$

$$u(r,\infty) = \frac{\phi\Delta\rho\omega^2}{6L}\left(r_2^3 + 3r_1^2 r - 3r_1^2 r_2 - r^3\right) \tag{6.77}$$

where A_n and $u(r,\infty)$ represent the transient and steady-state components of the displacement. The time $\tau_L = 4a^2/(\pi^2 D)$, where $a = r_2 - r_1$, is the longest characteristic time of the process.

Figure 6.32a displays the time evolution of the network/solvent interface position in a PAAm hydrogel under a centrifugal acceleration of $\alpha = 2.26 \times 10^4$ g. At early times, the interface moves at nearly constant velocity before reaching saturating. From Eq. (6.77), the initial velocity for $t << \tau_L$ is:

$$v = \frac{\phi\Delta\rho r_1\omega^2}{f} = \frac{\phi\Delta\rho g\alpha}{f} \tag{6.78}$$

where $\alpha = r_1\omega^2/g$ is the centrifugal acceleration in units of gravity. This expression indicates that the initial velocity is proportional to α, a relationship confirmed experimentally for a poly(vinyl alcohol) gel (Fig. 6.32b) [60]. Using Eq. (6.78), the friction coefficient f was estimated as 9.2×10^{13} Nsm^{-4}. From the steady-state solution of Eq. (6.77), the displacement at $r = r_1$ is given by:

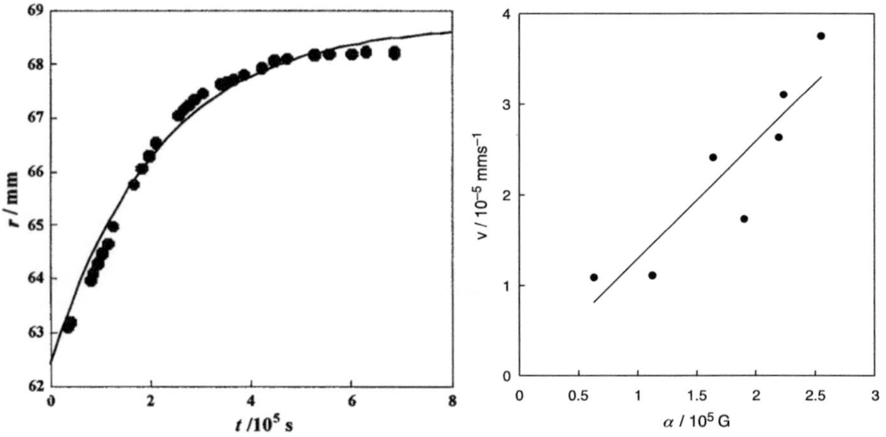

Fig. 6.32 **a** Position of the solvent–gel interface (the moving surface) of a PAAm hydrogel as a function of time at the centrifugal acceleration of 2.26×10^4 G. The solid line the best-fitted curve of Eq. 6.77. The fitted values of τL and L are 2.16×10^5 s and 1.50×10^4 Pa, respectively. Reprinted from Ref. [59] with the permission of AIP Publishing. **b** Initial velocity of the solvent-gel interface (v) at $t << \tau L$ as a function of centrifugal acceleration (α) for poly(vinyl alcohol) gels. Reprinted from Ref. [60] with the permission of Elsevier publishing

$$u(r_1, \infty) = \frac{\phi \Delta\rho\omega^2}{6L}\left(2r_1^3 - 3r_1^2 r_2 + r_2^3\right) \tag{6.79}$$

This equation shows that the magnitude of steady-state shrinking is determined by the balance between centrifugal force and the longitudinal modulus L. Using Eq. (6.79) and experimental data, L was evaluated to be 1.5×10^4 Pa. Furthermore, the solid line in Fig. 6.32a, corresponding to the first term ($n = 1$) in Eq. (6.77), closely fits the experimental data, capturing the time evolution of shrinkage. The diffusion coefficient D, computed from the experimentally obtained f and L, agrees with the value measured in free-swelling experiments, supporting the consistency of the diffusio-mechanical framework.

6.2.8 Other Examples of Diffusio-Mechanical Coupling

In addition to cases discussed so far, several other phenomena arise from diffusio-mechanical coupling—that is, the interplay between mechanical deformation and solvent transport in gels. The following examples illustrate how mass transfer influences the mechanical behavior of soft, solvent-containing materials.

When microgel suspensions are subjected to shear, the deformation induces solvent flow out of the microgels. This shear-induced deswelling reflects a coupling between shear stress and the internal solvent pressure gradient, leading to dynamic changes in microgel volume [61].

When a pressure gradient drives solvent flow through gel films, the solvent influx causes the film to swell. This behavior reflects how mechanical loading can alter solvent distribution, thereby modifying the gel volume and stress state [62].

Gels subjected to torsional deformation can exhibit shrinkage perpendicular to the torsion axis. Additionally, a negative first normal stress difference can emerge, a behavior contrary to the behavior of typical viscoelastic materials. This effect arises from the redistribution of solvent with the gel, altering internal pressure and leading to macroscopic volume contraction [63].

6.2.9 Summary of Diffusio-Mechanical Coupling

Table 6.1 summarizes the diverse phenomena arising from diffusio-mechanical coupling in gels. Each entry highlights the type of deformation, the primary coupling mechanism (i.e., how solvent migration is triggered by mechanical stimuli), characteristic mechanical outcomes, and key controlling parameters. Collectively, these examples demonstrate that mass transport significantly modulates the mechanical behavior of gels, depending on both geometry and mode of deformation.

Table 6.1 Summary of diffusio-mechanical coupling in gels

Deformation type	Primary diffusio-mechanical effect	Key observations/ outcomes	Controlling parameters	Sections
Uniaxial stretching (constant strain)	Strain-induced swelling (solvent inflow)	Stress decreases over time; Poisson's ratio changes from 1/2 → 1/4	$\tau_{sw} = d^2/D$, $\mu(t)$, D	6.2.2, 6.2.3
Uniaxial stretching (constant rate)	Time-dependent swelling	$\mu(t)$ varies with t/τ_{sw}; strain-rate dependent stress-strain behavior	$\dot{\varepsilon}$, τ_{sw}	6.2.4
Sinusoidal strain	Frequency-dependent swelling	Dynamic Poisson's ratio shifts from 0.5 (high ω) to 0.25 (low ω)	μ, ω	6.2.5
Uniaxial compression (constant speed)	Strain-induced shrinking (solvent outflow)	Ultra-slow compression leads to toughening and high compressibility without fracture	Compression speed v, τ_{sw}	6.2.6
Ultracentrifugal field	Centrifugal force-induced shrinking	Interface displacement ~ ω/f; steady-state shrinkage reflects modulus L	ω, $\Delta\rho$, f, L	6.2.7
Shear flow on microgels	Solvent discharge under flow	Microgels dynamically deswell under shear flow	Shear rate, microgel size	6.2.8
Solvent flow through gel films	Flow-driven swelling	Solvent influx expands gel film	Pressure gradient, permeability	6.2.8
Torsional deformation	Shrinking and negative normal stress	Volume reduction orthogonal to shear; emergence of negative normal stress	Shear angle, torsion rate	6.2.8

6.3 Rheology of Sol-Gel Transition

6.3.1 Sol-Gel Transition

Gelation refers to the process by which monomers or polymers form an infinite three-dimensional network. Macroscopically, this transition is observed as a shift from a fluid-like state to a solid-like state, where the material loses its ability to flow. This phenomenon is commonly termed the "sol-gel transition," with "sol" denoting the fluid state. In this context, a sol is defined as a system composed of clusters with finite size, whereas a gel contains at least one cluster that spans the entire system—effectively an infinite cluster. To illustrate, consider gelation via the polymerization of tetra-functional monomers. Let p be the fraction of reacted functional groups. As p increase, clusters grow. At a critical value $p = p_c$, an infinite cluster forms, connecting one side of the system to the other. This point is called the "gel point," and the system just beyond it is referred to as a "critical gel." The regions $p < p_c$ and $p > p_c$ correspond the sol and gel states, respectively. Notably, even for $p > p_c$, many finite clusters still coexist with the infinite cluster, unless p becomes sufficiently large.

Gel formation has long been a classical issue in polymer science. Early theoretical work by Flory and Stockmayer employed a tree-like model to describe cluster growth and the onset of gelation (Fig. 6.33a) [2, 64, 65]. This model conceptualizes gelation as a branching process where connections grow outward from a single point, assuming that all reactions occur between different molecules and neglecting intramolecular cyclization. Later theoretical developments incorporated insights from critical phenomena and percolation theory, providing a more comprehensive framework. Percolation theory based on lattice models implicitly considers excluded volume effects and intramolecular cyclization—factors neglected in the original tree model [66–68]. For example, Fig. 6.33b depicts a bond percolation model on a two-dimensional square lattice where all grid points (sites) are occupied, and the probability p governs whether neighboring sites are connected by bonds. The critical percolation threshold p_c depends on lattice type (e.g., square, cubic, diamond), but the critical behavior near the gel point is governed by the spatial dimension d (typically $d = 3$ in real systems), not the lattice geometry. The tree model corresponds to the Bethe lattice, effectively a mean-field model representing $d \geq 6$, where loop formation is negligible. The incorporation of fractal geometry into percolation theory further advanced the theoretical understanding of gelation. The critical gel, with its complex branched structure, is regarded as a self-similar fractal object lacking a characteristic length scale. Its structural complexity and sparsity are characterized by the "fractal dimension" [69–71].

At the gel point, pronounced changes in rheological properties occur due to the emergence of an infinite cluster. The zero-shear viscosity (η_0) diverges, and the equilibrium shear modulus (G_e) begins to emerge (Fig. 6.34). In the sol state ($p < p_c$), $G_e = 0$ because the system exhibits fluidity. As p approaches p_c, the average cluster size grows significantly, leading to a sharp rise in η_0. Beyond the gel point (p

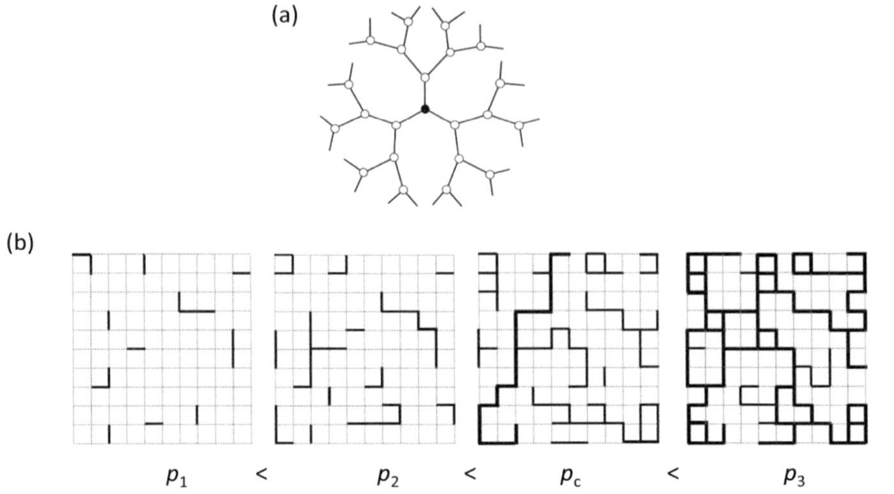

Fig. 6.33 **a** Bethe lattice employed in tree-like model. **b** Two-dimensional bond percolation model at various values of bond probability (p)

$> p_c$), the material behaves as a solid, with $\eta_0 \to \infty$ and $G_e > 0$. Further increases in p raise G_e due to the growing density of cross-links (mesh density). When all functional groups are bonded ($p = 1$), G_e reaches its maximum.

In this section, we describe how rheological measurements can be employed to accurately determine the gelation point, analyze the fractal nature of the critical gel, and evaluate the relevance of percolation theory to the sol-gel transition.

Fig. 6.34 Divergence of zero-shear viscosity and emergence of equilibrium elastic modulus at the gel point

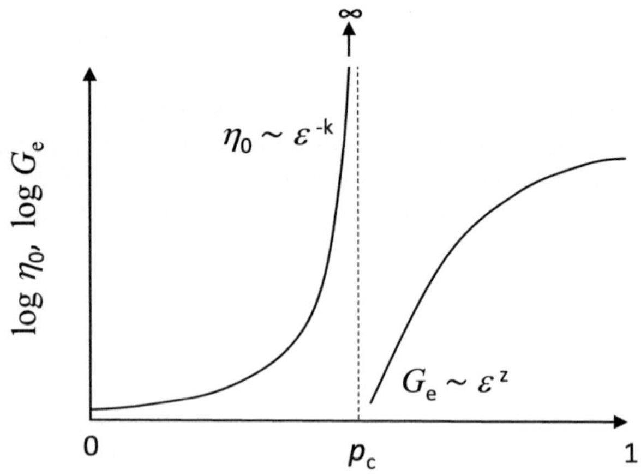

6.3.2 Sol-Gel Transition Based on Percolation Theory

As the system approaches the gel point, many physical quantities exhibit critical behavior —they either diverge to infinity or converge to zero. This behavior can be described by power-laws characterized by critical exponents, using the relative distance from the gel point, denoted by ε, defined as:

$$\varepsilon = \frac{|p - p_c|}{p_c} \tag{6.80}$$

where p represents the bond or site occupation probability in a percolation model, and corresponds to variables such as reaction conversion or precursor concentration in actual gelation reactions. In experiments, p is often indirectly controlled by parameters such as reaction time or temperature.

The weight average molecular weight (M_w) of finite clusters in the sol state increases with p and diverges at the gel point p_c, reflecting the emergence of an infinite cluster. This divergence is characterized by the critical exponent γ:

$$M_w \sim \varepsilon^{-\gamma} (p < p_c) \tag{6.81}$$

Another key parameter is the correlation length (ξ), which represents the average size of finite clusters (excluding the infinite clusters when $p > p_c$). Near the gel point, ξ behaves as:

$$\xi \sim \varepsilon^{-\nu} \tag{6.82}$$

Furthermore, the mass M of the largest cluster scales with ξ via the fractal dimension D:

$$M \sim \xi^D (p \approx p_c) \tag{6.83}$$

For three-dimensional systems ($d = 3$), percolation theory predicts $D = 2.5$, indicating the sparse, fractal structure of the infinite cluster. Although not detailed here, percolation theory provides critical exponents such as ν, γ, and β (regarding the gel fraction in the gel state), and these exponents exhibit universality—they depend only on the spatial dimension d, not on specific system details. For $d = 3$, typical values are $\nu = 0.89$, $\gamma = 1.74$. These exponents are interrelated through scaling laws [67, 68].

6.3.3 *Theoretical Background of the Rheology of the Sol-Gel Transition*

Rheological behavior is sensitive to the formation and growth of the network structure during gelation. As mentioned before, the zero-shear viscosity η_0 diverges, while a finite equilibrium shear modulus G_e emerges at the gel point. Their critical behaviors near p_c (Fig. 6.34) follow power laws with respect to the distance from the gel point, ε, as:

$$\eta_0 \sim \varepsilon^{-k} (p < p_c) \tag{6.84}$$

$$G_e \sim \varepsilon^{z} (p < p_c) \tag{6.85}$$

where k and z denote the critical exponents for η_0 and G_e, respectively. However, determining p_c directly from η_0 and G_e is difficult, as these quantities diverge or vanish at p_c. Instead, time-dependent moduli such as the relaxation modulus $G(t)$ and the dynamic modulus $G^*(\omega)$ are more practical and measurable both below and above p_c.

In viscoelastic measurements, the material is typically subjected to a step strain $[\gamma(t) = \gamma_0]$ or a sinusoidal strain $[\gamma(t) = \gamma_0 \sin(\omega t)]$. The resulting stress response defines the modulus:

$$G(t) = \sigma(t)/\gamma_0 \tag{6.86}$$

Under sinusoidal deformation, the stress response $\sigma(t)$ can be decomposed into in-phase (elastic) and out-of-phase (viscous) components:

$$\begin{aligned}\sigma(t) &= \sigma_0 \sin(\omega t + \delta) \\ &= G'\gamma_0 \sin \omega t + G''\gamma_0 \sin(\omega t + \pi/2)\end{aligned} \tag{6.87}$$

$$G'(\omega) = \sigma_0 \cos \delta/\gamma_0 \tag{6.87a}$$

$$G''(\omega) = \sigma_0 \sin \delta/\gamma_0 \tag{6.87b}$$

$$\left|G^*(\omega)\right| = \left[G'(\omega)^2 + G''(\omega)^2\right]^{1/2} \tag{6.87c}$$

where G' and G'' are the storage and loss moduli, respectively.

The long-time behavior of $G(t)$ and the low-frequency limit of $G'(\omega)$ reflect the emergence of G_e. In the sol state, $G_e = 0$; in the gel state, $G_e > 0$. (Fig. 6.35). At the gel point ($p = p_c$), the material exhibits intermediate behavior between a liquid

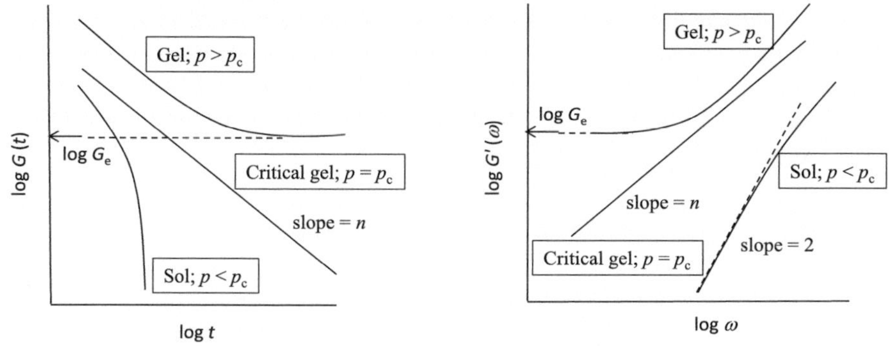

Fig. 6.35 **a** Time dependence of relaxation modulus and **b** frequency dependence of storage modulus in the regions of $p < p_c$ (sol), $p = p_c$ (critical gel), and $p > p_c$ (gel)

and a solid. Winter and Chamber [72, 73] demonstrated that a critical gel exhibits a power-law relaxation over a wide t range:

$$G(t) = St^{-n}(p = p_c; \lambda_0 < t < \infty) \tag{6.88}$$

where S is the gel strength, and n is the viscoelastic exponent or relaxation exponent. In general, $G(t)$ can be expressed as the Laplace transform of the relaxation spectrum $H(\tau)$:

$$G(t) = \int_0^{\tau_{max}} H(\tau)\exp(-t/\tau)\frac{d\tau}{\tau} + G_e \tag{6.89}$$

where τ_{max} is the longest relaxation time, and G_e is nonzero for a gel, zero for a sol. For a critical gel, the spectrum is:

$$H(\tau) = \frac{S}{\Gamma(n)}\tau^{-n}(p = p_c; \lambda_0 < t < \infty) \tag{6.90}$$

where $\Gamma(n)$ represents Gamma function. Using this, the storage and loss moduli take the form:

$$G'(\omega) = C_1\omega^n(p = p_c; 0 < \omega < 1/\lambda_0) \tag{6.91a}$$

$$G''(\omega) = C_2\omega^n(p = p_c; 0 < \omega < 1/\lambda_0) \tag{6.91b}$$

where C_1 and C_2 are the constants. The power-law behavior over wide t and ω ranges indicates the absence of a characteristic relaxation time at the gel point ($\tau_{max} \rightarrow \infty$), reflecting the self-similar or fractal structure which lacks a characteristic length scale. This fractal structure is composed of hierarchical elements that span multiple length and time scales. The cutoff time λ_0 corresponds to the relaxation time of the smallest unit within the self-similar structure, potentially associated with the glass transition or local segmental motion. Applying the Kramers-Kronig relation to Eq. (6.91) yields an important result: the loss tangent $\tan\delta$ remains constant over a broad frequency range at the gel point:

$$\tan\delta(\omega) = \frac{G''(\omega)}{G'(\omega)} = \tan\left(\frac{n\pi}{2}\right) = \text{Constant}(p = p_c;\, 0 < \omega < 1/\lambda_0) \quad (6.92)$$

This constancy of $\tan\delta$ signifies that both elastic and viscous components scale identically with frequency (ω^n), a hallmark of the critical gel state. Moreover, this ratio depends solely on the viscoelastic exponent n, making it a universal and experimentally accessible criterion for identifying the gel point.

Another key feature near p_c is the divergence of the longest relaxation time τ_{max} (Fig. 6.36a) [73]:

$$\tau_{max} \sim \varepsilon^{-\alpha^-} (p < p_c) \quad (6.93a)$$

$$\tau_{max} \sim \varepsilon^{-\alpha^+} (p < p_c) \quad (6.93b)$$

where α^- and α^+ denote the critical exponents in the sol and gel regions, respectively. In the sol region, τ_{max} reflects the relaxation of the largest finite cluster; in the gel region, it characterizes the time required for $G(t)$ to reach G_e (Eq. 6.89) (Fig. 6.36b). From Eq. (6.88), τ_{max} in the gel region is obtained using $t = \tau_{max}$:

$$\tau_{max} = \left(\frac{S}{G_e}\right)^{1/n} (p > p_c) \quad (6.94)$$

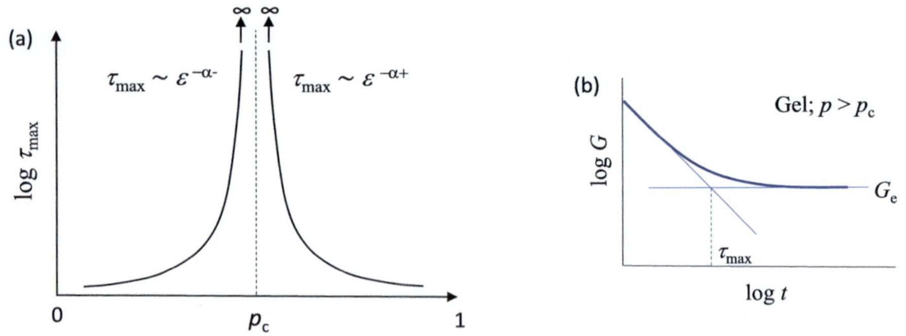

Fig. 6.36 **a** Divergence of the longest relaxation time at the gelation point. **b** Evaluation of the longest relaxation time beyond the gelation point

The τ_{max} in the gel region reflects the relaxation of dangling chains in infinite network and residual finite clusters. As p increases, these relaxable units are gradually incorporated into the rigid network, thereby reducing τ_{max}.

The zero-shear viscosity η_0 is related to $H(\tau)$ as:

$$\eta_0 = \int_0^{\tau_{max}} H(\tau)\mathrm{d}\tau \tag{6.95}$$

$$\eta_0 \sim \tau_{max}^{1-n}(p < p_c) \tag{6.96}$$

From these relations, the critical exponents k and z are expressed as:

$$k = (1-n)\alpha^{-}(p < p_c) \tag{6.97}$$

$$z = n\alpha^{+}(p > p_c) \tag{6.98}$$

Assuming symmetric divergence for τ_{max} on both sides of p_c, a scaling relation emerges [74]:

$$n = \frac{z}{k+z} \tag{6.99}$$

Thus, the rheological exponents n, k, and z are interrelated. Percolation theory links these rheological exponents to the fractal dimension D. For systems with excluded volume effects but no hydrodynamic interaction [69, 70]:

$$k = \nu(D+2-d) \tag{6.100}$$

$$z = d\nu \tag{6.101}$$

$$n = \frac{d}{D+2} \tag{6.102}$$

Using $d = 3$ and $D = 2.5$, we obtain $n = 2/3$. If hydrodynamic interactions are also screened, an alternative relation applies [75]:

$$n = \frac{d(d+2-2D)}{2(d+2-D)} \tag{6.103}$$

These theoretical relations enable the estimation of D from experimentally determined values of n (Fig. 6.37). For example, in the 3D percolation model, Eqs. (6.100) and (6.101) yield $k = 1.33$ and $z = 2.67$. An alternative conductivity-based approach

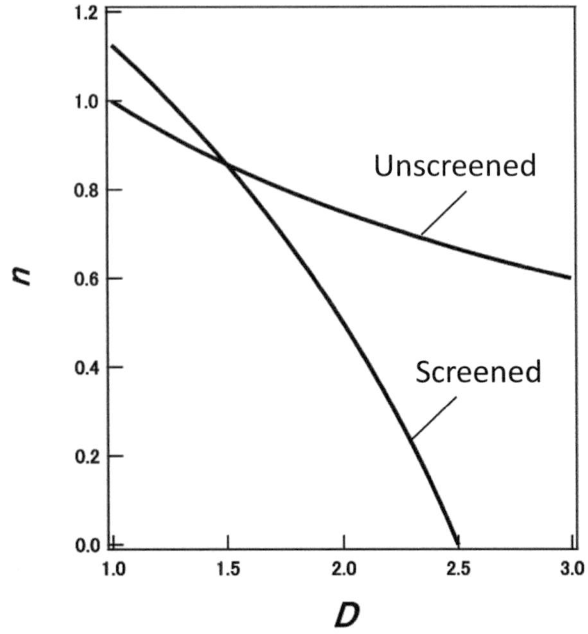

Fig. 6.37 Relation between relaxation exponent n and fractal dimension D for a three-dimensional network. Reprinted with permission from Ref. [75]. Copyright 1989 American Chemical Society

suggests $k = 0.75$ and $z = 1.94$, highlighting an analogy between gelation and electrical conduction in heterogeneous media near p_c [76, 77]. In this analogy, the divergence of η_0 and the emergence of G_e near the gel point correspond to the behavior of conductivity in conductor-superconductor and conductor-insulator mixtures near p_c.

6.3.4 Experiments for Sol-Gel Transition

This section describes how the gel point can be determined and how the critical behavior of viscoelastic properties can be analyzed using rheological data obtained during the gelation process.

6.3.4.1 Gel Point

Figure 6.38a presents rheological measurements of network formation in a polybutadiene system with difunctional end groups and a trifunctional cross-linker [78]. For reaction times $t < 120$ min, $G'' > G'$ is observed over the entire frequency (ω) range, indicating a sol state with no infinite network. For $t > 200$ min, $G' > G''$ holds across all frequencies, and G' shows a clear plateau at low ω, signifying the formation of an infinite gel network. At $t = 160$ min, both the log-log plots of $G'(\omega)$ and $G''(\omega)$

exhibit linear behavior with identical slopes across the entire ω range. This satisfies Eq. (6.91) and corresponds to the gel point criterion.

Figure 6.38b shows the evolution of the loss tangent, tan δ, as a function of reaction time t at various ω, using the same dataset [78]. In the sol state, tan$\delta > 1$ (i.e., $G'' > G'$), while in the gel state, tan$\delta < 1$ (i.e., $G' > G''$). Although the G' and G'' values vary with ω in both states, a specific reaction time, $t_c = 160.8$ min, exists at which all tan $\delta(t)$ curves intersect at a single point (tan$\delta_c = 1.75$), independent of ω. This ω-independent crossover point marks the gelation time (t_c), in accordance with Eq. (6.92). Power-law behavior of $G'(\omega)$ and $G''(\omega)$ at the gel point has been widely observed [79]. Across a wide range of precursor concentrations ($w_f = 0.25$ ~ 0.96), all samples exhibit power-law scaling at the gel point [79] (Fig. 6.39).

The power-law form of $G(t)$ (Eq. 6.88) is mathematically equivalent to that of $G^*(\omega)$ (Eq. 6.91), and both provide valid criteria for identifying the gel point. Noted that t in $G(t)$ refers to the time during rheological test, not reaction time. Because accurately capturing the power-law behavior in $G(t)$ requires long-duration measurements, experiments often involve quenching the reaction. In contrast, $G^*(\omega)$ can typically be obtained during gelation and within shorter timescales, making it a more practical tool for characterizing the sol-gel transition. As a result, $G^*(\omega)$ is more commonly used in experiments, while accurate time-domain tracking of $G(t)$ during gelation is relatively rare. However, since $G(t)$ and $G^*(\omega)$ are mathematically related through Fourier transform, $G(t)$ can be computed from measured $G^*(\omega)$ data. Figure 6.40a shows $G^*(\omega)$ data for the gelation of polydimethylsiloxane (PDMS) undergoing terminal cross-linking [80]. Figure 6.40b illustrates the corresponding $G(t)$ obtained via Fourier transformation [80]. At the gel point, $G(t)$ follows a power law with $n \approx 0.5$ over six decades of time. In the long-time regime, flow behavior (in

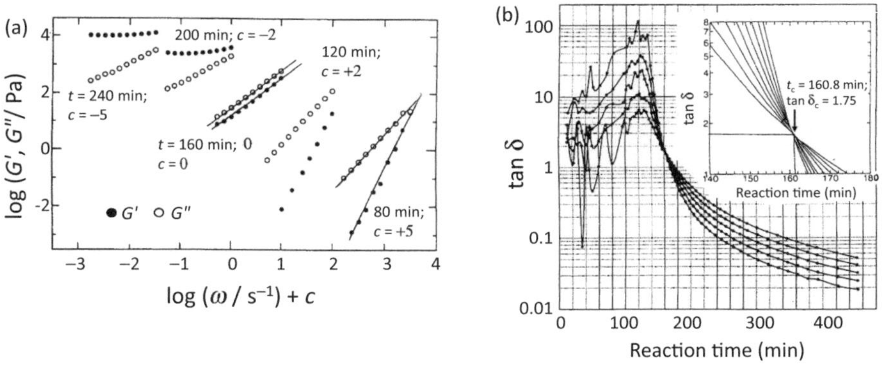

Fig. 6.38 **a** Storage and loss modulus as a function of angular frequency (ω) at various elapsed times for the gelation of hydroxy-terminated polybutadience and trifunctional isocyanate. **b** Loss tangent as a function of elapsed time at various ω: $\omega = 7.7$, 14, 28, 52 and 100 s^{-1}. The inset shows the magnification of the region near t_c. The five curves meet at one point corresponding to t_c. The data are reproduced from Ref. [78]

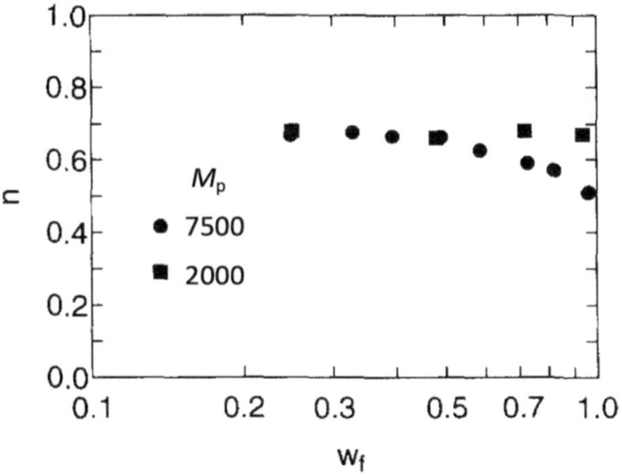

Fig. 6.39 (a) Angular frequency dependence of storage and loss modulus at the gelation point for five solutions with various precursor concentrations (w_f) for end-reactive dimethyl silyl poly(propylene oxide).. **b** Dependence of n on concentration (w_f) and molecular weight (M_p) of precursor chains. Reprinted from Ref. [79] with the permission of Elsevier publishing

the sol state) and plateau behavior (in the gel state) are clearly observed, consistent with trends shown in Fig. 6.35a.

The relaxation index n can be determined from the value of tan δ at the gel point or from the slope of Eq. (6.91). This value ($n \approx 0.7$) is largely independent of precursor molecular weight M_p and weight fraction w_f, except at high w_f with long-chain precursors ($M_p = 7500$) [79]. Systems exhibiting $n < 0.7$ show significant chain entanglement prior to gelation, as confirmed by the rheological measurements in the initial state [79]. This trend—where entanglement reduces n—has been observed in other gelation systems as well [81].[87] In the cases shown in Fig. 6.39, n tends to fall below 0.7, when w_f exceeds a certain threshold, likely due to hydrogen bonding interactions between chain ends. The result $n \approx 0.7$ is consistent with the theoretical prediction of $n = 2/3$ when excluded volume effects are considered. However, care is needed when interpreting this agreement, particularly for highly concentrated

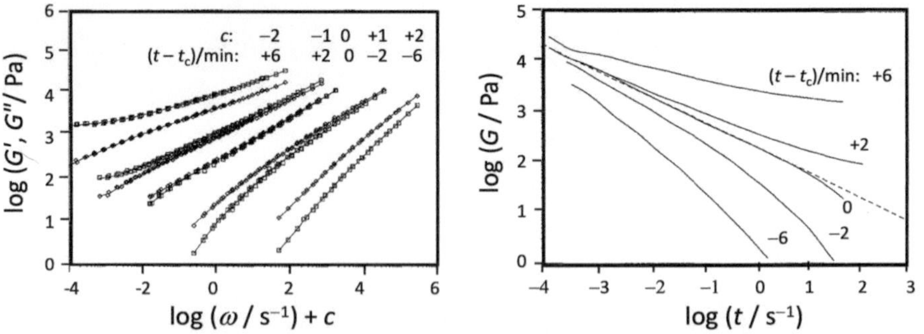

Fig. 6.40 **a** Storage and loss modulus as a function of angular frequency (w) at various elapsed times in the gelation of end-reactive polydimethylsiloxane. The curves are horizontally shifted by a factor of A to avoid the overlapping. **b** Relaxation modulus at various elapsed times calculated from the data in (**a**). Data are reproduced from Ref. [80]

systems where excluded volume effects are suppressed. For instance, substituting $n = 0.7$ into Eq. (6.103), which assumes full screening of excluded volume effects, yields a fractal dimension $D \approx 1.8$.

Reported values of n span a wide range (0.3 ~ 0.8), suggesting that n is highly sensitive to molecular-level interactions within the system. Nevertheless, the power-law behavior of moduli over wide time or frequency ranges (Eq. (6.91)) remains a robust and broadly applicable criterion for identifying the gel point. This power-law scaling reflects the emergence of a self-similar fractal network and offers a physical foundation for defining the gelation threshold [69, 82].

In some studies, the condition of $G' = G''$ (i.e., $\tan\delta = 1$) at a single frequency is used as a simple criterion for the gel point. However, frequency sweep data (Figs. 6.38 and 6.39) often show that $\tan\delta$ at the gel point is greater than unity. This suggests that the point where $\tan\delta = 1$ depends on the frequency (i.e., the observation timescale), as shown in the inset of Fig. 6.38. In contrast, the gel point defined by Eq. (6.92) is independent of frequency (timescale), reinforcing the conclusion that the condition that $\tan\delta = 1$ at a single frequency cannot serve as a universal criterion for identifying the gel point or phase transition.

6.3.4.2 Critical Behavior of Viscosity and Elastic Modulus

Figure 6.41 shows double-logarithmic plots of η_0 versus the reduced distance from the gel point (ε), where ε is calculated using reaction time as a proxy for p, and the gelation time t_c (Fig. 6.38a) as p_c [79]. In the region of sufficiently small ε (i.e., close to the gel point), a clear power-law relationship is observed, with the slope corresponding to the critical exponent k. Across various precursor concentrations w_f, the k value is consistently close to 1.0, excepting at high w_f where pronounced entanglement effects lead to deviations from this trend. Similarly, the critical exponent z, characterizing the growth of the equilibrium shear modulus (G_e) beyond the gel point, is obtained from the slope of the log-log plots of G_e versus ε (Fig. 6.42) [79]. The z value is approximately 1.9 and remains relatively constant across different w_f, again with deviations observed at high w_f where entanglement effects become significant.

Substituting $k = 1.0$ and $z = 1.9$ into the scaling relation Eq. (6.99) yields $n = 0.66$. This n value agrees with the independently determined $n \approx 0.66$ obtained from the ω dependence of G' and G'', thereby validating the scaling law (Eq. 6.99) that links the viscoelastic critical exponents. In particular, the consistency supports the assumption of symmetric divergence of the longest relaxation time τ_{max} in both sol and gel states, i.e., $\alpha = \alpha^- = \alpha^+$.

Reported values of k and z in the literatures span broad ranges, depending on the specific gelation system: k typically varies from 0.7 to 1.7, and z from 1.7 to 3.2. Much like the relaxation exponent n, the k and z values appear to be influenced by molecular-level factors, such as chain entanglements, cross-linking chemistry, and specific intermolecular interactions. For instance, Figs. 6.41 and 6.42 demonstrate that both k and n tend to increase at high precursor concentrations, where entanglement effects are more pronounced. Despite such variability, considerable effort has

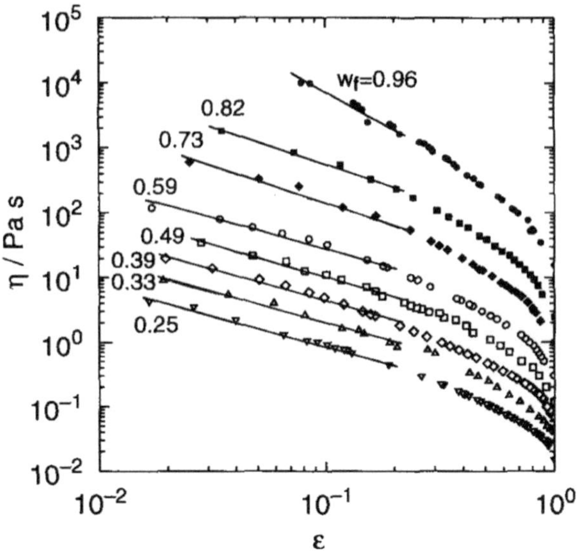

Fig. 6.41 Double-logarithmic plots of η_0 versus ε for the solutions of end-reactive dimethyl silyl poly(propylene oxide) with various precursor concentrations (w_f). The slope corresponds to k in the relation of $\eta_0 \sim \varepsilon^{-\sigma\sigma k}$. Reprinted from Ref. [79] with the permission of Elsevier publishing

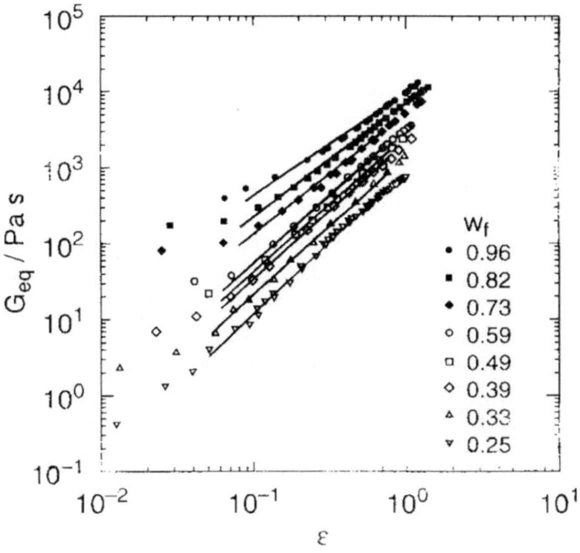

Fig. 6.42 Double-logarithmic plots of G_e versus ε for the solutions of end-reactive dimethyl silyl poly(propylene oxide) with various precursor concentrations (w_f). The slope corresponds to z in the relation of $G_e \sim \varepsilon^z$. Reprinted from Ref. [79] with the permission of Elsevier publishing

been devoted to exploring the universality of these critical exponents, through systematic studies that account for the effects of entanglement, cross-linking mechanism, and other system-specific characteristics [83].

References

1. M. Doi, *Soft Matter Physics* (Oxford University Press, Oxford, 2013)
2. P.J. Flory, *Principles of Polymer Chemistry* (Cornell University Press, Ithaca, NY, 1953)
3. L.R.G. Treloar, *The Physics of Rubber Elasticity* (Clarendon Press, Oxford, 1975)
4. W. Kuhn, Kolloidzschr **76**, 258 (1936)
5. F.T. Wall, Statistical thermodynamics of rubber II. J. Chem. Phys. **10**(7), 485–488 (1942)
6. H.M. James, E. Guth, Theory of the increase in rigidity of rubber during cure. J. Chem. Phys. **15**(9), 669–683 (1947)
7. J.D. Ferry, *Viscoelastic Properties of Polymers*, 3rd edn. (Wiley, New York, 1980)
8. M. Doi, S.F. Edwards, *The Theory of Polymer Dynamics* (Oxford Unviversity Press, Oxford, 1986)
9. M. Gottlieb, C.W. Macosko, G.S. Benjamin, K.O. Meyers, E.W. Merrill, Equilibrium modulus of model poly(dimethylsiloxane) networks. Macromolecules **14**(4), 1039–1046 (1981)
10. T. Sakai, T. Matsunaga, Y. Yamamoto, C. Ito, R. Yoshida, S. Suzuki, N. Sasaki, M. Shibayama, U.I. Chung, Design and fabrication of a high-strength hydrogel with ideally homogeneous network structure from tetrahedron-like macromonomers. Macromolecules **41**(14), 5379–5384 (2008)
11. D.R. Miller, C.W. Macosko, New derivation of post gel properties of network polymers. Macromolecules **9**(2), 206–211 (1976)
12. S.K. Patel, S. Malone, C. Cohen, J.R. Gillmor, R.H. Colby, Elastic modulus and equilibrium swelling of poly (dimethylsiloxane) networks. Macromolecules **25**(20), 5241–5251 (1992)
13. R. Everaers, Entanglement effects in defect-free model polymer networks. New J. Phys. **1**(1), 12 (1999)
14. K. Urayama, T. Kawamura, S. Kohjiya, Elastic modulus and equilibrium swelling of networks crosslinked by end-linking oligodimethylsiloxane at solution state. J. Chem. Phys. **105**(11), 4833–4840 (1996)
15. Y. Akagi, J.P. Gong, U. Chung, T. Sakai, Transition between phantom and affine network model observed in polymer gels with controlled network structure. Macromolecules **46**(3), 1035–1040 (2013)
16. L.M. Dossin, W.W. Graessley, Rubber elasticity of well-characterized polybutadiene networks. Macromolecules **12**(1), 123–130 (1979)
17. N.R. Langley, Elastically effective strand density in polymer networks. Macromolecules **1**, 348 (1968)
18. M. Zhong, R. Wang, K. Kawamoto, B.D. Olsen, J.A. Johnson, Quantifying the impact of molecular defects on polymer network elasticity. Science **353**(6305), 1264–1268 (2016)
19. B. Yohsuke, K. Urayama, T. Takigawa, K. Ito, Biaxial strain testing of extremely soft polymer gels. Soft Matter **7**(6), 2632 (2011)
20. P.J. Flory, J. Rehner, Statistical mechanics of cross-linked polymer networks I. Rubberlike elasticity. J. Chem. Phys. **11**(11), 512–520 (1943)
21. E.M. Arruda, M.C. Boyce, A 3-dimensional constitutive model for the large stretch behavior of rubber elastic-materials. J. Mech. Phys. Solids **41**(2), 389–412 (1993)
22. M. Mooney, A theory of large elastic deformation. J. Appl. Phys. **11**(9), 582–592 (1940)
23. R.S. Rivlin, R.E. Keightley, Large elastic deformations of isotropic materials IV: further developments of the general theory. Philos. Trans. R. Soc. London. Ser. A Math. Phys. Sci. **241**(835), 379–397 (1948)
24. T. Kawamura, K. Urayama, S. Kohjiya, Multiaxial deformations of end-linked poly(dimethylsiloxane) networks. 1. Phenomenological approach to strain energy density function. Macromolecules **34**(23), 8252–8260 (2001)
25. K. Urayama, An experimentalist's view of the physics of rubber elasticity. J. Polym. Sci. Part B Polym. Phys. **44**(24), 3440–3444 (2006)
26. A. Kloczkowski, J.E. Mark, B. Erman, A diffused-constraint theory for the elasticity of amorphous polymer networks. 1. Fundamentals and stress-strain isotherms in elongation. Macromolecules **28**(14), 5089–5096 (1995)

27. R.J. Gaylord, J.F. Douglas, The localization model of rubber elasticity. 2. Polym. Bull. **23**(5), 529–533 (1990)
28. S.F. Edwards, T. Vilgis, The effect of entanglements in rubber elasticity. Polymer (Guildf). **27**(4), 483–492 (1986)
29. M. Kaliske, G. Heinrich, An extended tube-model for rubber elasticity: statistical-mechanical theory and finite element implementation. Rubber Chem. Technol. **72**(4), 602–632 (1999)
30. M. Rubinstein, S. Panyukov, Nonaffine deformation and elasticity of polymer networks. Macromolecules **30**(25), 8036–8044 (1997)
31. X.J. Xing, P.M. Goldbart, L. Radzihovsky, Thermal fluctuations and rubber elasticity. Phys. Rev. Lett. **98**(7), 75502 (2007)
32. M. Gottlieb, R.J. Gaylord, Experimental tests of entanglement models of rubber elasticity. 1. Uniaxial extension-compression. Polymer (Guildf). **24**(12), 1644–1646 (1983)
33. G.B. McKenna, L.J. Zapas, Experiments on the small strain behavior of crosslinked natural rubber. Part I. Torsion. Polymer (Guildf). **24**, 1495–1501 (1983)
34. K. Urayama, T. Kawamura, S. Kohjiya, Multiaxial deformations of end-linked poly(dimethylsiloxane) networks. 2. Experimental tests of molecular entanglement models of rubber elasticity. Macromolecules **34**(23), 8261–8269 (2001)
35. K. Urayama, T. Kawamura, S. Kohjiya, Multiaxial deformations of end-linked poly(dimethylsiloxane) networks. 4. Further assessment of the slip-link model for chain-entanglement effect on rubber elasticity. J. Chem. Phys. **118**(12), 5658–5664 (2003)
36. Y. Kondo, K. Urayama, M. Kidowaki, K. Mayumi, T. Takigawa, K. Ito, Applicability of a particularly simple model to nonlinear elasticity of slide-ring gels with movable cross-links as revealed by unequal biaxial deformation. J. Chem. Phys. **141**(13) (2014)
37. T. Katashima, K. Urayama, U.-I. Chung, T. Sakai, Strain energy density function of a near-ideal polymer network estimated by biaxial deformation of tetra-PEG Gel. Soft Matter **8**(31), 8217–8222 (2012)
38. T. Katashima, K. Urayama, U.-I. Chung, T. Sakai, Probing the cross-effect of strains in non-linear elasticity of nearly regular polymer networks by pure shear deformation. J. Chem. Phys. **142**(17) (2015)
39. A.N. Gent, A new constitutive relation for rubber. Rubber Chem. Technol. **69**(1), 59–61 (1996)
40. P. Bladon, M. Warner, Elasticity of nematic networks and nematic effects in conventional rubbers. Macromolecules **26**(5), 1078–1085 (1993)
41. J. Gao, J.H. Weiner, Chain force concept in systems of interacting chains. Macromolecules **24**(18), 5179–5191 (1991)
42. K. Iwata, S.F. Edwards, New model of polymer entanglement—localized gauss integral model—plateau modulus Gn, topological 2nd virial-coefficient A2-theta and physical foundation of the tube model. J. Chem. Phys. **90**(8), 4567–4581 (1989)
43. N. Hirayama, K. Tsurusaki, Topological interaction between loop structures in polymer networks and the nonlinear rubber elasticity. Nihon Reoroji Gakkaishi **39**(1–2), 65–73 (2011)
44. Y. Okumura, K. Ito, The polyrotaxane gel: a topological gel by figure-of-eight cross-links. Adv. Mater. **13**(7), 485–487 (2001)
45. K. Urayama, T. Takigawa, Volume of polymer gels coupled to deformation. Soft Matter 8(31) (2012)
46. T. Takigawa, K. Urayama, Y. Morino, T. Masuda, Simultaneous swelling and stress relaxation behavior of uniaxially stretched polymer gels. Polym. J. **25**(9) (1993)
47. A. Konda, K. Urayama, T. Takigawa, Strain-rate-dependent poisson's ratio and stress of polymer gels in solvents revealed by ultraslow stretching. Macromolecules 44(8) (2011)
48. A.M. Hecht, F. Horkay, E. Geissler, M. Zrinyi, Experiments on swollen networks under anisotropic deformation. Polym. Commun. **31**(2), 53–55 (1990)
49. M. Fujine, T. Takigawa, K. Urayama, Strain-driven swelling and accompanying stress reduction in polymer gels under biaxial stretching. Macromolecules **48**(11), 3622–3628 (2015)
50. F. Horkay, E. Geissler, A.M. Hecht, M. Zrinyi, Osmotic swelling of poly(vinyl acetate) gels at the.THETA. temperature. Macromolecules **21**(8), 2589–2594 (1988)

51. N. Murata, A. Konda, K. Urayama, T. Takigawa, M. Kidowaki, K. Ito, Anomaly in stretching-induced swelling of slide-ring gels with movable cross-links. Macromolecules **42**(21), 8485–8491 (2009)
52. E. Geissler, A.M. Hecht, F. Horkay, M. Zrinyi, Compressional modulus of swollen polyacrylamide networks. Macromolecules **21**(8), 2594–2599 (1988)
53. T. Tanaka, D.J. Fillmore, Kinetics of swelling of gels. J. Chem. Phys. **70**(3), 1214–1218 (1979)
54. T. Takigawa, K. Urayama, T. Masuda, Time dependent poisson's ratio of polymer gels in solvent. Polym. J. **26**(2), 225–227 (1994)
55. T. Takigawa, S. Nosaka, Y. Takakura, K. Urayama, Dynamic swelling properties of a poly(N-isopropylacrylamide) gel measured by a magnetic force-driven rheometer. Polym. J. **35**(10), 819–822 (2003)
56. S. Nosaka, K. Urayama, T. Takigawa, Static and dynamic swelling properties of poly(N-isopropylacrylamide) gels in the swollen state. Polym. J. **37**(9), 694–699 (2005)
57. K. Nakamura, E. Shinoda, M. Tokita, The influence of compression velocity on strength and structure for gellan gels. Food Hydrocoll. **15**(3), 247–252 (2001)
58. K. Urayama, Y. Taoka, K. Nakamura, T. Takigawa, Markedly compressible behaviors of gellan hydrogels in a constrained geometry at ultraslow strain rates. Polymer (Guildf). **49**(15), 3295–3300 (2008)
59. K. Urayama, S. Okada, S. Nosaka, H. Watanabe, T. Takigawa, Kinetics of shrinking of polymer gels induced by ultracentrifugal fields. J. Chem. Phys. **122**(2) (2005)
60. S. Nosaka, S. Okada, Y. Takayama, K. Urayama, H. Watanabe, T. Takigawa, Compression of poly(vinyl alcohol) gels by ultracentrifugal forces. Polymer (Guildf). **46**(26), 12607–12611 (2005)
61. A. Zanina, T. Budtova, Hydrogel under shear: a rheo-optical study of the particle deformation and solvent release. Macromolecules **35**(5), 1973–1975 (2002)
62. T. Takigawa, K. Uchida, K. Takahashi, T. Masuda, Flow-induced swelling of poly(vinyl alcohol) gel. J. Chem. Phys. **111**(5), 2295–2300 (1999)
63. T. Yamamoto, Y. Masubuchi, M. Doi, Large network swelling and solvent redistribution are necessary for polymer gels to show negative normal stress. ACS Macro Lett. **6**(5), 512–514 (2017)
64. W.H. Stockmayer, Theory of molecular size distribution and gel formation in branched-chain polymers. J. Chem. Phys. **11**(2), 45–55 (1943)
65. W.H. Stockmayer, Theory of molecular size distribution and gel formation in branched polymers II. General Cross Linking. J. Chem. Phys. **12**(4), 125–131 (1944)
66. P.G. de Gennes, *Scaling Concepts in Polymer Physics* (Cornell University Press, New York, 1979)
67. D. Stauffer, A. Coniglio, M. Adam, *Gelation and Critical Phenomena BT—Polymer Networks*, K. Dušek (Springer Berlin Heidelberg, Berlin, Heidelberg, 1982) pp. 103–158
68. D. Stauffer, *Introduction to Percolation Theory* (Taylor & Francis, London, 1985)
69. M. Muthukumar, Dynamics of polymeric fractals. J. Chem. Phys. **83**(6), 3161–3168 (1985)
70. J.E. Martin, D. Adolf, J.P. Wilcoxon, Viscoelasticity near the sol-gel transition. Phys. Rev. A **39**(3), 1325–1332 (1989)
71. M. Rubinstein, R.H. Colby, *Polymer Physics* (Oxford University Press, Oxford, 2003)
72. F. Chambon, H.H. Winter, Linear viscoelasticity at the gel point of a crosslinking PDMS with imbalanced stoichiometry. J. Rheol. **31**(8), 683–697 (1987)
73. H.H. Winter, M. Mours, Adv. Polym. Sci. **134**, 165 (1997)
74. J.E. Martin, D. Adolf, J.P. Wilcoxon, Viscoelasticity of near-critical gels. Phys. Rev. Lett. **61**(22), 2620–2623 (1988)
75. M. Muthukumar, Screening effect on viscoelasticity near the gel point. Macromolecules **22**(12), 4656–4658 (1989)
76. P.G. de Gennes, On a relation between percolation theory and the elasticity of gels. J. Phys. Lett. **37**, 1–2 (1976)
77. M. Adam, M. Delsanti, D. Durand, Mechanical measurements in the reaction bath during the polycondensation reaction, near the gelation threshold. Macromolecules **18**(11), 2285–2290 (1985)

78. A. Koike, N. Nemoto, Y. Watanabe, K. Osaki, Dynamic viscoelasticity and FT-IR measurements of end-crosslinking α, ω-dihydroxyl polybutadiene solutions near the gel point in the gelation process. Polym. J. **28**, 942 (1996)
79. A. Koike, N. Nemoto, M. Takahashi, K. Osaki, Dynamic viscoelasticity of end-linking α, ω-dimethyl silyl poly(propylene oxide) solutions near the gel point. Polymer (Guildf). **35**(14), 3005–3010 (1994)
80. M. Baumgaertel, H.H. Winter, Determination of discrete relaxation and retardation time spectra from dynamic mechanical data. Rheol. Acta **28**(6), 511–519 (1989)
81. H. Henning Winter, A. Izuka, M.E. De Rosa, Experimental observation of the molecular weight dependence of the critical exponents for the rheology near the gel point. Polym. Gels Networks **2**(3), 239–245 (1994)
82. M.E. Cates, Brownian dynamics of self-similar macromolecules. J. Phys. Fr. **46**(7), 1059–1077 (1985)
83. V. Adibnia, R.J. Hill, Universal aspects of hydrogel gelation kinetics, percolation and viscoelasticity from PA-hydrogel rheology. J. Rheol. **60**(4), 541–548 (2016)
84. T. Kawamura, K. Urayama, S. Kohjiya, Multiaxial deformations of end-linked poly(dimethylsiloxane) networks. III. Effect of entanglement density on strain-energy density function. J. Polym. Sci. Part B Polym. Phys. **40**(24), 2780–2790 (2002)
85. M. Doi, Gel dynamics. J. Phys. Soc. Japan **78**(5), 52001 (2009)
86. T. Yamaue, M. Doi, The stress diffusion coupling in the swelling dynamics of cylindrical gels. J. Chem. Phys. **122**(8), 84703 (2005)
87. M. Takahashi, K. Yokoyama, T. Masuda, T. Takigawa, Dynamic viscoelasticity and critical exponents in sol-gel transition of an end-linking polymer. J. Chem. Phys. **101**(1), 798–804 (1994)

Chapter 7
Structure Analysis of Gels

Mitsuhiro Shibayama

Abstract Most gels are polymer networks formed by long polymer chains and cross-linking agents that swell in a solvent. The cross-linkers bond the polymer chains together and prevent them from dissolving. Characteristic properties of gels, such as swelling/shrinking, elasticity, rheology/viscoelasticity, friction, chemical/physical sorption/desorption, and optical properties, are important from both academic and industrial perspectives, and have been the subject of much research. Since these gel properties are closely related to the gel structure, the structural analysis of gels is particularly important in the study of gels. However, due to the inherently heterogeneous structure of gels, structural analysis of gels has not been easy and has not been able to go beyond the realm of phenomenology. Finally, from the late 1990s to the beginning of this century, a systematic understanding of gels has been developed. In this chapter, we first describe the "static" structural analysis of gels by scattering methods such as X-ray/neutron/light scattering, morphological methods such as microscopy, and NMR, one of the spectroscopic methods. Then, the dynamic properties of gels by dynamic light scattering, i.e., "dynamic" structural analysis, is described, including theory and experiment, and various applications, such as gelation point determination methods.

Keywords Inhomogeneity · Scattering · NMR · Microscopy · DLS · High-strength gel

Most gels are polymer networks formed by long polymer chains and cross-linking agents that swell in a solvent. The cross-linkers bond the polymer chains together and prevent them from dissolving. Characteristic properties of gels, such as swelling/shrinking, elasticity, rheology/viscoelasticity, friction, chemical/physical sorption/desorption, and optical properties, are important from both academic and industrial

M. Shibayama (✉)
Comprehensive Research Organization for Science and Society (CROSS), Neutron Science and Technology Center, Ibaraki, Japan
e-mail: m_shibayama@cross.or.jp

Y. Osada (ed.), *Science of Gel*, https://doi.org/10.1007/978-981-95-0301-8_7

perspectives, and have been the subject of much research [1]. Since these gel properties are closely related to the gel structure, the structural analysis of gels is particularly important in the study of gels. However, due to the inherently heterogeneous structure of gels, structural analysis of gels has not been easy and has not been able to go beyond the realm of phenomenology. Finally, from the late 1990s to the beginning of this century, a systematic understanding of gels has been developed [2–4].

In this chapter, we first describe the "static" structural analysis of gels by scattering methods such as X-ray/neutron/light scattering, morphological methods such as microscopy and NMR, one of the spectroscopic methods. Then, the dynamic properties of gels by dynamic light scattering, i.e., "dynamic" structural analysis, is described, including theory and experiment, and various applications, such as gelation point determination methods.

7.1 Methodology of Structural Analyses and Gel Inhomogeneity

7.1.1 Methods of Structural Analyses

There are various methods for structural analysis of polymer gels, including scattering, microscopy, imaging, and spectroscopy. Table 7.1 lists the major methods. Scattering is a commonly used method for structural analysis of many materials, including gels. Scattering is a method for structural analyses using electromagnetic waves such as (visible) light and X-rays, and particle beams such as neutrons [5, 6]. The spatial extent of the object can be chosen according to the wavelength of the "wave" used as a probe, and the object of structural analysis can be selected according to the type of "wave." For example, X-ray and neutron scattering methods cover structures from several nm to several hundred nm, while light scattering covers tenths of micrometers to several micrometers. On the other hand, electron scattering is not suitable for bulk structure analysis because of the strong interaction between electrons and materials (including gas molecules in the air), but are rather useful for surface characterization of materials in high vacuum and as a source for electron microscopy. Infrared and ultraviolet rays are also strongly absorbed by materials, making them suitable for "absorption spectroscopy" to study characteristic functional groups in materials rather than for scattering.

Microscopy is a very powerful technique because it allows direct observation of materials as images in real space. However, optical microscopy has too low a magnification to study polymer gels, and electron microscopy, while powerful for nanometer-order characterization, requires a high vacuum and is not suitable for wet gels containing solvents. On the other hand, cryo-TEM (cryo-TEM) [7], which has recently been widely used in biological systems, is a very promising technique, but the problems of shape change due to rapid quenching and subsequent microtoming (ultra-thin section preparation) and electron beam damage due to observation under

Table 7.1 Structural Characterization of polymer gels

Classification	Methodology	Size[a]	Characteristics/Remarks
Scattering	Light scattering (LS)	100 nm–1 μm	Microgel size
	Dynamic light scattering (DLS)		Network size
	Small-angle X-ray scattering (SAXS)	1 nm–100 nm	Microgel size, network size
	Small-angle neutron scattering (SANS)	1 nm–100 nm	Microgel size, network size
	Neutron spin echo (NSE)	1 nm–10 nm	Network dynamics
Microscopy	Transmission microscopy (TEM)	10 nm–1 μm	Dried gel structure, feeze-dried fracture surface
	Cryo-TEM	10 nm–1 μm	Gel structure
	Scanning microscopy (SEM)	10 nm–1 μm	Dried gel structure, feeze-dried fracture surface
	Atomic force microscopy (AFM)	1 nm–100 nm	Surface structure
	Confocal microscopy (CLMS)	1 nm–100 nm	Three-dimensional structure
Imaging	NMR imaging	~ mm	Concentration distribution
Spectroscoopy	infrared (IR)	< 1 nm	Network structure, cross-link density, hydrogen bond, hydrophobic interaction
	Raman	< 1 nm	Crystal structure
	Circular dichroism (CD)	< 1 nm	Helix
	NMR	< 1 nm	Network structure, cross-link density, hydrogen bond, hydrophobic interaction

[a] Typical size

electrons cannot be avoided. On the other hand, atomic force microscopy (AFM) and laser-scanning confocal microscopy (LSCM) are suitable for gel studies because they do not require a high vacuum and do not cause electron damage.

Spectroscopy is a technique for studying absorption and resonance signal changes in materials by scanning wavelengths or magnetic fields, from detection of functional groups to analysis of chirality and crystal structure. Spectroscopic methods include infrared (IR) absorption spectroscopy, Raman scattering spectroscopy, circular dichroism (CD), nuclear magnetic resonance spectroscopy (NMR), and fluorescence spectroscopy. These spectroscopic methods are widely used to identify and quantify the functional groups that make up polymer networks. For example, the concentration of cross-linkers and the degree of hydrogen bonding can be evaluated by spectroscopy. Though spectroscopy is limited as a structural analysis method because it does not provide spatial information, it provides insight into the concentration, state, and local interactions (< 1 nm) of functional groups and can be used for structural analysis when necessary.

7.1.2 Gel Inhomogeneity

The most important feature of gels is the cross-linked structure. When cross-linking is introduced into a polymer solution, the polymer chains become topologically fixed, and as a result, the polymer solution loses fluidity and becomes a gel. When cross-linkers are present in a polymerization reaction system from monomers, the cross-linkers are incorporated into polymer chains during the reaction, forming a branched structure, which eventually becomes a gel. In this process, a sparse and dense network structure is formed. Such sparseness and entanglement of the mesh structure are called "gel inhomogeneity." Particularly in the case of chemical gels, the mesh structure once formed will not unravel or recombine, so inhomogeneity is a major issue in the physical properties of gels. It should be noted, here, there are many possible cases of inhomogeneity.

Figure 7.1 classifies the inhomogeneity of polymer gels [8]. (a) is spatial inhomogeneity, (b) is topological inhomogeneity, (c) is bonding inhomogeneity, and (d) is mobility inhomogeneity. (a) can be investigated mainly by static scattering. (b) and (c) are closely related to macroscopic properties such as gel dynamics and swelling/shrinking behavior, and NMR plays an important role from the viewpoint of structural analysis. (d) is evaluated by dynamic light scattering and position-scanning static light scattering. In the following, the static structure analysis of gels by scattering, microscopy, and NMR is described. Then, the dynamic light scattering method, which is very useful for the dynamic structure of gels, will be discussed.

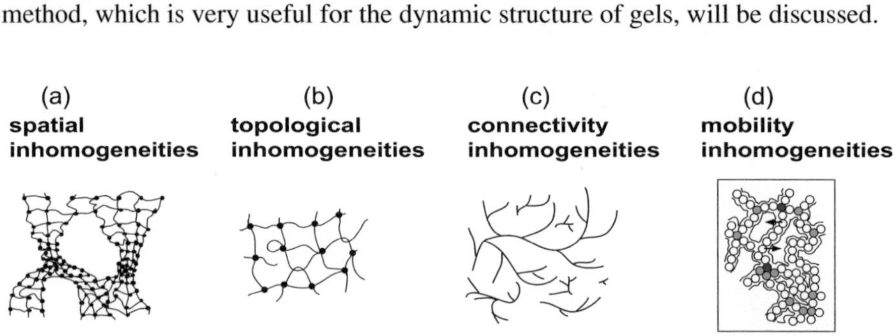

Fig. 7.1 Schematic diagram of gel heterogeneity. **a** Spatial heterogeneity, **b** topological heterogeneity, **c** binding heterogeneity, **d** mobility heterogeneity

7.2 Scattering Theory

The general theory of scattering and the scattering theory of polymer solutions are discussed in detail in many reference books [5, 6]. In this section, we will focus on polymer gels.

7.2.1 Scattering Functions of Polymer Gels

What happens when cross-linking is introduced into a polymer solution? Will the conformation of the polymer chain change significantly? First, let us try a thought experiment. Figure 7.2 shows a schematic diagram of a semi-dilute polymer solution and a polymer gel of the same polymer concentration. In a polymer solution, the polymer chains are perturbed with time, and this figure is a snapshot of a certain moment in time. Since this is a semi-dilute solution, the blob concept can be used. The dotted line in the figure represents a blob. Long polymer chains are divided into blobs within the range of the excluded volume effect. In the case of photocross-linking or gamma-ray cross-linking, cross-links are randomly introduced and gels are formed. Since the concentration of the polymer does not change, the gel is expected to be in the same state as a polymer solution of the same concentration. If this is the case, cross-linking does not change the conformation of the polymer chains, so the scattering function is almost the same for the solution and the gel. Here, we assume that the scattering from the cross-linking point is negligible. Is this thought experiment correct? We will discuss this issue below.

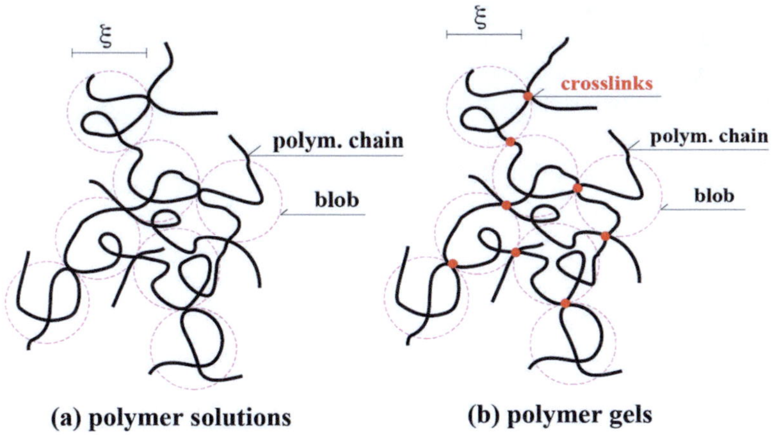

Fig. 7.2 Schematic of **a** a semi-dilute polymer solution and **b** a polymer gel of the same polymer concentration. The (red) dotted line in the figure represents a blob

7.2.1.1 Blob Concept of Polymer Gel

De Gennes proposed the famous C^* theorem, which states that "the polymer concentration in a polymer gel is automatically adjusted to be the same as that in the corresponding polymer solution" [9]. Figure 7.2 shows this schematically. If this is the case, it can be assumed that the scattering intensity, which reflects the structure, is the same for gels and solutions. However, it is known that the scattering intensity of the gel is significantly increased by the introduction of cross-linking. In the following, we will discuss the cross-linking effect in scattering in detail, starting with a discussion of scattering theory based on the C^* theorem.

7.2.1.2 Scattering Function of Gels Predicted from Thermodynamic Theory

When a gel is prepared by introducing cross-linking into a semi-dilute solution, an elastic term associated with the cross-linking appears in the free energy. The free energy of an uncharged gel system is given by the following equation [10, 11].

$$\frac{F}{k_B T} = (1-\phi)\ln(1-\phi) + \chi\phi(1-\phi) + \frac{3\phi}{2N_X}\left[\left(\frac{\phi_0}{\phi}\right)^{2/3} - 1 - \ln\left(\frac{\phi_0}{\phi}\right)\right] \tag{7.1}$$

where ϕ and ϕ_0 is the volume fraction of the polymer constituting the gel at observation and in the reference state, i.e., the volume fraction at the time of gel preparation. χ is the interaction parameter, and N_X is the number of monomers between adjacent cross-linking points. On the other hand, the last term on the right side of Eq. (7.1) indicates the change in elastic energy of the partial chain consisting of N_X segments due to swelling from volume ϕ_0 to ϕ[12, 13].

From Eq. (7.1), the osmotic pressure of the system is

$$\begin{aligned}\frac{a^3\Pi}{k_B T} &= \phi^2\frac{\partial}{\partial\phi}\left(\frac{F_{\text{site}}}{k_B T\phi}\right) = \frac{\phi_0}{N_X}\left[\frac{1}{2}\left(\frac{\phi}{\phi_0}\right) - \left(\frac{\phi}{\phi_0}\right)^{1/3}\right] - \phi - \ln(1-\phi) - \chi\phi^2 \\ &\cong \frac{\phi_0}{N_X}\left[\frac{1}{2}\left(\frac{\phi}{\phi_0}\right) - \left(\frac{\phi}{\phi_0}\right)^{1/3}\right] + \left(\frac{1}{2} - \chi\right)\phi^2 + \frac{\phi^3}{3} + \cdots\end{aligned} \tag{7.2}$$

In scattering theory, the measure of length is the absolute value of the scattering vector $\mathbf{q}$, $q \equiv (4\pi/\lambda)\sin\theta$ where λ is the wavelength of the wave used and 2θ is the scattering angle. The scattering intensity in the thermodynamic limit $q = 0$ is

$$I(q=0) = K_N k_B T\left(\frac{\partial^2 F}{\partial\phi^2}\right)^{-1} = K_N\frac{k_B T\phi^2}{a^3 M_{os}} \tag{7.3}$$

Here, M_{os} is the longitudinal modulus of elasticity of the gel [14, 15] and

$$M_{os} = K_{os}^{gel} + \frac{4}{3}\mu = \frac{k_B T}{a^3}\left\{\frac{\phi_0}{N_X}\left[\frac{1}{2}\left(\frac{\phi}{\phi_0}\right) - \frac{1}{3}\left(\frac{\phi}{\phi_0}\right)^{1/3}\right] + (1 - 2\chi + \phi)\phi^2 + \cdots\right\} + \frac{4}{3}\mu \quad (7.4)$$

where $K_{os}{}^{gel}$ is the osmotic modulus and μ is the shear modulus. Since it is known that the scattering function of a semi-dilute polymer solution is represented by the Ornstein–Zernike function (Lorentz function), the scattering intensity function of the gel is

$$I(q) = K_N \frac{k_B T \phi^2}{a^3 M_{os}(1 + \xi^2 q^2)} \quad (7.5)$$

Here, $M_{os} > K_{os}$ in general since $K_{os}{}^{gel}$ and μ are both positive. Thus, theoretically, one would expect the scattering from the gel to be weaker than that from the solution. However, the scattering intensity from gels observed in practice is often much stronger than that from the corresponding solutions. In the next section, we discuss the phenomenology of scattering from gels.

7.2.2 Gel Inhomogeneity in Scattering

In the 1980s, when scattering studies became popular, the existence of excess scattering from gels was often pointed out. Scattering was phenomenologically defined as the sum of the solution component of the gel and the excess scattering due to the introduction of cross-linking [16, 17]. However, a controversy arose over the nature of this excess scattering, whether it is "annealed scattering" [18] or "frozen inhomogeneities" [19]. After some controversy, a change in the speckle pattern of light scattering during gel stretching was the deciding factor, and the excess scattering from gels was concluded to be "frozen inhomogeneities" associated with the introduction of cross-linking [8, 20, 21].

Figure 7.3 shows a schematic of gel concentration fluctuations [16]. (a) is a reactor batch gel (gel at the time of preparation) and (b) is a gel at the time of swelling. When cross-links (indicated by dots) are introduced into a polymer solution, the cross-linking points are generally introduced randomly (Fig. 7.3a). As discussed in Fig. 7.2, however, merely introducing cross-linking into a polymer solution does not create inhomogeneity in concentration fluctuations. Note that, when the gel is swollen or stretched, as shown schematically in Fig. 7.3b, the dense and sparse areas of the cross-link distribution swell (deform) differently, and the inhomogeneity of the cross-link distribution becomes apparent as concentration fluctuations. As a result, excess scattering appears in the scattering intensity.

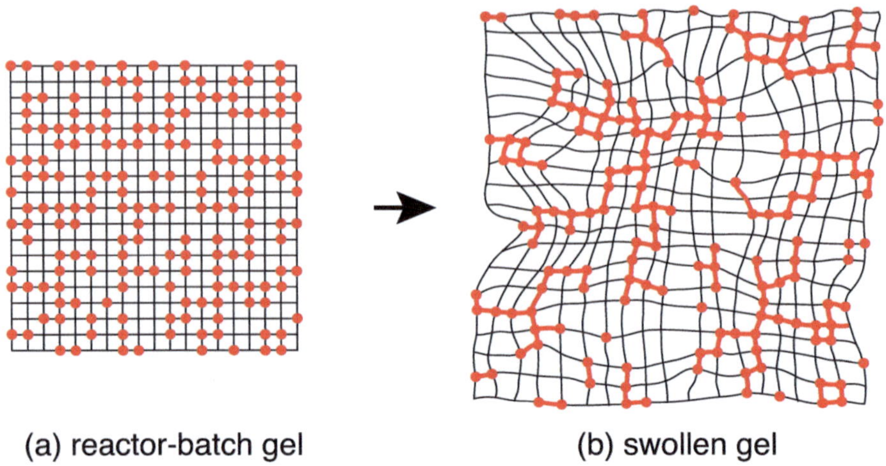

Fig. 7.3 Model of gel concentration fluctuations [16]. **a** Reactor batch gel (gel at preparation), **b** gel at swelling. Red circles represent cross-linking points and red lines represent clusters connected by dense cross-linking (Bastide and Leibler [16]; reproduced with permission, © American Chemical Society)

7.2.3 Phenomenology of Scattering Functions of Polymer Gels

As discussed in the previous section, the observed scattering intensity $I(q)$ from the gel is stronger than the corresponding scattering intensity $I_{\mathrm{soln}}(q)$ from the polymer solution. Therefore, the scattering intensity $I(q)$ from the gel is often expressed as the sum of the scattering $I_{\mathrm{soln}}(q)$ from the solution and the excess scattering $I_{\mathrm{ex}}(q)$, as in the following equation.

$$I(q) = I_{\mathrm{soln}}(q) + I_{\mathrm{ex}}(q) \tag{7.6}$$

where q is the absolute value of the scattering vector defined by $q = (4\pi/\lambda)\sin(\theta)$ and is a "measure of reciprocal space" with the dimension of the reciprocal of the length. λ is the wavelength of the wave used and 2θ is the scattering angle. In a swollen gel, the polymer concentration is in the semi-dilute range. Therefore, at large wavenumber vectors ($q \to \infty$), the effect of cross-linking disappears and the scattering $I(q)$ of the gel should be equal to the scattering $I_{\mathrm{soln}}(q)$ of the solution. In other words, the scattering $I(q)$ of the gel

$$I(q) \underset{q=\infty}{\longrightarrow} I_{\mathrm{soln}}(q) = \frac{I_{\mathrm{soln}}(0)}{1 + \xi^2 q^2} \tag{7.7}$$

where ξ is the correlation length (or blob size) already defined. On the other hand, at small values of q, the effect of cross-linking appears and the scattering intensity of the gel is larger than that of the solution. Various expressions have been proposed for this excess scattering $I_{\mathrm{ex}}(q)$,

(i) Lorentz function [17, 22]

$$I_{\mathrm{ex}}(q) = \frac{I_{\mathrm{ex}}(0)}{1 + \Xi^2 q^2} \quad (\Xi \gg \xi) \tag{7.8}$$

(ii) stretched exponential function [23]

$$I_{\mathrm{ex}}(q) = I_{\mathrm{ex}}(0) \exp\left[-(q\Xi)^{\alpha}\right] \tag{7.9}$$

(iii) Debye-Bueche type squared Lorentz function [24]

$$I_{\mathrm{ex}}(q) = \frac{I_{\mathrm{ex}}(0)}{(1 + \Xi^2 q^2)^2} \tag{7.10}$$

where Ξ is a quantity that indicates the magnitude of inhomogeneity of the gel. Also, α is the stretched exponent, which has been reported to take values between 0.7 and 2 in many cases [25, 26]. These functions for $I_{\mathrm{ex}}(q)$ are used to represent excess fluctuations or solid-like inhomogeneities in the gel. For example, Eq. (7.8) considers the existence of another correlation length $\Xi(\gg \xi)$ in the gel, and Eq. (7.9) is an extended form of the Guinier function ($\alpha = 2$), which is often used in the scattering of dilute systems and assumes randomly distributed components that correspond to independent "islands" (e.g., domains with high cross-link density) with respect to each other [27]. It has been reported that $\alpha = 2$ for end-cross-linked polydimethylsiloxane swollen with toluene and $\alpha = 0.725$ for polyvinylacetate gel). Equation (7.10), on the other hand, is a function called the Debye-Bueche function [28], which represents the scattering from a random two-phase system with a sharp interface. The validity of these functions is discussed below.

7.2.4 *Statistical Mechanics Theory for Scattering Functions of Polymer Gels*

It is not easy to describe the inhomogeneity of gels theoretically. A statistical mechanics theory of polymer gels was proposed by Panyukov and Rabin (PR theory) [29], assuming Gaussian statistics at the cross-linking points. According to this theory, the two types of concentration fluctuations present in gels are

$$\rho(r) = \rho^{\mathrm{th}}(r) + \rho^{\mathrm{eq}}(r) \tag{7.11}$$

The meaning of this expression is shown schematically in Fig. 7.4. (a) is the thermal fluctuation $\rho^{\mathrm{th}}(r)$, which is a fluctuation that exists in polymer solutions and is zero when averaged over time. (b) is the frozen fluctuation $\rho^{\mathrm{eq}}(r)$, which does not fluctuate with time and does not become zero when averaged over time, and is a concentration fluctuation created by the introduction of cross-linking. The Fourier transform of $\rho^{\mathrm{eq}}(r)$ is

$$\rho^{\mathrm{eq}}(q) = \frac{n(q)}{1 + wg(q)} \tag{7.12}$$

$n(q)$ is the Fourier transform of the inhomogeneous equilibrium concentration fluctuation when the exclusion parameter w is zero ($w = 0$). In the case of charged gels, where the gel mesh is charged, the exclusion parameter w is a function of distance, so its Fourier transform shows a q-dependence. In other words

$$w_0(q) = \phi_0^{5/4}N + \frac{\hat{l}_{\mathrm{B}} f_0^2 \phi_0^{5/4} N^2}{Q^2(\phi_0/\phi)^{2/3} + \hat{l}_{\mathrm{B}} f_0 \phi_0^{5/4} N} \tag{7.13}$$

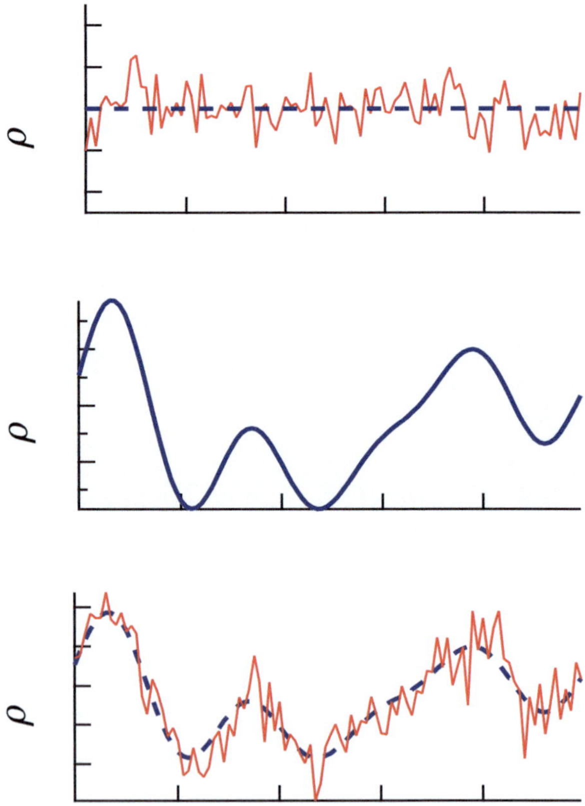

Fig. 7.4 **a** Thermal fluctuation $\rho^{\mathrm{th}}(r)$ in polymer solution, **b** frozen fluctuation $\rho^{\mathrm{eq}}(r)$ created by cross-linking, and **c** the sum of the two fluctuations (**a**) and (**b**) in the gel [29]

where χ is the interaction parameter, l_B is the Bjerrum length (about 7 Å for water at 25 °C), f_0 is the dissociation degree and C_s is the salt concentration. Q is the reduced scattering vector normalized by the fluctuation radius of the monomer between adjacent cross-linking points, and

$$Q \equiv (aN_X^{1/2}/\sqrt{6})q \tag{7.14}$$

As the gel swells or deswells and the concentration changes, the exclusion parameter w also changes, and

$$w = (1 - 2\chi + \phi)\phi N + \frac{l_B f^2 \phi N^2}{Q^2 + l_B(f + 2C_s)\phi N} \tag{7.15}$$

When discussing the statistical dynamics of gels, it is necessary to understand that there are two types of averages: thermal average $\langle X \rangle$ and ensemble average $\overline{X}$. By definition, the thermal and ensemble averages of $\rho^{eq}(r)$ are $\langle \rho^{eq}(q) \rangle \neq 0$ and $\overline{\rho^{eq}(q)} = \overline{\langle \rho^{eq}(q) \rangle} = 0$. The structure factor due to the thermal fluctuation $\rho^{th}(r)$ is called the thermal correlator $G(q)$ and is defined as

$$G(q) \equiv \left\langle \rho^{th}(q)\rho^{th}(-q) \right\rangle = \frac{g(q)}{1 + wg(q)} \tag{7.16}$$

On the other hand, the static correlator $C(q)$, which represents the frozen fluctuations, can be obtained by taking the ensemble average of the auto-convolution of $\rho^{eq}(q)$

$$C(q) \equiv \overline{\rho^{eq}(q)\rho^{eq}(-q)} = \frac{\nu(q)}{\left[1 + wg(q)\right]^2} \tag{7.17}$$

The following is an example of a case in which the value of the where

$$\nu(q) = \langle n(q)n(-q) \rangle \tag{7.18}$$

From the above, the structure factor $S(q)$ is given by

$$S(q) \equiv \overline{\langle \rho(q)\rho(-q) \rangle} = G(q) + C(q) = \frac{g(q)}{1 + wg(q)} + \frac{\nu(q)}{\left[1 + wg(q)\right]^2} \tag{7.19}$$

Note that Eq. (7.19) is given by the sum of two contributions, $G(q)$ and $C(q)$. From this, the scattering intensity is given by

$$I(q) = K_N \phi N_X S(q) \tag{7.20}$$

The following is an example of the use of a cross-linked point. Where N_X is the number of segments between the bridging points. In addition, $g(q)$ in Eq. (7.16) is given by

$$g(q) = \frac{1}{Q^2/2 + (4Q^2)^{-1} + 1} + \frac{2}{(1 + Q^2)^2(\phi_0/\phi)^{2/3}} \tag{7.21}$$

The $(4Q^2)^{-1}$ in the denominator of the first term on the right-hand side of Eq. (7.21) corresponds to the term[30] first introduced by de Gennes in the scattering function of cross-linked heteropolymers. On the other hand, the static correlator for the frozen fluctuations is

$$C(q) = \frac{\phi N_X}{\left[1 + wg(q)\right]^2(1 + Q^2)^2}\left[6 + \frac{9}{w_0 - 1 + (1/2)Q^2(\phi_0/\phi)^{2/3}}\right] \tag{7.22}$$

where the excluded volume parameter w_0 at sample preparation is given by Eq. (7.13). Since $C(q) = 0$ for the polymer solution, the cross-linking term, $(4Q^2)^{-1}$, and the second term in the first term of Eq. (7.21) are not necessary. Therefore, Eq. (7.19) is equivalent to the scattering function of the polymer solution, and the correlation length ξ is given by

$$\xi = a[12(1 - 2\chi + \phi)\phi N_X]^{-1/2} \tag{7.23}$$

The PR theory has been used to describe the scattering functions of deformed and ionized gels [31]. As one example, for gels prepared at cross-linking concentrations far enough from the critical cross-linking state, the structure factor is a simple sum of the Lorentz function (L) and the squared Lorentz function (SL), which is similar to the sum of Eq. (7.7) and (7.10) [32], *i.e.*,

$$S(q) \cong \frac{S_L(0)}{1 + \xi^2 q^2} + \frac{S_{SL}(0)}{(1 + \xi^2 q^2)^2} \tag{7.24}$$

For isotropically swollen gels, the following scattering functions have been proposed by Onuki [15, 19].

$$I(q) \cong \frac{K_N}{a^3}\frac{k_B T\phi^2}{M_{os}}\left[\frac{1}{1 + \xi^2 q^2} + p(q)\left(\frac{\phi_0}{\phi}\right)^{2/3}\frac{\mu}{M_{os}}\frac{1}{(1 + \xi^2 q^2)^2}\right] \tag{7.25}$$

where $p(q)$ is the Fourier transform of the autocorrelation function of the cross-link density fluctuations [15]. When the cross-linking points are spatially randomly distributed according to a Poisson distribution, $p(q)$ may be assumed to be constant $p(q) = p_{\text{int}}$, and Eq. (7.25) becomes

$$I(q) = \frac{I_L(0)}{1 + \xi^2 q^2} + \frac{I_{SL}(0)}{(1 + \xi^2 q^2)^2} \tag{7.26}$$

where $I_L(0)$ and $I_{SL}(0)$ are the coefficients at $q = 0$ of the dynamic and static components, respectively, that contribute to the scattering intensity $I(q)$. It should be noted here that Eqs. (7.24) through (7.26) differ from the scattering functions discussed in Eqs. (7.6) through (7.10) in that both static and dynamic fluctuations are represented by the same correlation length ξ. More recently, analytical methods (Eqs. (7.6) and (7.10)) that evaluate the dynamic fluctuation (shielding length) and static inhomogeneity measures independently of ξ and Ξ, respectively, are more commonly used.

7.3 Scattering Functions of Gels Under Various Experimental Conditions

7.3.1 Cross-Linking Effect

As mentioned many times before, cross-linking increases the scattering intensity, especially on the small-angle side. This phenomenon has been known since the 1970s as cross-link inhomogeneity. For example, the observation of inhomogeneity of cross-linked rubber by light scattering [33] and the increase in scattering intensity $I(q)$ with increasing cross-link density in polyacrylamide (PAAm) gels with different cross-link densities [23] have been reported. Initially, this increase in scattering intensity was explained by the thickening of the tufted network chains. However, as discussed in the previous section, this increase in scattering intensity is now understood as an increase in static inhomogeneity.

Figure 7.5 shows the dependence of the light scattering (LS) and neutron scattering (SANS) intensity distribution $I(q)$ of poly(*N*-isopropylacrylamide) (PNIPAM) gel on cross-linking density. The scattering intensity increases as the concentration of methylenebisacrylamide (BIS), the cross-linking agent, increases with the amount of C_{BIS} added. This trend extends not only to small-angle neutron scattering (SANS), but also to light scattering (LS) with smaller q, and the dependence on cross-linking density is even more pronounced in this small-angle region. At high cross-link densities, the inhomogeneity in the gel becomes large enough to be observed by light scattering, and the gel eventually becomes cloudy.

It should be noted here that the inhomogeneity of the gel is highly dependent on the conditions and environment under which the gel was prepared. Figure 7.6 shows the dependence of the SANS curves of PNIPAM gels prepared at (a) 10 °C, (b) 20 °C, (c) 25 °C, and (d) 27 °C on the cross-linker concentration C_X [34],

$$C_X = \frac{2C_{BIS}}{2C_{BIS} + C_{NIPA}} \times 100 \tag{7.27}$$

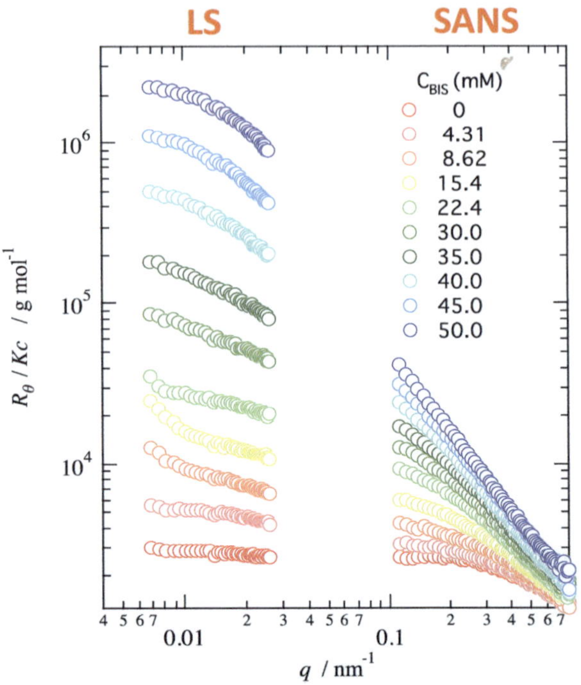

Fig. 7.5 Light scattering (LS) and neutron scattering intensity distribution (SANS) $I(q)$ of poly(N-isopropylacrylamide) (PNIPA) gel as a function of cross-linking density. Concentration C_{BIS} is the concentration of the cross-linking agent methylenebisacrylamide (BIS) (Takata et al. [34]; reproduced with permission, © American Chemical Society)

From this series of figures, it can be seen that the inhomogeneity of the gel depends not only on the cross-linker concentration but also on the gel preparation temperature T_{prep}, and that the scattering intensity increases with an increase in T_{prep}. The solid lines in the figures are the results of calculations based on the PR theory (Eqs. (7.11)–(7.22)) [29] and reproduce the experimental results well.

Figure 7.7 shows interesting comparative results. (a) is a SANS curve of a BIS cross-linked gel (chemically cross-linked gel, monomer-cross-linked gel), which is

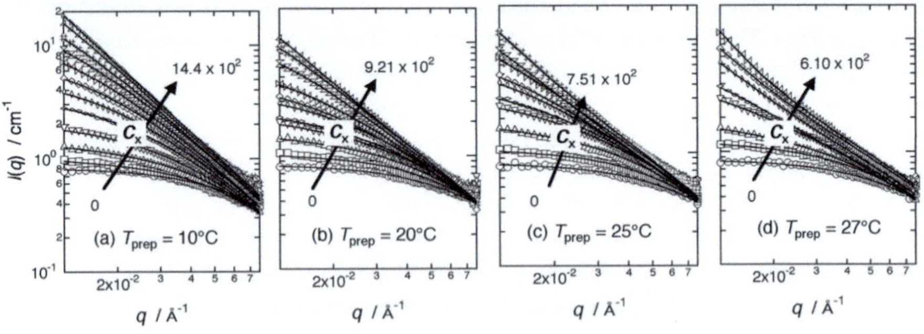

Fig. 7.6 Cross-linker concentration C_{x} dependence of SANS curves of PNIPA gels. Preparation temperatures: **a** 10 °C, **b** 20, **c** 25, **d** 27 °C [34] (Takata et al. [34]; reproduced with permission, © American Chemical Society)

obtained by polymerizing *N*-isopropylacrylamide (NIPAM) monomer in water in the presence of BIS, as is usually done. On the other hand, (b) is a SANS curve of a gamma-ray cross-linked gel (polymer-cross-linked gel) obtained by irradiating an aqueous solution of PNIPAM with various doses of gamma radiation [35]. While the scattering intensity function of the monomer-cross-linked gel shows a strong dependence on the concentration of the cross-linking agent, that of the polymer-cross-linked gel is not so strong. In other words, the introduction of cross-linking in monomer-cross-linked gels causes a marked increase in inhomogeneity, whereas the inhomogeneity in polymer-cross-linked gels does not change much. This experimental fact indicates that the inhomogeneity of the gels is not necessarily caused by the introduction of cross-linking, but is highly dependent on the difference in cross-linking methods. That is, in the polymerization and cross-linking reactions of monomers, inhomogeneity is formed in the polymerization process as described in Sect. 3.2.1. On the other hand, in the case of gamma-ray cross-linking (polymer-cross-linking), cross-linking proceeds by linking polymer chains existing in the polymer solution, which is explained by the fact that inhomogeneity is difficult to form. Again, the solid line in the figure is the result of fitting based on PR theory [35]. The scale of inhomogeneity evaluated from the fitting Ξ is given by

$$\Xi = \sqrt{Na^2/12(w_0 - 1)} \tag{7.28}$$

Ξ is plotted in Fig. 7.8 as a function of the cross-linker concentration (C_{BIS}: lower axis) and of the irradiation dose (dose: upper axis). The differences in the occurrence of inhomogeneity due to the method of cross-linking are very clearly shown.

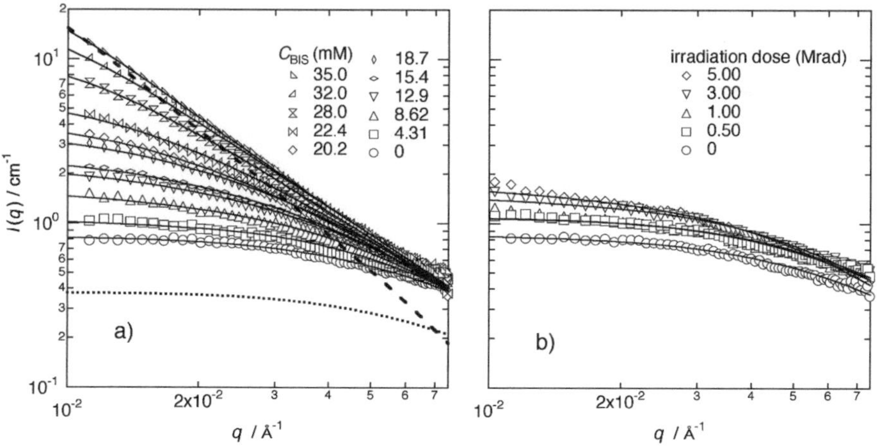

Fig. 7.7 SANS curves for **a** BIS cross-linked gel (chemical cross-linked gel; monomer-cross-linked gel) and **b** gamma-ray cross-linked gel (polymer-cross-linked gel) [35] (Norisuye et al. [35]; reproduced with permission, © Elsevier)

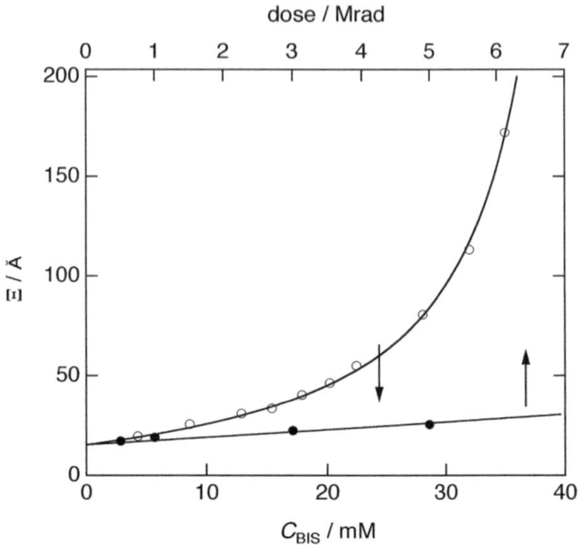

Fig. 7.8 Cross-linker concentration (C_{BIS}; lower axis) and irradiation dose (dose; upper axis) dependence of the measure of inhomogeneity Ξ [35] (Norisuye et al. [35]; reproduced with permission, © Elsevier)

7.3.2 Swelling and De-swelling Gels

Do polymer chains in gels expand or contract as the gel swells or shrinks? The spreading of PDMS-labeled chains in polydimethylsiloxane (PDMS) networks under equilibrium swelling conditions has been studied [36]. Four- and six-branched PDMS networks were prepared by end-cross-linking unlabeled PDMS (lightly hydrogenated PDMS: 80%) and deuterated PDMS (20%), and the z-average radius of gyration of the PDMS mesh chains, R_z, and the corresponding linear PDMS chains were evaluated by SANS as a function of the PDMS molecular weight, M_z.

$$R_z = 0.17 M_z^{0.58} \tag{7.29}$$

This confirms de Gennes' theoretical prediction (C^* theorem) that the spreading of network chains in a good solvent automatically adjusts to that of the corresponding polymer solution, and the power exponent of 0.58 strongly supports that the system is a good solvent system.

The spreading of deuterated label PS chains (radius of gyration R_g) in a network of end-cross-linked polystyrene (PS) in benzene has also been investigated by SANS as a function of swelling degree $\Phi = V/V_{dry}$ [37]. The results are shown in Fig. 7.9. The solid, dashed, and dotted lines are theoretically predicted curves based on the affine model, the terminal elongation model (four branches), and the phantom network model [38, 39], respectively.

$$\left\langle R_g^2 \right\rangle^{1/2} = \left\langle R_{g,dry}^2 \right\rangle^{1/2} \Phi^{1/3} \quad \text{(Affine model)} \tag{7.30}$$

$$\left\langle R_g^2 \right\rangle^{1/2} = \left\langle R_{g,dry}^2 \right\rangle^{1/2} \left(\frac{\Phi^{2/3} + 1}{2} \right)^{1/2} \quad \text{(Terminal elongation model)} \tag{7.31}$$

$$\left\langle R_g^2 \right\rangle^{1/2} = \left\langle R_{g,dry}^2 \right\rangle^{1/2} \left(\frac{\Phi^{2/3} + 3}{4} \right)^{1/2} \quad \text{(Phantom mesh model)} \tag{7.32}$$

Surprisingly, as shown in Fig. 7.9, the Φ-dependence of the measured R_g was much smaller than either theoretical prediction. They conclude that the deformation of network molecules is small compared to macroscopic swelling/de-swelling [40].

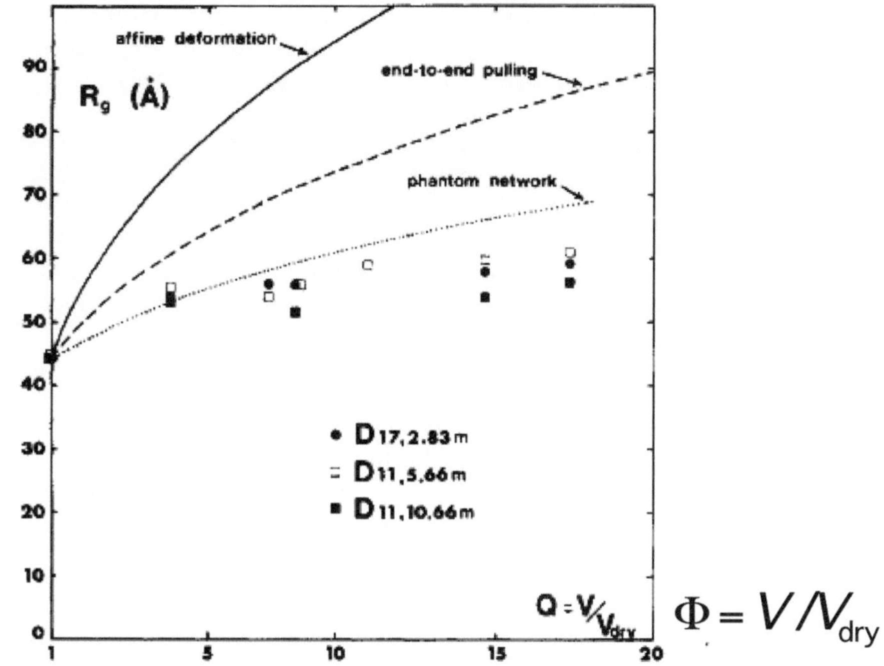

Figure 2. Radius of gyration of the labeled chains of the network as a function of the swelling degree.

exhibits long-time fluctuations, which may produce an error in the determination of the flat background *B* of the autocorrelation function. Therefore, instead of using the general procedure based on the mean unclipped and clipped count rates per sample time,[20] we have considered *B* as an adjustable parameter in the fitting procedure. For the investigated gels the ratio between the fitted and computed values of *B* ranges from 1.002 to 1.02.

Fig. 7.9 Swelling ratio $\Phi = V/V_{dry}$ dependence of the spreading (radius of gyration R_g) of deuterated labeled PS chains in end-cross-linked polystyrene (PS) networks in benzene [37] (Bastide et al. [37]; reproduced with permission, © American Chemical Society)

However, these experimental results and their interpretation remain questionable. Recently, it has been reported that there is an affine correspondence between the correlation lengths of tetra-PEG hydrogels obtained by end-cross-linking of tetra-polyethylene glycol macromers at preparation and at equilibrium swelling, respectively [41]. This is discussed in detail in Sect. 7.3.8.4.

7.3.3 Stretching Gel

It has been observed that the scattering intensity increases in the direction of stretching when the gel is uniaxially stretched [42]. The deformation behavior of polystyrene (PS) gel by uniaxial stretching was investigated by SANS. The cross-linked PS gels in 1,2-dichloroethane solution were prepared by introducing cross-links randomly by Friedel–Crafts reaction, the sol component was removed, and dried. Then, the gel was immersed in deuterated toluene, stretched uniaxially at a stretching ratio of λ, and SANS was performed [43]. Figure 7.10 shows the SANS contour plot. As shown in the figure, the contour plot is an ellipse elongated in the direction of stretching for small values of q, and an ellipse elongated perpendicular to the direction of stretching for large values of q. In particular, for $\lambda \geq 1.0$, the contour plot is an ellipse elongated in the direction of stretching. In particular, the elliptical scattering pattern seen on the small-angle side for $\lambda \geq 1.33$ is called an abnormal butterfly pattern [44]. The reason is that such a scattering pattern is opposite to the theoretical prediction (which predicts an elliptical pattern elongated in the direction perpendicular to the stretching direction). Though this abnormal scattering pattern was explained as thermal concentration fluctuations [18], it was re-interpreted as quasi-static fluctuations (also called frozen or quenched fluctuations), which are fluctuations intrinsic to the gel. The elliptic scattering intensity function is given by [19]

$$I(q;\beta) = \frac{K_N \phi^2 k_B T}{a^3} \left\{ \frac{1}{M_{os}^*(\beta)[1+\xi^2(\beta)q^2]} + p \frac{\left[J(\beta) - (\phi_0/\phi)^{-2/3}\right]^2 (\phi_0/\phi)^{2/3} \mu}{M_{os}^{*2}(\beta)[1+\xi^2(\beta)q^2]^2} \right\} \quad (7.33)$$

where β is the angle between the scattering vector q and the stretching direction, ϕ_0 is the volume fraction of the polymer in the reference state, $p(0 < p \ll 1)$ is the network inhomogeneity defined in Eq. (7.25), $J(\beta)$ is the modified term of the shear modulus

$$J(\beta) = \left(\lambda^2 - \frac{1}{\lambda}\right)\cos^2\beta + \frac{1}{\lambda} \quad (7.34)$$

The equation is given as where $J(\beta) = \lambda^2$ in the stretched direction ($\beta = 0°$) and $J(\beta) = \lambda^{-1}$ in the vertical direction ($\beta = 90°$). $M^*_{os}(\beta)$ is the direction-dependent longitudinal modulus of elasticity of stretching, and is given by

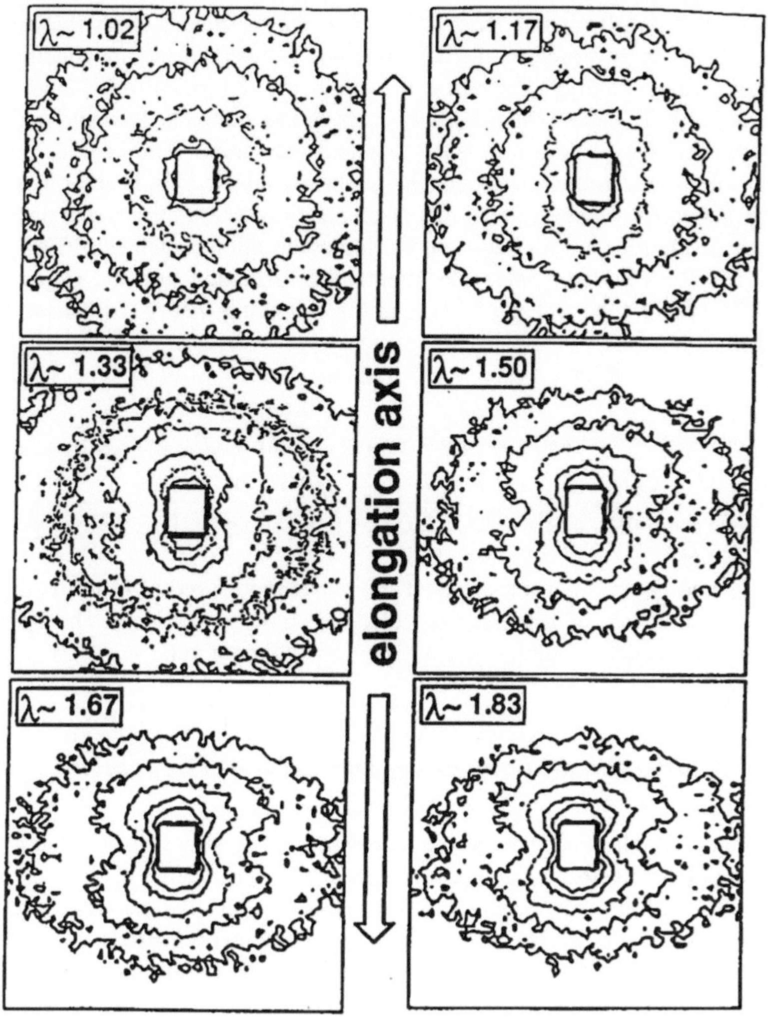

Fig. 7.10 SANS contour plot of the deformation behavior of polystyrene (PS) gel by uniaxial stretching [43]. λ is the stretching ratio (Mendes et al. [43]; reproduced with permission, © American Chemical Society)

$$M_{\text{os}}^{*}(\beta) = K_{\text{os}} + \left[\frac{1}{3} + J(\beta)\right]\mu \tag{7.35}$$

The first term on the right-hand side represents thermal fluctuations and the second term represents quasi-static fluctuations.

In the study of swollen and stretched PS networks [43], the observed SANS patterns were compared with (i) the cluster model [45], (ii) the Onuki theory [19], and (iii) the Rabin-Bruinsma theory [18]. The cluster model states that the scattering intensity is governed by the deformation of clustered inhomogeneities [45]. However, neither of these theories can explain the deformation behavior. In the studies of butterfly patterns in randomly cross-linked networks [46] and end-cross-linked PS networks [47], experimental results do not agree with theoretical predictions. Furthermore, strong inhomogeneity was also found to exist in end-cross-linked polymer networks. The reason for this discrepancy between experiment and theory is that these theories are all based on ad hoc assumptions, such as p (Eq. (7.33)). In contrast, the PR theory puts the effect of deformation into the $g(q)$ function in Eq. (7.21) and $C(q)$ in Eq. (7.17), and the following theory for deformation is proposed [29].

$$g = \frac{1}{Q^2/2 + (4Q^2)^{-1} + 1} + \frac{2(\phi/\phi_0)^{2/3}\phi_0^{1/4}}{(1 + Q^2)^2} \tag{7.36}$$

$$C(q) = \frac{\phi N_X}{\left[1 + wg(q)\right]^2 (1 + Q^2)^2} \times \left[6 + \frac{9}{w_0 - 1 + \left\{(Q_p^2\lambda^2 + Q_t^2/\lambda^2)\right\}(\phi_0/\phi)^{5/12}/2}\right] \tag{7.37}$$

where Q_p and Q_t are the projections of the wavenumber vector $\mathbf{Q}$ in the direction of the stretching axis and its perpendicular direction, respectively. The PR theory explains well the deformation of weakly charged polymer gels.

7.3.4 Volume Phase Transitions and Critical Phenomena

In a polyacrylamide (PAAm) gel in a mixed solvent system of water and acetone, a volume phase transition was discovered in which the volume of the gel changes dramatically and discontinuously with slight changes in solvent composition [10]. First, it was found that this phenomenon does not occur in PAAm gels, but in P(AAm/AAc) gels in which a portion of the PAAm chain in the PAAm gel was hydrolyzed to acrylic acid (AAc), and it was necessary to introduce a charged group to the gel to cause a large discontinuous phase transition [48]. Subsequently, by replacing the acrylamide (AAm) component with a temperature-responsive *N*-isopropylacrylamide, a poly(*N*-isopropylacrylamide/acrylic acid) copolymer gel (P(NIPAM/AAc) gel, which exhibits a volume phase transition. The volume phase transition of P(NIPAM/AAc) was then investigated in detail from the viewpoint of thermodynamics [49]. The volume phase transition was found to occur not only in P(NIPAM/AAc) gels but also in gels made from nucleic acids, polysaccharides, and proteins that represent natural polymers, indicating that the volume phase transition is a universal phenomenon [50]. Starting from this research, various studies have been conducted.

In PNIPAM homopolymer gels, we found that light scattering intensity and relaxation time diverge when approaching LCST [51]. It was concluded that this is a critical phenomenon [52] observed in many materials, including two-component systems and magnetic materials. These are classified into several universal classes.

Critical phenomena have also been observed in polymer gels. Based on the temperature dependence of the degree of swelling of a series of PNIPAM gels prepared by varying the concentration of cross-linking agent in the equilibrium swelling state ($\Pi = 0$: isobaric gel), it has been classified as a three-dimensional icing system [53]. On the other hand, there is a study that investigated temperature-induced critical phenomena by performing SANS experiments on PNIPAM gels prepared in heavy water ($\phi = 0.083$: isobaric gel) and the corresponding PNIPAM heavy water solution, where the volume was kept constant. The PNIPAM gel continuously transitioned to the contracted phase at 34.6 °C, and the SANS intensity curves diverged at the critical temperature according to the Ornstein–Zernike type scattering function [54].

In P(NIPAM/AAc) gels undergoing a discontinuous volume phase transition, a characteristic scattering peak is observed in the SANS curve [55]. Figure 7.11 shows a photograph of a P(NIPAM/AAc) gel undergoing a volume phase transition (right) and the SANS curve (left) together with a schematic diagram of the gel network (superimposed to the photograph) [56]. As the temperature increased, the scattering peaks became larger and sharper. When the volume phase transition began, two phases appeared: a swollen phase (left) and a shrunken phase (right), as if in a gas–liquid transition. The boundary between the two phases shifted to the left and right when the temperature was slightly increased or decreased, respectively. The difference from the gas–liquid transition is that the two phases are continuously connected because the gel is a continuum (this is discussed in detail in Chap. 5).

Figure 7.12 compares the phase behavior of (a) a neutral gel (PNIPAM homopolymer gel: uncharged gel) and (b) a weakly charged P(NIPAM/AAc) gel [56]. In both cases, the horizontal axis is the gel concentration ϕ and the vertical

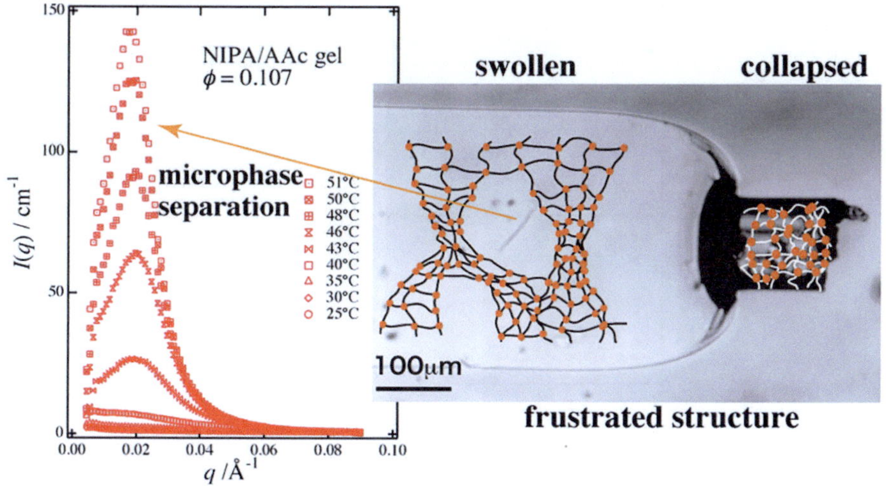

Fig. 7.11 Photograph of (PNIPA/AAc) gel undergoing a volumetric phase transition and SANS curve with a schematic of the gel meshwork [56] (Shibayama [56]; reproduced with permission, © The Society of Polymer Science, Japan)

axis is the temperature. In (a), the gel was initially in an equilibrium swollen state (lower left photo), but as the temperature increased, the gel concentration increased (collapsed) along the isobaric curve and eventually reached a transparent shrunken state (photo on the right). However, when the temperature is suddenly increased, the gel moves along the isovolumetric curve and phase separates without much change in size because there is no time margin for shrinking (upper left photo). On the other hand, in (b) charged gel, the gel shrinks along the isobaric curve as the temperature is slowly increased, but it suddenly shrinks when it reaches T_c and enters the shrunken phase (as shown by the photograph in the case of swollen-shrunken coexisting state).

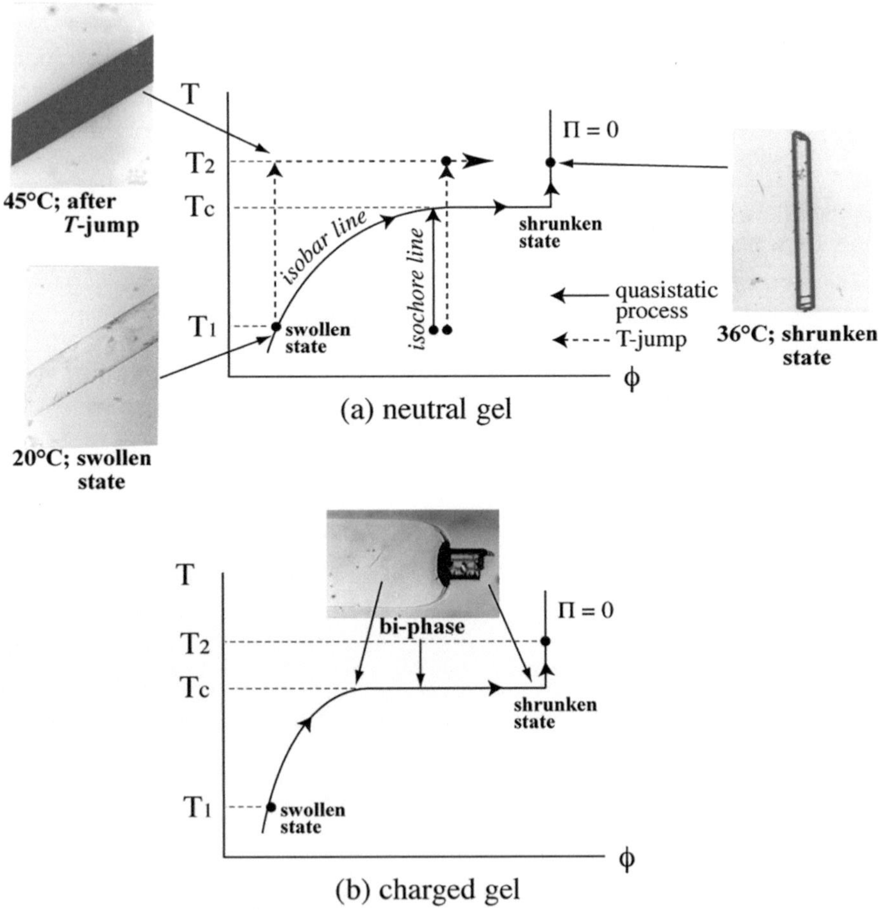

Fig. 7.12 Comparison of phase behavior of **a** neutral gel (PNIPA homopolymer gel; uncharged gel) and **b** weakly charged PNIPA/AAc gel [56] (Shibayama [56]; reproduced with permission, © The Society of Polymer Science, Japan)

7.3.5 Charged Gels and Microphase Separation

As described in the previous section, introducing a charge into an uncharged gel causes a greater degree of swelling of the gel. When charge is introduced into the gel while the gel volume is kept constant, static inhomogeneity in the gel is suppressed due to the strong Donnan potential [57]. However, this situation changes drastically when the solvent becomes poor. This phenomenon is observed not only in weak electrolyte gels but also in weak electrolyte solutions. The homogenization that tries to maintain electrical neutrality and the phase separation caused by poor solvents compete with each other, and the gel undergoes microphase separation [55]. This phenomenon was first theorized for weakly charged polyelectrolyte solutions (Borue-Erukhimovich theory, BE theory) [58]. On the experimental side, a scattering peak was reported to appear in the SANS curve of a partially neutralized polyacrylic acid (PAAc) gel [59].

A more vivid example of this phenomenon is the SANS study of P(NIPAM/AAc) gel, a copolymer gel of *N*-isopropylacrylamide and acrylic acid [55]. In P(NIPAM/AAc) gel copolymerized with approximately 5% acrylic acid, the gel underwent a discontinuous volume phase transition with increasing temperature. The SANS curve was measured as a function of temperature, and a sharp scattering peak appeared above about 40 °C and increased significantly with increasing temperature. The gel underwent a volumetric phase transition and shrank at around 50 °C. This has already been shown in Fig. 7.11.

The BE theory was used to explain SANS curves because it explains the phase behavior and scattering function of weakly charged polymer solutions and weakly charged polymer gels well [55]. However, since this theory was originally proposed for weakly charged polymer solutions, there were limitations in its application to gels. Later, a new theory was proposed that extended the PR theory [29, 60] to charged systems [31]. From this, the scattering function of a weakly charged polymer gel can be described by using Eq. (7.22) from Eqs. (7.13) and (7.19), and a comparison of theory and experiment was made [61, 62].

7.3.6 Physical Gels

Since cross-linking in gels is usually introduced by polymerizing monomers in the presence of a cross-linking agent, the cross-linking agent is unevenly distributed and frozen inhomogeneity is formed during polymerization reaction. As a result, as discussed in Sect. 7.3.2, the inhomogeneity increases with increasing cross-linker concentration [23, 26, 63, 64]. However, this argument is questionable in the case of physical gels, where cross-linking is removable. This is because the introduction of cross-links in physical gels is reversible under chemical equilibrium or physical aggregation/dissociation equilibrium. In this case, cross-links are expected to be randomly distributed in the gel to minimize the free energy of the system. If this is the case, physical gels are expected to have very low inhomogeneity compared to chemical gels.

In the following, we present examples of studies on biopolymer gels, which make up the majority of physical gels. The structure of physical gels has been studied for many years. The sol–gel transition, the structure of not only gels but also pre-gels, and cross-linked structures have been discussed. These are described in detail in the book by Guenet [65].

7.3.6.1 Biopolymer Gels

Most biopolymer gels are physical gels. The structure and sol–gel transition of such physical gels have long been studied by many researchers using small-angle X-ray scattering (SAXS), SANS, and other techniques [65]. For example, in a SANS study of the thermoreversible sol–gel transition of agarose in water/DMSO mixtures [66], it was reported that agarose gels consist of very rigid molecules with a helical structure. Such a structure can be represented by the worm chain model and is known to undergo a transition from q^{-2} to q^{-1} with q near q^*. de Cloizeaux [67]

$$q^2 I(q) = 6\mu_L / l_p \quad (q < q^*) \tag{7.38}$$

$$q^2 I(q) = \mu_L \left(\pi q + \frac{2}{3 l_p} \right) \quad (q > q^*) \tag{7.39}$$

where μ_L is the mass per unit length and l_p is the persistence length of the chain. The crossover q^* is given by

$$q^* = 16/3\pi l_p \approx 1.7 l_p^{-1} \tag{7.40}$$

However, when the polymer concentration is high, a q^{-2} flat region appears due to the polymer shielding effect, and the correlation length ξ is obtained by the following equation [68].

$$I(q) \sim (q^2 + \xi^{-2})^{-1} (q < q^*) \tag{7.41}$$

$$I(q) \sim q^{-1}(1 + 2q^{-1}\xi^{-1})(q > q^*) \tag{7.42}$$

This type of crossover is observed for $l_p > \xi$, but when the correlation length is sufficiently large compared to the persistence length ($l_p > \xi$), it behaves as a worm chain. They conclude that in agarose gels, the structure is a single helix rather than a double helix model.

7.3.6.2 Physical Gels Consisting of Microcrystals

Many physical gels are formed with microcrystals as cross-linking points, and the crystallization/melting of the microcrystals corresponds to the sol–gel transition. In the study of syndiotactic polystyrene (SPS) gels formed by SANS in various solvents [69], the structure of SPS/chloroform gel is very different from that of SPS/*o*-dichlorobenzene gel, and the SPS/chloroform gel was reported to be a more loosely aggregated gel.

On the other hand, poly(vinyl alcohol) (PVA) gels prepared in a mixture of water and DMSO have been structurally analyzed in a wide q-range ($10^{-4} \leq q \leq 10$ Å^{-1}) using wide-angle neutron scattering (WANS), SANS, ultrasmall-angle neutron scattering (U-SANS), and light scattering [70]. It has been reported that (1) PVA gel is composed of PVA microcrystals and (2) the distance between the microcrystals is about 200 Å. It has also been reported that phase separation proceeds even after the time-resolved U-SANS experiment reaches the macroscopic gelation point [71].

7.3.6.3 Block Copolymer Gels

Physical gels consisting of block copolymer solutions have been studied for more than 30 years, mainly using SAXS. The cross-links consist of microphase-separated structures composed of hydrophobic solvent components. Many of these gels have been introduced in technical books on block copolymers [72], and studies on micelle formation, microphase-separated structures, and rheological behavior of physical gels using SANS have also been reported [73, 74]. In addition, structural changes from block copolymer solutions to hard gels via soft gels [75], gels with body-centered or face-centered lattice structures [76], highly sensitive thermo-induced gels [77], and solvent-induced microphase separation transitions [78] have also been reported.

7.3.6.4 Ion Complex Gels

PVA ion complex hydrogels, or slimes, have also been studied [24, 79]. Aqueous solutions of PVA and Congo red (CR) have also been reported to gel (PVA/CR gel) [80], and PVA/CR has been shown to (i) gel without hysteresis, (ii) undergo sol–gel transition at moderate temperatures near room temperature (T_{gel} ~ 43 °C), and (iii) to be stable at room temperature [81]. The structures of PVA/CR gel have also been investigated in detail [80–82]. In these studies, the gelation point has also been identified by examining static inhomogeneity in the appearance of speckle images of light scattering and significant changes in dynamics at the gelation point [83].

7.3.7 Oil Gelators (Oil Gelling Agents)

Certain low-molecular-weight organic compounds such as fatty acids and their derivatives, steroids, and amino acids gel in organic solvents. These compounds are called gelators and have attracted much attention since the 1990s because of their wide range of applications and unique structures and properties [84]. The unanswered questions are (1) what is the necessary structure of the oil gelatos, (2) how is the gel formed and what is the structure after gelation, and (3) how does the packing of the oil gelling agent molecules affect the physical properties of the gel? It goes without saying that SANS and other scattering methods are very effective in answering these questions. If one were to ask what is the most effective structure for space-filling, the answer would be fibrous fibrils (one-dimensional objects).

Oil gelators assume a fibrillar structure in the gel, as evidenced by many microscopic observations and scattering techniques. In the scattering method, rod-shaped molecules are

$$I(q) \sim K_{\mathrm{N}} M_{\mathrm{L}} \left[2\frac{J_1(qR)}{qR} \right]^2 \tag{7.43}$$

is expressed as where $J_1(x)$ is a Bessel function of the first kind. M_{L} is the mass per unit length of fibrils of radius. In gels composed of oligosiloxanes, hydrogen bonds between amide groups are reported to be responsible for intermolecular association [85].

7.3.8 Various High-Strength Gels and Their Deformation Behavior

Since the beginning of this century, gels developed based on completely new designs have been produced one after another. Figure 7.13 shows their structural schematic diagrams [86]. (a) is a slide-ring gel in which the cross-linking points are replaced with cyclodextrins so that the cross-linking points can slide along the molecular chain [87], (b) is a nanocomposite gel (NC gel) [88] consisting of organic and inorganic components, (c) is a double network gel (DN gel) [89], which is made by making a gel once and then immersing it in a second monomer solution and polymerizing its components, and (d) is a tetra-PEG gel [90] made by cross-linking two nearly monodisperse tetrabranched polymers with complementary functional groups at their mutual ends. Their structural features and deformation mechanisms obtained by the scattering method are described below.

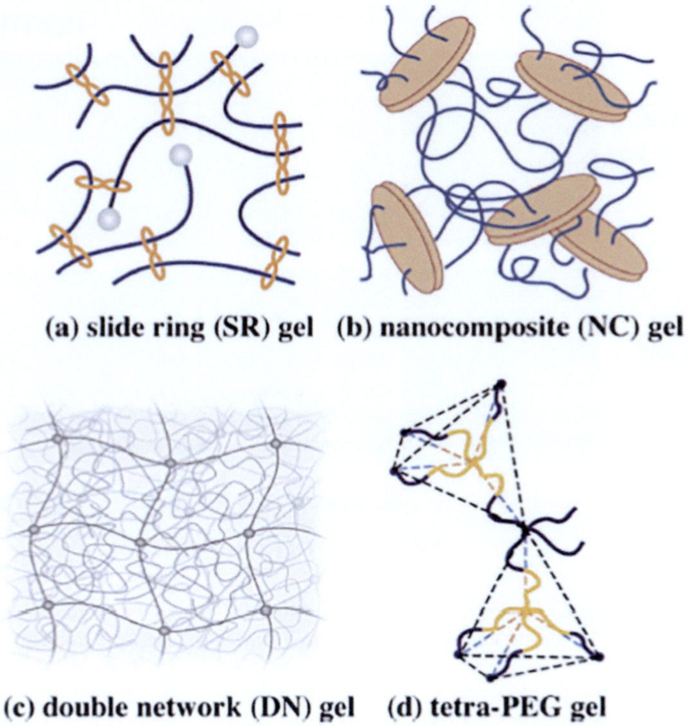

Fig. 7.13 Structural schematic of various high-strength gels [86]: **a** slide-ring gel, **b** nanocomposite gel, **c** double network gel (DN gel), and **d** tetra-PEG gel

7.3.8.1 Slide-Ring Gels

Slide-ring gels have the ability to automatically disperse the strain in the gels by sliding cross-linking points on the polymer chain during deformation [87]. Therefore, the structure is resistant to deformation. Figure 7.14 shows a series of small-angle neutron scattering (SANS) patterns of a slide-ring gel stretched uniaxially [91]. The λ in the figure indicates the stretching ratio. The gel with a cross-linker concentration of 0.7% (upper panel) showed a longitudinal scattering pattern (normal pattern: homogeneous deformation) as the gel was stretched in the lateral direction, while the gel with a cross-linker concentration of 2.0% (lower panel) showed a stretched pattern in the lateral direction (abnormal pattern: gel-specific deformation). This indicates that at high cross-linker concentrations, the abnormal pattern appears due to increased inhomogeneity as a result of stretching, as in conventional gels (see Sect. 7.3.3), while at low cross-linker concentrations, the slide-ring gel chains are stretched in the direction of stretching, as in polymers in an elongated flow field, indicating that the movable cross-linking points work effectively during deformation.

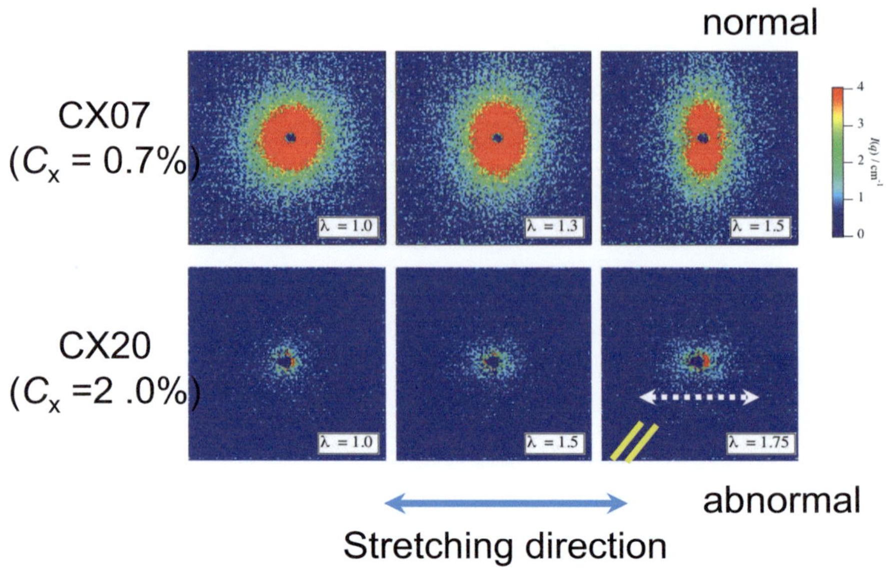

Fig. 7.14 Small-angle neutron scattering (SANS) pattern of uniaxially stretched slide-ring gels [91] (Karino et al. [91]; reproduced with permission, © American Chemical Society)

7.3.8.2 Nanocomposite Gel (NC Gel)

NC gels do not break even when stretched about 10 times, and when the load is removed, it almost recovers, although some residual strain remains [88]. Since NC gels are a three-component system consisting of polymer (P), clay (C), and water (w), contrast variation SANS is effective. Contrast variation SANS is a method of evaluating the scattering intensity distribution $I(q)$ by isolating and extracting each pair of components that make up the scattering intensity distribution $I(q)$, by continuously changing the scattering length density of the solvent (w) in a SANS experiment for a multicomponent system [92]. As shown in Fig. 7.15, the scattering intensity $I(q)$ is decomposed into its pairwise components $S_{ij}(q)$. In the case of an aqueous system, the scattering length density can be varied continuously by changing the solvent composition from light water to heavy water. In the case of a three-component system, six different $S_{ij}(q)$ are obtained, but if the system is incompressible, they are not independent of each other, yielding two $S_{ii}(q)$ and one $S_{ij}(q)$ (i $\neq$ j). Figure 7.16 shows the SANS pattern of a uniaxially stretched NC gel [93]. S_{CC} and S_{PP} in the upper and lower panels are the partial scattering functions of the clay and polymer, respectively. These pieces of information can be obtained by the contrast matching method, in which the scattering length density of one of the components is matched to that of the medium, but the S_{CP} in the middle panel can only be obtained by contrast variation SANS. In both patterns, stretching deforms the scattering into a longitudinally elongated shape, indicating orientation of the clay in the stretching

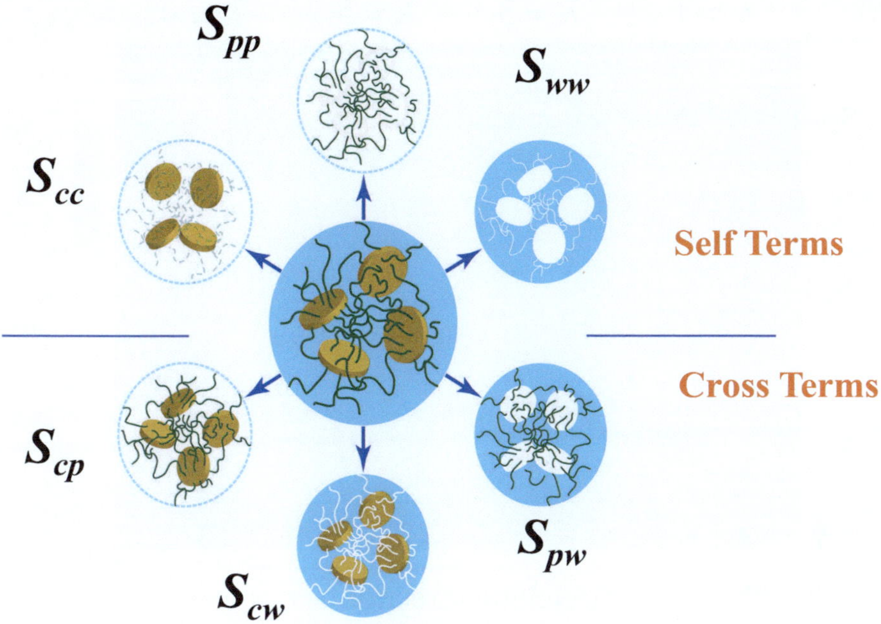

Fig. 7.15 Schematic of contrast variation SANS for nanocomposite gels. The scattering intensity $I(q)$ is decomposed into its respective pairwise components $S_{ij}(q)$

direction and stretching of the polymer chains in the stretching direction. The S_{CP}s show that a considerable amount of polymer chains are adsorbed on the clay surface and that these adsorbed molecules are desorbed by stretching.

7.3.8.3 Double Network Gels

In the case of double network gels (DN gels) [89], it is believed that the fragile components in the gel are destroyed by deformation, absorbing energy and producing excellent mechanical properties [94].

7.3.8.4 Tetra-PEG Gels

In order to suppress the development of heterogeneity due to cross-linking, tetra-PEG gels are prepared by preparing a hyperbranched symmetric prepolymer solution near the C^* concentration and cross-linking the ends [90]. The resulting gel exhibits almost no inhomogeneity characteristic of gels, and is described only by an Ornstein–Zernike type scattering function similar to that of semi-dilute polymer solutions [95,

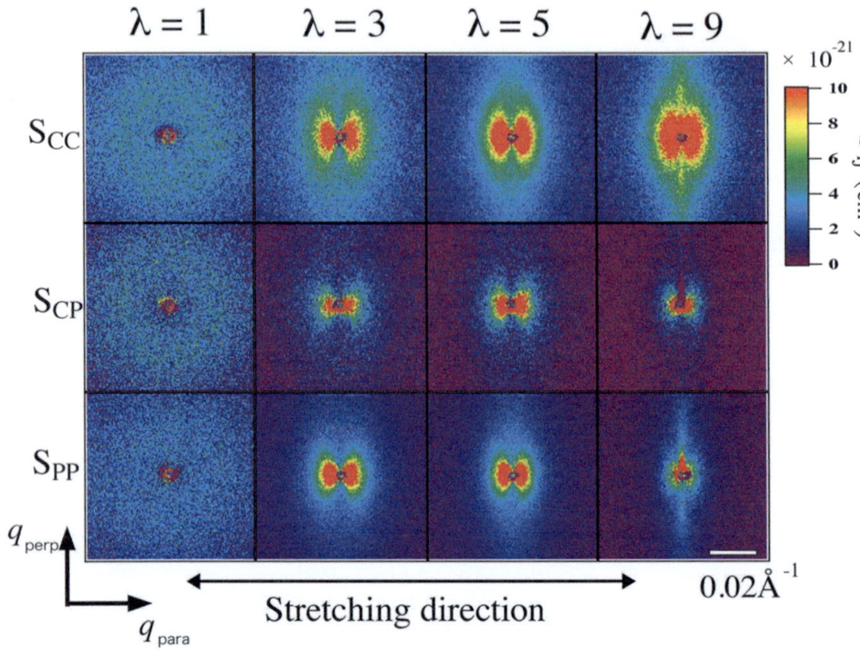

Fig. 7.16 SANS patterns of uniaxially stretched NC gels. [93] The upper panel represents the clay-clay S_{CC}, the middle panel the clay-polymer S_{CP}, and the lower panel the polymer–polymer S_{PP} partial scattering function (Nishida et al. [93]; reproduced with permission, © American Physical Society)

96]. This indicates that a uniform network structure is formed without inhomogeneous distribution of cross-linking points upon introduction of cross-linking. In other words, it can be thought of as an infinite network formed by the three-dimensional association of four-arm polymer modules (elastic blob) with entropy-elastic properties. Therefore, viscous components are rarely observed, and it has been found that the gel exhibits ideal spring-like behavior. Furthermore, even when the gel is stretched, no inhomogeneity at all appears in the scattering pattern [97].

Initially, tetra-PEG gel was developed as one of the high-strength gels, but its uniformity of network structure and good biocompatibility, as well as its ability to prepare imperfect network gels by controlling cross-linking defects, have been used as model polymer networks for quantitative evaluation of gel scattering functions [98], rubber elasticity theory [99], and percolation theory [100]. Figure 7.17 shows the results of a systematic study of the osmotic pressure and SANS scattering intensity as a function of polymer concentration and molecular weight for a tetra-PEG hydrogel and the corresponding tetra-PEG solution. This indicates that the longitudinal modulus calculated from osmotic pressure and that evaluated from SANS are the same within experimental error [98]. This is an experimental proof of de Gennes' C^* theorem discussed in Sect. 7.2.1, and it is another new fact about the nature of gels.

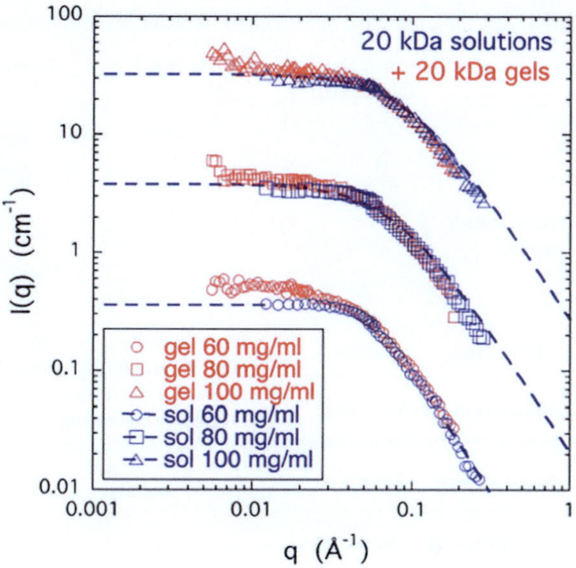

Fig. 7.17 SANS scattering intensity distribution function for aqueous tetra-PEG solution and tetra-PEG gel [98] The dashed lines are fitting results with Ornstein–Zernike function (Horkay et al. [98]; reproduced with permission, © American Institute of Physics)

Furthermore, tetra-PEG gels have attracted much attention not only from gel researchers but also from researchers in materials science and medical materials [101]. For example, recently, tetra-PEG gels have been used as copolymers with completely different properties, which are end-cross-linked to form co-networks to provide functions such as temperature response and solvent selectivity [102, 103], and gels with ultra-low osmotic pressure similar to biological tissue have been fabricated and applied to medical applications [104]. Scattering analysis of these tetra-PEG gel derivatives has also become an indispensable tool for structural analysis [100, 105].

7.3.8.5 Deformation Mechanism of High-Strength Gels

The four types of high-strength gels described above were all obtained by different molecular designs, and their characteristics are also very different. Small-angle neutron scattering (SANS) techniques have been very effective in elucidating the structure and deformation mechanisms of these gels. Table 7.2 shows a summary [56]. (a) is a conventional chemically cross-linked gel. The network structure is non-uniform, and deformation causes stress concentration, resulting in immediate gel rupture. SANS experiments have shown that in the cyclic gel (b), the stress concentration in the mesh during deformation is dispersed by the movement of movable cross-linking points. In the NC gel of (c), it was clarified that not only do the clay and polymer chains orient and stretch in the stretching direction as the gel is stretched, but also the polymer chains adsorbed on the clay act as a reservoir, enabling the gel to be highly stretchable. In the case of tetra-PEG gel (d), it was found that the elastic

Table 7.2 Structures and deformation mechanisms of four types of gels Revealed by SANS

	(a) Conv. Gel (chem. Gel)	(b) Slide ring gel	(c) Nanocomposite gel	(d) Tetra-PEG gel
Deformatior model	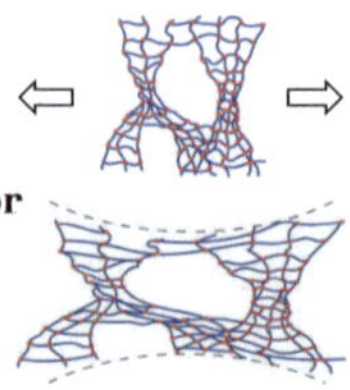	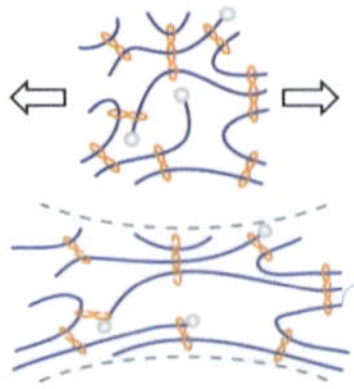	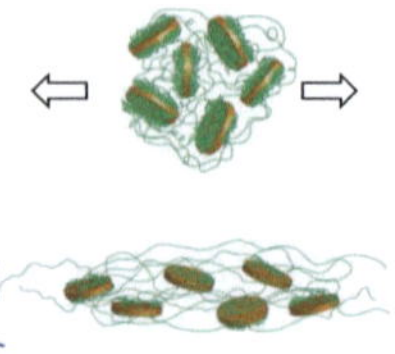	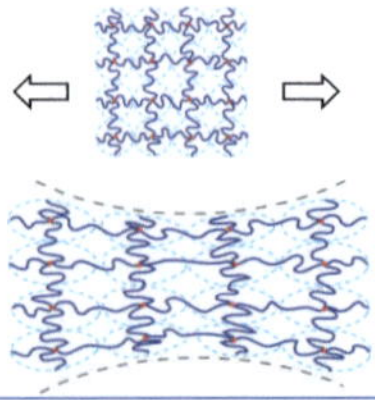
Mech. Prop.	brittle	High elasticity High deformability	High elasticity High deformability	High elasticity High deformability
Inhomogeneities	Very large	Low C_x : decreasing High C_x : increasing	Small	None
Scatt. pattern	Abnormal	Normal / abnormal liq. like/gel like	Normal liq. like/clay scatt.	Isotropic
Charac. feature	Inhomog. str.	Movable cross-links	Plane crosslink, large molec. wt.	Elastic blob

blobs form a three-dimensional network structure as a module with few defects and behave almost as an ideal elastic body.

7.4 Microscopy

7.4.1 Various Microscopy Methods

Electron microscopes use a beam of electrons instead of visible light, and images are obtained by scanning or focusing the beam. Figure 7.18a and b show an overview of a scanning electron microscope (SEM) and a transmission electron microscope (TEM), respectively. The intensity of the secondary electron beam emitted from the sample is measured by a detector to produce an image. For samples that do not conduct electricity, such as polymers, the conductivity of the sample surface is increased by depositing carbon or other materials, or a replica of the sample is taken for observation. The structure of a TEM is similar to that of an optical microscope, using a focusing lens or objective lens made of electromagnets instead of an optical lens to focus electron beams and form an image. The image obtained on the projection surface is photographed or converted into an electronic image using an imaging device such as a CCD. Since both of these methods require operation in a high vacuum, the sample cannot be observed in the form of a gel containing moisture, and must be prepared as a dried gel or freeze-fractured. Recently, cryo-TEM technology has been rapidly developed, in which a sample containing water

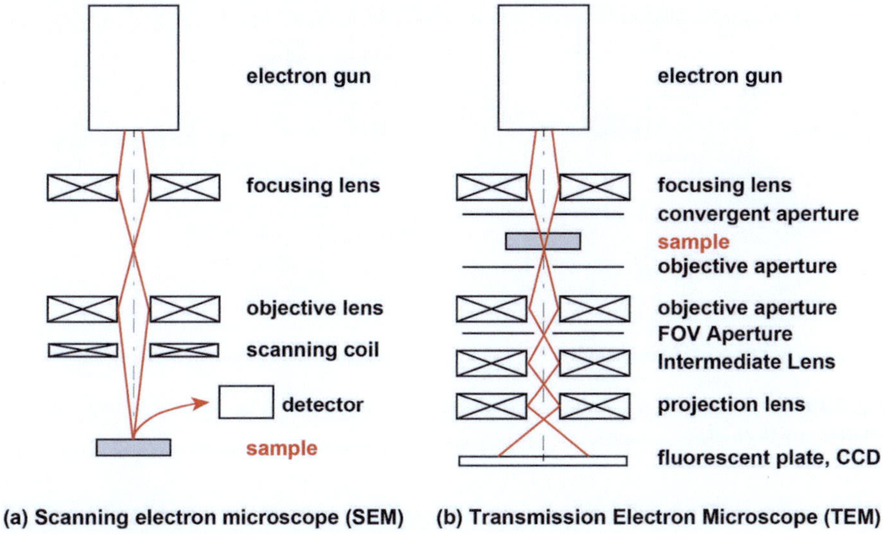

Fig. 7.18 Schematic of **a** scanning electron microscope (SEM) and **b** transmission electron microscope (TEM)

is rapidly placed in liquefied ethane chilled with liquid nitrogen to vitrify the water, which is then observed. For samples with a regular structure (crystal structure) such as proteins, it is now possible to obtain TEM images with very high spatial resolution and even three-dimensional images by taking many cryo-TEM images and processing them with a computer. Therefore, cryo-TEM has become an indispensable tool for observing protein crystal structures. On the other hand, the resolution of cryo-TEM is still limited for the observation of samples such as gels that do not have a well-defined structure.

Figure 7.19a and b show an overview of atomic force microscopy (AFM) and laser scanning confocal microscopy (LSCM), respectively. In AFM, a cantilever (cantilever spring) is attached to the tip of a probe, which is brought into contact with the sample (Z-scan), and the amount of deflection of the cantilever is optically read by an optical sensor as displacement or force generated. The amount of deflection of the cantilever at that time is read optically by an optical sensor as displacement or force generated. The sample position (X, Y) information is obtained by scanning the sample with a piezoelectric element. On the other hand, the LSCM irradiates a laser beam onto the sample and uses a beam splitter to guide the reflected light from the sample to the detector for imaging. By installing a confocal system that combines a lens (L) and a pinhole on the light emitting side and the light receiving (detecting) side, respectively, image information can be obtained only at the points that satisfy the confocal condition. Then, by scanning the sample in the z-axis direction, image information in the depth direction of the sample can be obtained continuously. There are two types of scanning of the sample in the x–y direction: the laser scanning method (Fig. 7.19b) and the sample scanning method.

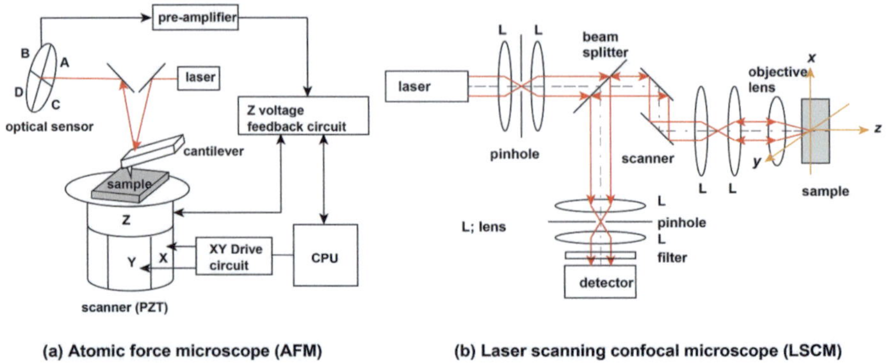

Fig. 7.19 Schematic of **s** atomic force microscopy (AFM) and **b** laser scanning confocal microscopy (LSCM)

Table 7.3 shows the statistics of gel research papers by SEM, TEM, AFM, LSCM, and cryo-TEM retrieved from Web of Science. The search was performed for each microscopy method (e.g., "scanning electron microscopy") and "polymer gel." It can be seen that a large number of gel studies are conducted using microscopy. Among these, SEM and TEM have by far the highest total number of search results and citations. The difference in statistics between SEM and TEM is partly due to the fact that SEM is relatively easy to prepare (direct observation of dried gel, replica observation, etc.) compared to TEM. In the case of TEM, the preparation of ultra-thin sections of several tens of nanometers, which is necessary for electron microscopic observation, is a little more difficult than that of SEM. AFM and LSCM, the next most cited techniques, have the highest average number of citations, even though they are relatively new techniques (gel research started in 1990). This is probably due to the ease of sample preparation, the fact that they do not require a high vacuum, and the fact that high-resolution images can be obtained relatively easily. In contrast, cryo-TEM has the fewest number of studies, but has the highest average number of citations, indicating that it has a large impact on many studies. Many of the studies used multiple microscopy methods in combination, and some examples of studies by microscopy method are presented below.

Table 7.3 The statistics of gel research papers

	SEM	TEM	AFM	LSCM	Cryo-TEM
Search results	**1517**	**825**	**535**	**435**	**21**
Total number of citations	**30,956**	**20,750**	**15,571**	**12,997**	**823**
Average number of citations	**20.41**	**25.15**	**29.1**	**29.88**	**39.19**
h-index	**76**	**67**	**63**	**58**	**17**

7.4.2 *Examples of Gel Studies by Microscopy*

7.4.2.1 SEM and TEM Methods

The structure of hydrogels composed of pH-responsive *N*-acetylgalactosamine (GalNAc)-loaded glutamine esters has been studied by LSCM, TEM, SEM, and X-ray diffraction and Fourier transform infrared spectroscopy (FT-IR) [106]. Fibril structures grown by self-assembly were observed, and under acidic conditions, the fibrils aggregated as a result of the neutralization of carboxylic acids. As a macroscopic pH response, a release of biologically active substances was also observed depending on pH.

There are also reports of synthesis and characterization of nanocomposites of microgels, nanogels, and hydrogels with a view to biomedical applications [107]. For the development of viral and non-viral vectors, microgels, nanogels, and semi-interpenetrating nanocomposites composed of quaternary ammonium salts of cationic monomers (3-acrylamidopropyl)- trimethylammonium chloride (APTMACl) Polymer network (semi-IPN) hydrogels were prepared, and their SEM, TEM, and AFM observations were performed. Bulk gels were prepared by conventional methods, microgels in lecithin solution in cyclohexane, nanogels in reversed-phase microemulsion, and semi-IPN gels in acrylamide (AAm), hydroxyethyl methacrylate (HEMA), or APTMACl gels with APTMACl as a second component. The semi-IPN was prepared by gelation of APTMACl as the second component in an APTMACl gel. Figure 7.20 shows SEM images of lyophilized cationic gels (3-acrylamidopropyl-trimethylammonium chloride, APTMACl) composed of quaternary ammonium salts (a) loaded with lecithin (phospholipid), (b) loaded with lecithin (phospholipid), and (b) after removal of lecithin. (c) and (d) are TEM images of the gel after filtering and partial removal of phosphate. As can be seen from the SEM images in Fig. 7.20(a) and (b), lecithin covers the APTMACl gel. However, as shown in Fig. 7.20c, d, the nanohydrogels are found to be properly cleaned by partially removing the lecithin coating with lecithin.

7.4.2.2 AFM Method

Friction coefficients have been measured for various gels on different substrates (glass and Teflon plates) and in air and water, and the relationship between the friction coefficient and the chemical structure of the gel, the type of substrate, and the presence of water has been discussed [108]. AFM has been used to investigate microscopic frictional forces in these studies. The gels used were poly(vinyl alcohol) (PVA) gel, an uncharged gel; gellan gel, a partially dissociated ionic gel; and poly(2-acrylamido-2-methylpropanesulfonic acid) gel (PHAMPS), a strong electrolyte gel, and its salts (PNaAMPS). The friction force was measured as a function of load using a friction meter to evaluate the macroscopic coefficient of friction, and the cohesive strength between the gel and the glass sphere was examined by AFM using

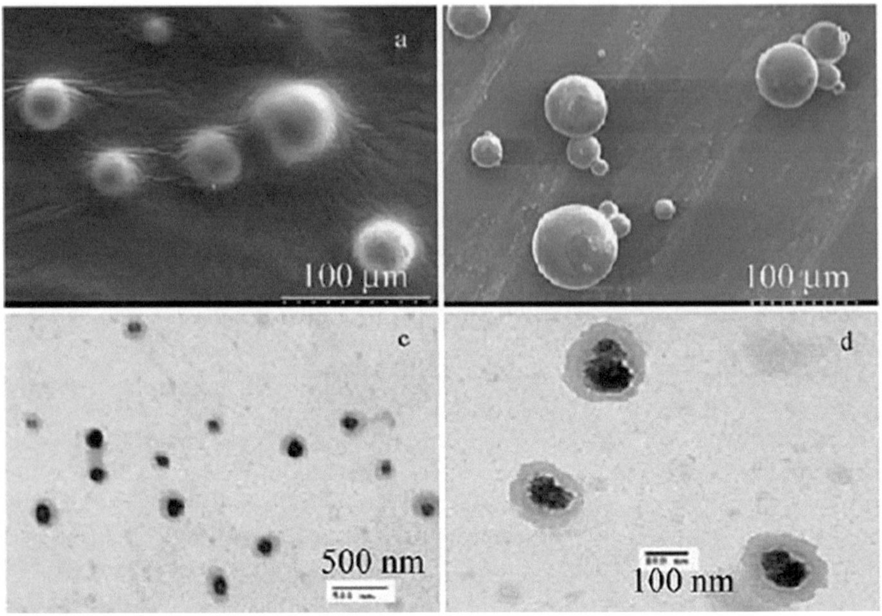

Fig. 7.20 SEM images of freeze-dried cationic gels (3-acrylamidopropyl trimethylammonium chloride; APTMACl) composed of quaternary ammonium salts. **a** With phospholipid, **b** after removal of phospholipid. **c** and **d** are TEM images of the gel after filtering and partial removal of phosphate [107] (Sahiner et al. [107]; reproduced with permission, © Springer)

a glass sphere as a probe to compare the relationship between the two. The study reported that the friction of the gel does not follow the simple Ammonton law ($F = \mu W$: the friction force F, the friction coefficient μ, and the vertical drag force W) that generally acts on surfaces in contact with objects. It has also been reported that the magnitude of friction is highly dependent on the chemical structure of the gel, with the order of PNaAMPS < gellan < PVA, and that this relationship has a good correlation with the strength of cohesion evaluated by AFM. In this example, AFM is used as a mechanical testing machine of nano-Newton (nN) order.

On the other hand, porous PNIPAM gels have been prepared using polyethylene glycol (PEG) as a pore-forming material, and gels with a fast response have been prepared [109]. The morphological observation using AFM and SEM clearly shows the porous structure in the gel. The root-mean-square surface roughness of the porous gel was determined to be 25 Å by AFM, which is larger than that of the conventional PNIPAM gel (20.8 Å). This porosity has been reported to provide much faster temperature response and greater swelling capacity than conventional PNIPAM gels, even in swelling and shrinking across the LCST. This contrasts favorably with the swelling and shrinking behavior of conventional PNIPAM gels discussed in Sect. 5.5.2, demonstrating that porosity is one effective method to achieve a fast gel response. AFM observations of microgels, nanogels, and semi-interpenetrating polymer network (semi-IPN) hydrogels [107] composed of APTMACl described above (Fig. 7.21) show that the nanogel particles are 20–50 nm in size, larger than expected. This is explained by the fact that the swollen gel is expanding on the mica substrate.

Mechanical responsiveness is essential to all living systems, and the mechanisms of mechanical responsiveness down to the tissue and cellular level are governed by a series of proteins. Microtubules, actin, intermediate filaments, and collagen are examples of such proteins, which self-assemble in a helical fashion. Using this motivation, studies of biomimetic gels consisting of mechanoresponsive polyisocyanopeptide hydrogels have been reported [110]. In this gel, isolated polymers and monolayers of bundled polymers were observed by AFM, and their height distribution was evaluated.

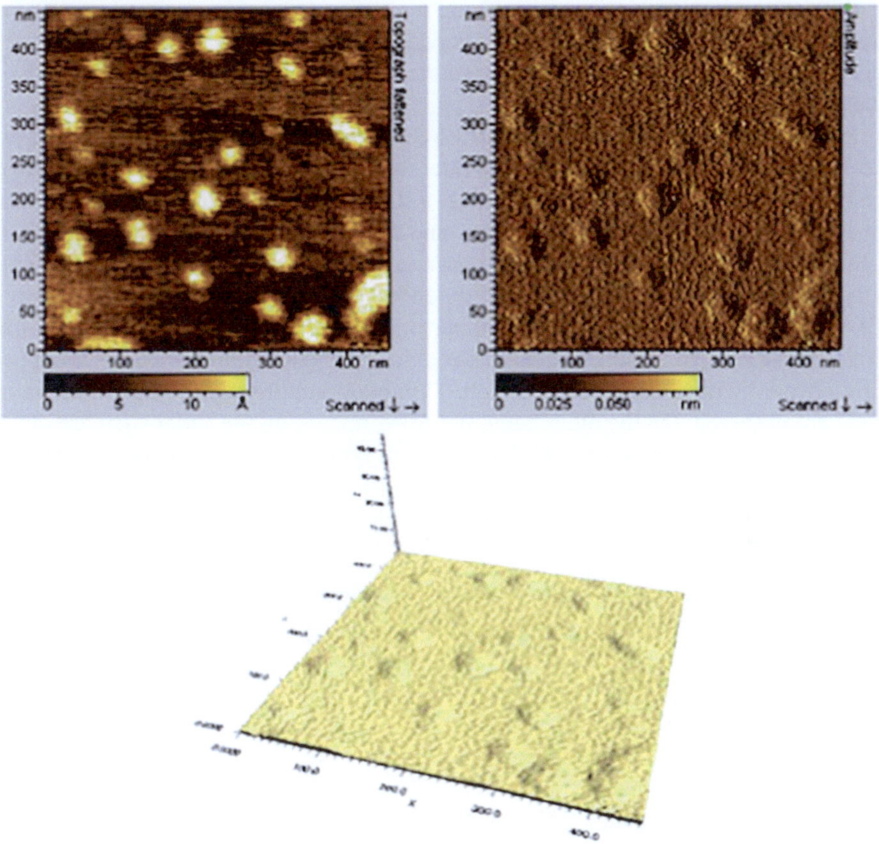

Fig. 7.21 AFM image of a cation gel (3-acrylamidopropyl trimethylammonium chloride; APTMACl) composed of freeze-dried quaternary ammonium salts [107] (Sahiner et al. [107]; reproduced with permission, © Springer)

7.4.2.3 LSCM Method

Laser scanning confocal microscopy (LSCM) can observe a two-dimensional image of a three-dimensional object at any depth by changing the focus. By continuously changing the depth and obtaining microscopic images, a three-dimensional image can be reconstructed [111]. The internal structure of PNIPAM gel was observed by LSCM, and a three-dimensional depiction of phase separation was drawn [111]. Figure 7.22 shows the reconstructed three-dimensional LSCM image of PNIPAM gel at 27 °C. (a) shows the dense domains of the polymer network brightly, (b) shows the boundary between the sparse and dense domains, one interface brightly and the other darkly, and (c) shows the sparse domains brightly. The reconstructed images of the three-dimensional internal structure in the gels revealed that the gels have a co(bicontinuous) domain structure. Thus, three-dimensional images are a powerful tool for the structural analysis of gels because they provide visual information not only on the size of the sparse structure of the gel, but also on the shape and depth of the gel.

X-ray diffraction, TEM, SEM, and LSCM observations have been made of *N*-acetylgalactosamine (GalNAc) glutamate supramolecular hydrogels that change volume in response to pH [112]. In Fig. 7.23, the image on the left is an LSCM image, the upper right is a TEM image, and the lower right is an SEM image. The LSCM image, in which the gel can be observed in its raw state without drying, shows long fibers with green fluorescence stained with a nitrobenzoxadiazole derivative (HANBD). The TEM images on the upper right show (a) swollen gel (under neutral conditions) and (b) shrunken gel (under acidic pH conditions) dried on a copper grid covered with a carbon film for 6 h. Thin hydrogel fibers (10–100 nm in diameter) are observed. On the other hand, the SEM images on the lower right are of (c) swollen gel and (d) shrunken gel, but frozen in liquid nitrogen and freeze-dried for one day. These results infer that the macroscopic opacity of the hydrogel is due to pH-induced fibril self-assembly caused by the densification of the gel fibers. Thus, it

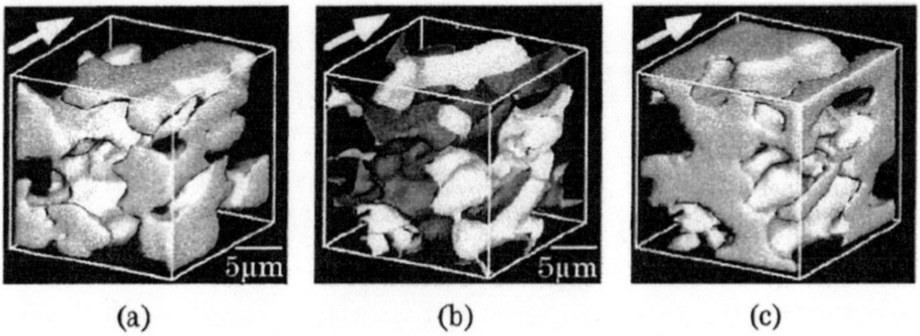

Fig. 7.22 3D LSCM images of PNIPAM gels at 27 °C. **a** Shows the dense domains of the polymer network brightly, **b** shows the boundary between the sparse and dense domains, one interface bright and the other dark, and **b** shows the sparse domains brightly [111] (Hirokawa et al. [111]; reproduced with permission, © American Chemical Society)

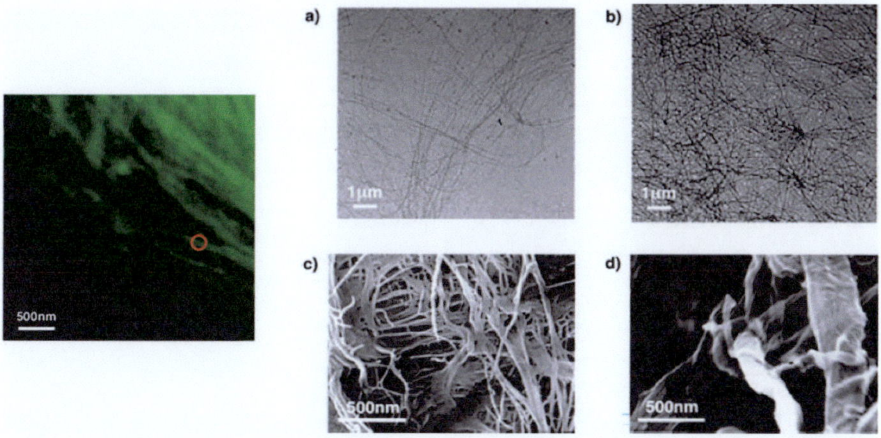

Fig. 7.23 Shrinking/swelling behavior of pH-responsive *N*-acetylgalactosamine (GalNAc) glutamate supramolecular hydrogels composed of two amphiphilic molecules [112]. Left, LSCM image of GalNAc gel stained with nitrobenzoxadiazole derivative (HANBD); top right, **a** swollen gel (under neutral conditions), **b** TEM image of a shrunken gel (under acidic pH conditions), and bottom right, SEM images of **c** swollen and **d** shrunken gels (Zhou et al. 1999; reproduced with permission, © Wiley)

can be seen that the combination of different microscopy techniques allows a more detailed structural analysis of the gels.

7.4.2.4 Cryo-TEM Method

Stimuli-responsive hydrogels play an important role in biomaterials such as regenerative medicine and drug delivery, and have been widely studied. As mentioned above, cryo-TEM has been applied to structural analysis of hydrogels in recent years. Here, we introduce an example.

Dual-response, highly absorbent thermally cross-linked hydrogels composed of methacrylate-substituted polyphosphazene (PMAPhos) were synthesized in response to temperature and pH, and the structural changes associated with swelling and contraction were studied by cryo-TEM [113]. Dried samples of PMAPhos gel were microtomed into 200 nm thick sections and pipetted onto pH 5 and pH 8 buffer grids. After a few minutes, cryo-samples were prepared by rapid cooling in a cryo-TEM sample preparation machine Vitrobot (EFI) and sublimation of water without icing. Figure 7.24 shows freeze-dried cryo-TEM images of PMAPhos hydrogels at (a) pH 5 (shrunken gel) and (b) pH 8 (swollen gel). In both figures, a polymer network is observed, but at pH 8, the morphology reflects the maximum swelling degree, while at pH 5, a more compact network structure can be seen. The ability to obtain microscopic images that reflect the true state of the gel (swollen/shrunken state) is

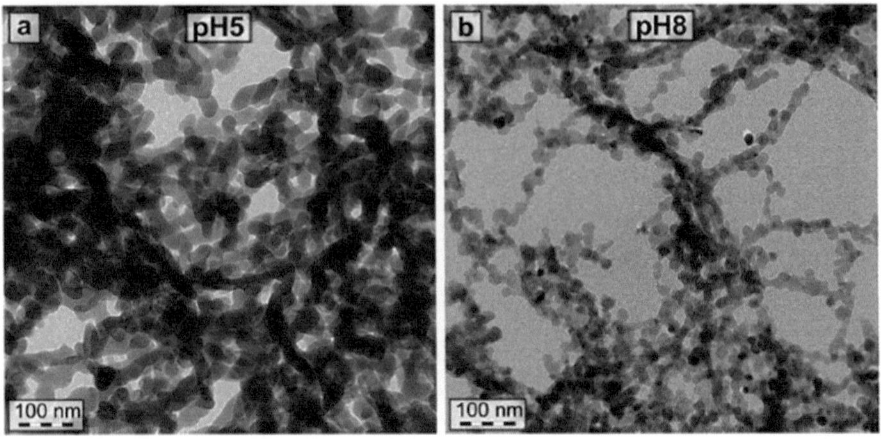

Fig. 7.24 Cryo-TEM images of freeze-dried PMAPhos hydrogels. **a** pH 5 (shrunken gel), **b** pH 8 (swollen gel) [113] (Nykänen et al. [113]; reproduced with permission, © Royal Society of Chemistry)

a feature of cryo-TEM, and its applications are expected to expand further in the future.

7.5 NMR Method

7.5.1 Principles of the NMR Method

NMR is an abbreviation for nuclear magnetic resonance, a technique for analyzing the molecular structure of materials at the atomic level by placing atomic nuclei in a magnetic field and observing the nuclear spin resonance phenomena. NMR is used in a wide range of fields such as life science, chemistry, drug and food development, and materials science because it provides useful information not only on molecular structure but also on inter- and intramolecular interactions, molecular mobility, etc. The structural analysis of polymer gels by NMR includes (i) studies of cross-link structures and molecular orientation during deformation, (ii) volume phase transitions, (iii) NMR imaging, (iv) probe-induced solvent diffusion, (v) inhomogeneity, and (vi) local binding inhomogeneity. The Fourier transform NMR (FT-NMR) method is used to obtain an NMR spectrum by placing a sample in a magnetic field and irradiating it with pulsed radio waves to excite a wide frequency range at once and then Fourier transforming the detected signal. When the static magnetic field $\mathbf{B}_0$ and the macroscopic magnetization vector $\mathbf{M}$ of the sample are arranged as shown in Fig. 7.25, the spin–spin relaxation constant T_2 and the spin–lattice relaxation constant T_1 are obtained using the following equations.

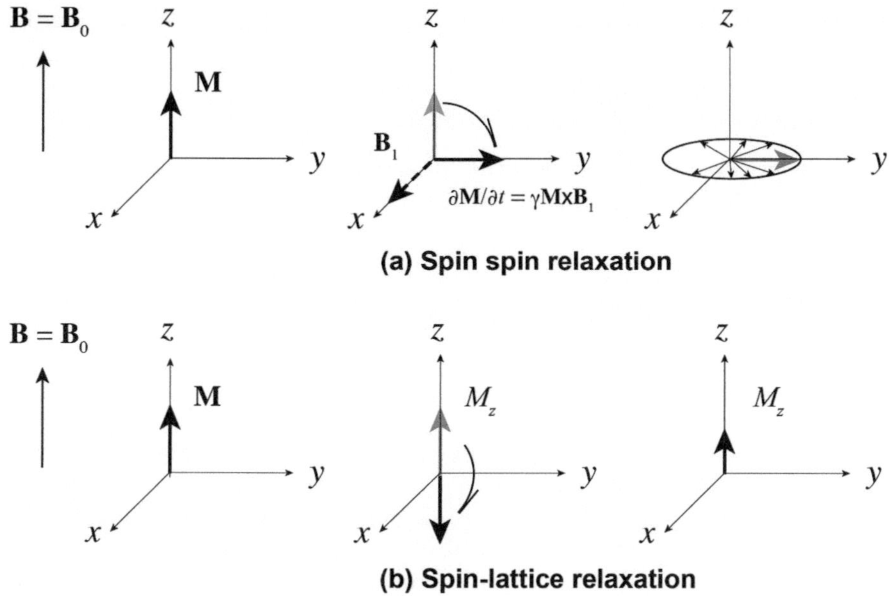

Fig. 7.25 Arrangement of static magnetic field and macroscopic magnetization vectors for measuring spin–spin relaxation (**a**) and spin–lattice relaxation (**b**)

Spin–spin relaxation (transverse relaxation)

$$M_{xy} = M_{xy}(0)\exp(-t/T_2) \tag{7.44}$$

Spin–lattice relaxation (longitudinal relaxation)

$$M_z = M_0\left[1 - 2\exp(-t/T_1)\right] \tag{7.45}$$

The characteristics of the pulse NMR method can be summarized as follows.

(1) The method is useful for obtaining information on the thermal motion state of molecular chains (molecular mobility) and the aggregation state of molecular chains.
(2) It does not provide detailed information on the chemical structure, but it does provide information on the system as a whole. The component amount can be evaluated from the difference in relaxation time.
(3) Free induction decay (FID) signal is obtained as a response to the pulse. The initial value of the signal is proportional to the number of protons in the material. If there are multiple components, the signal is the sum of the signals of each component.
(4) If there is a difference in mobility among the components, the rate of decay (transverse relaxation time: T_2) is different, and the relaxation time and the volume ratio of each component can be determined separately.

(5) The solid component has a shorter T_2, while the liquid or soft component has a longer T_2.
(6) The longitudinal relaxation time T_1 is measured by the inversion recovery method or the saturation recovery method.

For the principle and theory of NMR, please refer to reference books [114, 115], etc. Here, we describe its application to the structural analysis of gels.

7.5.2 *Examples of NMR Studies of Cross-linked Structures*

7.5.2.1 Gelation and Cross-Linking Structures

In NMR time-resolved experiments on the gelation process of acrylamide, it has been reported that T_2 suddenly decreases at the gelation point [116]. Figure 7.26 shows the time variation of T_2 for acrylamide prepared under three different conditions (AA, BAA, APS, and TEMED), where AA, BAA, APS, and TEMED are the amounts of acrylamide (monomer), bisacrylamide (cross-linker), ammonium persulfate, and tetramethylenediamine in 100 mL aqueous solution. Samples A, B, and C are conditions in which the amount of monomer is reduced to 1/5 in A based on B, and the amount of initiator and polymerization accelerator is doubled in C based on B. From the experiment, it can be seen that T_2 increased significantly when the amount of cross-linking agent was reduced (from B to A). Also, when the amount of initiator and polymerization accelerator is doubled (from B to C), T_2 increased and gelation occurred more quickly. Thus, real-time observation of gelation can be obtained by measuring the time variation of T_2.

In addition, a pioneering study on molecular orientation in polymer networks under strain was conducted by ^{1}H NMR [117], and the cross-linking mechanism of poly(vinyl alcohol)/borax gel (slime) was studied by ^{11}B NMR [118].

7.5.2.2 Volume Phase Transition

The structure of uncharged and charged poly(*N*-isopropylacrylamide) (PNIPAM) gels has been investigated near the volume phase transition temperature based on the line width and relaxation time variation of ^{1}H NMR chemical shifts [119]. In this study, it was found that (i) a broad peak corresponding to the shrunken phase appears near the volume phase transition temperature, while the sharp peak observed in the swelling phase disappears, (ii) the charged PNIPAM gel is composed of a hard component and a mobile component, while the uncharged PNIPAM gel is homogeneous, and (iii) the (iii) the methylene groups and methylene protons in the main chain do not change significantly before and after the volume phase transition and remain mobile across the transition, while (iv) the mobility of the methyl groups in the *N*-isopropylmethyl group decreases significantly as it enters the collapsed

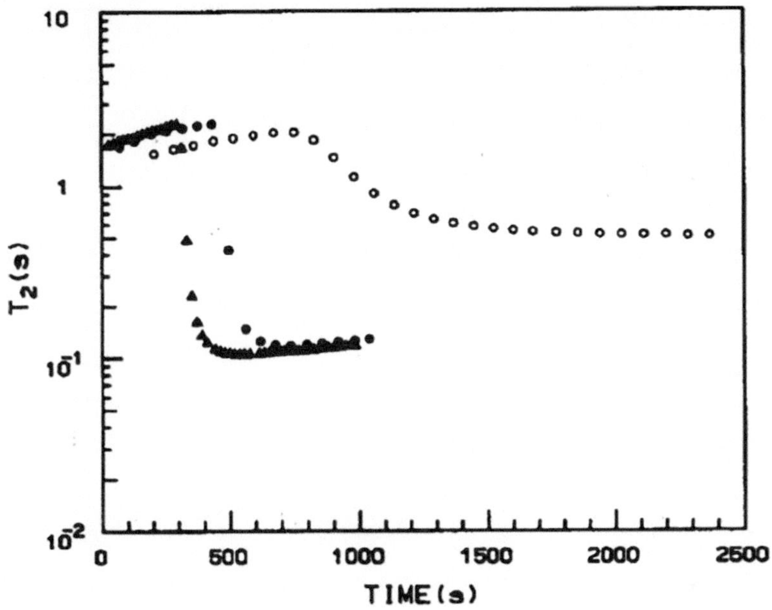

Fig. 7.26 Time variation of T_2 in NMR time-resolved experiments of the gelation process of acrylamide [116]. AA, BAA, APS, and TEMED are the amounts of acrylamide (monomer), bisacrylamide (cross-linking agent), ammonium persulfate, and tetramethylenediamine in 100 mL of water solution. The conditions for A are (AA, BAA, APS, TEMED) = (6 g, 0.2 g, 20 mg, 80 mL); for B, BAA is five times the amount of A (6 g, 1 g, 20 mg, 80 mL); and for C, APS and TEMED are twice the amount of B (6 g, 0.2 g, 40 mg, 160 mL) (Tanaka et al. [116]; reproduced with permission, © American Institute of Physics)

phase due to the volume phase transition. (iv) The methyl group in the *N*-isopropyl group shows a significant decrease in mobility as it enters the collapsed phase due to the volume phase transition. Figure 7.27 shows a study of the temperature dependence of PNIPAM aqueous solution [120]. The T_2 of side chains and main chain ^{1}H in a PNIPAM aqueous solution changes significantly across the lower critical solution temperature (LCST) of PNIPAM, indicating that the T_2 of side chain ^{1}H increases (increase in mobility) and that of main chain [1] H decreases (suppression in mobility).

7.5.2.3 NMR Imaging

Nuclear magnetic resonance imaging (MRI) has rapidly developed as a diagnostic technique for observing the distribution of water and fat in living organisms, especially as tomographic images. NMR imaging was actively studied in the 1990s [121]. ^{1}H NMR imaging has been used to image the ^{1}H spin density and spin–spin relaxation time of polymethacrylate (PMMA) hydrogels, which are caused by strain when

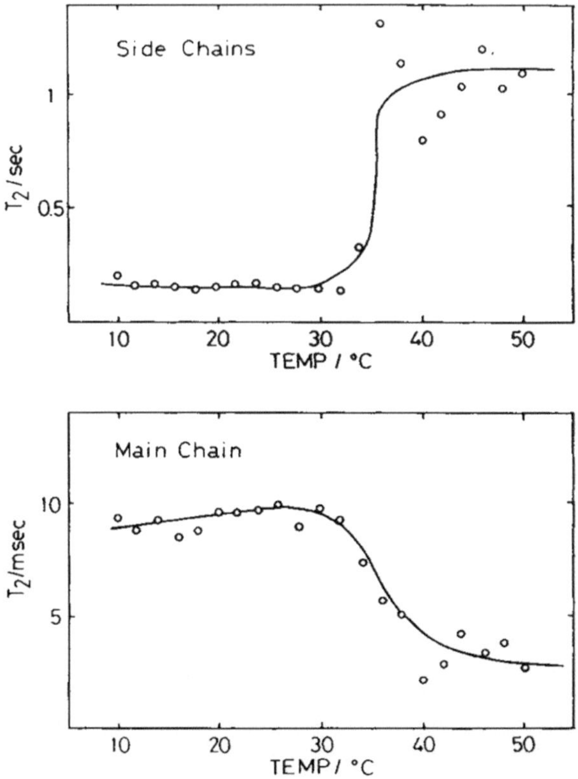

Fig. 7.27 Temperature dependence of T_2 of side chain ^{1}H and main chain ^{1}H of poly(*N*-isopropylacrylamide) (PNIPA) aqueous solution [120] (Ohta et al. [120]; reproduced with permission, © Wiley)

stress is applied to the hydrogel [122]. Fig. 7.28 shows a ^{1}H spin density distribution diagram of a disk-shaped PMMA hydrogel placed in a glass cell and compressed by a glass piston with the center axis shifted. The differences in color indicate the stress distribution applied to the gel, with darker colors indicating more stress and compression. When the same experiment was performed in heavy water, no NMR signal was obtained. Thus, the ^{1}H NMR experiment does not give a polymer signal due to the broadening of the dipole, but does give a water signal trapped in the gel.

7.5.2.4 Inhomogeneity

A series of studies have been conducted on gel inhomogeneity by NMR using small molecule compounds and polymers as probes. The inhomogeneity of polystyrene and polyethoxyacrylate gels in dimethylformamide (DMF) has been studied by time-dependent diffusion NMR using amino acids as probes. In this study, it was reported that in the short time region, the diffusion of amino acids has two modes reflecting the inhomogeneity of the gel network, but in the long-time region, it becomes a single mode because the inhomogeneity is averaged out [123]. The dependence of

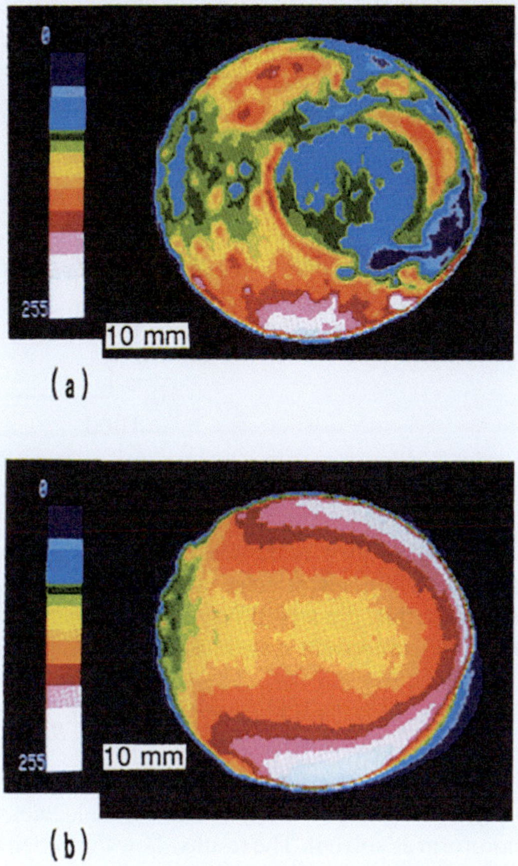

Fig. 7.28 ^{1}H-NMR imaging of a polymethacrylate (PMMA) hydrogel under stress [122]. **a** and **b** show the [1]H-spin density distribution at 1.0 mm and 1.9 mm from the top of the gel, respectively. (Yasunaga et al., 1992; reproduced with permission, © American Chemical Society)

inhomogeneity of polyacrylic acid gels on cross-link density and degree of swelling has also been investigated based on changes in the diffusion coefficient of probes [124].

The PS molecular weight dependence of the diffusion behavior of PS in polymethyl methacrylate (PMMA) gels with deuterated chloroform as a solvent has been studied [125]. The diffusion behavior of the probe PS was investigated by time-dependent diffusion ^{1}H NMR in the presence of PS molecules with molecular weights ranging from 4 to 400k. The following relationship exists between NMR echo signal intensity and magnetic field gradient parameters.

$$\ln\left[\frac{A(g)}{A(0)}\right] = -\gamma^2 g^2 D\delta^2\left(\Delta - \frac{\delta}{3}\right) \tag{7.46}$$

where g is the magnetic field gradient intensity, $A(g)$ is the echo signal intensity at $t = 2\tau$, $A(0)$ is the echo signal intensity when no magnetic field is applied, τ is

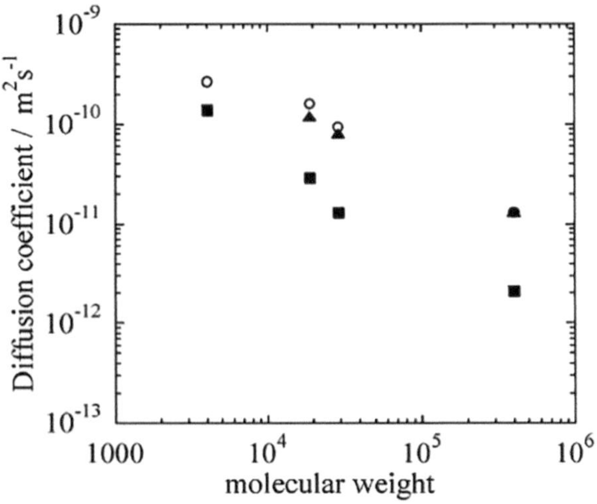

Fig. 7.29 Diffusion coefficient of probe PS molecules in PMMA gel with deuterated chloroform as solvent as a function of probe PS molecular weight [125] (Kamiguchi et al. [125]; reproduced with permission, © American Chemical Society)

the pulse interval, γ is the magnetic rotation ratio in ^{1}H, δ is the duration of the magnetic field gradient pulse, and Δ is the gradient pulse interval called diffusion time. From this, the self-diffusion coefficient D of the probe molecule is evaluated. For a multicomponent system, Eq. (7.46) can be generalized to

$$\frac{A(g)}{A(0)} = \sum_i f_i \exp\left[-\gamma^2 g^2 D_i \delta^2 \left(\Delta - \frac{\delta}{3}\right)\right] \tag{7.47}$$

Equation (7.47) was applied to PS molecules in a PMMA gel with deuterated chloroform as solvent. The results show that when the molecular weight of the probe PS molecule is small (4k: the radius of gyration $R_H \approx 1.5$ nm), the system can be treated as a single-component system, but as the molecular weight increases, the system gradually becomes a two-component system, consisting of a fast and slow components. Figure 7.29 plots the diffusion coefficients of the fast and slow components versus the molecular weight of the probe PS molecules. As shown in the figure, when the molecular weight is small enough, the diffusion coefficient of the fast component is almost equal to that of the pure solvent (white circles), but when the molecular weight is large, the behavior of the system becomes two-component.

7.5.2.5 Bond Inhomogeneity

Recently, an NMR method, multiple quantum NMR, has been proposed that is very suitable for studying the motion of polymer chains in elastomers and entanglement in polymer melts [126]. In these materials, anisotropy exists on a quasi-local scale about the entanglement point. In other words, topological constraints such as cross-linking suppress fast motions such as Rouse modes. As a result, correlations with

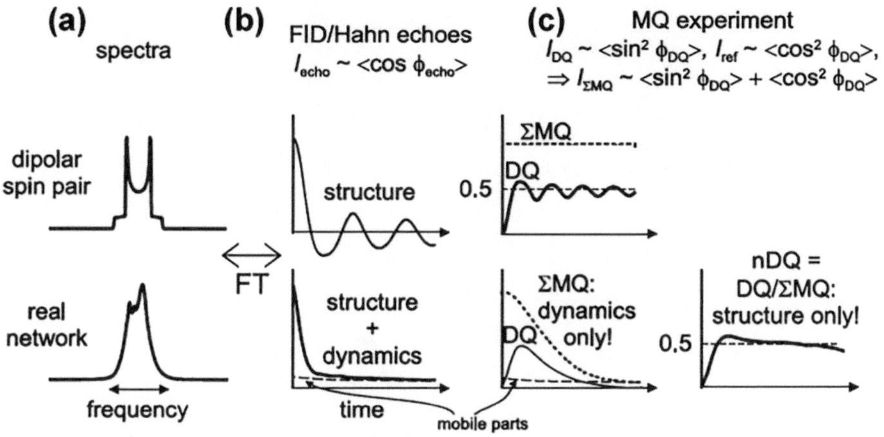

Fig. 7.30 Comparison of multiple quantum NMR and magnetic field gradient NMR methods [126] (Saalwächter et al. [126]; open access, Elsevier)

finite relaxation times associated with these constraints appear. The signal intensity is directly related to the cross-link density and the entanglement length, which can be studied in detail to obtain information on the local structure of rubbers and gels. Figure 7.30 compares the principle of multiple quantum NMR with that of conventional magnetic field gradient NMR. In the case of magnetic field gradient NMR (b), the signal from a homogeneous object is observed as dipole spin pairs (upper left in (a)), and structural information is obtained from the Fourier transform. In the case of a polymer network such as rubber ((a), bottom left), the signal is broadened and the information becomes a mixture of structure and dynamics. In the case of multiple quantum NMR (c), the DQ can be evaluated separately from the dynamics and structural information. Here, DQ is the spin-pair double-quantum interference quantity, which represents the orientation correlations of the various pairs. This method, combined with Monte Carlo simulations, has been applied to a detailed analysis of the cross-linked structure of a tetra-PEG gel (Sect. 7.3.8.4) [127]. As a result, as shown in Fig. 7.31, the existence of non-ideal cross-link structures such as (a) single bonds, (b) double bonds, and (c) double-double and triple bonds has been reported.

7.6 Dynamic Light Scattering

The study of gels by dynamic light scattering (DLS) began in the early 1970s and is now an important method for the structural analysis and dynamics of gels [128]. Compared to rheology, which is a classical gelation analysis, dynamic light scattering is a non-destructive and non-contact measurement method, and is therefore increasingly being used for development research and quality control as well as basic research as a simple and important method. The features of dynamic light scattering

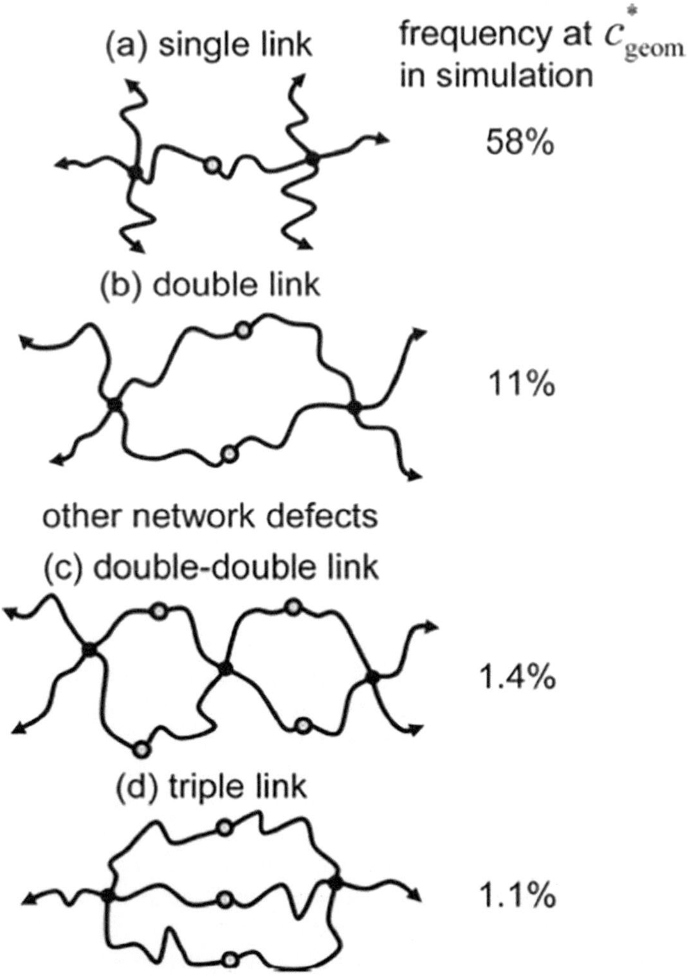

Fig. 7.31 Non-ideal cross-linked structure of tetra-PEG gels: **a** single bond, **b** double bond, **c** double-dual or triple bond [127] (Lange et al. [127]; reproduced with permission, © American Chemical Society)

include: (i) it is a non-contact measurement using light and does not damage the sample, (ii) the measurement time is short, taking only a few tens of seconds, (iii) the range of sample sizes that can be measured is very wide (1 nm to 5 μm), and (iv) only the refractive index and viscosity of the solvent are needed for data analysis, with no other external constants.

In dynamic light scattering measurements of gels, the structure of the gel is examined using light, a probe with a very high frequency, so depending on the type of gel, the gel appears to be either liquid or solid. Another characteristic of gels is that the molecules form an infinite network structure, so the network itself cannot move in translation and is in a state of violent oscillation. This is called "collective diffusion" or "cooperative diffusion" of gels (networks), and was theorized in detail by Tanaka et al. [14]. On the other hand, dynamic light scattering measurements during the gelation process provide detailed information on the gelation process and are used to determine the gelation point [128].

7.6.1 Gel Structure Analysis and Gelation Point Evaluation

As discussed in Sect. 5.2, Tanaka et al. treated gel as a continuum, considered the relationship between the strain tensor and stress tensor acting on the mesh elements of gel, and proposed the following diffusion equation for strain in gel [14].

$$\frac{\partial u}{\partial t} = \frac{K + 4\mu/3}{f} \frac{\partial^2 u}{\partial z^2} \tag{7.48}$$

where u is the displacement, t is the time, K is the bulk modulus, μ is the shear modulus, and f is the friction coefficient. The direction of displacement is along the z-axis of the laboratory coordinate system, and the coefficient on the right-hand side is the diffusion coefficient

$$D = \frac{K + 4\mu/3}{f} \tag{7.49}$$

Considering that the displacement of the network element is caused by an external electric field with an electric field vector along the z-axis and that the scattered light electric field is proportional to the magnitude of the concentration fluctuation, i.e., displacement vector, the correlation function of the scattered light electric field from the scattering elements of the gel is given by

$$\langle E^*(q,t)E(q,0)\rangle \sim \frac{k_B T}{K + 4\mu/3} \exp\left[-\frac{(K + 4\mu/3)q^2 t}{f}\right] \tag{7.50}$$

where the scattering vector $\mathbf{q}$, i.e., the direction of observation, is $\mathbf{q} = (0, 0, q)$ along the z-axis (longitudinal mode scattering). If the scattered electric field is observed in the x- or y-axis (transverse mode scattering), depolarized scattering is observed, and the correlation function is given by

$$\left\langle E^*_{\mathrm{dep}}(q,t)E_{\mathrm{dep}}(q,0)\right\rangle \sim \frac{k_B T}{\mu} \exp\left(-\frac{\mu q^2 t}{f}\right) \tag{7.51}$$

Equations (7.50) and (7.51) show that gel diffusion can be expressed as a single exponential function.

From the dynamic light scattering measurement of the gel, the correlation length ξ is obtained as the length corresponding to the hydrodynamic radius R_H of the dispersed system. According to Tanaka, D is obtained as [129]

$$D = \frac{\int_0^\infty g(\mathbf{r})[k_B T/(6\pi\eta\mathbf{r})]\mathrm{d}\mathbf{r}}{\int_0^\infty g(\mathbf{r})\mathrm{d}\mathbf{r}} \tag{7.52}$$

This equation indicates that the cooperative diffusion coefficient is the average of $k_B T/(6\pi\eta r)$ with the spatial correlation function $g(r)$ of the polymer segments as a weight function, where η is the solvent viscosity.

On the other hand, the spatial correlation function for a quasi-dilute solution of a polymer in a good solvent is given by the number density of monomers as c and the blob size as ξ [9]

$$g(r) = c\frac{\xi}{r}\exp\left(-\frac{r}{\xi}\right) \tag{7.53}$$

Thus, from Eqs. (7.52) and (7.53), we have

$$D = \frac{k_B T}{6\pi\eta\xi} \tag{7.54}$$

According to de Gennes' C^* theorem, ξ is equal for a quasi-dilute solution of a polymer and for an equilibrium swollen gel of the same concentration [9]. Hence, ξ is understood as the distance between adjacent cross-linking points in the network, i.e., the mesh size of the gel. However, it should be noted that this is true only for gels in good solvents. ξ should be taken as the correlation length of the concentration fluctuation, not as the gel mesh size. In any case, the cooperative diffusion theory makes a significant contribution to the evaluation of the diffusion coefficient of gels and to the evaluation of the correlation length (mesh size) of gels.

7.6.2 Non-ergodicity and Static/Dynamic Component Separation of Scattered Light Intensity

When a laser beam is shined on a white paper, a shimmering speckle appears. This is called speckle, which is caused by interference of reflected light on the surface of the paper, resulting in modulation of intensity in the reflected light. Speckles occur not only in reflected light but also in scattered light. Figure 7.32 shows the results of the measurement of scattered light from (a) a polymer solution and (b) a polymer gel while shifting the measurement position. In the case of the polymer solution, the scattered light intensity does not change significantly when the measurement position is scanned, whereas in the case of the gel, there is a marked dependence of the scattered light intensity on the measurement position. This is due to the fact that the gel is a non-ergodic system in which the equivalence between the time average and the ensemble average (position average) does not hold [64].

$$\langle I\rangle_{\mathrm{E}} \neq \langle I\rangle_T \tag{7.55}$$

Therefore, to describe the scattered light intensity from a gel, it is necessary to define two types of intensity: the time-averaged scattered light intensity $\langle I\rangle_T$ and the ensemble-averaged scattered light intensity $\langle I\rangle_E$. $\langle I\rangle_T$ is the long-time average of the scattered light intensity at a measurement point, and $\langle I\rangle_E$ is the ensemble average of the scattered light intensity at many measurement points. Hence,

$$\langle I\rangle_{\mathrm{E}} \equiv \left\langle I_p(\tau)\right\rangle_{\mathrm{E}} = \frac{1}{N}\sum_{p}^{N}\langle I_p\rangle_T \tag{7.56}$$

where p is the sample position and N is the number of measurement points.

Non-ergodic systems are found not only in gels, but also in glassy materials and other systems in which fluctuations of the refractive index (or polarizability) in the material are "frozen." Just as there are time-averaged and ensemble-averaged scattered light intensities, there are time-averaged and ensemble-averaged correlation functions, which are defined by the following equations.

$$g_{T,p}^{(2)}(\tau) \equiv \frac{\left\langle I_p(0)I_p(\tau)\right\rangle_T}{\left\langle I_p(0)\right\rangle_T^2} \tag{7.57}$$

$$g_{\mathrm{E}}^{(2)}(\tau) \equiv \frac{\left\langle\left\langle I_p(0)I_p(\tau)\right\rangle_T\right\rangle_E}{\left\langle\left\langle I_p(0)\right\rangle_T^2\right\rangle_E} \tag{7.58}$$

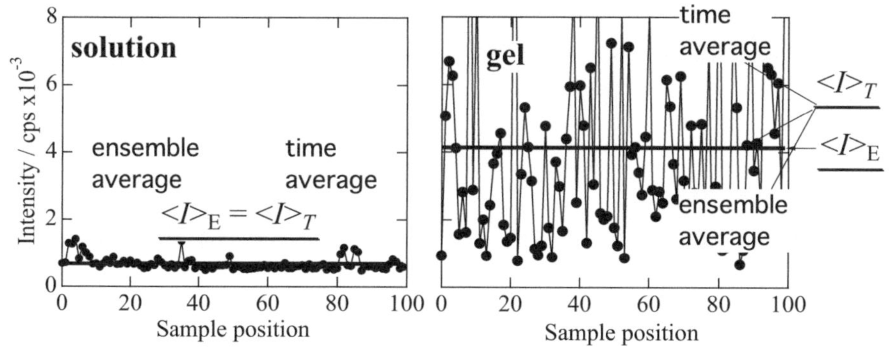

Fig. 7.32 Difference in scattering between polymer solution (**a**) and polymer gel (**b**). In the polymer solution, the scattering intensity is almost the same regardless of the measurement position, whereas the scattering from the gel depends strongly on the measurement position

In a gel, the scattering elements can be considered to have two parts: one that is mobile, such as a polymer chain segment in a polymer solution, and one that is immobile with respect to time, such as a cross-linking point or a region with high cross-linking density. In this case, the scattered electric field can be considered to consist of two contributions, a dynamic component and a static component. The static component is considered to depend on the sample position p because of the spatial distribution of the cross-link density.

$$E_p(t) = E(t)_{\mathrm{F}} + E_{C,p} \tag{7.59}$$

The autocorrelation functions of the dynamic and static components correspond to the scattered light intensities I_{F} and I_{C}, respectively. Since the autocorrelation functions of the dynamic and static components correspond to the scattered light intensities I_{F} and I_{C}, respectively, the following equations can be written

$$\left\langle E^*(t)_{\mathrm{F}} E(t)_{\mathrm{F}} \right\rangle_T = \left\langle E^*(t)_{\mathrm{F}} E(t)_{\mathrm{F}} \right\rangle_{\mathrm{E}} \equiv I_{\mathrm{F}} \tag{7.60}$$

$$\left\langle E^*_{\mathrm{C},p} E_{\mathrm{C},p} \right\rangle_T \equiv I_{\mathrm{C},p}, \quad \left\langle I_{C,p} \right\rangle_{\mathrm{E}} \equiv I_{\mathrm{C}} \tag{7.61}$$

Here, the superscript asterisk denotes complex conjugate. The correlation function $g_{\mathrm{F}}{}^{(1)}(q, \tau)$ of the electric field for the dynamic component is

$$g_{\mathrm{F}}^{(1)}(q,\tau) \equiv \frac{\left\langle E^*_{\mathrm{F}}(q,0) E_{\mathrm{F}}(q,\tau) \right\rangle_{\mathrm{E}}}{\langle I_{\mathrm{F}}(q,0) \rangle_{\mathrm{E}}^2} = \frac{\left\langle E^*_{\mathrm{F}}(q,0) E_{\mathrm{F}}(q,\tau) \right\rangle_T}{\langle I_{\mathrm{F}}(q,0) \rangle_T^2} \tag{7.62}$$

On the other hand, for the correlation function of scattered light intensity is given by [130]

$$\left\langle g_p^{(2)}(q,\tau) \right\rangle_T - 1 = X_p^2 \left| g_{\mathrm{F}}^{(1)}(q,\tau) \right|^2 + 2X_p(1 - X_p) \left| g_{\mathrm{F}}^{(1)}(q,\tau) \right| \tag{7.63}$$

where X_{p} is the ratio of the dynamic component of the scattered light intensity to the total scattered light intensity as is given by

$$X_p = \frac{I_{\mathrm{F}}}{\langle I \rangle_{T,p}} \tag{7.64}$$

Unlike the heterodyne method ($I_{\mathrm{F}} \ll I_{\mathrm{C}} = I_{\mathrm{LO}}$) described above, in the case of gels, $I_{\mathrm{F}} \approx I_{\mathrm{C}}$ in many cases, so it is called the partial heterodyne method. The partial heterodyne scattering from the gel is illustrated in Fig. 7.33. The interference of scattering between moving parts results in scattering that can be observed by the homodyne method, but if one of the scatterers does not vary with time (e.g., cross-linking points), it becomes a local oscillator, and the scattered light becomes

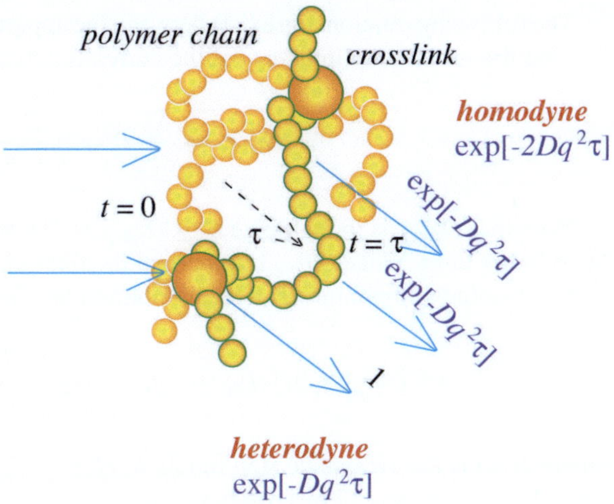

Fig. 7.33 Schematic diagram showing that the scattering from the gel is composed of homodyne and heterodyne components

scattering that can be observed by the heterodyne method. Thus, the scattering from the entire system is the sum of these two types of scattering.

Now, defining ${\sigma_{\mathrm{I},p}}^2$ as the initial amplitude of $\left\langle g_p^{(2)}(q,\tau)\right\rangle_T$, and assuming $\tau = 0$ in Eq. (7.63), we obtain

$$\sigma_{\mathrm{I},p}^2 \equiv g_{T,p}^{(2)}(q,0) = X_p(2 - X_p) \tag{7.65}$$

For ergodic systems, ${\sigma_{\mathrm{I},p}}^2 = 1$ because $X_p = 1$, while for non-ergodic systems, ${\sigma_{\mathrm{I},p}}^2 < 1$ because $0 < X_p < 1$, and it is position-dependent. On the other hand, ${g_{\mathrm{F}}}^{(1)}(q,\tau)$ is, from cooperative diffusion theory (Eq. (7.50)), a single relaxation function is given by

$$g_{\mathrm{F}}^{(1)}(q,\tau) = \exp(-Dq^2\tau) \tag{7.66}$$

In the analysis of the measured data, the correlation function $\left\langle g_p^{(2)}(q,\tau)\right\rangle_T$ of the measured scattered light intensity can be fitted directly using Eq. (7.63), but since the difference in the exponential terms, i.e., D and $2D$, has little effect on the correlation function, $\left\langle g_p^{(2)}(q,\tau)\right\rangle_T$ can be approximated to

$$\left\langle g_p^{(2)}(q,\tau)\right\rangle_T - 1 = \sigma_{\mathrm{I},p}^2 \exp(-2D_{\mathrm{app},p}q^2\tau) \tag{7.67}$$

and the apparent diffusion coefficient $D_{\mathrm{app},p}$ is obtained from the limit of $\tau = 0$ in this equation as

$$D_{\mathrm{app},p} = -\frac{1}{2q^2}\lim_{\tau\to 0}\frac{\mathrm{d}}{\mathrm{d}\tau}\ln\left\{\left\langle g_p^{(2)}(q,\tau)\right\rangle_T - 1\right\} \tag{7.68}$$

The following relationship exists between the apparent diffusion coefficient $D_{\mathrm{app},p}$, X_p, and the initial amplitude $\sigma_{\mathrm{I},p}$ of the correlation function,

$$D_{\mathrm{app},p} = \frac{X_p}{\sigma_{\mathrm{I},p}^2} D = \frac{D}{2 - X_p} \tag{7.69}$$

Note that $D_{\mathrm{app},p}$ is in the range $D \le D_{\mathrm{app},p} \le 2D$, and the scattered light intensity has a lower limit. Since $\langle I \rangle_{T,p}$ is a thermal fluctuation, its lower limit is proportional to the absolute temperature T. The distribution of $\langle I \rangle_{T,p}$ is given by

$$P\left(\langle I \rangle_{T,p}\right) \approx H\left[\langle I \rangle_{\mathrm{T},p} - \langle I_{\mathrm{F}} \rangle_T\right] \exp\left(-\frac{\langle I_{\mathrm{F}} \rangle_{T,p} - \langle I_{\mathrm{F}} \rangle_T}{\langle I \rangle_{\mathrm{E}} - \langle I_{\mathrm{F}} \rangle_T}\right) \tag{7.70}$$

where $H(x)$ is the Heaviside step function, $H(x) = 1$ $(x \ge 0)$ and $H(x) = 0$ $(x < 0)$. From Eqs. (7.64) and (7.69), we have [64]

$$\frac{\langle I \rangle_{T,p}}{D_{\mathrm{app},p}} = \frac{2}{D} \langle I \rangle_{T,p} - \frac{\langle I_{\mathrm{F}} \rangle_T}{D} \tag{7.71}$$

This equation can be used to separate the static and dynamic components of the scattered light intensity (static/dynamic component separation) and to evaluate the cooperative diffusion coefficient. Next, we will show how this relationship can be used to separate the static and dynamic components of the scattered light intensity and dynamics from the gel.

7.6.3 Static/Dynamic Component Separation of Scattered Light Intensity and Evaluation of Cooperative Diffusion Coefficient

Figure 7.34 illustrates how to evaluate the key parameters that characterize the dynamics of a gel from dynamic light scattering measurements. First, dynamic light scattering measurements are performed on each speckle point of the gel while scanning the measurement position (Fig. 7.34a), and a position-dependent correlation function is obtained (Fig. 7.34b). By curve fitting this with a single exponential relaxation function and plotting the obtained apparent diffusion coefficient D_{app} against the scattered light intensity according to Eq. (7.71), the dynamic component of the scattered light intensity, the true cooperative diffusion coefficient D, and the ensemble average scattered light intensity can be obtained (Fig. 7.34d). This is called the static/dynamic separation of the scattered light intensity. Figure 7.34c shows that, according to Eq. (7.70), is an exponential function with a lower limit.

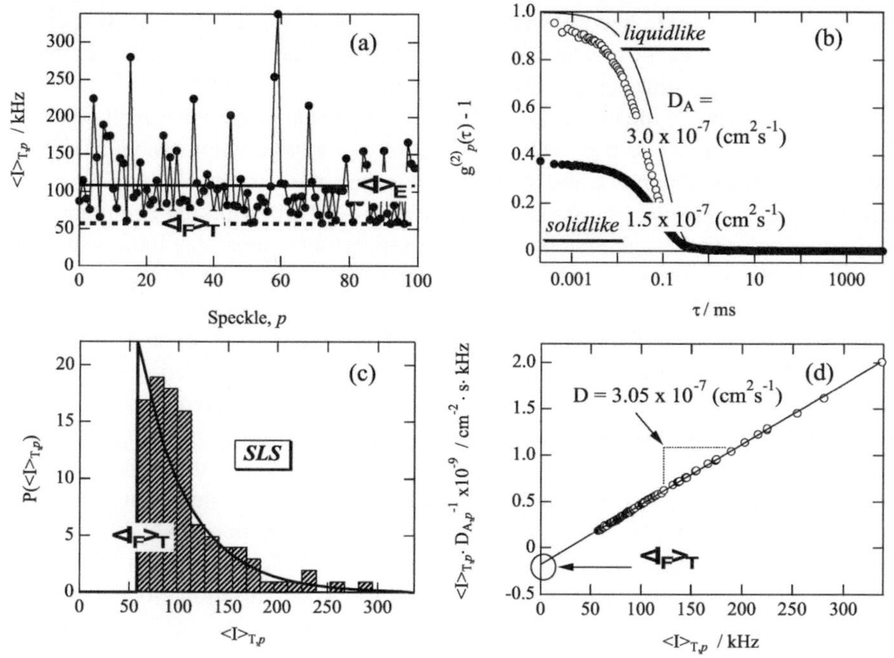

Fig. 7.34 Methods to separate and extract dynamic and static components and diffusion coefficients from a dynamic light scattering experiment [128]

7.6.4 *Gelation Analysis by Light Scattering*

Controlling gelation is a critical issue in synthetic processes involving cross-linking reactions. The gelation point is usually determined by visual observation of the flow of the sample during the gelation process or by measuring the viscosity from time to time. Another classical method is the characterization of precursors extracted from the reaction system (molecular weight measurement, end group determination, etc.) (precursor method). The ESR method, which relies on radical capture, and NMR and IR methods based on the quantification of chemical species are methods to estimate the reaction rate by quantifying the concentration of radicals and chemical species of interest. However, neither of these methods do not directly determine the gelation point. Static light scattering and small-angle X-ray and neutron scattering are not suitable for real-time measurements because the size and structure of individual clusters cannot be determined without dilution of the system. On the other hand, time-resolved dynamic light scattering has been proposed as a non-contact, non-destructive, "in situ" method for observing the dynamics of cross-linking reaction systems, and has been demonstrated in many systems. An overview of the method is given below.

The monomer solution scatters little light because the scatterers are small. However, as the polymerization reaction proceeds, the molecular weight increases and the molecular size becomes larger, which gradually scatters light. If a cross-linking agent is present together with the monomer, the polymerization is accelerated by the cross-linking reaction, and eventually a gel with infinite molecular weight is formed. On the other hand, in a polymerization system in which no cross-linking agent is added, chain extension reactions (polymerization reactions) occur, but molecular development and growth throughout the system by branching, etc. do not occur. Therefore, the presence or absence of a cross-linking agent causes a large difference in the scattered bodies formed (polymer chains or polymer clusters). Using this phenomenon, the gelation point in a gelation reaction system can be determined contactlessly and non-destructively by simply monitoring the scattered light intensity $I(t_p)$ (t_p: reaction time) from the reaction system [131]. In other words, the gelation process can be determined in detail by using time-resolved scattered light intensity and time-resolved dynamic light scattering methods. Interestingly, once the gelation point is passed in the reaction system, the scattering intensity decreases because the gel (infinite network) occupies the whole or most of the system. In other words, the scattering is no longer from a dispersed cluster system, but from local concentration fluctuations in a uniform medium of infinite networks. This process corresponds to the gas–liquid transition of a small molecule one-component system, and is analogous to the critical opalescence during condensation from a gas to a liquid.

7.6.5 Cluster Distribution Analysis of Gels and Time-Resolved Dynamic Light Scattering

We have discussed that the presence or absence of cross-links shows a significant difference in the change in scattered light intensity during the polymerization process. In this session, we show that the effect of gelation also appears in the correlation function itself and in the distribution function obtained by inverse Laplace transforming the correlation function. Figure 7.35 shows (a) the time-correlation function $g^{(2)}(\tau) - 1$ and (b) their distribution functions $P(\Gamma^{-1})$ [132]. Here, Γ^{-1} is the characteristic relaxation time of the system. From the figure, it can be seen that $g^{(2)}(\tau) - 1$ is a function with a certain characteristic relaxation time. This characteristic relaxation time shifts toward the long side as c increases. However, in the case of this system, the relaxation time shifts to the short time side significantly above $c \approx 100$ mM. At the same time, the initial amplitude of the correlation function tends to decrease from 1.

The distribution function $P(\Gamma^{-1})$ in Fig. 7.35b shows that as c increases, the peak broadens and shifts toward the long-time side until $c \approx 100$ mM, after which the peak shifts toward the short time side and narrows. This indicates that the formation of finite clusters (nanogel formation) occurs at lower concentrations and gel (bulk gel) formation occurs at higher concentrations, with $c \approx 100$ mM as the boundary. Furthermore, this boundary is the critical gelation concentration, at which very large

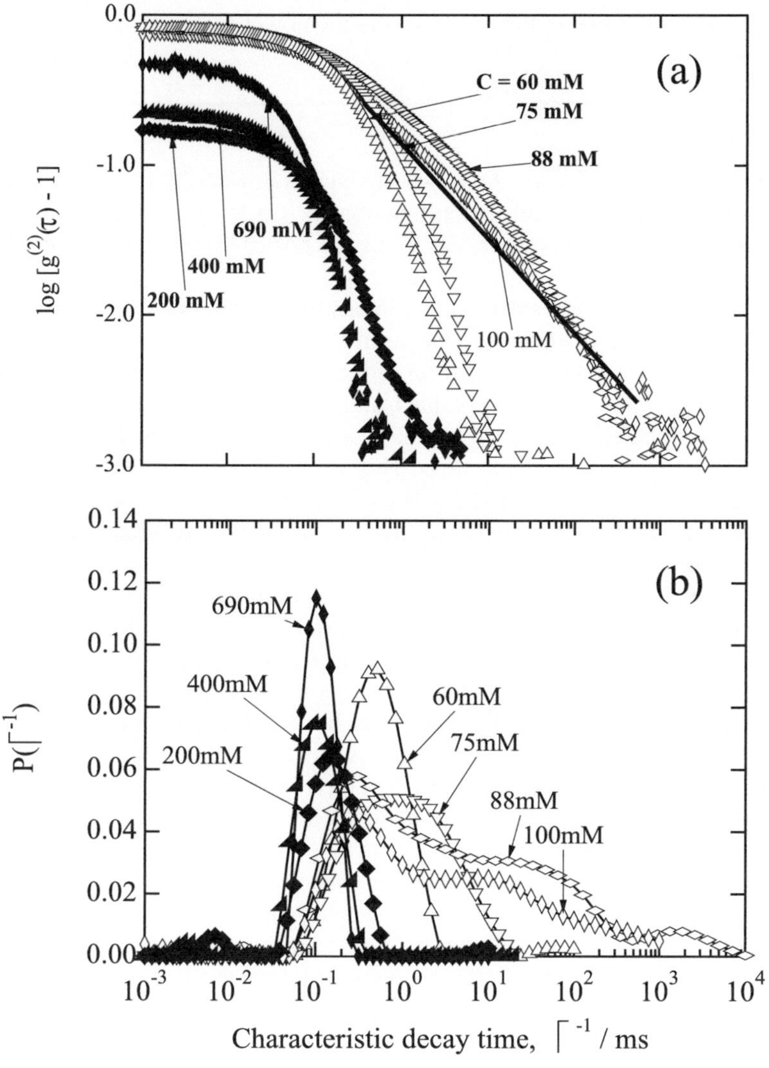

Fig. 7.35 Dynamic light scattering results of *N*-isopropylacrylamide (NIPA) polymerization systems prepared at various monomer concentrations with a constant monomer/cross-linker ratio. **a** Correlation function, **b** distribution function [132] (Norisuye et al. [132]; reproduced with permission, © American Chemical Society)

clusters are formed. According to percolation theory, it is known that just below the gelation point, a small number of large clusters are continuously distributed among numerous smaller clusters. These clusters are self-similar, and the cluster size distribution follows a power law. The distribution curve for $c \approx 100$ mM in the figure shows just such a situation, indicating that clusters are distributed in a power-law manner with a lower limit of small clusters where Γ^{-1} corresponds to 10^{-1} ms

= 100 ns. The solid line in Fig. 7.35a shows the power function

$$g^{(2)}(\tau) - 1 = \sigma_{\mathrm{I}}^{2}\left\{A\exp(-Dq^{2}\tau) + \frac{1-A}{[1+(\tau/\tau^{*})]^{(D_{\mathrm{p}}-1)/2}}\right\} \tag{7.72}$$

The results are fitted using the following equation. where A is the fraction of the diffuse component in the correlation function ($0 \leq A \leq 1$), τ^* is the lower limit of the characteristic relaxation time, and D_{p} is the power exponent for the correlation function of the intensity.

Combining the time-resolved light intensity measurement method described above with the cluster distribution analysis of gels (dynamic light scattering method), a time-correlation function $g^{(2)}$ (q,τ; t_{p}) of the scattered light intensity that changes from moment to moment can be obtained, and the dynamics of the gelation and reaction processes can be determined by analyzing it in a non-contact, non-destructive manner. Here, t_{p} is the reaction time. This is called time-resolved dynamic light scattering, and it is applied not only to the tracking of gelation reactions but also to the evaluation of changes in dynamics during bulk polymerization without solvent and to the tracking of vitrification.

Real-time observation of $g^{(2)}$ (τ) in many gelation reaction systems reveals four characteristic behaviors at the gelation point. An example of the gelation process of tetramethoxysilane (TMOS) is shown in Fig. 7.36 [133]. At the gelation point, (a) the intensity of scattered light increases rapidly, (b) $g^{(2)}$ (τ) – 1 becomes a power function, (c) the distribution function $P(\Gamma^{-1})$ becomes broad, and (d) the initial amplitude of the correlation function, $g^{(2)}$ (0) – 1, becomes $\ll 1$ after the gelation. The fact that the gelation point can be determined non-destructively and without contact using these four methods is of great significance and highly applicable [131, 134]. The exponent D_{p} is less than 1, and in most cases takes a value between 0.3 and 0.73, from which the degree of gel branching, etc., can be estimated [135]. The power-law behavior at the gelation point and the broadening of the distribution function $P(\Gamma^{-1})$ correspond to a fractal-like (self-similar) cluster distribution, corresponding to the conventional percolation model [136]. In contrast, the decrease in $g^{(2)}$ (0) – 1 after the gelation point indicates that the system changed to a non-ergodic system by becoming a gel [137].

7.6.6 Probe Dynamic Light Scattering

Dynamic light scattering experiments in which a small amount of polystyrene particles, gold nanoparticles, or silica particles are mixed with a sample are used to investigate the viscosity and mobility of the sample. This is called probe dynamic light scattering. In general, by selecting probes that are monodisperse in size, have a large scattering capacity, and have no (or little) interaction with the sample (medium), the dynamics of the medium can be investigated in detail. Since the scattering intensity from the probe particles is much larger than that from the gel itself, the dynamics of

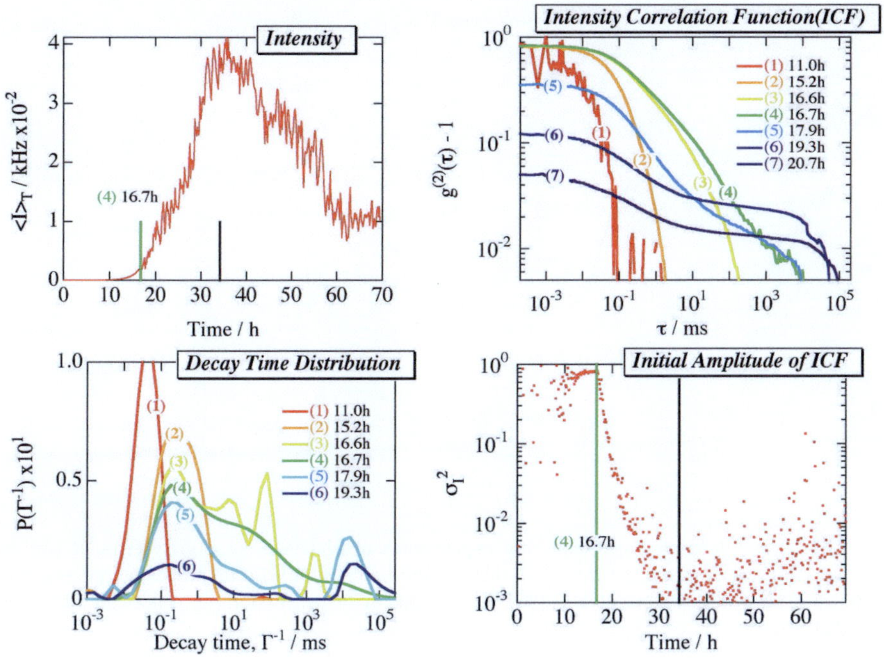

Fig. 7.36 Four methods to determine the gelation point by light scattering. **a** Scattering intensity, **b** correlation function, **c** distribution function, and **d** initial amplitude of the correlation function

the medium can be investigated in detail through the movement of the probe particles by performing a probe scattering experiment.

Figures 7.37 and 7.38 show examples of polystyrene (PS) microparticles probed into poly(*N*-isopropylacrylamide) (PNIPAM) solution (or gel) [138]. Two systems, with and without PS Latex (PS microparticle dispersion), were prepared during gel synthesis, and the dependence on polymer concentration c Fig. (7.37) and on cross-linker concentration c_{BIS} Fig. (7.38) were examined. From the DLS experiments, both PNIPAM and probe dynamics were observed in the with Latex system, and a comparison with the without Latex system shows that the diffusion of PS slows down as c increases. (long-time shift of the peak due to PS) can be seen as c increases. On the other hand, when the polymer concentration is increased to $c = 200$ mM (above the gelation point) and the c_{BIS} dependence is examined Fig. (7.38), PS motion is clearly observed at the beginning ($c_{\text{BIS}} = 0$ mM, i.e., in solution), but as cross-linking is introduced, the PS-derived peak does not change its position, becomes smaller and eventually, it disappears. This indicates that the PS particles are trapped in the mesh and become immobile due to the increase in mesh density. Thus, probe-diffuse dynamic light scattering allows us to observe the dynamics of both the probe and the medium.

In probe dynamic light scattering, it is sometimes desirable to observe only the dynamics of the probe. In such cases, the isorefractive index probe dynamic

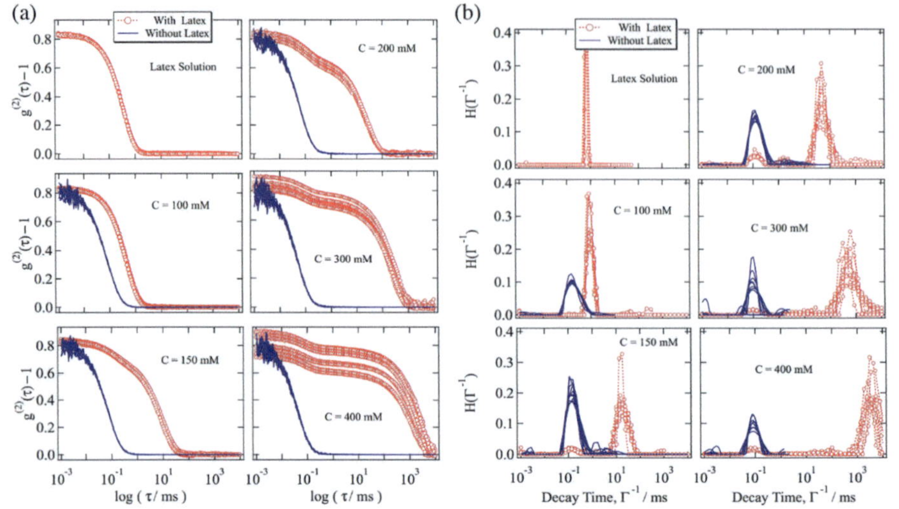

Fig. 7.37 Probe dynamic light scattering experiments of PNIPAM solution systems at various concentrations. **a** Correlation function, **b** distribution function [138]. The probes are polystyrene particles (latex) (Shibayama et al. [138]; reproduced with permission, © American Chemical Society)

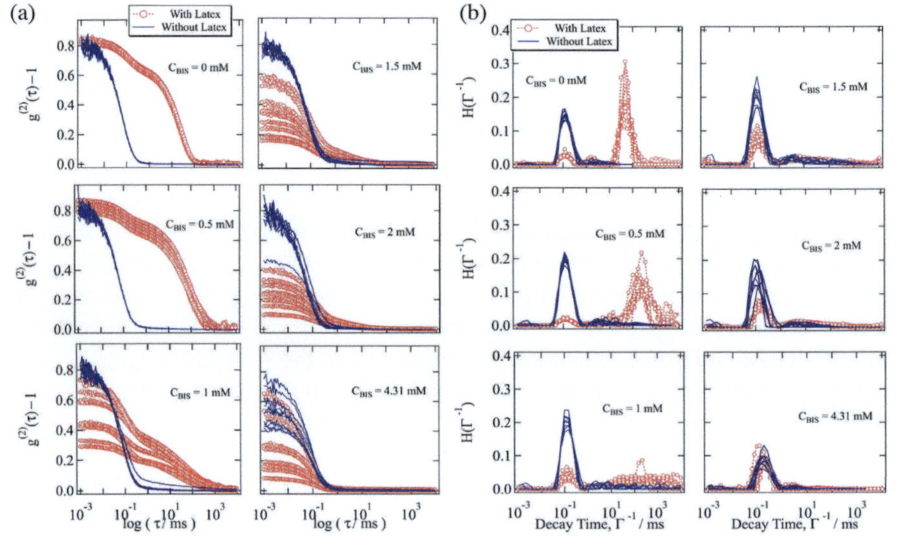

Fig. 7.38 Probe dynamic light scattering experiments of PNIPAM systems with various cross-linker concentrations. **a** Correlation function, **b** distribution function [138]. The probe is a polystyrene particle (latex) (Shibayama et al. [138]; reproduced with permission, © American Chemical Society)

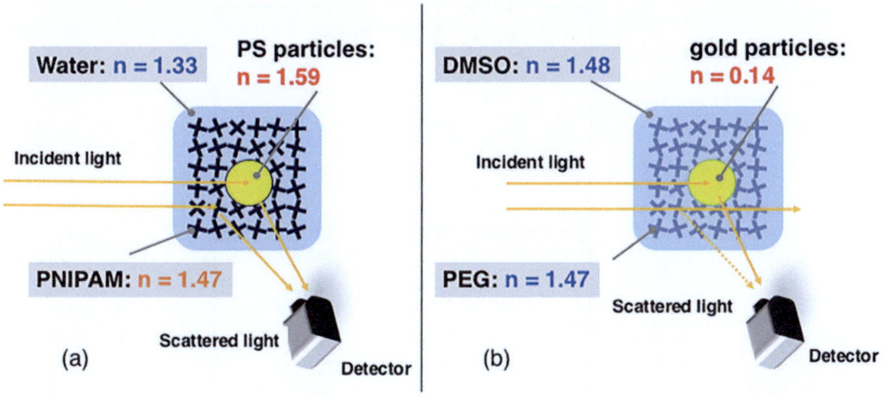

Fig. 7.39 **a** Probe dynamic light scattering and **b** refractive index matched dynamic light scattering [139] (Shibayama et al. [139]; reproduced with permission, © American Chemical Society)

light scattering method, in which the refractive indices of the media are equal, has been reported. Figure 7.39 shows a schematic representation of two types of probe dynamic light scattering methods: normal probe dynamic light scattering and isorefractive index probe dynamic light scattering [139]. (a) shows an example of using polystyrene particles (PS) as probes in a poly(*N*-isopropylacrylamide) (PNIPAM) solution (or gel). Since there is a significant difference in the refractive index between the solvent water, PNIPAM, and PS, the dynamics of both PNIPAM and the probe are observed in the DLS experiment. On the other hand, in case (b), by using DMSO, which has almost the same refractive index as that of polyethylene glycol (PEG), as the solvent and using gold nanoparticles as the probe, only the dynamics of the probe particles can be selectively studied [140].

Figure 7.40 shows an example using gold microparticles as probe particles and DMSO as solvent [140]. In this case, since the refractive indices of the polymer and solvent are almost equal, there is almost no scattering from the polymer, and only the dynamics of the gold microparticles can be observed. Hence, the gelation process can be observed through the dynamics of the gold particles, and the change in the "local viscosity" felt by the probe particles can be studied in detail. Interestingly, the figure shows that the probe particles maintain some mobility for a while after gelation. This confirms the presence of a sol component in the system for a while after gelation (infinity of the largest cluster).

7.6.7 *Microscopic Dynamic Light Scattering*

As mentioned at the beginning of this section, dynamic light scattering is not suitable for measuring opaque, turbid, or colored systems. This is because of the difficulty of measurement due to the lack of light penetration and multiple scattering. Various

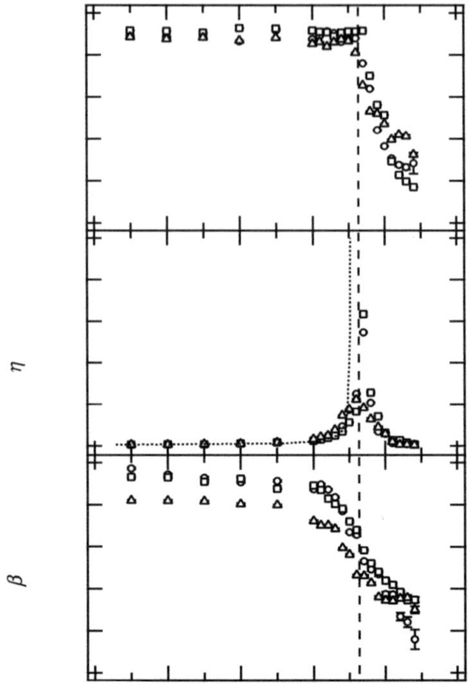

Fig. 7.40 Dynamics of tetra-PEG gels evaluated by refractive index matched dynamic light scattering. The dashed line represents the reaction rate at the gelation point. The dynamics of the probe particles can be evaluated even after gelation [140] (Li et al. [140]; reproduced with permission, © American Chemical Society)

methods have been devised to overcome this problem. Here, we introduce a microscopic dynamic light scattering method that combines 180-degree scattering and confocal measurement. The configuration is shown in Fig. 7.41 [141]. Using the microscopic dynamic light scattering system, it is possible to observe the dynamics of colloidal particles in milk (multiple scattering system) and colloidal particles in ink (strong light-absorbing medium system) without the need for dilution, even in systems that do not allow light to pass through at all, such as milk and ink, respectively.

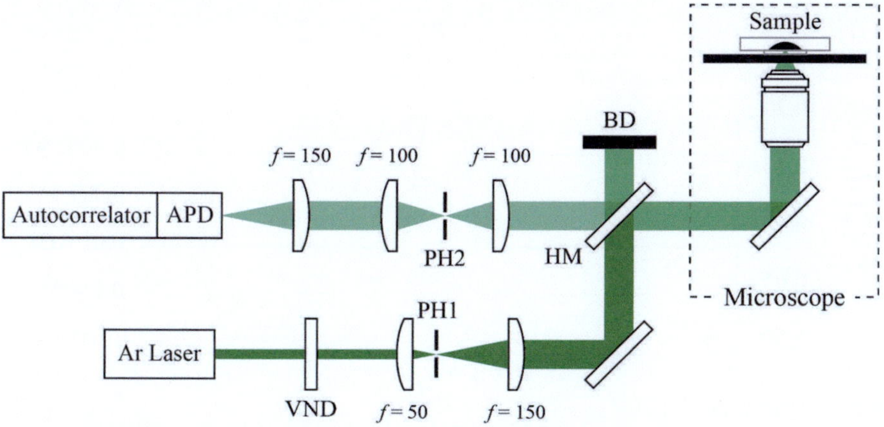

Fig. 7.41 Optics of the microscopic dynamic light scattering system [141] (Hiroi and Shibayama [141]; open access, Optica Pub. G.)

Figure 7.42 shows the dynamic scattering measurement results of (a) polystyrene latex and (b) India ink [141]. The solid and dashed lines indicate the measurement results obtained with a microscopic dynamic light scattering apparatus and a conventional dynamic light scattering apparatus, respectively. Taking (a) as an example, the ordinary dynamic light scattering system requires 100-fold (0.01 wt%) to 10,000-fold (0.0001 wt%) dilution of the sample for measurement, whereas the microscopic dynamic light scattering system can measure the undiluted solution (1 wt%) itself. In (b), the dynamic light scattering data can be obtained even from a 10 wt% undiluted solution of India ink, which is a completely black liquid that is unlikely to allow light to penetrate through it.

Another feature of the microscopic dynamic light scattering system is that it is a confocal system, so changes in the dynamics of particles in the depth direction from the liquid surface can also be investigated. Therefore, microscopic dynamic light scattering is expected to be further developed as a method to improve the shortcomings and limitations of conventional dynamic light scattering methods.

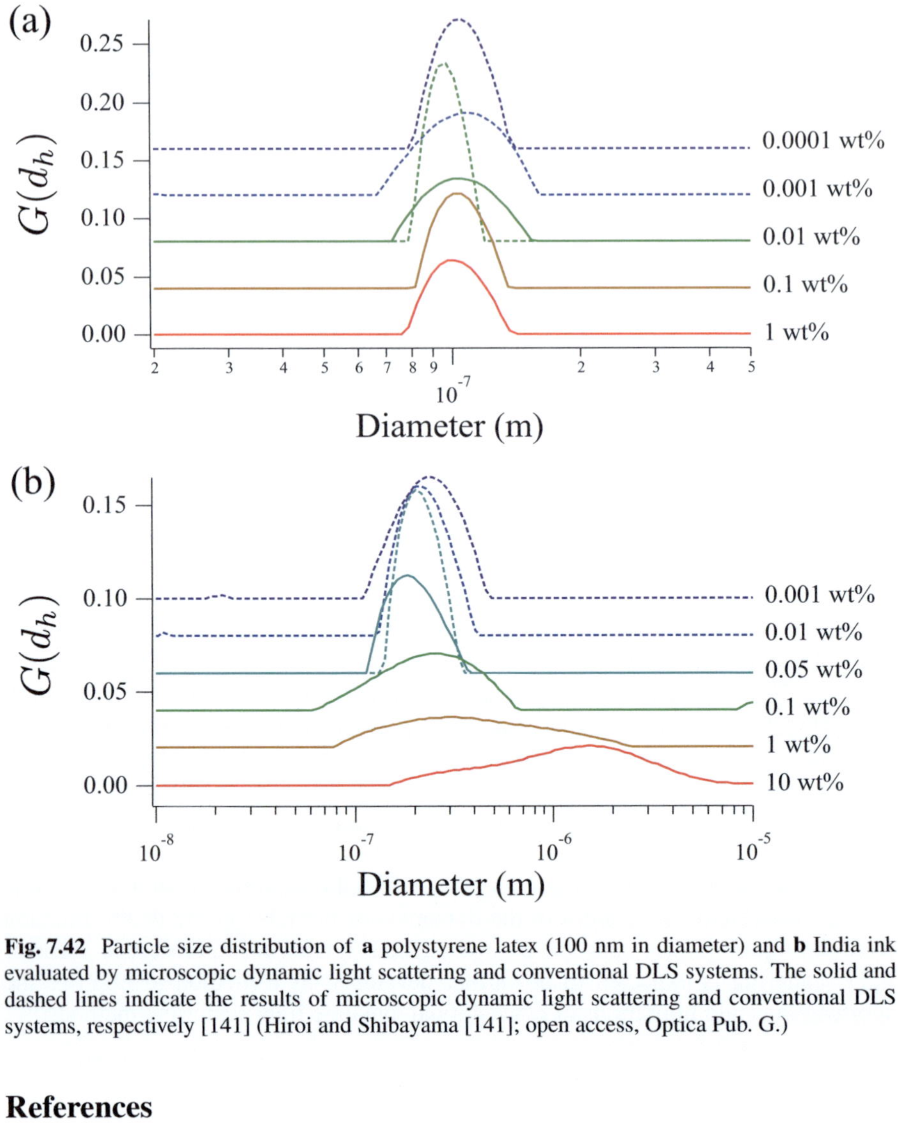

Fig. 7.42 Particle size distribution of **a** polystyrene latex (100 nm in diameter) and **b** India ink evaluated by microscopic dynamic light scattering and conventional DLS systems. The solid and dashed lines indicate the results of microscopic dynamic light scattering and conventional DLS systems, respectively [141] (Hiroi and Shibayama [141]; open access, Optica Pub. G.)

References

1. Y. Osada, K. Kajiwara, *Gels Handbook* (Academic Press, New York, 2000)
2. J.P. Cohen Addad, *Physical Properties of Polymer Gels* (Wiley, New York, 1996)
3. M. Shibayama, K. Kajiwara (Supervisors), *Recent Trends in Polymer Gels* (CMC Publishing, 2004) (in Japanese).
4. The Chemical Society of Japan (ed.), Amazing Soft Materials—Latest Functional Gel Research (CSJ Current Review) (Kagaku-dojin, 2010), p.192
5. M. Shibayama, T. Sato, T. Iwai, Y. Kimura, *Fundamentals and Application of Light Scattering Method* (Kodansha, 2014)
6. T. Hashimoto, *Principles and Applications of X-ray, Light, and Neutron Scattering* (Kodansha, 2017)

7. Y. Talmon, H.T. Davis, L.E. Scriven, E.L. Thomas, Cold-stage microscopy system for fast-frozen liquids. Rev. Sci. Instrum. **50**, 698–704 (1979)
8. M. Shibayama, Spatial inhomogeneity and dynamic fluctuations of polymer gels. Macromol. Chem. Phys. **199**, 1–30 (1998)
9. P.G. de Gennes, *Scaling Concepts in Polymer Physics* (Cornell University Press, Ithaca, 1979)
10. T. Tanaka, Collapse of gels and the critical endpoint. Phys. Rev. Lett. **40**, 820–823 (1978)
11. M. Shibayama, T. Tanaka, Volume phase transition and related phenomena of polymer gels. Adv. Polym. Sci. **109**, 1–62 (1993)
12. P.J. Flory, J. Rehner Jr., Statistical mechanics of cross-linked polymer networks II. Swelling. J. Chem. Phys. **11**, 521–526 (1943)
13. L.R.G. Treloar, *The Physics of Rubber Elasticity* (Clarendon Press, Oxford, 1975)
14. T. Tanaka, L.O. Hocker, G.B. Benedek, Spectrum of light scattered from a viscoelastic gel. J. Chem. Phys. **59**, 5151–5159 (1973)
15. A. Onuki, Theory of phase transition in polymer gels. Adv. Polym. Sci. **109**, 63–121 (1993)
16. J. Bastide, L. Leibler, Large-scale heterogeneities in randomly cross-linked networks. Macromolecules **21**, 2647–2649 (1988)
17. A. Baumgartner, C.E. Picot, *Molecular Basis of Polymer Networks* (Springer, Berlin, 1989)
18. Y. Rabin, R. Bruinsma, Strain effects on thermal fluctuations in soft solids: butterfly patterns. Europhys. Lett. **20**, 79 (1992)
19. A. Onuki, Scattering from deformed swollen gels with *heterogeneities*. J. Phys. II France **2**, 45–61 (1992)
20. C. Rouf, J. Bastide, J.M. Pujol, F. Schosseler, J.P. Munch, Strain effect on quasistatic fluctuations in a polymer gel". Phys. Rev. Lett. **73**, 830–833 (1994)
21. C. Rouf-George, J.P. Munch, F. Schosseler, A. Pouchelon, G. Beinert, F. Boué, J. Bastide, Thermal and quenched fluctuations of polymer concentration in poly(dimethylsiloxane) gels. Macromolecules **30**, 8344–8359 (1997)
22. S. Candau, J. Bastide, M. Delsanti, Structural, elastic, and dynamic properties of swollen polymer networks. Adv. Polym. Sci. **44**, 27–71 (1982)
23. S. Mallam, F. Horkay, A.M. Hecht, E. Geissler, Scattering and swelling properties of inhomogeneous polyacrylamide gels. Macromolecules **22**, 3356–3361 (1989)
24. W. Wu, M. Shibayama, S. Roy, H. Kurokawa, L.D. Coyen, S. Nomura, R.S. Stein, Physical gels of aqueous poly(vinyl alcohol) solutions: a small-angle neutron-scattering study. Macromolecules **23**, 2245–2251 (1990)
25. F. Horkay, A.M. Hecht, S. Mallam, E. Geissler, A.R. Rennie, Macroscopic and microscopic thermodynamic observations in swollen poly(vinyl acetate) networks. Macromolecules **24**, 2896–2902 (1991)
26. Y. Cohen, O. Ramon, I.J. Kopelman, S. Mizrahi, Characterization of inhomogeneous polyacrylamide hydrogels. J. Polym. Sci. Polym. Phys. Ed. **30**, 1055–1067 (1992)
27. A. Guinier, G. Fournet, *Small-Angle Scattering of X-rays* (Wiley, New York, 1955)
28. P. Debye, A.M. Bueche, Scattering by an inhomogeneous solid. J. Appl. Phys. **20**, 518–525 (1949)
29. S. Panyukov, Y. Rabin, Statistical physics of polymer gels. Physics Report **269**, 1–132 (1996)
30. P.G. de Gennes, Effect of cross-links on a mixture of polymers. J. Phys. Lett. **40**, L69 (1979)
31. Y. Rabin, S. Panyukov, Scattering profiles of charged gels: frozen inhomogeneities, thermal fluctuaitons, and microphase separation. Macromolecules **30**, 301–312 (1997)
32. M. Shibayama (under the supervision of M. Shibayama and K. Kajiwara), *Recent Trends in Polymer Gels, Neutron Scattering* (CMC Publishing, 2004), pp. 297–309
33. R.S. Stein, The determination of the inhomogeneity of crosslinking of a rubber by light scattering. J. Polym. Sci. Part B: Polym. Lett. **7**, 657–660 (1969)
34. S. Takata, T. Norisuye, M. Shibayama, Small-angle neutron scattering study on preparation temperature dependence of thermosensitive gels. Macromolecules **35**, 4779–4784 (2002)
35. T. Norisuye, N. Masui, Y. Kida, M. Shibayama, D. Ikuta, E. Kokufuta, S. Ito, S. Panyukov, Small angle neutron scattering studies on structural inhomogeneities in polymer gels : radiation cross-linked gels vs chemically cross-linked gels. Polymer **43**, 5289–5297 (2002)

36. M. Beltzung, J. Herz, C. Picot, Investigation by small-angle neutron scattering of the chain conformation in equilibrium-swollen poly(dimethylsiloxane) networks. Macromolecules **16**, 580–584 (1983)
37. J. Bastide, R. Duplessix, C. Picot, Small-angle neutron scattering and light spectroscopy investigation of polystyrene gels under osmotic deswelling. Macromolecules **17**, 83–93 (1984)
38. D.S. Pearson, Scattered intensity from a chain in a rubber network. Macromolecules **10**, 696–701 (1977)
39. M. Warner, S.F. Edwards, Neutron-scattering from strained polymer networks. J. Phys. A **11**, 1649 (1978)
40. J. Bastide, C. Picot, S. Candau, Some comments on the swelling of polymeric networks in relation to their structure. J. Macromol. Sci., Part B: Phys. **19**, 13 (1981)
41. Y. Akagi, J.P. Gong, U.I. Chung, T. Sakai, Transition between phantom and affine network model observed in polymer gels with controlled network structure. Macromolecules **46**, 1035–1040 (2013)
42. F. Boue, J. Bastide, M. Buzier, SANS probing of statics and dynamics of polymer networks under deformation: relation between the scatterings of labeled paths and labeled free chains dissolved in the network, in *Molecular Basis of Polymer Networks*, ed. by A. Baumgartner, C.E. Picot (Springer, Berlin, 1989), pp.65–81
43. E. Mendes, R. Oeser, C. Hayes, F. Boue, J. Bastide, Small-angle neutron scattering study of swollen elongated gels: butterfly patterns. Macromolecules **29**, 5574–5584 (1996)
44. E.J. Mendes, P. Lindner, M. Buzier, F. Boue, J. Bastide, Experimental evidence for inhomogeneous swelling and deformation in statistical gels. Phys. Rev. Lett. **66**, 1595–1598 (1991)
45. J. Bastide, L. Leibler, J. Prost, Scattering by deformed swollen gels: butterfly isointensity patterns. Macromolecules **23**, 1821–1825 (1990)
46. A. Ramzi, F. Zielinski, J. Bastide, F. Boue, Butterfly patterns: Small-angle neutron scattering from deuterated mobile chains in a randomly cross-linked polystyrene network. Macromolecules **28**, 3570–3587 (1995)
47. A. Ramzi, A. Hakiki, J. Bastide, F. Boue, Uniaxial extension of end-linked polystyrene networks containing deuterated free chains studied by small-angle neutron scattering: effect of the network chains and the size of the free chains. Macromolecules **30**, 2963–2977 (1997)
48. T. Tanaka, Gels. Sci. Am. **244**, 110–138 (1981)
49. S. Hirotsu, Y. Hirokawa, T. Tanaka, Volume-phase transitions of ionized N-isopropylacrylamide gels. J. Chem. Phys. **87**, 1392–1395 (1987)
50. T. Amiya, T. Tanaka, Phase transitions in cross-linked gels of natural polymers. Macromolecules **20**, 1162–1164 (1987)
51. E. Sato-Matsuo, T. Tanaka, Kinetics of discontinuous volume-phase transition of gels. J. Chem. Phys. **89**, 1695 (1988)
52. H.E. Stanley, *Introduction to Phase Transition and Critical Phenomena* (Oxford University Press, New York, 1971)
53. Y. Li, T. Tanaka, Study of the universality class of the gel network system. J. Chem. Phys. **90**, 5161–5166 (1989)
54. M. Shibayama, T. Tanaka, C.C. Han, Small-angle neutron-scattering study on poly(*N*-isopropyl acrylamide) gels near their volume-phase transition-temperature. J. Chem. Phys. **97**, 6829–6841 (1992)
55. M. Shibayama, T. Tanaka, C.C. Han, Small-angle neutron scattering study on weakly charged temperature sensitive polymer gels. J. Chem. Phys. **97**, 6842–6854 (1992)
56. M. Shibayama, Small-angle neutron scattering on polymer gels: phase behavior, inhomogeneities, and deformation mechanisms. Polym. J. **43**, 18–34 (2011)
57. M. Shibayama, Y. Fujikawa, S. Nomura, Dynamic light scattering study of poly(*N*-isopropylacrylamide-*co*-acrylic acid) gels. Macromolecules **29**, 6535–6540 (1996)
58. V. Borue, I. Erukhimovich, A statistical theory of weakly charged polyelectrolytes: fluctuations, equation of state, and microphase separation. Macromolecules **21**, 3240–3249 (1988)

59. F. Schosseler, F. Ilmain, S.J. Candau, Structure and properties of partially neutralized poly(acrylic acid) gels. Macromolecules **24**, 225–234 (1991)
60. S. Panyukov, Y. Rabin, Polymer gels: frozen inhomogeneities and density fluctuations. Macromolecules **29**, 7960–7975 (1996)
61. M. Shibayama, K. Kawakubo, T. Norisuye, Comparison of the experimental and theoretical structure factors of temperature sensitive polymer gels. Macromolecules **31**, 1608–1614 (1998)
62. F. Ikkai, O. Iritani, M. Shibayama, C.C. Han, Effect of degree of cross-linking on spatial inhomogeneity in charged gels. 3. Ionized effect. Macromolecules **31**, 8526–8530 (1998)
63. F. Schosseler, R. Skouri, J.P. Munch, S.J. Candau, Swelling and cross-linking density effects on the structure of partially ionized gels. J. Phys. **II**(4), 1221–1239 (1994)
64. M. Shibayama, T. Norisuye, S. Nomura, Cross-link density dependence of spatial inhomogeneities and dynamic fluctuations of poly(*N*-isopropylacrylamide) gels. Macromolecules **29**, 8746–8750 (1996)
65. J.M. Guenet, *Thermoreversible Gelation of Polymers and Biopolymers* (Academic, New York, 1992)
66. C. Rochas, A. Brulet, J.M. Guenet, Thermoreversible gelation of agarose in water/dimethyl sulfoxide mixtures. Macromolecules **27**, 3830–3835 (1994)
67. J. de Cloizeaux, G. Jannink, *Polymer in Solution: their Modeliing and Structure* (Oxford University Press, Oxford, 1990)
68. S.F. Edwards, The theory of polymer solutions at intermediate concentration. Proc. Phys. Soc. **88**, 265–280 (1966)
69. M. Kobayashi, T. Yoshioka, T. Kozasa, K. Tashiro, J. Suzuki, S. Funahashi, Y. Izumi, Structure of physical gels formed in syndiotactic polystyrene solvent systems studied by small-angle neutron-scattering. Macromolecules **27**, 1349–1354 (1994)
70. T. Kanaya, H. Takeshita, Y. Nishikoji, M. Ohkura, K. Nishida, K. Kaji, Micro- and mesoscopic structure of poly(vinyl alcohol) gels determined by neutron and light scattering. Supuramol. Sci. **5**, 215–221 (1998)
71. H. Takeshita, T. Kanaya, K. Nishida, K. Kaji, Ultra-small-angle neutron scattering studies on phase separation of poly(vinyl alcohol) gels. Phys. Rev. E **61**, 2125–2128 (2000)
72. I.W. Hamley, *The Physics of Block Copolymers* (Oxford University Press, Oxford, 1998)
73. H. Li, G.E. Yu, C. Booth, J.P.A. Fairchough, A. Ryan, K. Mortensen, Mesophase behavior of aqueous micellar solutions of triblock copolymers of ethylene oxide and 1,2-butylene oxide (type EmBnEm). Langmuir **19**, 1075–1081 (2003)
74. M.J. Park, K. Char, Gelatin of PEO-PLGA-PEO triblock copolymers induced by macroscopic phase separation. Langmuir **20**, 2456–2465 (2004)
75. V. Castelletto, I.W. Hamley, R.J. English, W. Mingvanish, SANS and rheology study of aqueous solutions and gels containing highly swollen diblock copolymer micelles. Langmuir **19**, 3229–3235 (2003)
76. G.A. McConnell, A.P. Gast, J.S. Huang, S.D. Smith, Disorder-order transitions in soft sphere polymer micelles. Phys. Rev. Lett. **71**, 2102 (1993)
77. S. Okabe, S. Sugihara, S. Aoshima, M. Shibayama, Heat-induced self-assembling of thermosensitive block copolymer I. Small-angle neutron scattering study. Macromolecules **35**, 8139–8146 (2002)
78. C. Fuse, S. Okabe, S. Sugihara, S. Aoshima, M. Shibayama, Water-induced self-assembling of solvent-sensitive block copolymer. Macromolecules **37**, 7791–7798 (2004)
79. M. Shibayama, H. Kurokawa, S. Nomura, M. Muthukumar, R.S. Stein, S. Roy, Small-angle neutron scattering from poly(vinyl alcohol)-borate gels. Polymer **33**, 2883–2890 (1992)
80. M. Shibayama, F. Ikkai, R. Moriwaki, S. Nomura, Complexation of poly(vinyl alcohol)-congo red aqueous solutions 1. viscosity behavior and gelation mechanism. Macromolecules **27**, 1738–1743 (1994)
81. M. Shibayama, F. Ikkai, S. Nomura, Microscopic structure and viscosity behaviours of ion-complexed polymer gels. Macromol. Symp. **93**, 277–281 (1994)

82. M. Shibayama, Poly(vinyl alcohol)-ion complex gels, in *The Polymeric Materials Encyclopedia: Synthesis, Properties, and Applications* (CRC Press, Florida, 1996), pp. 7020–7026
83. F. Ikkai, M. Shibayama, Static inhomogeneities in thermoreversible gels. Phys. Rev. Lett. **82**, 4946–4949 (1999)
84. P. Terech, R.G. Weiss, Low molecular mass gelators of organic liquids and the properties of their gels. Chem. Rev. **97**, 3133–3159 (1997)
85. S. Okabe, K. Ando, K. Hanabusa, M. Shibayama, Dynamic light scattering and small-angle neutron scattering studies on organogels formed with a gelator. J. Polym. Sci. Part B: Polym. Phys. **42**, 1841–1848 (2004)
86. M. Shibayama, Structure of high-strength gels studied by neutron scattering. Polymer **59**, 701–704 (2010)
87. Y. Okumura, K. Ito, The polyrotaxane gel: a topological gelby figure-of-eight cross-links. Adv. Mater. **13**, 485–487 (2001)
88. K. Haraguchi, T. Takehisa, Nanocomposite hydrogels: a unique organic-inorganic network structure with extraordinary mechanical, optical, and swelling/de-swelling properties. Adv. Mater. **14**, 1120–1124 (2002)
89. J.P. Gong, Y. Katsuyama, T. Kurokawa, Y. Osada, Double-network hydrogels with extremely high mechanical strength. Adv. Mater. **15**, 1155–1158 (2003)
90. T. Sakai, T. Matsunaga, Y. Yamamoto, C. Ito, R. Yoshida, S. Suzuki, N. Sasaki, M. Shibayama, U. Chung, Design and fabrication of a high-strength hydrogel with ideally homogeneous network structure from tetrahedron-like macromonomers. Macromolecules **41**, 5379–5384 (2008)
91. T. Karino, Y. Okumura, C. Zhao, T. Kataoka, K. Ito, M. Shibayama, SANS studies on deformation mechanism of slide-ring gel. Macromolecules **38**, 6161–6167 (2005)
92. H. Endo, J. Allgaier, G. Gompper, B. Jakobs, M. Monkenbusch, D. Richter, T. Sottmann, R. Strey, Membrane decoration by amphiphilic block copolymers in bicontinuous microemulsions. Phys. Rev. Lett. **85**, 102–105 (2000)
93. T. Nishida, H. Endo, N. Osaka, H. Li, K. Haraguchi, M. Shibayama, Deformation mechanism of nanocomposite gels studied by contrast variation small-angle neutron scattering. Phys. Rev. E **80**, 030801 (2009)
94. J.P. Gong, Why are double network hydrogels so tough? Soft Matter **6**, 2583–2590 (2010)
95. T. Matsunaga, T. Sakai, Y. Akagi, U. Chung, M. Shibayama, Structure characterization of tetra-peg gel by small-angle neutron scattering. Macromolecules **42**, 1344–1351 (2009)
96. T. Matsunaga, T. Sakai, Y. Akagi, U. Chung, M. Shibayama, SANS and SLS studies on tetra-arm PEG gels in as-prepared and swollen states. Macromolecules **42**, 6245–6252 (2009)
97. T. Matsunaga, H. Asai, Y. Akagi, T. Sakai, U. Chung, M. Shibayama, SANS studies on tetra-PEG gel under uniaxial deformation. Macromolecules **44**, 1203–1210 (2011)
98. F. Horkay, K. Nishi, M. Shibayama, Decisive test of the ideal behavior of tetra-PEG gels. J. Chem. Phys. **146**, 164905 (2017)
99. K. Nishi, F. Fujii, U. Chung, M. Shibayama, T. Sakai, Experimental observation of two features unexpected from the classical theories of rubber elasticity. Phys. Rev. Lett. **119**(1–7), 26801 (2017)
100. X. Li, K. Hirosawa, T. Sakai, E.P. Gilbert, M. Shibayama, SANS study on critical polymer clusters of tetra-functional polymers. Macromolecules **50**, 3655–3661 (2017)
101. M. Shibayama, X. Li, T. Sakai, Gels: from soft matter to biomatter. Ind. Eng. Chem. Res. **57**, 1121–1128 (2018)
102. H. Kamata, Y. Akagi, Y. Kayasuga-Kariya, U.-I. Chung, T. Sakai, "Nonswellable" hydrogel without mechanical hysteresis. Science **343**, 873–875 (2014)
103. H. Kamata, X. Li, U. Chung, T. Sakai, Design of hydrogels for biomedical applications. Adv. Healthcare Mater. **4**, 2360–2374 (2015)
104. K. Hayashi, F. Okamoto, S. Hoshi, T. Katashima, D.C. Zujur, X. Li, M. Shibayama, E.P. Gilbert, U. Chung, S. Ohba, T. Oshika, T. Sakai, Fast-forming hydrogel with ultralow polymeric content as an artificial vitreous body. Nat. Biomed. Eng. **1**, 0044(1)–0044(7) (2017)

105. S. Nakagawa, X. Li, H. Kamata, T. Sakai, E.P. Gilbert, M. Shibayama, Microscopic structure of the "nonswellable" thermoresponsive amphiphilic conetwork. Macromolecules **50**, 3388–3395 (2017)
106. S.-L. Zhou, S. Matsumoto, H.-D. Tian, H. Yamane, A. Ojida, S. Kiyonaka, I. Hamachi, PH-Responsive shrinkage/swelling of a supramolecular hydrogel composed of two small amphiphilic molecules. Chem. Eur. J. **11**, 1130–1136 (2005)
107. N. Sahiner, W.T. Godbey, G.L. McPherson, V.T. John, Microgel, nanogel and hydrogel-hydrogel semi-IPN composites for biomedical applications: synthesis and characterization. Colloid Polym. Sci. **284**, 1121–1129 (2006)
108. J. Gong, Y. Iwasaki, Y. Osada, K. Kurihara, Y. Hamai, Friction of Gels. 3. Friction on solid surfaces. J. Phys. Chem. B **103**, 6001–6006 (1999)
109. Y. Zhang, J.F. Douglas, B.D. Ermi, E.J. Amis, Influence of counterion valency on the scattering properties of highly charged polyelectrolyte solutions. J. Chem. Phys. **114**, 3299–3313 (2001)
110. P.H.J. Kouwer, M. Koepf, V.A.A. Le Sage, M. Jaspers, A.M. van Buul, Z.H. Eksteen-Akeroyd, T. Woltinge, E. Schwartz, H.J. Kitto, R. Hoogenboom, S.J. Picken, R.J.M. Nolte, E. Mendes, A.E. Rowan, Responsive biomimetic networks from polyisocyanopeptide hydrogels. Nature **493**, 651–655 (2013)
111. Y. Hirokawa, H. Jinnai, Y. Nishikawa, T. Okamoto, T. Hashimoto, Direct observation of internal structures in poly(*N*-isopropylacrylamide) chemical gels. Macromolecules **32**, 7093–7099 (1999)
112. Z.B. Zhou, H. Matsumoto, K. Tatsumi, Low-melting, low-viscous, hydrophobic ionic liquids: aliphatic quaternary ammonium salts with perfluoroalkyltrifluoroborates. Chem. Eur. J. **11**, 752–766 (2005)
113. V.P.S. Nykänen, A. Nykänen, M.A. Puska, G.G. Silva, J. Ruokolainen, Dual-responsive and super absorbing thermally cross-linked hydrogel based on methacrylate substituted polyphosphazene. Soft Matter **7**, 4414–4424 (2011)
114. I. Ando, M. Kobayashi, M. Kanekiyo, S. Kuroki, S. Ando, S. Matsukawa, H. Kurosu, H. Yasunaga, S. Amiya, NMR spectroscopy in polymer science, in *Experimental Methods in Polymer Science*, ed. by T. Tanaka (Academic, San Diego, 2000), pp. 261–494
115. A. Asano, Review of NMR phenomena from the basics. Bunseki **6**, 266–274 (2014)
116. H. Tanaka, K. Fukumori, T. Nishi, Study of chemiment. J. Chem. Phys. **89**, 3363–3372 (1988)
117. B. Deloche, Orientational order induced in strained polymer networks as revealed by ^{2}H-NMR, in *Molecular Basis of Polymer Network*, ed. by A. Braumgartner, C. Picot (Springer, Berlin, 1989)
118. M. Shibayama, M. Sato, Y. Kimura, H. Fujiwara, S. Nomura, ^{11}B NMR study on the reaction of poly(vinyl alcohol) with boric acid. Polymer **29**, 336–340 (1988)
119. T. Tokuhiro, T. Amiya, A. Mamada, T. Tanaka, NMR study of poly(*N*-isopropylacrylamide) gels near phase transition. Macromolecules **24**, 2936–2943 (1991)
120. H. Ohta, I. Ando, S. Fujishige, K. Kubota, Molecular motion and ^{1}H NMR relaxation of aqueous poly(*N*-isopropylacrylamide) solution under high pressure. J. Polym. Sci. Part B: Polym. Phys. **29**, 963 (1991)
121. P.T. Callaghan, *Principles of nuclear magnetic resononce microscopy* (Oxford Science Publications, Oxford, 1991)
122. H. Yasunaga, H. Kurosu, I. Ando, Spatial information on a polymer gel as studied by ^{1}H NMR imaging. 1. Image analysis of stress-strain. Macromolecules **25**, 6505–6509 (1992)
123. Y. Yamane, M. Matsui, H. Kimura, S. Kuroki, I. Ando, Diffusional inhomogeneity of probe molecules in chemically cross-linked polymer gels as studied by time-dependent diffusion NMR. Macromolecules **36**, 5655–5660 (2003)
124. Y. Yamane, I. Ando, F.L. Buchholz, A.R. Reinhardt, S. Schlick, Detection of spatial inhomogeneity in poly(acrylic acid) gels by measuring time-dependent diffusion coefficients of a probe in NMR experiments: effect of the degree of cross-linking and degree of swelling. Macromolecules **37**, 9841–9849 (2004)
125. K. Kamiguchi, S. Kuroki, M. Satoh, I. Ando, Diffusional behaviors of polystyrenes with different molecular weights in the same PMMA gel network elucidated by time-dependent diffusion NMR spectroscopy. Macromolecules **41**, 1318–1322 (2008)

126. K. Saalwächter, Proton multiple-quantum NMR for the study of chain dynamics and structural constraints in polymeric soft materials. Prog. Nucl. Magn. Reson. Spectrosc. **51**, 1–35 (2007)
127. F. Lange, K. Schwenke, M. Kurakazu, Y. Akagi, U. Chung, M. Lang, J.-U. Sommer, T. Sakai, K. Saalwaechter, Connectivity and structural defects in model hydrogels: a combined proton NMR and Monte Carlo simulation study. Macromolecules **44**, 9666–9674 (2011)
128. M. Shibayama, Physical properties analysis of gels by dynamic light scattering, in *Fundamentals and Applications of Light Scattering*, ed. by M. Shibayama, N. Sato, T. Iwai, Y. Kimura (Kodansha, 2014), p. 209)
129. T. Tanaka, Light scattering from polymer gels, in *Dynamic Light Scattering*, ed. by R. Pecora (Plenum Press, New York, 1985), pp. 347–362
130. J.G.H. Joosten, J.L. McCarthy, P.N. Pusey, Dynamic and static light scattering by aqueous polyacrylamide gels. Macromolecules **24**, 6690–6699 (1991)
131. M. Shibayama (ed.), *New approaches to characterization of organic-inorganic composites* (Catching up with Polymer Science and Inorganic Chemistry, NTS, 2001), pp.41–73
132. T. Norisuye, M. Takeda, M. Shibayama, Cluster-size distribution of cross-linked polymer chains across the gelation threshold. Macromolecules **31**, 5316–5322 (1998)
133. M. Shibayama, T. Norisuye, Dynamics of silica gels in gelation process. AIP Conf. Proc. **469**, 85–92 (1999)
134. M. Shibayama, T. Norisuye, Gel formation analyses by dynamic light scattering. Bull. Chem. Soc. Japan. **75**, 641–659 (2002)
135. M. Shibayama, Structural analysis of polymer gels. Funct. Mater. **19**, 7–15 (1999)
136. M. Adam, D. Lairez, Sol-gel transition, in *Physical Properties of Polymeric Gels*, ed. by J.P. Cohen Addad (Wiley, New York, 1996), p. 87
137. P.N. Pusey, W. van Megen, Dynamic light scattering by non-ergodic media. Physica A **157**, 705–741 (1989)
138. M. Shibayama, Y. Isaka, Y. Shiwa, Dynamics of probe particles in polymer solutions and gels. Macromolecules **32**, 7086–7092 (1999)
139. M. Shibayama, X. Li, Probe diffusion dynamic light scattering of polymer solutions and gels, in *Gels and Other Soft Amorphous Solids*, ACS Symposium Series Vol. 51, ed. by F. Horkay (American Chemical Society, Washington, 2018), pp. 51–69
140. X. Li, N. Watanabe, T. Sakai, M. Shibayama, Probe diffusion of sol-gel transition in an isorefractive polymer solution. Macromolecules **50**, 2916–2922 (2017)
141. T. Hiroi, M. Shibayama, Dynamic light scattering microscope: accessing opaque samples with high spatial resolution. Opt. Express **21**, 20260–20267 (2013)

Chapter 8
Gels in Living Organism

Yoshihito Osada

Abstract Living organism largely consists of hydrogels with highly hierarchical orders and builds up organs and tissues. This chapter describes the structure and properties of extracellular matrix which is the hydrogel laterally existing through soft tissues. This will be followed by soft tissues, cells, and cytoskeletal protein assemblies along the hierarchical order. Specific viscoelastic functions of biological gels which are essential for vital activities will be emphasized.

Keywords Extracellular matrix · Proteoglycan aggregate · Connective tissue · Collagen · Tendon · Ligament · Articular cartilage · Friction · Stress-stiffen · Stress-soften · Cytoskeletal protein · Actin · Microtubule · Persistent length

8.1 Soft Tissue of Human Body is Hydrogel

Human body consists of hard tissue [bone, tooth, nails, and the outer layers of skin) and soft tissue. Soft tissue can further be divided into connective tissue which forms ligaments, tendons, cartilage, corneal stroma, adipose (fat) and organs such as vein, nerve, and muscle. Apart from the question whether bone is the gel state or not [1], a substantial part of the volume of the soft tissue is essentially made of hydrogel consisting of the macromolecular network called extracellular matrix (ECM) which is the filamentous proteins and polysaccharides networks filled with water. There, various kinds of proteins and polysaccharides are penetrated, aggregated and cross-linked each other to form the network connecting cells embedded there and determine the physical and biological properties of the tissue as whole.

Tissues widely vary their stiffness in several orders of magnitude [2]. The rigidity of the ECM is commonly determined by Young's modulus which is significantly different across cell types. For example, it ranges from less than 10^3 Pa of human brain or liver, 10^4 Pa of muscle, 10^6 Pa of articular cartilage to 10^7 kPa of rigid bone. This difference is a result of the quantitative and qualitative biochemical properties

Y. Osada (✉)
RIKEN, Wako, Saitama, Japan
e-mail: osadayoshi@riken.jp

Y. Osada (ed.), *Science of Gel*, https://doi.org/10.1007/978-981-95-0301-8_8

of various biomacromolecules consisting of ECM network, i.e., their concentration, structure and the categories. Mechanical properties of typical tissues and organs are summarized in Table 8.1.

ECM network consists of mainly polysaccharide (proteoglycan) and proteins such as collagen, elastin, fibrous proteins (fibronectin, laminin). Among them collagen and elastin the most influentially determine the mechanical properties of the ECM by changing the relative ratio of these two proteins. In addition, the network of these biological tissues is in highly-charged state and their mechanical properties are substantially attributed to the strong polyelectrolyte nature. Articular cartilage, which contains anionic proteoglycan-rich ECM, for example, has remarkable elasticity, low surface friction, and ability to resist enormous physical forces. These features are directly related to the high-water content obtained by negatively charges of aggrecan/hyaluronic acid complexes which often bound proteins. This distinctive feature also enables the organism to transport solvated molecules easily and effectively in highly swollen state while keeping its mechanically tough solidity as gel state. Thus, the role of negative charges on tissues plays critical roles.

Cells have to organize themselves spatially within ECM in required form and respond mechanically and physiologically with their environment. They have to be correctly shaped, physically robust, and properly structured internally. Many of them have to be able to change their shape and move from place to place. They are required to rearrange their internal components as they grow, divide, and adapt to changing outside circumstances where the responses with ECM play important role.

The mechanical functions developed to a very high degree in eukaryotic cells are driven by a filamentous network system called cytoskeleton. Cytoskeleton is composed of three types of self-assembled protein filaments–actin, tubulin and intermediate. In vivo the mechanical properties of actin cytoskeleton are regulated predominantly by actin binding proteins which enable to align the filament, form bundles, cross-link the filaments, and branched dendritic networks and consequently form mesh-like actin hydrogels [3]. Owing to the functions of actin binding protein, cytoplasmic actin hydrogels obtain the storage modulus G' of order 10^{-1}–10^{2} Pa [4] which is required to maintain the shape of the cell.

This chapter describes fundamental aspects, structure and properties of highly hierarchical biological tissues viewed as gel state. Description will be made first along with the ECM, the hydrogel which laterally exists through every soft tissue and then along the hierarchically ordered structure of the living organism—organ, cells, and molecular assemblies.

Table 8.1 Elastic properties of organs under tension (human, unless otherwise specified)

	Maximum tensile strength (MPa)	Maximum strain at breaking (%)	Modulus (MPa)
Soft tissue			
Liver (rabbit)	0.024	46	0.02
Liver (human)			0.6
Skeletal muscle (rectus abdominis)	0.11	61	0.02
Cardiac muscle	0.11	64	0.08
Smooth muscle (under relaxed) Smooth muscle (under contracted)		300 300	0.006 0.01
Small intestine (longitudinal direction)	0.56	43	0.2
Stomach (longitudinal direction)	0.56	93	0.015
Large intestine {longitudinal direction)	0.69	117	0.02
Coronary arteries	1.1	64	0.1
Wet spongy bone (vertebrae)	1.2	0.6	200
Umbilical cord	1.5	59	0.7
Vena cava (longitudinal direction)	1.5	100	0.04
Ligaments (cattle)	2.1	130	0.5
Gall bladder (rabbit)	2.1	53	0.05
Cardiac valves	2.5	15	1
Intervertebral disc	2.8	57	2
Cerebral vein	2.8	80	6.9
Hyaline cartilage (synovial joints)	2.9	18	24
Elastic cartilage (external ear)	3.1	26	4.5
Vertebrae	3.5	0.8	410
Skin (face)	3.8	58	0.3
Intervertebral disc (compression)	11	32	6
Nerves	13	18	10
ACL ligament	36	15	350
Tendons (calcaneal = Achilles)	54	9	250

(continued)

Table 8.1 (continued)

	Maximum tensile strength (MPa)	Maximum strain at breaking (%)	Modulus (MPa)
Hair (head)	197	40	12,000
Hard tissue			
Dentin (wet teeth) (compression)	162	4.2	6000
Femoral compact bone (compression)	162	1.8	10,600
Femoral compact bone	109	1.4	10,600
Cortical bone	50–150	1–3	7000–30,000
Nail	18	141	160

Modified using data by Yamada [8]

8.2 Extracellular Matrix (ECM)—Main Component of Soft Tissue

The role of ECM which occupies a main part of the volume of the soft tissue was once considered to serve mainly to sustain the mechanical toughness of tissues, but it is proved that ECM is physiologically active in controlling the behaviors of the cells embedded there.

ECM is the gel state filled with water and network constituted by two main and highly organized biopolymers: polysaccharide and proteins (Fig. 8.1). There, various kinds of polysaccharides and proteins are penetrated each other to form interwoven and cross-linked networks and determine the physical and biological properties of the tissue. They are produced mainly by the cells in the EMC, called fibroblasts, but EMC itself can regulate various functions of the cell as development, survival, migration, proliferation, and others. Since these activities of cells are closely related to the stiffness of ECM, their rigidity is very important. For example, cell migrates under the guidance of rigidity gradients called durotaxis. Most normal cells migrate up rigidity gradients in the direction of greater stiffness which arise from differential structural properties of the ECM.

One of main component of ECM is the polysaccharide chain called glycosaminoglycan (GAG). GAG is linear polysaccharide composed of repeating disaccharide, in most cases with sulfated amino group (*N*-acetylglucoseamine or *N*-acetylgalactoseamine) and uronic acid (glucuronic or iduronic) [5]. GAG is rigid macromolecules resulted from its chain backbone structure and highly-charged state due to the ionized sulfate and carboxylate groups. This highly negative charges attract positive cations such as sodium, potassium and calcium ions to give high osmotic pressure. The rigid structure with full of charges brings GAG highly extended chain conformations and occupies a huge volume relative to the mass. The high osmotic pressure produced by negative charges is the reason why attracts a huge amount

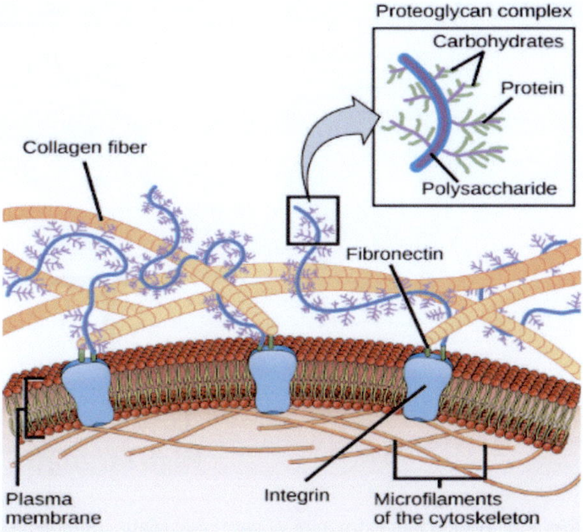

Fig. 8.1 Schematic illustration of extra cellular matrix (from Wikipedia)

of water to form highly water swollen hydrogel. This also creates swelling pressure and enables tissues to resist compressive forces and reduce the sliding friction. Hyaluronan (or hyaluronic acid) is the simplest GAG found in all tissues and especially is an important substance of joint fluid in the cartilage as described below.

GAG is the major component of the connective tissue and often covalently bound to proteins to form proteoglycans with a molecular weight of $1 \sim 4 \times 10^6$. The proteoglycans are complex aggregate of two-stage assembly. The smallest unit of GAG is two ionizing groups: chondroitin sulfate, with a molecular weight of typically 2×10^4, and keratin sulfate, with a rather smaller molecule, both being unbranched chains of disaccharide units. About 100 molecules of chondroitin sulfate and about 50 molecules of keratin sulfate are attached with a protein with a molecular mass about 20,000(first assembly). Tens of these proteoglycans further bind with a hyaluronic acid to form proteoglycan aggregates which has giant molecular mass as large as 50×10^6 which length goes up to 1.2 μm. This aggregate provides the tissue with unique gel-like property and resistance to deformation through water absorption. As described, the origin of mechanical toughness of the cartilage matrix located at the knee joint that sustains the pressure of hundreds of atmospheres is mainly attributed to this high osmotic pressure. In fact, the large quantities of fully ionized chondroitin sulfate and keratan sulfate in aggrecan play an important role in the hydration of cartilage (Fig. 8.2) Aggrecan is one of the most important proteoglycans forming very large aggregates; more than 4 μm long. Aggrecan molecules are non-covalently bound to hyaluronan through linking proteins like a bottlebrush (Fig. 8.2). Weight bearing and lubrication properties of cartilage are associated primarily with this matrix and its large water content (ca. 75–80 wt%).

On the other hands their structural rigidity and the stretched 3D network structure enable the cells to allow effective migration and transportation within the EMC.

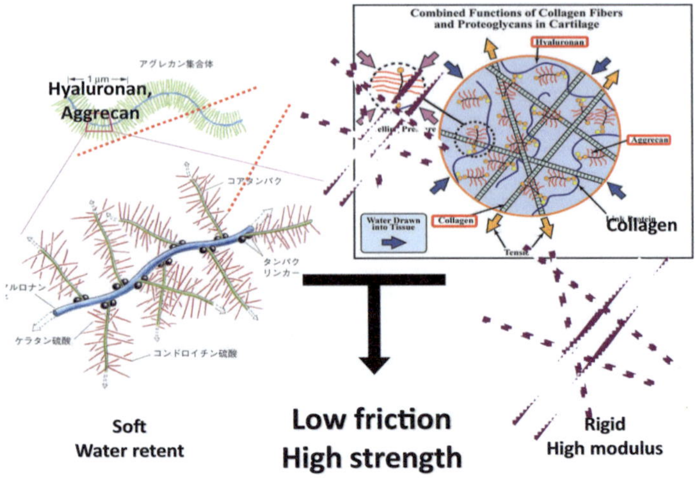

Fig. 8.2 Hierarchical structure of cartilage. Main component of cartilage is collagen and proteoglycan aggregates (from Wikipedia)

This function is best illustrated by the most abundant proteoglycan in cartilage. Proteoglycans may also help to trap and store growth factors within the ECM. The effect of size and shape on the transport process of some proteins in the gel is described in 2.4

Fibronectin is another important glycoprotein. They connect cells with collagen fibers in the ECM. Fibronectins also facilitate cell movement to heal from the injury by binding to platelets during blood clotting as will be described in detail next.

Proteins such as collagen, elastin, and some others are another main component of ECM. Collagens are the most abundant protein in the ECM, i.e., in the human body. In contrast to GAG which resists to compressive pressure, collagen fibers aggregate each other to give triple-stranded coiled-coil structure and form collagen fibrils that resist tensile forces. These collagen fibrils form highly-ordered network structure in the tissue. The size and the shape are strongly dependent on the species of tissue but they are well fabricated to fit the role and the function of tissues. In mammalian skin, for example, the collagen fibrils are highly interwoven network structure to resist tensile stress from every direction. In contrast, collagen fibrils in tendons form network oriented each in parallel bundles aligned along the axis of tension to resist to one-dimensional stress. Collagen fibrils in cornea arrange in parallel on the layer, but form the network structure penetrating with different directions one another at nearly right angles on each layer level.

Elastin is another important protein which constitutes the ECM. Elasticity is an important property in many tissues, especially, in blood vessels, lungs, and skin, since they are required to have strong and elastic functions and in fact, these tissues contain high amounts of elastin. Artery and aorta are also organs which require high

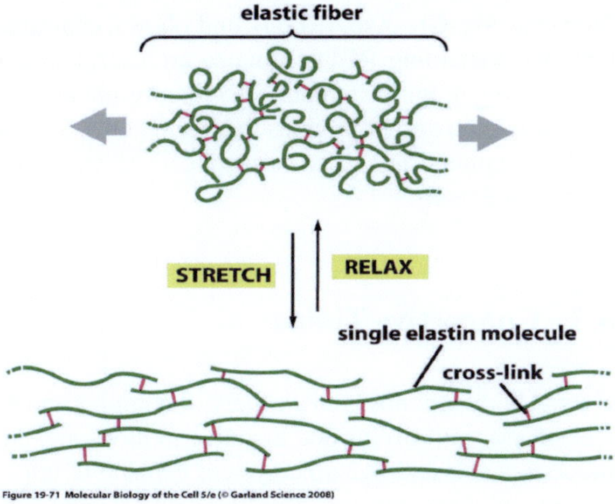

Fig. 8.3 Network of elastin fibers are flexible. It deforms easily and recovers quickly (from Wikipedia)

elasticity and elastin is their dominant protein comprising 50% of the dry weight of the largest.

Elastin network, in contrast to collagens, gives the ECM the high elasticity, the recovering force, and toughness through its strong elastic nature. This originated from its very flexible conformational structure which allows reversible extending—recoiling property of the polymer chains. Here, the elastin polypeptide chain, like the rubbers, can take a loose "random coil" conformation that allows the network to stretch and recover the random coil conformation (Fig. 8.3).

Elastin.

However, the robust elasticity of the tissue is not obtained solely by elastin, but the triple-stranded rigid collagen fibrils form the interwoven structure with elastin to give the double structured network, thus limit the extent of stretching and prevent the tissue from tearing the tissue by restricting unnecessary large extension. The cross-linking of elastin through lysine are also found. The elastin core is covered with a sheath of microfibrils composed of number of distinct glycoproteins.

In addition to collagen and elastin ECM contains number of proteins that have specific binding sites for other component in ECM and for receptors of the cells to physiologically and mechanically functionalize them. Fibronectin is a large glycoprotein and typical one. Fibronectin plays important roles for cell adhesion to the matrix and for guiding cell migrations in vertebrate embryos.

It should note that there are two types of fibronectin: one is soluble fibril and the other is insoluble fibril. Soluble fibronectin is found in the blood plasma and circulates in the body to enhance blood clotting when required, wound healing, and phagocytosis. Fibrous fibronectin is composed of fibrous aggregates connected by disulfide bonding on the cells surface. This fibronectin forms fibrils only on certain cells in the presence of fibronectin-binding proteins called integrin. The fibronectin fibrils on the cell surface keep highly stretched conformation under tension associated with actin filaments. It is well known that fibrinogen contains a specific tripeptide sequence Arg-Gly-Asp, (RGD), and plays a central role of cell binding. Fibrinogen peptides containing RGD sequence are useful to develop biocompatible materials anti-clotting drugs to implant cells that mimic these peptides on their surfaces.

Laminins are other proteins that form network structures to resist tensile forces in the basal lamina. They also assist in cell adhesion and bind other ECM components such as collagens.

8.3 Connective Tissue

Human body is exposed to a variety of complex and diverse environment of mechanical stresses and strains due to active motions and passive deformation. Connective tissue is a soft and dynamic system capable to appropriately respond against each of stresses. One can find beautiful and efficient geometrical network structure of connective tissues.

8.3.1 Tendon

Tendon is the connective tissue consisting of dense network of flexible protein fibers and contains with few cells and little ground substance. Tendon links muscle with skeleton. Tendon contains 75–85% dry collagen (95% is type I collagen), nearly 3% elastin, and 1–2% proteoglycans [5]. It has highly one-directional oriented structural as shown in Fig. 8.4, similarly to that of ligaments, but fibers are well arranged with highly hierarchical structure into packets called fascicles, and the bundles of collagen fibers are more parallel oriented than in ligaments. The highly-ordered hierarchical structure of tendon in Fig. 8.4 is due to serve one-directional stretching force behaving as a passive transmitter. The hierarchy of essentially unidirectional arrangement of fibers starts from the single collagen macromolecule to the tropo-collagen triple helix, then to the microfibril of five tropocollagens, followed by the tetragonal lattice of microfibrils forming sub-fibrils and finally, the collagen fibril [6]. The network collagen fibrils are aligned in parallel to form bundles along the axis of tension to resist tensile stress. There, chemical covalent cross-linking and various kinds of inter-macromolecular interaction such as, hydrogen bonding, electrostatic force, and van der Waals' forces are involved to mechanically strengthen each hierarchical component. In addition, highly water containing proteoglycans bind and occlude these collagen fibrils to form 3D architecture to form the proteoglycan hydrogel as whole.

The numbers and the properties of these chemical and physical bridges are sustained to change with age. Figure 8.5 shows the stress–strain relationship of rat tail tendon as a function of age [7]. As shown in this figure, mechanical responses of tendon deformation have a strong age-dependence. The elastic module which is the slope of stress strain relation is constant, while the strain becomes smaller with the

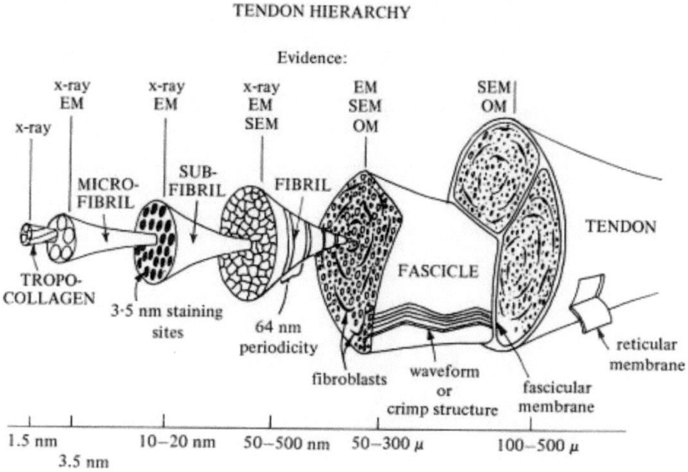

Fig. 8.4 Hierarchical organization of tendon [6] (from Wikipedia)

age. The dependence is believed to have arisen due to changes in number and structure of bridging interactions [7]: When the bridging density increases with age, the tendon, in general, loses the flexibility and the Young's modulus increases. The mean tensile strength of the normal patellar tendon is ranged between 60 and 100 MPa, while the elastic modulus ranges from 400 to 1000 MPa Table 8.1. The wide diversity of the data comes from different age, species, and type of specimens.

8.3.2 Ligament

Ligament is another connective tissue consisting of dense network of flexible fibers and contains few cells. It binds bones and supports the organs in place. Ligaments contain 55–65% water, 35–45% dry substances containing 70–80% collagen (mostly type I), 10–15% elastin, and 1–3% proteoglycans [6]. Like tendon, the collagen fiber in the ligament forms multi-scale hierarchy through its self-assembly process and gives helices assembles starting from microfibrils (4 nm in diameter), sub-fibrils (20 nm in diameter), assemble of fibrils (50–500 nm in diameter), and finally to collagen fibers (100–300 μm in diameter). There, fibroblast cells are embedded to produce the collagen. The elasticity is more important in ligaments than in tendons. The average tensile strength and elastic modulus of (cattle) ligament are 2.1 and 130 MPa, while those of Achilles tendon are 54 and 9.0 MPa Table 8.1. The average tensile strength and elastic modulus of human ligaments have been reported as 35 and 316 MPa. Thus, tensile strength of ligament is smaller than that of tendon, while the maximum strain of the ligament is large enough, i.e., more flexible.

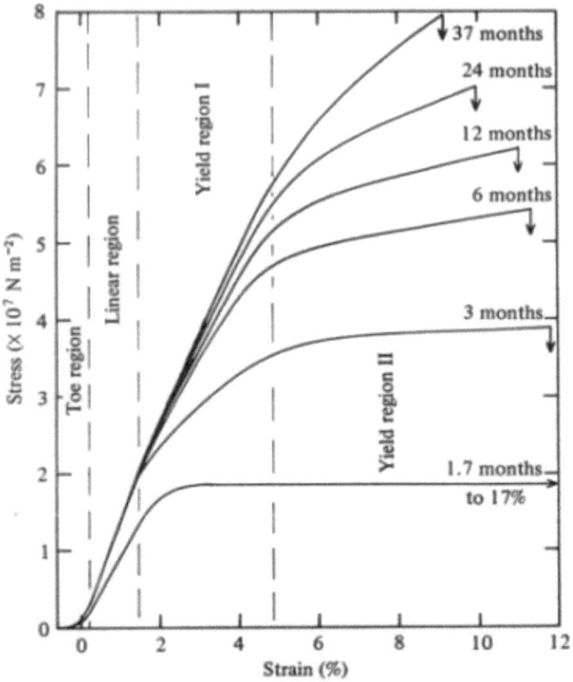

Fig. 8.5 Stress–strain behavior of rat tail tendon as a function of age. Strain rate: 8% per minute [7]

Finally it should be noted that besides tendon and ligament, other tensile load-bearing tissues such as muscle and keratin aggregates have similar uniaxial fibrous hierarchical structure of collagen.

8.3.3 Articular Cartilage

Articular cartilage covers the bones at synovial joints and performs two important mechanical functions: one is shock absorbing function to buffer the outside forces transmitted through the joint as to minimize the stresses. The articular cartilage can sustain mechanical stresses cyclically which exceed 10 MPa. Articular cartilage is the gel state and therefore, orders of magnitude deformable than the bone of the articular joint. Another important function is to reduce friction by using well-lubricating surface and to prevent wearing. The frictional coefficient of cartilages of human and animal joints are low as low as 0.001 - 0.03 which are two orders of magnitude lower than that of solid friction.

Cartilage constitutes of solid matrix about 30% by mass in which collagen is a 40–70% (~80% type II collagen), 15–40% proteoglycan, and 70% water containing inorganic salts and chondrocyte cells that produce the cartilage. Cartilage itself is viscoelastic and porous material with c.a. 5 nm voids filled with water. The collagen network has gradient-arranged interwoven network structure which enables to sustain most of the load in tension by deforming it shape, while both the solid and liquid carry the load under the compression (Fig. 8.2).

The network of cartilage has a pertinent structure. The layer of cartilage next to the articular surface (up to about 200 μm deep) contains more collagen than the deeper layers, and the dominant direction of the collagen fibers is parallel to the surface to resist and minimize the frictional force, while in the deepest zone the dominant direction is perpendicular to the surface to sustain the vertical stress. The majority of water in the cartilage is bound with the proteoglycans to form a swollen gel state and minimize the vertical stress but some is excluded from the gel through the voids under compression, thus dissipates energy as it flows and minimize the friction. The viscoelasticity of cartilage is controlled by the exudation of fluid through the pores in this biphasic gel structure thus obtain a low frictional coefficient.

Despite the importance of the cartilage, it is not well understood up to now why the cartilage friction of the joints is so low, even under such high pressure between the bone surfaces which reaches up to 3–18 MPa and the sliding velocity is so low as a few cm s-1. Under these conditions, the hydrodynamic lubrication of solid friction does not work since the lubricating liquid layer cannot be sustained between two solid surfaces. It would be certain that the synovial fluid plays crucial role for frictional behavior of cartilages. Synovial fluid is essentially a dialysate of blood plasma in which contains proteoglycan aggregate (hyaluronan aggrecan), and allows the cartilage swelling. Proteoglycan aggregate is a highly-charged hyaluronate-protein complex which have a roughly spherical shape of c.a. 0.5 μm and can effectively occupy large volumes in the synovial fluid (Fig. 8.2). Due to this composition, synovial fluid can exhibit pronounced non-Newtonian behavior which allows to sensitively decrease its viscosity with increasing shear rate from 10^{-1}–10^{2} N s m^{-2} at a shear rate of 10^{-1} s^{-1} down to 10^{-2}–10^{-1} N s m^{-2} at a shear rate of 10^{3} s^{-1}.

There is another polysaccharide that plays a tribological role on the surfaces of tissue and organ. That is mucins. Mucins consist of high molecular weight, extensively glycosylated glycoproteins and make smooth lubrication of tissues and organs and protect cell surface from damage. It was suggested that gel formation of gastric mucous glycoprotein is markedly different from those of typical polysaccharides such as agarose and carrageenan which form ordinary regular macro-structures. Mucin is considered to form giant 3D complicated architecture by the strong intermolecular interactions between oligosaccharide chains that brings about the extraordinary viscous property even at low concentration as 20–30 g/ℓ.

8.3.4 Vitreous Body and Stroma of Cornea

The vitreous body is a transparent, colorless hydrogel that fills the space in the eye. It is located between retina and lens and keeps the place by pressing it against the choroid. The major macromolecular components of the vitreous body are hyaluronic acid (hyaluronan) (HA) and collagens. There, hyaluronic acid is in highly hydrated state and occupies a large specific volume of 2000–3000 ml/g. It is arranged in a linear helix conformation which allow to undergo both a compacted and extended conformations. The molecular weight of hyaluronic acid ranges from 3 to 4.5×10^6 and the concentration of hyaluronic acid is highest in the cortical gel which makes the hyaluronic acid form a highly entangled open coil.

Vitreous collagen consists of fibrils 7–28 nm in diameter and they are covalently linked each other. The fibrils consist of a mixture of various types of collagens of II, V, and IX. The assembly of these collagens enhances the stability and strength of the fibrils. Transmission electron microscopy reveals that the fibrils have uniform with alternative layers of different density originated from a filament consisting of a succession of linked 20-nm beads with a spacing ranging from 35 to 85 nm (Fig. 8.6). There, an interaction between the collagen fibrils and HA exists that provides the vitreous gel with its unique mechanical and optical properties. The large domains of HA stabilize and spread apart the collagen fibrils, thus, minimizing light scattering. The viscoelastic properties of vitreous are the result of complex molecular interactions between the collagen fibrils and HA [9]. The spaces between collagen fibrils are filled by HA and other GAGs such as chondroitin 6-sulfate. The GAGs link together as aggregates via hydrophobic and hydrogen bonding, resulting in a three-dimensional honeycomb structure. This ordered aggregation of GAGs is fragile and has clinical implications Fig. 8.6.

The stroma exhibits highly transparent property originated from well-organized architecture of collagen fibers. There, the collagen fibrils form special lamella structure with 1.5–2.5 μm thick where about 200 flat plates are superimposed one after another. These fibrils are interwoven with different angles: the fibers of each lamella are parallel with one another, but with a certain angle to those of next lamella layers. These lamellas are produced by keratocytes (fibroblasts) located between the lamella

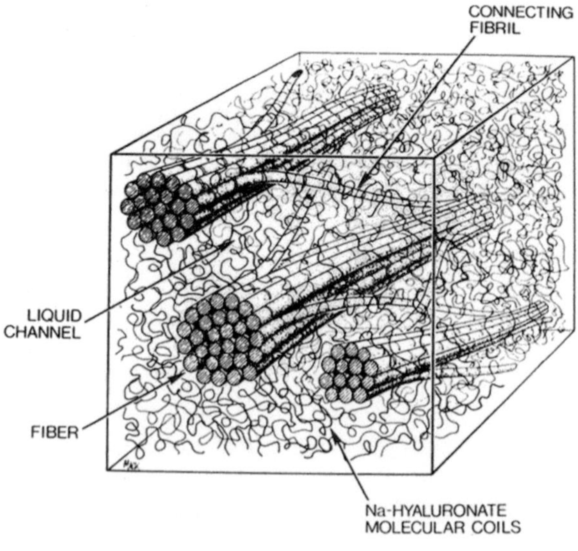

Fig. 8.6 Schematic illustration of vitreous ultrastructure showing the interaction of collagen fibrils with hyaluronic acid to form the gel of vitreous bodyl (Sebag et al. [10] from Wikipedia)

Table 8.2 Structure and components of the eye

The cornea	The majority (real layers) consists of collagen fibers and mucopolysaccharide. The fiber aligns parallel and the transparency is attained by the size smaller than the wavelength
The lens	The lens is surrounded by a membrane containing collagen. The structure of the cells at the front surface is fibrous and ordered structure of α、β, and γ crystalline forms. It becomes higher density and higher refractive index when approaches to the center of the lens. When this ordered structure is disrupted, the system becomes cloudy resulting in cataracts
The vitreous	The vitreous is a composite structure consisting of collagen fibers and hyaluronic acid. The water content is as high as 99%. It maintains the shape of the eyeball, absorbs mechanical impact, maintains the balance of the aqueous humor in the eyeball, and regulates eye pressure. Details are not known

collagen fibers. The main components and structure of the eye are summarized in Table 8.2

8.4 Cells

8.4.1 Viscoelastic Behaviors of Cells

Cells of living organisms are considered as an active soft and wet material with fascinating mechanical properties. Cells, regardless of a part of organs or isolated single cell, live under frequently and irregularly changing mechanical stress from external environments but maintain their vitality. These mechanical properties are presumably obtained under various external mechanical circumstances through long time of their evolution that they were born, developed and adapted. A variety of biological processes such as cell growth, stem cell differentiation, cell crawling, wound healing, protein regulation, cell malignancy, and even apoptosis have obviously obtained mechanically responding to environmental changes of the external mechanical conditions. In fact, these cell functions including cellular homeostasis, cell–cell communication, tissue formation, and locomotive functions are closely associated with their cellular viscoelasticity. Such adaptive mechanical responses and functions in some cases follow the classical mechanical law which is well expressed by creep and stress relaxation. But in many case they don't follow the classical theory of mechanics and follow unique nonlinear power-law responses.

As described in Chap. 1 briefly, cells are complex bi-phase systems constituting of an elastic solid with a certain rigidity and a viscous fluid and show viscoelastic property as typical hydrogels. For example, animal cells are covered with a plasma membrane which is supported with relatively inextensible thin cortical layer of $\approx$ 100 nm. This cortical network is composed of actin filaments bridged with actin binding proteins and encloses a viscous liquid called cytoplasm that fills the cell. Such viscoelastic behavior of the cell allowed us to treat with systems similarly to cross-linked polymer gels containing the cytoplasm inside. However, scientists knew soon after that it is too simplified and recognized that cells are made of a highly nonlinear material that change the stiffness and fluidity frequently under biological requirement, quickly and constantly remodeling to adapt to environmental changes. Thus, cells show highly complex dynamic behaviors far from those from static non-living materials. They are often heterogeneous spatially within cell and change with time. An important fact is that all these mechanical behaviors are mainly induced by the contractile forces produced by the cytoskeletal network in cells called cytoskeleton. The cellular cytoskeleton is a dynamic network of filamentous proteins consisting of three types of protein filaments–actin, tubulin, and intermediate (refer to 8.5). Thus, cytoskeleton and cytoplasm determine and control the viscoelastic properties and mechanical strength of the cell through cascade processes based on their complex interactions. Nowadays these mechanical responses are considered regulated through a direct physical connection between ECM and the nucleus across the plasma membrane at the genome. Indeed, cytoskeletal connections between the ECM and the nucleus via integrin receptors have recently been experimentally proved.

Therefore, the measurement of the progressive spatial–temporal changes of viscoelastic properties of living cells under the physiological conditions is crucially important to understand the comprehensive cellular processes. However, measurements of the mechanical change responding to external signals in cells in vitro are still difficult tasks. The reason is that cells have heterogeneous compositions and their various constituents have diverse mechanical properties which give extremely complicated signals. Thus, the mechanical responses of cell are an complex of all these individual components.

Nevertheless, comprehensive methods have recently been developed to measure the mechanical and rheological properties of single cell. These include magnetic twisting cytometry (MTC), micropipette aspiration, microplate cell manipulation, optical stretchers, particle-tracking micro-rheology, optical traps, and atomic force microscopy (AFM). Important point is that the response depends on the manner of measurement technique and timescale of the experiment. The micropipette aspiration, microplate rheometry, and optical stretching are used to investigate whole cell mechanics, while probe- or particle-based techniques such as magnetic probes, optical tweezers and particle-tracking cell rheology are used to measure the mechanical properties on local level. AFM is now becoming popular techniques to measure the local force as well as mapping of whole cells across the cell surface with sub-10-nm resolution. Conventional technique to evaluate the viscoelastic properties of living cells is to obtain force-distance curves by applying a rectangular stress-relaxation measurement. Frequency-dependence of shear modulus of cells is also typical method to investigate either from strain response to an applied stress or torque or from the Brownian motion of attached tracer particles.

Although the results obtained by these methods give strongly fluctuating data depending on the origin of samples, they give valuable data to estimate overall mechanical properties of the cells. Many studies agree that cells respond stiffer when probed at higher frequencies as typical viscoelastic materials. However, many of them show that the cell stiffness G^*(complex modulus) does not depend linear on the frequency ω but obeys a weak power law as $G^*(\omega) \sim \omega^{\beta}$, with small exponents. Note that the β of physical gel is two and that of viscous flow is one (refer to 2–2), while the value of the exponent β of cells lies in the range 0.1–0.35, and depends on the part of the cell that is probed. Therefore, β value makes a useful to mechanically identify different structures of the cell. At higher indentations, viscous effects led to an increase of the apparent elastic modulus. The experimental results show that viscous contribution that followed a weak power law increases at larger cell indentations.

More recently, cytoplasm elasticity and viscosity of many different types of cells, including fibroblasts, mouse embryonic fibroblasts, human umbilical vein endothelial cells, human induced pluripotent stem cells, and *human embryonic stem cells,* have been studied under different conditions by particle-tracking micro-rheology. The average viscous modulus and elastic modulus (both at the frequency of 1 Hz) were determined to be in the range of 1–50 Pa and 1–80 Pa, respectively. Theoretically, these values of cellular viscoelasticity obtained by micro-rheology are valid only under condition that the cell can be regarded as an equilibrium system. However, even though live cells are far from equilibrium, the contribution of non-thermal force to particle fluctuation has been proved to be much lower than that of the thermal force at the frequency higher than 10 Hz (or equivalently, for time scale shorter than 0.1 s). Under such conditions the cells can be regarded as in quasi-equilibrium state.

The cytoskeleton undergoes dynamic structure changes to lead the viscoelastic property to support the cell division. Specifically, in the beginning of cell division, microtubules release tubulin, and the free tubulins assemble to form spindles to orient the chromosomes' arrangement. As early as 1965, the viscosity of fibroblasts in prophase, metaphase, and anaphase had been measured and reported as 0.26, 0.35, and 0.24 Pa, respectively. Later, the same method was used to measure the intracellular viscosity of the spermatocytes and was determined to be in the range of 0.04–0.36 Pa in the temperature range of 30 to 5 °C. An increase in intracellular viscoelasticity was attributed to an increase in microtubule density and cross-linking.

The intracellular viscoelasticity of HeLa cells at different phases during cell division was also measured by injecting 100 nm (diameter) carboxylated polystyrene beads into the cytoplasm. During the process of cell division, both intracellular elasticity and viscosity was found to increase from 10^{-1} to 10^{0} Pa in the frequency range of 10^{-1}–10^{1} Hz in the transition from the metaphase to the anaphase. The result is presumably due to the remodeling and redistribution of cytoskeleton. The change of cortical stiffness during cell division was also measured by AFM and found that the elasticity of the cortex increased from 1.7 to 3.9 and 10.1 kPa, respectively.

8.4.2 Fibroblasts

Optical trapping method is useful for a low stress and for a small deformation to measure. The viscoelastic property of single fibroblasts cell was measured by using

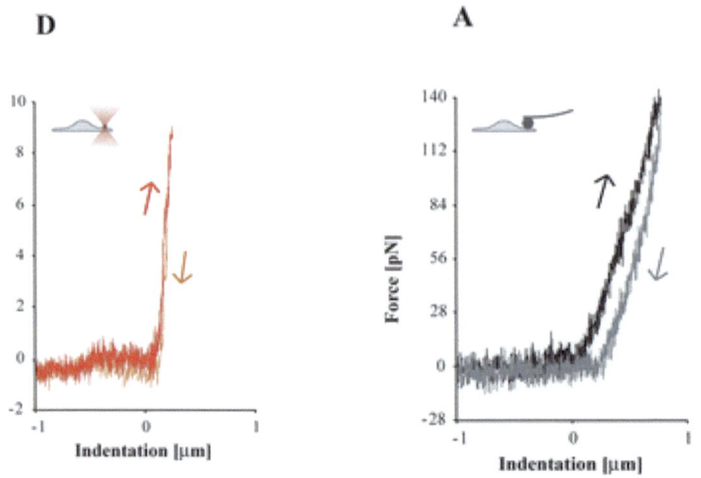

Fig. 8.7 Sub-micron cell indentation with the optical trapping method for a low stress a small deformation (**a**), and AFM for a higher stress and a higher deformation (**b**). Of fibroblasts cell [11]

an optical trap in vertical direction, normal to the plane of the coverslip in the range of forces below 30 pN and deformations below 200 nm. Under such small deformations, the cell showed a reversible elastic response of ~ 100 Pa, independent on the rate of deformation and indicated the dissipated energy is negligible (Fig. 8.7a) [11]. This means that the elastic modulus of fibroblast is mainly determined by the actin cortex which Young's modulus has c.a.100 Pa. This value was well coincided with the data obtained by AFM measurements performed at different indentation rates and at forces up to 30–75 pN. Dependence of elasticity on deformation rates was measured using AFM at the rates from 0.3 to14 μm/s and found that the Young's modulus within a force of 7–30 pN does not depend on these indentation rates. This means that the cells respond elastic for small deformations [11].

However, there appears considerable hysteresis when a cell is indented with a force of 140 pN or more (Fig. 8.7b). This indicates that when the force becomes large, the response is not elastic and the energy is dissipated by the viscous components during indentation. When forces applied to the cell increase higher than 150 pN, the Young's modulus increased up to 330 Pa with increase with deformation rate. This result indicates viscous effects increasingly contribute to the cell stiffness. The apparent modulus was found to increase with a power law of an exponent $\beta = 0.31$ in an indentation range of 0.8 and 1.2 μm, while decreased to $\beta = 0.17$. At smaller indentations between 0.15 and 0.45 μm. This indicates that when the cell deformation becomes larger, the viscous effects have a stronger effect on the apparent modulus.

The transition from elastic at small deformations to viscous at higher deformation shows that the elastic and viscous components are not homogeneously distributed in the cell: at low forces the probe mainly deforms the surface layer whereas at higher forces the probe penetrates deeper into the cell and its viscous interior is measured. Predominant structure at the cell surface is the lipid bilayer and that of the interior cytoplasma is mainly the actin cortex. Thus, the obtained modulus is an “apparent value” consisting of the complexes of actual Young’s modulus and a viscous component Fig. 8.7.

8.4.3 *Red Blood Cell (RBC)*

Viscoelastic behavior of the red blood cell (RBC) is particularly important from the circulatory pathology viewpoint. RBC is often larger than capillary diameter, nevertheless it is strictly required to flow and transit smoothly through capillary. Therefore, RBC is required to have extremely high spatial–temporal deformability with microcirculatory function as to quickly respond to the size, shape of capillary and blood flowing mode. The complex modulus of the isolated RBC was measured using ferromagnetic microbead by directly attaching to the membrane of healthy RBC. An oscillatory magnetic field was subjected to cause cell deformation from 0.1 to 100 Hz [12]. The apparent storage stiffness G' of the RBC was found as c.a.10^{-3} Pa/nm which was insensitive to frequency up to 30 Hz, while the loss modulus G'' ranged from 10^{-4} up to 10^{-2} Pa/nm which increased with increase in frequency from10^{-1} to 10^{2} Hz with a power law of the exponent $\beta = 0.64$. When the load was removed, the bead quickly returned to its original position and experimentally demonstrated entire reversible elastic deformability.

8.4.4 *Plant Cell*

In contrast to animal cells, cells in very firm plant cellulosic fibers have very high and tough property. The primary cell wall of plant, for example is made of two, sometimes three structurally independent but interacting networks and form gels [13]. In the plant cell cellulose does not exist as a single fiber, but as the microfibrils which are paracrystalline assembles of dozens of (1-4)β D-glycan cellulose chains. They strongly connect each other with cooperative hydrogen bonding along the chain length and give mechanical toughness. Cross-linking glycans bind with cellulose microfibrils through hydrogen bonding. These cellulose microfibrils and cross-linking glycans lie embedded in the second network of matrix consisting of pectic polysaccharides (pectin is a mixture of heterogeneous, branched, and highly hydrated polysaccharides rich in D-galacturonic acid.). The third independent network consists of structural proteins or a phenylpropanoid network. They bind tightly to the glycan chains in the cellulose microfibrils and fix the microfibrils into the proper spatial arrangement. Here again, water is the main component of cell membrane and accordingly the assembly of these polysaccharides occurs in an aqueous environment to form hydrogel. Water is critically important to maintain the networks in the proper conformations. Water is the medium that allows the permeation of ions and molecules through their interstitial space and it is critical to sustain the living organism. There, some molecules as large as 4 nm can permeate through the pores of primary wall due to extended conformations formed by the rod-like network of cellulose.

8.4.5 Cell Nucleus

Mechanical stress gives an important influence not only on the cells, but on the regulation of gene expression through a direct physical connection from the ECM to the nucleus across the plasma membrane. The evidence for this hypothesis is now becoming evident and it is considered that mechanical perturbations of nuclear matrix proteins can influence gene expression by exposing or obscuring molecular binding sites during mRNA transcription or through changes in DNA conformation. Therefore, in order to protect genome integrity the structural stability of nuclei is essential. Despite extensive knowledge on nuclear architecture and components, the mechanical response or deformation behavior of the nucleus remains largely unknown.

Earlier studies showed that nuclei in a condensed form of chromatin exhibits substantial mechanical resistance when aspirated by a micropipette or indented by an AFM cantilever. Another micropipette aspiration experiment showed that chondrocytes and their nuclei exhibit substantial deformation followed by a slow flowing into the micropipette when applied stress and indicated that chondrocytes and nucleus are characterized as a viscoelastic solid rather than that of a viscous fluid with cortical tension. Genetic manipulation showed that mechanical properties of the nucleus changes with the state of compaction and spatial localization of chromatin and leads to decreased ability to withstand mechanical force.

Another experiment showed that isolated nuclei behave as viscoelastic solid materials similarly to the cytoplasm, and was 3–4 times stiffer and nearly twice as viscous as the cytoplasm. There is another report that the equilibrium Young's modulus of the nucleus is a order of 1 kPa, which is consistent with an elastic modulus of 1–5 kPa of isolated eukaryote chromosomes and neutrophil nuclei. A thin microneedle (diameter; ~1 μm) based force measurement combined with biochemical perturbations of chromatin of human cell nuclei allowed quantitative determination of the mechanical response of isolated nuclei to localized forces [12]. In the presence of 5 mM Mg^{2+} (Mg^{2+} ions are abundant more than 1 mM in nuclei) at which chromatin is assembled into highly compacted domains, nuclei showed sub-micrometer- scale deformation under a small stretching force (~20 nN) and largely retained their round shape. The deformation was restored as soon as the force was removed and showed predominantly elastic property. On the other hand, when a larger force is applied (~100 nN), the nuclei exhibited more remarkable deformation which persisted the deformed shape after the force is removed and showed more viscous response. The repeated stretch–release cycles with different force magnitudes revealed that nuclei increased a resisting force in proportion to the increase of deformation and yielded at around 50 nN. The rigidity of the nuclei within the elastic deformation was determined as 55.4

± 21.9 n N/μm and showed that nuclei in the condensed chromatin have significant elastic rigidity. However, it was revealed that these mechanical responses of chromatin are totally disrupted when the condensed state is changed by decreasing cation concentration, by acetylating histone tails, or by digesting linker DNA and results in a loss of nuclear rigidity. In contrast, chromatin assembly factors, such as cohesin, condensin II, and a major nuclear envelope protein lamin were not affected. This suggests that linker DNA and inter-nucleosomal interaction via histone tails play a predominant role in providing nuclear rigidity. Thus, rigidity of nuclei depends on the organization properties of chromatin, especially on the continuity of the nucleosomal fiber and interactions between nucleosomes.

8.5 Cytoskeletal Network in the Cell

As described in the previous section, the cell keeps internal mechanical tension to maintain the shape and to resist deformation in response to the external force. It was also emphasized that the mechanical responses of the cell are inherently coupled with such biological activities as cell division, cell movement and others. All these mechanical responses are involved by the filamentous cross-linked biopolymer network inside eukaryotic cells called cytoskeleton. These filaments have spatially well-organized and dynamic 3D structure which enable not only to maintain the shape, but to give the motion and necessary deformation upon biological requirements under the frequently changing mechanical environment. Cytoskeleton is also a major contributor as a scaffold to support the 3D morphology of cell interior.

The cellular cytoskeleton consists of three types of self-assembled protein filaments–actin, tubulin, and intermediate [3] (Fig. 8.8). Actin filaments determine the shape of the cell's surface and are responsible for whole cell locomotion as muscle contraction, cell motility, cell division cytokinesis. Microtubules determine the positions of membrane-enclosed organelles, direct intracellular transportation of organelle, and chromosomal segregation. Intermediate filaments provide mechanical strength to resist compression and shear stress.

These networks exhibit highly nonlinear elasticity and ATP-fueled non-thermal motilities. Although recent technological advances enabled precise measurements

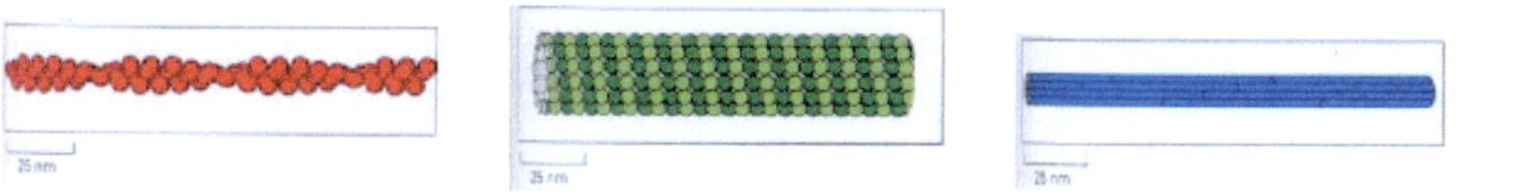

a) b) c)

Fig. 8.8 cytoskeletal proteins, (red): actin filament, (green): microtubules, (blue):intermediate filament (From Wikimedia Commons, the free media repository)

of mechanical response of cytoskeletal protein networks over a wide range of length scales, a complete understanding of the mechanical behaviors of cytoskeleton is still far from complete. One possible approach to solve this problem is certainly to measure dynamic mechanical responses of the cytoskeleton using living cells, but the mechanical responses of living cytoskeleton in the living cell is so complicated and extremely difficult to elucidate the physical origins of this cytoskeletal elasticity as described in previous section.

One possible approach is to study the behaviors of reconstituted network of cytoskeletal proteins in vitro. In fact, these measurements enabled to some extent to elucidate various mechanical responses and revealed the mechano-chemical responses on suitable physical models. It was found that the mechanical response of cytoskeletal networks is highly nonlinear critically dependent on the amount of the stress applied and this was, at least a part of the origin of the nonlinear responses of the cell as described in the previous section. When the elasticity increases with increasing applied stress or strain, the materials are called "stress-stiffen" or "strain-stiffen". On the contrary, when the elasticity decreases with increased stress or strain, the material is called "stress-soften" or "strain-soften". Many cytoskeletal networks, for example, F-actin networks and intermediate filament networks show such stress-stiffening behaviors. In addition, cytoskeletal networks prepared in vitro show the length-scale dependences at which the measurement is taken due to the inherent rigidity of cytoskeletal polymers which makes the study more complicated. However, through these measurements of reconstructed cytoplasmic network in vitro, a wide range of unique behaviors of cytoskeletal proteins within these networks has been revealed. Study using these reconstructed "non-living" materials has made possible to build predictive models to illustrate mechano-biological behaviors which is becoming new area of modern life science. Methodology and viscoelastic understanding of gel science could greatly contribute as view from materials science (refer to cytoskeletal gel in 9-5).

8.5.1 Actin Filament

Actin is the most abundant protein found in eukaryotic cells. It comprises 10% of the total protein mass in muscle cells and up to 5% in non-muscle cells. More than 100 globular protein monomers (G-actin) are polymerized to form fibrous actin (F-actin) with a diameter of 5 nm, thickness of ~ 8 nm and contour lengths up to 20 mm.

F-actin is a polar fiber with helical structure in the same direction which three single F-actin convolute each other to form coiled-coil stranded structure and forms F-actin fibers. They form a cortex beneath the cytoplasmic membrane to support cell movement.

The extensional modulus or Young's modulus of F-actin is very high showing approximately 10^9 Pa and this value is almost the same with that of some plastics. However, since the filament diameter of F-actin is so thin, bending modulus is very low and thermal energy actively causes transverse fluctuations resulting in substantial shortening of the end-to-end length to give rather extended coil-like conformation. The ratio of bending modulus to thermal energy kT is defined as the persistence length. This is the length over which vectors tangent to the filament contour does not correlate due to the thermal bending fluctuations. The persistence length of F-actin is 8–17 μm, which is larger than that of DNA 0.05 μm, but smaller than microtubules 1000 μm and categorized as semi-flexible chain.

When F-actin is extended, it exhibits viscoelasticity with linear and nonlinear responses depending on the extent of elongation. In the linear regime, tensile force applied to the end of the filament results in proportional extension of the filament and the ratio of the tensile force to the displacement is constant (Fookian). However, at high extension, the tensile force diverges nonlinear and the force-extension relationship becomes to depend sensitively on the magnitude of extension. When the extension of single F-actin approaches the contour length, the nonlinearity in the force–extension relationship becomes more complex and diverse. Elucidation to describe the full model of the mechanical response of cytoskeletal networks is still not completed.

In response to cell signaling, F-actin fibers continuously perform polymerization and depolymerization processes called tread milling and form bundles as well as networks. In vivo the mechanical properties and functions of actin cytoskeleton are predominantly regulated by actin binding protein. This is the protein which aligns the filament, form bundles, bind covalently individual actin filament to give bundles or cross-linked mesh-like hydrogels. There are two types of actin binding protein and each displays the required arrangement to form actin network in the cell: one is bundling proteins which cross-link actin filaments to form a parallel array. *α-Actinin and fimbrin* are categorized in this type of actin binding protein. The other is gel-forming proteins which hold two actin filaments together at a large angle to each other to form a network structure. S*pectrin and filamin* are categorized in this type. Due to functions of these actin binding proteins, cytoplasmic actin hydrogels obtain the storage modulus G' of order 10^{-1}–10^{2} Pa which exceeds the modulus required to maintain the shape of the cell.

Since the viscoelastic study of live cytoskeleton is very difficult, F-actins networks was reconstituted in vitro to quantitatively evaluate the contribution of cytoskeleton mechanics. The reconstituted F-actin forms a gel-like network showing an entropic elastic response where the mesh size depended on the F-actin concentration. Similar to F-actin fiber, the semi-flexible networks deformed in proportional to the force within the low force range, while they show strain-stiffening with change in contour length of F-actin filament under a larger force. If the concentrations of F-actin and actin cross-linking protein are both increased, the obtaining actin filaments become stiff and the entropic elasticity is reduced. Since such filaments are cross-linked by the protein, they can sustain much larger stresses compared to pure F-actin networks, providing a reasonable explanation that cells can resist a broad range of external stresses or internal tension. The nonlinear strain-stiffening response of these F-actin networks at macroscopic scales is the emergent cooperative result of the nonlinear stiffening of individual actin filaments. These nonlinear mechanics of in vitro cross-linked F-actin networks also imply the autonomous regulation of cell stiffness.

8.5.2 Microtubule

Microtubule (MT) is another important protein constituting of the cytoskeletal network of eukaryotic cells that plays various important biological roles. MT has the largest fibrous structure which is composed of protein heterodimers: α and β tubulin as the building subunits. There, α and β tubulins alternately bind to form protofilaments and laterally arrange to form very rigid cylindrical hollow tubes of 25 nm in diameter. Because of its unique cylindrical structure, MT have a high bending rigidity that originated from the large diameter of 25 nm which give very rigid structure with a large persistence length as 1000 μm which is about160 times larger than actin filaments. As such, they constitute an example of rod-like polymers, whose mechanical and rheological properties are not sufficiently understood.

As a part of cytoskeletal network, MT displays unique and important mechanical functions which are very different from those of F-actin network. They form networks that guide active intracellular transport and support the overall cell structure. MT mechanically pulls the chromosomes apart to split the dividing cell into two at mitosis stage. MT transports substances and drugs from one part in the cell to another. It mechanically supports the plasma membrane and stabilizes the cell to sustain the stress and strain. Besides, MT enables sperm to swim, fibroblasts to crawl across surfaces in the white blood cells and an axon to extend in the neuron.

MT has high dynamic nature. Depending on the condition, it can perform growth and de-composition cyclically known as dynamic instability. This dynamic behavior allows MT to rapidly reconstruct the structure into different functional network architectures. MT is able to change its shape from specialized mitotic spindle in the dividing cells to the radial network that controls migration of polarized interphase cells. All these functions are made possible by the mechanical toughness of tubular MT and the cross-linked network. MT has numbers of regulatory accessory proteins called microtubule-associated proteins in its network and due to this, MT network as a cytoskeleton behaves as chemomechanical network and the mechanical property of the network is controlled as whole by the MT-associated proteins. For instance, the MT of axon has a stiff and straight form and keeps the extended structure since it must carry out of a long-distance axonal transportation. In contrast, MT in a proliferating cell is quite dynamic and changes its form easily to enable rapid redistribution during transitions between interphase and mitosis. A fact that mesh size of an in vitro MT network is orders of magnitude smaller than its persistence length makes thermal fluctuations minimize. This gives the MT subsequently different behaviors from those of actin networks, especially at high strain and a high-frequency ranges.

Mechanical properties of MT have been studied by using a variety of techniques such as optical tweezers, hydrodynamic flows, osmotic pressure and AFM. Elastic modulus of MT wall was found to be very high showing almost 1 GPa.

The linear and nonlinear viscoelastic properties of purified MT solution as well as network cross-linked by biotin–avidin was measured. Both solutions and networks showed elastic moduli on the order of a few Pa in the linear regime. The elastic moduli show a power-law dependence on tubulin concentration C with $G' \sim C^{\nu}$, where $\nu \approx 1.4$ for solutions and $\nu \approx 1.6$–1.8 for networks. Under larger deformations a yield stress is observed which depends on the concentration. It should be noted that the rheology of MT solutions could not be explained by the Doi–Edwards model, where specific interaction between rigid rods is absent. This means pure MT solution undergoes effective cross-linking behavior very similar to the permanently cross-linked networks [14].

8.5.3 Intermediate Filaments

Intermediate filaments (IFs) are the third cytoskeletal network which is not sufficiently investigated. It contains various proteins such as keratins, vimentin, desmin, neurofilaments, and lamins and plays dynamic and mechanical contributions in the cell. Lamins, for example, sustain the shape and mechanical stability of the nucleus. Vimentin forms a composite network with actin fibers and MT in the cytoplasm to give cellular mechanical integrity.

IF is the filamentous network composed of these proteins and have approximately 10-nm-wide. Each filament of IF network consists of α-helical fibrous proteins which are bound laterally and longitudinally each other to form coiled-coil bundles. Unlike F-actin or microtubules, these filaments are nonpolar. The diameter and the length of the filament depend on the protein and the conditions. The persistence length for vimentin filaments is on the order of 1 μm which is one order of magnitude smaller than that of F-actin and three orders of magnitude smaller than that of MT.

Bending modulus, Young's modulus, and a shear modulus of IF filament were measured using AFM, both native and glutaraldehyde-stabilized vimentin filaments and were found as 300 MPa, 900 MPa, and 2 MPa, respectively. The very low value of shear modulus suggests that individual subunits in the filament may slide relative to each other. Filamentous proteins in IF such as neuro filaments, desmin, and keratin are considered as flexible and stretched several times their original length before breaking. The full stress–strain curve for individual IFs is not yet measured, but large bundles of IFs are found to be highly extensible and can sustain under very large stresses before breaking. This remarkable extensibility of IFs is a strong contrast to actin or MT, which are broken under relatively small extensive force. This is the unique mechanical function that characterizes a role of Ifs in cell, i.e., it provides mechanical stability to cells under the large deformations.

An elastic modulus of vimentin networks has been determined as ranged between 10 and 100 Pa which increases with increase in IF concentration. IF networks are highly dynamic and exhibit strain-stiffen property like F-actin network. This property and the extensive nature of the individual filaments allow IF network to withstand large stresses and strains before rupturing. These strain-stiffening behaviors show a unique dependence of the modulus on stress which are different from those of rigid actin cross-linked networks. Another interesting mechanical feature of IF network is when IF network consisted of large number of bundles shows much increased persistence lengths and increased bending than those with a lower number of bundles. This means bundle number of IF is important on elasticity of the cell. However, the mechanism that regulates the dynamics and mechanical transmission in the IF networks remains unsolved. Exploring the mechanical responses of composite networks of IFs with F-actin and MT is also another future subject to elucidate the mechanical response of live cells.

References

1. The author considers the bone is in the gel state because it largely consists of inter-woven composite made of hydroxyapatite and collagen and the total water content of the bone is around 11%
2. H. Abe, K. Hayashi, M. Sato (eds) *Data Book on mechanical Properties of Living Cells, Tissues, and Organs* (Springer, Tokyo 1996), *The Mechanical Properties of Biological Materials* (Cambridge University Press, 1980)
3. B. Alberts, B. *Molecular Biology of the Cell*, 4th edn. (Garland Science, New York, 2002)
4. B. Fabry et al., Scaling the microrheology of living cells. Phys. Rev. Lett. **87**, 148102 (2001)
5. I.P. Herman, *Physics of the Human Body* (Springer, 2007) (Chapter 4)
6. J. Kastelic, A. Galeski, E. Baer, The multicomposite structure of tendon. Connective Tissue Res. **5**, 11–23 (1978)
7. S. Torp, E. Baer, C.D. Armeniades, E. Baer, in *Structure in Fibrous Biopolymers*, vol. 223, ed. by E.D.T. Atkins, A. Keller (Butterworth, London)
8. H. Yamada, *Strength of Biological Materials* (Williams & Wilkins, Baltimore 1970)
9. S. Nawaz, P. Sánchez, K.L.S. Bodensiek, M. Simons et al., Cell visco-elasticity measured with afm and optical trapping at sub-micrometer deformations. **7**(9), e45297 (2012)
10. J. Sebag, E.A. Balazx, Morphology and ultrastructure of human vitreous fibers. Invest. Ophthalmol. Vis. Sci. **30**, 1871 (1989)
11. M. Puig-de-Morales-Marinkovic, K.T. Turner, J.P. Butler, J.J. Freidberb, S. Suresh, Am. J. Physiol. Cell Physiol. **293**, 597 (2007)
12. Y. Shimamoto, S. Tamura, H. Masumoto, K. Maeshima, Mol. Biol. Cell **28**, 1580 (2017)
13. B. Buchanan, W. Gruissem, R.L. Jones, *Biochemistry & Molecular Biology of Plants* (American Society of Plant Biologists, Maryland, 2000)
14. Y.-C. Lin[†], G.H. Koenderink[†‡], F.C. MacKintosh[§], D.A. Weitz*[†], Macromolecules **40**(21), 7714–7720 (2007)

Chapter 9
Biomimetic Functions of Hydrogel

Yoshihito Osada

Abstract Recent progress to obtain hydrogels with biomimetic functions is introduced. Recently developed functions of hydrogels which possess tissue-like functions as tendon, cartilage, muscle-like motility, and ECM-like scaffold for regenerated medicine will be introduced. Reversible sol–gel functions of highly hierarchical cytoskeletal protein gels and on–off drug delivery gels are also focused.

Keywords Topological gel · Nanocomposite gel · Double network gel · Tetra-PEG gel · Chemomechanical gel · Artificial muscle · Gel actuator · Shape memory gel · Scaffold · Drug delivery gel · Cytoskeletal gel · Hierarchical gel · Self-assembly

Scientists of polymer hydrogels have made a great effort over the many years trying to mimic the biological functions of organs, soft tissues, cells, and others using gel. As explained in Chap. 8, soft tissues such as tendon, ligament, cartilage, and others are the hydrogels largely made of fibrous proteins and polysaccharides. They are mainly composed of extra cellular matrix (ECM), three-dimensional highly organized and inter-woven hydrogel. Therefore, it is rather understandable that the gel scientists have made a tremendous effort to obtain the artificial tissues using polymer hydrogels.

ECM in tissues has three symbolic roles.

(1) Pertinent and excellent mechanical performance (tendon, ligament, and skin)
(2) Extremely low and robust frictional behaviors (articular cartilage)
(3) Effective and selective transportation of solute molecules for cell viability (every tissues).

Considering these excellent functions of tissues, scientists easily find that there exist several problems to be solved and shortages to be improved for polymer hydrogels at present to replace the tissue with synthetic hydrogels. They are:

Y. Osada (✉)
RIKEN, Wako, Saitama, Japan
e-mail: osadayoshi@riken.jp

Y. Osada (ed.), *Science of Gel*, https://doi.org/10.1007/978-981-95-0301-8_9

(1) Improving the mechanical toughness of the hydrogel to sustain under a severe mechanical dynamic environment as tendons and ligaments.
(2) Controlling the surface properties of hydrogels with low sliding resistance against other tissues. The articular cartilage sustains a daily compression of 10 MPa, while showing a very low resistance against sliding motion.
(3) Controlling the transportation of ions, hormones, drugs, and nutrition and improving the permeability to them. ECM controls the transportation of them and also cell migration by changing the local stiffness.

Gel scientists have been striving to design hydrogels with these critical parameters by using synthetic as well as natural polymer gels. It is the ultimate goal for gel scientists to design hydrogels with these critical parameters which natural tissues possess and to replace them with artificial ECM made of hydrogel.

In addition, hydrogels are highly expected to make 3D scaffolds for future regenerative medicine to cultivate induced pluripotent stem (iPS) cells for repairing and regenerating organs. Hydrogels are also expected to be a pertinent matrix which delivers drug and other biological substances at required time to required place (DDS).

In this chapter, recent progress to obtain hydrogels for biomedical uses is described.

9.1 Mechanically Tough Gel for Soft Tissue

The first problem to be solved for their practical applications is lack in their mechanical strength. Common hydrogels are very weak and brittle and easily broken even under a small amount of stress as a few kPa. The reason of lack of mechanical strength is attributed to its solution-like nature, i.e., low density of polymer chains and low friction between the chains. In addition, many synthetic hydrogels are highly inhomogeneous as described in Chaps. 1and 7 and brings about localized stress and additional decrease in mechanical strength. Thus, it is really a challenging problem to obtain hydrogels as strong as soft tissues.

This very important problem appears to have almost been solving mainly by Japanese scientists. Recently, four types of hydrogels with excellent mechanical performance have been developed [1–4]: They are "topological (TP) gel," "nanocomposite (NC) gel," and "double network (DN)" gel and "tetra-PEG gel" (Fig. 9.1). The first three cases still include inhomogeneity which is one of the reasons to makes hydrogel weak, but the problems seem to have been solved in the following different approaches; TP gel can spontaneously adjust the cross-linking distribution, NC gel reduced the cross-linking inhomogeneity effectively by using clay and DN gel by dissipating applied stress effectively. The tetra-PEG gel succeeded to minimize the inhomogeneity by using highly reactive coupling reaction.

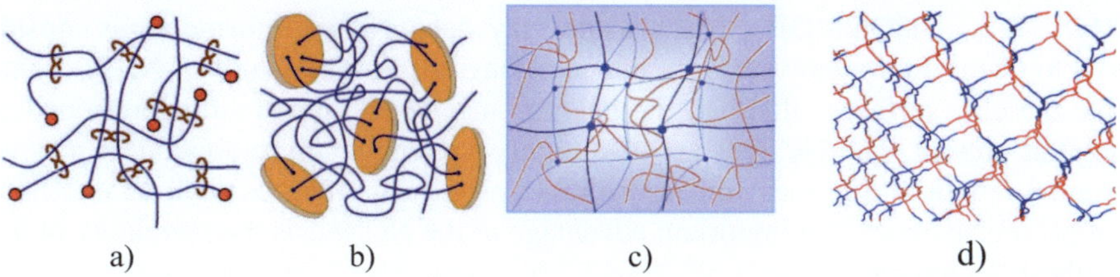

Fig. 9.1 Schematic network structures of mechanically tough hydrogels. **a** Topological gel, **b** nanocomposite gel, **c** double network gel, **d** tetra-PEG gel

9.1.1 Topological (TP) Gel

TP gels have figure-of-eight cross-linkers; they can slide along polymer chains (Fig. 9.1a). A typical example is the polyrotaxane gel synthesized using a supramolecular chemistry technique [1]; a polyrotaxane consists of a poly(ethylene glycol) (PEG) chain through which large end-groups, α-cyclodextrin (CD) circles are introduced at the chain ends and then chemically cross-led in an aqueous solution to give polyrotaxane gel. Point here is the cross-linking CD is able to slide along the network chain when stress is applied, thus can remove the local stress by making the conformational changes. Therefore, TP gel is highly swelling and stretchable. The gel swells to about 500 times its original weight and can be stretched to about 20 times its original length (in the as-prepared state). Another unique characteristic of the TP gels appears in their macroscopic and equilibrium mechanical behavior when the cross-linking density is low: the stress–strain (S–S) curve of the TP gel under uniaxial elongation exhibits lower convexity over all the range of strain. This is different from the usual chemical gels, for which the S–S curve exhibits an upper convexity in a small strain regime.

Cross-linkers account for the special feature of the TP gels, which distinguishes them from the usual chemical and physical gels. In the chemical gel, the tension of each polymer chain is heterogeneous and scission of the polymer chains gradually occurs, while in the TP gel, the relative sliding of the polymer chains by figure-of-eight cross-linkers occurs to equalize the tension of each polymer chain. The authors refer to such an adjustment of the tensions by sliding as a "pulley effect" [1]. The sliding mode (pulley effect) is obviously related to the large water absorbance mentioned above.

9.1.2 Nanocomposite (NC) Gel

In the NC gel, polymer chains are cross-linked via inorganic clay slabs on the scale of a several tens of nanometers (Fig. 9.1b) [2]. A typical combination of chemicals is N-isopropylacrylamide (NIPA) for the polymer chains and a hectorite

$[Mg5.34Li_{0.66}Si8O\ 20(OH)_4]Na_{0.66}$ as the clay slabs. Network formation is considered to occur on the clay surfaces due to strong adsorption nature of NIPA chains onto the clay surfaces, although the gelation mechanism is not sufficiently understood at present. The NC gels are also highly stretchable up to about 10 times its original length and possess other favorable physical properties such as excellent optical transparency. An important advantage of the NC gels is the simplicity of the synthesis technique.

9.1.3 Tetra-PEG Gel

Tetra-PEG gel is formed by the polycondensation method between mutually reactive tetra-armed prepolymers, namely (Tetra-PEG) having amine (Tetra-PEG-NH_2) and activated ester (Tetra-PEG-NHS) terminal groups (Fig. 9.1c) [3]. This type of cross-end coupling has advantages on suppressing the heterogeneity over conventional polycondensation and in fact, the reaction is considered to achieve up to 90% of reaction conversion. The homogeneity of small angle neutron scattering showed no excess scattering in the range below 200 nm and suggested the inhomogeneity which leads to decrease the mechanical performances was substantially suppressed. The models of elastic modulus shift from the phantom network model to the affine network model with an increase in polymer volume fraction.

9.1.4 Double Network (DN) Gel

The DN gel consists of two interpenetrated polymer networks: one is made of highly cross-linked rigid polymer and the other of loosely cross-linked flexible polymer (Fig. 9.1d) [4]. DN gel has a special interest since the mechanical enhancement reminds that of ECM which controls the hardness and softness by changing the rigid(collagen)–flexible(elastin) composition. Figure 9.2 shows stress–strain curves for the polyacrylamodo-2-methylpropane sulfonic acid (PAMPS)-poly acrylamide (PAAm) DN gel. The DN gel holds up to a stress of 17.2 MPa. where the vertical strain λ. is about 92%, while the single PAMPS gel and PAAm gel show breaking at the stresses of 0.4 MPa ($\lambda = 41\%$) and 0.8 MPa ($\lambda = 84\%$, respectively. The DN gels, containing about 90 wt. % water, possess both hardness (elastic modulus 0.3 MPa) and toughness (fracture stress ~10 MPa) and the fracture energy attains up to10^3 J m^{-2} [4] much exceeding the reported values of typical hydrogels that ranges within 0.1–1 J m^{-2} and those of the common rubbers. The reason of the mechanical toughness of DN gel is attributed to the loosely cross-linked second network effectively absorbs the elastic energy around the crack either by viscous dissipation or by large deformation, thus preventing the crack growth to a macroscopic level. Dynamic light scattering (DLS) analysis revealed that the first network contains inhomogeneous large "voids" which inherently appears by the nature of radical polymerization mechanism (refer to

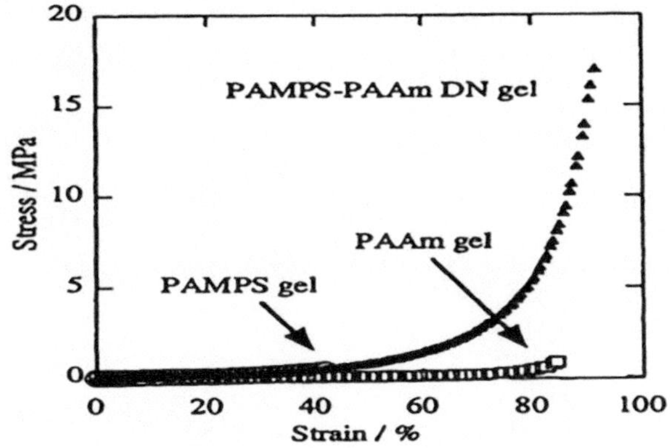

Fig. 9.2 Stress–strain curves for PAMPS—PAAm DN gel, PAMPS gel, and PAAm gel [4]

Chap. 3) to which the second polymer penetrates, entangles into the PAMPS network due to its ductile nature and fills the "voids" [4].

This situation is similar to the porous structure of animal bones which prevent crack propagation by suppressing stress concentrations as a "crack-stop." Since DN hydrogels can withstand from a few to tens of MPa mechanical strength and the value is comparable to that imposed on human articular cartilage (1.9–14.4 MPa), it seems to satisfy the requirement of mechanical property for an articular cartilage substitute. Under the same approach, biological gels composed of bacterial-cellulose (BC), gelatin, sodium alginate, gellan gum, ι-carrageenan [5], and electroconductive gels composed of polythiophene derivatives (E-DN-gels) which show both ionic and electronic conductivities were developed [6].

9.2 Low Surface Friction for Cartilage

Another very important problem to be solved is to reduce the frictional coefficient of hydrogel. This is a fundamental problem of mechanics as well as for making artificial joints and muscle system.

In contrast to the artificial lubrication systems of solid matters which show a frictional coefficient $\mu \approx 0.1$ in the presence of lubricant, bio-organs show fascinating surface frictional behavior; the cartilage of animal joints has a friction coefficient of 0.001–0.03 [7]. It is not understood why the synovial cartilage at the joints makes the friction so low even under high pressure between the bone surfaces as 3–18 MPa, and under the low sliding velocity as a few centimeters per second. Under such conditions, the lubricating liquid layer cannot be sustained between two solid surfaces and the hydrodynamic lubrication does not work. On the basis of fact that the cartilage is hydrogel, a study of friction on hydrogels has systematically been studied and found that the hydrogel can exhibit extremely low lubrication under certain conditions (Fig. 9.3) [8].

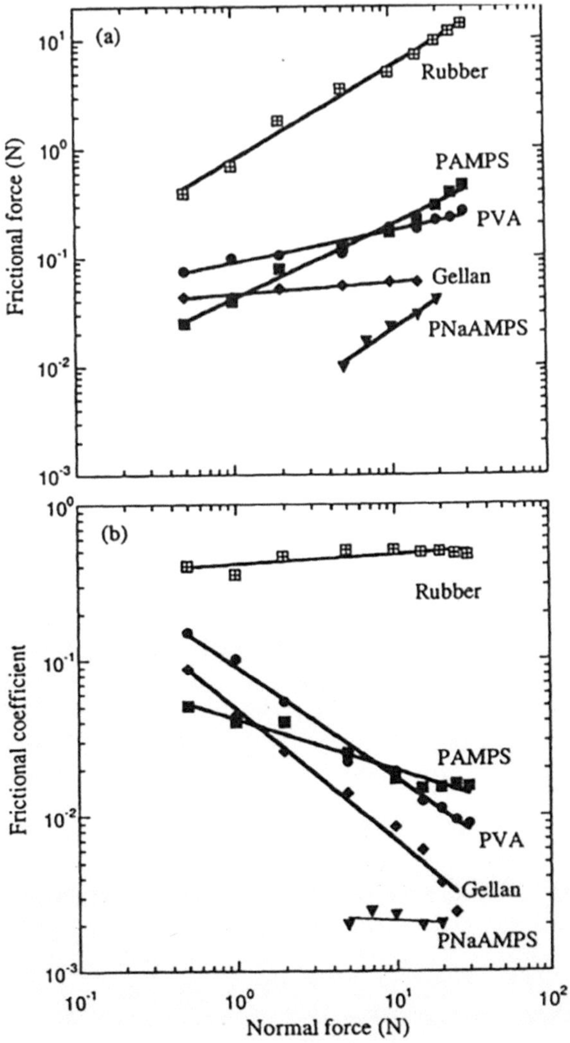

Fig. 9.3 Dependences of frictional force (**a**) and friction coefficient (**b**) on the load [8]. Hydrogels used were indicated in the figure. PNaMPS: sodium salt of AMPS, PVA:poly(vinyl-alcohol), The measurement was carried out on a glass substrate in air [8]

To understand the friction between two solids, one should go back to the Amonton's law proposed in 1699, $F = \mu W$ which says that the frictional force F is linearly proportional to load W and frictional coefficient μ, and does not depend on the apparent contact area A of two solid surfaces or sliding velocity *v*. It was found that the friction of gel does not simply obey neither the Amonton's law nor Newton's liquid lubrication law, but shows totally different and complicated features. It was found that the friction of gel strongly depends on the surface properties of opposing substrates and sliding conditions and Gong et al. proposed a "Repulsion-Adsorption Model" [9].

According to a repulsion-adsorption model [10], when the polymer network of the gel and the substrate is repulsive, the polymer network will be repelled from the surface and the friction is dominated by liquid lubrication of water layer surrounding the polymer chains. In contrast, when the polymer network has an attractive with the solid surface, the polymer network is adsorbed onto the substrate and the adsorbing polymer chains will be stretched while sliding and the elastic force increases with the deformation and eventually detaches the adsorbing polymer chain from the substrate, which in turn appears as the frictional force.

The model predicts that (1) when the substrate is repulsive and the compressive stress (normal load) is not very high, the frictional force is lower than in the attractive case and it increases linearly with the normal pressure and velocity (2) when the substrate is attractive, the frictional force increases with the attraction strength. For weak attraction, the pressure dependence of the frictional force is much weaker than in the repulsive case. It becomes stronger when the attraction strength increases.

In contrast to the case of solid friction, frictional force of the gel was depended on the apparent area A and accordingly, the power dependences of load (**W**) and apparent area (**A**) is expressed as

$$\mathrm{F} \propto \mathrm{W}^{\alpha}\mathrm{A}^{\beta} \tag{9.1}$$

α is obtained by the experiments varying W under constant A. Similarly, β is obtained by the experiments varying A under constant W. Both α and β varied from 0 to 1, but surprisingly, the sum was 1.0 (Fig. 9.4) [10].

$$\alpha + \beta = 1.0 \tag{9.2}$$

$$F \propto (W/A)^{\alpha} A \quad (0 < \alpha < 1) \tag{9.3}$$

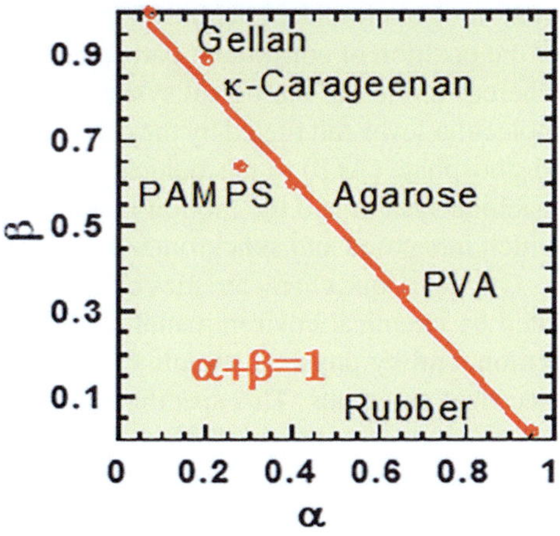

Fig. 9.4 Power dependences of load (α) and apparent area (β) [10]

Most importantly, the coefficient of friction of gels, μ, varies over a wide range and exhibits very low values as low as $\mu \approx 10^{-3}$–10^{-4}, which usually cannot be obtained from the friction between two solid materials.

It is crucially important to evaluate the wear properties to apply gels as artificial articular cartilage, because articular joints are subjected to high, rapid shear forces for millions of cycles over a lifetime. The pin-on-flat wear testing revealed that the maximum wear depth of DN gels under 10^6 of friction cycles, equivalent to 50 km of friction are similar to that of UHMWPE and demonstrated DN gel is potentially usable as a replacement material for artificial cartilage.

Although mechanism of the friction of the cartilage is not fully understood, the importance of polysaccharide molecules in the synovial fluid is pointed out. In this connection, it was found that the presence of solvated polyelectrolyte immersed in the gel as well as polyelectrolyte brushes on a hydrogel surface can drastically reduce the sliding-friction coefficient as low as 10^{-4}–10^{-5} [11, 12]. Thus, novel hydrogel materials, showing fracture strengths as high as ~ 10 MPa and frictional coefficients as low as 10^{-5} will find wide applications as *zero "μ" technology* in the biomedical field, for example, as substitutes for articular cartilage and bio-tissue.

It is still a question why a gel possessing solvated polymer chains and dangling polymer chains on its surface exhibits such a low frictional coefficient. However, study on friction using hydrogel is important to understand the mechanism of friction in biological systems which may lead to designing low-friction artificial systems.

9.3 Soft and Wet Actuator for Muscle

Man-made machines differ from biological power systems at least from two aspects. One is their materials of architecture and the other is their energy source. Man-made machines are basically made of hard and dry materials like metals and ceramics, whereas biological systems are of soft and wet molecular components. Secondary, motion of man-made machines is made by a consequence of relative displacement of the position of constituent parts and is driven by electrical or combustion energy, whereas motion of biological system as muscle is achieved by the deformation on a molecular level and fueled by the direct conversion of chemical energy of adenosine-tri-phosphate (ATP) to mechanical power. In other words, biology is the molecular machine system and the motion is made by the cooperative molecular deformation which integrated and synchronized and developed to a macroscopic level.

One of unique characteristics of a hydrogel is that the swelling state can be modulated by chemical environmental changes such as pH, ionic species, ionic concentration, and by physical stimuli such as temperature, UV light, magnetic field, and electrical potentials. This specific nature enables dynamic control of the expansion and contraction of the gel and lead to numerous gel applications as stimuli-responsive active materials. Electro-driven artificial muscles, magnetic-responsive actuators, temperature controlled shape memory gels, and on–off drug delivery systems which are introduced in this chapter are the pioneering examples of these soft and wet gel devices.

A chemomechanical system is defined as one that can produce macroscopic mechanical energy caused by microscopic deformation in response to changes in an external environment [13]. The chemomechanical systems are often called as artificial muscle or gel actuator performing as a new type of mechanical transducer. They are often accompanied with the physic-chemical transportation of ions, change in ionization state, change in chemical structure, and electrochemical reactions in the gel.

In the contrary, mechanical deformation of polyelectrolyte gels induces a spontaneous ionization which in turn produces electrical potential as large as a few millivolts [14]. This is the piezo-electric phenomena which reminds those of bones and collagen fibers and may have functions as a soft and wet touch-sensor or switching elements.

9.3.1 Chemomechanical Gel

When polyelectrolyte gel is interposed between a pair of plate electrodes and a DC voltage is applied, the counter ions in the fluid travel toward the electrode carrying surrounding water together, react on the electrode and release water. As a result, the gel contracts or bends accompanying with a repetitive current oscillation and entrainment phenomena [15]. The mechanism of contraction was interpreted as electrokinetic phenomenon and theoretically analyzed by one dimensional electrokinetic model [16].

On the base of this phenomenon, an electrically driven artificial worm-like muscle (gel looper) and a swinging pendulum have been made [17, 18] (Fig. 9.5a, b). The principle of these motions of the gel is associated with a reversible and cooperative complexation of the surfactant molecules with the polymer gel whereupon the electric field dominates and control the anisotropic complexation at one side of the gel surface. The response is inherently associated with the cooperativity of electrostatic binding of the surfactant molecules, whereupon the hydrophobic interaction of alkyl length of the surfactant molecules play crucial role.

The hydrogels with molecularly ordered structure such as copolymer of acrylic acid(AA) and stearyl acrylate (SA) [poly(AA-co-SA)] exhibits shape memory behaviors upon changing temperature: When the gel is warmed above order—disorder transition temperature, it becomes soft and can be freely deformed to give a new shape(write in). While the gel is kept below the transition temperature, it holds the new shape, but when the gel is heated again above the transition temperature, it spontaneously recovers the original shape (erasing) (Fig. 9.5c) [19]. Using this order—disorder transition in water—organic fluid, Chemical Motor which drive the gel rapid and prolonged rotation was developed [20]. The velocity of rotation ranges from 400 to 3000 r.p.m depending on its size (Fig. 9.5d). Driving force of the rotation comes from the surface spreading of the organic fluid squeezed out from the gel (Marangoni effect) in the process of recovery of ordered structure of the gel.

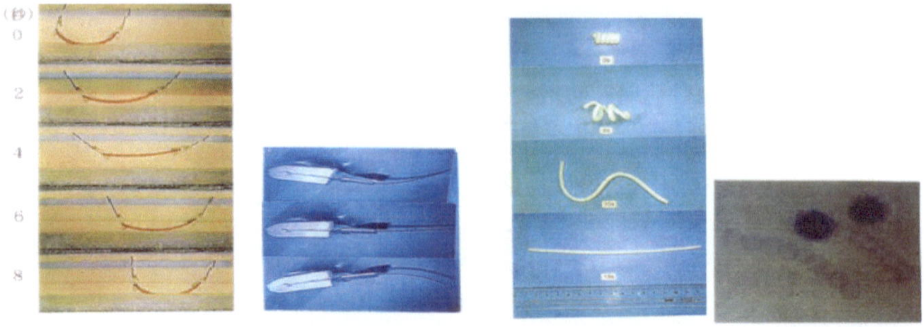

a) b) c) d)

Fig. 9.5 Chemomechanical gels; **a** gelooper, **b** gel fish, **c** shape memory gel, **d** chemical motor [18, 19]

9.3.2 Actuator of Electroconductive Gel

Electrically driven actuation can be made by using electroconductive polymer networks [21]. The principle is based on the reversible oxidation/reduction (doping/de-doping) reactions which accompany with concomitant production of charges on the conductive polymers as well as the counter ions in the solution. The electrochemical reactions followed the Faraday's laws shown in Eq. (9.4).

When the conductive polymer such as polypyrrole is oxidized through the cathode, it generates positive charges on the polymer chains due to electron extraction. The positive charges simultaneously generate negative counter anions $(A^{-1})^{-}$ in the solution to keep the film charge balanced.

$$\underset{\text{Neutral chains}}{(CP^0) + n(A^-)_{sol}} + m(S) \; \underset{\text{Oxidized chains}}{\left[(CP^{n+})(A^-)n(S)m\right]_{gel}} + n(e^-)_{metal} \tag{9.4}$$

The presence of positive charges in the film and the counter anions in the solution enhance the substantial solvation of water (or other solvent) to give the film highly swollen state and the film expands or stretches. The oxidized polymer film is then reduced through anode to neutralize the cationic polymer and the film loses the charges and the counter ions in the solution disappears. This makes the film de-solvation and the film shrinks.

Another type of conducting polymers is polymers of thiophene and its derivatives. In this case the polymers are reduced when electrons are injected to the chains via anode to give negative charges (*n*-doping) [21]. This must automatically be compensated by production of cations in the solution and this process results in an expansion of the conductive polymer by the same reason as polypyrrole. The reaction in this case can be summarized as Eq. (9.5).

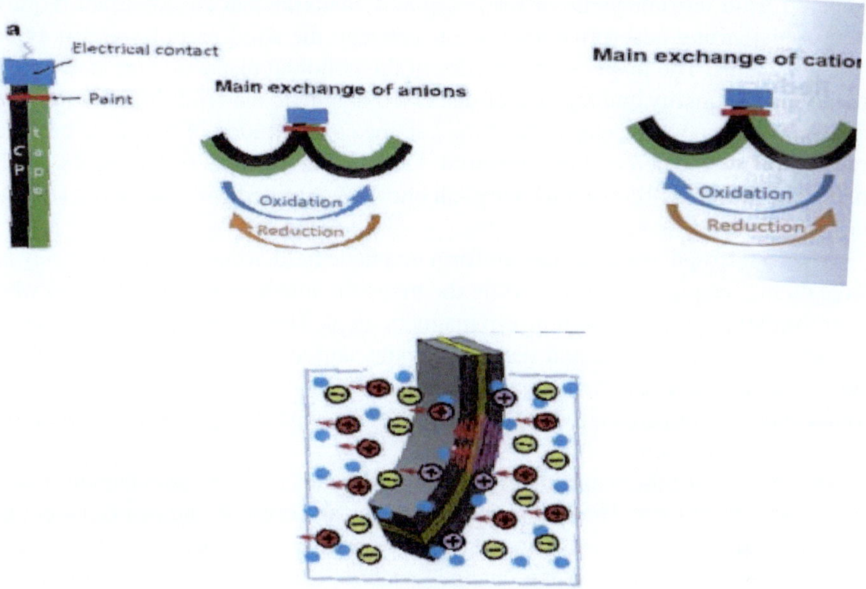

Fig. 9.6 Bilayered chemomechanical device composed of electroconductive polymer film (active) and a non-conductive film (inert) [21]

$$\underset{\text{Neutral chains}}{(\mathrm{CP}^0) + n(\mathrm{C}^+)_{\mathrm{sol}} + \mathrm{m(S)} + n(\mathrm{e}^-)_{\mathrm{metal}}} \rightarrow \underset{\text{Oxidized chains}}{\left[\left(\mathrm{CP}^{n^-}\right)(\mathrm{C}^+)n(\mathrm{S})_m\right]\mathrm{gel}} \tag{9.5}$$

One of the simplest ways to transduce volume changes of the conducting polymer film is through bilayer bending structure and an example of free-standing conducting polymer film is shown in Fig. 9.6. It is composed of an active layer made of electroconductive polymer capable of changing its volume during reaction and a passive layer which keeps its volume constant. Polyacetylens are another typical electroconductive polymers and various attempts to apply them have been made such as tweezers, micro-bulbs, and guides for optical fibers were proposed using [22].

Generally, chemomechanical actuators based on the electroactive conductive polymers can be operated under relatively low voltage. Although the strain is not large enough at present, they can exhibit large shrinking and bending forces.

9.3.3 Actuator of Magnetic-Responsive Gel

Magnetic field-sensitive gel, or simply ferro-gels is the chemically cross-linked polymer networks swollen with a colloidal dispersion of mono-domain magnetic particles with a typical size of about 10 nm [23]. The magnetic filler particles can be prepared from ferromagnetic or ferri-magnetic materials but an important requirement is the strong adsorptive interaction between the solid particles and polymer network chains. The magnetic properties of the colloidal dispersion are determined mainly by the quality and the size of dispersed magnetic material. In order to maintain the highest possible elasticity of the gel, the concentration of magnetic particles must be below the percolation threshold. One typical example of ferro-gels can be made from the cross-linked poly(vinyl-alcohol) (PVA) filled with magnetite (Fe_3O_4) particles [23].

When a ferro-gel is placed in a uniform magnetic field, it does not produce any net force on the magnetic particles, while the magnetic interactions are enhanced when placed under a spatially non-uniform magnetic field. The magnetic force depends on the magnitude of the field and the particle size and above a certain magnetic field the interaction induces changes in network conformation. The change of network conformation is integrated to lead to macroscopic shape changes of the whole ferro-gel to give the motion.

The principle of the shape changes and motility lies in an unique magneto-elastic properties of ferro-gel. The magnetic field drives and controls the displacement but the overall shape change is dominated by the combination of magnetic with elastic interactions. A variety of shape changes such as bending, elongation, stretching, contraction, and a combination of these basic distortions can be obtained repeatedly and many times without damaging the gel. The response time to obtain the new equilibrium shape is less than a second, and this seems to be independent of the size of the gel. All the shape changes are completely reversible. The magneto-elastic behaviors of ferro-gel cylinders suspended vertically between plan-parallel poles of electromagnets is shown in Fig. 9.7 [23] The relative displacement of the lower end of the ferro-gel plotted against the steady current intensity does not show notable increases in a low current region. However, when a current intensity increased more, an abrupt large elongation displacement occurs at a certain current intensity and this abrupt change in the size occurs within an extremely small increase of the current intensity. A further increase in the current intensity results in only small and gradual displacement. An elongation caused by a non-uniform magnetic field attains as large as 40% of the initial length.

Decrease of the current contracts the ferro-gel and similarly to the case of extension, an abrupt contraction occurs at the same current intensity. The discrete shape transition occurred within a time interval of 1 s, regardless of the gel size. It should be noted that in parallel to the relative displacement, forces that appear at the elongation and contraction also abruptly produced.

9.3.4 ATP-Fueled Biological Gel

As described in Chap. 8, actin-myosin and microtubule (MT)-kinesin are the constituent components of cytoskeletal proteins, and also the main biological power systems that can provide mechanical work. Motor proteins can be categorized into

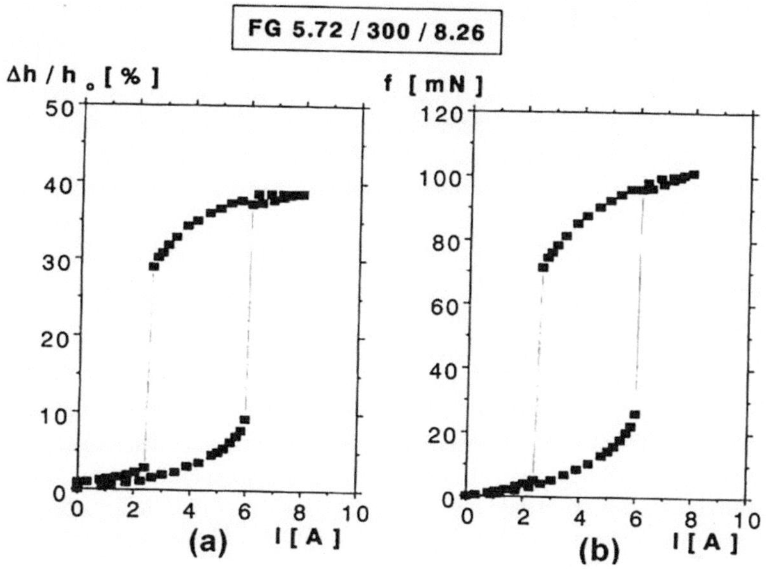

Fig. 9.7 Magnet-elastic behavior of a ferro-gel. **a** Relative displacement and **b** force generated as a function of steady current [23]

two groups: rotating motors and linear motors. Actin-myosin and MT-kinesin systems are linear motor proteins and are found in relatively larger living organisms, while rotating motors are observed in bacterial flagellum and mitochondrial.

Since biological motors driven by the direct conversion of chemical energy to mechanical energy are much more efficient than the man-made machines. Since they possesses of immense potential in biotechnology and biomaterial science, a variety of ATP-fueled biological gels of actin-myosin and MT-kinesin as building block have been studied.

F-actin fibers have been conjugated with various type of synthetic cationic polymers under share stress to form gels of polyelectrolyte complex. This large and aligned protein-polymer fibrous gels were then chemically cross-linked. When the oriented F-actin- complex gel was placed on the surface of oriented fibrous myosin gel, and added a drop of ATP, it moved along the myosin gel by coupling to ATP hydrolysis **(Nano-biomachine)** [24]. The mean velocity on the non-oriented myosin gel was $0.69 \mu ms^{-1}$, while that on the oriented myosin gel was $0.83 \mu ms^{-1}$ and both exceeded the velocity of native F-actin. The velocity of the motion of F-actin gel was proportional to the polarity of F-actin fiber and it was substantially dominated by the chemical structure of polymeric cations. The F-actin bundles exhibited a synchronized motion with periodical oscillating waves meaning that the self-assembled and covalently bound actin gel exert a highly cooperative and synchronized motion coupling to the oriented myosin fiber gel **(emergence).** Chemically cross-linked filamentous microtubules (MT) gel also exhibited an enhanced and prolonged motility on the kinesin-immobilized surface than that of un-cross-linked MT [25].

9.4 Scaffolds for ECM

The goal of tissue engineering is to restore diseased or damaged tissue by delivering functional cells, and signal molecules such as growth factors to the affected area. For this purpose, the scaffolds where the cell survive and grow should have functions to release of the signal molecules and uptake nutrients and oxygen to keep the implanted cells alive.

In living tissue, many cellular processes necessary for keeping the vital activity are regulated by spatial–temporal interactions with the surrounding extracellular matrix (ECM) in the tissue. Cells exist in 3D microenvironments of ECM with intricate cell–cell and cell–matrix interactions and complex transport dynamics for nutrients and cells. Cells bind to the ECM through cell surface receptors, secrete, assemble, and continuously regenerate the ECM, through which various cellular activities such as migration, proliferation and differentiation are maintained. Thus the functions of tissue of the human body owes to a close temporal and spatial cooperation of cell/ECM interactions. As described in Chap. 8, ECM is the porous and water containing 3D gel mainly composed of fiber-forming proteins and polysaccharides. Therefore, designing and fabrication of the ECM-like materials using hydrogels are highly interesting both from basic science and medical applications, since hydrogels, like ECM, are also able to permeate oxygen and transport ion water-soluble nutrients to keep vital activities [26].

However, there are various requirements to use polymer gels for biomedicals. Biocompatibility is one of them. Medical polymers must function immediately upon contact with internal organs and other systems without rejection reactions. In many common natural and synthetic gels, however, a number of platelet adsorption, as well as formation of blood clots occur. These phenomena are obviously due to chemical and physical characteristics of constituent polymers. The mechanism of blood clotting process is considered to be extremely complex and how this clotting phenomenon occurs is yet far from a full understanding. Recently, a notable progress has been made in this area and demonstrated the high potential of hydrogels as scaffolds for tissue engineering applications.

9.4.1 ECM Originated Materials

Direct usage of natural ECM components from organ or tissue may be the simplest way to develop the ECM-like materials. The main benefit is the presence of 3D fibrous and porous topographies, as well as macrostructures like the vasculature. However, they are required perfect decellularization (remove cells from their original ECM) to minimize potential antigens since they can cause any inflammatory responses or immune-mediated implant rejection. In addition, the decellularization process may give damages on the micro-structure of ECM constituents—proteoglycans (PG) and glycosaminoglycans (GAG). As described in Chap. 8, the structural damage of

PGs and GAGs may cause serious problems on the performance and durability of ECM since PG and GAG strongly connect cells through collagen bundles to keep various cellular responses. However, the technique of decellularization to maintain the natural ECM texture have been greatly improved and hydrogels using natural ECM or ECM components have been transferred toward clinical trials for human skin or keratinocytes. Since they already contain native biochemical information, whole ECMs matrix made from tissues such as omentum or myocardium were used to form a hydrogel after decellularization and used for several preclinical feasibility studies. It promoted functional improvement of the post-infarcted heart. Recently xenogenic decellularized small intestinal submucosa ECM was used as a biomaterial for Achilles tendon repair.

In some cases, hydrogels prepared from proteins in the ECM potentially capable to generate functional scaffolds for regenerative medicine application. A common way to obtain native proteins is to isolate them from human or animal tissues. These so-called artificial or ECM-like proteins are of great interest in biomaterial scaffold production.

Single ECM protein, for example, tropoelastin is non-thrombogenic, and therefore, it was used to form stable hydrogels to induce several cell responses. Tropoelastin is able to self-organize under physiological conditions to give spherical nanoparticles which coalesce and form a porous hydrogel in the presence of glutaraldehyde. Thus, tropoelastin is a very promising biomaterial for regenerative medicine purposes.

Collagen type I is widely known to form hydrogels and in situ polymerizing injectable hydrogels is commercially available. It is often used to replace cartilage defects. The tyramine-modified hyaluronan forms a hydrogel when interacts with peroxidase and enables the growth of embryonic stem cells. Collagen and some other natural and synthetic polymers are regarded as the useful polymeric materials that are used, for example, as artificial skin. Artificial skin requires soft proper moisture permeation and should absorb bodily fluid and adheres well to scar tissue without causing inflammation or rejection. It also accelerates the growth of fiber cells. Collagen fulfills many of these requirements and used widely as biomedical polymers.

However, once again, it should note that ECM derived biopolymers have good biocompatibility, but have concerns potential immunogenic reactions and virus contamination. Animal derived materials such as collagen gels are widely used as scaffolds for tissue engineering because of their general compatibility with living tissues, while they often cause allergic reactions and carry dangerous pathogens including prions and cause a variety of neurodegenerative diseases in humans and animals. Other viruses might also be carried as pathogens in animal derived scaffolds.

9.4.2 Natural and Synthetic Gels

There is a need for alternative sources of animal derived scaffolds. One possible approach to obtain synthetic ECM is to use well-established 3D in vitro models, for example, gels from synthetic polymers and bio-macromolecules. Synthetic hydrogels have various advantages over these natural sources: a well-defined and controlled chemical composition and structure. They are infection-free, stable under high-temperature sterilization, easy handling, and are relatively inexpensive.

Since the self-assembling peptides are a 100% chemically synthesized material, the use of self-assembling peptide hydrogels can minimize the risk of biological contamination and the influence from the undefined factors in EHS gels. Self-assembling peptides which are characterized by a stableβ-sheet structure are another candidates to solve these problems. They undergo self-assembly to give nanofibers, then inter-woven matrices and further form a hydrogel scaffold [27, 28]. They are widely used in engineering studies, drug delivery systems and hemostatic applications.

To elicit the implanted cell functions, the scaffolds should transmit several mechanical stimulations and the effects have been proven in several cell culture systems. In order to provide a scaffold to prove the effect of mechanical stimulation on three-dimensionally cultured cells, an artificial self-assembling peptide: RADA16 (RADARADARADARADA; R. arginine, A. alanine, and D. aspartic acid) hydrogel scaffold was developed [29]. In this hydrogel, mouse skeletal muscle cells proliferated successfully. When the scaffold was statically stretched, a rapid phosphorylation of extracellular signal-regulated kinase (ERK) was observed and demonstrated that the developed self-assembling peptide gel is noncytotoxic and is a suitable tool for investigating the effect of mechanical stimulation on 3D cell culture.

Biomimetic hydrogels are also versatile for microfabrication for cell encapsulated biochips and tissue chips. Recent advances in the field of "smart" synthetic polymers have made possible to mimic some of the key bio- and physic-chemical functions as close as an ECM. Artificial ECM-like polymer hybrids are made by using natural polymers such as alginate, chitosan, and hyaluronic acid. Since these gels have ECM-like bioactive functions of protein binding and downstream signal pathway to make cell adhesion, proteolytic degradation and growth factor-binding, a variety of bioactive molecules derived from ECM: fibronectin, collagen, laminin, peptides, heparin, and growth factors have been incorporated and conjugated into synthetic hydrogels. Effect of surface modification of cell adhesive proteins such as laminin, fibronectin collagen and RGD sequence for the cell cultivation were also studied using synthetic hydrogels like PEG, PVA, and poly-lactic acid. Attempts mimicking the dermis and epidermis of human skin and human corneal limbal crypts have been made in vitro. Biomimetic engineering of synthetic hydrogels is an attractive strategy to develop scaffolds for tissue engineering, which provides fundamental knowledge to understand cell/scaffold interactions, cellular response and tissue formation.

It should note, however, that current biomimetic hydrogel scaffolds are mostly still lack of the hierarchical structure and multiple bio-functions which ECM inherently possesses. An important future work is to mimic the structure, morphology, and bioactivity of the ECM as closely as possible to design and produce hydrogels with an ideal biomimetic microenvironment to support cell growth and tissue regeneration.

9.4.3 Three-Dimensional (3D) Cell Culture

3D cell culture that allows cells to make 3D grow like ECM in vivo has a special scientific importance in terms of development of artificial ECM. 3D scaffold should allow the cells to move in the matrix in a similar way as in living tissue to ensure migration, differentiation, survival, and growth. 3D scaffold should also sustain solute diffusion and binding to effector proteins (like growth factors and enzymes) for cell viability. In fact, regenerative strategies using iPS cell and stem cell as tissue engineering therapies are being very actively developed with the aim to replace or regenerate damaged tissues and organs. For this purpose, the design of appropriate 3D scaffolds is crucially important for the generation of bio-inspired replacement tissues and they are often made of degradable or non-degradable hydrogels. Nowadays various types of hydrogels have been prepared from natural (proteins, peptide, polysaccharides) and synthetic polymers are available for tissue engineering scaffolds to cultivate cells, all attempting to mimic natural ECM structure.

To obtain an acellular human-based ECM-scaffold, fibroblasts were incorporated into gel matrix to create a reconstituted skin and transplanted into a mouse model. Also fibroblast-derived cell sheet was prepared, which was then rolled up to form a blood vessel, decellularized and served as a tissue-engineered blood vessel. There are examples of cell-produced materials that have been clinically utilized to support damaged cornea, heart muscle or cartilage. Development of organs using multiple layered cell sheets have also been attempted, where cell sheets as well as various types of hydrogels are studied intensely in vitro and some have already been successfully applied in clinical trials.

Successful designing of 3D gels would promote scientific understanding of the cellular behavior in ECM in vivo. Some reports indicate that cell adherence, spreading and proliferation are correlated with the charge of the gels, but detailed correlation between the surface charge (and possibly many other factors) and cell adherence and proliferation has so far, little obtained experimentally. Therefore, designing of 3D scaffolds with the natural ECM functions are one of the goals for bioengineers.

9.5 Hierarchical Cytoskeletal Gels

Self-organization is the process by which isolated components organize autonomously and spontaneously into the ordered structure. As described in Chap. 8, ECM is characterized by their ability to form hierarchical inter-woven network architectures through self-organization of the components keeping their second- and third-ordered structure. That is nature's preferred way of building ECM to give pertinent vital functions on various scales.

An important fact here is that the highly ordered hierarchical architecture is crucial to exert emergent function of living system. In another words, the highly hierarchical structure is the required origin to display the emergence in nature. Emergent function is often arisen at newly organized macroscopic structures formed in the given environment as result of a very large nonlinear ensemble of the interactions at a lower hierarchical or constituent molecular level. Therefore, a building up the integrated hierarchical 3D architecture is critically important to mimic the gels with emergent functions of living system. The great difficulty lies in how to integrate them to form a highly ordered structure keeping the original functions as nature have.

Cytoskeleton is the filaments system composed of three types of self-assembled protein filaments: actin, tubulin, and intermediate fibers. These filaments are formed by the self-assembly process and form spatially organized multi-step hierarchical structure. They are physically robust and are responsible not only to maintain the shape of the cell, but to give pertinent shape change and the motion upon changing the environment.

Chemically cross-linked 3 D gels of fibrous actin (F-actin) have recently been made by using tetra-PEG-maleimide which used as a cross-linker of actin fibers [30]. The storage modulus G' of the F-actin gel showed 10^3 Pa which exceeded those of cytoplasmic actin gels formed by actin-binding proteins. This suggests the F-actin gel is robust and mechanically stable, although it is formed by self-assembly processes. The important point is the cross-linked F-actin gel undergoes reversible sol–gel transition in vitro by changes in the ionic concentration or by applying a small oscillatory share stress. Figure 9.8 shows the time profile of storage modulus (G') and loss modulus (G'') by alternative applying and removing share strain. An abrupt decrease of G' indicates the decomposition of the gel (transition to sol), and an abrupt increase of G' indicates self-repairing of actin gel (transition to gel). Regardless of duration of strain applied, an abrupt increase of storage modulus (G') demonstrates the rapid and autonomous self-repairing of F-actin gel [30]

Interesting feature here is notable periodical oscillations in G' and G'' is observed during the reversible sol–gel transition which continued for more than ten hours on the chemically cross-linked F-actin gel [31] (Fig. 9.9a). From the power spectrum analysis, it was revealed that storage modulus of the actin hydrogel oscillates as a typical reverse-sawtooth waveform with 0.0006 Hz. The amplitude of oscillation decreased with decreasing degree of cross-linkage of the gel, and no oscillatory waveform was observe at all on the F-actin gel solution (Fig. 9.9b). This indicates that the chemical cross-linking is critical to induce the mechanical oscillation. It

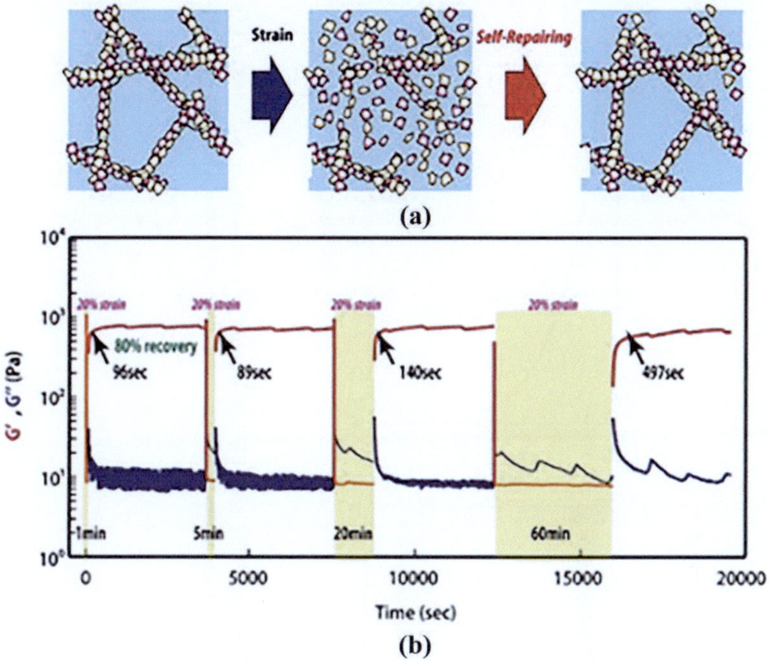

Fig. 9.8 **a** Schematic illustration of decomposition and self-repairing processes of F-actin gel. **b** Time profile of storage modulus (G') and loss modulus (G'') by repeated application and removal of share strain. An abrupt increase of G' indicates the self-repairing of actin gel. Duration of 200% oscillatory strain does not influences on the recovery time

is known that F-actin undergoes polymerization and depolymerization dynamics (treadmilling) in the presence of ATP to ensure the shape change and the motility of the cell. The oscillation of G' was entirely suppressed when phalloidin which inhibits the depolymerization of F-actin is present (Fig. 9.9c). These results demonstrate that polymerization and depolymerization dynamics (treadmilling) of chemically cross-linked actin hydrogel causes the autonomous oscillation of storage modulus. Extremely important point here is that the polymerization and depolymerization dynamics, which is a chemical process is transferred to mechanical oscillation in the cross-linked gel which synchronized and amplified to give the large-scale mechanical oscillation of G' with periodical frequency. This is the emergence experimentally obtained by the hierarchical soft and wet matter.

Tubulin is another cytoskeletal protein that forms a cylindrical assembly called microtubule (MT) by the spontaneous polymerization. A thermo-responsive 3D MT gel was obtained by simultaneous polymerization and subsequent chemical cross-linking of tubulins [32]. The MT gel exhibited a reversible so-gel transition by temperature changes between 4 and 37 °C. Apart from the typical synthetic polymer gels where the constituent polymer network is made of covalent bonding, these

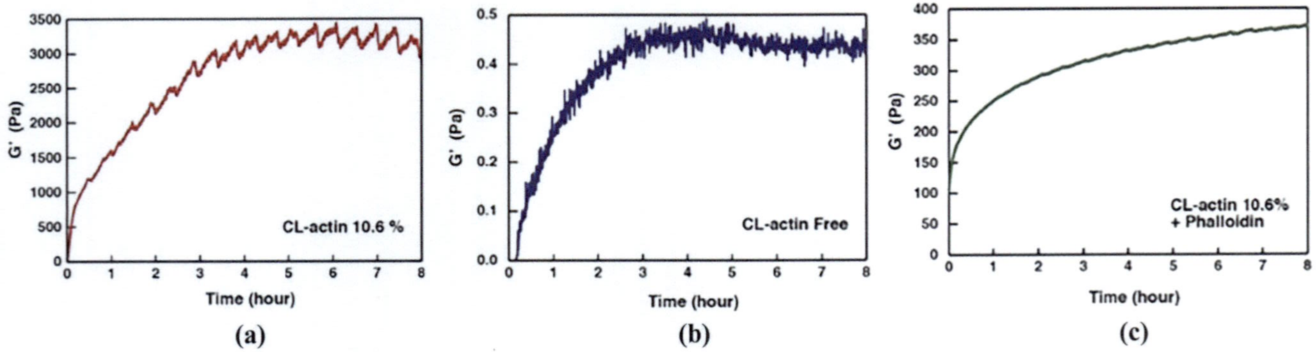

Fig. 9.9 Time profiles of G' of F-actin gel prepared under various conditions. **a** Contains 10.6% tetra-PEG-maleimide as a cross-linker, **b** no tetra-PEG-maleimide contains, **c** contains 10.6% tetra-PEG-maleimide and 200 mM phalloidin. (note the scale on vertical axis); measured under 1% strain and 1 Hz frequency

cytoskeletal hydrogels consist of self-organized assembly of the globular protein–G-actin or tubulin to form high-ordered hierarchy and polymerizes and depolymerizes reversibly to undergo sol–gel transition. Both F-actin and MT gels with hierarchical orders showed highly cooperative and synchronized responses.

9.6 Drug Delivery Gels

One of important subject in pharmaceutical science is to minimize the side effects of drugs and it is possible if the right amount of drug is administered at the right time and to the right place. This is a much sought-after form of drug delivery system (DDS) and many researches in the areas of medicine, pharmacology, and engineering have devoted in this field.

Research on DDS began with the development of sustained release drugs that are able to maintain drug concentration in the body within the effective (desired) range for an extended period of time. In particular, techniques to release drugs at a constant rate for prolonged period of time (zero order release) from various matrices have been made using permeable membranes under controlled osmotic pressure. However, more important is how to effectively deliver the drug to areas of illness (targeting) at required time whenever necessary (pulse delivery system). Drug delivery that employs hydrogels is promising because they can control both concentration and distribution of the drug in the body with the aid of stimuli-responsive polymers, i.e., delivery systems that responds to external stimuli such as pH, temperature, electric fields, magnetic fields, and others.

9.6.1 Diffusion-Controlled Delivery

If the gel containing drug does not show any specific changes in structure while delivering, drug delivery process is dominated diffusion. Diffusion-controlled delivery is classified by two: one is reservoir-type devices where the process is controlled by polymer barriers through which the entrapped drug permeates. When the external concentration is zero and the drug in the container is over-saturated, the delivery rate is constant regardless of the time (zero order delivery). However, if the drug concentration is less than the saturation, the drug concentration distribution changes with time and delivery rate follows Fick's first law of diffusion.

The other is a monolithic device that contains drug over the saturation concentration (monolithic dispersion type), or lower the saturation concentration (monolithic dissolution type). The drug delivery behavior of the monolithic dissolution follows Fick's second law of diffusion. When the swelling state of the gel changes with drug delivery process, the profile of drug release is influenced by the swelling state and does not simply follow the diffusion-controlled type but is a function of both swelling state and drug diffusion rate. If the swelling (water absorption) rate is high and drug diffusion is limited, the delivery follows the diffusion type release, while when the swelling rate is much less than the diffusion rate, the delivery follows the swelling profile.

9.6.2 On–off Controlled Delivery

9.6.2.1 Thermo-Responsive Gel

On–off drug delivery with a small temperature variation near body temperature was made by adjusting the phase transition temperature of a thermos-responsive copolymer gel composed of *N*-isopropyl acrylamide (NIPAAm), dimethylacrylamide (DMAAm), and butyl methacrylate(BMA) [33] (Fig. 9.10).

Figure 9.10 shows on and off delivery of indomethacin upon repeated temperature changes near the body temperature between 36 and 38 °C. Under low temperature the gel keeps swollen state and releases the drug while under elevated temperature the release substantially decreased since the gel shrinks and becomes hydrophobic. When temperature is raised to 38 °C, a sharp delivering peak is observed in Fig. 9.10. This was attributed to a squeezing effect by abrupt and large shrinkage of the gel volume. This abrupt shrinkage subsequently forms a non-permeable hydrophobic skin layer and the release is substantially inhibited.

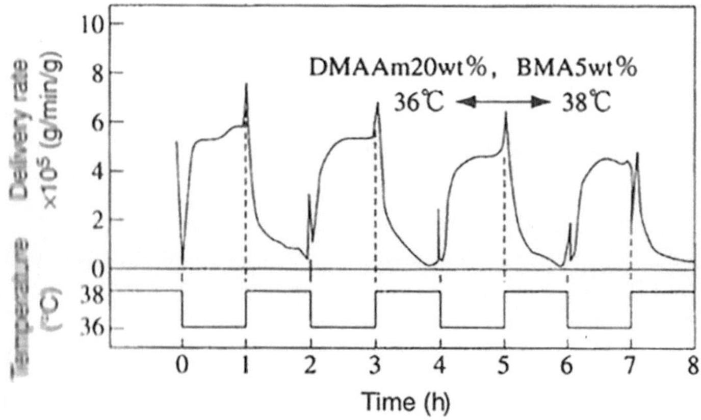

Fig. 9.10 On–off drug delivery of indometacine by temperature change. Gel: cross-linked copolymer consisting of *N*-isopropyl acrylamide-dimethylacrylarnide-butyl methacrylate [33]

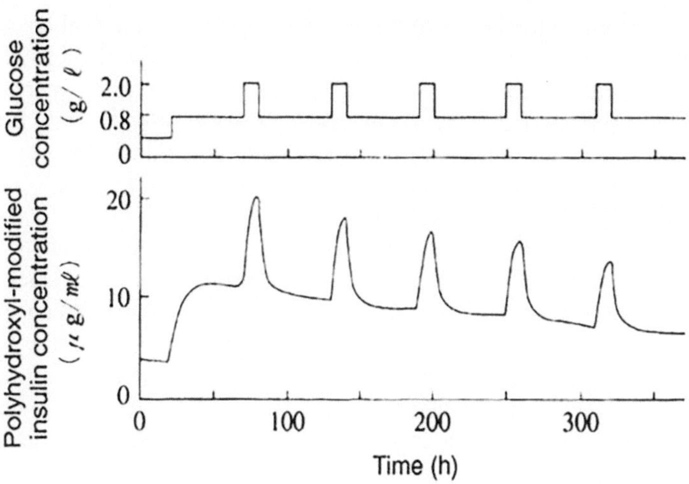

Fig. 9.11 On–off delivery of insulin by responding glucose [35]

9.6.2.2 Chemically Responsive Gel

An attempt to control glucose permeability in response to pH has been made by using various polyelectrolyte polymer gels. A chemically cross-linked polyacrylic acid gel was interposed by two different pH solutions; one at pH 7.0 and the other pH was varied. When pH of the solution of one side is acidic, the gel keeps shrunken state and glucose permeation was restricted, while when pH is increased, the gel swelled and glucose permeation was enhanced.

A typical example that responds to a certain chemical compound is the delivering system of insulin which responds to blood sugar levels [34]. Glucose oxidase (GOD) was fixed in a copolymer gel of *N*,*N*-dimethylaminoethyl methacrylate and hydroxyethyl methacrylate. When GOD reacts with glucose, it induces pH decrease and the gel, responding to pH drop, swells and starts to deliver insulin. Another example of insulin release upon glucose is based on the exchanging reaction of insulin and glucose at the polymer complexes. A water-swelling copolymer gel containing phenyl boric acid was synthesized to which insulin was chemically bound via hydroxyl groups of phenyl boric acid [35]. When the gel contacts with glucose, an exchange reaction occurs since boric acid binds glucose more strongly, and releases insulin into the solution (Fig. 9.11).

9.6.2.3 Physically-Responsive Gel

Along with recent developments in communication technology, various physical stimuli, such as electric fields, magnetic fields, UV light, and ultrasound are used as signals of DDS. The on–off control of drug delivery using electrical signals was attempted for insulin, glucose and raffinose release [36]. Polyelectrolyte gels such as poly(acrylic acid), poly(methacrylic acid), poly(dimethylaminopropyl acrylamide) undergo substantial contraction upon applying electric field and the mechanism was interpreted as electrophoretic phenomenon[37]. When an electric field is applied at constant intervals to the polyelectrolytes physically containing drug, the polyelectrolyte gel shrinks and the drug delivery is enhanced. The rate of drug release was proportional to the voltage applied.

Pulse-like drug delivery was achieved by applying a magnetic field through a ring magnet located in the gel matrix. The device is hemispherical plastic shape coated with a non-permeable film with a delivery hole in the center. When a magnetic field is applied, a repeated pulsed release of insulin was obtained [38]. An external microwave irradiation also controlled drug delivery. Methotrexate, a chemotherapeutic drug, was covalently bound to biodegradable poly(hydroxypropyl glutamine) and injected into a tumor. An irradiation of external microwave of 430 MHz selectively heated the device and delivered the drug.

Appendix

Czech scientist O. Wichterle Who Discovered Soft Contact Lens

The most successful example of gels for biomedical application would be the soft contact lenses developed by Czech scientists Wichterle and Lím. They together. Succeeded in preparing a cross-linked and transparent gel—poly(2-hydroxyethyl methacrylate) (pHEMA)—which is able to absorb water up to 40% with suitable mechanical properties. Wichterle thought pHEMA might be a suitable material for contact lenses and patented in 1953 and first used as an orbital implant in 1954. In 1960 commercially produced appeared SCL (O. Wichterle and D. Lím, Nature 185, 117 (1960). Polymers from which soft lenses are manufactured improved over the next 25 years, primarily in terms of increasing oxygen permeability, by varying the ingredients. In 1998, the first silicone hydrogel contact lenses were released which has extremely high oxygen permeability that had been used for the previous 30 years. Later on, various gels made of polyvinyl alcohol, polyvinylpyrrolidone, poly hydroxyl methacrylate, polyacrylamide, polydimethylsiloxane, and others have been proposed for soft contact lenses.

The vitreous is a gel composed of network of collagen, hyaluronic acid, and a small amount of soluble protein and glycoproteins containing 99% water in which hyaluronic acid is entangled with collagen fibers according to the electron microscopy observation. The network has a homogeneous structure to keep the transparency.

Attempts to replace vitreous materials have been made, while materials practically usable have not been developed mainly, due to lack of durable biocompatibility and transparency. Gels designed for transparency in bioorganism are the cornea, lens, and vitreous materials of eyes. The requirements for vitreous replacement materials include: (1)transparency; (2) similar refractive index and density to the vitreous; (3)

no germs and nontoxic; (4) no antigen effect; (5) nonflammable; (6) no enhancement of cell growth; (7) stability in the body; (8) slow absorption·(9) no solubility in water; and (1 0) ability to inject. However, as the vitreous replacement material is just a filler for a short period of time during retinal or lens surgery, it only needs to, satisfy conditions (1), (2), (3), (4), (5), and (10)'.

References

1. Y. Okumura, K. Ito, Adv. Mater. **13**, 485–487 (2001)
2. K. Haraguchi, T. Takehisa, Adv. Mater. **14**, 1120–1124 (2002)
3. T. Matsunaga, T. Sakai, Y. Akagi, U. Chung. M. Shibayama, Macromolecules **42**, 1344 (2009)
4. J.P. Gong, Y. Katsuyama, T. Kurokawa, Y. Osada, Adv. Mater. **15**, 1155–1158 (2003)
5. A. Nakayama, A. Kakugo, J.P. Gong, Y. Osada, M. Takai. T. Erata, S. Kawano, Adv. Funct. Mater. **14**, 1124–1128 (2004).
6. R. Kishi, K. Hiroki, T. Tominaga, K. Sano, H. Okuzaki, J. G. Martinez, T. F. Otero Y. Osada, J. Polym. Sci. Polym. Phys. **50**, 790 (2012)
7. C.W. McCutchen, *Lubrication of Joints, the Joints and Synovial Fluid* (Academic Press, New York, 1978)
8. J.P. Gong, M. Higa, Y. Iwasaki, Y. Katsuyarna, Y. Osada, J. Phys. Chem. B **101**, 5487–5489 (1997)
9. J.P. Gong, Y. Osada, Chem. Phys. **109**, 8062–8068 (1998)
10. J.P. Gong, Y. Iwasaki, Y. Osada, K. Kurihara, Y. Hamai, J. Phys. Chem. B **103**, 6001–6006 (1999)
11. J. Klein, E. Kumacheva, D. Mahalu, D. Perahia, L Fetters, Nature **370**, 634 (1994)
12. D. Kaneko, T. Tada, T. Kurokawa, J.P. Gong, Y. Osada, Adv. Mater. 535 (2005)
13. Y. Osada, *Conversion of Chemical into Mechanical Energy by Synthetic Polymers (Chemomechanical Systems) in Advances in Polymer Science*, vol. 82 (Springer, 1987)
14. K. Sawahata, J.P. Gong, Y. Osada, Macromol. Rapid Commun. Rapid Commun. **16**, 713 (1995)
15. Gong, J.P., Y. Osada, Chem. Lett. **449** (1995)
16. J.P. Gong, N. Komatsu, T. Nitta, Y. Osada, J. Phys. Chem. B **101**, 740 (1997)
17. Y. Osada, H. Okuzaki, H. Hori, Nature **355**, 242 (1992)
18. Y. Osada, S.B. Ross-Murphy, Sci. Am. **268**, 82 (1993)
19. Y. Osada, A. Matsuda, Nature **376**, 219 (1995)
20. T. Mitsumata, K. Ikeda, J.P. Gong, Y. Osada, Langmuir **16**, 307 (2000)
21. T. F. Otero, J.G. Martinez, in *Conducting Polymers as EAP: Fundamentals and Materials Chapter 11 in Electromechanically Active Polymers*, ed. by F. Carpi (Springer, 2016)
22. R.H. Berghman, Synth. Met. **78**, 339 (1996)
23. M. Zrinyi, D. Szabo, G. Filipcsei, J. Feher, in *Electrical and Magnetic Field-Sensitive smart Polymer, Chapter 11 in Polymer Gels and Networks*, ed. by Y. Osada, A.R. Khokhlov (Marcel Dekker 2002)
24. A. Kakugo, S. Sugimoto, J.P. Gong, Y. Osada, Adv. Mater. **14**, 1124 (2002)
25. R. Kawamura, K. Sano, K. Ijiro, Y. Osada, RSC Adv. **32953** (2014)
26. K.G. Cornwell, A. Landsman, K.S. James, Clin. Podiatr. Med. Surg. **26** (2009)
27. S. Zhang, T.C. Holems, C.M. DiPersio, R.O. Hynes, X. Su, A. Rich, Biomaterials **16**, 1385 (1995)
28. H. Yokoi, T. Kinoshita, S. Zhang, Proc. Natl. Acad. Sci. U.S.A. **102**, 8414 (2005)
29. Y. Nagai, H. Yokoi, K. Kaihara, K. Naruse, Biomaterials **33**, 1044 (2012)
30. K. Sano, R. Kawamura, Y. Osada, Biomacromolecules **4173** (2011)
31. Y. Osada, R. Kawamura, K. Sano, *Hydrogels of Cytoskeletal Proteins* (Springer, 2016)
32. K. Sano, R. Kawamura, Y. Osada, Biomacromolecules **1409** (2011)

33. R. Yoshida, K. Sakai, T. Okano, Y. Kimura, Y, J Biomater. Sci. Polym. Edn. **6**, 585 (1994)
34. J. Kost, T.A. Horbett, B.D. Ratner, M. Singh, 1 Biomed. Mater. Res. **19**, 1117 (1985)
35. D. Shiino, Y. Murata, K. Kataoka, Y. Koyama, M. Yokoyama, T. Okano, Y. Sakurai, Biomaterials **15**, 121 (1994)
36. K. Sawahata, M. Hara, H. Yasunaga, Y. Osada, J Control. Rel. **14**, 253 (1990)
37. J.P. Gong, T. Nitta, Y. Osada, J. Phys. Chem. **98**, 9583 (1994)
38. D.T. Hsieh, R. Langer, in *Controlled Release Delivery System*, ed. by T.J. Roseman, S.Z. Masdorf (Marcel Dekker, New York, 1982), p. 107

MIX
Papier aus verantwortungsvollen Quellen
Paper from responsible sources
FSC® C105338

If you have any concerns about our products,
you can contact us on
ProductSafety@springernature.com

In case Publisher is established outside the EU,
the EU authorized representative is:
Springer Nature Customer Service Center GmbH
Europaplatz 3, 69115 Heidelberg, Germany

Printed by Libri Plureos GmbH
in Hamburg, Germany